Integrated Analytical Systems

Series Editor:
Radislav A. Potyrailo
GE Global Research Center
Niskayuna, NY

For other titles published in the series, go to
www.springer.com/series/7427

Xudong Fan
Editor

Advanced Photonic Structures for Biological and Chemical Detection

Springer

Editor
Xudong Fan
Dept. Biological Engineering
University of Missouri
240D Bond Life Sciences Center
1201 E. Rollins St.
Columbia, MO 65211
USA

ISBN 978-1-4614-2442-0 e-ISBN 978-0-387-98063-8
DOI: 10.1007/978-0-387-98063-8
Springer Dordrecht Heidelberg London New York

Softcover reprint of the hardcover 1st edition 2009

Printed on acid-free paper

Springer is part of Springer Science+Business Media (www.springer.com)

Series Preface

In my career I've found that "thinking outside the box" works better if I know what's "inside the box."
Dave Grusin, composer and jazz musician

Different people think in different time frames: scientists think in decades, engineers think in years, and investors think in quarters.
Stan Williams, Director of Quantum Science Research, Hewlett Packard Laboratories

Everything can be made smaller, never mind physics;
Everything can be made more efficient, never mind thermodynamics;
Everything will be more expensive, never mind common sense.
Tomas Hirschfeld, pioneer of industrial spectroscopy

Integrated Analytical Systems

Series Editor: Dr. Radislav A. Potyrailo, GE Global Research, Niskayuna, NY

The book series *Integrated Analytical Systems* offers the most recent advances in all key aspects of development and applications of modern instrumentation for chemical and biological analysis. The key development aspects include (i) innovations in sample introduction through micro- and nanofluidic designs, (ii) new types and methods of fabrication of physical transducers and ion detectors, (iii) materials for sensors that became available due to the breakthroughs in biology, combinatorial materials science, and nanotechnology, and (iv) innovative data processing and mining methodologies that provide dramatically reduced rates of false alarms.

A multidisciplinary effort is required to design and build instruments with previously unavailable capabilities for demanding new applications. Instruments with more sensitivity are required today to analyze ultra-trace levels of environmental pollutants, pathogens in water, and low vapor pressure energetic materials in air. Sensor systems with faster response times are desired to monitor transient in vivo events and bedside patients. More selective instruments are sought to analyze specific proteins in vitro and analyze ambient urban or battlefield air. For these and many other applications, new analytical instrumentation is urgently needed. This book series is intended to be a primary source on both fundamental and practical information of where analytical instrumentation technologies are now and where they are headed in the future.

Looking back over peer-reviewed technical articles from several decades ago, one notices that the overwhelming majority of publications on chemical analysis has been related to chemical and biological sensors and has originated from Departments of Chemistry in universities and Divisions of Life Sciences of governmental laboratories. Since then, the number of disciplines has dramatically increased because of the ever-expanding needs for miniaturization (e.g., for in vivo cell analysis, embedding into soldier uniforms), lower power consumption (e.g., harvested power), and the ability to operate in complex environments (e.g., whole blood, industrial water, or battlefield air) for more selective, sensitive, and rapid determination of chemical and biological species. Compact analytical systems that have a sensor as one of the system components are becoming more important than individual sensors. Thus, in addition to traditional sensor approaches, a variety of new themes have been introduced to achieve an attractive goal of analyzing chemical and biological species on the micro- and nanoscale.

Foreword

Photonic biological and chemical sensors are one of the truly important areas in photonics today. They have evolved from the traditional cuvette-based apparatuses used a few decades ago and optical waveguide- and fiber-based devices developed in 1980s into the highly sophisticated photonic structures that are commonly used today. These include photonic crystals, fiber gratings, and optical ring resonators, etc., which are marked by significant improvement in sensing sensitivity, sample volumes, fluidic integration, and detection multiplexing capability.

Research and development of photonic bio/chemical sensors have seen rapid growth in recent years, driven mainly by existing and imminent needs in health care, defense and security, and environmental protection and by innovations in photonic technologies. However, for the most part, information concerning state-of-the-art photonic bio/chemical sensors can be found only by sifting through piles of scientific manuscripts. Therefore, a timely and concise survey of all different photonic bio/chemical sensing technologies is becoming increasingly important and desirable, especially in order to enable the education of up-and-coming scientists in this field.

This present book provides a broad overview of various modern photonic biological and chemical sensors that are still under intensive investigation worldwide. It emphasizes the photonic sensing structures being used in contemporary research and their underpinning photonic technologies. Meanwhile, the applications and the advantages of those photonic structures in sensing are also clearly elucidated. I believe that the book will be highly valuable to the bio/chemical sensing community and to graduate students in disciplines like biomedical engineering, electrical engineering, mechanical engineering, chemical engineering, and applied physics.

Sydney, Australia — Benjamin J. Eggleton

Preface

Photonics has been employed as a powerful tool in biological and chemical detection for many years. Compared to other sensing technologies, the sensing transduction signal from photonic sensors can be generated from optical parameters in both the spectral and time domains, including refractive index, optical absorption, fluorescence, polarization, lifetime, and even nonlinear optical processes such as lasing, Raman scattering, and multiphoton absorption and emission. These different types of signals are complementary to each other and provide vast amount of information regarding the presence and interaction of bio/chemical molecules.

In recent years, rapid advancements in photonic technologies have significantly enhanced the photonic bio/chemical sensor performance, especially in the areas of (1) interaction between the light and analyte, (2) device miniaturization and multiplexing, and (3) fluidic design and integration. This has led to drastic improvements in sensor sensitivity, enhanced detection limit, advanced fluidic handling capability, lower sample consumption, faster detection time, and lower overall detection cost per measurement.

This book consists of 19 chapters written by the worldwide experts in photonic sensors. It is intended to capture most of the important up-to-date fascinating research and applications of novel photonic structures in bio/chemical sensing. The subjects extend from optical ring resonators, distributed feedback resonators, Fabry-Pérot interferometers, Mach-Zehnder interferometers, Young interferometers, and spinning-disc interferometers to fiber gratings, photonic crystals, micro/nanofibers, photonic wires, antiresonant reflecting optical waveguides (ARROWs), slot waveguides, metal-clad waveguides, reverse symmetry waveguides, and folded waveguides.

The topics in this book are arranged into three sections dealing with (1) photonic chemical vapor sensing, (2) photonic biological sensing, and (3) integration of the photonic sensors with microfluidics. However, it should be emphasized that a number of photonic sensors presented in this book combine or integrate multiple photonic structures to achieve enhanced performance. Additionally, many sensors in this book are capable of performing both biological sensing and liquid/vapor chemical sensing. The above categorization is done simply out of consideration for balancing the book's content and giving readers some general thematic guidance.

While this book surveys a diversified range of photonic sensor structures, it is certainly impossible in one book volume to provide full coverage of all such structures known to science. Thus plasmonic photonic structures, photonic crystal fibers, and nanoparticles will be covered by other upcoming books in this Springer Series "Integrated Analytical Systems."

This book will be of interest to researchers from multiple disciplines that include physicists, chemists, engineers, nanotechnologists, and biologists, who work in the area of bio/chemical detection and analysis in university laboratories, the biotechnology industry, the healthcare industry, homeland security and defense, pharmaceuticals, the food industry, and environment protection. The book will also serve as an excellent resource for graduate students wanting an introduction to the exciting and fast-pacing research frontiers in photonic bio/chemical sensors.

I would like to thank all of the contributors for writing truly informative and stimulating chapters. I am also indebted to Dr. Radislav Potyrailo, the book series editor, who helped me choose this very intriguing topic on photonic bio/chemical sensing. My gratitude further goes to Dr. Siegfried Janz, Dr. Johannes S. Kanger, and Dr. David Erickson, who kindly provided the nice figures for the book cover and to Dr. Benjamin Eggleton for the foreword.

Columbia, Missouri Xudong Fan

Contents

Section II Photonic Structures for Biochemical Sensing

Section III Microfluidics Enabled Photonic Sensing Systems

Contributors

Matthew D. Butts
Chemistry Technologies and Material Characterization, General Electric Company, Global Research Center, Niskayuna, NY 12309, USA

Chung-Yen Chao
Department of Electrical Engineering and Computer Science, University of Michigan, Ann Arbor, MI 48109, USA

P. Cheben
Institute for Microstructural Sciences, National Research Council Canada, Ottawa, ON, Canada K1A 0R6

Antao Chen
Applied Physics Laboratory, University of Washington, Seattle, WA 98105, USA

Andrea Cusano
Optoelectronic Division, Engineering Department, University of Sannio, C.so Garibaldi 107, 82100 Benevento, Italy

Antonello Cutolo
Optoelectronic Division, Engineering Department, University of Sannio, C.so Garibaldi 107, 82100 Benevento, Italy

A. Delâge
Institute for Microstructural Sciences, National Research Council Canada, Ottawa, ON, Canada K1A 0R6

Tao Deng
Chemistry Technologies and Material Characterization, General Electric Company, Global Research Center, Niskayuna, NY 12309, USA

A. Densmore
Institute for Microstructural Sciences, National Research Council Canada, Ottawa, ON, Canada K1A 0R6

Zhebo Ding
Chemistry Technologies and Material Characterization, General Electric Company, Global Research Center, Niskayuna, NY 12309, USA

David Erickson
Sibley School of Mechanical and Aerospace Engineering, Cornell University, Ithaca, NY 14853, USA

Xudong Fan
Department of Biological Engineering, University of Missouri, Columbia, MO 65211, USA

Greg Frye-Mason
ICx Biodefense, Albuquerque, NM 87107, USA

Sarah E. Genovese
Chemistry Technologies and Material Characterization, General Electric Company, Global Research Center, Niskayuna, NY 12309, USA

Michele Giordano
Institute for Composite and Biomedical Materials (IMCB), National Research Council (CNR), P.le Enrico Fermi 1, 80055, Portici, Napoli

Julie Goddard
Sibley School of Mechanical and Aerospace Engineering, Cornell University, Ithaca, NY 14853, USA

Nicholas J. Goddard
School of Chemical Engineering and Analytical Science (CEAS), University of Manchester, Manchester, UK

L. Jay Guo
Department of Electrical Engineering and Computer Science, University of Michigan, Ann Arbor, MI 48109, USA

Aaron R. Hawkins
ECEn Department, Brigham Young University, 459 Clyde Building, Provo, UT 84602, USA

Robert Horvath
Research Institute for Technical Physics and Materials Science MTA-MFA (KFKI Campus), Konkoly Thege Miklós út 29–33, H-1121 Budapest, Hungary

S. Janz
Institute for Microstructural Sciences, National Research Council Canada, Ottawa, ON, Canada K1A 0R6

Johannes S. Kanger
Faculty of Science and Technology, University of Twente, 7500 AE Enschede, The Netherlands

Anders Kristensen
DTU Nanotech, Technical University of Denmark, 2800 Kongens Lyngby, Denmark

Scott Lacey
Department of Physics and Astronomy, Franklin & Marshall College, Lancaster, PA 17604, USA

B. Lamontagne
Institute for Microstructural Sciences, National Research Council Canada, Ottawa, ON, Canada K1A 0R6

J. Lapointe
Institute for Microstructural Sciences, National Research Council Canada, Ottawa, ON, Canada K1A 0R6

Tao Ling
Department of Electrical Engineering and Computer Science, University of Michigan, Ann Arbor, MI 48109, USA

G. Lopinski
Steacie Institute for Molecular Sciences, National Research Council Canada, Ottawa, ON, Canada K1A 0R6

Sudeep Mandal
Department of Applied and Engineering Physics, Cornell University, Ithaca, NY 14853, USA

R. McKinnon
Institute for Microstructural Sciences, National Research Council Canada, Ottawa, ON, Canada K1A 0R6

T. Mischki
Steacie Institute for Molecular Sciences, National Research Council Canada, Ottawa, ON, Canada K1A 0R6

Stephan Mohr
School of Chemical Engineering and Analytical Science (CEAS), University of Manchester, Manchester, UK

Kristian Mølhave
DTU Nanotech, Technical University of Denmark, 2800 Kongens Lyngby, Denmark

Niels Asger Mortensen
DTU Fotonik, Technical University of Denmark, 2800 Kongens Lyngby, Denmark

Paul H.J. Nederkoorn
Ostendum R&D BV, 7500 AE Enschede, The Netherlands

David D. Nolte
Department of Physics, Purdue University, West Lafayette, IN 47907, USA

Pierluigi Pilla
Optoelectronic Division, Engineering Department, University of Sannio, C.so Garibaldi 107, 82100 Benevento, Italy

Radislav A. Potyrailo
Chemistry Technologies and Material Characterization, General Electric Company, Global Research Center, Niskayuna, NY 12309, USA

A.T. Rosenberger
Department of Physics, Oklahoma State University, Stillwater, OK 74078, USA

Jacob Scheuer
Department of Physical Electronics, School of Electrical Engineering, Tel-Aviv University, Tel-Aviv 69978, Israel

J.H. Schmid
Institute for Microstructural Sciences, National Research Council Canada, Ottawa, ON, Canada K1A 0R6

Holger Schmidt
School of Engineering, University of California Santa Cruz, Santa Cruz, CA 95064, USA

Siyka I. Shopova
MicroParticle PhotoPhysics Lab, Polytechnic Institue of New York NY 11201, USA

Nina Skivesen
Inano Interdisciplinary Nanoscience Center, University of Aarhus, Denmark

Vinod Subramaniam
Faculty of Science and Technology, University of Twente, 7500 AE Enschede, The Netherlands

M. Sumetsky
OFS Laboratories, 19 Schoolhouse Road, Somerset, NJ 08873, USA

Yuze Sun
Department of Biological Engineering, University of Missouri, Columbia, MO 65211, USA

Jonathan D. Suter
Department of Biological Engineering, University of Missouri, Columbia, MO 65211, USA

P. Waldron
Institute for Microstructural Sciences, National Research Council Canada, Ottawa, ON, Canada K1A 0R6

Xuefeng Wang
Department of Physics, Purdue University, West Lafayette, IN 47907, USA

Tao Wei
Department of Electrical and Computer Engineering, Missouri University of Science and Technology, Rolla, MO 65409, USA

Ian M. White
Department of Bioengineering, University of Maryland, College Park, MD 20742, USA

Hai Xiao
Department of Electrical and Computer Engineering, Missouri University of Science and Technology, Rolla, MO 65409, USA

D.-X. Xu
Institute for Microstructural Sciences, National Research Council Canada, Ottawa, ON, Canada K1A 0R6

Aurel Ymeti
Ostendum R&D BV, 7500 AE Enschede, The Netherlands

Ming Zhao
College of Optical Sciences, University of Arizona, Tucson, AZ 85721, USA

Hongying Zhu
Department of Biological Engineering, University of Missouri, Columbia, MO 65211, USA

Mohammed Zourob
GROUPE GDG ENVIRONNEMENT LTÉE, 430 rue Saint-Laurent, Trois-Rivières (QC),Canada G8T 6H3

Chapter 1
General Introduction

Xudong Fan

Photonic sensors employ light to convert bio/chemical processes into a detectable signal, i.e., a sensing transduction signal. As compared to other sensing technologies, the strength and versatility of photonic sensors lie in the wide range of optical properties available in both the spectral and time domains that serve to generate the sensing transduction signal. These properties include, but are not limited to, refractive index, optical absorption, fluorescence, polarization, lifetime, and even nonlinear optical processes such as lasing, Raman scattering, and multiphoton absorption and emission. The sensing signals from these optical properties are complementary to each other. When used alone or in combination, they can provide vast amount of information regarding the presence and interaction of bio/chemical molecules. As a result, photonic bio/chemical sensors have broad applications in healthcare, defense, homeland security, the food industry, the biotechnology industry, pharmaceuticals, and environmental monitoring and protection.

For a photonic bio/chemical sensor, three characteristics, among others, are crucial in determining its performance. (1) Light–analyte interaction. This dictates the transduction signal. Stronger light–matter interaction usually results in a higher sensitivity and better (e.g., lower) detection limit. (2) Sensor miniaturization and multiplexing. These are directly related to sample consumption, device portability, detection time, and detection cost. (3) Integration of fluidics with photonic sensing elements. Effective and efficient fluidics not only reduces the sample consumption and hence the cost, but also enhances light–analyte interaction and expedites the detection processes.

In the past decade, rapid advancements in photonic technologies have significantly benefited photonic bio/chemical sensors. Numerous novel photonic structures have been invented and developed with tremendous improvement in the three

X. Fan
Department of Biological Engineering, University of Missouri, Columbia, MO 65211, USA
e-mail: fanxud@missouri.edu

X. Fan (ed.), *Advanced Photonic Structures for Biological and Chemical Detection*, Integrated Analytical Systems,
DOI 10.1007/978-0-387-98063-8_1,

important characteristics mentioned above. For example, through light interference, photonic crystals are capable of confining light within extremely small detection volumes for extremely high light–analyte interaction and thus, sensitive molecular detection (Chap. 16). Maximal light–analyte interaction is also achieved with the antiresonant reflecting optical waveguide (ARROW) structure, which is engineered so that both light and analyte can simultaneously be guided in the same liquid core (Chap. 18). In an optical ring resonator the light circulates along the resonator and repeatedly interacts with analyte deposited on the resonator surface (Chaps. 2, 5, 6, 8, 9, 12, 13, 14, 17, and 19), thus drastically improving the detection limit and lowering the sample consumption. In the BioCD platform (Chap. 11), thousands of samples can be analyzed quickly with very low cost.

The collection of chapters in this book represents the most recent global efforts in the research and development of photonic bio/chemical sensing structures. The photonic structures included in book are quite diversified, ranging from optical resonators and interferometers to photonic crystals and specially designed waveguides. For guidance, they are summarized as follows:

- Optical ring resonators (Chaps. 2, 5, 6, 8, 9, 12, 13, 14, 17, and 19)
- Distributed feedback resonators (Chap. 12)
- Fiber gratings (Chaps. 3 and 7)
- Fabry-Pérot interferometers (Chap. 7)
- Mach-Zehnder interferometers (Chaps. 7 and 9)
- Michelson interferometers (Chap. 7)
- Young interferometers (Chap. 10)
- Spinning-disc interferometers (Chap. 11)
- Photonic crystals (Chaps. 4 and 16)
- Micro/nanofibers and photonic wires (Chaps. 9 and 13)
- Slot waveguides (Chaps. 8 and 9)
- Folded waveguides (Chap. 9)
- Metal clad waveguides (Chap. 15)
- Reverse symmetry waveguides (Chap. 15)
- Liquid core antiresonant reflecting optical waveguides (Chap. 18)

The reader will notice that many photonic sensors presented in this book combine or integrate multiple photonic structures to achieve better sensing performance. For example, nanofibers in the form of a ring resonator provide highly sensitive bio/chemical detection (Chap. 13). Long period gratings in combination with a Mach-Zehnder or Michelson interferometer configuration perform sensitive chemical vapor sensing (Chap. 7). In Chap. 9, folded waveguide structures are integrated with the ring resonator or Mach-Zehnder interferometer design. While such ideas may make categorization difficult, they represent some of the most innovative designs that the field has ever seen.

The contributed chapters are divided into three sections. The first section is dedicated to chemical vapor sensing. In the majority of the photonic vapor sensors described here, a layer of vapor sensitive material (polymer, ceramics, or colloidal crystal film, etc.) is coated onto the photonic sensing structure (Chaps. 3, 6, and 7) or forms the sensing structure itself (Chaps. 2 and 4). The refractive index or the thickness of the vapor sensing material is changed when it interacts with vapor molecules, and these changes can be detected as the sensing transduction signal. For increased sensing specificity, in Chaps. 3 and 7, molecular sieves are used to selectively detect molecules of a particular size, and in Chap. 6, ring resonator-based microgas chromatography is employed to separate and identify analytes by their respective retention times. Chemical vapors can also be detected by their absorption band. Chapter 5 discusses cavity enhanced optical absorption using a high Q-factor ring resonator.

The second section focuses on biosensing and liquid chemical sensing. Most of the photonic biosensors in this section rely on so-called evanescent sensing or surface sensing. The light is typically highly confined in the close proximity of the solid sensing surface, which is coated with a layer of biorecognition molecules that have affinity for specific analytes. When analytes, i.e., target molecules bind to the surface, local refractive index changes, since biomolecules and cells generally have different refractive indices than buffer solutions. The decay length of the evanescent field is only a few tens of nanometers, which is advantageous in sensitively detecting the signal generated by the target molecules that bind to the sensing surface while rejecting the one by background molecules that do not bind to the surface but still exist in bulk solution.

Conventional evanescent sensing works exceedingly well for relatively small biomolecules such as proteins and DNA molecules whose size is much smaller than the decay length. However, it becomes less sensitive when detecting biospecies, such as cells, with dimensions over 1 μm. In Chap. 15, deep-probe waveguide sensors are developed to overcome this limitation, which have a decay length comparable to the size of the biospecies of interest.

The sensors in this section can also be utilized to detect chemicals in liquid through the bulk solution refractive index change induced by the presence of target chemicals. Since no recognition molecules are used, this type of chemical sensing may usually have low specificity. However, these sensors may perform excellently in conjunction with other technologies such as capillary electrophoresis, mass spectrometer, and liquid chromatography in chemical detection.

The third section discusses microfluidics-enabled photonic sensing systems. In these types of sensors, liquid not only carries analytes, but also defines photonic structures and provides photonic functions. In photonic crystals and ARROW sensing structures (Chaps.16 and 18), both light and liquid are confined within aqueous holes or channels. In Chaps. 17 and 19, liquid itself forms a part of ring resonators. In addition to the detection of surface-bound molecules (Chap. 16), molecules in free solution can also be analyzed (Chaps. 17–19). This is a very attractive and advantageous feature, which eliminates the necessity of immobilizing biorecognition molecules to sensor surfaces (thus simplifying the sensing protocol) and enables detection of molecules in their natural form without the interference from a solid state substrate. In the sensors presented in this section, sensing signals generated from refractive index, fluorescence, Raman emission, and laser emission are demonstrated.

Future research and development of photonic bio/chemical sensors will continue to be driven by technological advancements in photonics. Besides refractive index, absorption, and fluorescence, when the light-analyte interaction is sufficiently strong, sensing relying on nonlinear optical processes will be practically viable some day, which may lead to even lower detection limits and reveal more detailed information with regards to the interaction of bio/chemical molecules. Additionally, intensive investigation in micro/nanofabrication and nanophotonics will lead to further miniaturization and multiplexing of photonic sensors. Consequently, detection of a sample at the level of femtoliter or even subfemtoliter will become common practice. Finally, apart from fluidics, integration of light sources (tunable or fixed), detectors, optical signal processing components (such as filters and polarizers, etc.) will also be important in future photonic sensor designs.

Section I
Photonic Structures for Chemical Vapor Sensing

Chapter 2
Microresonator Sensors Made in Polymers with Functional Chromophore Dopants

Antao Chen

Abstract The optical properties such as index of refraction and optical absorption of many chromophore-doped polymers are sensitive to the physical and chemical environment to which the polymers are exposed. A variety of microring resonator sensors have been realized with chromophore-doped polymers. Detection sensitivity is further enhanced by optical microresonator structures such as waveguide microring resonator and fiber Bragg gratings. Chromophore-doped polymers also offer some desirable flexibility in device fabrication. Ultraviolet light and electron beam can reduce the index of refraction of the polymer. The photobleaching and electron beam bleaching methods form optical waveguides in a single fabrication step and do not involve solvents or wet chemicals, and can be applied to polymers that are not compatible with other waveguide fabrication techniques.

This chapter provides an overview of the basic principles and designs of such sensors. A chemical sensor to detect trace explosives and a broadband fiber optic electric-field sensor are presented as practical examples. The polymers used for the trace explosive sensor are unpoled and have chromophores randomly orientated in the polymer hosts. The electric field sensor uses a poled polymer with chromophores preferentially aligned through electrical poling, and the microring resonator is directly coupled to the core of optical fiber.

2.1 Introduction

Polymers with specific functionalities can be realized by incorporating various dopants, such as laser dyes, rare earth ions, quantum dots, and functional chromophores into the host polymer. Chromophores are molecules or chemical groups as part of a larger molecule, and they have characteristic absorption bands in the

A. Chen
Applied Physics Laboratory, University of Washington, Seattle, WA 98105, USA
e-mail: antaochen@apl.washington.edu

X. Fan (ed.), *Advanced Photonic Structures for Biological and Chemical Detection*, Integrated Analytical Systems,
DOI 10.1007/978-0-387-98063-8_2,

optical wavelengths, giving them a distinct color. The color of chromophores ranges from red to blue depending on the specific chromophore. Some chromophores have long been used as dyes in the textile industry. If chromophores are randomly oriented in the polymer matrix, the polymer can have strong third-order nonlinear optic (NLO) effect, two-photon absorption and fluorescence, and sensitivity to chemical analytes. Chromophores of electron donor–π electron charge transfer bridge-electron acceptor (D–π–A) structure can have a strong dipole moment and second-order NLO effect. Such chromophores can be preferentially orientated through electrical or optical poling. The poled polymers become electro-optic materials and they have been used for high speed electro-optic (EO) modulation and switching, generation of terahertz signals through optical rectification, and second harmonic generation[1–4]. High electronegative chemicals such as nitroaromatic explosives trinitrotoluene (TNT) and dinitrotoluene (DNT), have shown strong interaction with conjugated charge transfer chromophores and thus change the optical properties of chromophore-doped polymers. This makes such polymers useful in detecting trace explosives[5]. Decomposition of chromophores through exposure to ultraviolet (UV) and energetic electron beam permanently reduces the index of refraction of the polymer. This property can be used to fabricate optical waveguides and microring resonators.

A broad range of optical devices, including wavelength filters, wavelength division multiplexers and demultiplexers, lasers, switches, modulators, dispersion compensators, and polarization rotators can be realized with microring resonators[6–10]. Microring resonators are also a good platform of miniature optical sensors to measure temperature, strain, and stress, and to detect chemical and biological agents[11]. High *Q*-factor and long photon life-time enhance sensor response. The small size of microring resonators also requires smaller amount of analytes for bio- and chemical sensing. The overall length of conventional optical waveguide devices based on Mach-Zehnder interferometers, directional couplers, Y-branch switches, and multimode interference (MMI) technologies is on the order of a centimeter. In contrast, microring resonator devices have much smaller (several micrometers to a few hundred micrometers) sizes. This unique advantage makes microring resonators ideal for large scale integration to form an array of sensors on a single chip. It also makes them highly suitable for monolithic integration with silicon integrated circuits that interface with the sensors.

Microring resonators have been realized in many different materials, including silicon-on-insulator, GaAs and InP, glass, and polymers[1,6–11]. Polymers offer a number of unique advantages compared to the other materials. Through modern molecular design and organic synthesis, polymers can have tailored optical (index of refraction, birefringence, absorption spectrum), device fabrication processing, and tuning/sensing (electro-optic, thermo-optic, chemo-optic, and photoelastic) properties to meet the requirements of specific applications and device designs. Many fabrication options exist for polymer microring resonators. They can be fabricated by photolithography and reactive ion etching (RIE), imprinting, soft-lithography molding, two-photon polymerization, and electron-beam (e-beam) writing. Polymer optical waveguides can also be fabricated on flexible, curved, and conformal substrates. These advantages allow polymer microring resonators to have functionality, performance, design, and fabrication flexibilities not available in other materials.

By combining chromophore-doped polymers with optical resonator structures such as waveguide microring resonators and fiber Bragg gratings, a variety of sensors can be made. We have discovered that some polymers containing chromophores of D–π-A structure can have strong interaction with molecules of nitroaromatic compounds such as TNT and DNT[5]. As nitroaromatic compounds represent the major group of high explosives, this new discovery can potentially have applications in homeland security, counter-terrorism, and land mine detection. Unlike electro-optic polymers[12], poling[13] is not necessary for this application, and the chromophores are randomly oriented in the polymer matrix. This naturally avoids possible thermal stability issues associated with poled nonlinear optical polymers. Significant changes in the UV-VIS absorption spectra and indices of refraction have been observed when the polymer is exposed to part-per-billion (ppb) levels of DNT vapor in air. A compact explosive sensor based on microresonators made in chromophore-doped polymers by photobleaching is presented. When a polymer that contains second-order nonlinear optical chromophores is poled and the chromophores are preferentially oriented, the polymer exhibits macroscopic second-order nonlinear optical properties and becomes an electro-optic polymer, and the polymer can be used for electro-optic sensors. The index of refraction of an EO polymer varies with external electric field. A broadband all-dielectric electric field sensor based on electrically poled chromophore-doped polymer[14] is described.

This chapter covers the recent progress in optical sensors based on microresonators made with chromophore-doped polymers. The chapter begins with basic theory of optical microring resonator sensors and a description of a simple approach that has been found effective in optimizing the design of microring resonators. Photobleaching and electron-beam bleaching fabrication of microring resonators with submicrometer feature size are described. These techniques are unique to chromophore-doped polymers. Finally two practical examples, a trace explosive sensor using an unpoled polymer and a broadband electric field sensor using a poled polymer will be discussed in detail.

2.2 Microring Resonator Theory and a Design Method

2.2.1 Basic Theory of Microring Resonators

Basic microring resonators consist of a ring waveguide as a resonant cavity and one or two bus waveguides, which are coupled to the ring waveguide and provide input and output ports for the device. When the round-trip phase shift of the ring waveguide is equal to multiples of 2π, circulating lights of different cycles in the ring interfere constructively with each other and build up resonance modes. As resonance occurs, light from the input port is trapped in the ring and the optical power of these resonance wavelengths are minimized at the through port, as shown in Fig. 2.1a, b.

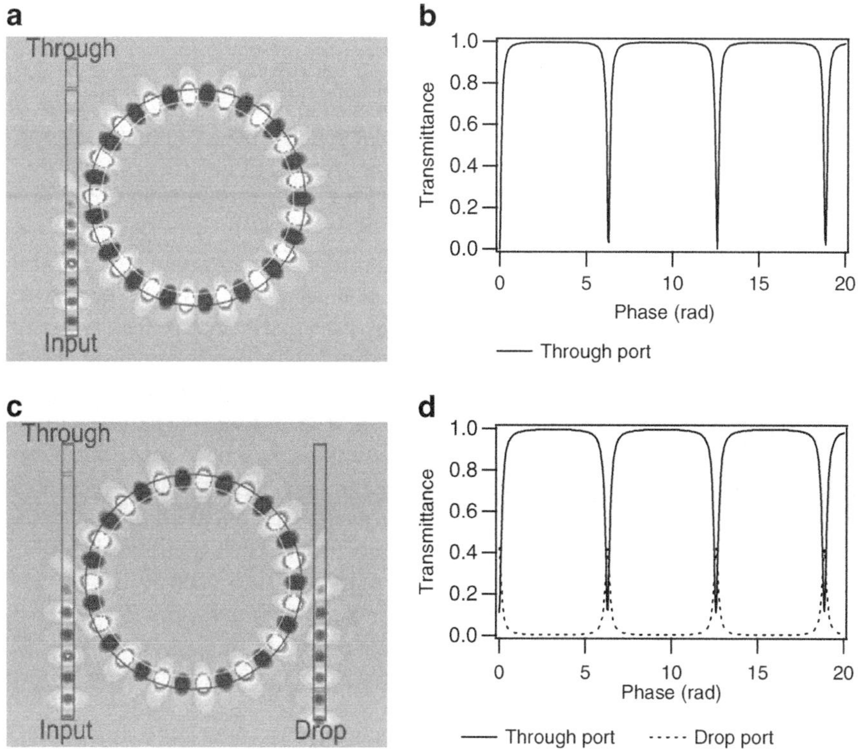

Fig. 2.1 Basic microring resonator structures and their output characteristics. (**a**) One ring resonator and one bus waveguide and (**b**) the corresponding resonance spectrum at the through port. (**c**) One ring resonator and two bus waveguides and (**d**) the corresponding resonance spectrum at the through port and drop port. Reprinted from Ref. 15 with permission. © 2008 Institute of Electrical and Electronics Engineers

If there is a second bus waveguide coupled to the ring, the light trapped in the ring will be coupled out to the second bus waveguide and resonance peaks will be present in the spectrum at the drop port (Fig. 2.1c, d). Because the round-trip phase shift of the ring waveguide is a function of both wavelength and the index of refraction of the waveguide material, the output power of the microring resonator varies with both wavelength and index of refraction. Resonance in the wavelength domain is the basis of channel filters and lasers, and the resonance shift due to the change of the index of refraction, which can be induced by external electric field or chemical analytes, forms the basis of optical switches, modulators, and sensors.

2.2.2 Sensitivity of Microring Resonator Sensors

The sensitivity in detecting the index of refraction change can be estimated using the maximum slope of the resonance and a minimum detectable change in the power at the optical output end. The minimum detectable change is usually limited by the noise equivalent power of the optical detector. For slow-varying signals such as outputs from chemical, biological, and temperature sensors, a relative intensity change of 0.2% is generally accepted as a detection limit criterion for optical sensors such as surface plasmon resonance (SPR) sensors[16–18].

The sensors are intended to be built with microresonators of high extinction for a large dynamic range; in such case the normalized optical power at the output of the fiber can be express as[19]

$$I_{\text{out}} = \frac{2\alpha^2(1-\cos(\theta))}{1+\alpha^4-2\alpha^2\cos(\theta)}, \tag{2.1}$$

where α is the round-trip loss factor ($0 \leq \alpha \leq 1$). θ is the round-trip phase shift of the resonator

$$\theta = \frac{2\pi^2 n_{\text{eff}} d}{\lambda_0}, \tag{2.2}$$

where n_{eff} is the effective index of the resonant mode, d is the diameter of the resonator, and λ_0 is the free-space wavelength.

Using a free-space wavelength of 1.55 μm (at which fixed and tunable wavelength telecom diode lasers are available at relatively low cost, and also in the low loss window of optical fiber), an effective index of refraction of 1.45, and a ring diameter of 50 μm, the full width half maximum $\Delta\theta_{\text{FWHM}}$ (with coupling taken into account) can be calculated numerically using (2.1). From $\Delta\theta_{\text{FWHM}}$ the loaded Q factor and the finess as a function of the total optical loss of the resonator are calculated and plotted in Fig. 2.2a. The Q factor is the ratio of the resonant wavelength over the wavelength full width half maximum of the resonance. The finess is the ratio of the separation of adjacent resonant wavelengths over the wavelength full width half maximum. Finess is an important factor when building sensor arrays with large number of microresonators on the same bus waveguide. Figure 2.2a shows that for a typical waveguide loss (due to absorption, surface scattering, and bending) of 1 dB/cm, a loaded Q of 10^5 can be reached. A Q of 10^6 requires a waveguide loss of 0.1 dB/cm. This level of low loss is challenging, yet still possible since optical waveguides with a low loss of 0.03 dB/cm (at 1.55 μm. 0.01 dB/cm at 830 nm, including the absorption and scattering losses) have been demonstrated previously[20–22].

The steepest slope can be found from the derivative $dI_{\text{out}}/d\theta$. Using the steepest slope and the criterion of 0.2% relative change of output intensity[16–18], the minimum detectable refractive index changes at various Q are calculated and given in Fig. 2.2b. With a Q factor of 10^6 (total loss of 0.1 dB/cm), an index of refraction change as small as 1.75×10^{-9} RIU (refractive index units) can be detectable, which is about 100 times better than that of typical SPR sensors. If the total loss could be kept below 0.04 dB/cm (based on the 0.03 dB/cm loss demonstrated from straight waveguides[20,21] plus 0.01 dB/cm for waveguide bending

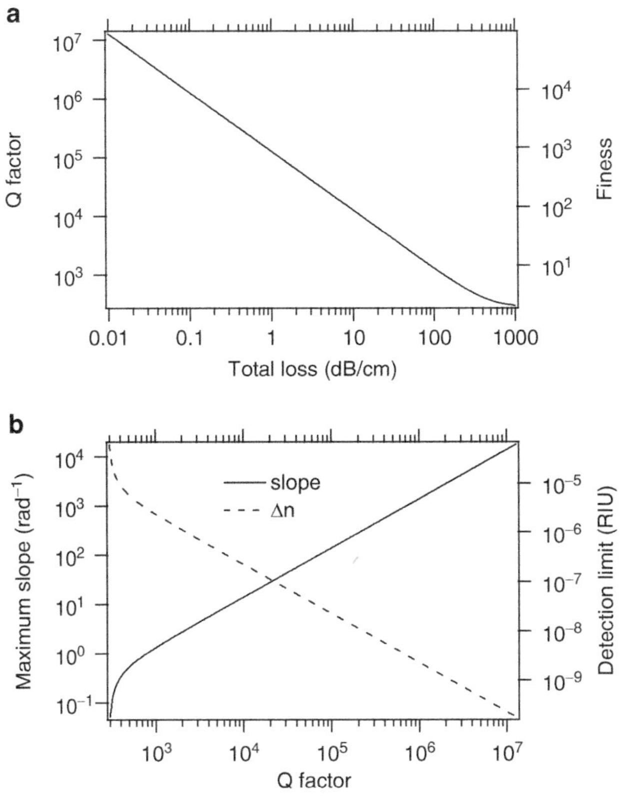

Fig. 2.2 (**a**) The loaded Q and finess of a 50 μm diameter ring resonator as a function of the total waveguide loss. The coupling loss is taken into account in the calculation (critical coupling is assumed). The total waveguide loss includes material absorption loss, waveguide bending loss, and scattering loss due to surface roughness and index inhomogeneity. (**b**) Maximum slope in rad^{-1} and sensitivity (minimum detectable change in refractive index) as functions of Q. The data are calculated using the 0.2% intensity variation criterion, a 50 μm ring diameter, and 1.55 μm wavelength

loss), the theoretical detection limit would be 7×10^{-10} RIU. For a given microring resonator, sensitivity scales with $1/\lambda_0^2$. Using a shorter wavelength could also significantly improve the sensitivity. As a side benefit of using shorter wavelengths, the optical losses of polymers are lower at shorter wavelengths than at telecom wavelengths. Polymer waveguides with loss as low as 0.01 dB/cm has been reported at the wavelength of 830 nm[22].

The above estimation is based on slow varying quasi-DC signals. For fast oscillating AC signals such as acoustic, ultrasound, and radio frequency (RF) electromagnetic waves, advanced detection techniques (low noise amplification, filtering, heterodyne, phase locked loop, etc.) developed in these technical areas and in optical telecom could be used to reach sensitivities much higher than the sensitivity based on the 0.2% DC detection criteria[23].

For an RF electric field sensor, bandwidth (speed) is another important requirement. The speed of a resonator is determined by the cavity ring down time $\tau = Q\lambda/c$. As in any devices using resonating structures, a trade off between sensitivity (which increases with Q) and speed (which decreases when Q increases) has to be made. At the wavelength of 1.55 μm, Q factors of 10^4 and 10^6 translates to a sensor bandwidths of 20 GHz and 200 MHz, respectively.

2.2.3 A Simplified Approach to Designing Microring Resonators

Q-factor and extinction ratio are two most important figures of merit of resonators. The Q is mainly affected by the scattering and bending loss of the ring waveguide, and the maximum extinction ratio is achieved when the critical coupling condition[19] is met. To vigorously model the complete microring resonator device the finite-difference time-domain (FDTD) algorithm needs to be used. FDTD solves the Maxwell's equations without making major approximations or assumptions to compromise the accuracy. However, FDTD algorithm is very computationally intensive and cluster computers are normally required to realistically model a microring resonator. The outputs from FDTD codes also require a great deal of postprocessing to extract device performance information such as resonant wavelengths, the Q-factor, the resonance extinction ratio, and the free spectral range (FSR).

A time-saving and much simpler approach is to divide the microring resonator into its building blocks and to consider the microring resonator as an optical circuit made of directional couplers and curved waveguides, as illustrated in Fig. 2.3. Beam propagation method (BPM) algorithm can separately calculate the coupling

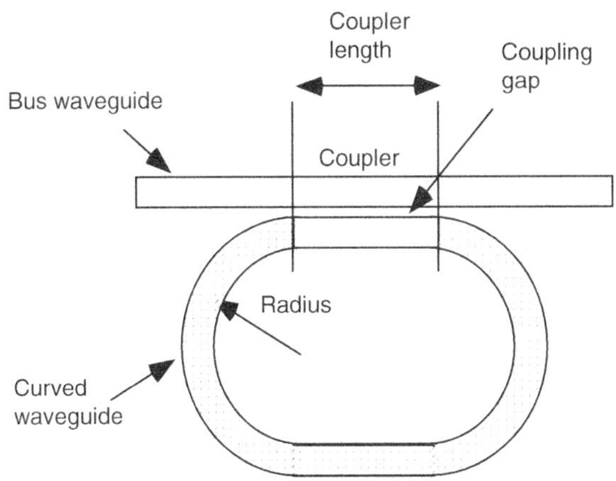

Fig. 2.3 By regarding a basic microring resonator as an optical circuit composed of a directional coupler and a curved waveguide (the cross hatched section), the characteristics of the microring resonator can be analyzed using simpler BPM code and transfer matrices instead of computation intensive FDTD simulation. Reprinted from Ref. 15 with permission. © 2008 Institute of Electrical and Electronics Engineers

ratio of directional couplers, the propagation constant, and loss of curved waveguide with satisfactory accuracy. A BPM code is much faster and requires much less computer memory than an FDTD code and the BPM simulation can be easily carried out on a standard personal computer. The results from the BPM simulation are then used to construct transfer matrices [24,25] that represent couplers and the ring waveguide. Transmission properties of every specified output port can be obtained by using the transfer matrices. For example, the calculated round-trip loss factor $|\alpha|^2$ for various index contrast and radius of curvature and the coupler attenuation $|t|^2$ vs. coupler length and gap, are plotted in Fig. 2.4a, b, respectively. From Fig. 2.4a one can find the possible combinations of index contrast and radius of curvature to maximize the $|\alpha|^2$ and Q. The figure clearly shows that smaller radius requires larger index contrast. For any given index contrast, there exists a peak of $|\alpha|^2$ and an optimal radius. Below the optimal radius the bending loss increases sharply, and above the optimal radius the perimeter of the ring increases and the round trip loss increases with it. After the peak $|\alpha|^2$ is determined the next step is to match $|t|^2$ with $|\alpha|^2$ using Fig. 2.4b to find the possible combinations of coupler length and gap to meet the critical coupling condition for maximum extinction ratio. Wider (>1 μm) gap or zero gap (MMI) couplers are often preferable because they have relatively large critical feature size and are easier to fabricate. By using this simple procedure one can quickly determine the approximate values of the key design parameters. From there an optimal design can be found by experimenting a number of design variations around these values.

2.3 Waveguide Fabrication Techniques Unique to Chromophore-Doped Polymers

2.3.1 Photobleaching

Traditional methods to fabricate polymer waveguide devices is patterning a photoresist etch mask on the polymer film, and then using RIE to transfer the photoresist pattern to polymer film, and finally removing the remaining photoresist on the waveguide. It is a process of multiple fabrication steps. The solvent in the photoresist often reduces the surface quality of the polymer and in some cases can dissolve the polymer and destroy the film. In these cases a method to fabricate polymer waveguides without using solvent or wet chemistry is highly desirable. For both poled and unpoled chromophore-doped polymers, optical waveguide can be patterned in a single step using ultraviolet light or electron-beam.

Conjugated charge transfer chromophores are more polarizable than the polymer host and polymers doped with chromophores have index of refraction higher than that of the host polymer alone. High-energy ultraviolet photons can break chemical bonds and destroy the conjugated-charge transfer system of the organic chromophores. Consequently, the index of refraction of the chromophore-doped optical polymer is reduced. Photobleaching has been used to make single mode channel

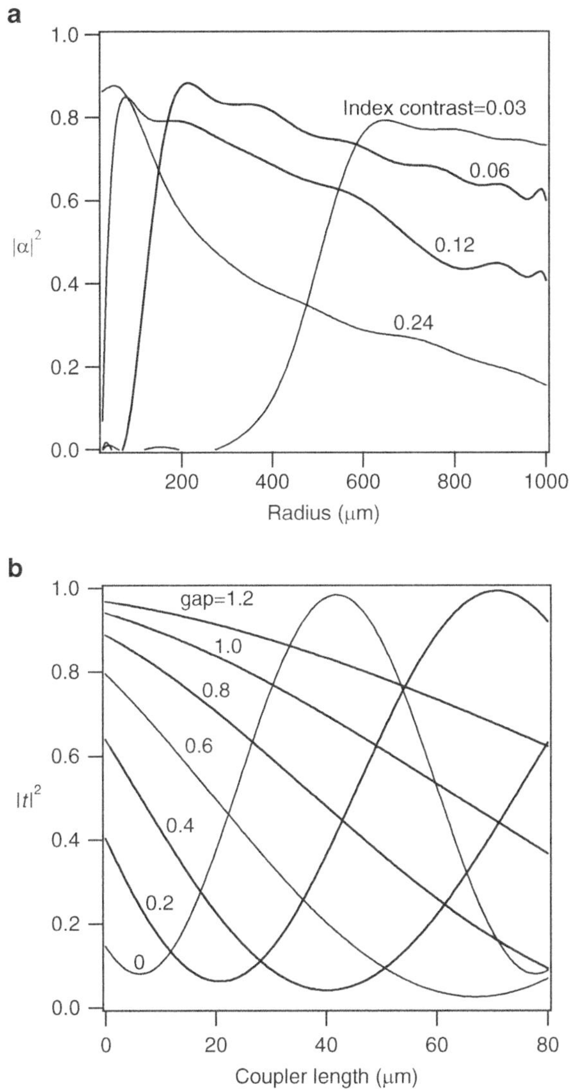

Fig. 2.4 BPM simulation results for polymer microring resonators: (**a**) Round trip attenuation vs. the core-cladding index contrast and radius of curvature. (**b**) Coupling induced attenuation vs. the length and gap of the coupler. The width and height of the waveguide is 2 μm. The wavelength used in the calculations is 1.55 μm. Reprinted from Ref. 15 with permission. © 2008 Institute of Electrical and Electronics Engineers

waveguides[26] and to perform postfabrication trimming of polymer waveguide devices[27]. Photobleaching through a photomask forms optical waveguides in a single step. It not only simplifies the device fabrication, but also reduces the sources of error from multiple fabrication steps. Photobleaching is a dry process and does not use solvents and wet chemicals.

The effectiveness of photobleaching in fabricating microring resonators[28] has been demonstrated in an amorphous polycarbonate (APC) doped with 25 wt.% of AJL8 chromophore (AJL8/APC). The chemical structure of AJL8 is given in Fig. 2.5. Figure 2.6 shows the change of the absorption spectra and refractive index of the AJL8/APC polymer after various degrees of exposure to UV light of 365 nm wavelength, indicating a permanent change of the material. In Fig. 2.6a, the absorption peak at 714 nm is the characteristic peak of the conjugated charge transfer system of the π-orbital electrons of the chromphore. The disappearance of the characteristic peak clearly indicates an irreversible photochemical decomposition of the chromophores. A decrease of refractive index as large as 0.08 RIU was observed. According to Fig. 2.4a, such index contrast is sufficient to make microring resonators of radius as small as 200 μm.

Photobleaching also reduces the thickness of the EO polymer by 80 nm, or 6%. The original thickness of the polymer film was 1.3 μm. Atomic force microscopy (AFM) picture (Fig. 2.7) of a photobleached waveguide shows that the corresponding cross-section profile matches the near-field diffraction pattern of the opaque line on the photomask. The decrease of thickness in the photobleached area is because some photodecomposition products are gaseous and diffuse out of the polymer thin film, and the volume of the polymer is reduced. AFM scan indicates that the photobleached region has the same level of surface smoothness as the unbleached region. Single mode waveguides were obtained for waveguides of width smaller than 6 μm. Waveguide losses were measured by cutback method. The propagation loss was found to be 2.3 dB/cm and 2.0 dB/cm for the horizontal (TE) and vertical (TM) polarizations, respectively. These loss values are similar to the losses of the ridge waveguides made from the same material by photolithography and RIE. Coupling loss between the fiber and the waveguide was 4 dB, which is also typical for the polymer waveguides with same mode size.

Based on the measurement and BPM simulation results, microring resonators were designed using the method described in the previous section. The ring waveguides are race track shaped and have 80 μm coupling length and 200 μm radius. The width of the waveguide is 4 μm. The AJL8/APC was spin cast from solution to form a film about 1 μm in thickness on a silicon wafer with 4 μm silicon dioxide. The silicon dioxide functions as a lower cladding for the polymer waveguides. Devices with coupling gaps ranging from 0.6 to 1.2 μm were fabricated and tested. A photomask made by electron-beam lithography was used to define the waveguides. The highest extinction ratio was obtained from resonators with a coupling gap of 1.1 μm. The resonance extinction ratio is 16 dB for TE and 7 dB for TM polarization, as shown in Fig. 2.8. The FSR between adjacent resonances was approximately 1.08 nm, which is in good agreement with the theoretical calculation. The Q factor of the resonators was obtained by fitting the theory to the experimental data and the Q was found to be in the range of 6,500–8,100. The insertion loss at the peak transmission was 8.7 dB, which is mainly due to the fiber-to-waveguide coupling at the input and output ports of the device.

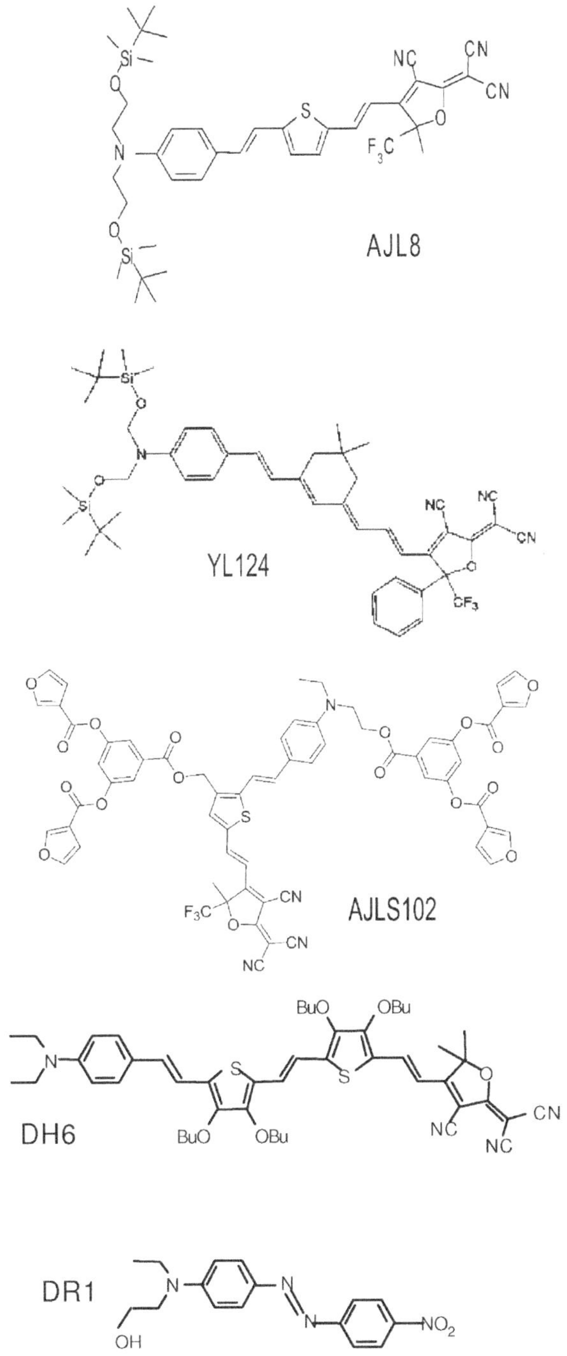

Fig. 2.5 Chemical structures of the chromophores used in this work. Reprinted from Ref. 15 with permission. © 2008 Institute of Electrical and Electronics Engineers

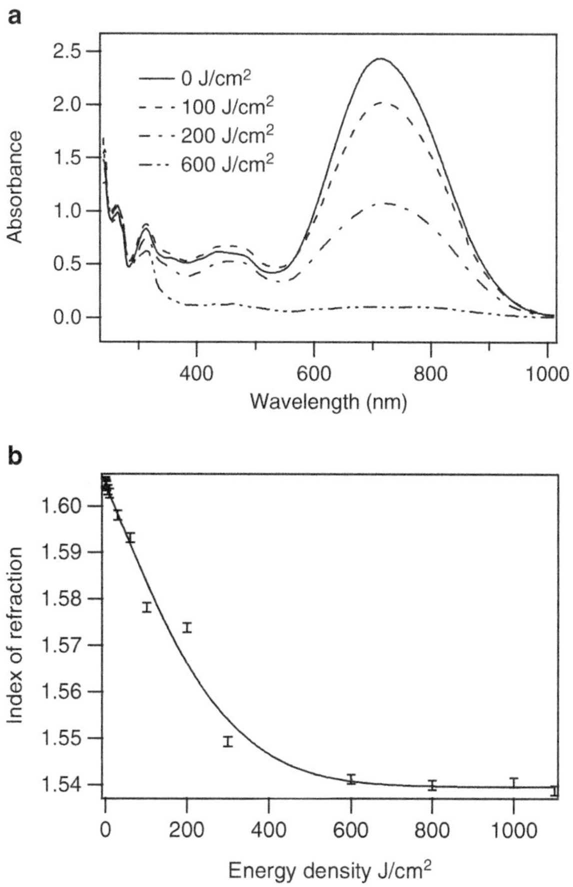

Fig. 2.6 The characteristic absorption peak disappears (**a**) and the refractive index decreases (**b**) as the photobleaching energy density increases. The refractive index was measured at the wavelength of 1.55 μm. Reprinted from Ref. 15 with permission. © 2008 Institute of Electrical and Electronics Engineers

2.3.2 Electron-Beam Bleaching

The resolution limit of photobleaching using ultraviolet light is about 1 μm. Sometimes the optimal design of microring resonator calls for a coupling gap much smaller than 1 μm. For instance, in order to achieve a large FSR the round-trip optical path needs to be minimized and therefore the coupler needs to have smaller gap in order to make the coupler shorter. Couplers with submicron coupling gaps can be made by electron-beam bleaching. Electron beam irradiation induced a refractive index decrease of about 0.06 RIU in polymethyl methacrylate (PMMA) polymer doped with YL124 chromophores[29] (Fig. 2.5). The degree of index change is similar to what is observed with photobleaching. Similarly, a decrease in the

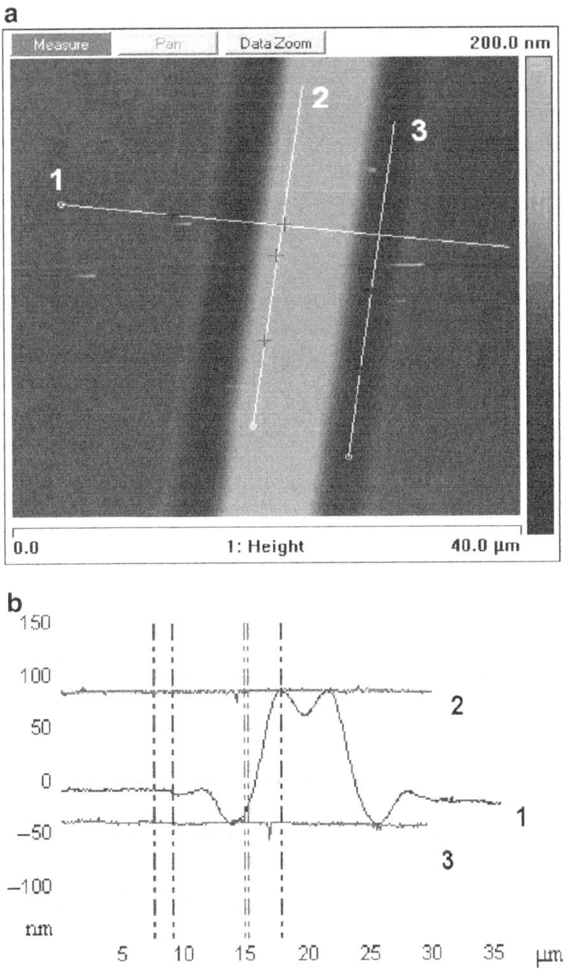

Fig. 2.7 The topographic view of a photobleached waveguide using atomic force microscopy (**a**), and the height profile along three scan lines (**b**). Reprinted from Ref. 15 with permission. © 2008 Institute of Electrical and Electronics Engineers

thickness of the polymer film after e-beam bleaching is also observed. Electron beam provides nanometer scale resolution, and microring resonators with critical feature size of 100–200 nm are readily achievable. E-beam bleaching can be carried out by a standard scanning electron microscope (SEM) with a pattern generator. Without the need of expensive high resolution photomasks, different designs of microring resonator devices can be generated easily with electron beam bleaching. This is especially effective for fast prototyping and design optimization. As with the photobleaching, electron beam bleaching is a single-step process and does not use wet chemicals. The waveguides are formed by exposing a strip of 20–50 µm wide on each side of the waveguide to create low index clad for lateral confinement,

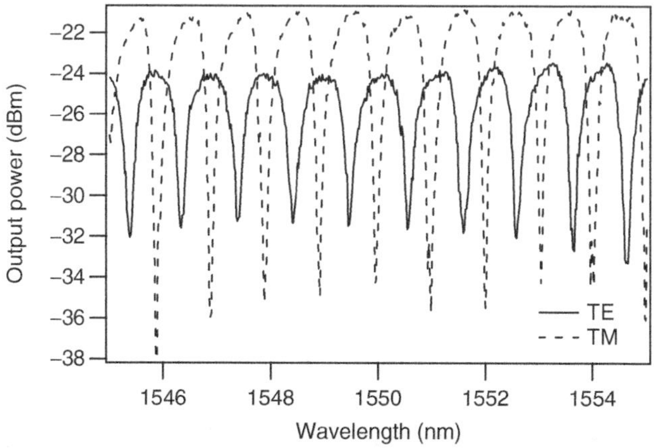

Fig. 2.8 Transmission spectrum of a microring resonator with 1.1 μm coupling gap. Reprinted from Ref. 15 with permission. © 2008 Institute of Electrical and Electronics Engineers

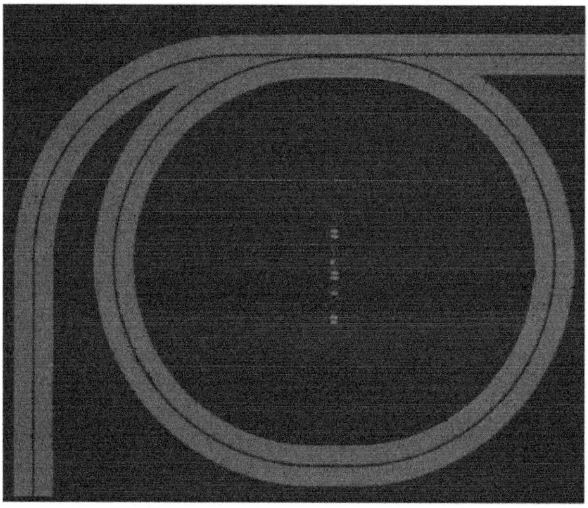

Fig. 2.9 A microscope image of a microring resonators fabricated with the electron beam bleaching method. The lighter areas in the image are bleached by electron beam. Reprinted from Ref. 15 with permission. © 2008 Institute of Electrical and Electronics Engineers

as shown in Fig. 2.9. It is found that it is not necessary to coat the polymer thin film with metal or conductive polymer during the e-beam exposure. This is different from SEM imaging of polymer samples, in which a conductive coating is usually required to prevent surface charging.

For a microring resonator with a waveguide width of 5 μm and coupling gap of 200 nm, a resonance extinction ratio of 12 dB for TE polarization and 9 dB for TM polarization has been achieved, as shown in Fig. 2.10. The ring resonator had a race track shape with circular sections of 500 μm radius and straight coupling section of 100 μm in length. The width of the two exposed strips is about 50 μm.

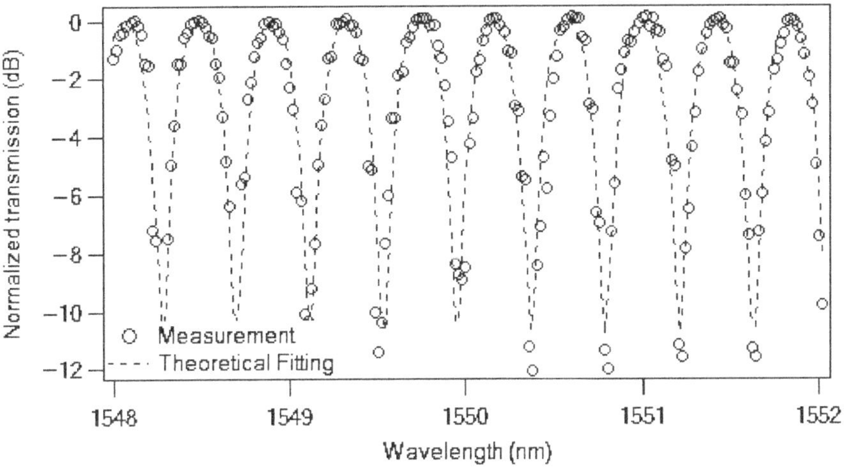

Fig. 2.10 Normalized transmission spectrum of a microring resonator made by e-beam bleaching. The input light is TE polarized. Reprinted from Ref. 15 with permission. © 2008 Institute of Electrical and Electronics Engineers

The polymer used is PMMA doped with YL124 chromophore at 20 wt% and the thickness of the polymer film is 2 μm. The film was spin coated on silicon substrate with 5 μm of thermal oxide as lower cladding. The microring resonator was patterned with a FEI Sirion SEM equipped with a nanometer pattern generation system. Accelerating voltage of 30 kV and beam current of 5 nA were used to reduce the e-beam writing time. For this polymer, the change of index saturates at e-beam dose above 500 $\mu C/cm^2$ and the dose of electron beam exposure used to pattern the microring resonator was 700 $\mu C/cm^2$.

2.4 Change of Linear Optical Properties Caused by Nitroaromatic Explosives

Recently we have observed that the trace vapors of electro-negative nitroaromatic explosives such as TNT and DNT cause significant changes in the linear optical properties, i.e., index of refraction and absorption spectrum, of certain chromophore-doped polymer thin films. Figure 2.11 shows the UV-VIS-NIR absorption spectra of a thin film of polymethylmethacrylate (PMMA) doped with chromophore DH6[30] (shown in Fig. 2.5) at 20 wt.% after various duration of exposure to DNT vapor in the air. The polymer films were spin coated from solution on glass slides, and dried overnight in vacuum at 60–70°C. The thickness of the film is 1 μm. The film is not electrically poled and the chromophores are randomly orientated in the matrix of the PMMA host. The main absorption peak originally at 630 nm shows a red shift as large as 70 nm and a reduced peak absorbance from 1.53 to 0.3 after

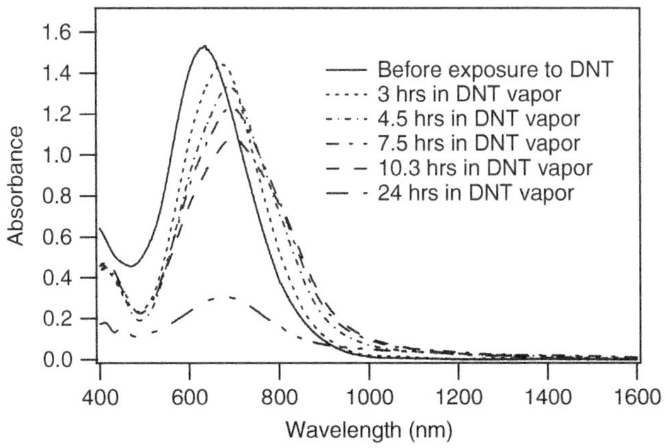

Fig. 2.11 Absorption spectrum of a DH6/PMMA thin film after it is exposed to DNT vapor at 65°C for various periods of time. The peak at about 630 nm is the characteristic absorption peak of the chromophore. Reprinted from Ref. 33 with permission. © 2008 American Chemical Society

the film is exposed to saturated DNT vapor in air at 65°C for 24 h. Similar changes have also been observed at room temperature although the same degree of change took place about ten times more slowly. There is also a noticeable change of color of the film after it is exposed to DNT vapor. The color of the polymer film changes from dark blue to light purple. The changes in the absorption spectrum and the corresponding color changes were observed in a number of host polymers doped at various levels with several different chromophores (DH6[30], AJL8[31], DR1[32]) of the D–π-A structure shown in Fig. 2.5. The type of host polymer (PMMA, APC, and bisphenol-A-polycarbonate) and solvent (chloroform and cyclopentanone) used to make the polymer solutions have only minor effect on the polymer thin film's response to DNT.

Accompanying the change in the absorption spectrum is a change in the index of refraction. Chromophores of D–π-A structure are highly polarizable and contribute to the index of refraction of the polymer. The polarizability, and therefore the index of refraction, can be affected by the electric field acting on the chromophore due to the local environment. The local electric field environment can, in turn, be affected by nitroaromatic explosive molecules that are highly electronegative and polar. To study the sensitivity and specificity of the polymer as trace explosive sensor materials, similarly prepared thin films of 20 wt.% AJL8 chromophore in APC host were, respectively, exposed to saturated vapors of DNT as well as other chemicals that are a common cause of false positives of explosives sensors. These chemicals include pollutants commonly found in the environment, e.g., salts, chemical fertilizers, and detergents. The thin film samples were individually sealed in separate containers with each of these chemicals, and thereby exposed to, at most, a saturated vapor concentration of these chemicals at ambient condition. The concentration of saturated DNT vapor in air at room temperature is known to be 100–120 parts-per-billion[34]. The index of refraction of these thin film samples was

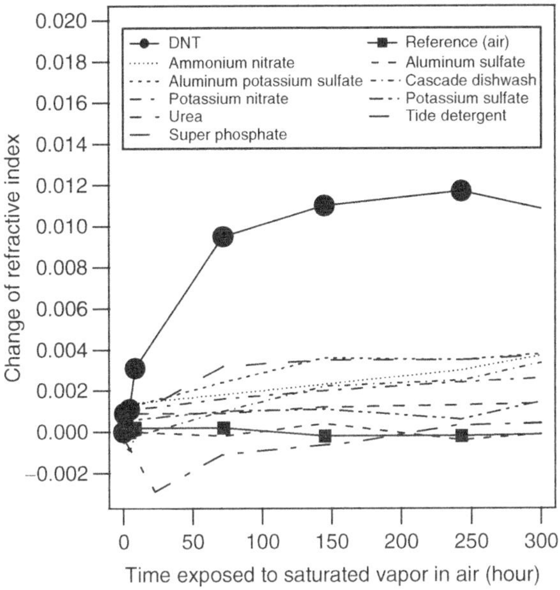

Fig. 2.12 The change of the index of refraction due to DNT is much larger than the change due to other chemicals found in the environment as pollutants. The polymer is 20 wt.% AJL8 chromophore in APC. Reprinted from Ref. 33 with permission. © 2008 American Chemical Society

measured at the wavelength of 1,550 nm with a prism coupler (Metricon 2010) after various time periods of the exposure to the chemical vapors. Figure 2.12 shows the index of refraction change of these samples over time. These results indicate that this polymer has good specificity to DNT. The maximum refractive index change of this polymer is 0.012 RIU at 100–120 ppb of DNT vapor concentration at room temperature. Considering that the index of refraction can be measured with a sensitivity of 10^{-7} RIU with well-engineered polymer microring resonators[11], a detection limit based on the index change of this polymer of 1 parts-per-trillion is practical. In general, the change in the index of refraction can be measured more accurately than the change in the absorption spectrum, for example, using microring resonators and fiber Bragg gratings. Therefore, sensors based on the refractive index change could be more sensitive than the sensors based on the absorption spectrum change.

The details of how nitroaromatic explosive molecules interact with the chromophores in the polymer matrix requires further study. Initial observations suggest that because nitroaromatic explosive molecules are highly electron-deficient, that chromophores have an electron-rich donor and bridge, and that both nitroaromatic explosives and chromophores are highly polar, explosive molecules and chromophores have a strong tendency to interact with each other. The interaction between explosives and the polymer takes place in two steps. In the initial step nitroaromatic explosive molecules create a more polar environment around the chromophores. The increased polar environment produces a solvatochromic red-shift of the absorption peak of chromophores, and a relatively small (on the order of 10^{-3} RIU) change in the index of refraction of the polymer. The effects of this step are reversible and the polymer recovers after it is removed from explosive vapors. If the polymer continues to be exposed to explosive vapors, more explosive molecules diffuse into the polymer, and the interaction enters the next step. Chromophores and explosive molecules start to form charge transfer complexes. Explosive molecules start to alter the band structure and disrupt the charge transfer bands of the chromophore, and the absorption peak begins to diminish, accompanied by a large (on the order of 10^{-2} RIU) change in the index of refraction. Both interactions are evident in Fig. 2.11, with an initial red-shift of the absorption peak followed by the subsequent reduction of the height of the absorption peak. The strong interaction of the second step was only observed in solid thin films, and the spectral change does not recover even after the film is subsequently baked at temperatures above 100°C for several days, indicating that the interaction between the chromophores and explosives is stronger than the thermal energy. Simply adding nitroaromatic explosives to the chromophore solution does not cause noticeable change of the chromophore absorption peak of the solution. Furthermore, if the polymer film that undergoes the absorption peak reduction after long exposure to explosives is redissolved in the solvent, the absorption spectrum returns to its original shape, as shown in Fig. 2.13. This recovery suggests that the interactions between chromophore and nitroaromatic explosives are physical in nature, rather than a chemical reaction. When the film is redissolved in the solvent, solvent molecules also interact with the chromophores and compete with explosives molecules. Since solvent

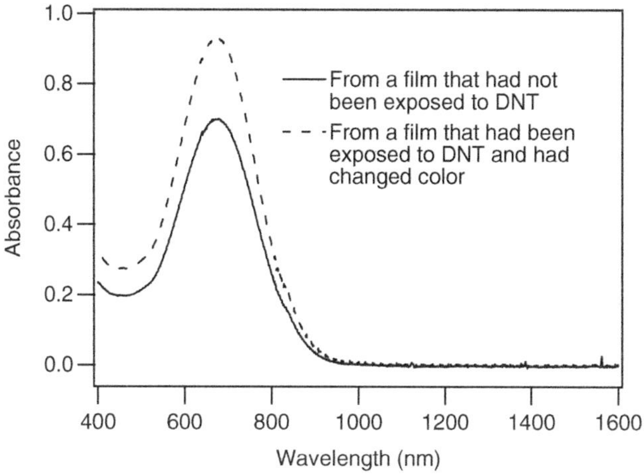

Fig. 2.13 Comparison of the absorption spectra of two DH6/PMMA thin films redissolved in cyclopentanone. One is a pristine film without being exposed to DNT and the other is a film that has been exposed to DNT and has changed color. After they are redissolved in solvent, the absorption spectra of both samples have the same peak wavelength and shape. The difference in the height of the peak is due to the difference in the concentration of the two polymer solutions, and is not important in the comparison. Reprinted from Ref. 33 with permission. © 2008 American Chemical Society

molecules in the polymer solution far outnumber explosive molecules, the solvent separates the explosive molecules from the chromophores.

For the AJL8/APC polymer, saturation was not achieved until 250 h of exposure to the DNT vapor. However, the sensing polymer produces a detectable amount of refractive index change within a few minutes of exposure to DNT vapor of 100 ppb concentration. The response rate is thought to be limited by the diffusion of DNT molecules into the polymer. In the initial experiments reported here PMMA and polycarbonate were used as the host polymers. These polymers are relatively less permeable to DNT. The response rate is expected to improve if a more permeable polymer host is used.

2.5 Poling and Electro-Optical Polymer

The electro-optic property of EO polymers comes from the NLO chromophores. When these chromophores are preferentially aligned to break the centrosymmetry of the material, the molecular level microscopic NLO effect of the molecules translates to the macroscopic second-order NLO effect of the polymer material. The poled material exhibits a strong macroscopic electro-optic effect.

An external electric field induces a change in the dipole moment known as polarization. However, the induced dipole moment has a saturation behavior which makes the polarization deviate for the linear response to the external E-field, and the total polarization p of the chromophore in an external electric field can be written as

$$p_i = \alpha_{ij}E_j + \beta_{ijk}E_jE_k + \gamma_{ijkl}E_jE_kE_l + \dots, \tag{2.3}$$

where α_{ij} is the linear polarizability, β_{ijk} is the first hyperpolarizability, and γ_{ijkl} is the second hyperpolarizability. $E_{i,j,k}$ are electric field components in the molecular coordinate system. The β term gives rise to the second-order NLO effect.

A polymer system with NLO chromophores randomly oriented does not allow even order nonlinear optical effect due to the central symmetry of the system. In EO polymers, these individual chromophores need to be aligned toward one direction by poling[35–37] to break the central symmetry of the polymer system. Poling is usually done by applying a strong DC field to the polymer. The poling field exerts a torque on these dipolar chromophores and rotates them so that they become preferentially orientated in the direction of poling field. The EO tensor of an EO polymer poled in the z direction is given by

$$\begin{pmatrix} 0 & 0 & r_{13} \\ 0 & 0 & r_{13} \\ 0 & 0 & r_{33} \\ 0 & r_{13} & 0 \\ r_{13} & 0 & 0 \\ 0 & 0 & 0 \end{pmatrix}. \tag{2.4}$$

r_{33} is typically several times greater than r_{13}. EO polymers in general have r_{33} in the range of 10–500 pm/V[38]. When an electric field E_z is applied to an EO polymer and the field is along the poling axis z, the refractive index of the EO polymer will change due to the linear electro-optic effect (the Pockels effect). The changes of ordinary index n_o (for light polarized perpendicular to the poling axis) and extraordinary index n_e are given by

$$\Delta n_o = -\frac{1}{2} n_o^3 r_{13} E_z \tag{2.5}$$

and

$$\Delta n_e = -\frac{1}{2} n_e^3 r_{33} E_z. \tag{2.6}$$

2.6 Microresonator Trace Explosive Sensor

The sensor is a microring resonator fabricated in the polymer thin film of 25 wt.% AJL8/APC polymer fabricated with photobleaching method. The width of the waveguide is 4 μm and the ring is in a race track shape. The straight sections are 80 μm long and the curved sections have a radius of 200 μm. The coupling gap is 1.1 μm. When the sensor is exposed to saturated DNT vapor at room temperature, a resonance shift toward longer wavelength was observed (Fig. 2.14a). The shift of resonance indicates an increase in the refractive index of the polymer consistent with the thin-film refractive index measurement results. After 1 h of exposure the shift was 1.2 nm and the maximum effect has not yet been reached. After removing the sensor from the DNT atmosphere, the sensor properties returned to their initial conditions (Fig. 2.14b). The sensor response was determined by measuring the wavelength of a resonance mode using an optical spectrum analyzer. A typical optical spectrum analyzer can measure the resonant wavelength to an accuracy of 0.05 nm. This accuracy corresponds to an accuracy of 0.00005 RIU for index of refraction measurements, and a sensitivity of ~5 ppb. Polymer microring resonators with optimized design and fabrication conditions and more accurate measurement instrument have demonstrated[11] an index of refraction measurement accuracy of 10^{-7} RIU, which corresponds to a detection sensitivity of a few parts-per-trillion (ppt).

In addition to polymer waveguide microring resonators, we also made trace explosives sensors by incorporating a fiber Bragg grating with the same polymer. The fabrication begins on a commercial fiber Bragg grating. The grating section of the fiber is first etched with hydrofluoric acid and the diameter of the fiber is reduced from 125 μm to about 30 μm. This allows the evanescent tail of the fiber

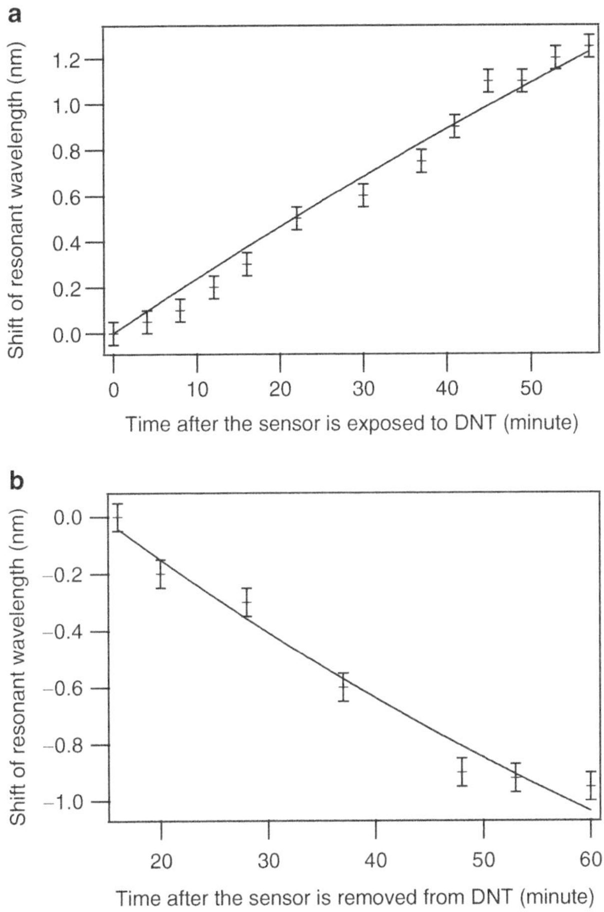

Fig. 2.14 The output of the sensor when it is exposed to (**a**) and removed from (**b**) saturated DNT vapor at room temperature. Reprinted from Ref. 33 with permission. © 2008 American Chemical Society

mode to extend into the polymer coating to probe the change of the refractive index of the polymer. The polymer is coated on the etched area of the fiber by dipping the fiber in the polymer solution and letting the coating dry with the fiber held in the vertical position. The thickness of the coating is about 1 μm. The change in the index of refraction of the polymer coating shifts the resonance of the fiber, as shown in Fig. 2.15. After the sensor fiber is exposed to saturated DNT vapor in air at room temperature for 16 h the change in the properties of the polymer become so large that not only is the resonance wavelength shifted, but the shape of the resonance is also greatly altered and distorted. Since both microring resonators and fiber Bragg gratings are high Q optical resonators, the microring resonators sensors and fiber Bragg grating sensors should have comparable sensitivity and response time if the sensor film of the same polymer and same thickness is used.

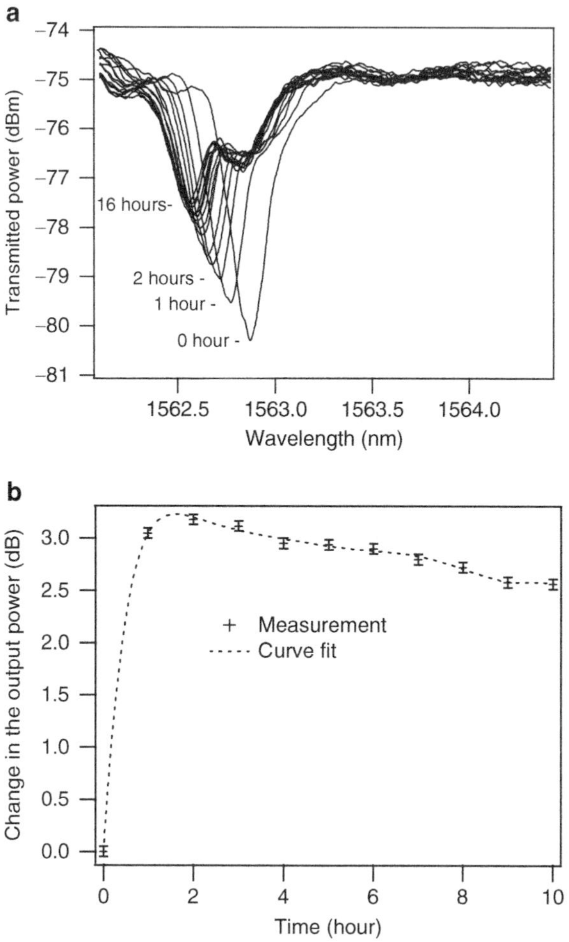

Fig. 2.15 (**a**) The change in the transmission spectrum of a fiber Bragg grating coated with DH6/PMMA in response to DNT. The sensor was exposed to saturated DNT vapor at room temperature. The measurements were taken at time intervals of one hour. The last measurement was taken after the sensor was exposed to DNT for 16 h. (**b**) The change of optical power at the output end of the fiber sensor for light of the wavelength of 1,562.9 nm. Reprinted from Ref. 33 with permission. © 2008 American Chemical Society

2.7 All-Dielectric Radio Frequency Electric Field Sensor

Optical sensors to measure electric field based on EO effect are useful for characterizing and testing high speed electronic devices and integrated circuits, electromagnetic compatibility measurements, electromagnetic imaging with millimeter and/or terahertz wave and radio frequency photonic links[39,40]. A number of optical electric field sensors based on Mach-Zehnder interferometers and polarization modulation have been previously reported[41,42]. Microring resonators of high Q factor could enable higher sensitivity with a compact device size. EO polymers have high EO coefficients, low dielectric constants, and femtosecond time scale EO responses. They have good potential for optical electric field sensors with higher sensitivity, lower invasiveness, and higher bandwidth.

A novel radio-frequency electric field sensor based on an EO polymer microring resonator directly coupled to the core of a side-polished optical fiber, shown in Fig. 2.16, has been successfully demonstrated[14]. A cross-linkable EO polymer that contains AJLS102 chromophore[43] (Fig. 2.5) was used to make the microring resonator. A $\sim$1.5 μm thin film was spin cast on a glass substrate coated with 100 nm of transparent conductive indium tin oxide (ITO). The film was subsequently electrically poled and cross-linked. The ITO was used as a poling electrode. The poled polymer film was then patterned with electron beam lithography and RIE. The patterned racetrack-shaped ring resonators had two semicircular sections of 35 μm radius and two straight sections of 50 μm in length. The waveguide width was 10 μm for better coupling with the fiber core. The transmission spectrum of the fiber was monitored while the EO polymer microring was being aligned to the fiber core at the polished region. After the best resonance extinction was reached the glass substrate of the ring was permanently attached to the fiber using a UV curable epoxy.

At a wavelength tuned at the slope of a resonance dip in the transmission spectrum, the intensity of the output light is modulated by the external radio frequency electric field. The variation of the output light intensity is proportional to the sharpness of the resonance, the intensity of the input light, and the strength of the electric field. The device showed a minimum detectable electric field intensity of 100 mV/m. This sensitivity was limited by the Q-factor of the resonator and the EO coefficient r_{33} of the EO polymer and the noise of the measurement system. The Q of the sensor is approximately 1,100, and therefore the theoretical bandwidth of the sensor is above 100 GHz. The device was tested at frequencies up to 550 MHz (Fig. 2.17), which is the upper limit of out test equipment. The Q is mostly limited by the absorption loss from ITO. EO polymer microring resonators without the ITO have demonstrated a Q-factor on the order of 10^5 [1], which translates to a sensitivity below 1 mV/m and bandwidth of 2 GHz. The excess loss of the fiber is on the order of 1 dB, due mostly to the roughness of the side-polish. Using the optical fiber itself as the bus waveguide and substrate avoids loss, reliability, and cost issues associated with the fiber-to-waveguide end coupling. It also eliminates high propagation loss coming from the long EO polymer waveguides in Mach-Zehnder type sensors. This structure allows the position of the ring to be adjusted to achieve optimal coupling and good resonance. The microring resonator can be either located in the middle of the fiber and operated in transmission configuration, or near one cleaved end of the fiber with a dielectric mirror coated on the end face and operate in reflection configuration like a needle-shaped EO probe. Using two-photon polymerization, it is also possible to make multiple rings directly on the side-polished flat surface of a free-standing fiber to form a sensor array, as shown in Fig. 2.16c.

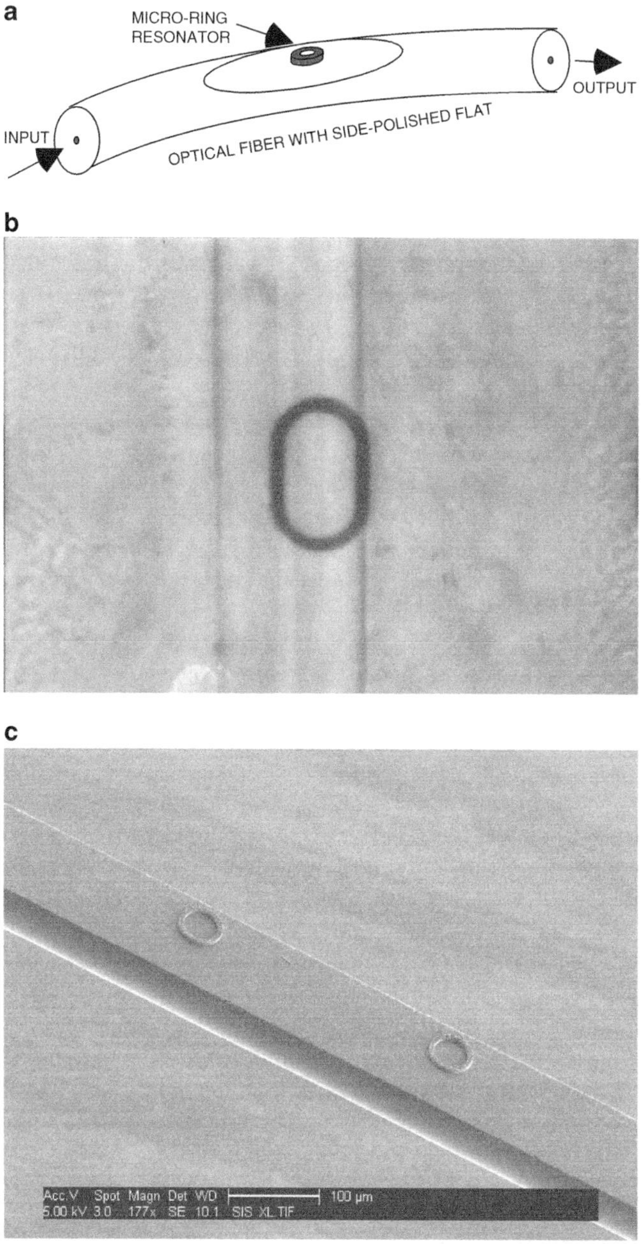

Fig. 2.16 (**a**) A schematic drawing of polymer microring resonator couple to a side-polished optical fiber. (**b**) A microscope image of the fabricated EO polymer electric field sensor. (**c**) SEM image of resonators fabricated on the polished flat of a free-standing fiber. The scale bar in the picture represents 100 μm. Reprinted from Ref. 15 with permission. © 2008 Institute of Electrical and Electronics Engineers

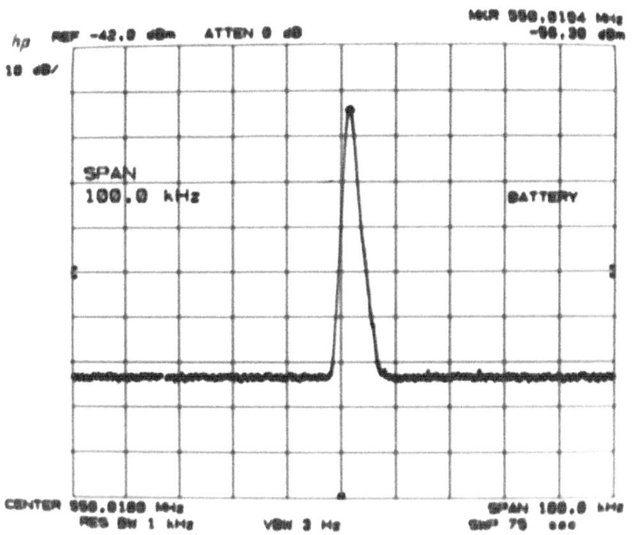

Fig. 2.17 The output of the RF electric field sensor at 550 MHz and 10 dBm RF input. Reprinted from Ref. 15 with permission. © 2008 Institute of Electrical and Electronics Engineers

2.8 Conclusion

Chromophore-doped polymers have great potential for many sensor applications. Key technologies of the design, fabrication, and electronic-photonic integration of polymer microring resonators and a number of novel microresonator sensors have been successfully developed. High Q factor of polymer microring resonators can be achieved through systematic design and optimization with fast BPM simulations and transfer matrix analysis. Chromophore-doped polymers also allow optical waveguides and microring resonators to be patterned using UV light and electron-beam in a single process step. Both methods are free of solvents, wet chemicals, and toxic gases. Microring resonators made with chromophore-doped polymer can be used to detect trace explosives. Detection sensitivity of DNT at ppb level with good specificity has been demonstrated, and the polymer has the potential of pushing the detection limit to the ppt level. An optical electric field sensor with an EO polymer microring coupled with a side-polished fiber was shown to have adequate sensitivity for many practical applications. This device structure uses a low loss optical fiber as a bus waveguide and greatly reduces the insertion loss from fiber-waveguide coupling and long EO polymer waveguides. Chromophores are man-made materials and as molecular design and organic synthesis continue to advance, new polymers with more desirable sensing properties will continue to emerge, enabling better sensors and new applications.

Acknowledgments The author thanks Profs. Larry R. Dalton and Alex K.-Y. Jen, and Lumera Corporation for providing polymer materials. The work present in this chapter is supported by

National Science Foundation Grant Number ECS-0437920, National Science Foundation Center on Materials and Devices for Information Technology Research (STC-CMDITR), Grant Number DMR-0120967, and Office of Naval Research Grant N00014–05–1–0843.

References

1 Rabiei, P.; Steier, W. H.; Zhang, Z.; Dalton, L. R., Polymer micro-ring filters and modulators, *J. Lightw. Technol.* **2002**, 20, 1968–1975

2 Dalton, L. R.; Robinson, B. H.; Jen, A. K.-Y.; Steier, W. H.; Nielsen, R., Systematic development of high bandwidth, low drive voltage organic electro-optic devices and their applications, *Opt. Mater.* **2003**, 21, 19–28

3 Enami, Y.; Peyghambarian, N.; Kawazu, M.; Jen, A. K. -Y., Hybrid electro-optic polymer and selectively buried sol-gel waveguides, *Appl. Phys. Lett.* **2003**, 82, 490–492

4 Sinyukov, A. M.; Hayden, L. M., Generation and detection of terahertz radiation with multilayered electro-optic polymer films, *Opt. Lett.* **2002**, 27, 55–57

5 Pyayt, A.; Zhang, X.; Luo, J.; Jen, A.; Dalton, L.; Chen, A., Optical micro-resonator chemical sensor, *Proc. SPIE* **2007**, 6556, 65561D-1–65561D-6

6 Little, B. E.; Chu, S. T.; Absil, P. P.; Hryniewicz, J. V.; Seiferth, F. G. F.; Gill, D.; Van, V.; King, O.; Trakalo, M., Very high-order microring resonator filters the WDM applications, *IEEE Photonics Technol. Lett.* **2004**, 16, 2263–2265

7 Xia, F.; Sekaric, L.; Vlasov, Y., Ultracompact optical buffers on a silicon chip, *Nat. Photonics* **2007**, 1, 65–71

8 Xu, Q.; Schmidt, B.; Pradhan, S.; Lipson, M., Micrometre-scale silicon electro-optic modulator, *Nature* **2005**, 435, 325–327

9 Dekker, R.; Klunder, D. J. W.; Borreman, A.; Diemeer, M. B. J.; Worhoff, K.; Driessen, A.; Stouwdam, J. W.; Van Veggel, F. C. J. M., Stimulated emission and optical gain in LaF_3:Nd nanoparticle-doped polymer-based waveguides, *Appl. Phys. Lett.* **2004**, 85, 6104–6106

10 Kuwata-Gonokami, M.; Jordan, R. H.; Dodabalapur, A.; Katz, H. E.; Schilling, M. L.; Slusher, R. E.; Ozawa, S., Polymer microdisk and microring lasers, *Opt. Lett.* **1995**, 20, 2093–2095

11 Chao, C. -Y.; Fung, W.; Guo, L. J., Polymer microring resonators for biochemical sensing applications, *IEEE J. Sel. Topics Quant. Electron.* **2006**, 12, 134–146

12 Steier, W. H.; Kalluri, S.; Chen, A.; Garner, S.; Chuyanov, V.; Ziari, M.; Shi, Y.; Fetterman, H.; Jalali, B.; Wang, W.; Chen, D.; Dalton, L. R., Applications of electro-optic polymers in photonics, *Mater. Res. Soc. Symp. Proc.* **1996**, 413, 147–157

13 Herminghaus, S.; Smith, B. A.; Swalen, J. D., Electro-optic coefficients in electric-field-poled polymer waveguides, *J. Opt. Soc. Am. B* **1991**, 8, 2311–2317

14 Sun, H.; Pyajt, A.; Luo, J.; Shi, Z.; Hau, S.; Jen, A. K. -Y.; Dalton, L.; Chen, A., All-dielectric electrooptic sensor based on a polymer microresonator coupled side-polished optical fiber, *IEEE Sens. J.* **2007**, 7, 515–524

15 Chen, A.; Sun, H.; Pyayt, A.; Dalton, L. R.; Luo, J.; Jen, A. K.-Y., Microring resonators made in poled and unpoled chromophore-containing polymers for optical communication and sensors, *IEEE J. Sel. Topics Quantum Electron.* **2008**, 14, 1281–1288

16 Homola, J.; Yee, S. S.; Gauglitz, G., Surface plasmon resonance sensors: review, *Sens. Actuators B* **1999**, 54, 3–15

17 Trouillet, A.; Ronot-Trioli, C.; Veillas, C.; Gagnaire, H., Chemical sensing by surface plasmon resonance in a multimode optical fibre, *Pure Appl. Opt.* **1996**, 5, 227–237

18 Ronot-Trioli, C.; Trouillet, A.; Veillas, C.; Gagnaire, H., Monochromatic excitation of surface plasmon resonance in an optical-fibre refractive-index sensor, *Sens. Actuators A* **1996**, 54, 589–593

19 Yariv, A., Universal relations for coupling of optical power between microresonators and dielectric waveguides, *Electron. Lett.* **2000**, 36, 321–322
20 Norwood, R. A.; Eldada, L.; Emo, S.; Gustus, J.; Rapoport, R.; Stengel, K. M. T.; Shacklette, L. W.; Wu, C.; Xu, C.; Yardley, J. T., Polymer optical interconnection technology: toward WDM applications, *Proc. SPIE* **1996**, 2690, 151–162
21 Eldada, L.; Xu, C.; Stengel, K. M. T.; Shacklette, L. W.; Yardley, J. T., Laser-fabricated low-loss single-mode raised-rib waveguiding devices in polymers, *J. Lightwave Technol.* **1996**, 14, 1704–1713
22 Eldada L.; Shacklette, L. W., Advances in polymer integrated optics, *IEEE J. Sel. Topics Quant. Electron.* **2000**, 6, 54–68
23 Wang, W. C.; Lin, W.; Marshall, H.; Skolnick, R.; Schaafsma, D., All-dielectric miniature wideband rf receive antenna, *Opt. Eng.* **2004**, 43, 673–677
24 Sun, H.; Dalton, L.; Chen, A., Systematic design and simulation of polymer microring resonators with the combination of beam propagation method and matrix model, In Digest of the IEEE LEOS Summer Topical Meetings, 2007, 217–218
25 Paloczi, G. T.; Huang, Y.; Yariv, A.; Mookherjea, S., Polymeric Mach-Zehnder interferometer using serially coupled microring resonators, *Opt. Express* **2003**, 11, 2666–2671
26 Ma, J.; Lin, S.; Feng, W.; Feuerstein, R. J.; Hooker, B.; Mickelson, A. R., Modeling photo-bleached optical polymer waveguides, *Appl. Opt.* **1995**, 34, 5352–5360
27 Chen, A.; Chuyanov, V.; Marti-Carrera, F. I.; Garner, S.; Steier, W. H.; Mao, S. S. H.; Ra, Y.; Dalton, L. R.; Shi, Y., Trimming of polymer waveguide Y-junction by rapid photobleaching for tuning the power splitting ratio, *Photonics Technol. Lett.* **1997**, 9, 1499–1501
28 Zhou, J.; Pyayt, A.; Dalton, L. R.; Luo, J.; Jen, A. K. Y.; Chen A., Photobleaching fabrication of microring resonator in a chromophore-containing polymer, *IEEE Photonics Technol. Lett.* **2006**, 18, 2221–2223
29 Liao, Y.; Anderson, C. Y.; Sullivan, P. A.; Akelaitis, A. J. P.; Robinson, B. H.; Dalton, L. R., Electro-optical properties of polymers containing alternating nonlinear optical chromophores and bulky spacers, *Chem. Mater.* **2006**, 18, 1062–1067
30 Jin, D.; Londergan, T.; Huang, D.; Wolf, N.; Condon, S.; Tolstedt, D.; Guan, H. W.; Cong, S.; Johnson, E.; Dinu, R., Achieving large electro-optic response: DH-type chromophores in both crosslinked systems and linear high-Tg systems, *Proc. SPIE* **2004**, 5351, 44–56
31 Luo, J.; Liu, S.; Haller, M.; Kang, J. -W.; Kim, T. -D.; Jang, S. -H.; Chen, B.; Tucker, N.; Li, H.; Tang, H. -Z.; Dalton, L. R.; Liao,Y.; Robinson, B. H.; Jen A. K. Y., Recent progress in developing highly efficient and thermally stable nonlinear optical polymers for electro-optics, *Proc. SPIE* **2004**, 5351, 36–43
32 Samoc, A.; Samoc, M.; Luther-Davies, B.; Kolev, V. Z.; Bagien, R. K.; Luo, X.; Zha, C., In situ second harmonic generation in disperse red 1 doped polymer and sol-gel films, *Molecular Cryst. Liq. Cryst.* **2006**, 446, 123–140
33 Chen, A.; Sun, H.; Pyayt, A.; Zhang, X.; Luo, J.; Jen, A.; Sullivan, P. A.; Elangovan, S.; Dalton, L. R.; Dinu, R.; Jin, D.; Huang, D., Chromophore-containing polymers for trace explosive sensors, *J. Phys. Chem. C* **2008**, 112, 8072–8078
34 Jose, A.; Zhu, Z.; Madigan, C. F.; Swager, T. M.; Bulovic, V., Sensitivity gains in chemosensing by lasing action in organic polymers, *Nature* **2005**, 434, 876–879
35 Bauer, W. S.; Yilmaz, S.; Wirges, W.; Gerhard-Multhaupt, R., Optimized poling of nonlinear optical polymers based on dipole-orientation and dipole-relaxation studies, *J. Appl. Phys.* **1994**, 75, 7211–7219
36 Hill, R. A.; Knoesen, A.; Mortazavi, M. A., Corona poling of nonlinear polymer thin films for electro-optic modulators, *Appl. Phys. Lett.* **1994**, 65, 1733–1735
37 Vorst, C. P. J. M. V. D.; Picken, S. J., Electric field poling of acceptor-donor molecules, *J. Opt. Soc. Am B* **1990**, 7, 320–325.
38 Dalton, L. R.; Robinson, B. H.; Jen, A.-K. Y.; Reid, P.; Eichinger, B.; Sullivan, P.; Akelaitis, A.; Bale, D.; Haller, M.; Luo, J.; Liu, S.; Liao, Y.; Firestone, K.; Bhatambrekar, N.; Bhattacharjee, S.; Sinncss, J.; Hammond, S.; Buker, N.; Snoeberger, R.;

Lingwood, M.; Rommel, H.; Amend, J.; Jang, S. H.; Chen, A.; Steier, W., Electro-optic coefficient of 500 pm/V and beyond for organic materials, *Proc. SPIE* **2005**, 5935, 1–12
39 Robertson, W. M. Optoelectronic Techniques for Microwave and Millimeter-Wave Engineering, Artech House, Norwood, MA, 1995
40 Ozaki, K.; Sekiguchi, H.; Wakana, S.; Goto, Y.; Umehara, Y.; Matsumoto, J., Novel optical probing system with submicron spatial resolution for internal diagnosis of VLSI circuits, *Proc. Int. Test. Conf.* **1996**, 269–275
41 Zhang, F.; Chen, F.; Qiu, K., An integrated electro-optic E-field sensor with segmented electrodes, *Microw. Opt. Tech. Lett.* **2004**, 40, 302–305
42 Yang, K.; Katehi, L. P. B.; Whitaker, J. F., Electric field mapping system using an optical-fiber-based electrooptic probe, *IEEE Microw. Wireless Comp. Lett.* **2001**, 11, 164–166
43 Shi, Z.; Hau, S.; Luo, J.; Kim, T.-D.; Tucker, N. M.; Ka, J.-W.; Sun, H.; Pyajt, A.; Dalton, L.; Chen, A.; Jen, A. K.-Y., Highly efficient diels-alder crosslinkable electro-optic dendrimers for electric-field sensors, *Adv. Funct. Mater.* **2007**, 17, 2557–2563

Chapter 3
Modal Transition in Nano-Coated Long Period Fiber Gratings: Principle and Applications to Chemical Sensing

Andrea Cusano, Pierluigi Pilla, Michele Giordano, and Antonello Cutolo

Abstract In this chapter, the numerical and experimental investigation of the phenomenon of the modal transition in long period fiber gratings (LPGs) coated with a polymeric overlay of nanometric thickness and of high refractive index (HRI) is reported. This layered structure shows a significant modification of the cladding mode distribution with respect to the bare device, which depends on the overlay features (refractive index and thickness) and on the surrounding medium refractive index (SRI). As a consequence, enhanced evanescent wave interaction with the surrounding environment and a powerful guided wave interaction with the overlay itself arc obtained. These effects were exploited to build very sensitive refractometers and chemical sensors. Moreover it was shown that the high SRI sensitivity region of the coated LPG can be tuned over the desired SRI range for specific applications by acting on the overlay thickness. The dip-coating technique was adopted to build the thin film coated devices and the syndiotactic polystyrene in its nanoporous form (δ form) was used as HRI material, and in particular as molecular sieve for the detection of trace amounts of chloroform in water. The recent applications to chemical sensing of the nano-coated LPGs are reviewed.

3.1 Introduction

Long Period Fiber Gratings (LPFGs or LPGs) are in-fiber photonic devices realized by inducing a periodic refractive index modulation of the core of a single mode optical fiber along an unjacketed portion of its length. They have periodicities ranging in hundreds of micrometers (typical values are 100–500 μm), refractive

---

A. Cusano (✉)
University of Sannio, Engineering Department - Optoelectronic Division, C.so Garibaldi 107, 82100, Benevento, Italy
e-mail: a.cusano@unisannio.it

X. Fan (ed.), *Advanced Photonic Structures for Biological and Chemical Detection*, Integrated Analytical Systems,
DOI 10.1007/978-0-387-98063-8_3,

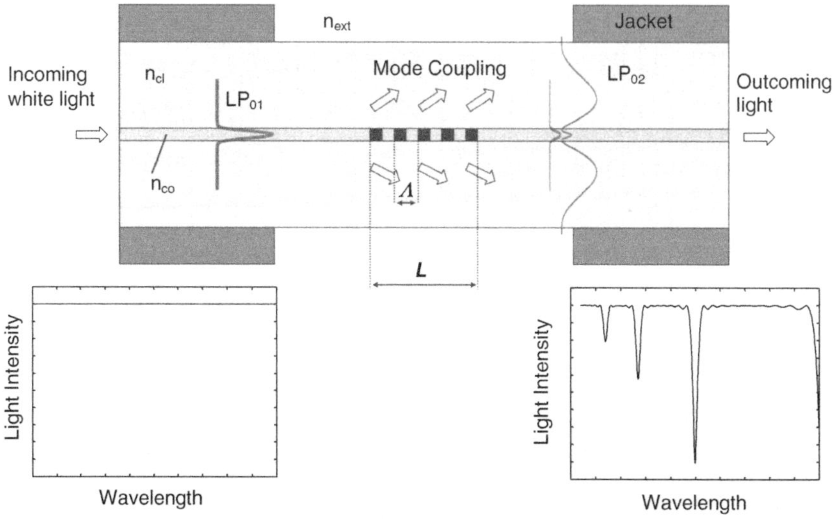

Fig. 3.1 Mode coupling in a LPG (not to scale). Dips are referred to as attenuation bands

index modulation strength of the order of 10^{-4} and lengths of few centimeters (typically 2–3 cm).

LPGs couples light from the fundamental guided core mode to discrete forward-propagating cladding modes and to each of them at a distinct wavelength where the so-called phase matching condition is satisfied[1]. Cladding modes are bounded in the optical fiber by the total internal reflection at the silica-air interface along the grating region. Just after the grating, cladding modes attenuate rapidly on propagation due to different loss mechanisms such as bends in the fiber, scattering losses and absorption by the fiber jacket. As a consequence attenuation bands are created in the transmission spectrum (see Fig. 3.1).

The shape of a LPG transmission spectrum in terms of attenuation bands central wavelength, depth and bandwidth depends on the effective refractive index difference between the core mode and the *j*th cladding mode, the overlap integral between the core mode field and the cladding mode field, the grating period, length and index modulation strength and on the physical parameters that characterize its status such as temperature, strain, bending and surrounding medium refractive index (SRI). In fact, whatever entity that is able to affect the difference of the effective indices of the core and cladding modes and/or its period and length, is able at the same time to affect the phase matching condition and the coupling strength[2].

In the past few years LPGs have been widely investigated for both sensing and communication applications[3,4]. In this chapter we focus on the application of LPGs to the measurement of the SRI and to chemical sensing.

An interesting feature of LPGs is that their sensitivity characteristics to a specific parameter depend drastically on the order of the coupled cladding mode and on the type of the fiber in which they are inscribed. This makes possible to discriminate between different types of perturbation that act simultaneously on the grating and offers the flexibility to design devices that are very sensitive or insensitive to a particular measurand[5–7].

In LPGs the way in which light is coupled out of the core mode depends on the refractive index of the medium, assuming infinitely extended in the radial direction that surrounds the LPG. If the SRI is lower than the cladding RI, the fiber supports bounded cladding modes that are maintained by total internal reflection at the cladding-surrounding medium interface. In this case, as the SRI increases the effective indices of the cladding modes increase with a consequent blue wavelength shift of the attenuation bands. The higher the cladding mode order, the higher the wavelength shift. This is because the higher order cladding modes extend further into the surrounding medium with their evanescent tail and hence are more sensitive to SRI changes. The spectral displacement of the attenuation bands for an increasing SRI has a nonlinear monotonic behavior with increasing sensitivity as the SRI approaches the cladding RI[8].

When the SRI is equal to the cladding one, the cladding layer becomes an infinitely extended medium and thus supports no discrete cladding modes. In this case, a broadband radiation mode coupling occurs with no distinct attenuation bands[8].

For even higher SRIs the attenuation bands start to reappear at wavelengths slightly higher than the case in which the external medium is air, with increased loss peaks but almost without wavelength shift. The presence of the attenuation bands in this situation is because the fiber partially confines the radiated field at the resonance wavelengths by the Fresnel reflection, rather than by total internal reflection, at the boundary between the cladding and the surrounding medium, forming leaky modes. As the SRI increases further, the leaky modes are better confined, increasing the resonance losses[9–12].

An approximate analytical expression for the LPGs sensitivity to SRI changes ($d\lambda_{res,j}/dn_{out}$) was reported by Chiang and co-workers[13]. They also found that the sensitivity may be significantly enhanced by reducing the cladding radius, being it inversely proportional to the cladding radius cubed. Reducing the cladding radius of a LPG has another effect, that is, the tuning of the spectral position of the attenuation bands toward higher wavelengths. However, both effects depend on the fact that thinning the fiber cladding reduces the effective refractive index of the cladding modes and as a consequence the resonance wavelengths increase as well as the cladding mode evanescent field interaction with the surrounding medium[14–16].

An immediate application suggested by the intrinsic SRI sensitivity of LPGs is the chemical sensing. However the first applications of LPGs in this field were not real chemical sensors but instead solution concentration sensors, since the chemical selectivity was missing[17–20]. In fact, a chemical sensor is usually composed of a sensitive layer which, interacting with the surrounding environment, selectively collects and concentrate molecules at or within the surface undergoing physical changes, and of an opportune transducer that converts into an interpretable and quantifiable term such modification of the sensitive layer[21]. To the best of our knowledge the first example of LPG-based (bio)chemical sensor was reported by DeLisa et al[22]. However if the refractive index of the chemo-sensitive overlay is lower than the cladding RI then the LPG interacts with it through the evanescent tail of the cladding modes. As we will see, there is a better way to exploit the cladding mode fields through a guided wave interaction with the sensitive overlay.

It was a common belief in the early studies of this topic that since the active materials have usually refractive indices higher than that of silica (in the following named high refractive index (HRI) materials), the LPG had to work in the leaky mode regime. This justifies the presence of many studies devoted to characterize experimentally and numerically the LPG behavior in this circumstance[9–12]. However this would have been true if the HRI surrounding medium had been considered indefinitely extended in the radial direction. The situation becomes very different when the LPG is coated with an HRI layer which has a thickness smaller than or comparable to the wavelength of the light used to feed it, as we will see in the rest of this chapter.

Rees and co-workers[23] first studied the effect of the deposition of thin HRI overlays onto LPGs by analyzing the wavelength shift of the attenuation bands. The Langmuir-Blodgett deposition technique was exploited to deposit an incrementally thicker overlay (up to few hundreds of nanometers) of an organic material whose refractive index was 1.58.

When azimuthally symmetric nano-scale overlays with higher refractive index than that of the cladding are deposited along the grating region, refraction-reflection regime at the cladding-overlay interface occurs while total internal reflection confines light rays to the overlay-surrounding medium interface, at least for SRIs lower than the cladding RI. The new cladding modes become bounded within the structure comprising the core, the cladding and the HRI overlay. In other words, the presence of the overlay determines the change in the power distribution of cladding modes within the transverse section of the coated optical fiber and in their effective refractive indices. In particular, the overlay deposition leads to a lowering of the cladding mode power bounded within the core and cladding layers, while part of the light power carried by the cladding modes moves towards the HRI overlay, which enhances the evanescent field in the surrounding medium and increases the cladding mode effective index. This significant modification of the cladding mode distribution depends on the layer features (refractive index and thickness) and on the SRI. As a consequence, there is improved sensitivity of the coated device in terms of wavelength shift and amplitude variations of the attenuation bands in response to changes of the aforementioned parameters. Moreover the thin HRI overlay constitutes a waveguide itself, therefore when the overlay parameters are properly chosen and the SRI is increased in a certain range, the transition of the lowest order cladding mode into an overlay mode occurs and almost simultaneously cladding mode reorganization takes place. In the rest of this chapter we will show that this phenomenon has wonderful benefits on the sensitivity of the device.

3.2 Theoretical Background

As aforementioned, the spectral characteristics of LPGs are determined by the coupling between the fundamental guided core mode and the co-propagating cladding modes when the core propagating light encounters a periodic perturbation of fiber refractive index. From the coupled mode theory the resonance wavelengths at which the transmission spectrum present the minima are approximately given by the well known first order Bragg condition[1]:

$$\lambda_{\mathrm{res},0j} = \left(n_{\mathrm{eff,co}} - n_{\mathrm{eff,cl}}^{0j}\right) \cdot \Lambda, \tag{3.1}$$

where $n_{\mathrm{eff,co}}$ and $n_{\mathrm{eff,cl}}^{0j}$ are the core and jth cladding mode effective indices respectively, Λ is the grating period.

It is worth noting that cladding modes excited in an LPG are usually numbered starting from the one with the highest effective index, so that higher order cladding modes have lower effective index. The effective index of the fundamental core mode is usually calculated by solving the scalar wave equation coming out from the weakly guidance approximation[24], so it is a linearly polarized mode (LP). The one with highest propagation constant and azimuthal symmetry is denoted as the LP_{01}. The effective index of the cladding modes are sometimes also calculated with the LP approximation so they are denoted as the LP_{0j} ($j = 2, 3\ldots$)[1,2]. However, while for the core mode the weakly guiding approximation is quite accurate due to the small refractive index difference between the core and cladding materials, for the cladding modes this approximation introduces bigger errors due to high index contrast between the cladding and the external medium that is usually air. For this reason, the cladding mode effective index is more accurately calculated by solving the full vector wave equation thus considering the exact modes of the waveguide that is the hybrid modes[25]. In this way, the cladding modes are denoted as HE_{1j} and EH_{1n} with j and n odd and even numbers, respectively, and j starting from 1. Finally, sometimes the core mode is also solved as an exact mode so it is denoted as the HE_{11}. In this case, the numbering of the cladding modes is with j and n even and odd numbers, respectively, and j starting from 2. Usually the refractive index modulation induced in the fiber core is azimuthally symmetric, so that only the cladding modes with azimuthal symmetry ($LP_{\nu j}$ and $HE_{\nu j}$ with $\nu = 0, 1$, respectively) are considered. In fact, the coupling mechanism is more efficient for cladding modes with the same symmetry of the fundamental core mode. However, there is no shortage of cases in the literature in which the writing method is intrinsically asymmetric thus introducing the coupling to asymmetric modes[26,27].

In this chapter, we will numerically analyze the effects of the HRI overlay on the cladding modes distribution with particular attention to its influence on the device sensitivity to the SRI changes and to the overlay refractive index changes. The structure to which we refer is depicted in Fig. 3.2.

The model makes a number of approximations. First, the modes of the four-layer (core, cladding, overlay, and surrounding medium) cylindrical waveguide are solved by exploiting the LP approximation[24]. Another approximation made in the following analysis is that the RI dispersion of the polymeric coating as well as those of the surrounding medium and of the optical fiber materials has not been taken into account. Furthermore, the overlay considered here is assumed to be optically transparent in the investigated wavelength range, or in other words the waveguide is considered lossless. Last, the presented analysis holds as long as the cladding

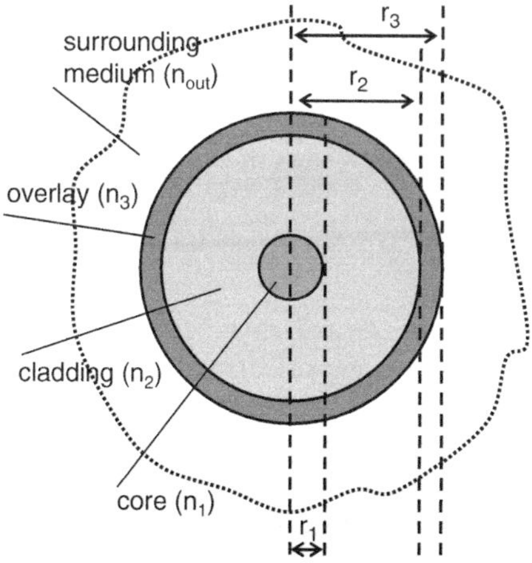

Fig. 3.2 Transversal section of the investigated structure (not to scale)

modes are supported, i.e., for SRI lower than the cladding RI. For higher SRIs, another model is required to take into account the coupling between the core mode and the radiation modes for the four-layer waveguide.

The calculation of the cladding mode effective index can be accomplished by an extension of the model for doubly clad fibers[28] or by following the transfer matrix method (TMM) proposed by Anemogiannis et al.[26] and successively widely adopted for the analysis of coated LPGs[29,30].

The transverse electric field component, with azimuthal order v, is given by:

$$\psi(r) = \begin{cases} A_0 Z_{v,1}\left(u_1 \frac{r}{r_1}\right) & \text{for} \quad r \leq r_1 \\ A_1 Z_{v,2}\left(u_2 \frac{r}{r_2}\right) + A_2 T_{v,2}\left(u_2 \frac{r}{r_2}\right) & \text{for} \quad r_1 < r \leq r_2 \\ A_3 Z_{v,3}\left(u_3 \frac{r}{r_3}\right) + A_4 T_{v,3}\left(u_3 \frac{r}{r_3}\right) & \text{for} \quad r_2 < r \leq r_3 \\ A_5 K_{v,4}\left(v \frac{r}{r_3}\right) & \text{for} \quad r > r_3 \end{cases}, \tag{3.2}$$

where:

$$Z_{v,i}(x) = \begin{cases} J_v(x) & \text{if } n_{\text{eff}} < n_i \\ I_v(x) & \text{if } n_{\text{eff}} > n_i \end{cases},$$

$$T_{v,i}(x) = \begin{cases} Y_v(x) & \text{if } n_{\text{eff}} < n_i \\ K_v(x) & \text{if } n_{\text{eff}} > n_i \end{cases},$$

$$u_i = r_i k_0 \sqrt{\left|n_i^2 - n_{\text{eff}}^2\right|} \quad \text{for} \quad i = 1, 2, 3,$$

$$v = r_3\, k_0 \sqrt{\left| n_{\mathrm{eff}}^2 - n_{\mathrm{out}}^2 \right|},$$

and r is the fiber radius, J_v and I_v are the ordinary Bessel functions of first and second kind of order v and Y_v and K_v are the modified Bessel functions of first and second kind of order v, respectively; n_1, n_2, and n_3 are the core, cladding, and overlay refractive indices, respectively, while n_{out} is the surrounding refractive index and n_{eff} is the effective refractive index; r_1 and r_2 are the core and cladding radius and $r_3 - r_2$ is the overlay thickness. In addition, A_0 is the normalization coefficient, while A_1, A_2, A_3, A_4, and A_5 can be obtained by imposing the continuity of the fields at the interface between core and cladding, cladding and overlay, and overlay and surrounding medium. Here, the effective refractive index of every cladding mode is achieved by numerical solution of the dispersion equation obtained by the continuity condition of the transverse fields.

According to the coupled mode theory[25], the interaction between optical modes is proportional to their coupling coefficient. Since the contribution of the longitudinal coupling coefficient is usually negligible, we will consider only the transversal coupling coefficient. Having adopted the LP approximation, in cylindrical coordinates the coupling coefficient between the two modes can be expressed as[26,29]:

$$K_{vj,mi} = \frac{\omega}{4P_0} \times \int_{\varphi=0}^{2\pi} \int_{r=0} \Delta\varepsilon(r,\varphi,z)\ \Psi_{vj}(r,\varphi)\ \Psi_{\mu i}(r,\varphi) r\ \mathrm{d}r\ \mathrm{d}\varphi, \tag{3.3}$$

where $\Psi(r,\varphi)$ is the transverse field for an LP mode, $\Delta\varepsilon(r,\varphi,z)$ is the permittivity variation, and P_0 is the power of each mode supposed to be the same. Under the assumptions of no azimuthal variation of the perturbed index profile and of weak guidance between the core and the cladding the permittivity can be expressed as:

$$\Delta\varepsilon(r,\varphi,z) \cong 2\,\varepsilon_0\, n_0(r)\, \Delta n(r,z), \tag{3.4}$$

where ε_0 is the free-space permittivity, $n_0(r)$ is the refractive index profile of the structure without the perturbation, and Δn is the refractive index modulation. This last function can be expressed as the product of three terms:

$$\Delta n(r,z) = p(r) \cdot \sigma(z) \cdot S(z), \tag{3.5}$$

where $p(r)$ is the transverse refractive index perturbation, which is usually assumed a nonzero constant only in the core:

$$p(r) = p_0 \cdot rect\left(\frac{r}{r_1}\right). \tag{3.6}$$

$\sigma(z)$ is the apodization factor, which we will consider a unitary constant, and $S(z)$ is the longitudinal refractive index modulation factor, which can be approximated by a Fourier series of two terms:

$$S(z) = s_0 + s_1 cos\left(\left(\frac{2\pi}{\Lambda}\right) z\right), \tag{3.7}$$

where Λ is the grating period. The coefficients of the Fourier series s_0 and s_1 depend on the exposure function of the LPG writing process, we will consider in the following numerical analysis $s_0 = s_1 = 1$. It is worth noting that $s_0 \times p_0$ is the average refractive index modulation of the core while $s_1 \times p_0$ is the refractive index modulation strength. Moreover, for the hypothesis of azimuthal symmetry of the refractive index modulation and of the deposited overlay, efficient coupling is achieved only between the core mode and azimuthally symmetric cladding modes, consequently only LP_{0j} cladding modes will been taken into account. After these approximations the coupling coefficients can be expressed as:

$$K_{0j,0i} = \left[s_0 + s_1 \cos\left(\left(\frac{2\pi}{\Lambda}\right) z\right)\right] \zeta_{0j,0i} \tag{3.8}$$

with:

$$\zeta_{0j,0i} = \frac{\omega\varepsilon_0}{2P_0} n_1 p_0 \int_0^{2\pi} \mathrm{d}\varphi \int_{r=0}^{r_1} R_{0j}(r) R_{0i}(r) r \, \mathrm{d}r, \tag{3.9}$$

where n_1 is the refractive index of the core.

From the coupled mode theory, it is possible to retrieve a modified phase matching condition, which is more accurate than (3.1)[26]:

$$\frac{2\pi}{\lambda}\left(n_{\text{eff},01}(\lambda) - n_{\text{eff},0j}(\lambda)\right) + s_0\left(\zeta_{01,01}(\lambda) - \zeta_{0j,0j}(\lambda)\right) = \frac{2\pi}{\Lambda}. \tag{3.10}$$

$\zeta_{01,01}$ and $\zeta_{0j,0j}$ are the self-coupling coefficients of the core and the jth cladding modes, respectively. They depend on the average refractive index change and on the overlap integral of the transverse field of a mode with itself evaluated in the region of the fiber where the perturbation is present (usually in the core). These coefficients are sometimes also called "dc" coupling coefficients because they account for the "dc" index change[2]. Moreover the normalized transmitted power by the fundamental guided core mode through the grating at the resonant wavelength regarding the jth cladding mode can be expressed as:

$$T_{0j} = \cos^2\left(\kappa_{01,0j} \cdot L\right), \tag{3.11}$$

where L is the grating length and $k_{01,0j}$ is the coupling coefficient between the fundamental core mode and the jth cladding mode. This coupling coefficient can be expressed as:

$$\kappa_{01,0j} = \frac{s_1}{2}\zeta_{01,0j}, \tag{3.12}$$

where s_1 is the second coefficient of the Fourier series coefficients and $\zeta_{01,0j}$ is the cross-coupling coefficient of the core and cladding modes. The coupling coefficient $\kappa_{01,0j}$ is sometimes called an "AC" coupling coefficient since it depends on the index modulation strength of the grating[2]. Complete power transfer between the core mode and the jth cladding mode occurs if $\kappa_{01,0j} \cdot L = \pi/2$.

3.3 Numerical Analysis of Coated LPGs

In this section, we will analyze the effects of a nano-sized HRI overlay on the distribution of the cladding modes (transversal fields, effective indices, coupling coefficients) to changes of its thickness and refractive index and to changes of the SRI.

With reference to Fig. 3.2, the analysis is performed with the standard Corning SMF-28 optical fiber parameters: numerical aperture 0.14, refractive index difference 0.36%, cladding and core diameter, 125 and 8.3 μm, respectively. The surrounding medium is assumed to be water ($n_{\text{out}} = 1.33$) unless specified differently. Every mode is characterized by the same power P_0 and the transversal fields are normalized to the maximum amplitude of the core mode. Figure 3.3a shows the fields related to the cladding modes LP_{03} and LP_{05} reported as a function of the radial coordinate within the transverse section of a bare LPG at a wavelength of 1310 nm. The same cladding mode fields are reported in Fig. 3.3b, this time for a HRI-coated LPG ($n_{\text{overlay}} = 1.578$) with overlay thicknesses of 200 and 250 nm.

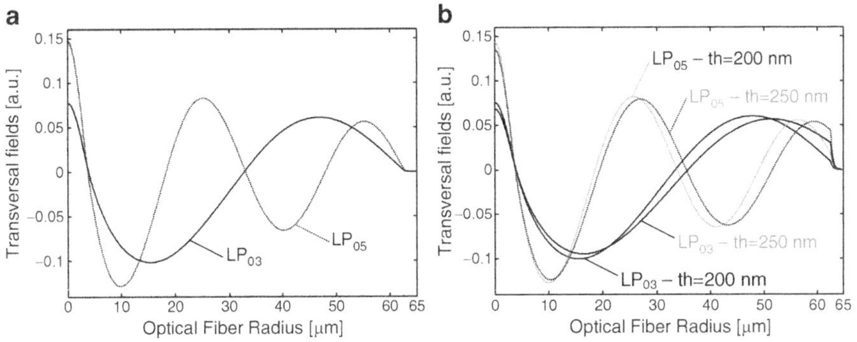

Fig. 3.3 LP_{03} and LP_{05} cladding modes in (**a**) bare fiber and (**b**) coated fiber with th = 200 nm and th = 250 nm thin overlay

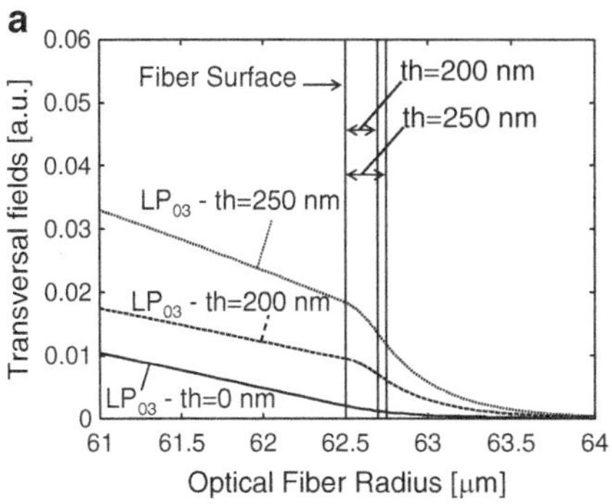

Fig. 3.4 (**a**) LP_{03} cladding mode in coated fibers with th = 200 nm and th = 250 nm; (**b**) LP_{03} and LP_{07} cladding modes in a coated fiber with th = 250 nm

From these figures it is apparent how the HRI overlay stretches the cladding mode fields toward itself for both modes and both overlay thicknesses. In addition, for the same cladding mode, the field content within the overlay increases as the overlay thickness is increased (see Fig. 3.4a). Finally, for a given overlay thickness, the field content within the overlay of higher order modes is higher compared with low order modes (see Fig. 3.4b)[31].

The same observations are true for the evanescent wave of the modes in the surrounding medium. Moreover, these changes in the field distributions are accompanied by an increase of the effective refractive indices of the cladding modes, as will be clarified later. For this reason and by virtue of (3.1), the resonance wavelengths of the coated LPG are expected to blue-shift in response to an overlay thickness change.

From these simulations the sensing mechanism and performances of the coated device can be outlined. If the HRI layer is a molecular sieve, or in other words, if it is able to absorb trace amounts of a chemical present in the surrounding medium and undergoes a subsequent change in its refractive index and/or thickness, then this change will cause strong modifications of the cladding mode fields and obviously of the effective refractive indices and overlap integrals. These effects in turn will be reflected in changes of the attenuation bands of the coupled cladding modes in terms of central wavelengths shift and loss peaks change. Moreover, due to the field enhancement within the overlay, as the thickness and the mode order are increased, higher sensitivity can be expected.

It is worth to stress that in contrast to bare LPGs, where the basic sensing principle is the evanescent interaction with the surrounding environment, here the sensing mechanism relies on a guided wave interaction with the overlay.

The same reasoning on the changes of the fields profile and effective indices is applicable if the overlay is not chemically interactive but the SRI is varied. In fact, the presence of the HRI overlay increases the extension of the evanescent tail of the modes in the surrounding medium making them more sensitive to the SRI changes compared with the bare device. What should be clear from these first observations is

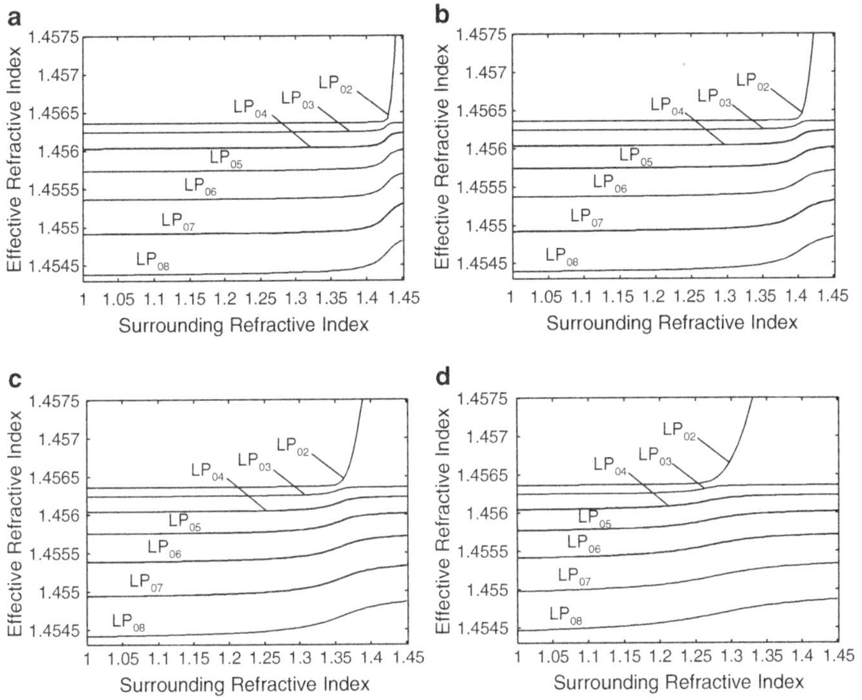

Fig. 3.5 Effective refractive index of the LP_{02}-LP_{08} cladding modes vs. the SRI in HRI coated fiber with: (**a**) 150 nm; (**b**) 200 nm; (**c**) 250 nm, and (**d**) 300 nm thin overlays

that the HRI-coated LPG can act as a real chemical sensor, which benefits from field enhancement in the sensitive layer or as a refractometer whose performances are expected to be better compared with the bare device.

Let us discuss now the effect of the SRI and of the overlay thickness on the effective refractive indices of the cladding modes. The effective indices of the first 7 cladding modes (LP_{02}–LP_{08}) are reported in Fig. 3.5a as a function of the SRI for HRI-coated LPG with overlay thickness of 150 nm and n_{overlay} of 1.578. As it can be observed, the effective refractive indices increase along with the SRI until a critical point is reached. At this point, a significant shift in the effective indices of the cladding modes occurs. For a fixed overlay thickness, there is a value of the SRI that leads the lowest order cladding mode (higher effective index) to be guided within the overlay. The transition from cladding mode to overlay mode is characterized by an effective index of the involved mode higher than the cladding RI. As the lowest order cladding mode moves to be guided within the overlay, a reorganization of the other cladding modes occurs. The immediate consequence is a simultaneous shift in the effective index of all the cladding modes to recover the original distribution. As a matter of fact, the effective index of the jth mode shifts to match the initial index of the $(j-1)$th mode, and so on[32]. Moreover, since the effective refractive index difference between consecutive cladding modes increases with the mode order, the higher the mode order, the greater the effective index change.

The same mechanism occurs for a fixed SRI by increasing the HRI overlay thickness and by varying the overlay refractive index as shown by Del Villar et al[29,33]. On these grounds and taking into account the direct relationship between effective indices and resonance wavelengths (see (3.1)), it is expected that by increasing the SRI, the attenuation band related to the jth cladding mode would span all the wavelength range to recover the initial position regarding the $(j-1)$th cladding mode. Moreover, the higher the cladding mode order, the greater the shift required to the jth mode to recover the initial position of the $(j-1)$th mode. If all this is true, an improved SRI sensitivity can be measured in HRI-coated LPGs near the transition point and in a wide SRI range compared with uncoated devices.

The mode transition is shown in Fig. 3.5b-d for different overlay thicknesses, namely 200, 250, and 300 nm. As observed in the figures, the transition region moves to lower SRIs as the overlay thickness increases, approaching the ambient index of 1.33 where bare LPGs demonstrate a significantly lower sensitivity. As a secondary effect, the SRI sensitivity becomes smaller as the overlay thickness increases. However, even if the absolute sensitivity decreases, the relative sensitivity in comparison to the bare device still increases as will be shown experimentally.

A fundamental feature of this mechanism is the possibility to tune the transition region in the desired SRI range by acting on the overlay thickness. This means that by using thick enough HRI coating, high sensitivity can be obtained also for SRIs approaching the air index, leading to the possibility to develop gas sensors.

The coupling coefficients of the first 7 cladding modes (LP_{02}–LP_{08}) are plotted in Fig. 3.6a as a function of the SRI when the overlay thickness is 150 nm. It is apparent from the figure that the coupling coefficients are a decreasing function of the SRI, thus a consequent decrease of the attenuation bands loss peak is expected according to (3.11), except for the case in which for some attenuation bands there is over-coupling ($\kappa_{01,0j} \cdot L > \pi/2$). In this last case an increase of the loss peak should be observed before the decrease. The trend of the coupling coefficients with the SRI can be attributed to the fact that by increasing the SRI the transverse field profile of the cladding modes is stretched toward the HRI coating and this causes a reduction of the overlap integral between cladding and core modes. When the first cladding mode becomes guided into the overlay, in correspondence of the transition point, its coupling coefficient vanishes rapidly. At the same point, the coupling coefficients of the higher order cladding modes shift to quasi recover the original configuration of the coupling coefficients. As already noticed for the effective refractive indices, the maximum sensitivity of the coupling coefficients to the SRI changes is shown in the transition region where the slope of the curves increases dramatically. But then, an opposite behavior is observed concerning the mode order. In fact, the lower the mode order, the greater the coupling coefficient change. This is clearly expected since the lowest order cladding modes are the first to experience the transition from cladding modes to overlay modes.

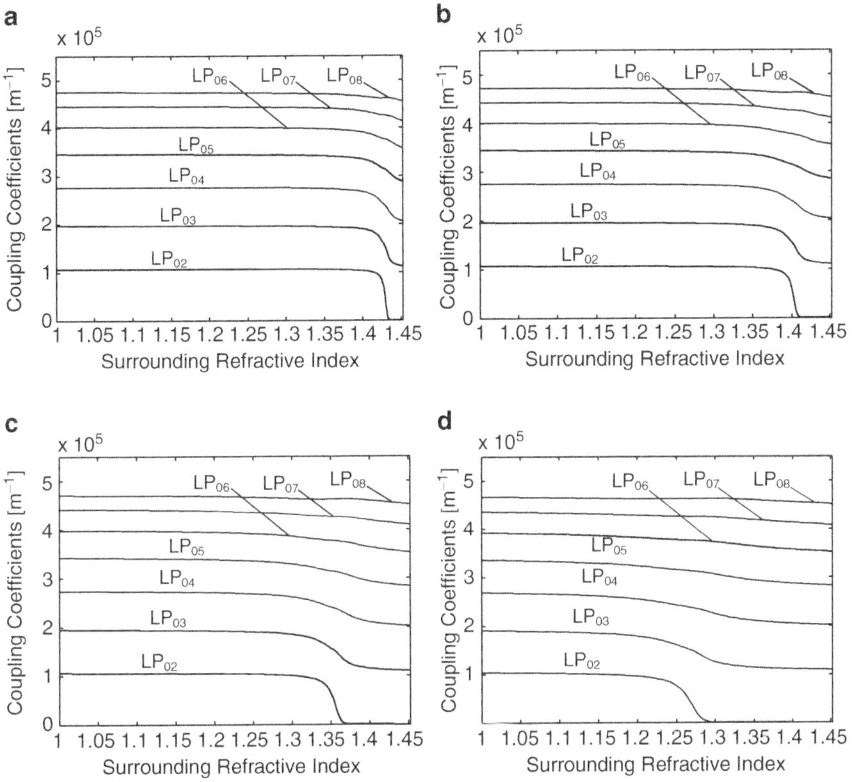

Fig. 3.6 Coupling coefficients of the LP_{02}-LP_{08} cladding modes vs. the surrounding refractive index in HRI coated fiber with: (**a**) 150 nm; (**b**) 200 nm; (**c**) 250 nm; and (**d**) 300 nm thin overlays

The coupling coefficients vs. the SRI are reported in Fig. 3.6 b-d for overlay thicknesses of 200, 250, 300 nm, respectively, always with the same coating index of 1.578. From these figures, it can be clearly inferred that increasing the overlay thickness the coupling coefficients curves shift toward lower SRIs as it happens for the effective refractive index curves.

A better understanding of the cladding mode reorganization as direct consequence of the SRI changes is provided in Fig. 3.7 in terms of cladding mode fields for a 200 nm coated LPG. In particular, the figure shows the transverse field profiles of the LP_{02}, LP_{03}, and LP_{04} cladding modes, respectively, before the transition (SRI = 1), during the transition (SRI = 1.40) and after the transition (SRI = 1.45). All the fields were normalized to the maximum amplitude of the LP_{01} core mode. As one can see, during the transition, the field in the overlay, and hence the interaction between the evanescent wave and the surrounding medium, reaches its maximum value. In addition, it can be clearly observed that the LP_{04} mode after the transition (dashed line in Fig. 3.7c) is qualitatively similar to the LP_{03} mode before the transition (solid line in Fig. 3.7b). In the same way, the LP_{03} mode after the

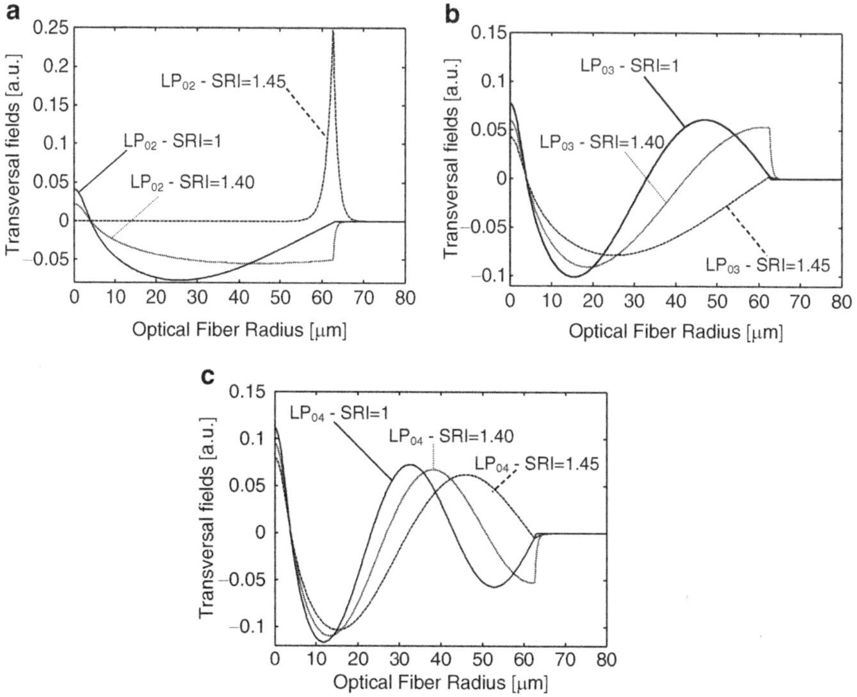

Fig. 3.7 Cladding mode fields in a 200 nm coated LPG, before transition (SRI = 1), in transition (SRI = 1.40) and after transition (SRI = 1.45): (**a**) LP_{02}; (**b**) LP_{03}; and (**c**) LP_{04}

transition (dashed line in Fig. 3.7b) is similar to the LP_{02} mode before the transition (solid line in Fig. 3.7a), while the LP_{02} mode after the transition (dashed line in Fig. 3.7a) is clearly guided within the overlay[34].

Finally, the resonance wavelengths are strongly related to the overlay refractive index through the effective index. What we have showed for the overlay thickness changes and SRI changes is similarly valid for the refractive index change of the overlay when the other parameters are kept constant. To conclude this numerical analysis, we plot in Fig. 3.8 the effective refractive index sensitivity to the overlay refractive index changes as function of the thickness for different cladding modes with water as surrounding medium.

From this figure, it can be derived that for a fixed mode order, a relevant sensitivity improvement can be obtained by increasing the overlay thickness until the transition region is approached[35]. Within the transition, the sensitivity of the lowest order cladding mode dramatically increases, because it starts to be guided within the overlay. Unfortunately, this high sensitivity cannot be exploited because the lowest order mode transition also leads to the vanishing of its related attenuation band, as we have just seen from the coupling coefficients analysis. For other modes, the sensitivity increases with the increased overlay thickness, reaching a maximum value very similar for each mode at the same coating thickness.

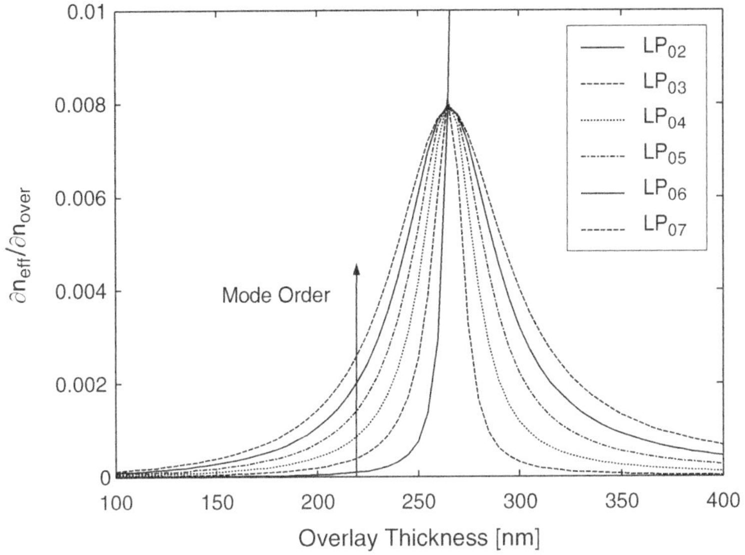

Fig. 3.8 Effective refractive index sensitivity to overlay RI changes vs. overlay thickness for different cladding modes

The highest sensitivity to the overlay refractive index changes occurs for an overlay thickness of 265 nm in correspondence of the mode transition, similarly to what happens with regard to the sensitivity to SRI changes, where a strong field enhancement within the overlay occurs (see Fig. 3.4). However, this is true only because the analysis is done at a fixed wavelength. If each cladding mode would be considered at the real wavelength at which it is coupled we would find that the higher the cladding mode order, the higher the sensitivity. Moreover, the thickness corresponding to the highest sensitivity would be slightly higher for higher order cladding modes. This is mainly a consequence of the fact that the higher order cladding modes are coupled at higher wavelengths. It is also interesting to note that the sensitivity characteristics of the cladding modes have a resonance-like shape. All the concepts here introduced will be experimentally investigated in a next section.

The simplifications of the model considered so far have been progressively removed in relatively recent works, which shined light on some not completely clear aspects related to the experimental results.

Del Villar et al. studied the phenomenon of the attenuation bands fading during the modal transition, which is not predicted by the numerical analysis presented here. They sophisticated the model first introducing a lossy overlay[36] and successively the hybrid modes instead of the LP modes[37]. They also found in this last case a double step transition as opposed to the one step transition seen in this section. The double step transition is better observed when there is high index contrast between the cladding and the overlay and if low radial order modes are considered.

More complex structures comprising multi-overlay geometries have been proposed to extend the degrees of freedom available for the design of the thin film coated LPGs sensors in terms of dynamic range, sensitivity, linearity[30]. Even the apparently adverse phenomenon of the attenuation bands vanishing has been dignified through the proposal of new coated LPG-based chemical sensor exploiting the absorbance change of the overlay on chemical interaction[38].

Finally, the presence of an HRI overlay on a LPG with a reduced cladding radius has been numerically investigated revealing a potentially higher sensitivity and an extended operational range[39,40].

3.4 Nano-Coated LPGs Fabrication

As we have seen in the previous section, the presence of a thin layer (few hundreds of nanometers) of a material with higher refractive index than the cladding is able to increase the sensitivity of the coated device to changes of the SRI and of the overlay parameters. This effect is particularly useful when the LPG is used as a transducer for a chemical sensor. The bare LPG, even though sensitive to SRI changes, is not able to work as a species-specific chemical sensor, or in other words is not able to discriminate among the causes of a RI change in its surrounding environment. Moreover, it is usually desirable to reveal trace amounts of pollutants, which do not produce any appreciable change of the environment RI. For this reason a chemical sensitive layer of material is used to act as a selective collector/concentrator. For sensitivity layers, they should optimize specific interactions with target analyte or narrow class of analytes; they should provide a fast and reversible diffusion of the penetrants and short recovery times; they should maintain the physical state over several cycles of use to avoid hysteresis effects, and thus to ensure the measurement reproducibility[21].

Polymers are among the most widely used materials for chemical sensing since they are able to collect and concentrate molecules on sensor surfaces by reversible sorption; they can be deposited on several substrates as thin adherent films with easy processing techniques; they are available in many kinds having different chemical and sorption properties, which can be used to enhance the selectivity of the sensors; last but not least, they are of low cost.

In this section, we will describe the properties of the material used in the experiment as well as the technique used to deposit it to form a thin overlay on the LPG.

3.4.1 Syndiotactic Polystyrene

Syndiotactic polystyrene (sPS) is a semicrystalline polymer, which shows four different forms, namely, α, β, γ, and δ, according to the nomenclature proposed by Guerra et al.[41] The first two forms, α and β, have a *trans*-planar conformation of the chains, while the other two, γ and δ, a s(2/1)2 helical conformation. Among these crystalline structures the δ form raises particular interest, as it is nanoporous[42]. This structure enables small organic molecules to be hosted between the helical chains in the crystalline phase of sPS to give rise to the corresponding δ-clathrated form. This can be obtained, for example, when the polymer is cast from solution. The compounds that can clathrate are aromatic molecules, such as benzene or toluene, small halogenated molecules, e.g. 1,2-dichloroethane, chloropropane or chloroform, or small oxygenated molecules such as tetrahydrofurane. The δ-clathrated form can be transformed into the "empty" δ form by a proper solvent treatment, by which the included molecules can be removed, leaving voids in the crystalline phase[43,44].

In this form sPS can absorb reversibly certain analytes, whose size and shape well fit the nanocavities establishing specific host guest interactions, when exposed to vapor or liquid environment where these compounds are present even in traces.

The nanoporous structure of the δ sPS is responsible of its higher sorption capability compared with other semicrystalline polymers since it allows the penetration not only in the amorphous phase but also within the crystalline domains that usually are more compact and then impervious to penetrants[45]. This is a consequence of the lower density of the crystalline phase (0.977 g/cm^3) with respect to that of the amorphous phase (1.05 g/cm^3). However, a previous study[46] demonstrated that the absorption in the amorphous phase is negligible compared with that in the crystalline phase at low concentration of analyte. Under this circumstance, it can be safely assumed that the sorption process concerns only those molecules of proper dimensions that are able to establish host-guest interaction with the nanocavities, like low molecular chlorinated or aromatic compounds. This allows δ sPS to be successfully used as polymeric coating for chemical sensing as well as for applications in chemical separation and water purification in membrane feature[47].

The sPS has been exploited as sensitive material for fiber optic chemical sensors based on reflectance measurements and aimed to detection of chloroform and toluene in water and air environments[48–50]. The refractive index of sPS thin films is estimated to be about 1.578. The effect of the analyte sorption in the crystalline domain was modeled as an increase in the material density, which in turn leads to an increase in the refractive index according to the Lorentz-Lorenz law:

$$\frac{n^2 - 1}{n^2 + 2} = \frac{N}{3M\varepsilon}\rho\beta, \tag{3.13}$$

where ρ is the density, β is the polarizability, N is the Avogadro number, M is the molecular weight of the repetitive polymer unit, and ε is the vacuum permittivity. It is worth noting that, since sorption occurs mainly in the crystalline nanocavities, at low concentration of analyte, no volume change of polymer layer occurs, while the average refractive index dramatically increases by 10^{-2} for few ppms of chloroform. The time constant τ of the analyte diffusion process through the sPS layer results to be $\tau \propto th^2/D$, where th is the overlay thickness and D is the concentration-averaged mutual diffusion coefficient[50,51].

3.4.2 Dip-Coating

In the dip coating, a substrate is immersed in a liquid solution and then withdrawn with a well-defined extraction speed under controlled temperature and atmospheric conditions. There are five essential steps involved in this process: immersion, extraction, wet layer formation, drainage, and solvent evaporation, as illustrated in Fig. 3.9.

The film thickness depends upon many parameters such as the withdrawal speed, the solid content, and the viscosity of the liquid. If the withdrawal speed is chosen such that the shear rates keep the system in the Newtonian regime, then the coating thickness depends upon the aforementioned parameters by the Landau-Levich equation[52]:

$$th = 0.94 \cdot \frac{(\eta \cdot v)^{2/3}}{\gamma_{\mathrm{LV}} \cdot (\rho \cdot g)^{1/6}}, \tag{3.14}$$

where th is the coating thickness, η is the solution viscosity, ρ is the density, γ_{LV} the liquid-vapor surface tension, g is the gravity, and v is the withdrawal speed.

It is worth to mention that there are other two techniques widely and successfully used to deposit nanoscale overlays onto LPGs, namely the Langmuir-Blodgett (LB) and the electrostatic self-assembly (ESA) techniques. Both techniques permit to deposit a thin organic film one molecular layer at a time and for this reason they permit a fine control of its thickness and architecture. On the other side, the dip-coating is relatively simple and less time consuming compared with ESA and LB and has been demonstrated to provide a lower roughness of the layer, which in turn determines its higher optical quality[36]. Moreover this technique can be realized

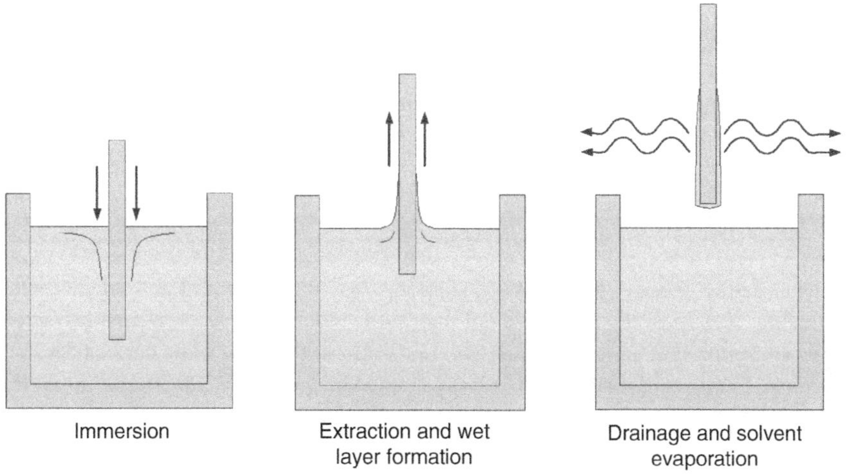

Fig. 3.9 Description of the LPG deposition and testing chamber (not to scale)

with much less expensive equipment than the sophisticated physical vapor deposition (PVD) or chemical vapor deposition (CVD). A drawback of the technique is lack of the accurate control of the thickness compared with ESA or LB techniques, since it depends on too many parameters, which are not easily controlled.

A good source of information about nano-structured coatings for fiber optic sensors, in particular, LPG-based sensors, can be found in a review paper by James et al[53] and the references therein.

3.4.3 Thin Film Deposition onto LPGs

A special holder was designed and realized to host the LPG straight without bending and under similar tensional conditions during both coating procedure and subsequent sensor testing to minimize the effects of the LPG cross-sensitivity to other environmental parameters.

The holder comprises two complementary moulds completely realized in Teflon, forming an about 20 mL (50-mm long) test chamber as shown in Fig. 3.10. Here, a V-groove geometry allows a good placement of the optical fiber. The optical fiber with the LPG can be arranged in the holder and put in a vertical position as illustrated in Fig. 3.11. A small weight (10 g) acting on the fiber by a pulley ensures a similar strain on the LPG in different experiments.

The optoelectronic setup involved for sensor fabrication monitoring comprises a broadband superluminescent diode (2 mW) operating at 1,310 nm and an optical spectrum analyzer for transmitted spectrum monitoring with a resolution of 0.05 nm (see Fig. 3.11). The sPS used for the realization of the coated devices was supplied by Dow Chemical Co. (Questra 101) presented a tacticity up to 98%, average molecular weight (GPC) $Mw = 3.2 \times 10^5$, polydispersity index $Mw/Mn = 3.9$. The LPGs used for the experiments were commercial 30-mm long LPG, with a nominal period of 340 μm, written in a standard Corning SMF-28 optical fiber.

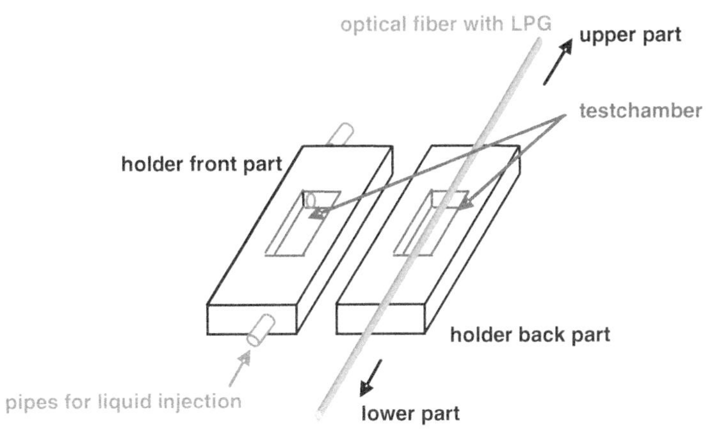

Fig. 3.10 Description of the dip-coating process

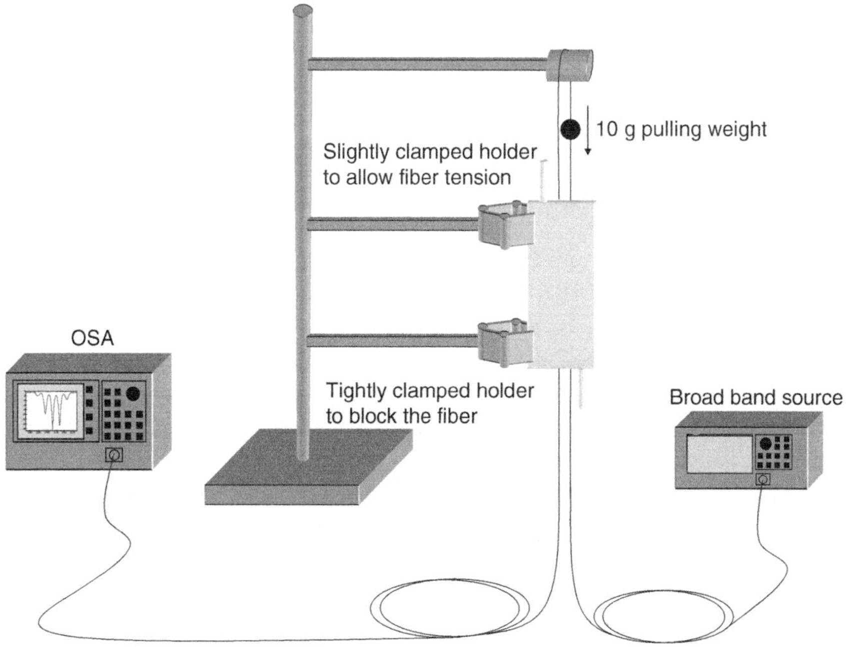

Fig. 3.11 Description of the experimental setup for sensor fabrication and testing (not to scale)

An overlay can be deposited onto the LPG filling up the chamber by means of a syringe with a solution of sPS in chloroform (typ. 2% by weight) and then emptying it out in few seconds. Different overlay thicknesses can be obtained by different extraction speeds and/or solution viscosities. The clathrate sPS film thus formed should be exposed to air for several hours and at a temperature not higher than 50°C, in order to extract chloroform and obtain the empty δ form sPS layer[35].

Figure 3.12 shows a comparison between the transmitted spectra of a bare and a 160 nm sPS coated LPG, with regard to the cladding mode LP_{06}.

Two effects are evident just after the dipping: a blue shift of λ_{res} of 2.45 nm and a decrease of the transmission loss peak of about 1 dB. After the complete desorption of the solvent molecules from the polymeric layer, the transmission spectrum demonstrated a final blue shift of 1.8 nm and a decrease of the transmission loss peak of 0.8 dB compared with the bare one.

The presence of the HRI coating induces an increase in the effective refractive index of the cladding modes and thus a decrease in the resonance wavelengths. In addition, the spatial shift of the cladding mode field profile toward the HRI overlay promotes a decrease of the overlap integral with the core mode and so of the transmission loss peak. The subsequent evaporation of the solvent molecules from the nano-cavities results in a reduction of its refractive index. This explains the partial

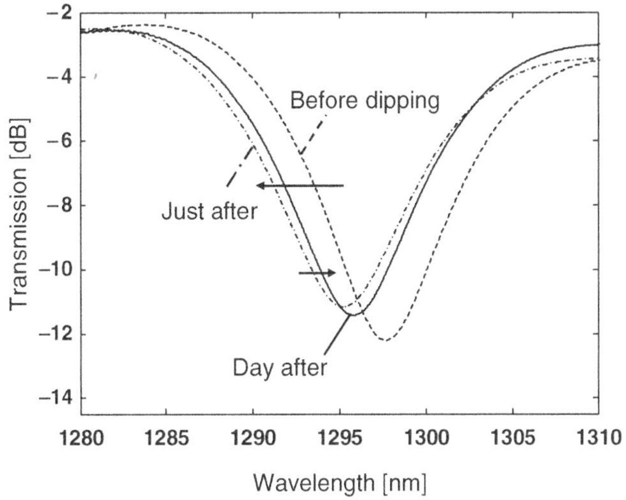

Fig. 3.12 Comparison of the attenuation band related to the cladding mode LP_{06} for a bare LPG and a 160 nm sPS coated LPG just after the dip coating and after solvent evaporation

recovery of the attenuation band initial position. Again a blue shift of the attenuation band is expected during exposure to analyte since chemical absorption within the polymeric layer induces an increase in its refractive index.

It is worth stressing that in the use of the coated LPG as a refractometer, and not as chemical sensor, the sPS layer acts just as a passive HRI layer and no chemical interaction with surrounding medium occurs. Therefore, any HRI overlay could be used for the refractometry, not necessarily a chemo-sensitive one as the sPS. However, in this work the same material has been used for both applications taking advantage of the experience ripened in its deposition.

The overlays thickness estimation is possible exploiting a twin fiber placed in the same deposition chamber and subject to the same coating procedure. The coated twin fiber can be cut by a precision cleaver and analyzed by scanning electron microscopy (SEM). One of the deposited thin overlay is clearly observable in the SEM image of the fiber section reported in Fig. 3.13.

Another way to directly measure the overlay thickness is to scratch the overlay itself by laser micromachining techniques, or simpler by a sharp blade, and then measuring the step by atomic force microscopy (AFM). An AFM topography image ($10 \times 10\ \mu m^2$) with cross section of a deposited sPS overlay is shown in Fig. 3.14, here can be clearly observed the uniformity of the layer along the curvature of fiber and its thickness. The amassment of material next to the trench wall is always present when the overlay is scratched. Moreover the fiber curvature is exaggerated due to the different scales of the axis. For this reason, a good measure of the overlay thickness is the difference of the ordinates just before and after the overshoot of the profile.

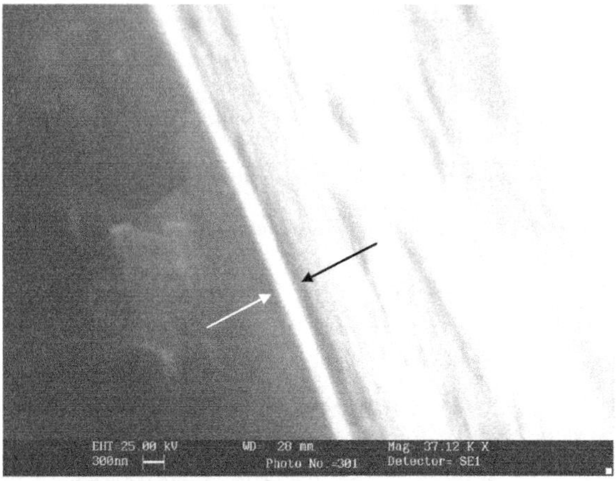

Fig. 3.13 SEM photogram reveals an overlay thickness of about 200 nm

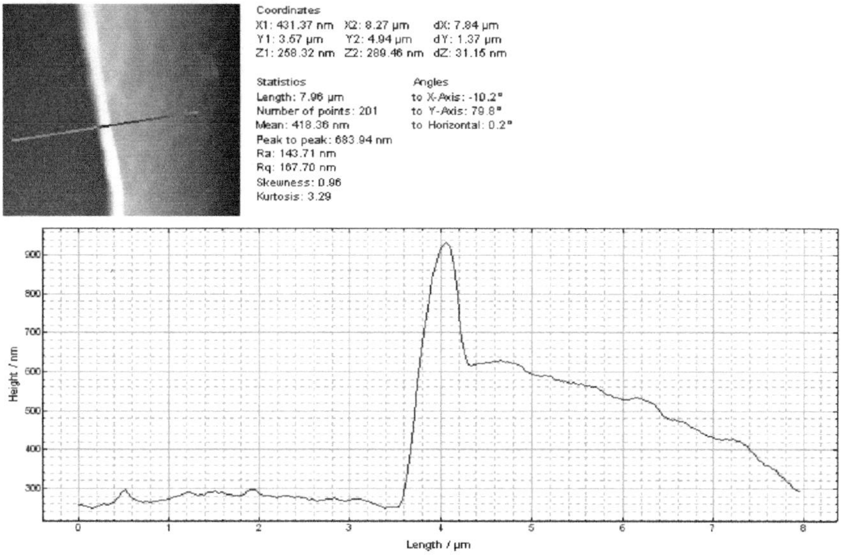

Fig. 3.14 AFM topography image with cross section of a deposited sPS overlay

3.5 Experimental Results

This section is devoted to the presentation of the experimental results concerning LPGs coated with nano-sized overlays of sPS. We will experimentally show the phenomenon of the modal transition and the consequent cladding mode reorganization by characterizing the spectral behavior of coated LPGs to SRI changes. The cladding mode transition is accompanied by a dramatic increase of the device sensitivity.

This consequence can be exploited to build high performances refractometers or chemical sensors. Moreover the transition region can be tuned over the desired SRI range by acting on the overlay thickness allowing the placement of the high sensitivity region also around SRIs where the bare LPG has a scarcely sensitive response. The wavelength shift of the attenuation bands, as a consequence of the mode transition, can be modeled by a simple analytic expression. The fitting of experimental determined transition curves (for different overlay thicknesses and mode orders) with the analytic form allows to derive some basic design criteria for the development of highly sensitive in fiber refractometers for specific SRI ranges.

The conclusion of the section deals with an application of the engineered HRI-coated LPG to chemical sensing. In particular, δ form sPS layers will be used on the LPG to detect trace amounts of chloroform in water revealing a sub-ppm detection capability.

3.5.1 Mode Transition: Effects on the SRI Sensitivity

To experimentally prove the phenomenon of the modal transition and its effects on the sensitivity of nano-coated LPGs to SRI changes, the dip coating, as deposition technique, and the sPS, as HRI material, were used. Here the LPG sensitivity (different for each cladding mode) is defined as the gradient of the wavelength shift of the attenuation bands minima with respect to the SRI ($\partial \lambda_{\text{res},j} / \partial \text{SRI}$).

The use of the dip-coating technique allows to obtain different overlay thicknesses by acting on the solution viscosity and extraction speed as stated by the Landau-Levich equation (see (3.14)). In particular, thicker overlays can be obtained by increasing the extraction speed and/or by increasing the solution viscosity.

The investigation of the thin film coated LPG under test involved recording its transmission spectra for different SRIs with an optical spectrum analyzer. A holder similar to that used for the deposition was used to host the coated LPG allowing its immersion in different liquids. The experimental setup was virtually identical to that depicted in Fig. 3.11 except for the light source, which was in this case a white light source of 400–1,800 nm wavelength range. The SRI was changed by using aqueous glycerol solutions whose refractive indices were measured by an Abbe refractometer at 589 nm.

Here, the sPS overlay was not used for its properties of species-specific sensitive overlay, since no suitable penetrant molecules are present in the used liquids, but just as a passive HRI coating.

As a preliminary test, the experimental setup just described was used to characterize the bare LPG. In Fig. 3.15 it is reported that the central wavelength and the transmission loss peak of the attenuation band centred around 1,300 nm (related to the cladding mode LP_{06}) as a function of the SRI. The refractive index sensitivity of the bare LPG was found to be −5.06 nm/RIU (refractive index units) and 3.59 dB/RIU, around the water refractive index, in terms of wavelength shift and amplitude

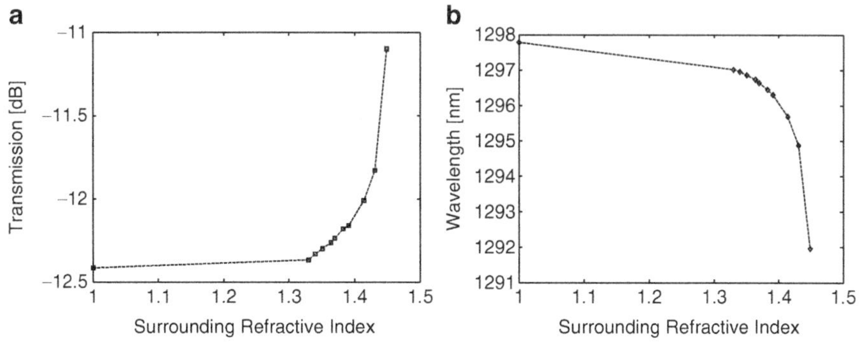

Fig. 3.15 Wavelength shift and loss peak change vs. SRI for the attenuation band around 1,300 nm of a bare LPG with period 340 μm

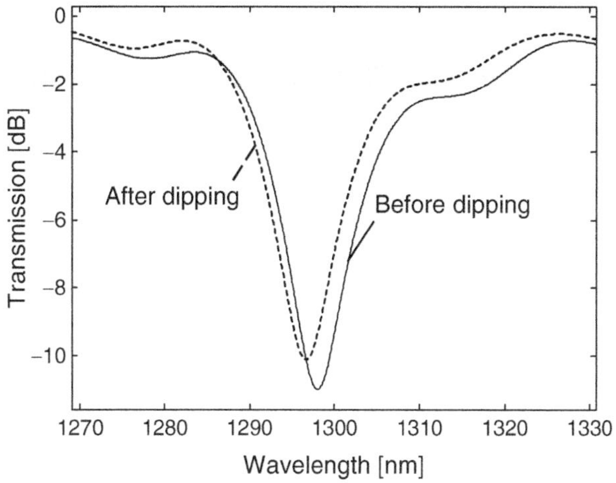

Fig. 3.16 Wavelength shift and loss peak change of the attenuation band related to the mode LP_{06} consequent to the deposition of a 150 nm sPS overlay

changes, respectively. It is worth noting the typical nonlinear monotonic behavior of the wavelength shift with maximum sensitivity for SRIs close to the cladding RI.

The transmission spectra, regarding the LP_{06} mode, of the bare 150 nm sPS coated LPG are reported in Fig. 3.16. As observed, a slight wavelength blue shift of about 1.4 nm is achieved due to the increase of the cladding mode effective refractive indices. The spectral shift is obviously greater for higher order cladding modes, even if not shown here. As predicted by the numerical analysis, the measured wavelength shift is small since for this overlay thickness and refractive index, the cladding modes are very far from the transition region when air is the surrounding medium.

Figure 3.17a reports the transmission spectra of the 150 nm coated LPG for different values of SRI in the range 1.33–1.472, while a zoom on the LP_{07} and LP_{08} modes in the wavelength range 1,350–1,680 nm is shown in Fig. 3.17b.

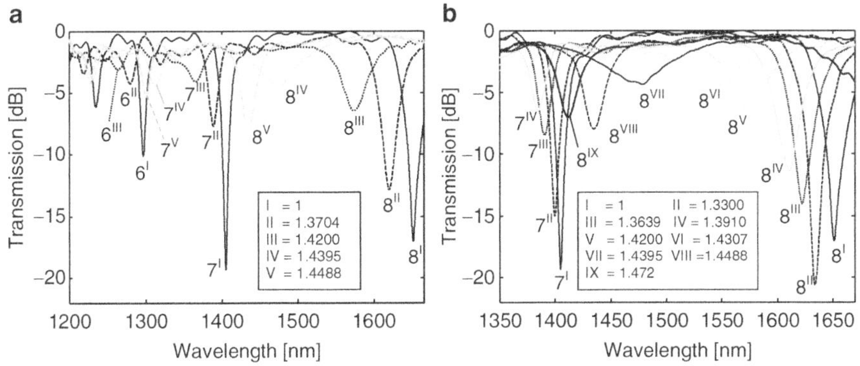

Fig. 3.17 (**a**) Transmission spectra of a 150 nm sPS coated LPG for different values of SRI in the range 1.33–1.472; (**b**) zoom on the LP_{07} and LP_{08} cladding modes

In both figures the attenuation band related to the LP_{0j} mode has been marked with j, j', and so on for the different SRIs. As observed, all the attenuation bands span the wavelength range, as a consequence of the changing SRI, to reach the next lower attenuation band position.

In agreement with the numerical analysis, as the SRI reaches approximately 1.42, a larger wavelength shift in all the attenuation bands is observed. With reference to Fig. 3.17b, it comes natural to name "transition index" the SRI value at which the maximum sensitivity with respect to the external medium occurs. When the lowest order cladding mode is subject to the transition from cladding mode to overlay mode, there is a consequent reorganization of all the other cladding modes. In particular, it is worth noting how at the end of the transition, i.e., for a SRI value approaching 1.45, all the attenuation bands related to the novel mode distribution exhibit a blue shift that is able to recover almost completely the spectral configuration before the transition. Instead, concerning the attenuation bands depth, the novel mode distribution is characterized by smaller loss peaks than those before the mode reorganization. A more clear understanding of the attenuation bands behavior is provided in Fig. 3.18a, where the wavelength shifts of the attenuation bands minima are reported.

Far from the transition region, the presence of the HRI coating leads to a slight increase in the SRI sensitivity with respect to the bare device, which can be roughly estimated to be a factor ranging from 2 to over 10 as the mode order decreases. Instead, within the transition region, a relevant enhancement in the SRI sensitivity is observed. As the mode order decreases, the sensitivity enhancement around 1.43 was estimated to be 8–70 times the sensitivity of the uncoated LPG at the same SRI. This is true if a linear fitting of the experimental data in the transition region is considered. Smaller, but still considerable, values for the sensitivity gains will be retrieved with a different fitting of the data in a next section. However that fitting will allow to model the phenomenon of the modal transition with a simple analytical formula, and for this reason will involve all the data and not just the points related to the transition region.

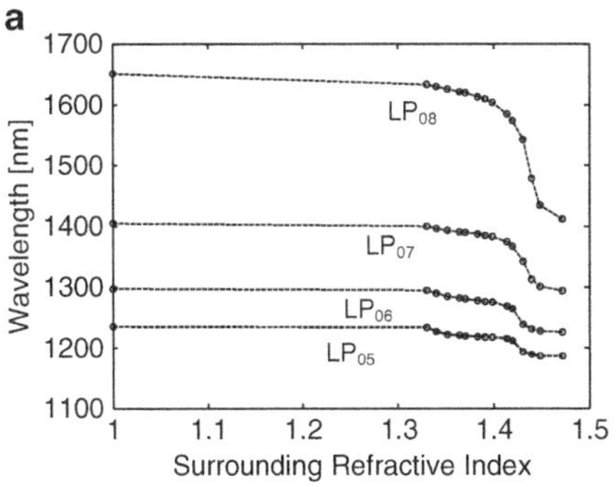

Fig. 3.18 (**a**) Wavelength shift of different cladding modes for the LPG coated with a 150 nm sPS overlay versus SRI; (**b**) loss peaks of different cladding modes versus SRI for the coated LPG with the same coating thickness

The behavior of the loss peaks related to different cladding modes vs. the SRI is reported in Fig. 3.18b. In particular, the attenuation band related to the cladding mode LP_{05}, during the transition, moves to shorter wavelengths and disappears from the spectral window under investigation. The attenuation band related to the cladding mode LP_{08} first increases its depth, thus demonstrating to be initially over-coupled, then follows the decreasing trend of the other attenuation bands.

All the attenuation bands during the transition demonstrated a decrease of the loss peak more relevant than that predicted by the first numerical analysis[29,34], but still significantly lower than that observed in other experiments where the deposition technique was different[54,55].

The reason of the attenuation bands vanishing during the transition was studied by Del Villar et al[36]. They showed that if an imaginary part in the overlay refractive index is considered, then during the modal transition there is a maximum of the imaginary part of the cladding mode effective index. As a consequence, the coupling efficiency is reduced. Another additional reason of the attenuation bands fading was then found in the consideration of the coupling mechanism involving the hybrid modes instead of the LP modes[56]. The physical origin of the imaginary part of the overlay refractive index was identified mainly in its roughness, which is a consequence of the overlay deposition technique (ESA or LB). On the contrary, the dip-coating technique permits to deposit low absorption overlays thus preserving rather well the attenuation bands in the transition region.

To demonstrate that the sensitivity gain region can be tuned over the desired SRI range by acting on the overlay thickness, another sPS overlay was deposited on a LPG nominally identical to the previous one. In this case, a blue wavelength shift of 2.8 nm was recorded (for the LP_{06}) after the deposition of the film and the overlay thickness was found to be about 180 nm. Then the SRI characterization of the coated device was performed and it is reported in Fig. 3.19 as wavelength shift for the mode LP_{08}.

There, a large shift in the central wavelength can be observed as the SRI approaches the value of 1.35. In correspondence of the transition, a wavelength

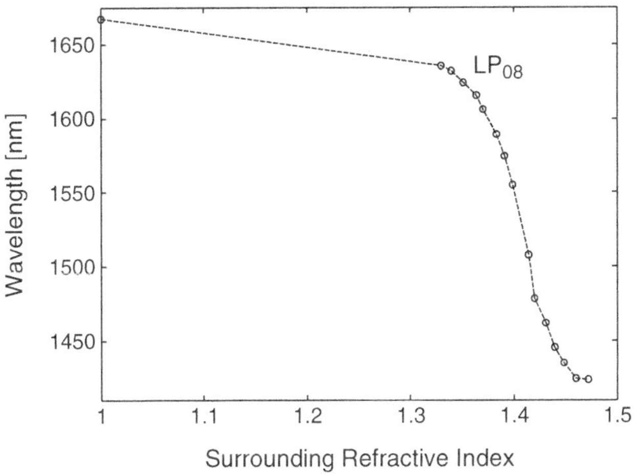

Fig. 3.19 Wavelength shift of the LP_{08} cladding mode for the LPG coated with a 180 nm sPS overlay vs. SRI

shift of over 200 nm was recorded for a refractive index change of ~0.1. This means that low cost and commercially available spectrometer with a wavelength resolution of 0.1 nm would ensure a SRI resolution of approximately 5×10^{-5}.

It is worth noting that in good agreement with the numerical analysis, the transition is slower and less marked as the overlay thickness increases. However, much higher sensitivities than those of the bare device can be obtained in correspondence of air ($n = 1$), water ($n = 1.33$), or other solutions as surrounding media, by using the suitable overlay thickness.

For the purpose of achieving a full understanding of the mode transition phenomenon, other sPS overlays with different thicknesses have been successively deposited on the same LPG. Each deposition was followed by the SRI characterization of the coated device. The experimental results, presented in the next section, have been analyzed to obtain the sensitivity characteristics of the HRI-coated LPG.

3.5.2 Sensitivity Characteristics of Nano-Coated LPGs

In this section, the sensitivity characteristics of HRI-coated LPGs have been investigated to outline their dependence on the overlay thickness and mode order. In addition, the experimental results here presented provide the basic design criteria for the development of highly sensitive in fiber refractometers and chemical sensors for specific SRI ranges.

The behavior of the central wavelengths of the attenuation bands for a 140 nm coated LPG as a function of the SRI is reported in Fig. 3.20. As observed, the resonance wavelengths of the investigated cladding modes (λ_{0j} for $j = 6, 7, 8$) decrease by increasing the SRI until a critical point is reached. In a well defined and

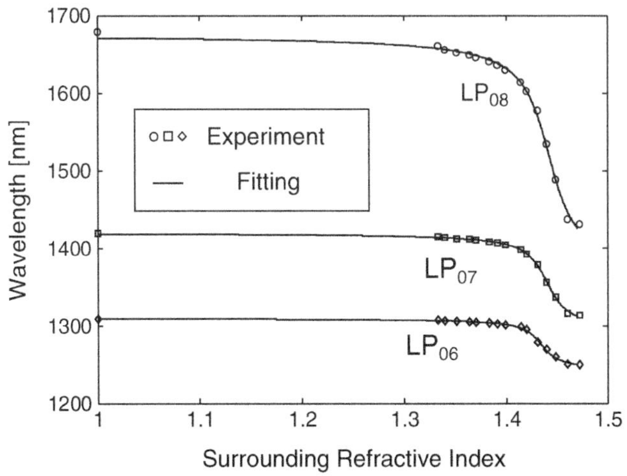

Fig. 3.20 Attenuation bands shift vs. SRI for a 140 nm sPS coated LPG

restricted SRI range, a significant increase of the SRI sensitivity occurs quasi simultaneously for all the attenuation bands.

The mechanism that induces such a change in the resonance wavelength behavior is the transition of the lowest order cladding mode into an overlay guided mode and the consequent mode redistribution. Here, the attention is focused on the analysis of the transition curves to determine the novel sensitivity characteristics of HRI-coated LPGs and to outline the dominant role of the overlay thickness and the mode order.

To this aim, an analytical form was proposed to model the transition curves. The Lorentzian-Cumulative function was found to be the best fit of the experimentally determined transition curves. This means that each resonance wavelength λ_{0j} related to the cladding mode LP_{0j} can be expressed as:

$$\lambda_{0j} = \lambda_{0j\_\mathrm{offset}} + \frac{S_{0j\_\max} \cdot \Delta \mathrm{SRI}_{3\mathrm{dB}\_0j}}{2} \cdot \left[\mathrm{arctg}\left(2 \frac{\mathrm{SRI}_{\mathrm{tr}\_0j} - \mathrm{SRI}}{\Delta \mathrm{SRI}_{3\mathrm{dB}\_0j}} \right) + \frac{\pi}{2} \right], \quad (3.15)$$

where $\lambda_{0j\_\mathrm{offset}}$ is an offset wavelength for the transition curve; $\mathrm{SRI}_{\mathrm{tr}\_0j}$ is the transition index, or the SRI that identifies the middle of the transition region; $S_{0j\_\max}$ is the maximum sensitivity; finally, $\Delta\mathrm{SRI}_{3\mathrm{db}\_0j}$ is the sensitivity bandwidth, or the SRI range in which the sensitivity is not less than half maximum. All these parameters, as we are going to see, depend on the cladding mode order and on the overlay thickness.

In Fig. 3.20, the experimental transition curves (circle, square, and diamond markers for LP_{08}, LP_{07}, and LP_{06} modes, respectively) are plotted together with the Lorentzian fitting, while the fitting parameters are shown in Table 3.1.

An excellent agreement between the experiment and the model was observed for all the attenuation bands. Moreover, the proposed fitting has been successfully

Table 3.1 Fitting parameters of the Lorentian-Cumulative function for the transition curves of a 140 nm coated LPG

140 nm coated LPG	λ_{0j_offset} (nm)	S_{0j_max} (nm)	SRI_{tr_0j}	ΔSRI_{3dB_0j}
LP_{06}	1,243	1,596	1.4344	2.7×10^{-2}
LP_{07}	1,293	2,575	1.4399	3.1×10^{-2}
LP_{08}	1,365	4,989	1.4423	4.0×10^{-2}

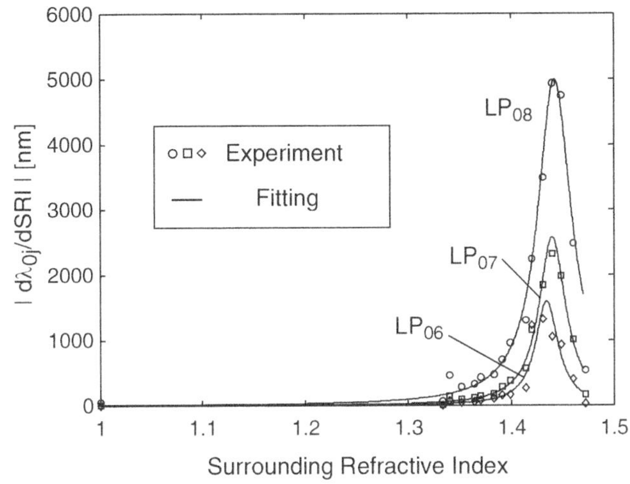

Fig. 3.21 SRI Sensitivity characteristics in 140 nm coated LPG for different cladding modes

tested with transition curves related to different thickness values and for different coated LPGs, as we will see soon.

Once modeled the transition curves, it is possible to retrieve an analytical form for the sensitivity characteristics in nano-sized coated LPGs. By using the SRI-derivative of the resonance wavelengths, the SRI sensitivity can be expressed as:

$$S_{0j} = \frac{\partial \lambda_{0j}}{\partial \mathrm{SRI}} = -\frac{S_{0j\_\max}}{1 + \left(2\frac{\mathrm{SRI}_{\mathrm{tr}\_0j} - \mathrm{SRI}}{\Delta \mathrm{SRI}_{3\mathrm{dB}\_0j}}\right)^2}, \qquad (3.16)$$

From (3.16) it is now clear the physical meaning of $S_{0j\_\max}$ and ΔSRI_{3db_0j}, since they represent the maximum sensitivity at the transition point and the sensitivity bandwidth evaluated at half maximum (3 dB). Figure 3.21 shows the absolute sensitivity characteristics for a 140 nm coated LPG and for the LP_{06}, LP_{07} and LP_{08} cladding modes, obtained by using (3.16). In the same figure, the experimentally estimated sensitivity is also reported. Once again good agreement has been observed between the experimental results and the analytical form used, thus demonstrating its suitability to model the transition curves in such devices.

As can be inferred by the results obtained, the effect of nano-sized HRI coatings on LPGs is a drastic modification of their sensitivity characteristics, which ranges from the nonlinear monotonic behavior of the bare device, showing maximum sensitivity at SRI close to the cladding RI, to a resonant-like behavior of the coated one. The sensitivity curves demonstrate a selective behavior, exhibiting the maximum value in correspondence of the transition index for each cladding mode[57].

The key parameters in determining the final sensing properties of this class of devices can be identified in maximum sensitivity, transition index, and bandwidth. All these parameters are an increasing function of the mode order. In particular, the transition SRI slightly increases for higher order modes passing from ~1.4344 for the LP_{06} to 1.4423 for the LP_{08} (see Table 3.1). The maximum sensitivity ranges from 1,596 nm/RIU for the LP_{06} to 2,575 nm/RIU and 4,989 nm/RIU for the LP_{07} and LP_{08}, respectively. Finally, also the sensitivity bandwidth increases with the mode order passing from $\sim 2.7 \times 10^{-2}$ for the LP_{06} to $\sim 4.0 \times 10^{-2}$ for the LP_{08}.

As we have seen in the last paragraph, the innovative feature of the nano-sized HRI-coated LPGs is not only the enhanced SRI sensitivity, but also the possibility to tune the high sensitivity region over the desired SRI range by acting on the overlay thickness. For this reason, the influence of the coating thickness on the sensitivity characteristics of the coated device was more deeply investigated.

In this regard, Fig. 3.22 shows the sensitivity characteristics of the LP_{08} cladding mode for three different values of the coating thickness: 140, 180, and 215 nm. The fitting parameters are reported in Table 3.2.

Also in this case a very good fitting of the experimental results was achieved by using (3.16), demonstrating once more the validity of the empirically deduced function in describing the phenomenon of the modal transition.

Here, it should be stressed that at the transition point there is the maximum attainable field content of the cladding modes in the overlay as well as the maximum evanescent field in the surrounding medium (see Sect. 3.3). As a consequence, one can tune the high sensitivity region in the desired SRI range by acting

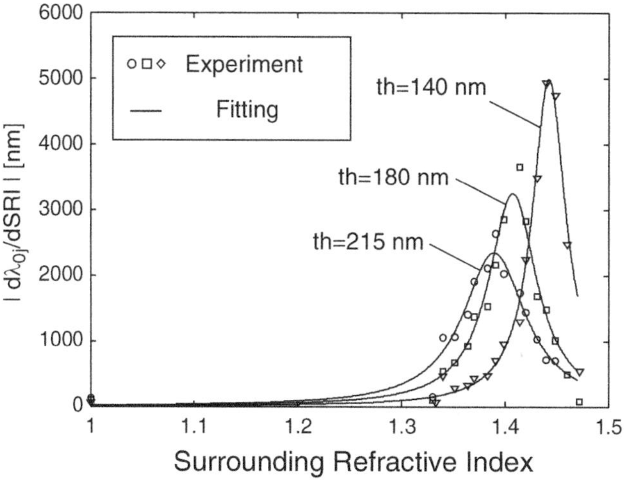

Fig. 3.22 Sensitivity characteristics of the mode LP_{08} for different overlay thickness values

Table 3.2 Fitting parameters of the Lorentian-Cumulative function for the transition curves of the mode LP_{08} for different overlay thickness values

LP_{08} cladding mode	S_{0j_max} (nm)	SRI_{tr_0j}	ΔSRI_{3dB_0j}
140 nm LPG	4,989	1.4423	4.0×10^{-2}
180 nm LPG	3,249	1.4073	5.7×10^{-2}
215 nm LPG	2,350	1.3894	7.4×10^{-2}

on the overlay thickness. Successively, the coated device can be exploited as a refractometer if the SRI is changed, and there is no chemical interaction between the surrounding environment and the overlay, or as a chemical sensor if in the surrounding environment is injected a chemical whose presence leads to a negligible SRI change, but which in turn determines a change in the overlay parameters (thickness or RI).

From Fig. 3.22, it is evident how the sensitivity characteristics can be tailored for chemical sensing in different environments (gaseous, aqueous) by acting on the coating thickness. In fact, by increasing the overlay thickness, the transition value of the SRI moves toward lower values, and thus the maximum sensitivity can be located in the desired SRI range. In particular, the SRI_{tr_0j} shifts from 1.4423 to 1.3894, changing the coating thickness from 140 to 215 nm. The maximum sensitivity is strongly affected by the coating thickness and decreases by increasing the overlay thickness, passing from 4,989 nm/RIU for a 140 nm coating thickness to 2,350 nm/RIU for 215 nm, with regard to the LP_{08} mode (see Table 3.2). Concerning the sensitivity bandwidth, an increasing trend can be observed as the coating thickness is increased passing from 4.0×10^{-2} for 140 nm of coating thickness to 7.4×10^{-2} for 215 nm.

It is worth noting that, although the maximum sensitivity decreases for thicker overlays, the sensitivity gain with respect to the bare device increases. In fact, the sensitivity gain functions are obtained as the ratio between the sensitivity characteristics of coated and bare LPGs. They are reported for different mode orders and overlay thickness values in Fig. 3.23.

A sensitivity gain higher than 20 was provided by a 140 nm coated LPG by selecting the LP_{06} mode for SRI approaching 1.43. Nevertheless, a sensitivity gain up to one order of magnitude for SRI approaching the water refractive index can be obtained by using a 215 nm LPG and by selecting the LP_{08} cladding mode.

On the basis of the reported results, it is evident how the selection of the overlay thickness and of the specific cladding mode offers a certain degree of flexibility to design chemical sensors with optimized sensitivity in the desired SRI range.

3.5.3 Coated LPGs as Highly Sensitive Optochemical Sensors

In this section, we present an application to chemical sensing of the nano-sized HRI-coated LPGs. This time the sPS overlay is used both as HRI material to enhance the sensitivity of the LPG and as a selective molecular sieve to collect

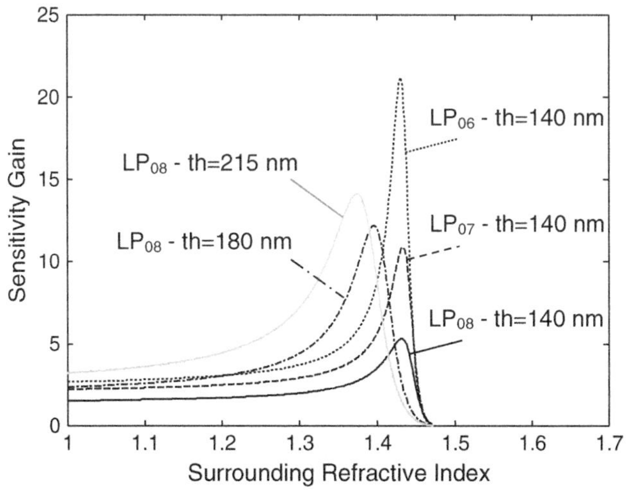

Fig. 3.23 Sensitivity gain between coated and bare LPG for different mode orders and overlay thickness values

and concentrate specific analyte molecules from the surrounding environment. It is important to remark that the sPS does not swell in response to chemical absorption at low concentrations, but rather it changes its refractive index as the analyte molecules fill the nano-voids of its crystalline domains (see Sect. 3.4).

In this experiment, we optimize the sensitivity of the coated LPG to the overlay RI changes for an SRI = 1.33 by acting on the overlay thickness.

We have shown that if the overlay features are properly chosen for a specific SRI, a strong field enhancement within the overlay occurs, leading to an excellent sensitivity of the cladding mode distribution to the coating parameters (thickness and RI) changes. In this case, as the sensitive material is sPS, the initial refractive index of the material is fixed at 1.58. Hence the tuning of the coated LPG in the high sensitivity region can be performed by changing the overlay thickness. In particular, here we are interested in chemical sensing in water, so the SRI at which maximize the performances of the sensor is 1.33.

The experimental setup was virtually identical to that depicted in Fig. 3.11 with two broadband superluminescent diodes (2 mW) operating at 1,310 and 1,550 nm, respectively, and an optical spectrum analyzer for transmitted spectrum monitoring with a resolution of 0.05 nm. A holder similar to that used for the deposition was used to host the coated LPG allowing also the conveying of pure distilled water or polluted water as the case. The temperature was held constant at 20°C.

3.5.3.1 Far from the Transition Region

The dip-coating method was used to deposit the δ form sPS layer on the LPG. The optical fiber with the LPG was arranged in the holder and a sPS chloroform solution (2% by weight) was added. A first overlay was then deposited emptying out the test chamber in about 3 s. The clathrate thus formed was finally exposed to air, for about 15 h at temperature of about 35°C, to extract chloroform. In particular, the effect that the overlay deposition has on the transmission spectrum of the LPG has been discussed in Sect. 4.3. The overlay thickness was found to be approximately 160 nm (th_1).

For this thickness value and for an SRI = 1.33 the cladding modes are far from the transition region, as it can be clearly inferred from Fig. 3.18 where the wavelength shift of the attenuation bands is reported as a function of the SRI for an overlay thickness (150 nm) very close to the one considered here. There, with reference to the attenuation band initially centred around 1,300 nm, the total blue wavelength shift of the considered mode is 71 nm. The transition region starts for an external refractive index of 1.34 (10% of the total wavelength shift), while the maximum sensitivity to SRI changes is shown for 1.42, where a wavelength shift of 25 nm was observed for a surrounding refractive index change of 0.011. At the end of the transition for a SRI of 1.439 (90% of the total wavelength shift), the attenuation band is centred around 1232.7 nm.

To test the sensor response to overlay refractive index changes induced by chemical sorption, chloroform was used as analyte. Measurements consisted in recording the transmission spectra of the sensing grating as the sorption of the analyte in the nanocavities promoted an increase of the polymer layer refractive index. The spectra were recorded every 40 s. The holder, with the sPS coated LPG, was connected with a thermostated beaker (20°C), containing initially 1 L of pure distilled water. In this way, the holder was filled up with pure water. Chloroform was then added by successive steps of 10 ppm (μL/L). The solutions were always magnetically stirred to ensure the maximum dispersion of the analyte in water and then fluxed to the sensor holder. Figure 3.24a shows the changes of the transmitted spectrum and Fig. 3.24b shows the time responses of the sensor in terms of wavelength shift and amplitude changes of the considered attenuation band to two successive 10 ppm chloroform exposures.

At these concentration levels, negligible effects on the SRI occur, thus changes in the attenuation bands reported here can be attributable only to the chemical sorption within the sensitive overlay. The 10 and 20 ppm chloroform concentrations induced a blue wavelength shift of 0.96 and 1.26 nm, respectively, and a decrease of the loss peak of 1.20 and 1.57 dB.

Here, a response time (10–90%) of about 21 min (τ_1) for 10 ppm chloroform concentration was measured. As the dynamic of the chemical sorption relies on the diffusion of the analyte through the sensitive overlay, the overlay thickness determines the time response of the sensor. Moreover, the dependence of the response times on the analyte concentrations can be attributed to the dependence of the diffusivity on the concentration of the absorbed analyte.

When chloroform was added, the equilibrium response of the sensor progressively decreased. This is probably related to the combination of the nonlinear behavior of the effective refractive index of the coupled cladding mode on the overlay refractive index with the nonlinear relationship between adsorbed mass of

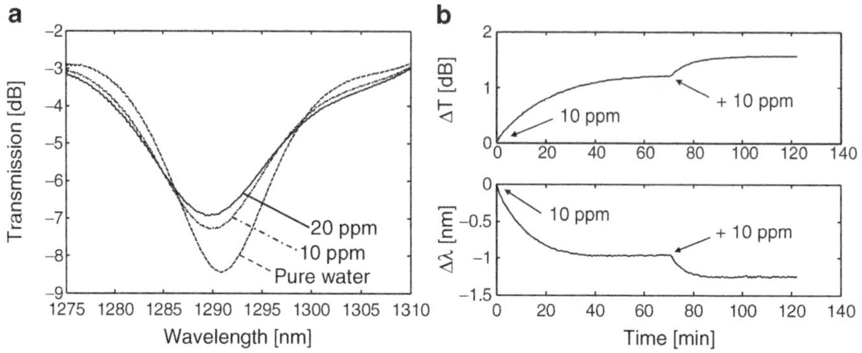

Fig. 3.24 (**a**) Changes in the transmission spectrum due to the analyte exposures; (**b**) time response of the attenuation band around 1,300 nm for a 160 nm sPS coated LPG exposed to chloroform

chloroform and its concentration in the liquid phase. Nonlinear sorption equilibrium was, in fact, reported in literature for sorption in the vapour and liquid phase[56–58].

3.5.3.2 Within the Transition Region

To verify the sensitivity enhancement when the transition region is approached, the first overlay was removed and a new thicker overlay was deposited using the same technique. Here, the measured thickness was approximately 260 nm (th_2) and an attenuation band was found in the 1,550 nm spectral region with water as surrounding medium. This attenuation band was originally located around 1,700 nm when the overlay thickness was 160 nm, but now the overlay thickness was big enough to tune the attenuation band around 1,550 nm. Moreover, from Fig. 3.19, for example, results are apparent that for the considered attenuation band the 1,550 nm spectral region is the half-way between its initial position and that of the next lower attenuation band, or in other words it represents the middle of the transition region. Therefore, we have a higher order cladding mode perfectly tuned in the middle of the transition region.

To prove the sensitivity enhancement when the attenuation band works within the transition region, a procedure similar to that of the previous paragraph by using chloroform exposure in water at room temperature was carried out.

Figure 3.25 shows the time response of the sensor in terms of wavelength shift and loss peak changes due to four successive 5 ppm chloroform exposures. A blue wavelength shift of 3.95, 6.45, 8.03, 8.79 nm and a decrease of the transmission loss 1.20, 1.91, 2.34, 2.53 dB were measured. In this case, the response time (10–90%) was estimated to be about 62 min (τ_2) for a single 10 ppm step (not reported here). When chloroform was removed by a continuous flux of distilled water, an

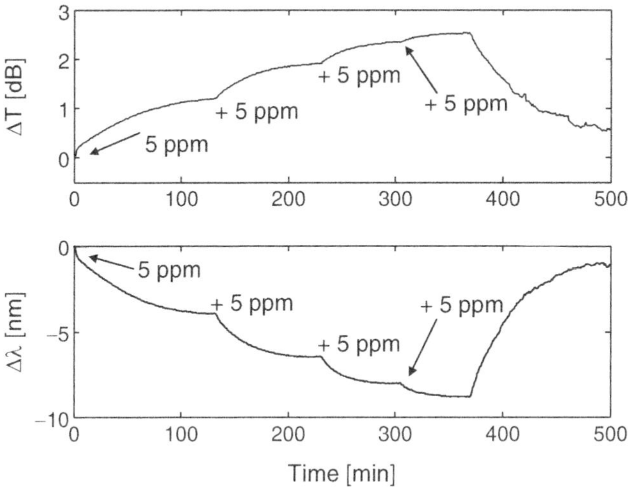

Fig. 3.25 Time response of the attenuation band around 1,550 nm for the 260 nm sPS coated LPG exposed to chloroform

excellent recovery was observed demonstrating the reversibility of the proposed configuration.

With regard to the response times, according to the diffusion theory and the experimental results previously reported a quadratic rule can be assumed[50,51]:

$$th_1/th_2 = (\tau_1/\tau_2)^{0.5}, \tag{3.17}$$

where th_l and th_2 are overlay thicknesses and τ_l and τ_2 are response times. In our case, with $th_l \approx 160$ nm $th_2 \approx 260$ nm, $\tau_l = 21$ min and $\tau_2 = 62$ min, we obtain 0.615 for the left side and 0.586 for the right one demonstrating good agreement with the data previously reported.

With regard to sensor sensitivity, Fig. 3.26 shows the wavelength shift and amplitude changes vs. chloroform concentration with quadratic interpolation for the 160 and 260 nm coated LPG compared with the uncoated one. As expected, without the sensing overlay negligible variations have been observed in both the measured parameters. Sensitivities of −0.130 nm/ppm and 0.163 dB/ppm were observed for the thinner in the range of 0–10 ppm. Sensitivities of −0.85 nm/ppm and 0.26 dB/ppm were measured for the thicker overlay, in the same range.

On the basis of these results, it is evident how the presence of the polymer layer is the uniquely responsible for the significant changes observed in the attenuation band due to the addition of few ppm in water. Moreover, a sensitivity improvement of more than six times in terms of wavelength shift was obtained in the range of 0–10 ppm by considering an overlay with a higher thickness and a higher order cladding mode.

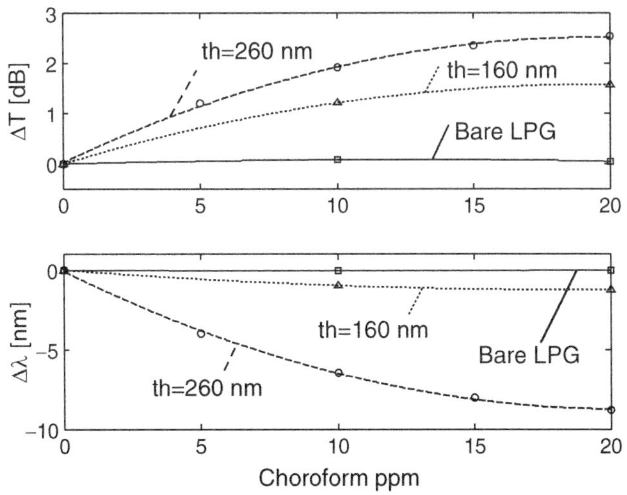

Fig. 3.26 Sensor sensitivity for bare LPG, coated LPG with 160 nm and 260 nm sPS overlay

The latter results are in good agreement with the numerical analysis previously reported and demonstrate the excellent performance of HRI-coated LPGs as opto-chemical sensors.

A great feature of the proposed configuration is the possibility to select the overlay thickness and mode order to meet specific requirements in terms of sensitivity and response time. In addition, with commercially available and low cost spectrometers, a wavelength shift resolution of 0.05 nm is easily available, allowing sub-ppm chemical detection.

3.6 Applications of (Nano)-Coated LPGs

The number of chemical sensors based on LPGs coated with a chemo-sensitive overlay is still quite limited and even lower is the number of those sensors for which the phenomenon of the modal transition is knowingly used to enhance the performances of the devices. Some of them are based on overlays with low refractive index and/or on thick overlays, which have the drawback of slow diffusion limited response times. One application that seems to attract particular interest is the measurement of the relative humidity (RH).

Changes in the transmission loss peak, instead of the attenuation bands wavelength shift, of a 297 nm gelatin-coated LPG were used for RH measurements[58]. The reported results showed a sensitivity of 0.833%RH/dBm in a high RH range (90–99%). They seemed to be attributable mainly to the swelling of the gelatin layer than to its refractive index change.

Relative humidity changes were measured with a nano-scale HRI-coated LPG in a wide range from 38.9% to 100% RH with a sensitivity of 0.2 nm/%RH, and an accuracy of 2.3% RH[59]. The material used was a hydro-gel layer composed by, among other chemicals, acrylic acid and vinyl pyridine. Fine adjust of the components proportion made possible to obtain a refractive index of the gel of about 1.55. The overlay thickness was estimated to be in that case 600 nm.

A very fast (response times of few hundreds of milliseconds) RH LPG-based sensor was proposed, making use of a thin layer of poly(ethylene oxide)/cobalt chloride[60]. The complex spectral response of the sensor was explained by referring to different parameters such as the refractive index change, absorption, swelling of the coating. Good resolution and dynamic range were also demonstrated.

Long period gratings modified by the deposition of a self-assembled colloid monolayer of gold have been shown to enhance SRI sensitivity, which was attributed to the localized surface plasmon resonance (LSPR) of gold nanoparticles[61]. The system was used after functionalization of the gold nanoparticles with dinitrophenyl (DNP) antigen, for the detection of anti-(DNP) with a limit of detection as low as 9.5×10^{-10} M.

A zeolite-coated LPG sensor for the detection of organic vapors was also proposed[62] (also see Chap. 7). The sensor exploited a thick (10 μm) film whose refractive index was estimated to be around 1.33. It is therefore apparent in that case the device probed the overlay refractive index changes through the evanescent tail of the cladding modes. The overlay just shifted the working point of the bare LPG while acting as a collector/concentrator for the vapors. However, the system showed a good sensitivity with a shift of 1.45 nm for an isopropanol vapor concentration of 5.5 ppm and shifted as much as 0.7 nm for a toluene vapor concentration of 0.22 ppm.

Preliminary results were reported about the deposition of a copper sensitive coating deposited onto a LPG by means of layer-by-layer electrostatic adsorption[63].

Corres and coworkers demonstrated a couple pH sensors realized by ESA deposition of nanostructured polymeric materials. Those sensors represented a truly engineered design exploiting the rules of the modal transition[64].

However, the applications of the phenomenon of the modal transition are not limited to the chemical sensing field. Indeed, cladding etching and an HRI layer of sputtered indium tin oxide were exploited to achieve an LPG-based tunable filter with a large tuning range and with limited reduction in the loss peak[65].

Also the use of tapered long period gratings (TLPGs) is a step toward the use of structures with reduced cladding radius, which provide an additional sensitivity gain with respect to UV-written LPGs and are more robust than etched UV-LPGs[66].

Finally, nano-coated cascaded TLPGs or coated LPGs with an ad-hoc laser generated defect in the overlay, if properly designed, could be a useful platform for chemical sensing especially thanks to the presence of finer-scale spectral features and longer interaction lengths, which offer potentially higher sensitivity and resolution in the measurement of changes in environmental parameters[67,68].

References

1 Vengsarkar, A. M.; Lemaire, P. J.; Judkins, J. B.; Bhatia, V.; Erdogan, T.; Sipe, J. E., Long-period fiber gratings as Band Rejection Filters, *J. Lightwave Technol.* **1996**, 14, 58–65

2 Erdogan, T., Fiber grating spectra, *J. Lightwave Technol.* **1997**, 15, 1277–1294

3 Chiang, K. S.; Liu, Q., Long-period gratings for application in optical communications, In Proceedings of the 5th International Conference on Optical Communications and Networks and 2nd International Symposium on Advances and Trends in Fiber Optics and Applications (ICOCN/ATFO 2006), Chengdu, China, Sept **2006**, 128–133

4 James, S. W.; Tatam, R. P., Optical fibre long-period grating sensors: characteristics and application, *Meas. Sci. Technol.* **2003**, 14, R49–R61

5 Bhatia, V., Applications of long-period gratings to single and multi-parameter sensing, *Opt. Express* **1999**, 4, 457–466

6 Shu, X.; Zhang, L.; Bennion, I., Sensitivity characteristics of long-period fiber gratings, *J. Lightwave Technol.* **2002**, 20, 255

7 Allsop, T.; Webb, D. J.; Bennion, I., A comparison of the sensing characteristics of long period gratings written in three different types of fiber, *Opt. Fiber Technol.* **2003**, 9, 210–223

8 Patrick, H. J.; Kersey, A. D.; Bucholtz, F., Analysis of the response of long period fiber gratings to external index of refraction, *J. Lightwave Technol.* **1998**, 16, 1606–1612

9 Duhem, O.; Henninot, J.-F.; Warenghem, M.; Douay, M., Demonstration of long-period grating efficient couplings with an external medium of a refractive index higher than that of silica, *Appl. Opt.* **1998**, 37, 7223–7228

10 Stegall, D. B.; Erdogan, T., Leaky cladding mode propagation in long-period fiber grating devices, IEEE Photon. *Technol. Lett.* **1999**, 11, 343–345

11 Hou, R.; Ghassemlooy, Z.; Hassan, A.; Lu, C.; Dowker, K. P., Modelling of long-period fibre grating response to refractive index higher than that of cladding, *Meas. Sci. Technol.* **2001**, 12, 1709–1713

12 Koyamada, Y., Numerical analysis of core-mode to radiation-mode coupling in long-period fiber gratings, *IEEE Photon. Technol. Lett.* **2001**, 13, 308–310

13 Chiang, K. S.; Liu, Y.; Ng, M. N.; Dong, X., Analysis of etched long-period fiber grating and its response to external refractive index, *Electron. Lett.* **2000**, 36, 966–967

14 Vasiliev, S. A.; Dianov, E. M.; Varelas, D.; Limberger, H. G.; Salathe, R. P., Postfabrication resonance peak positioning of long-period cladding-mode-coupled gratings, *Opt. Lett.* **1996**, 21, 1830–1832

15 Kim, S.; Jeong, Y.; Kim, S.; Kwon, J.; Park, N.; Lee, B., Control of the characteristics of a long-period grating by cladding etching, *Appl. Opt.* **2000**, 39, 2038–2042

16 Zhou, K.; Liu, H.; Hu, X., Tuning the resonant wavelength of long period fiber gratings by etching the fiber's cladding, *Opt. Commun.* **2001**, 197, 295–297

17 Patrick, H. J.; Kersey, A. D.; Bucholtz, F.; Ewing, K. J.; Judkins, J. B.; Vengsarkar, A. M., Chemical sensor based on long-period fibre grating response to index of refraction, *Proc. Conf. Lasers Electro Opt.* (CLEO '97) **1997**, 11, 420–421

18 Falciai, R.; Mignani, A. G.; Vannini, A., Long period gratings as solution concentration sensors, *Sens. Actuators B* **2001**, 74, 74–77

19 Allsop, T.; Zhang, L.; Bennion, I., Detection of organic aromatic compounds in paraffin by a long-period grating optical sensor with optimized sensitivity, *Opt. Comm.* **2001**, 191, 181–190

20 Falate, R.; Kamikawachi, R. C.; Müller, M.; Kalinowski, H. J.; Fabris, J. L., Fiber optic sensors for hydrocarbon detection, *Sens. Actuators B* **2005**, 105, 430–436

21 Grate, J. W.; Abraham, M. H., Solubility interaction and design of chemically selective sorbent coatings for chemical sensor and arrays, *Sens. Actuators B* **1991**, 3, 85–111

22 DeLisa, M. P.; Zhang, Z.; Shiloach, M.; Pilevar, S.; Davis, C. C.; Sirkis, J. S.; Bentley, W. E., Evanescent wave long-period fiber bragg grating as an immobilized antibody biosensor, *Anal. Chem.* **2000**, 72, 2895–2900

23 Rees, N. D.; James, S. W.; Tatam, R. P.; Ashwell, G. J., Optical fiber long-period gratings with Langmuir-Blodgett thin-film overlays, *Opt. Lett.* **2002**, 27, 686–688
24 Gloge, D., Weakly guiding fibers, *Appl. Opt.* **1971**, 10, 2252–2258
25 Erdogan, T., Cladding mode resonances in short and long period fibre grating filters, *J. Opt. Soc. Am. A* **1997**, 14, 1760–1773
26 Anemogiannis, E.; Glytsis, E. N.; Gaylord, T. K., Transmission characteristics of long- period fiber gratings having arbitrary azimutal/radial refractive index variation, *J. Lightwave Technol.* **2003**, 21, 218–227
27 Rego, G.; Ivanov, O. V.; Marques, P. V. S., Demonstration of coupling to symmetric and antisymmetric cladding modes in arc-induced long-period fiber gratings, *Opt. Express* **2006**, 14, 9594–9599
28 Monerie, M., Propagation in doubly clad single-mode fibers, *IEEE J. Quantum Electron.* **1982**, 18, 535–542
29 Del Villar, I.; Matías, I.; Arregui, F.; Lalanne, P. Optimization of sensitivity in long period fiber gratings with overlay deposition, *Opt. Express* **2005**, 13, 56–69
30 Hochreiner, H.; Cada, M.; Wentzell, P. D., Tuning the Response of Long-Period Fiber Gratings for Chemical Sensing Applications, in Next-Generation Spectroscopic Technologies, Brown, C. D.; Druy, M. A.; Coates, J. P., Eds.; *Proc. SPIE* **2007**, 6765, 676504
31 Cusano, A.; Pilla, P.; Contessa, L.; Iadicicco, A.; Campopiano, S.; Cutolo, A.; Giordano, M.; Guerra, G., High sensitivity optical chemo-sensor based on coated long period gratings for sub ppm chemical detection in water, *Appl. Phys. Lett.* **2005**, 87, 234105
32 Cusano, A.; Iadicicco, A.; Pilla, P.; Contessa, L.; Campopiano, S.; Cutolo, A.; Giordano, M., Cladding modes re-organization in high refractive index coated Long Period Gratings: effects on the refractive index sensitivity, *Opt. Lett.* **2005**, 30, 2536–2538
33 Del Villar, I.; Matias, I. R.; Arregui, F. J., Long-period fiber gratings with overlay of variable refractive index, *IEEE Photon. Technol. Lett.* **2005**, 17, 1893–1895
34 Cusano, A.; Iadicicco, A.; Pilla, P.; Contessa, L.; Campopiano, S.; A. Cutolo, M. Giordano, Mode transition in high refractive index coated long period gratings, *Opt. Express* **2006**, 14, 19–34
35 Cusano, A., Iadicicco, A.; Pilla, P.; Contessa, L.; Campopiano, S.; Cutolo, A.; Giordano, M.; Coated long period fiber gratings as high sensitivity opto-chemical sensors, *J. Lightwave Technol.* **2006**, 24, 1776–1786
36 Del Villar, I.; Matias, I. R.; Arregui, F. J.; Achaerandio, M., Nanodeposition of materials with complex refractive index in long-period fiber gratings, *J. Lightwave Technol.* **2005**, 23, 4192
37 Del Villar, I.; Matias, I. R.; Arregui, F. J. Influence on cladding mode distribution of overlay deposition on long-period fiber gratings, *J. Opt. Soc. Am. A* **2006**, 23, 651–658
38 Hochreiner, H.; Wentzell, P. D., Modelling new applications of long-period fibre gratings: Absorbance measurements, Proceedings Europtrode IX, 30 March–2 April 2008, Dublin, Ireland
39 Yang, J.; Yang, L.; Xu, C.; Huang, W.; Li, Y., Long-period grating refractive index sensor with a modified cladding structure for large operational range and high sensitivity, *Appl. Opt.* **2006**, 45, 6142–6147
40 Yang, J.; Yang, L.; Xu, C.; Li, Y., Optimization of cladding-structure-modified long-period-grating refractive-index sensors, *J. Lightwave Technol.* **2007**, 25, 372–380
41 Guerra, G.; Vitagliano, V. M.; De Rosa, C.; Corradini, P., Polymorphism in melt crystallized syndiotactic polystyrene samples, *Macromolecules* **1990**, 23, 1539–1544
42 Petraccone, V.; Auriemma, F.; Dal Poggetto, F.; De Rosa, C.; Guerra, G.; Corradini, P., On the structure of the mesomorphic form of syndiotactic polystyrene, *Makromolekulare Chemie* **1993**, 194, 1335–1345
43 De Rosa, C.; Guerra, G.; Petraccone, V.; Pirozzi, B., Crystal structure of emptied clathrate form (δe form) of syndiotactic polystyrene, *Macromolecules* **1997**, 30, 4147–4152
44 Milano, G.; Venditto, V.; Guerra, G.; Cavallo, L.; Ciambelli, P.; Sannino, D., Shape and volume of cavities in thermoplastic molecular sieves based on syndiotactic polystyrene, *Chem. Mater.* **2001**, 13, 506–1511

45 Guerra, G.; Manfredi, C.; Musto, P.; Tavone, S., Guest conformation and diffusion into amorphous and emptied clathrate phases of syndiotactic polystyrene, *Macromolecules* **1998**, 31, 1329–1334
46 Mensitieri, G.; Venditto, V.; Guerra, G., Polymeric sensing films absorbing organic guests into a nanoporous host crystalline phase, *Sens. Actuators B* **2003**, 92, 255–261
47 Guerra, G.; Mensitieri, G.; Venditto, V., Method for the detection of organic pollutants using syndiotactic polystyrene polymers as sensing elements, US Patent 20020073764: **2002**
48 Giordano, M.; Russo, M.; Cusano, A.; Cutolo, A.; Mensitieri, G.; Nicolais, L., Optical sensor based on ultrathin films of δ-form syndiotactic polystyrene for fast and high resolution detection of chloroform, *Appl. Phys. Lett.* **2004**, 85, 5349
49 Giordano, M.; Russo, M.; Cusano, A.; Mensitieri, G.; Guerra, G., Syndiotactic polystyrene thin film as sensitive layer for an optoelectronic chemical sensing device, *Sens. Actuators B* **2005**, 109, 177–184
50 Giordano, M.; Russo, M.; Cusano, A.; Mensitieri, G., An high sensitivity optical sensor for chloroform vapours detection based on nanometric film of δ-form syndiotactic polystyrene, *Sens. Actuators B* **2005**, 107, 140–147
51 Crank, J., The Mathematics of Diffusion, Clarendon Press, Oxford, 1975
52 Scriven, L. E., Physics and applications of dip coating and spin coating, *Mater. Res. Soc. Symp. Proc.* **1988**, 121, 717–729
53 James, S. W.; Tatam, R. P., Fibre optic sensors with nano-structured coatings, *J. Opt. A* **2006**, 8, S430–S444
54 Ishaq, I. M.; Quintela, A. S.; James, W.; Ashwell, G. J.; Lopez-Higuera, J. M.; Tatam, R. P., Modification of the refractive index response of long period gratings using thin film overlays, *Sens. Actuators B* **2005**, 107, 738–741
55 Del Villar, I.; Achaerandio, M.; Matías, I. R.; Arregui, F. J., Deposition of overlays by electrostatic self-assembly in long-period fiber gratings, *Opt. Lett.* **2005**, 30, 720–722
56 Del Villar, I.; Corres, J. M.; Achaerandio, M.; Arregui, F. J.; Matias, I. R., Spectral evolution with incremental nanocoating of long period fiber gratings, *Opt. Express* **2006**, 14, 11972–11981
57 Cusano, A.; Iadicicco, A.; Pilla, P.; Cutolo, A.; Giordano, M.; Campopiano, S., Sensitivity characteristics in nanosized coated long period gratings, *Appl. Phys. Lett.* **2006**, 89, 201116
58 Tan, K. M.; Tay, C. M.; Tjin, S. C.; Chan, C. C.; Rahardjo H., High relative humidity measurements using gelatin coated long-period grating sensors, *Sens. Actuators B* **2005**, 110, 335–341
59 Liu, Y.; Wang, L.; Zhang, M.; Tu, D.; Mao, X.; Liao, Y., Long-period grating relative humidity sensor with hydrogel coating, *IEEE Photon. Technol. Lett.* **2007**, 19, 880–882
60 Konstantaki, M.; Pissadakis, S.; Pispas, S.; Madamopoulos, N.; Vainos, N. A., Optical fiber long-period grating humidity sensor with poly(ethylene oxide)/cobalt chloride coating, *Appl. Opt.* **2006**, 45, 4567–4571
61 Tang, J. L.; Cheng, S. F.; Hsu, W. T.; Chiang, T.Y.; Chau, L. K., Fiber optic biochemical sensing with a colloidal gold-modified long period fiber grating, *Sens. Actuators B* **2006**, 119, 105–109
62 Zhang, J.; Tang, X.; Dong, J.; Wei, T.; Xiao, H., Zeolite thin film-coated long period fiber grating sensor for measuring trace chemical, *Opt. Express* **2008**, 16, 8317–8323
63 Keith, J.; Hess, L. C.; Spendel, W. U.; Cox, J. A.; Pacey, G. E., The investigation of the behavior of a long period grating sensor with a copper sensitive coating fabricated by layer-by-layer electrostatic adsorption, *Talanta* **2006**, 70, 818–822
64 Corres, J. M.; del Villar, I.; Matias, I. R.; Arregui, F. J., Fiber-optic pH-sensors in long-period fiber gratings using electrostatic self-assembly, *Opt. Lett.* **2007**, 32, 29–31
65 Lee, J.; Chen, Q.; Zhang, Q.; Reichard, K.; Ditto, D.; Mazurowski, J.; Hackert, M.; Yin, S., Enhancing the tuning range of a single resonant band long period grating while maintaining the resonant peak depth by using an optimized high index indium tin oxide overlay, *Appl. Opt.* **2007**, 46, 6984–6989

66 Pilla, P.; Giordano, M.; Korwin-Pawlowski, M. L.; Bock, W. J.; Cusano, A., Sensitivity characteristics tuning in tapered long period gratings by nanocoatings, *IEEE Photon. Technol. Lett.* **2007**, 19, 1517–1519
67 Del Villar, I.; Arregui, F. J.; Matias, I. R.; Cusano, A.; Paladino, D.; Cutolo, A., Fringe generation with non-uniformly coated long-period fiber gratings, *Opt. Express* **2007**, 15, 9326–9340
68 Pilla, P.; Foglia Manzillo, P.; Giordano, M.; Korwin-Pawlowski, M. L.; Bock, W. J.; Cusano, A., Spectral behavior of thin film coated cascaded tapered long period gratings in multiple configurations, *Opt. Express* **2008**, 16, 9765–9780

Chapter 4
New Approach for Selective Vapor Sensing Using Structurally Colored Self-Assembled Films

Radislav A. Potyrailo, Zhebo Ding, Matthew D. Butts, Sarah E. Genovese, and Tao Deng

Abstract We demonstrate that selective sensing of multiple vapors can be accomplished using a structurally colored colloidal crystal film formed from composite core/shell nanospheres and multivariate spectral analysis of vapor response. To improve the detection of color changes of the sensing colloidal crystal film at relatively low vapor partial pressures ($P/P_0 \leq 0.1$, where P is the partial pressure of vapor and P_0 is the saturation vapor pressure), we apply a differential spectroscopy measurement approach. The vapor-sensing selectivity is provided by the combination of the composite nature of the colloidal nanospheres in the film with the multivariate analysis of the spectral changes of the film reflectivity upon exposure to different vapors. The multianalyte sensing was demonstrated using a colloidal crystal film composed of 326-nm diameter core polystyrene spheres coated with a 20-nm thick silica shell. Discrimination of water, acetonitrile, toluene, and dichloromethane vapors using a single sensing colloidal crystal film was evaluated applying principal components analysis of the reflectivity spectra.

4.1 Introduction

Chemical sensing based on responsive materials goes back to times when the Romans used papyrus impregnated with an extract of acorns for selective colorimetric determinations of iron sulfate and copper sulfate and times when Lewis used a litmus paper for detection of acids and alkalis in the late eighteenth century[1,2]. In modern times, colorimetric detection involves organic molecules with color transitions[3,4], and it has expanded to the use of nanofabricated sensing materials.

R.A. Potyrailo (✉)
Chemistry Technologies and Material Characterization, General Electric Company, Global Research Center, Niskayuna, NY 12309, USA
e-mail: potyrailo@crd.ge.com

X. Fan (ed.), *Advanced Photonic Structures for Biological and Chemical Detection*, Integrated Analytical Systems,
DOI 10.1007/978-0-387-98063-8_4,

Table 4.1 Examples of mechanisms of color generation in nanostructured photonic sensing materials

Response type	Mechanism	Selected references
Localized plasmon resonance on noble metal nanostructures	Noble metal nanostructures exhibit a strong UV-visible extinction band with its peak position affected by the dielectric constant and thickness of the material surrounding the nanostructures	7,11–13
Bragg diffraction on crystalline colloidal arrays	Photonic crystal material is composed of a crystalline colloidal array that diffracts light at wavelengths determined by the optical lattice spacing, which is affected by the presence of analyte	5,14,15
Bragg mirrors on periodic stacks of layers	Periodic stacks of metal nanoparticles or dielectric layers with alternating high and low refractive index produce a desired reflectance of the mirror that depends on the thickness and the refractive index of the layers in the stack	16,17
Interferometry on porous silicon	The average refractive index of the porous silicon layer is affected by analyte adsorption, resulting in a shift of the Fabry-Pérot fringes	6,18
Combined light diffraction and interference on hierarchical nanostructures	Combined action of adsorption and capillary condensation of vapors in the domains of ordered hierarchical nanostructures	19

The creation of photonic structures for chemical and biological detection has been possible through nanofabrication[5–10].

There are several mechanisms of color generation in existing nanofabricated photonic sensing materials (see Table 4.1)[5–7,11–19]. An advantage of these photonic structures over organic dyes is in the elimination of photobleaching processes. Unfortunately, the main limitation of existing nanofabricated photonic sensing materials is their low response selectivity to different analytes. Thus, their selectivity is conventionally enhanced by using chemically selective moieties or layers[5,20–22]. For monitoring of multiple analytes, individual nanostructured sensing films are required to be combined in an array where each film will be responsive to a certain class of analytes[23].

Recently, we have developed an approach for selective multianalyte sensing based on the structurally colored colloidal crystal film formed from monodisperse submicron composite core/shell spheres[15]. Because color changes of sensing colloidal crystal films are negligible at relatively low vapor partial pressures ($P/P_0 \leq 0.1$), a straightforward detection of color changes cannot be applied. To overcome this limitation, we apply a differential spectroscopy measurement approach coupled with a multivariate analysis of differential reflectance spectra. The vapor-sensing selectivity is provided by the combination of the composite nature of the colloidal nanospheres in the film with the multivariate spectral analysis of the film reflectivity upon exposure to different vapors. Multivariate spectral analysis correlates spectral changes with the effects caused by different chemical species and aids in establishing the identity and concentration of chemical species.

It was previously shown that spherical composite and core/shell submicron particles can be attractive in optical switching and optical memory storage[24], separations[25], thermosensing applications[26], electrochemical sensing[27], and as integrated chromatographic and detection elements[28] and superparamagnetic photonic crystals[29]. Also, shell layers in core/shell submicron particles were shown to serve other functions such as to increase colloid stability, aid dispersion in various media, alter the optical and electrical properties of the core, add robustness or permselectivity, and provide a continuous framework from which inverse lattices may be generated[30,31]. Our work demonstrates that selective chemical sensing can be achieved with spherical composite core/shell submicron particles without the need for doping of core and shell with organic dyes.

4.2 Detection Concept

The method for selective chemical sensing using colloidal crystal films is depicted in Fig. 4.1. The steps for the fabrication of the colloidal crystal films and determination of selective chemical sensing response are summarized in Table 4.2.

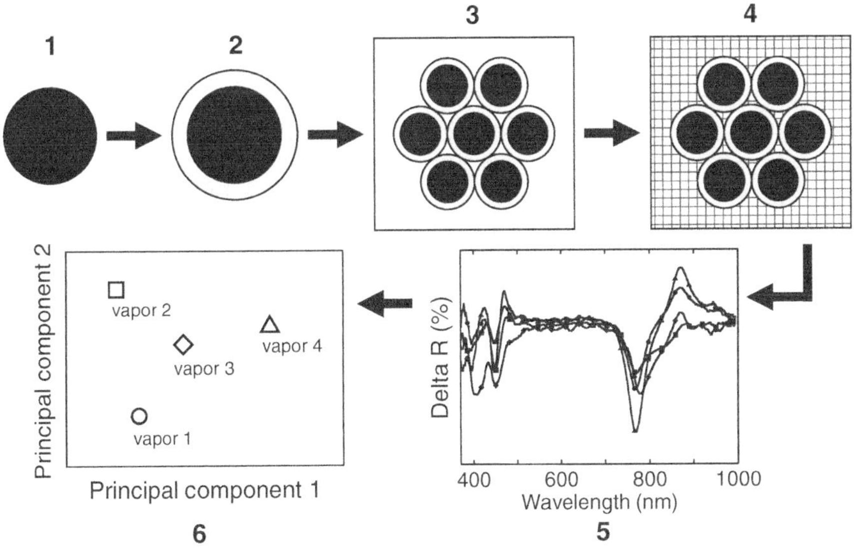

Fig. 4.1 Diagram of the method for selective vapor detection that includes fabrication of core (1) and core-shell (2) materials, their assembly into a colloidal crystal film (3), exposure of the film to different vapors (4), measurements of the spectral response of the film (5), and multivariate analysis of the spectra (6) to obtain a vapor-selective response of the colloidal crystal film

Table 4.2 Steps for the fabrication of the colloidal crystal films and for determination of selective chemical sensing response

Step	Description
1	Fabricate submicron spherical core of material X responsive to one class of vapors
2	Deposit shell of material Y responsive to another class of vapors
3	Assemble colloidal crystal film with core/shell X/Y submicron spheres responsive to both classes of vapors
4	Expose colloidal crystal film to both classes of vapors
5	Measure differential reflectivity spectral response of the colloidal crystal film
6	Perform multivariate spectral analysis to obtain selective multianalyte response

Fabrication of composite colloidal spheres involves two steps: submicron particles are fabricated from a material preferentially responsive to one class of chemicals followed by a step in which the submicron spheres are coated with a shell that is preferentially responsive to another class of chemicals. A colloidal crystal array is subsequently self-assembled into a 3D ordered film.

As an example of composite core/shell submicron particles, we made colloidal spheres with a polystyrene core and a silica shell. The polar vapors preferentially affect the silica shell of the composite nanospheres by sorbing into the mesoscale pores of the shell surface. This vapor sorption follows two mechanisms: physical adsorption and capillary condensation of condensable vapors[17]. Similar vapor adsorption mechanisms have been observed in porous silicon[20] and colloidal crystal films fabricated from silica submicron particles[32], however, with lack of selectivity in vapor response. The nonpolar vapors preferentially affect the properties of the polystyrene core. Sorption of vapors of good solvents for a glassy polymer leads to the increase in polymer free volume and polymer plasticization[32].

Differentiation of vapor responses of the colloidal crystal film was accomplished with spectral measurements of the shape changes of the diffraction peak. Selectivity of response was obtained by applying multivariate data analysis to correlate these spectral changes to the effects of species of different chemical nature and to establish the identity and concentration of species.

Typically, significant changes in color of colloidal crystal films can be achieved upon an expansion or contraction of the crystalline lattice, leading to red or blue shifts of the Bragg diffraction[33]. The lattice constant can be changed by incorporating the light-diffracting submicron particles into polymeric materials swellable in water[5,14] and air[33,34]. Color changes can also be accomplished through changes in the refractive index of the material. These changes can be caused by adsorption of chemicals onto nanostructured materials such as porous silicon[8], colloidal crystal films assembled from mesoporous silica spheres[35], colloidal photonic crystals with engineered defects[34], hierarchical nanostructures that utilize combined effects of diffraction and interference of light[19], etc.

However, optical nanostructured sensors exhibit a significant color change only at relatively high partial pressures of detected vapors, while at relative vapor partial pressures $P/P_0 \leq 0.1$, color shifts are very small[8,19,34,35]. Thus, for applications that require detection of trace concentrations of vapors, monitoring of the *color change* is not practical. Nevertheless, determinations of low vapor concentrations ($P/P_0 \leq 0.1$) of a variety of toxic vapors are of significant practical importance.

To overcome the limitation of detecting only a color change in the sensing colloidal crystal films, we apply a differential spectroscopy measurement approach coupled with the multivariate analysis of differential reflectance spectra. In differential spectroscopy, the differential spectrum accentuates the subtle differences between two spectra. Thus, in optical sensing, when the spectral shifts are relatively small, it is well accepted to perform measurements of the differential spectral response of sensing films before and after analyte exposure[6,19]. Therefore, the common features in two spectra of a sensing film before and after analyte exposure cancel and the differential spectrum accentuates the subtle differences due to analyte response.

To evaluate the vapor responses of the colloidal crystal film, we measured differential reflectance spectra ΔR defined as:

$$\Delta R = 100\% \times (R/R_0 - 1), \qquad (4.1)$$

where R is a spectrum collected from the colloidal crystal film upon an exposure to a vapor of interest and R_0 is a spectrum collected upon exposure to a carrier gas (dry N_2). Other approaches for the improvement of the evaluation of small spectral differences can be also applied[36].

It is possible to tune the sensor response through the manipulation of the particle chemical composition. Individual composite colloidal spheres that have the desired sensing response multifunctionality can be fabricated using a variety of approaches. Monodisperse polystyrene core submicron particles can be synthesized by the controlled persulfate-initiated polymerization of styrene and a silica shell can be deposited onto the core through controlled tetraalkoxysilane hydrolysis[37]. To tailor the sensor response of the composite submicron particles, the compositions of the core and the shell can be varied depending on the analytes of interest. In addition to polystyrene, copolymers of styrene and substituted styrenes (e.g., alpha-methyl-styrene) can be formed in submicron particle synthesis. Other types of polymers (e.g., polymethyl-methacrylate) can be easily employed for fabrication of cores of composite submicron particles. When silica shells are deposited through the use of tetraalkoxysilane precursors, the shells are hydrophilic in nature. Silica shell surfaces can also subsequently be functionalized with hydrophilic or hydrophobic groups. The surface of the spheres can be rendered hydrophobic by, for example, a hexamethyldisilazane surface treatment. In addition to self-assembly, various methods such as Langmuir-Blodgett techniques, sedimentation, template-assisted assembly, dielectrophoresis, etc. can be used to form colloidal crystals from monodisperse submicron particles.

4.3 Fabrication of Core-Shell Colloidal Crystal Films for Selective Chemical Sensing

Core-shell colloidal crystal films were prepared in three steps as outlined in Table 4.2. *First*, spherical submicron polystyrene particles were prepared by known methods[38,39]. The size of isolated polystyrene beads was 326 ± 5 nm as determined by analysis of scanning electron microscopy (SEM) images using standard techniques.

Second, the polystyrene submicron particles were coated with a silica shell[15]. The shell thickness was determined by transmission electron microscopy (TEM) to be 20 ± 2.4 nm. To minimize "bulk" condensation of silane and to drive the condensation of tetraethoxysilane to the surface of the polystyrene submicron beads, the submicron particles were first treated with phenyltrimethoxysilane. The phenyltrimethoxysilane functions as a compatibilizer, with the phenyl group being associated with the submicron particle surface and the hydrolyzed silane providing anchor sites with which the hydrolysis products of tetraethoxysilane can react. The formed core/shell submicron particles had a hydrophilic surface, consistent with the fact that they readily dispersed in water. Hydrophobic core/shell submicron particles, which were prepared by treating the hydrophilic product with hexamethyldisilazane, do not disperse in water. There was invariably some degree of soluble siloxane oligomer and solid silica formation, but most of these byproducts were effectively removed upon core/shell product isolation by centrifugation. The oligomers and silica particles, being much smaller than the desired core/shell spheres, remained suspended while the core/shell spheres settled out when centrifuged. Figure 4.2a illustrates a TEM image of the core/shell submicron particles after their fabrication. The surface of these submicron particles contained only a negligible amount of residual nano-sized silica. This residual silica was substantially removed during the preparation of the

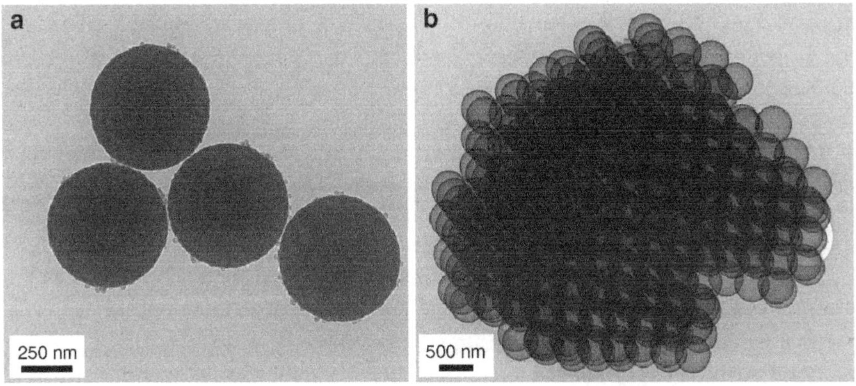

Fig. 4.2 TEM images of fabricated nanoparticles. (**a**) Isolated composite core/shell submicron particles. (**b**) Hollow silica submicron particles prepared by removing the polystyrene core to demonstrate the high quality of the formed sol-gel shell of the composite nanospheres employed to prepare sensing colloidal crystal films

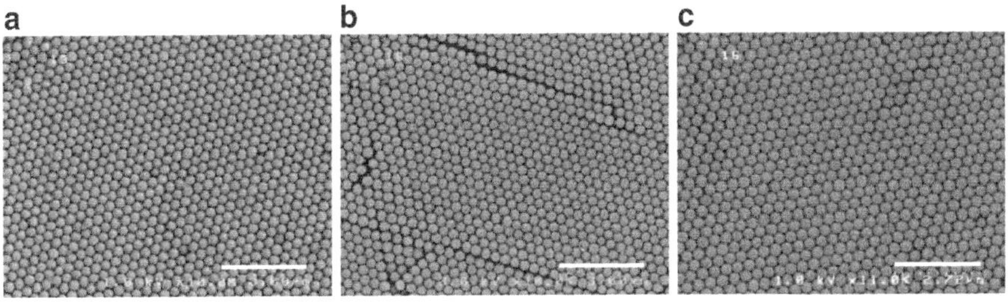

Fig. 4.3 SEM images of the long-range packing assembly of composite colloidal crystal array films assembled on different substrates: (**a**) polycarbonate sheet, (**b**) Teflon sheet, and (**c**) Teflon film. The submicron particles were composed of 326-nm diameter polystyrene beads coated with a 20-nm thick SiO_2 shell. Scale bars in (**a–c**) are 3.00, 3.00, and 2.72 μm, respectively

Fig. 4.4 SEM image of a cross section of the beginning of the assembled colloidal crystal film to illustrate the growth of multiple layers. Reprinted from Ref. 15 with permission. © 2008 Institute of Electrical and Electronics Engineers

colloidal crystal films. To visualize the core/shell nature of these nanospheres, we removed the core of the nanospheres by a heat treatment and performed their further characterization by TEM as shown in Fig. 4.2b. This image is typical of those obtained from multiple synthetic batches.

Third, an evaporation-assisted self-assembly technique[40] was adapted to form colloidal crystal films onto several types of substrates such as polycarbonate and Teflon. The fluid-driven self-assembly of colloidal spheres initiated by the solvent evaporation[41] generated highly ordered 3D structures. The assembled colloidal crystal film has a close-packed 3D face-centered cubic (fcc) ordering as has been determined previously[42,43]. Figure 4.3 shows the top view images of the hexagonally packed layers of the (111) face of the fcc lattice of the colloidal crystal films self-assembled on a polycarbonate sheet, a Teflon sheet, and a Teflon film. The negligible residual silica that was visible on the surface of individual submicron particles (Fig. 4.2a) was fully removed by washing with anhydrous ethanol via repeated centrifugation and resuspension and was not detectable. A cross section of the beginning of the assembled colloidal crystal film (Fig. 4.4) illustrates the growth of multiple layers. Such thickness gradients in colloidal crystal sensing films may be useful as an additional parameter to control sensor response[2]. The photograph of the self-assembled sensing film on a Teflon film under white light illumination is illustrated in Fig. 4.5. The film displayed a uniform color when observed at a constant angle. A reflected light spectrum of the film is illustrated in Fig. 4.6. Reflectance measurements perpendicular to the (111) plane of the composite colloidal crystal film were performed using a portable halogen light source and a portable fiber-optic spectrograph. The common arm of the fiber-optic probe

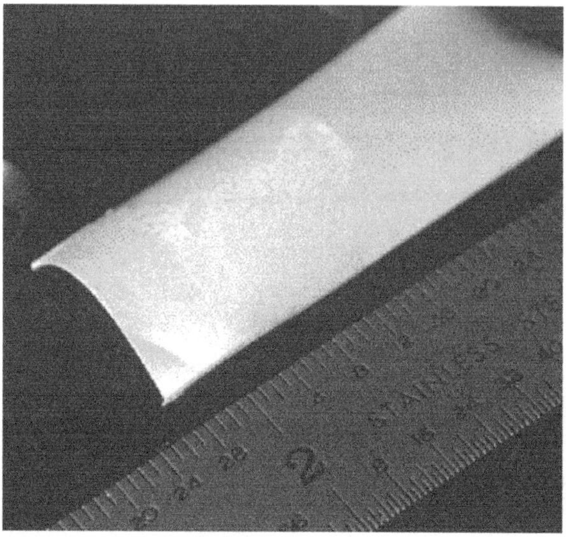

Fig. 4.5 Optical photograph of the composite colloidal crystal array film self-assembled on a Teflon film. The submicron particles were composed of 326-nm diameter polystyrene beads coated with a 20-nm thick SiO_2 shell. Reprinted from Ref. 15 with permission. © 2008 Institute of Electrical and Electronics Engineers

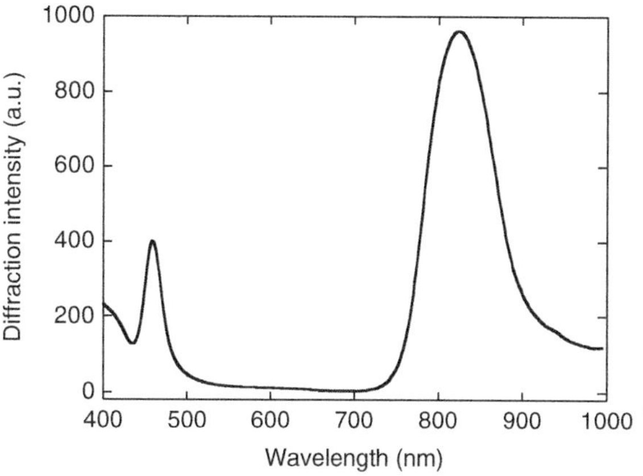

Fig. 4.6 Reflectance spectrum of the composite colloidal crystal array film self-assembled on a Teflon film measured over a 400–1,000 nm range. The submicron particles were composed of 326 nm diameter polystyrene beads coated with a 20-nm thick SiO_2 shell. Reprinted from Ref. 15 with permission. © 2008 Institute of Electrical and Electronics Engineers

illuminated the sensing film at a small angle relative to the normal to the surface with a ~1-mm diameter illumination spot. The reflection of the colloidal crystal film resulted in a pronounced photonic stop band of the fundamental Bragg diffraction peak at 824 nm. Also, there is a weaker diffraction peak at 459 nm, which was likely to be from higher order planes[44].

Table 4.3 Solvent polarity and refractive index of tested vapors[46,47]

Tested vapor	Solvent polarity E_T (kcal mol^{-1})	Relative solvent polarity E_T^N	Refractive index
Water	63.1	1.000	1.333
Acetonitrile	45.68	0.460	1.342
Dichloromethane	40.78	0.309	1.424
Toluene	33.9	0.099	1.494

4.4 Spectral Response to Vapors

For chemical sensing with the colloidal crystal film, we used measurements of the reflectivity spectra. It is known that the resolution of optical resonances is better in reflectance rather than in transmission mode[45]. Possible packing defects in the film are more pronounced in transmission measurements contributing to diffuse scattering. In reflection mode, the top (111) planes contribute the most to the spectrum, reducing the contributions from packing defects. Different vapor concentrations were generated by bubbling dry nitrogen through pure solvents and diluting the resulting vapors with dry nitrogen at different ratios, keeping the total flow constant at 400 cm^3/min at a constant laboratory temperature. Thus, the concentrations of vapors were different fractions of the saturated (maximum) concentrations of vapors P_0 under lab conditions.

For the evaluation of the response of the sensor, we selected several vapors of different polarity. The vapors included water (H_2O), acetonitrile (ACN), toluene, and dichloromethane (DCM). Solvent polarity and refractive index of tested vapors are listed in Table 4.3[46,47]. The spectral range for the evaluation of the vapor responses of the colloidal crystal film was selected as 700–995 nm, which covered only the fundamental Bragg diffraction peak on the (111) planes of the colloidal crystal film to further reduce effects from possible stacking defects in the film as suggested in the literature[44].

The spectral reflectivity of the sensing film before and after the exposure to different vapors (all at $P/P_0 = 0.1$) is illustrated in Figs. 4.7 and 4.8. Similar to other photonic nanostructured sensors[8,19,34,35], the spectral shifts upon response to low vapor concentrations are relatively small. Thus, to accentuate the subtle differences due to vapor response, we measured the differential reflectance spectra ΔR as described by equation (4.1).

Changes in the differential reflectance spectra ΔR of the sensing film upon exposure to different vapors at various concentrations are presented in Fig. 4.9. These spectra illustrate several important findings. For polar vapors such as water and ACN (see Fig. 4.9a, b respectively), the differential reflectance spectra have a stable baseline and consistent well-behaved changes in the reflectivity as a function of analyte concentration. The response of the colloidal crystal film to nonpolar vapors such as DCM and toluene (see Fig. 4.9c, d respectively) is quite different compared with the response to polar vapors. There are pronounced analyte concentration-dependent baseline offsets that are likely due to

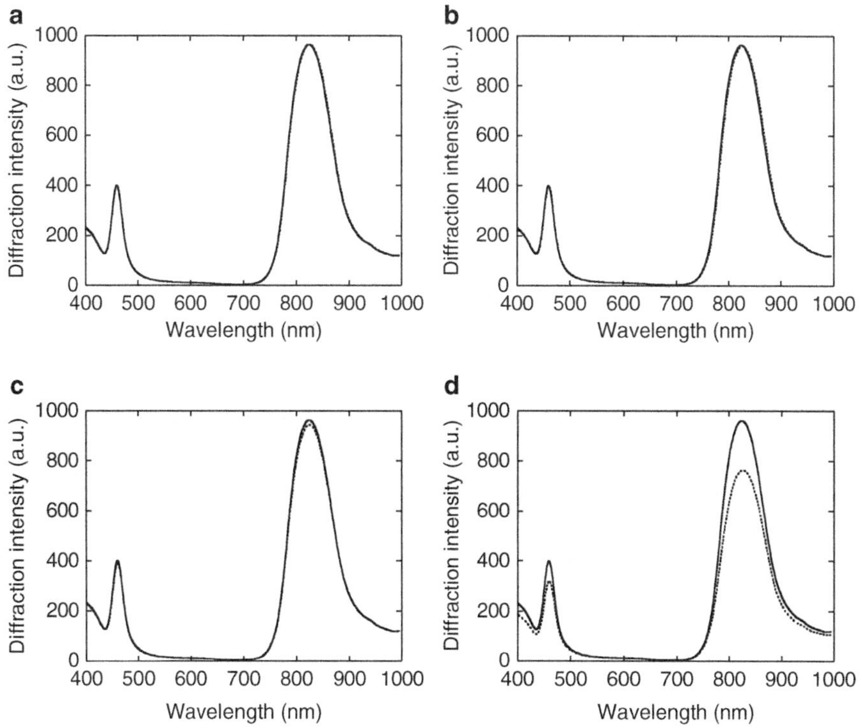

Fig. 4.7 The spectral reflectivity of the sensing film measured over the range 400–1,000 nm before (*solid line*) and after (*dotted line*) the exposure to different vapors (all at $P/P_0 = 0.1$): (**a**) water, (**b**) ACN, (**c**) DCM, and (**d**) toluene

the significant distortions of the optical lattice upon exposure to nonpolar vapors. These distortions are likely to be caused by the reversible slight swelling of the core of the submicron spheres by the vapors at their high tested concentrations.

4.5 Dynamic Response to Vapors

The reversibility of interactions of the composite colloidal crystal film with different vapors was also evaluated. Figure 4.10 illustrates the dynamic response of the sensing film at several wavelengths (770, 835, and 870 nm) upon triplicate exposures to water vapor at four concentrations (0.02, 0.04, 0.07, and 0.1 P/P_0). The response and recovery kinetics upon exposure to water vapor were fully reversible and rapid (under 5 s). A comparison of dynamic response of the sensing film at five different wavelengths upon exposures to water and toluene vapors at four concentrations is presented in Fig. 4.11. These data illustrate that the direction, magnitude, and kinetics of the responses to these vapors were quite different.

A comparison of the dynamic response profiles plotted at several wavelengths (770, 835, and 870 nm) for all four tested vapors at 0.02 P/P_0 is shown in Fig. 4.12.

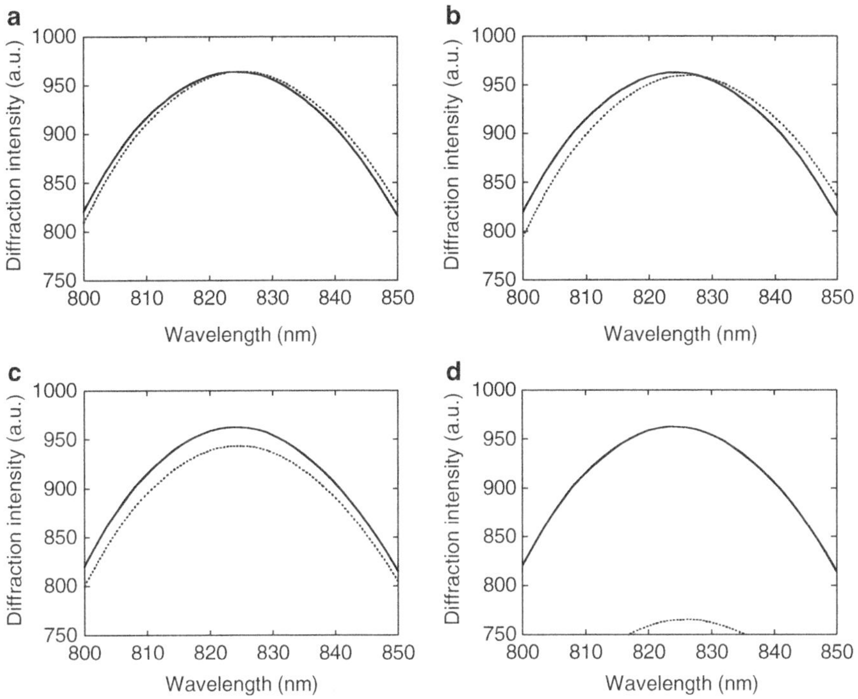

Fig. 4.8 Changes in the peak position and intensity of the spectral reflectivity of the sensing film measured over the range 800–850 nm before (*solid line*) and after (*dotted line*) the exposure to different vapors (all at $P/P_0 = 0.1$): (**a**) water, (**b**) ACN, (**c**) DCM, and (**d**) toluene

ACN vapor had the most pronounced permutation of the relatively rapid and relatively slow response and recovery kinetics. Such behavior could be due to the combination of physical and chemical adsorption. Physical adsorption effects are typically pronounced with rapid response and recovery kinetics because of the relatively low energies of physical interactions between vapors and the sensing surface. Chemical adsorption effects have much slower recovery kinetics because of the relatively high energies of chemical interactions between vapors and the sensing surface. The recovery from all tested vapors was reversible with the slowest recovery after the exposure to ACN on the order of several hours from the highest tested vapor concentration of 0.1 P/P_0.

4.6 Quantitation of Vapor Concentrations

To evaluate the quantitation capabilities of the experimental design, univariate calibration curves were constructed at 770 nm for four tested vapors as shown in Fig. 4.13. Upon exposure to the highest tested concentration of DCM and toluene

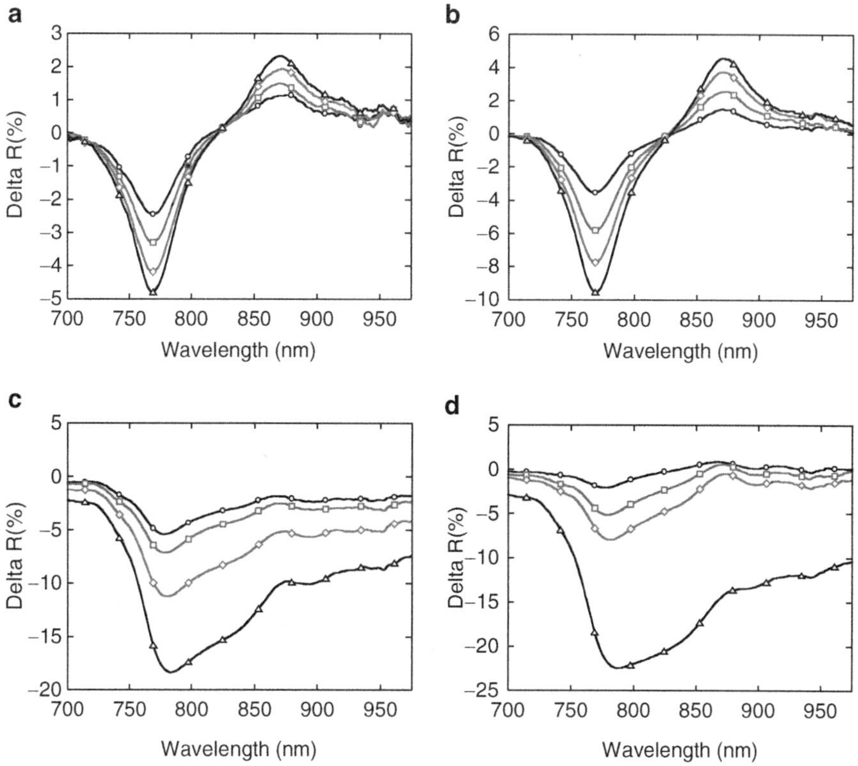

Fig. 4.9 Differential reflectance spectra for four different concentrations of vapors: (**a**) water, (**b**) ACN, (**c**) DCM, and (**d**) toluene. Vapor concentrations: 0.02 (*circles*), 0.04 (*squares*), 0.07 (*diamonds*), 0.1 (*triangles*) P/P_0. Reprinted from Ref. 15 with permission. © 2008 Institute of Electrical and Electronics Engineers

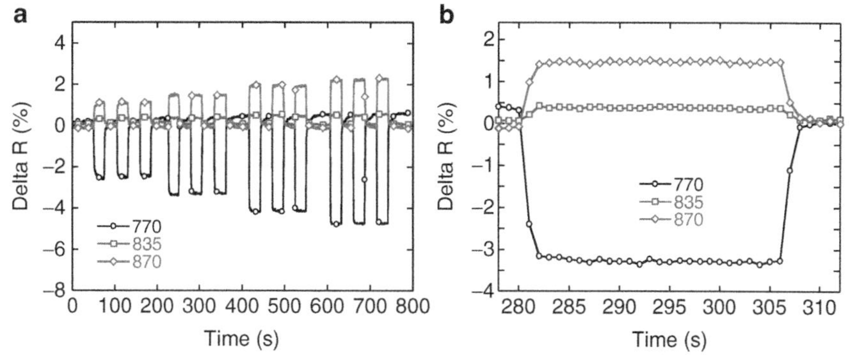

Fig. 4.10 Dynamic response of the sensing film to water vapor at three different wavelengths. (**a**) Response upon replicate ($n = 3$) exposures to water vapor at four concentrations (0.02, 0.04, 0.07, and 0.1 P/P_0). (**b**) Portion of response from (**a**) to illustrate the response and recovery kinetics. Wavelengths: 770 nm (*circles*), 835 nm (*squares*), and 870 nm (*diamonds*)

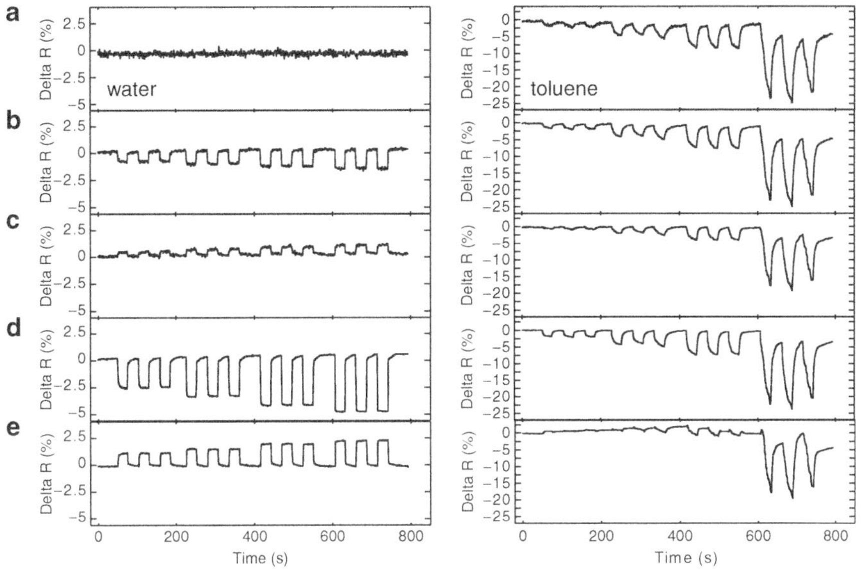

Fig. 4.11 Comparison of the dynamic response of the sensing film at five different wavelengths upon replicate ($n = 3$) exposures to water and toluene vapors at four concentrations. Wavelengths: (**a**) 410 nm, (**b**) 450 nm, (**c**) 470 nm, (**d**) 770 nm, (**e**) 870 nm. *Left column* is the response to water vapor, *right column* is the response to toluene vapor. Concentrations of vapors: 0.02, 0.04, 0.07, and 0.1 P/P_0

(0.1 P/P_0), the colloidal crystal film exhibited an additional increase in the response sensitivity likely due to the swelling of the core of the core/shell submicron particles. This effect was more pronounced upon exposure to toluene.

Detection limits for all tested vapors were further calculated according to a widely accepted approach[36]. According to this approach, the detection limit was calculated at the signal-to-noise value of three ($S/N = 3$) from measurements of the spectral response in presence of a blank gas (dry N_2) and 0.02 P/P_0 vapor concentration. The calculated detection limits were 11, 62, 15, and 20 ppm for water, ACN, DCM, and toluene vapors, respectively. The achieved detection limits were within the range of interest or better than those needed for their determinations for industrial applications. The permissible exposure limits as time-weighted average for general industry are 40, 50, and 100 ppm for ACN, DCM, and toluene, respectively[48].

ACN and DCM are also known to be chemical warfare agent (CWA) simulants. ACN is a known simulant for blood CWAs, while DCM is a simulant for choking CWAs[49,50]. For determinations of ACN and DCM as CWA simulants, the detection limits need to be improved[51]. Such improvements may be possible by the application of more stable light source and detector to reduce noise in the measured signal.

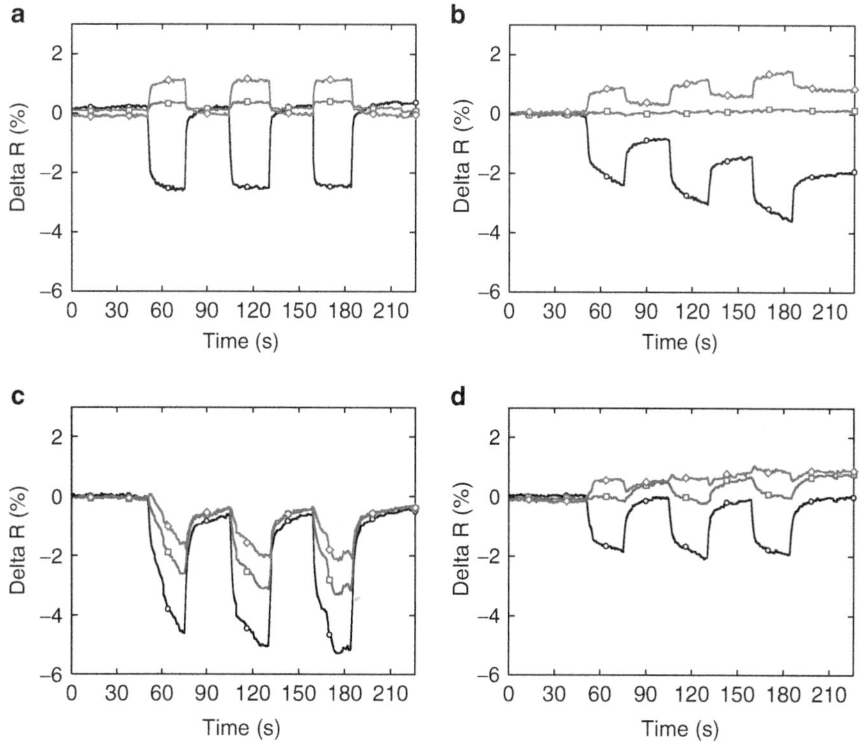

Fig. 4.12 Dynamic response profiles at different wavelengths upon replicate ($n = 3$) exposures to vapors at 0.02 P/P_0 concentration: (**a**) water, (**b**) ACN, (**c**) DCM, and (**d**) toluene. Wavelengths: 770 nm (*circles*), 835 nm (*squares*), and 870 nm (*diamonds*). Reprinted from Ref. 15 with permission. © 2008 Institute of Electrical and Electronics Engineers

Discrimination of vapors using a single sensing colloidal crystal film was done using a principal components analysis (PCA) model generated with mean-centered differential reflectance ΔR spectra over the 700–995 nm range. Mean centering of each ΔR spectrum was accomplished by subtracting the mean of the spectrum from the respective original spectrum. PCA is a multivariate data analysis technique that was used to project the ΔR spectra onto a subspace of lower dimensionality with removed collinearity. PCA achieves this objective by explaining the variance of the data matrix **X** in terms of the weighted sums of the original variables with no significant loss of information. These weighted sums of the original variables are called principal components (PCs). Upon applying the PCA, the data matrix **X** is expressed as a linear combination of orthogonal vectors along the directions of the PCs[52]:

$$\mathbf{X} = \mathbf{t}_1\mathbf{p}_1^{\mathrm{T}} + \mathbf{t}_2\mathbf{p}_2^{\mathrm{T}} + \cdots + \mathbf{t}_A\mathbf{p}_K^{\mathrm{T}} + \mathbf{E}, \tag{4.2}$$

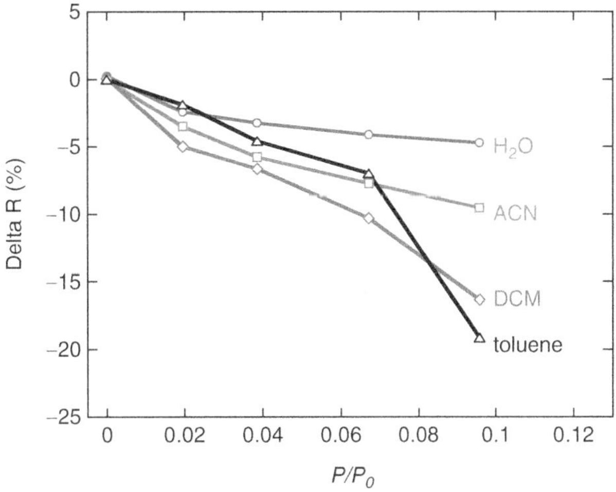

Fig. 4.13 Univariate calibration curves constructed at 770 nm for four tested vapors over 0–0.1 P/P_0 concentration range. Reprinted from Ref. 15 with permission. © 2008 Institute of Electrical and Electronics Engineers

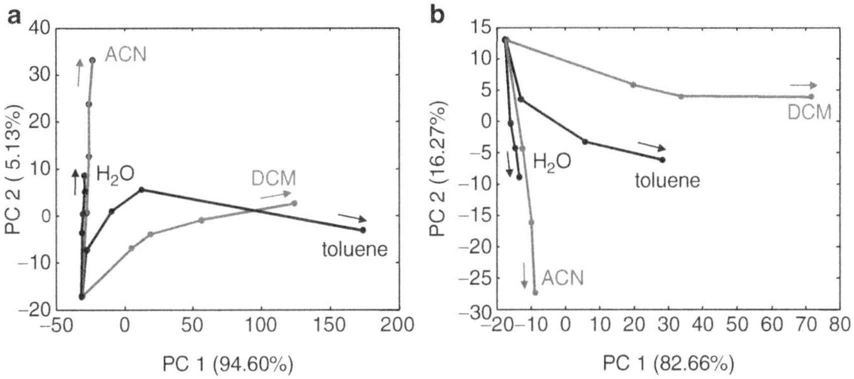

Fig. 4.14 Multivariate response of the colloidal sensor film for four tested vapors over (a) 0–0.07 and (b) over 0–0.1 P/P_0 concentration ranges. Multivariate analysis was done using PCA tools by processing ΔR spectra over 700–995 nm range. *Arrows* indicate the increase of concentrations of each vapor. Reprinted from Ref. 15 with permission. © 2008 Institute of Electrical and Electronics Engineers

where $\mathbf{t}_i$ and $\mathbf{p}_i$ are the score and loading vectors, respectively, K is the number of PCs, $\mathbf{E}$ is a residual matrix that represents random error, and T is the transpose of the matrix.

Results of the multivariate analysis of the ΔR spectra as scores plots are presented in Fig. 4.14. Initially, PCA of the ΔR spectra was performed for all tested concentrations, including the highest tested concentration of 0.1 P/P_0 (see Fig. 4.14a).

The polar and nonpolar vapors at different concentrations were separated in PCA space. The best selectivity was obtained between toluene and DCM vapors at low concentrations, while water and ACN vapors were almost unresolved. At the highest tested vapor concentration of 0.1 P/P_0, toluene and DCM vapors likely induced film swelling that resulted in the decrease in the vapor response selectivity. Overall, two first PCs described more than 98% of total variance in the spectral data with PC_2 contribution of only 5% from the original spectral data. This relatively small contribution of PC_2 was likely due to the film-swelling effects at high vapor concentration.

Thus, we further evaluated the vapor response selectivity of the composite colloidal crystal film when the largest tested concentration of vapors (0.1 P/P_0) was not considered in the PCA model (see Fig. 4.14b). Again, two first PCs described more than 98% of total variance in the spectral data. However, at low vapor concentrations, PC_2 had a very significant contribution of 16% from the original spectral data. This large contribution conclusively demonstrates the high efficiency of the PCA model to discriminate low concentrations of different vapors using this single sensing film.

Upon exposure to high concentrations of nonpolar solvent vapors characterized by a refractive index less than that of polystyrene, the vapor induced swelling of the polystyrene can cause an increase in the lattice spacing in the colloidal crystal film. This effect could be attenuated through the use of a core material having a refractive index less than that of polystyrene. Such a situation should lead to the reduction in the measured signal changes when measurements are done at a single wavelength. However, upon vapor exposure, the composite colloidal crystal film acts as a mixed phase where the refractive index distribution is affected by both the core and shell of each colloidal sphere. The full spectrum measurements of vapor-induced distortions of the diffraction peak shape significantly reduce this effect. Indeed, in the past, full-spectrum measurements of vapor interactions with thin polymeric films[53] have been successfully employed to determine separately vapor-induced changes in polymer thickness and refractive index.

4.7 Conclusions

We have shown a new concept for selective chemical sensing based on composite core/shell polymer/silica colloidal crystal films. The vapor response selectivity is provided via the multivariate spectral analysis of the fundamental diffraction peak from the colloidal crystal film. Of course, as with any other analytical device, care should be taken not to irreversibly "poison" this sensor. For example, a prolonged exposure to high concentrations of nonpolar vapors will likely to irreversibly destroy the composite colloidal crystal film. Nevertheless, sensor materials based on the colloidal crystal films promise to have an improved long-term stability over the sensor materials based on organic colorimetric reagents incorporated into polymer films due to the elimination of photobleaching effects. In the experiments reported here, vapor discrimination was achieved with individual vapors. Our immediate future experiments are planned with vapor mixtures keeping in mind that existing sensors and even sensor arrays cannot quantify multiple individual components in very complex mixtures[54] without additional time-domain data processing.

Acknowledgments This work has been supported by GE Corporate long-term research funds. We thank Prof. Sanford Asher from the University of Pittsburgh for useful discussions.

References

1 Zolotov, Y. A.; Ivanov, V. M.; Amelin, V. G., Chemical test methods of analysis, In Wilson & Wilson's Comprehensive Analytical Chemistry; Barcelo, D., Ed.; Elsevier, Amsterdam, **2002**
2 Potyrailo, R. A.; Mirsky, V. M., Combinatorial and high-throughput development of sensing materials: The first ten years, *Chem. Rev.* **2008**, 108, 770–813
3 Charych, D. H.; Nagy, J. O.; Spevak, W.; Bednarski, M. D., Direct colorimetric detection of a receptor-ligand interaction by a polymerized bilayer assembly, *Science*. **1993**, 261, 585–588
4 Rakow, N. A.; Suslick, K. S., A colorimetric sensor array for odour visualization, *Nature*. **2000**, 406, 710–713
5 Holtz, J. H.; Asher, S. A., Polymerized colloidal crystal hydrogel films as intelligent chemical sensing materials, *Nature*. **1997**, 389, 829–832
6 Lin, V. S.-Y.; Motesharei, K.; Dancil, K.-P. S.; Sailor, M. J.; Ghadiri, M. R., A porous silicon-based optical interferometric biosensor, *Science*. **1997**, 278, 840–843
7 Elghanian, R.; Storhoff, J. J.; Mucic, R. C.; Letsinger, R. L.; Mirkin, C. A., Selective colorimetric detection of polynucleotides based on the distance-dependent optical properties of gold nanoparticles, *Science*. **1997**, 277, 1078–1081
8 Li, Y. Y.; Cunin, F.; Link, J. R.; Gao, T.; Betts, R. E.; Reiver, S. H.; Chin, V.; Bhatia, S. N.; Sailor, M. J., Polymer replicas of photonic porous silicon for sensing and drug delivery applications, *Science*. **2003**, 299, 2045–2047
9 Wehrspohn, R. B.; Schweizer, S. L.; Schilling, J.; Geppert, T.; Jamois, C.; Glatthaar, R.; Hahn, P.; Feisst, A.; Lambrecht, A., Application of photonic crystals for gas detection and sensing, In Photonic Crystals; Busch, K., Ed.; Wiley-VCH Verlag, Weinheim, Germany, **2004**, 238–246
10 Benabid, F.; Couny, F.; Knight, J. C.; Birks, T. A.; Russell, P. S. J., Compact, stable and efficient all-fibre gas cells using hollow-core photonic crystal fibres, *Nature*. **2005**, 434, 488–491
11 Amarie, D.; Onuta, T.-D.; Potyrailo, R.; Dragnea, B., Submicrometer cavity surface plasmon sensors, *J. Phys. Chem. B*. **2005**, 109, 15515–15519
12 Dovidenko, K.; Potyrailo, R. A.; Grande, J., Focused ion beam microscope as an analytical tool for nanoscale characterization of gradient-formulated polymeric sensor materials, In Combinatorial Methods and Informatics in Materials Science. Materials Research Society Symposium Proceedings; Fasolka, M.; Wang, Q.; Potyrailo, R. A.; Chikyow, T.; Schubert, U. S.; Korkin, A., Eds.; Materials Research Society, Warrendale, PA, **2006**, 894, 231–236
13 Stewart, M. E.; Anderton, C. R.; Thompson, L. B.; Maria, J.; Gray, S. K.; Rogers, J. A.; Nuzzo, R. G., Nanostructured plasmonic sensors, *Chem. Rev.* **2008**, 108, 494–521
14 Miyata, T.; Asami, N.; Uragami, T., A reversibly antigen-responsive hydrogel, *Nature*. **1999**, 399, 766–769
15 Potyrailo, R. A.; Ding, Z.; Butts, M. D.; Genovese, S. E.; Deng, T., Selective chemical sensing using structurally colored core-shell colloidal crystal films, *IEEE Sensors J.* **2008**, 8, 815–822

16 Convertino, A.; Capobianchi, A.; Valentini, A.; Cirillo, E. N. M., A new approach to organic solvent detection: High-reflectivity bragg reflectors based on a gold nanoparticle/teflon-like composite material, *Adv. Mater.* **2003**, 15, 1103–1105

17 Snow, P. A.; Squire, E. K.; Russell, P. S. J.; Canham, L. T., Vapor sensing using the optical properties of porous silicon bragg mirrors, *J. Appl. Phys.* **1999**, 86, 1781–1784

18 Gao, J.; Gao, T.; Sailor, M. J., Porous-silicon vapor sensor based on laser interferometry, *Appl. Phys. Lett.* **2000**, 77, 901–903

19 Potyrailo, R. A.; Ghiradella, H.; Vertiatchikh, A.; Dovidenko, K.; Cournoyer, J. R.; Olson, E., Morpho butterfly wing scales demonstrate highly selective vapour response, *Nature Photonics.* **2007**, 1, 123–128

20 Gao, T.; Gao, J.; Sailor, M. J., Tuning the response and stability of thin film mesoporous silicon vapor sensors by surface modification, *Langmuir.* **2002**, 18, 9953–9957

21 Potyrailo, R. A., Polymeric sensor materials: Toward an alliance of combinatorial and rational design tools ?, *Angew. Chem. Int. Ed.* **2006**, 45, 702–723

22 Bailey, R. C.; Hupp, J. T., Large-scale resonance amplification of optical sensing of volatile compounds with chemoresponsive visible-region diffraction gratings, *J. Am. Chem. Soc.* **2002**, 124, 6767–6774

23 Asher, S. A., Crystalline colloidal array chemical sensing devices, In ACS PRF summer school on nanoparticle materials, June 6–18, 2004. Eastern Michigan University, Ypsilanti, MI, **2004**

24 Zhang, J.; Coombs, N.; Kumacheva, E., A new approach to hybrid nanocomposite materials with periodic structures, *J. Am. Chem. Soc.* **2002**, 124, 14512–14513

25 Fong, B.; Turksen, S.; Russo, P. S.; Stryjewski, W., Colloidal crystals of silica-homopolypeptide composite particles, *Langmuir.* **2004**, 20, 266–269

26 Suzuki, D.; Kawaguchi, H., Modification of gold nanoparticle composite nanostructures using thermosensitive core-shell particles as a template, *Langmuir.* **2005**, 21, 8175–8179

27 Osterloh, F.; Hiramatsu, H.; Porter, R.; Guo, T., Alkanethiol-induced structural rearrangements in silica-gold core-shell-type nanoparticle clusters: An opportunity for chemical sensor engineering, *Langmuir.* **2004**, 20, 5553–5558

28 Goodey, A. P.; McDevitt, J. T., Multishell microspheres with integrated chromatographic and detection layers for use in array sensors, *J. Am. Chem. Soc.* **2003**, 125, 2870–2871

29 Xu, X.; Friedman, G.; Humfeld, K.; Majetich, S.; Asher, S. A., Superparamagnetic photonic crystals, *Adv. Mater.* **2001**, 13, 1681–1684

30 Liz-Marzan, L. M.; Mulvaney, P., The assembly of coated nanocrystals, *J. Phys. Chem. B.* **2003**, 107, 7312–7326

31 Nayak, S.; Lyon, L. A., Ligand-functionalized core/shell microgels with permselective shells, *Angew. Chem. Int. Ed.* **2004**, 43, 6706–6709

32 Zhang, R.; Graf, K.; Berger, R., Swelling of cross-linked polystyrene spheres in toluene vapor, *Appl. Phys. Lett.* **2006**, 89, 223114

33 Arsenault, A. C.; Kitaev, V.; Manners, I.; Ozin, G. A.; Mihi, A.; Míguez, H., Vapor swellable colloidal photonic crystals with pressure tunability, *J. Mater. Chem.* **2005**, 15, 133–138

34 Fleischhaker, F.; Arsenault, A. C.; Tétreault, N.; Wang, Z.; Kitaev, V.; Peiris, F.; Mihi, A.; Miguez, H.; von Freymann, G.; Manners, I.; Zentel, R.; Ozin, G. A., "Smart" defects in colloidal photonic crystals, In Materials Research Society Symposium Proceedings, Materials Research Society, Warrendale, PA, **2006**, Vol. 901E; paper # 0901-Ra22-27-Rb22-27

35 Yamada, Y.; Nakamura, T.; Ishi, M.; Yano, K., Reversible control of light reflection of a colloidal crystal film fabricated from monodisperse mesoporous silica spheres, *Langmuir.* **2006**, 22, 2444–2446

36 Ingle, J. D., Jr.; Crouch, S. R. Spectrochemical Analysis, Prentice Hall, Englewood Cliffs, NJ, **1988**

37 Butts, M. D.; Genovese, S. E.; Glaser, P. B.; Williams, D. S., Hollow silica particles and methods for making same, US Patent Application 20070036705: **2007**

38 Goodwin, J. W.; Ottewill, R. H.; Pelton, R., Studies on the preparation and characterization of monodisperse polystyrene lattices V: The preparation of cationic latices, *Colloid Polymer Sci.* **1979**, 257, 61–69

39 Ali, S. A.; Sengupta, M. J., Preparation and characterization of monodisperse polystyrene latexes of varying particle sizes without the use of surfactants, *Polym. Mater. Sci. Eng.* **1991**, 8, 243–250
40 Jiang, P.; Bertone, J. F.; Hwang, K. S.; Colvin, V. L., Single-crystal colloidal multilayers of controlled thickness, *Chem. Mater.* **1999**, 11, 2132–2140
41 Denkov, N.; Velev, O.; Kralchevski, P.; Ivanov, I.; Yoshimura, H.; Nagayama, K., Two-dimensional crystallization, *Nature.* **1993**, 361, 26–26
42 Ye, Y.-H.; Mayer, T. S.; Khoo, I.-C.; Divliansky, I. B.; Abrams, N.; Mallouk, T. E., Self-assembly of three-dimensional photonic-crystals with air-core line defects, *J. Mater. Chem.* **2002**, 12, 3637–3639
43 Xu, X.; Asher, S. A., Synthesis and utilization of monodisperse hollow polymeric particles in photonic crystals, *J. Am. Chem. Soc.* **2004**, 126, 7940–7945
44 Asher, S. A.; Weissman, J. M.; Tikhonov, A.; Coalson, R. D.; Kesavamoorthy, R., Diffraction in crystalline colloidal-array photonic crystals, *Phys. Rev. E.* **2004**, 69, 066619
45 Egen, M.; Voss, R.; Griesebock, B.; Zentel, R.; Romanov, S.; Torres, C. S., Heterostructures of polymer photonic crystal films, *Chem. Mater.* **2003**, 15, 3786–3792
46 Reichardt, C., Solvatochromic dyes as solvent polarity indicators, *Chem. Rev.* **1994**, 94, 2319–2358
47 Lide, D. R., Ed., CRC Handbook of Chemistry and Physics, 74th edn.; CRC Press, Boca Raton, FL, **1993**
48 Sittig, M., Handbook of Toxic and Hazardous Chemicals and Carcinogens, Noyes Publications, Park Ridge, NJ, **1991**
49 Choi, N.-J.; Lee, Y.-S.; Kwak, J.-H.; Park, J.-S.; Park, K.-B.; Shin, K.-S.; Park, H.-D.; Kim, J.-C.; Huh, J.-S.; Lee, D.-D., Chemical warfare agent sensor using mems structure and thick film fabrication method, *Sens. Actuators B.* **2005**, 108, 177–183
50 Tomchenko, A. A.; Harmer, G. P.; Marquis, B. T., Detection of chemical warfare agents using nanostructured metal oxide sensors, *Sens. Actuators B.* **2005**, 108, 41–55
51 Bevelacqua, A.; Stilp, R., Terrorism handbook for operational responders, Delmar Publishers, Albany, NY, **1998**
52 Wise, B. M.; Gallagher, N. B., PLS_Toolbox version 2.1 for use with MATLAB, Eigenvector Research, Inc., Manson, WA, **2000**
53 Spaeth, K.; Gauglitz, G., Characterisation of the optical properties of thin polymer films for their application in detection of volatile organic compounds, *Mat. Sci. Eng. C.* **1998**, 5, 187–191
54 Lewis, N. S., Comparisons between mammalian and artificial olfaction based on arrays of carbon black-polymer composite vapor detectors, *Acc. Chem. Res.* **2004**, 37, 663–672

Chapter 5
Methods of Cavity-Enhanced Laser Absorption Spectroscopy Using Microresonator Whispering-Gallery Modes

A. T. Rosenberger

Abstract Theoretical analysis of, and experimental results using, chemical sensing techniques based on microcavity-enhanced optical absorption are presented. Two methods are described in detail, and several extensions and enhancements of these methods are discussed briefly. Both techniques involve novel applications of tunable diode laser absorption spectroscopy in which cavity enhancement is provided by a dielectric microresonator (<1 mm in diameter) with whispering-gallery modes (WGMs) excited by tapered-fiber coupling. The evanescent component of a WGM allows for interaction with the analyte. The first method is used for the detection of trace gases in the ambient air by measuring the coupling-fiber throughput as the laser scans in frequency. Centimeter effective absorption path lengths are measured, in agreement with theory. The second method employs the observation of thermal bistability to enable measurement of absorption due to a coating applied to, or molecules adsorbed on, the microresonator's surface. Absorption by the water layer on a fused-silica surface agrees with theory, and results for thermal accommodation coefficients and thin-film absorption are also presented.

5.1 Introduction

Dielectric microresonators supporting whispering-gallery modes (WGMs) are becoming increasingly useful for numerous applications in optics. These resonators, often made of fused silica because of its extremely low absorption and scattering losses, can be spherical, cylindrical, disk-shaped, or toroidal; they are typically a few hundred μm in diameter. Light in a WGM circulates around the circumference of the resonator, localized near the surface (for example, at the equator of a sphere) by total internal reflection. The impact of these devices is based on certain attractive

A.T. Rosenberger
Department of Physics, Oklahoma State University, Stillwater, OK 74078-3072, USA
e-mail: atr@okstate.edu

X. Fan (ed.), *Advanced Photonic Structures for Biological and Chemical Detection*, Integrated Analytical Systems,
DOI 10.1007/978-0-387-98063-8_5,

properties of WGMs, such as high quality factor (Q) and low mode volume, and on the possibility of efficient coupling of light into and out of these modes[1]. The high Q of a WGM means that light makes many round trips in the resonator before being lost from the mode by outcoupling, scattering, or absorption. This feature, combined with small mode volume and efficient coupling using prisms, angle-polished fibers, or tapered fibers, makes high intracavity power enhancement easy to achieve. Various applications such as filtering, lasing, modulation, nonlinear optics, sensing, and spectroscopy are enabled by these properties of WGMs[2].

Part of the WGM field extends outside the dielectric, and it is this evanescent portion that interacts with the surrounding medium or with a surface layer, making the WGM a sensitive probe of the microresonator's environment. Two sensing methods will be described here. The first method is microcavity-enhanced laser absorption spectroscopy of molecules in the ambient gas or liquid[3,4], to be discussed in the following three sections. Section 5.2 will give a theoretical analysis of the operation of a chemical sensor based on this technique; Section 5.3 will present results for its use in detection of atmospheric trace gases; and a discussion of possible extensions and enhancements will follow in Sect. 5.4. The second method is based on using the temperature sensitivity of WGM resonance frequencies to characterize surfaces and thin films[5]. Section 5.5 is devoted to this technique, and in Sect. 5.6 the chapter concludes with a summary and outlook.

5.2 Theoretical Analysis of Microcavity-Enhanced Absorption Sensors

In the application area of label-free chemical detection and spectroscopy, several different methods of using WGM microresonators have previously been reported. Analyte molecules in the ambient gas or liquid or on the resonator's surface can be detected by measuring the frequency shift of a WGM caused by the analyte's perturbation of the ambient's index of refraction[6,7]. In addition, analyte absorption will change the effective Q of the WGM, and the associated modification of mode linewidth can be measured[8]. One can also envision using cavity-ringdown spectroscopy[9]. The method described in this section is more closely related to other earlier work[10] in which what is measured is the analyte absorption effect, for continuous-wave input, on either the depth of the throughput dip or the strength of the drop signal. As explained below, dip-depth detection can be more sensitive than other tunable single-frequency methods, and drop-signal detection lends itself well to broadband spectroscopy.

5.2.1 *Tunable Single-Frequency Operation*

Consider a microcavity in which light is coupled into a WGM from an adiabatically tapered fiber tangentially in contact with the resonator. Tunable single-frequency

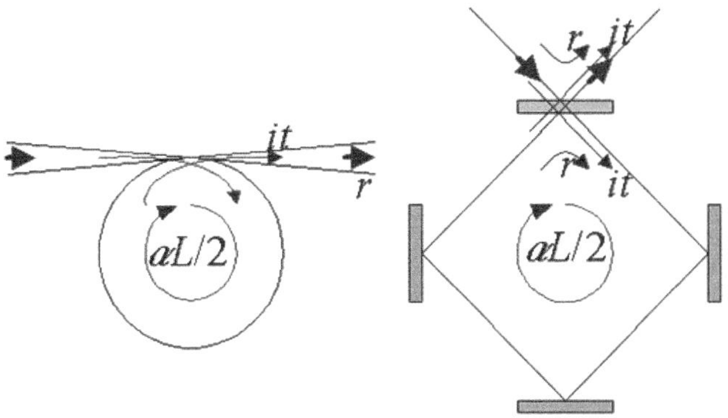

Fig. 5.1 Four-mirror ring cavity model. *Left*, microcavity and tapered fiber in contact. Light can couple from the fiber into the resonator and back into the fiber. *Right*, the four-mirror ring cavity equivalent. The top mirror is partially transmitting; all others have 100% reflectivity. Reprinted from Ref. 3 with permission. © 2008 Optical Society of America

light is injected at one end of the tapered fiber and the throughput spectrum is detected at the other end. The throughput will display a Lorentzian dip for each WGM resonance that is excited, because some power is lost to scattering and absorption in the resonator's WGM.

The microcavity can be modeled as a four-mirror ring cavity, as depicted in Fig. 5.1, in which one of the mirrors is partially transmitting while all the others are assumed to be perfect reflectors. The ring cavity model is a valid analog to a fiber-coupled microresonator when only a single fiber mode is excited. For light incident on a microresonator, this is accomplished by making the first fiber taper transition adiabatic. However, light coupling out of the microresonator can couple into many fiber modes, and any light coupled into higher-order modes will be lost if the second taper transition is also adiabatic. Fortunately, it is possible to make this lossy coupling into higher-order fiber modes negligibly small by properly choosing the diameter of the tapered fiber[11].

A round-trip power loss αL, where α is the loss coefficient and L is the microresonator circumference, is assumed; it models the intrinsic loss (primarily surface scattering) of a WGM microresonator. The reflection and transmission coefficients r and it, relating to the field mode amplitudes, are taken to be real and imaginary, respectively, without loss of generality; this corresponds to the usual choice for fiber or prism coupling[10,12,13]. The assumption of negligible power loss into radiation modes requires that these coefficients satisfy the relation $t^2 = 1 - r^2$. Because Q is so high, even under conditions of loading by the coupler, the intrinsic round-trip loss and the coupling loss (mirror transmissivity or probability of photon tunneling between fiber and microresonator) will always be small, so $\alpha L << 1$ and $T = t^2 << 1$ will be assumed throughout this chapter.

By summing the field over round trips, the net reflected power fraction (fiber throughput fraction) is found to be

$$R_\delta = \left| \frac{r - \mathrm{e}^{-\alpha L/2}\mathrm{e}^{\mathrm{i}\delta}}{1 - r\,\mathrm{e}^{-\alpha L/2}\mathrm{e}^{\mathrm{i}\delta}} \right|^2, \tag{5.1}$$

where $\delta = 2\pi nL(\nu - \nu_0)/c$ is the round-trip phase accumulation due to detuning of the input frequency ν from resonance ν_0 (c is the speed of light and n is the resonator's index of refraction). The deviation of the throughput fraction from unity gives the dip profile:

$$M_\delta = 1 - R_\delta = \frac{T\alpha L}{\frac{1}{4}(T + \alpha L)^2 + 4\sin^2\frac{1}{2}\delta} \cong \frac{T\alpha L}{\frac{1}{4}(T + \alpha L)^2 + \delta^2}. \tag{5.2}$$

This function is the usual Airy profile, with maxima at $\delta = p2\pi, p = 0, 1, 2, \ldots$. The spacing of 2π between adjacent modes is the free spectral range (FSR) in phase. (Recall that the FSR in frequency is given to good approximation by $c/(nL)$. Because many higher-order WGMs can be excited, the actual spacing between adjacent modes is much less than the FSR.) The full width at half-maximum of a WGM resonance is thus seen to be given by $\Delta\delta = T + \alpha L$, the total round-trip loss. The final expression in (5.2), valid when $\delta << 1$, shows the Lorentzian profile of a dip.

The dip depth on resonance ($\delta = 0$) is determined by the ratio of the coupling loss to intrinsic loss, $x = T/\alpha L$, and is given by

$$M_0 = \frac{4x}{(1 + x)^2}. \tag{5.3}$$

If $x = 1$, critical coupling is obtained and the dip depth attains its maximum value of 100%; the microresonator is said to be undercoupled if $x < 1$ and overcoupled for $x > 1$. While the coupling loss remains constant, the effective intrinsic loss can be changed by interaction of the evanescent fraction (f) of the WGM with the surrounding medium. The effective loss coefficient can then be written as $\alpha = \alpha_\mathrm{i} + f\alpha_\mathrm{a} + f\alpha_\mathrm{s}$, where the three terms denote true intrinsic loss, absorption (and perhaps also scattering) by the analyte, and absorption in the solvent (or ambient).

Absorption by the analyte causes a change in dip depth that, when small (weak analyte absorption), is proportional to the change in analyte absorption coefficient, in analogy with Beer's law. A theoretical effective absorption path length $L_\mathrm{eff}^\mathrm{t}$ can then be obtained from the dip-depth dependence on the absorption coefficient of the analyte

$$\frac{\mathrm{d}M_0}{M_0} \cong -\left(\frac{1 - x}{1 + x}\right)\frac{f}{\alpha_\mathrm{i} + f\alpha_\mathrm{s}}\mathrm{d}\alpha_\mathrm{a} \cong -L_\mathrm{eff}^\mathrm{t}\mathrm{d}\alpha_\mathrm{a}. \tag{5.4}$$

In the absence of solvent absorption, this effective absorption path length can be expressed as

$$L_{\mathrm{eff}}^{\mathrm{t}} = \left(\frac{1-x}{1+x}\right)\frac{f}{\alpha_{\mathrm{i}}}. \tag{5.5}$$

This effective length is defined in the low-analyte-absorption limit, $\alpha_{\mathrm{a}}L_{\mathrm{eff}}^{\mathrm{t}} << 1$ (or $f\alpha_{\mathrm{a}} << \alpha_{\mathrm{i}}$), which is the condition for (5.4) to hold. Note that in the strongly undercoupled or overcoupled limits ($x << 1$ or $x >> 1$) the relative detection sensitivity is determined by the intrinsic loss only. This can be advantageous, since having the tapered fiber in contact with the microresonator tends to produce overcoupling, especially when the system is immersed in a liquid. Thus in the strongly overcoupled case, this method has greater relative sensitivity (here, fractional change in dip depth) than the frequency-shift, mode-width, or ringdown methods, all of whose relative sensitivities are determined by the total loss. Recall that the total loss determines the linewidth and the cavity lifetime, and in the overcoupled limit coupling loss dominates. The relative frequency-shift sensitivity is measured as a fraction of the full linewidth, the change in mode width is relative to the full width, and ringdown measures the fractional change in the overall lifetime; however, the relative change in dip depth depends on the ratio of analyte absorption to intrinsic loss only.

An experimental effective absorption path length $L_{\mathrm{eff}}^{\mathrm{e}}$ can be found by measuring the dip depth in the absence of analyte (M_0) and in the presence of analyte ($M_0 + \Delta M_0$)

$$L_{\mathrm{eff}}^{\mathrm{e}} \cong \frac{1}{\alpha_{\mathrm{a}}}\ln\frac{M_0}{M_0 + \Delta M_0}. \tag{5.6}$$

As with (5.4), this holds for small changes in dip depth or $\alpha_{\mathrm{a}}L_{\mathrm{eff}}^{\mathrm{e}} << 1$. Comparison of experimental and theoretical effective absorption path lengths for detection of atmospheric trace gases shows good agreement[4], as discussed in Sect. 5.3.

This method also allows for sensitive detection of an analyte in a strongly absorbing solvent. In (5.4), if $f\alpha_{\mathrm{s}} >> \alpha_{\mathrm{i}}$, the effective absorption path length can still be as large as $1/\alpha_{\mathrm{s}}$. In effect, solvent absorption can shift the sensor from one sensitive regime to another – from strongly overcoupled to strongly undercoupled, enabling absolute analyte sensitivity, i.e., actual signal amplitudes, that would be difficult to achieve in single-pass direct absorption through the same effective path length.

5.2.2 *Broadband Operation*

In modeling this mode of operation, a microsphere, with a prism coupler used to excite precessing modes[14], is taken to be analogous to a ring cavity with two identical partially transmitting mirrors. Here we will concentrate on the transmitted signal (drop power output, enabled by precession and detected by a spectrometer) rather than the reflected signal (throughput) (see Fig. 5.2).

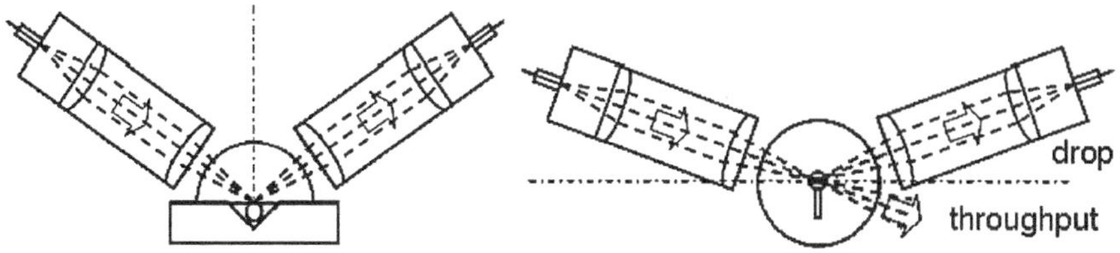

Fig. 5.2 Hemispherical prism coupling scheme. The microsphere is in a V-groove channel for the solvent and analyte. *Left*, polar view showing input and output coupling. *Right*, equatorial view illustrating how precessed light is collected at the drop port. Reprinted from Ref. 3 with permission © 2008 Optical Society of America

First consider single-frequency response. Summing over round trips and arbitrarily taking the second partially transmitting mirror to be at half the round-trip distance from the input mirror (the final result will not depend on this choice), gives the drop power fraction

$$D_\delta = \left| \frac{T\,\mathrm{e}^{-\alpha L/4}\mathrm{e}^{\mathrm{i}\delta/2}}{1 - r^2\mathrm{e}^{-\alpha L/2}\mathrm{e}^{\mathrm{i}\delta}} \right|^2 = \frac{T^2}{\left(T + \frac{1}{2}\alpha L\right)^2 + 4\sin^2\frac{1}{2}\delta} = D_0 \frac{\left(\frac{1}{2}\Delta\delta\right)^2}{\left(\frac{1}{2}\Delta\delta\right)^2 + \left(2\sin\frac{1}{2}\delta\right)^2}. \quad (5.7)$$

In this system, the mode width is $\Delta\delta = 2T + \alpha L$ and the resonant drop fraction is D_0.

Now, consider the case where the input is broadband and the drop signal is detected by a spectrometer, which integrates over a frequency interval that we will take to be equal to the FSR (for convenience; again, the final result will not depend on this choice, and the spectrometer resolution interval may be greater than or less than the FSR, as long the conditions noted below are satisfied). Assuming the input power, the reflection and transmission coefficients, and the intrinsic loss all to be independent of frequency over this interval, the integrated drop fraction D_I, the fraction of the incident power in the integration interval that is transmitted out the drop port, is given by

$$D_\mathrm{I} = \frac{D_0}{\sqrt{1 + \left(\frac{4}{\Delta\delta}\right)^2}} \rightarrow \begin{cases} D_0 = 4T^2/(\Delta\delta)^2 & \text{for } \Delta\delta >> 1 \\ D_0\Delta\delta/4 = T^2/\Delta\delta & \text{for } \Delta\delta << 1 \end{cases}. \quad (5.8)$$

The spectrometer signal will be proportional to this integrated fraction times the incident power in the detection interval. Note the functional dependence of the detected signal on the linewidth (or total loss) $\Delta\delta = 2T + \alpha L$ in the two limiting cases of large and small linewidth. The first limiting case says that when the linewidth fills the integration interval, the drop fraction equals what would be found by using a single-frequency source tuned to WGM resonance; this holds for either low Q or many overlapping modes in the integration interval, so that integration introduces no additional linewidth dependence. The second limiting case is the usual one for well-separated modes, and is the same as what results from approximating each WGM's narrow transmitted lineshape as a Lorentzian; this holds as long as the integration interval is wide compared to the WGM linewidth.

Let the effective loss coefficient now include, in addition to intrinsic loss, a contribution due to absorption (and scattering) by the analyte contained in the evanescent fraction f of the WGM: $\alpha = \alpha_i + f\alpha_a$. The effect of the analyte on the resonant (single-frequency) drop signal D_{0a}, when analyte absorption is a small fraction of the total loss, can be written in terms of an approximate effective absorption path length L_{eff} as defined below:

$$D_{0a} = \frac{4T^2}{(2T + \alpha_i L + f\alpha_a L)^2} = \frac{D_0}{\left(1 + \frac{1}{2}\alpha_a L_{eff}\right)^2}, \quad \text{where } L_{eff} = \frac{2fL}{2T + \alpha_i L}. \tag{5.9}$$

Here, L_{eff} is the effective absorption path length as defined in the low-analyte-absorption limit. However, the last expression for D_{0a} is valid even for large analyte absorption, that is, there are no restrictions on the size of $\alpha_a L_{eff}$ as long as $f\alpha_a L \ll 1$.

Then, for broadband input, the integrated drop fraction in the presence of analyte, D_{Ia}, will also be given by (5.9) in the large linewidth limit. For precessing-mode drop-port output collected without being spatially filtered by an aperture[15], this limit applies for the following reason. The incident focused light (Fig. 5.2) incorporates a bundle of wavevectors and so excites precessing modes over a range of angles with respect to the equatorial plane. The round-trip distance depends on this angle in a spheroid, so the precessing modes are frequency shifted by amounts depending on angle. The result is many modes (not only a range of different angles, but also ranges of different radial orders and different polar orders) that overlap to fill the integration interval, so when the analyte absorption broadens the linewidth and reduces the transmission of each mode, the broadening is not noticed, because the integration interval remains filled. Thus only the decrease in amplitude is observed, just as in the single-frequency resonant case. The overlapping of modes also means that the exact value of the spectrometer resolution interval does not matter. This large-linewidth-limit functional dependence of the integrated drop signal on $\alpha_a L_{eff}$ as given in (5.9) has been tested in recent experimental work[15], described briefly below in Sect. 5.4.

5.3 Microcavity-Enhanced Detection of Atmospheric Trace Gases

So far, in label-free chemical sensing applications, the interaction of a WGM's evanescent component with an analyte in the ambient or adsorbed on the microresonator's surface has led to the development of two sensing methods. These are monitoring of the WGM resonance frequency shift due to the analyte's change of the effective index of refraction[6,7] and measurement of the amount of Q spoiling (increase in WGM linewidth) resulting from analyte absorption[8].

Another method that we introduced earlier[16] and described in Sect. 5.2 is based on recording changes in the resonant dip depth observed in the coupling-fiber throughput[10]. In this implementation[4], we use a cylindrical microcavity in which a WGM is excited around its circumference using an adiabatically tapered fiber tangentially in contact with the cylinder. Tunable diode laser (TDL) light is injected at one end of the tapered fiber and the throughput at the other end displays a Lorentzian dip every time a WGM resonance is excited. The analysis of such an efficiently coupled microcavity is well known[10–13]. Absorption by a molecular trace gas in the air surrounding the microresonator will change the depth of a resonant dip, and this change is what we measure[3] This method enables TDL absorption spectroscopy experiments in which the long effective absorption path lengths that result are confined within less than a cubic millimeter.

Although the cylindrical microresonators provide a lower sensitivity than the microspheroids that we used earlier[16] (however, see further discussion in Sect. 5.3.3), they offer some very attractive advantages. First, they are far easier to mount in a strain tuner, and are much less susceptible to mechanical failure than microspheroids tuned by compression. Second, alignment is easier since there is only one degree of freedom to adjust (making the taper perpendicular to the cylinder axis) vs. two for the spheroids. Third, the mode spectrum is sparser, making locking of a WGM to the laser easier. And, finally, because the microcylinder is just a standard optical fiber, different resonators are nearly identical in diameter, providing better repeatability of results.

For chemical sensing applications, we model the microresonator as described in Sect. 5.2.1, and monitor the depth of a resonant dip in the fiber throughput. This dip depth is given by (5.3). As described in Sect. 5.2.1, absorption by the analyte causes a small change in dip depth that is proportional to the analyte absorption coefficient[3], as in (5.4). When a WGM is frequency locked[17] to the input laser, the WGM stays on resonance with the laser and, as the laser scans in frequency, the throughput of the bitapered fiber displays the locus of the bottom of the WGM resonance dip as it follows the laser scan. When an absorption line of the surrounding gas is present within the scan, there is a change in the WGM dip depth that traces the contour of the respective absorption line. Since in such a setup the microresonator is always on resonance, light interacting with the gas executes many round trips around the circumference, resulting in a long effective interaction path length.

The theoretical effective absorption path length was given in (5.5); the experimental effective absorption path length takes the form shown in (5.6) and is found by measuring the dip depth in the absence of analyte (M_0) and in the presence of analyte ($M_0 + \Delta M_0$). Equation (5.6) thus provides a way to express the relative sensitivity (fractional change in dip depth) of the WGM gas sensor based on the experimental effective absorption path length, the length of a hypothetical absorption cell that has the same sensitivity. The model is tested by comparing experimental and theoretical effective path lengths.

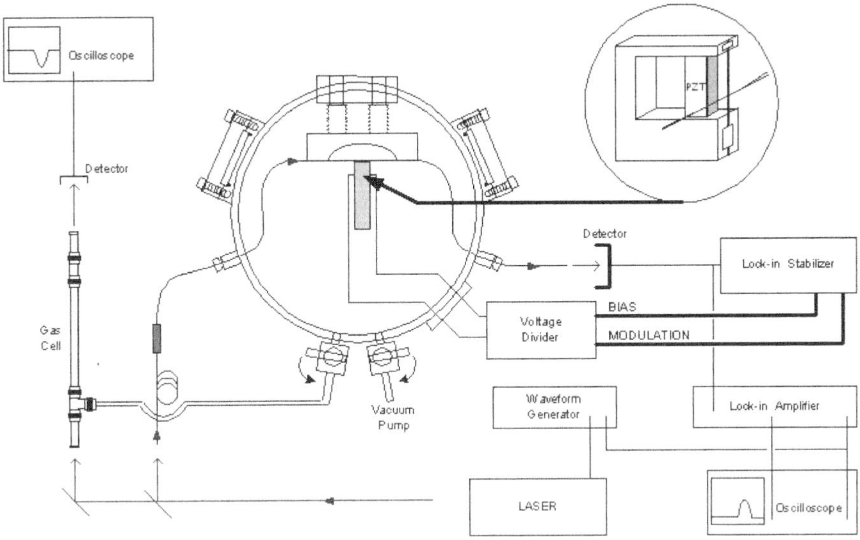

Fig. 5.3 Experimental setup. Light from a frequency-scanned cw diode laser is launched into a bitapered fiber to excite WGMs of the microresonator. The microresonator is held in a PZT fixture for tuning and locking purposes (*inset*). Reprinted from Ref. 4 with permission. © 2008 Optical Society of America

5.3.1 Experimental Setup

Figure 5.3 shows the experimental setup[4]. A cylindrical vacuum chamber with a transparent top plate and side windows for viewing the optical setup is connected to a reference cell that is used for calibration. The gas mixes and pressures in the chamber and cell can be varied independently. Typically, a low partial pressure of the molecular gas of interest is added to air at atmospheric pressure in the main chamber. Light from a cw TDL is coupled into the reference cell and into a single-mode optical fiber. The fiber passes through a polarization controller that is adjusted to ensure that WGMs of a single polarization (TE or TM) are excited. Then the fiber is fed into the vacuum chamber where its adiabatic bitaper is brought into contact with the microresonator. The position and the orientation of the coupling fiber are controlled, via bellows-sealed feedthroughs, by a positioner located outside the chamber. The 125-μm-diameter cylindrical fused-silica microresonator is mounted in a fixture (inset, Fig. 5.3) that allows tuning of the WGMs by stretching the resonator when a voltage is applied to the piezoelectric transducer (PZT). The microcylinder is just a standard optical fiber whose jacket has been softened by soaking in acetone and then pulled off. The fiber is then soaked in clean acetone and wiped with lens paper and methanol. This preserves the optical quality of the microresonator's surface. After the coupling fiber exits the chamber, its output is collected onto a detector. The signal from the detector is split, part of it being fed into a lock-in stabilizer in order to lock individual WGMs to the laser.

The locking is done as follows[17]. A WGM is tuned to be in resonance with the TDL by adjusting the PZT bias voltage, and the lock-in stabilizer is switched on. The lock-in stabilizer applies a 518-Hz dither voltage to the PZT and peak-locks the WGM to the laser. Then as the laser scans in frequency the lock-in stabilizer tunes the dc bias voltage on the PZT, maintaining the WGM on resonance with the laser over the entire scan range. When locked, the maximum excursion from true resonance is less than 5% of the WGM's linewidth. This tuning jitter results in unfiltered noise on the signal that is less than 1% of the dip depth.

The rest of the detector signal is noise filtered and amplified by a lock-in amplifier. The output of the lock-in amplifier is monitored by an oscilloscope, and recorded as the laser scans across the gas's absorption line. The result is a spectral profile of the gas absorption, impressed on the depth of the locked resonance dip. This is then analyzed using (5.6) to find an experimental effective absorption path length.

5.3.2 Results

Experimental results are reported here for a few trace gases, having absorption features in the vicinity of 1.65 μm, in air at atmospheric pressure. The gases investigated were methane (CH_4)[18,19], methyl chloride (CH_3Cl)[18], and ethylene (C_2H_4)[19]. The transitions involved belong to overtone and combination bands involving the C-H bonds. For example, the absorption lines of methane in this wavelength range correspond to the C-H asymmetric stretch vibrational overtone $2\nu_3$[18–20]. There are three lines of approximately equal strength that are roughly equally spaced at intervals of about 400 MHz and pressure broadened by air to about 4 GHz each[20]. The methyl chloride lines probably belong to the perpendicular component of the $2\nu_4$ band[18]; however, the assignment for ethylene is less certain[19].

These results are presented in Fig. 5.4 and the details are summarized in Table 5.1. For each gas in Fig. 5.4, the dip-depth variation (change in M_0, at 50× amplification) shows the detected absorption profile. These are not to scale, in the sense that the full dip is far too large to show at this amplification. Straight lines are added to make the variation clear; the slope of these lines (again, amplified 50×) may be an effect of the differential tuning of adjacent WGMs resulting in slightly varying mode overlap, because there is no variation in coupling (or in x) over the 10-GHz frequency scan range. When the gas compositions and pressures are the same in the test chamber and in the reference cell, both give the same lineshape. For example, the top traces in Fig. 5.4 (methane) can be fitted well by a sum of three unresolved Lorentzians.

In Table 5.1, the theoretical effective absorption path length is calculated from (5.5) using, for example, in the case of methane, $x = 0.28$ (calculated from the dip depth M_0; the dip gets shallower with analyte absorption, so the WGM is under-coupled), $f = 1.6\%$ (estimated from a computation of the field distributions at the same wavelength in a microsphere of the same diameter), and $\alpha_i = 0.0061\ \text{cm}^{-1}$

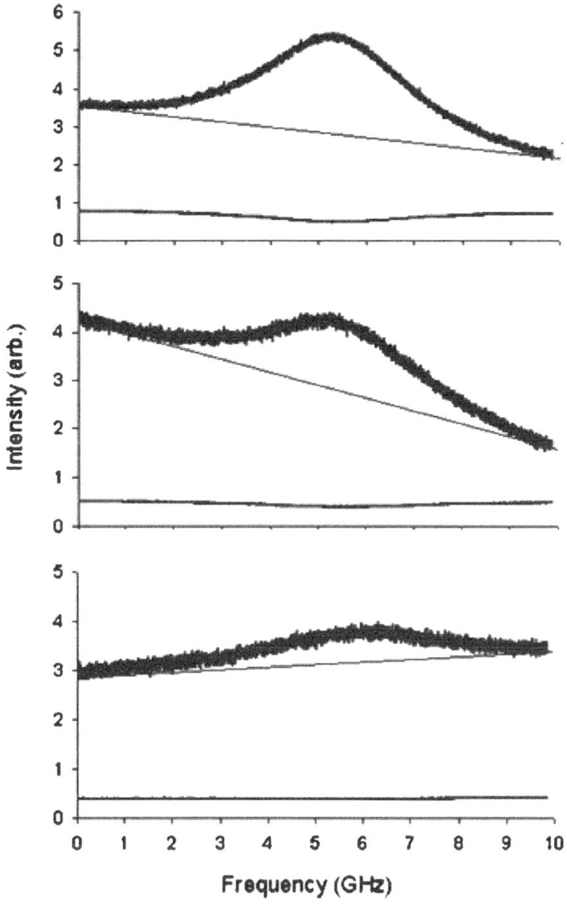

Fig. 5.4 From top down: the measured absorption profiles of methane, methyl chloride, and ethylene obtained using a WGM locked to the laser. In each case, the top trace shows the amplified variation in dip depth and the bottom trace is the transmission profile of the gas in a 16-cm absorption cell. The frequency axis shows the tuning range. Reprinted from Ref. 4 with permission. © 2008 Optical Society of America

Table 5.1 Summary of results from Fig. 4, for gases in air at atmospheric pressure. Reprinted from Ref. 4 with permission. © 2008 Optical Society of America

Gas	L^{t}_{eff} (mm)	L^{e}_{eff} (mm)	WGM Q	λ (nm)	Partial pressure (atm)
CH_4	15.0 ± 3.0	17.7 ± 1.7	7×10^6	1,653.722	0.01
CH_3Cl	7.2 ± 1.5	6.3 ± 1.1	3×10^6	1,651.59	0.01
C_2H_4	13.6 ± 2.7	13.2 ± 2.6	5×10^6	1,654.23	0.02

(found from the values of Q and x). The experimental effective absorption path length was obtained using (5.6) in which, again for methane, $M_0 = 68.4\%$ and $\Delta M_0 = -0.7\%$. The value of the methane absorption coefficient that was used for this concentration, $\alpha_a = 0.0058\ \mathrm{cm}^{-1}$, was extracted from the reference cell transmission measurements, and is in reasonable agreement with the value expected from parameters found in the HITRAN database[20]. Similar calculations were done for the other gases.

For each trace gas in Table 5.1, there is good agreement between the theoretical and the experimental effective absorption path lengths. For the theoretical values, the main source of error is in calculating the evanescent fraction, because it takes on different values for different modes, and the WGM that is used is not definitively identified. Uncertainty in the measured value of α_a and the residual noise on the traces corresponding to the locked WGMs are responsible for the error in determining the experimental effective absorption path length.

5.3.3 Summary

Cavity-enhanced laser absorption spectroscopy using microresonator WGMs has been demonstrated by locking the WGM resonances to the laser. Atmospheric trace gases in the evanescent fraction of the locked WGM are detected. The gas molecules remain in the ambient air and are not adsorbed onto the microresonator surface. The relative detection sensitivity is characterized by an effective absorption path length, and its experimentally determined value is in good agreement with the theoretical prediction. The major advantages of our setup are miniaturization and relatively low cost. Although the detection volume is less than a cubic millimeter, relatively large effective absorption path lengths on the order of centimeters are obtained.

From Fig. 5.4, the detection limit can be estimated to be about one-tenth of the concentrations shown or about 1,000 parts per million. This is an order of magnitude higher than that estimated for a microsphere in Ref. 16, because the Q is lower by a factor of 20 – 5×10^6 vs. 10^8. The other advantages of the microcylinder mentioned in the introduction to this section outweigh this sensitivity deficit and, in addition, we now have very good agreement with theory. Nevertheless, other microresonator geometries can also be envisioned for enhancing the sensitivity of the setup. Using a fiber fusion splicer, for example, a microspheroidal resonator with two stems can be obtained. Mounting it into the same PZT device would enable tuning of the WGMs in the same fashion as the microcylinder[21]. However, the curved surfaces of the microspheroid should, in principle, result in higher-Q WGMs with greater detection sensitivities, i.e., longer effective absorption path lengths.

5.4 Extensions of Method and Further Enhancements

The cavity-enhanced evanescent-wave sensing method described in Sect. 5.2 and applied to gases in Sect. 5.3 can also be employed for the detection of chemicals in liquid solution. In addition, the intracavity enhancement provided by the high-Q

microresonator can be further enhanced by other techniques, thereby increasing the sensitivity. Some of these extensions and enhancements are briefly described in this section.

5.4.1 Sensing in Liquids: Laser and Broadband

Both the modes of operation described in Sect. 5.2 may be used for the detection of chemicals in liquid solution. Because the analyte's absorption linewidth is very broad, overlapping several (or many) WGMs, no tuning of the microresonator, or locking of a WGM to the scanning laser is necessary. In fact, a broadband source such as a light-emitting diode (LED) may be used.

Microcavity-enhanced absorption sensing experiments were performed on two dyes dissolved in methanol, using a cw tunable Ti:sapphire laser at 800 nm and a TDL at 1,550 nm[22]. In both cases, the experimental sensitivity was such that 100-pM changes in concentration could be detected. This is especially remarkable at the longer wavelength, because at 1,550 nm methanol is strongly absorbing, with an absorption coefficient of about 8.8 cm^{-1}. This provides further confirmation of the theory of Sect. 5.2.1[3]. However, the experimental sensitivities are approximately two orders of magnitude greater than predicted by the theory. This might be a result of adsorption of dye molecules onto the surface of the microresonator, which could be tested by using certain surface treatments. Another possible explanation is multiple-fiber-mode excitation, described in Sect. 5.4.2.

The second implementation, described in Sect. 5.2.2, uses a broadband source and spectral detection, using a spectrometer with resolution of the order of the resonator's FSR (the FSR is about 200 GHz for a 330-μm diameter microsphere, or 0.2 nm for wavelengths around 550 nm). Precessing modes in a microsphere are employed to give a drop spectrum that is modified by analyte absorption. In this case, the total loss determines the sensitivity, but with prism coupling it is more likely that the system will be in the undercoupled limit. The setup shown in Fig. 5.2 has been used for the detection of Lissamine Green B dye in pH 5 citrate buffer using a 635-nm LED light source[15]. Effective path lengths of about 1 cm were observed, and the behavior of the drop signal as a function of concentration (hence α_a) was in most cases exactly as predicted by (5.9).

5.4.2 Prospective Methods for Sensing Enhancement

Two methods of further enhancing detection sensitivity rely on the use of multiple resonators or multiple fiber modes. The first will just be mentioned briefly, because although it is absorption based it uses a frequency shift. When two microresonators have resonances that are coincident in frequency, and the second resonator is brought near to the first resonator, which is in contact with the coupling fiber, the coupling between resonators will split the mode and produce the analog of electromagnetically induced transparency. This coupled-resonator-induced transparency (CRIT) response[23] is sensitive to absorption in the ambient, and the presence of absorbing molecules will cause the frequency splitting to change[24]. This provides another sensing method that is enhanced by the CRIT response and benefits from the fact that the splitting is independent of any overall frequency shift.

The second enhancement method involves multimode excitation of a single WGM. If one uses a tapered coupling fiber whose taper-down transition is nonadiabatic, but whose taper-up transition is adiabatic, then as light propagates from the single-mode untapered fiber, where it is core guided, into the taper waist region, where it is cladding guided, higher-order fiber modes can be excited, most likely the HE_{12} mode in addition to the fundamental HE_{11}. These modes have the same frequency but different propagation constants. They couple to the same WGM, but their relative input phase will depend on the position of the microresonator along the taper waist region. Light will couple out of the WGM back into the taper modes, but only the HE_{11} will then couple into the core-guided mode of the untapered fiber for detection, because the second taper transition is adiabatic. For the proper combination of relative phases and coupling strengths, we have shown that the absorption sensitivity can be enhanced by about two orders of magnitude over the usual single-fiber-mode case. The throughput dip depth is no longer a good indicator of intracavity power, and thus a small change in effective intrinsic loss can make a large change in the dip signal. It may be that the experimental disagreement from theory for liquid sensing in Sect. 5.4.1 is a result of this multiple-fiber-mode coupling effect.

5.4.3 Gold-Nanorod-Enhanced Sensing

We have developed a technique for growing gold nanorods directly on the surface of a microresonator, using semiconductor (HgTe) nanoparticles as seeds[25]. This method has proven to give a better nanorod yield than the use of gold nanoparticle seeds. When the nanorods have an aspect ratio (length to width) of about 5 or more, their longitudinal surface plasmon resonance is shifted from the visible into the infrared (~800 nm). This enhances the coupling of light from a tapered fiber into WGMs[26,27]. The strong surface field enhancement provided by the longitudinal plasmon resonance effectively increases the strength of the WGM's evanescent component, giving a large increase in fiber-microsphere coupling; the strength of this coupling is measured by the probability of tunneling into the WGM experienced by a photon as it travels down the fiber past the sphere. Remarkably, the coupling can be increased by a factor that is typically between 100 and 1,000, while the intrinsic loss (surface scattering) increases by less than a factor of three.

The enhanced evanescent field, as confirmed by the coupling enhancement results noted above, has been shown to significantly enhance detection sensitivity in preliminary experiments using dye in water solution at 800 nm. In addition, a nanorod-coated microresonator can be used for surface-enhanced Raman scattering (SERS). SERS greatly increases the number of molecular species that can be detected, because the molecule no longer has to be resonant with the driving laser.

5.4.4 *Summary*

The enhancement techniques briefly described in this section are just a few of those that can be used to make cavity-enhanced evanescent-wave chemical detection even more sensitive. The use of dielectric microresonators with tapered-fiber coupling enables performance that is difficult to achieve in other configurations, while minimizing the sensor size.

5.5 Surface and Thin-Film Characterization

It is well known that the WGMs of dielectric microresonators can serve as sensitive probes of their environment, through the environment's interaction with the WGMs' evanescent components. In addition to microcavity-enhanced laser absorption spectroscopy of molecules in the ambient gas or liquid, another effect that can be used as an environmental probe is the temperature sensitivity of WGM resonance frequencies[28–30]. A high-Q microresonator can easily measure mK temperature changes; while this can be a complication for some applications, it can also be put to use in detecting heat absorbed and heat lost by the microresonator. At wavelengths around 1,570 nm, the resonance frequency will shift down by approximately 1.6 GHz for each Kelvin increase in temperature, due primarily to the change in refractive index[31], to which is added the effect of thermal expansion.

The experiments reported here are extensions of some of our earlier work[32]. First, a microsphere is heated by focusing a laser beam onto it; upon turnoff of the heating beam, the microsphere relaxes back to room temperature. From its relaxation rate, observed via the WGM frequency shift using a weak TDL, we can calculate the thermal conductivity of the surrounding air. Doing this for a range of pressures allows us to determine the thermal accommodation coefficient of air on the surface (roughly the probability that a gas molecule will equilibrate with the surface after one collision). The surface in these experiments is either bare fused silica or a polymer film coating. This is a novel method for measuring accommodation coefficients. In a second experiment, use of the diode laser at higher power (without an external heating beam) produces thermal bistability; since the experiment with external heating determined the heat loss, the bistable response gives us the heat supplied by the fraction of total optical power lost due to absorption. This absorption results primarily from a water layer on the surface of a bare sphere, and for a coated sphere it is due to the polymer film. The precision of these

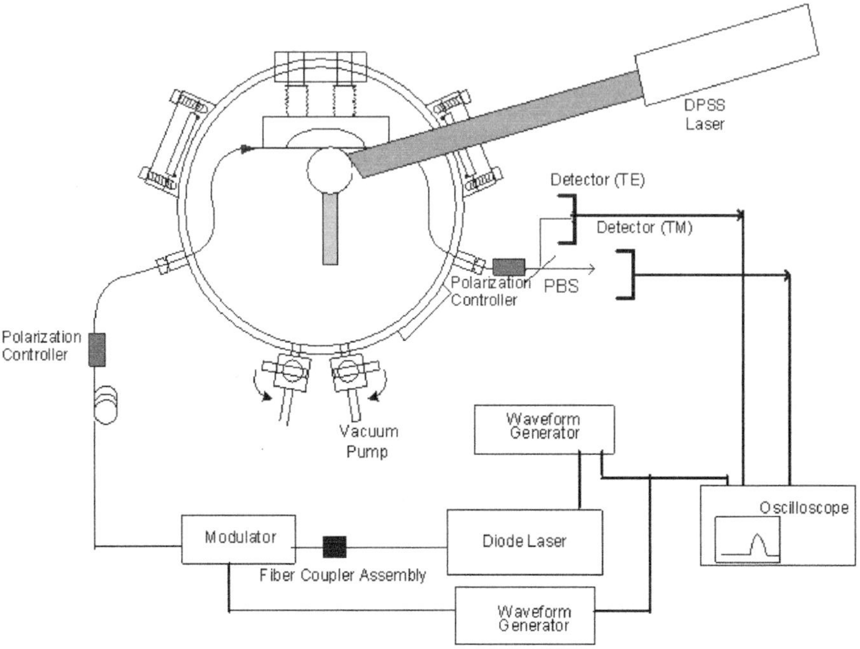

Fig. 5.5 Experimental setup. The diode laser is frequency scanned by one waveform generator, while the other controls the modulation. The light couples from a tapered fiber into and back out of microsphere WGMs, and the throughput is detected. A polarizing beamsplitter (PBS) separates throughput of the two polarizations. A diode-pumped solid-state laser can be used as an external heat source for the microsphere, and the vacuum chamber allows control over the ambient pressure. Reprinted from Ref. 5 with permission. © 2008 International Society for Optical Engineering

measurements is quite good (~5%), so the prospect of using these techniques for thin-film characterization is promising.

5.5.1 Experiments

The experimental setup is shown in Fig. 5.5. The microsphere is mounted in a cylindrical vacuum chamber that has a transparent top plate and side windows for viewing. The pressure in the chamber can be varied from above atmospheric to about 2 mTorr. Light from a cw TDL is coupled into a modulator, from which it exits into a single-mode optical fiber that passes through a polarization controller before being fed into the chamber, where its adiabatic bitapered region is brought into contact with the microsphere. The position and orientation of the coupling fiber are controlled, via bellows-sealed feedthroughs, by a positioner located outside the chamber. The polarization controller is adjusted to ensure that WGMs of a single polarization (TE or TM) are excited, and after the fiber is fed out of the chamber the two polarizations in the throughput can be detected separately. The modulator is normally off, but can be turned on to provide a square-wave intensity modulation; the throughput response to the square-wave input tells us whether a WGM is undercoupled or overcoupled (intrinsic or coupling loss dominant, respectively) without changing the geometry of the system. The microsphere is actually a prolate spheroid of average radius about 300 μm, fabricated from optical fiber in such a way as to have very thin stems both above and below. The double stem helps to make the mounting more rigid, since the stems are so thin (tapering to just a few μm). The 532-nm heating beam of about 1 W comes from a frequency-doubled $Nd{:}YVO_4$ laser, and is focused through a side window onto the microsphere.

The diode laser is scanned up and down in frequency by a triangle wave, so that the scan should be linear in time and have the same rate in both directions. In the thermal accommodation coefficient experiments, the external beam heats the microsphere to a few K above room temperature and is then turned off. The diode laser is kept at fairly low power ($\sim$7 μW) so that it does not appreciably heat the microsphere. Displacement of a WGM's throughput dip from one scan trace to the next is analyzed to find the relaxation time constant as the microsphere returns to room temperature. Results from the two scan directions are averaged to reduce error due to residual scan nonlinearity. This is done over a wide range of pressures (about four orders of magnitude). The time constant provides the measured thermal conductivity of the surrounding air, and fitting the thermal conductivity vs. pressure curve determines the thermal accommodation coefficient, as described in Sect. 5.5.2.

The thermal bistability experiments use the same system, minus the external heating laser. Now the diode laser power incident on the microsphere is on the order of 150 μW, so that the diode laser will heat the microsphere if it is scanned slowly enough across a WGM resonance. Since the relaxation time constant is known from the previous experiment, fitting the bistable throughput trace to a simple model will determine the single free parameter, which is the fraction of the total power loss that is due to absorption (assuming the absorbed power is converted to heat rather than reradiated). This is described in Sect. 5.5.3. Because the absorption in fused silica is so low in this wavelength range, we assume that all the absorption takes place in the water layer on the surface of the bare sphere or in the polymer layer on the coated sphere. Then from the known absorption coefficient for bulk water, we can estimate the thickness of the water layer. Likewise, knowing the thickness of the polymer film, we can find its absorption coefficient.

5.5.2 Accommodation Coefficients

Analysis of these experiments is simplified by the fact that the thermal conductivity of fused silica is much greater than that of air. Therefore, the microsphere can be treated as being in thermal equilibrium (uniform internal temperature distribution) at all times because of its very fast internal relaxation[33]. We assume that heat loss through the stems is negligible because of their small masses and long conduction paths. Because the microsphere is so small, convection is negligible[34], and so heat loss occurs through conduction by the surrounding air and by radiation. Define T to be the deviation of the microsphere's temperature above room temperature T_R (in this section, T_c will be used to denote coupling loss), and the relaxation equation can be written as follows for $T << T_R$:

$$\frac{dT}{dt} = -\frac{1}{mc}\left(4\pi a k_{air} + 16\pi a^2 \varepsilon \sigma T_R^3\right)T = -\frac{1}{\tau}T, \tag{5.10}$$

where m is the microsphere's mass, c is the specific heat of fused silica, a is the microsphere's radius, k_{air} is the measured thermal conductivity of air, ε is the emissivity of fused silica, σ is the Stefan–Boltzmann constant, and τ is the thermal relaxation time constant. Because the temperature variation is small, any temperature dependence of these parameters can be neglected. However, k_{air} will depend on pressure; the measured value will begin to decrease as the molecular mean free path becomes comparable to the size of the microsphere. The pressures used are such that our data are predominantly in the slip-flow or temperature-jump regime,[35] that is, the pressure is higher than the free-molecular-flow regime. This means that the pressure dependence of k_{air} is given by

$$k_{air}(p) = \frac{k_{atm}}{1 + \left(\frac{2-\alpha}{\alpha}\right)\frac{k_{atm}\sqrt{2\pi R T_R}}{(c_p + c_v)pa}}. \tag{5.11}$$

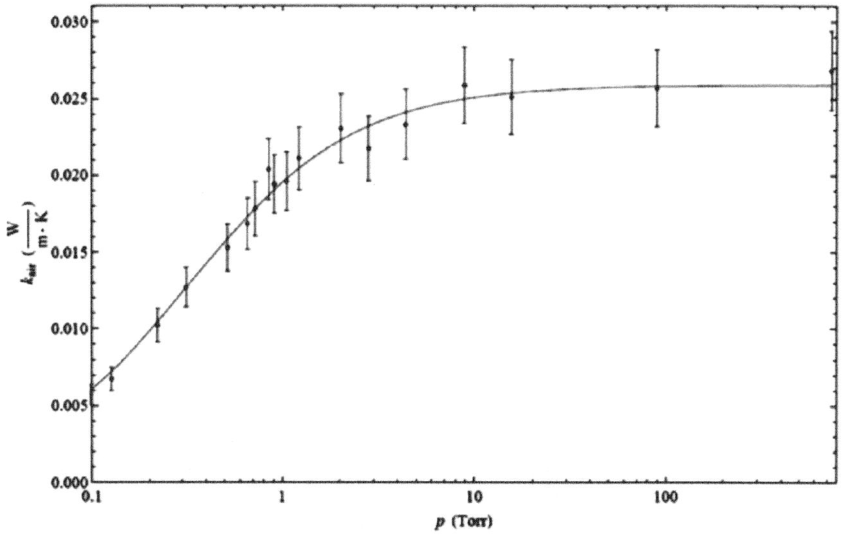

Fig. 5.6 Pressure dependence of thermal conductivity of air, measured using a PDDA-coated microsphere of effective radius 298 μm. The fit to (5.11), shown as the *curve*, gives a thermal accommodation coefficient of 0.92 for air on PDDA. Reprinted from Ref. 5 with permission. © 2008 International Society for Optical Engineering

In (5.11), k_{atm} is the thermal conductivity of air at atmospheric pressure; R is the gas constant per unit mass of air; c_p and c_v are the heat capacities of air at constant pressure and volume, respectively; p is the air pressure; and α is the thermal accommodation coefficient for air on the surface of interest. Fitting a plot of k_{air} vs. p then determines the value of α, the single free parameter.

Experimental values for k_{air} vs. p were found for two fused-silica microspheres, one bare and the other coated with a polyelectrolyte layer (Fig. 5.6). The second microsphere was dipped in a 0.5% solution of poly(dimethyldiallylammonium) chloride (PDDA) for about 30 min to produce a polyelectrolyte surface layer approximately 1-nm thick[36]. Because the microresonators were so prolate, an effective radius was chosen that gave the best fit of (5.11) to the data. For example, the bare sphere had a minor radius of 295 μm, a major radius of 375 μm, and an effective radius of $a = 325$ μm. Similarly, the PDDA-coated sphere had an effective radius of $a = 298$ μm. For the accommodation coefficient of air on fused silica, we find $\alpha = 0.84 \pm 0.03$, in good agreement with previous measurements for nitrogen and oxygen on glass[37]. Our value found for the accommodation coefficient of air on PDDA is $\alpha = 0.92 \pm 0.03$, slightly larger than on fused silica, as might be expected from the more porous structure of the PDDA surface.

For the PDDA-coated sphere, since the coating thickness is much less than the effective radius, we use the specific heat of fused silica in (5.10). The emissivity of PDDA is not known, so the value for fused silica was used ($\varepsilon = 0.87$); the results are not sensitive to the exact value, because the radiation loss is smaller than the heat loss by conduction. Looking at the data in Fig. 5.6, no obvious effect of heat loss from the stems is seen. If stem loss were significant, the measured value of k_{air} would begin to fall off at a higher pressure than it does. Perhaps, however, the somewhat larger scatter and error in the data at high pressures may be a consequence of some stem loss, as well as residual heating by the diode laser. The fit of (5.11) to the data is quite sensitive to the value of α, so its error is only on the order of 3% despite the larger errors in k_{air}. The fit remains good even to the lowest pressures of 0.1 Torr, indicating that the transition to free-molecule behavior has not yet been seen. At even lower pressures, estimated for these spheres to be on the order of 10 mTorr[35], the data points should begin to fall below the curve of (5.11). This appears to be an entirely new method for measuring the accommodation coefficient; nothing similar is mentioned in the literature[37].

5.5.3 Thermal Bistability

In these experiments, power absorbed from the WGM-exciting diode laser, operating at higher power, heats the sphere. The laser is scanned slowly in frequency across a resonance. The scan rate is slow enough that the assumption of internal thermal equilibrium in the microsphere holds, except perhaps during the fast throughput power jumps (see Figs. 5.7 and 5.8). In equilibrium, the power absorbed depends on the detuning from the temperature-dependent resonance, the depth of the resonance dip, and the incident power. Thus a heating term can be added to the right-hand side of (5.10) to describe the microsphere's temperature as a function of time[32]:

$$\frac{dT}{dt} = \frac{\beta M_0 P_{\text{inc}}}{mc} \frac{\left(\frac{\Delta\nu}{2}\right)^2}{(\nu(t) - \nu_0 + bT)^2 + \left(\frac{\Delta\nu}{2}\right)^2} - \frac{T}{\tau}, \tag{5.12}$$

where M_0 is the fractional dip depth[3], P_{inc} is the incident power (equal to throughput power when off resonance), $\Delta\nu$ is the linewidth of the Lorentzian WGM at $\nu_0 - bT$ where b is the 1.6 GHz K^{-1} temperature shift, and the only free parameter, β, is the fraction of the total power loss ($M_0 P_{\text{inc}}$) that is due to absorption. Fitting a throughput trace to a tuning curve derived from (5.12) gives the value of β. Now from (5.3), where the loss ratio is $x = T_c / 2\pi a \alpha_i$, and the measured (loaded) quality factor

$$Q = \frac{\nu_0}{\Delta\nu} = \frac{2\pi n}{\lambda(1 + x)\alpha_i}, \tag{5.13}$$

where n is the index of refraction of fused silica and λ is the wavelength, the intrinsic loss coefficient α_i can be found. There are two solutions, corresponding to the undercoupled ($x < 1$) and overcoupled ($x > 1$) cases, which is why the modulator is needed in the experiment to determine the coupling regime. Then the effective absorption coefficient is $\alpha_{\text{abs}} = \beta\alpha_i$. An approximate expression for the relation between the effective absorption coefficient of the water layer or thin film, of thickness δ, and its bulk absorption coefficient α_b has been given[38]:

$$\delta \cong \left(\frac{\lambda a}{\pi n^5}\right)^{1/2} \frac{\alpha_{\text{abs}}}{4\alpha_b}. \tag{5.14}$$

Using (5.14) and the determined value of α_{abs}, we can estimate δ if α_b is known, and vice versa. Two examples of thermal bistability data, fit to a calculated tuning curve based on (5.12), are shown below. Figure 5.7 is for the bare sphere, and Fig. 5.8 is for the PDDA-coated sphere. In the figures, the laser scans slowly across a TM-polarized WGM dip (taking several thermal relaxation times to scan $\Delta\nu$), first down in frequency, then reversing at the vertical dashed line, and scanning back up in frequency across the same mode. The continuous smooth lines are the theoretical fits.

The fitting to the data in Figs. 5.7 and 5.8 gives the values of α_{abs} to about 5% precision. The value found for the bare sphere is consistent with the values found at other pressures and for overcoupled modes. Using (5.14) and the known bulk absorption coefficient of water (at these wavelengths, water has $\alpha_b \approx 800$ m^{-1})[14] we estimate $\delta \approx 6.7$ pm from Fig. 5.7. This corresponds to water coverage on the

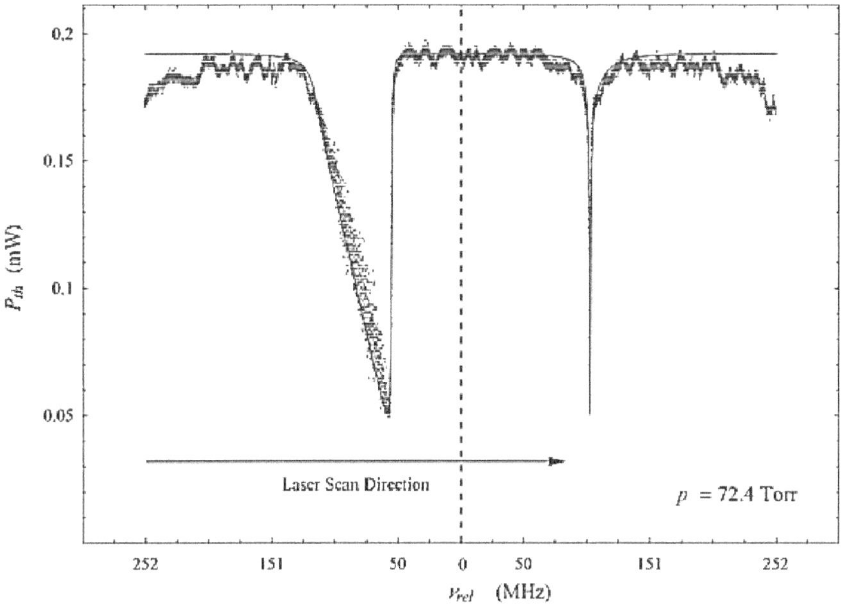

Fig. 5.7 WGM resonance dip showing thermal bistability. For this undercoupled mode in the bare sphere with $Q = 3.14 \times 10^7$, the fit gives $\alpha_{abs} = 0.00419\ m^{-1}$. Reprinted from Ref. 5 with permission. © 2008 International Society for Optical Engineering

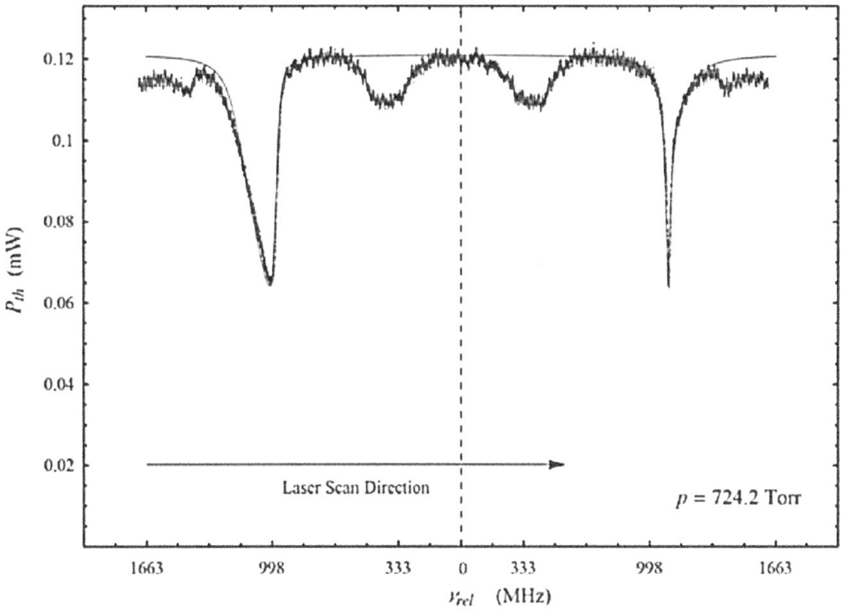

Fig. 5.8 WGM resonance dip showing thermal bistability. For this undercoupled mode in the PDDA-coated sphere with $Q = 2.25 \times 10^6$, the fit gives $\alpha_{abs} = 0.428\ m^{-1}$. Reprinted from Ref. 5 with permission. © 2008 International Society for Optical Engineering

fused-silica surface of about 5–6% of a monolayer. A typical value is about one monolayer; our values may be lower than this because the sphere had been held at a low pressure for some time before the measurements were made. Applying the same analysis to Fig. 5.8, we estimate $\delta \approx 0.66$ nm. This value indicates that the absorption coefficient of the PDDA film is of the same order of magnitude as that of bulk water.

5.5.4 Summary

The temperature sensitivity of the resonant frequencies of WGMs in a microsphere has been employed to measure the thermal conductivity of the ambient air as a function of pressure, and these results have enabled a quantitative explanation of thermal optical bistability observed in the microsphere under conditions of larger incident power. Measuring the thermal relaxation time to determine k_{air} vs. p, and fitting the results to the temperature-jump model of (5.11), determines the value of the thermal accommodation coefficient of air on the microsphere surface. Since these measurements can easily be done with various gases, using different surface coatings (thin compared to the wavelength), a large number of gas-surface interactions can be studied. This optical technique using microresonators is very well suited to making these measurements, and so adds a new method for the study of the interaction of gases with surfaces.

Measuring thermal bistability and fitting the observed mode profiles to the model, combined with the intensity modulation that permits determining the coupling regime without having to move any part of the setup, allow us to characterize the system losses. The coupling loss and intrinsic loss can be determined separately, and the intrinsic loss separated into its scattering and absorption components[39]. The absorption component is predominantly due to the surface water layer on a bare sphere, or to the coating on a sphere to which a thin film has been applied. This technique then also permits the measurement of the absorption coefficient of the film, and it can be applied to films of any refractive index and absorption coefficient, provided only that the film is thin enough.

5.6 Conclusions

The methods and results presented in this chapter describe the utility of chemical sensing techniques based on microcavity-enhanced optical absorption. In Sect. 5.2, a theoretical analysis was given for both tunable single-frequency operation and broadband operation. Using a single tunable frequency and measuring the effect of analyte absorption on the depth of the throughput dip was shown to give better relative sensitivity in some cases than methods such as frequency shift or cavity ringdown. It was further shown that this method makes it possible to measure weak analyte absorption against the background of a strong solvent absorption. It was also shown that using a broadband source and spectral detection of the drop-port output gives nearly as great a sensitivity as the tunable single-frequency method. For both methods, it was shown how experimental results can be easily compared to the theory.

The tunable single-frequency method was used, with the added modification of locking the WGM to the laser, for the detection of atmospheric trace gases. In this case, good agreement was found between experiment and theory. The low-Q cylindrical microresonators used in Sect. 5.3 provide centimeter effective absorption path lengths, but high-Q microspheres used in similar experiments have given effective path lengths as large as 10 m[32]. Tunable single-frequency sensing experiments in liquids, described in Sect. 5.4.1, have confirmed the ability to detect an analyte in a strongly absorbing solvent. In liquids, effective path lengths of several meters were found; this is nearly two orders of magnitude greater than the theoretical expectation, perhaps because of an inadvertent enhancement by one of the methods mentioned in Sect. 5.4.2. There are several ways to further enhance the sensitivity of microcavity-enhanced absorption sensing, and some of these will be quite easy to implement, such as growing a sparse distribution of gold nanorods on the microresonator surface.

The broadband analysis was confirmed by the experimental results mentioned in Sect. 5.4.1. This method can also be further enhanced by some of the techniques described in Sects. 5.4.2 and 5.4.3. The conclusion is that these methods of microcavity-enhanced optical absorption sensing provide compact, inexpensive, and sensitive detectors for molecular species in the ambient gas or liquid, and that further increases in sensitivity can be implemented to make them even more competitive. The molecular-transition specificity that is implicit in absorption spectroscopy is a limiting restriction, but the surface-enhanced Raman sensing that is enabled by metallic nanoparticles on the microresonator surface can significantly increase the number of molecular species that could be detected.

The sensing methods summarized thus far are intended for absorption detection of molecules in the ambient, but molecules (or indeed thin films) on the microresonator surface can also be detected. In particular, if the surface is covered to such an extent that the optical energy absorbed heats the microresonator, the resulting thermal bistability in the frequency-scan response can be used to determine the absorption and/or thickness of the thin-film coating. This and surface characterization by measurement of the thermal accommodation coefficient were described in Sect. 5.5. These methods offer quite precise measurement, provided that certain reasonable and easily implemented assumptions are satisfied.

In conclusion, when a WGM is excited in a dielectric microresonator, its evanescent component provides a convenient probe of the microresonator's surroundings. Various ways to implement evanescent-wave sensing have been devised, but the emphasis of this chapter has been on microcavity-enhanced absorption spectroscopy. The techniques described here have broad applicability, can even be used with broadband sources, and lend themselves well to further enhancement methods. We are looking forward to continuing our development of these sensors.

Acknowledgments The following former and current students have made significant contributions to this work: Jeromy Rezac, George Farca, Siyka Shopova, Elijah Dale, and Deepak Ganta. Also contributing were these students and colleagues: Chuck Blackledge, Razvan Stoian, Michael Humphrey, Sarah Bates, Seth Koterba, Jiangquan Zhang, Brian Strecker, Donna Bandy, Bret Flanders, Lee Elizondo, Sitong Yuan, and Mike Lucas. This work was supported by the National Science Foundation under award numbers 0329924 and 0601362, by the Oklahoma Center for the Advancement of Science and Technology under award numbers AR022–052 and AR072–066, by the Oklahoma State Regents for Higher Education, and by ICx Nomadics, Inc.

References

1 Matsko, A. B.; Ilchenko, V. S., Optical resonators with whispering-gallery modes – Part I: Basics, *IEEE J. Sel. Top. Quantum Electron.* **2006**, 12, 3–14
2 Ilchenko, V. S.; Matsko, A. B., Optical resonators with whispering-gallery modes – Part II: Applications, *IEEE J. Sel. Top. Quantum Electron.* **2006**, 12, 15–32
3 Rosenberger, A. T., Analysis of whispering-gallery microcavity-enhanced chemical absorption sensors, *Opt. Express* **2007**, 15, 12959–12964
4 Farca, G.; Shopova, S. I.; Rosenberger, A. T., Cavity-enhanced laser absorption spectroscopy using microresonator whispering-gallery modes, *Opt. Express* **2007**, 15, 17443–17448
5 Rosenberger, A. T.; Dale, E. B.; Ganta, D.; Rezac, J. P., Investigating properties of surfaces and thin films using microsphere whispering-gallery modes, In Laser Resonators and Beam Control X; Kudryashov, A. V.; Paxton, A. H.; Ilchenko, V. S., Eds.; *Proc. SPIE* **2008**, 6872, 68720U
6 Teraoka, I.; Arnold, S.; Vollmer, F., Perturbation approach to resonance shifts of whispering-gallery modes in a dielectric microsphere as a probe of a surrounding medium, *J. Opt. Soc. Am. B* **2003**, 20, 1937–1946
7 Hanumegowda, N. M.; Stica, C. J.; Patel, B. C.; White, I.; Fan, X., Refractometric sensors based on microsphere resonators, *Appl. Phys. Lett.* **2005**, 87, 201107
8 Armani, A. M.; Vahala, K. J., Heavy water detection using ultra-high-*Q* microcavities, *Opt. Lett.* **2006**, 31, 1896–1898
9 Savchenkov, A. A.; Matsko, A. B.; Mohageg, M.; Maleki, L., Ringdown spectroscopy of stimulated Raman scattering in a whispering gallery mode resonator, *Opt. Lett.* **2007**, 32, 497–499
10 Boyd, R. W.; Heebner, J. E., Sensitive disk resonator photonic biosensor, *Appl. Opt.* **2001**, 40, 5742–5747
11 Humphrey, M. J.; Dale, E.; Rosenberger, A. T.; Bandy, D. K., Calculation of optimal fiber radius and whispering-gallery mode spectra for a fiber-coupled microsphere, *Opt. Commun.* **2007**, 271, 124–131
12 Gorodetsky, M. L.; Ilchenko, V. S., Optical microsphere resonators: Optimal coupling to high-*Q* whispering-gallery modes, *J. Opt. Soc. Am. B* **1999**, 16, 147–154
13 Cai, M.; Painter, O.; Vahala, K. J., Observation of critical coupling in a fiber taper to a silica-microsphere whispering-gallery mode system, *Phys. Rev. Lett.* **2000**, 85, 74–77
14 Gorodetsky, M. L.; Ilchenko, V. S., High-Q optical whispering-gallery microresonators: Precession approach for spherical mode analysis and emission patterns with prism couplers, *Opt. Commun.* **1994**, 113, 133–143
15 Westcott, S. L.; Zhang, J.; Shelton, R. K.; Bruce, N. M. K.; Gupta, S.; Keen, S. L.; Tillman, J. W.; Wald, L. B.; Strecker, B. N.; Rosenberger, A. T.; Davidson, R. R.; Chen, W.; Donovan, K. G.; Hryniewicz, J. V., Broadband optical absorbance spectroscopy using a whispering gallery mode microsphere resonator, *Rev. Sci. Instrum.* **2008**, 79, 033106
16 Rosenberger, A. T.; Rezac, J. P., Whispering-gallery-mode evanescent-wave microsensor for trace-gas detection, In Biomedical Instrumentation Based on Micro- and Nanotechnology; Mariella, R. P., Jr.; Nicolau, D. V., Eds.; *Proc. SPIE* **2001**, 4265, 102–112

17 Rezac, J. P.; Rosenberger, A. T., Locking a microsphere whispering-gallery mode to a laser, *Opt. Express* **2001**, 8, 605–610

18 Chou, S.-I; Baer, D. S.; Hanson, R. K., Diode laser absorption measurements of CH_3Cl and CH_4 near 1.65 μm, *Appl. Opt.* **1997**, 36, 3288–3293

19 Boschetti, A.; Bassi, D.; Iacob, E.; Iannotta, S.; Ricci, L.; Scotoni, M., Resonant photoacoustic simultaneous detection of methane and ethylene by means of a 1.63-μm diode laser, *Appl. Phys. B* **2002**, 74, 273–278

20 Rothman, L. S.; Rinsland, C. P.; Goldman, A.; Massie, S. T.; Edwards, D. P.; Flaud, J.-M.; Perrin, A.; Camy-Peyret, C.; Dana, V.; Mandin, J.-Y.; Schroeder, J.; McCann, A.; Gamache, R. R.; Wattson, R. B.; Yoshino, K.; Chance, K. V.; Jucks, K. W.; Brown, L. R.; Nemtchinov, V.; Varanasi, P., The HITRAN molecular spectroscopic database and HAWKS (HITRAN atmospheric workstation): 1996 edition, *J. Quant. Spectrosc. Radiat. Transfer* **1998**, 60, 665–710

21 von Klitzing, W.; Long, R.; Ilchenko, V. S.; Hare, J.; Lefèvre-Seguin, V., Tunable whispering gallery modes for spectroscopy and CQED experiments, *New J. Phys.* **2001**, 3, 14.1–14.14

22 Farca, G., Cavity-Enhanced Evanescent-Wave Chemical Sensing Using Microresonators, PhD dissertation, Oklahoma State University, **2006**

23 Smith, D. D.; Chang, H.; Fuller, K. A.; Rosenberger, A. T.; Boyd, R. W., Coupled-resonator-induced transparency, *Phys. Rev. A.* **2004**, 69, 063804

24 Naweed, A.; Farca, G.; Shopova, S. I.; Rosenberger, A. T., Induced transparency and absorption in coupled whispering-gallery microresonators, *Phys. Rev. A* **2005**, 71, 043804

25 Shopova, S. I.; Blackledge, C. W.; Rosenberger, A. T.; Materer, N. F., Gold nanorods grown from HgTe nanoparticles directly on various surfaces, *Appl. Phys. Lett.* **2006**, 89, 023120

26 Shopova, S. I., Nanoparticle-Coated Optical Microresonators for Whispering-Gallery Lasing and Other Applications, PhD dissertation, Oklahoma State University, **2007**

27 Shopova, S. I.; Blackledge, C. W.; Rosenberger, A. T., Enhanced evanescent coupling to whispering-gallery modes due to gold nanorods grown on the microresonator surface, *Appl. Phys. B* **2008**, 93, 183–187

28 Il'chenko, V. S.; Gorodetskii, M. L., Thermal nonlinear effects in optical whispering gallery microresonators, *Laser Phys.* **1992**, 2, 1004–1009

29 Collot, L.; Lefèvre-Seguin, V.; Brune, M.; Raimond, J.-M.; Haroche, S., Very high-Q whispering-gallery mode resonances observed on fused silica microspheres, *Europhys. Lett.* **1993**, 23, 327–334

30 Carmon, T.; Yang, L.; Vahala, K. J., Dynamical thermal behavior and thermal self-stability of microcavities, *Opt. Express* **2004**, 12, 4742–4750

31 Malitson, I. H., Interspecimen comparison of the refractive index of fused silica, *J. Opt. Soc. Am.* **1965**, 55, 1205–1209

32 Rezac, J. P., Properties and Applications of Whispering-Gallery Mode Resonances in Fused Silica Microspheres, PhD dissertation, Oklahoma State University, **2002**

33 Carslaw, H. S.; Jaeger, J. C., Conduction of Heat in Solids, Clarendon, Oxford, **1959**, Chap. IX, 230–254

34 Foss, W. R.; Davis, E. J., Transient laser heating of single solid microspheres, *Chem. Eng. Commun.* **1996**, 152–153, 113–138

35 Kennard, E. H., Kinetic Theory of Gases, McGraw-Hill, New York, NY, **1938**, Chap. VIII, 291–337

36 Rogach, A. L.; Koktysh, D. S.; Harrison, M.; Kotov, N. A., Layer-by-layer assembled films of HgTe nanocrystals with strong infrared emission, *Chem. Mater.* **2000**, 12, 1526–1528

37 Saxena, S. C.; Joshi, R. K., Thermal accommodation and adsorption coefficients of gases, In Vol. II-1 of McGraw-Hill/CINDAS Data Series on Material Properties; Touloukian, Y. S.; Ho, C. Y., Eds.; McGraw-Hill, New York, NY, 1981

38 Vernooy, D. W.; Ilchenko, V. S.; Mabuchi, H.; Streed, E. W.; Kimble, H. J., High-Q measurements of fused-silica microspheres in the near infrared, *Opt. Lett.* **1998**, 23, 247–249

39 Rokhsari, H.; Spillane, S. M.; Vahala, K. J., Loss characterization in microcavities using the thermal bistability effect, *Appl. Phys. Lett.* **2004**, 85, 3029–3031

Chapter 6
Rapid Chemical Vapor Detection Using Optofluidic Ring Resonators

Yuze Sun, Siyka I. Shopova, Ian M. White, Greg Frye-Mason, and Xudong Fan

Abstract The optofluidic ring resonator (OFRR) is a novel gas sensing technology platform. In an OFRR gas sensor, the OFRR interior surface is coated with a layer of vapor-sensitive polymer. The interaction between the polymer and the gas molecules flowing through the OFRR results in a change in polymer refractive index and thickness, which can be detected by the circulating waveguide modes supported by the circular cross section of the OFRR. Due to the excellent fluidics of a capillary, the OFRR is capable of detecting chemical vapors rapidly with very low sample volume. In addition, the OFRR is highly compatible with gas chromatography (GC) and is a promising platform for development of micro-GC (μGC) with unique multipoint, on-column detection capability. In this chapter, we will discuss the fundamental operational principles of the OFRR gas sensor, followed by examples of rapid detection of several representative vapor analytes. The development of an OFRR-based μGC system and its applications in explosive separation and detection will also be presented.

6.1 Introduction

Chemical vapor sensors play an ever-increasing role in the environmental monitoring, homeland security, defense, and health care. The desirable characteristics of a chemical vapor sensor include ultrahigh sensitivity, specific and rapid response to certain vapor molecules, as well as the ability for on-the-spot chemical analysis, which usually requires the sensor to be small, portable, reusable, stable, robust, and cost effective. Toward this end, various sensing techniques have been studied

---

X. Fan (✉)
Department of Biological Engineering, University of Missouri, Columbia, MO 65211, USA
e-mail: fanxud@missouri.edu

X. Fan (ed.), *Advanced Photonic Structures for Biological and Chemical Detection*, Integrated Analytical Systems,
DOI 10.1007/978-0-387-98063-8_6,

extensively, including electrochemical gas sensors[1], microcantilever-based sensors[2], and surface acoustic wave sensors[3].

Optical sensors are another powerful tool in gas detection and analysis. Various optical techniques have been employed, which include optical absorption measurement (such as direct absorption measurement[4,5], Fourier transform infrared spectroscopy[6,7], photoacoustic spectroscopy[8], and cavity ring down spectroscopy[9], also see Chap. 5), photoluminescence measurement[10], Raman spectroscopy[11] and surface-enhanced Raman spectroscopy[12], and refractive index (RI) measurement.

The RI measurement is advantageous over other detection methods such as absorption and fluorescence in that the sensing signal depends on the sample concentration rather than the total amount of the sample. Therefore, it can be employed in a number of miniaturized vapor sensors. The RI measurement relies on the RI changes when the polymer interacts with gas molecules. A variety of optical structures have been developed to detect the RI changes, such as waveguides[13], fiber gratings[14–17] (also see Chaps. 3 and 7), Fabry–Pérot cavities[18,19] (also see Chap. 7), interferometers[20,21], and surface plasmon resonance[22].

Recently, optical ring resonators are under investigation as a promising chemical vapor sensing technology[23–29]. One of the major advantages of the ring resonator is its compact size. In a ring resonator, the light propagates in the form of whispering-gallery modes (WGMs) or circulating waveguide modes[30]. Although the ring resonator is only a few tens to a few hundreds of micrometers in diameter, the effective detection length can significantly be enhanced due to the high Q-factors of the WGMs. Therefore, the ring resonator technology enables large density of sensor arrays for portable devices with multiplexed detection capability. Like many other optical chemical vapor sensors, the ring resonator vapor sensor relies on the polymer to provide selectivity toward analytes. In the presence of vapor molecules, the polymer undergoes RI and/or thickness change[22,23], resulting in a WGM spectral shift. Therefore, by directly or indirectly monitoring the WGM spectral position in real time, both quantitative and kinetic information regarding the interaction between vapor molecules and the polymer can be extracted.

So far, there are four basic ring resonator configurations that can potentially be used as a vapor sensor, the first three of which are shown in Fig. 6.1. They include (1) chip-based planar ring resonators made of solid dielectric materials such as SiON and coated with a vapor-sensitive polymer as the cladding (Fig. 6.1a)[23,24]; (2) chip-based polymer ring resonators, in which polymer is used as the building block of the ring resonator and also as the sensing material (Fig. 6.1b)[25,26]; and (3) free-standing microspheres or cylinders whose exterior surface is coated with a layer of polymer (Fig. 6.1c). While these ring resonators have been investigated for sensitive detection of alcohol[23,25], ammonia[24], and explosives[26] (also see Chap. 2), they suffer from cumbersome and ineffective gas fluidics, which is also commonly seen in many other optical vapor sensors. In some cases, these sensors are placed in a large chamber with a gas flow rate as high as tens of liters per minute[14,31]. This not only requires a large amount of gas samples, but also significantly slows down the sensing response and recovery time (a few to a few tens of minutes may be needed[23,25]). Both may deteriorate the sensor performance, especially in the

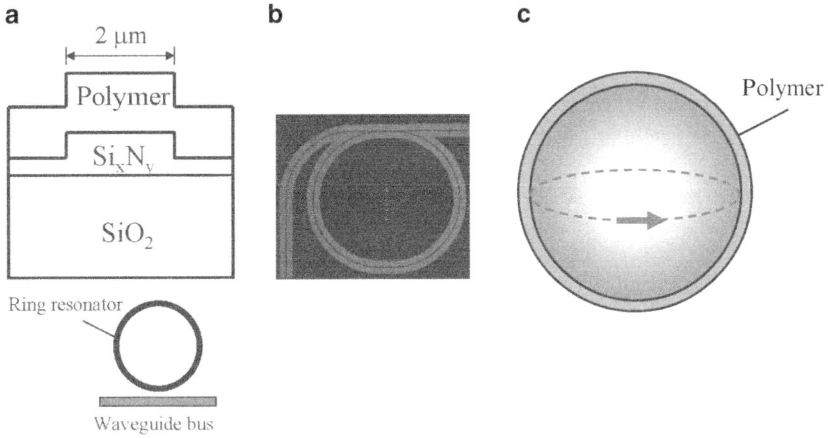

Fig. 6.1 Various configurations of ring resonator vapor sensors

situation where fast response and short recovery time are needed and where only low sample quantity is available.

Very recently, we developed the fourth ring resonator configuration for rapid chemical vapor sensing based on an optofluidic ring resonator (OFRR)[27–29]. As illustrated in Fig. 6.2, the OFRR utilizes a thin-walled glass capillary whose circular cross section forms a ring resonator to support WGMs. The OFRR serves as both a microfluidic channel to deliver samples and a ring resonator to detect the gaseous analyte passing through the capillary. In contrast to OFRRs used as label-free biosensors and microfluidic lasers (see Chaps. 14 and 19), when used for chemical vapor sensing, the OFRR interior surface is coated with a thin layer of vapor-sensitive polymer. The WGM has significant optical presence in the polymer layer and detects the RI and/or thickness change in the polymer when it interacts with gas molecules.

The OFRR has several distinctive advantages in comparison with the other three types of ring-resonator-based vapor sensors. Since the OFRR achieves dual use of the capillary as a fluidic channel and as a sensing head, it requires only a few μL sample volume, in contrast to a typical level of a few liters used in planar-ring-resonator-based vapor sensors that use a chamber and extra delivery channels[28]. Additionally, the sensing events take place on the inner surface of the capillary. The circular nature of the capillary ensures the most efficient interactions between the vapor molecules and the polymer layer. Consequently, rapid detection becomes possible[27,28]. Moreover, the OFRR is highly compatible with GC technology that is widely used in analytical chemistry for sample separation and identification, which can be explored for the development of μGC system[27]. In turn, many GC technologies, such as coating methods[32], can be adapted for OFRR vapor sensor studies. In this chapter, we will first discuss the operation principles of the OFRR chemical vapor sensors, followed by examples of rapid gas detection and OFRR-based μGC development.

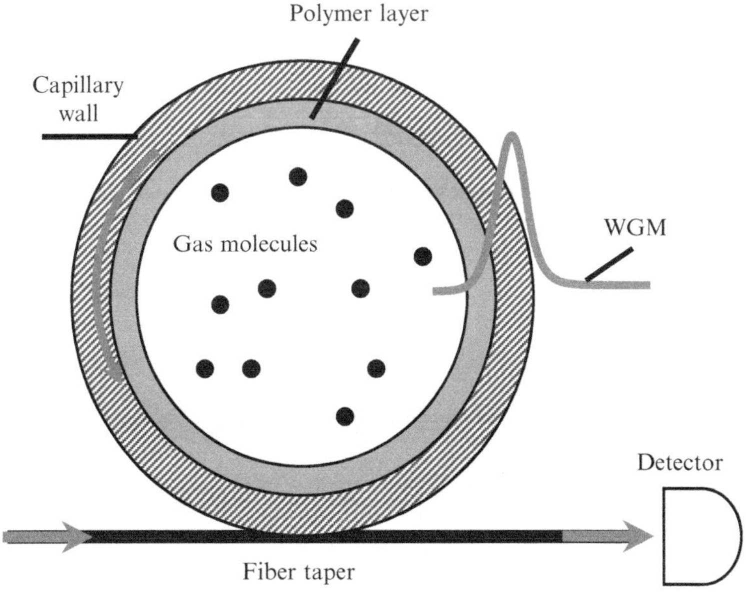

Fig. 6.2 Cross-sectional view of the OFRR chemical vapor sensor

6.2 Theoretical Model of the OFRR Vapor Sensor

The OFRR has a layer of polymer on its interior surface. When interacting with vapor molecules, the polymer undergoes RI changes and/or thickness changes[20,22–25,28,33,34], which result in a spectral shift in the WGM. This detection scheme is similar to what is used in OFRR biosensing, where the WGM spectral position is monitored (see Chap. 14). The WGM spectral position can be solved by considering the four-layer Mie model, as illustrated in Fig. 6.3, which is different from the OFRR used for biosensing, where only a three-layer Mie model is needed. The radial distribution of the WGM for the four-layered OFRR is given by[35]

$$E_{m,l}(r) = \begin{cases} AJ_m(kn_1r) & (r \leq \mathrm{OD}/2 - d - t) \\ BJ_m(kn_2r) + CH_m^{(1)}(kn_2r) & (\mathrm{OD}/2 - d - t \leq r \leq \mathrm{OD}/2 - d) \\ DJ_m(kn_3r) + EH_m^{(1)}(kn_3r) & (\mathrm{OD}/2 - d \leq r \leq \mathrm{OD}/2) \\ FH_m^{(1)}(kn_4r) & (r \geq \mathrm{OD}/2) \end{cases}, \quad (6.1)$$

where J_m and $H_m^{(1)}$ are the mth Bessel function and the mth Hankel function of the first kind, respectively. The RI of the core, the polymer, the ring resonator wall, and the surrounding medium are described by n_1, n_2, n_3, and n_4. $k = 2\pi/\lambda$, where λ is the WGM wavelength in vacuum. The WGMs have two polarizations, a-mode and

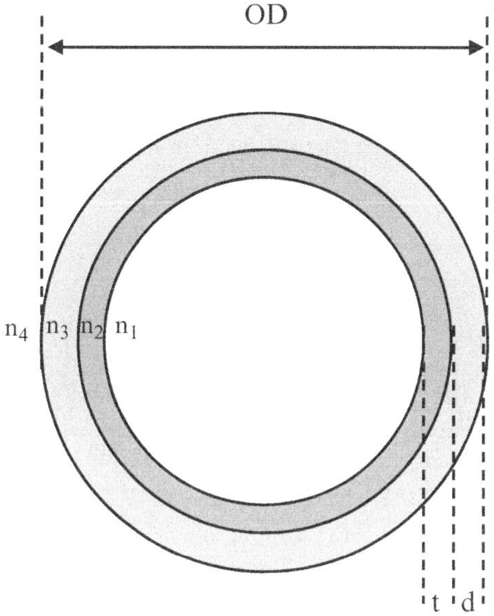

Fig. 6.3 Four-layer model for the OFRR vapor sensor. *OD*: ring resonator outer diameter; *t*: polymer thickness; *d*: ring resonator wall thickness; n_1, n_2, n_3, and n_4 are the refractive indices for the medium inside (air), polymer, silica ring resonator, and medium outside (air), respectively

b-mode, with the magnetic field and the electric field being along the cylinder longitudinal direction, respectively.

The OFRR sensor sensitivity can be described by

$$S = \frac{\mathrm{d}\lambda}{\mathrm{d}\rho} = \frac{\partial\lambda}{\partial t}\cdot\frac{\partial t}{\partial\rho} + \frac{\partial\lambda}{\partial n_2}\cdot\frac{\partial n_2}{\partial t}\cdot\frac{\partial t}{\partial\rho} + \frac{\partial\lambda}{\partial n_2}\cdot\frac{\partial n_2}{\partial\rho}, \tag{6.2}$$

where $\mathrm{d}\lambda/\mathrm{d}\rho$ is the WGM spectral shift due to the change of the vapor molecule density in the polymer matrix, ρ. $\partial\lambda/\partial t$ and $\partial\lambda/\partial n_2$ refer to the WGM thickness sensitivity (S_t) and RI sensitivity (S_{RI}), respectively, which are the intrinsic properties associated with the optical modes of the coated ring resonator. $\partial t/\partial\rho$ and $\partial n_2/\partial\rho$ are the polymer swelling/shrinkage and the RI change due to the vapor molecule absorption, which depend on the polymer–analyte interaction. Note that the RI change can be caused by either the polymer volume change induced by vapor molecules, or by the doping effect due to the presence of vapor molecules in the polymer matrix[22], as described, respectively, by the second and the third term on the right-hand side of (6.2).

RI sensitivity for the polymer RI change, S_{RI}, is related to the fraction of light in the polymer, η, by[36–38]

$$S_{RI} = \frac{\partial\lambda}{\partial n_2} = \frac{\lambda}{n_{eff}}\eta, \tag{6.3}$$

where $n_{eff} = m\lambda/2\pi R$ (R, ring resonator radius) is the WGM effective RI. However, the thickness sensitivity is more complicated, which will be discussed later in Sect. 6.3. While the polymer thickness change due to vapor molecules adsorption and the thickness-induced RI change may have to be determined experimentally[20,33,34], the RI change due to the doping effect can be modeled by the Lorentz–Lorenz equation[22]

$$\delta n_2 = \frac{\left(n_2^2 + 2\right)^2}{6n_2} \frac{1}{3\varepsilon_0} (\delta\rho)\alpha, \tag{6.4}$$

where α is the vapor molecule polarizability. Equation (6.4) relates the RI change in the polymer layer to the vapor molecule density in the polymer matrix, which is further related to the vapor concentration in free space, ρ_0, by $K = \rho/\rho_0$, where K is the partition coefficient ranging from a thousand to hundreds of thousand[39,40]. It should be noted that the polymer RI change and thickness change may not work additively. For example, while the doping-induced polymer RI change is always positive, its thickness-related RI decreases when the polymer swells upon interaction with the analyte.

6.3 Theoretical Analysis of the OFRR Chemical Vapor Sensor

Using the model presented in Sect. 6.2, we systematically investigate two cases where the polymer RI is lower and higher than that of the wall.

6.3.1 Low RI Polymer Coating

When the polymer RI is smaller than or close to that of the ring resonator, the polymer layer can be regarded as an extension of the ring resonator wall, regardless of the polymer thickness. Initially, when the polymer layer is thin, only the evanescent field exists in the polymer layer and the RI sensitivity is low. With the increased polymer thickness, higher order modes start to move inward, resulting in a higher RI sensitivity for those modes while the lower order modes are not affected much. This behavior is shown in the insets of Fig. 6.4a. The thickness sensitivity, S_t, can be deduced from the slope of the curves in Fig. 6.4a. S_t is always positive. However, it drops gradually to zero when polymer thickness increases, meaning that the sensor is insensitive to any polymer thickness change for the thick polymer layer. The RI sensitivity, S_{RI}, is plotted in Fig. 6.4b, which increases monotonically with the increased polymer thickness. The maximum of S_{RI} depends highly on the mode order. The higher order the mode, the higher S_{RI} it can achieve. This is a

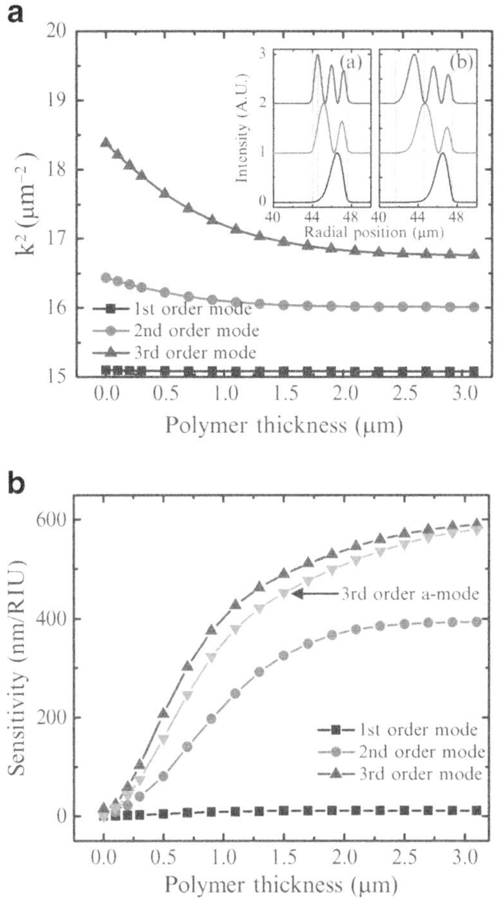

Fig. 6.4 (**a**) k^2 as a function of polymer thickness for the first three WGMs. The relevant parameters are OD = 95 μm, $d = 3$ μm, $n_1 = 1$, $n_2 = 1.47$, $n_3 = 1.45$, $n_4 = 1$, $m = 257$. All for b-mode. *Inset*: Intensity radial distribution for the first three modes when the polymer thickness is 0.5 μm (**a**) and 2.9 μm (**b**). *Vertical lines* indicate the boundaries of the ring resonator and the polymer layer. (**b**) The corresponding RI sensitivity. The RI sensitivity for the third WGM of different polarization (a-mode) is also plotted. Reprinted from Ref. 29 with permission. © 2008 Optical Society of America

result of (6.3), as higher order modes have more fraction of light in the polymer. Polarization-dependent S_{RI} is also given in Fig. 6.4b, showing that the b-mode has a slightly higher sensitivity than the a-mode, opposite to the OFRR biosensors where the evanescent field in the core is used for sensing (see Chap. 14)[41].

For the OFRR-based vapor sensor, the ring resonator wall thickness has a significant impact on the sensor performance. Since the polymer layer is treated as the extension of the ring resonator, the relative thickness between the wall and the polymer determines the radial intensity distribution of the WGMs. As a result,

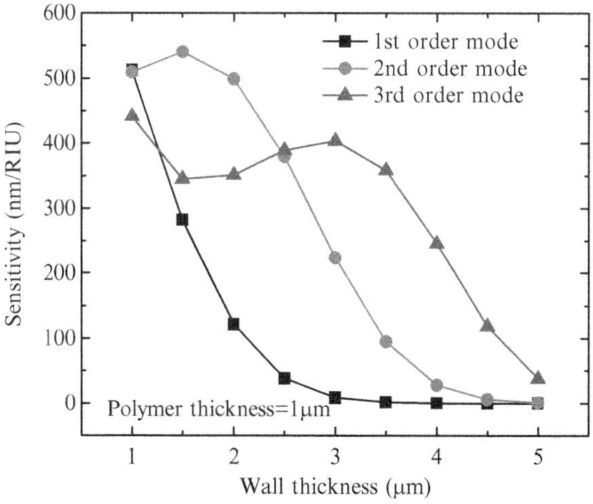

Fig. 6.5 RI sensitivity as a function of the OFRR wall thickness for the first three WGMs. The polymer thickness is fixed at 1 μm. Other parameters are the same as in Fig. 6.4. Reprinted from Ref. 29 with permission. © 2008 Optical Society of America

the fraction of light and hence the RI sensitivity is lower with a thicker-walled ring resonator, as shown in Fig. 6.5. Eventually, the RI sensitivity drops to zero when the wall is sufficiently thick, in which case, the WGM is predominantly confined within the wall.

6.3.2 *High RI Polymer Coating*

When a polymer of sufficiently high RI is used as the coating material, the WGMs exhibit completely different behavior. Initially, the light is confined within the ring resonator wall as a wall mode. With the increased polymer thickness, the polymer starts to support a new set of WGMs that interact with the wall modes to form a so-called photonic molecular mode that has characteristics from both wall mode and polymer layer mode[42,43]. Figure 6.6a shows the dispersion for the three lowest modes in such a polymer-coated ring resonator. The energy splitting occurs when the first-order wall mode (shown as the dashed line in Fig. 6.6a) intersects the polymer modes. Eventually, when the polymer is sufficiently thick, polymer modes and wall modes are decoupled. This is equivalent to creating a new polymer ring resonator in a capillary and the glass capillary simply acts as the physical substrate. Figure 6.6b shows such a transition for the second-order mode.

The RI sensitivity, S_{RI}, of the above sensor structure is given in Fig. 6.7. Whereas the sensitivity for the first-order mode increases monotonically with the increased wall thickness, the sensitivity for the second and third order modes oscillates significantly. In particular, S_{RI} becomes nearly zero at certain regions that

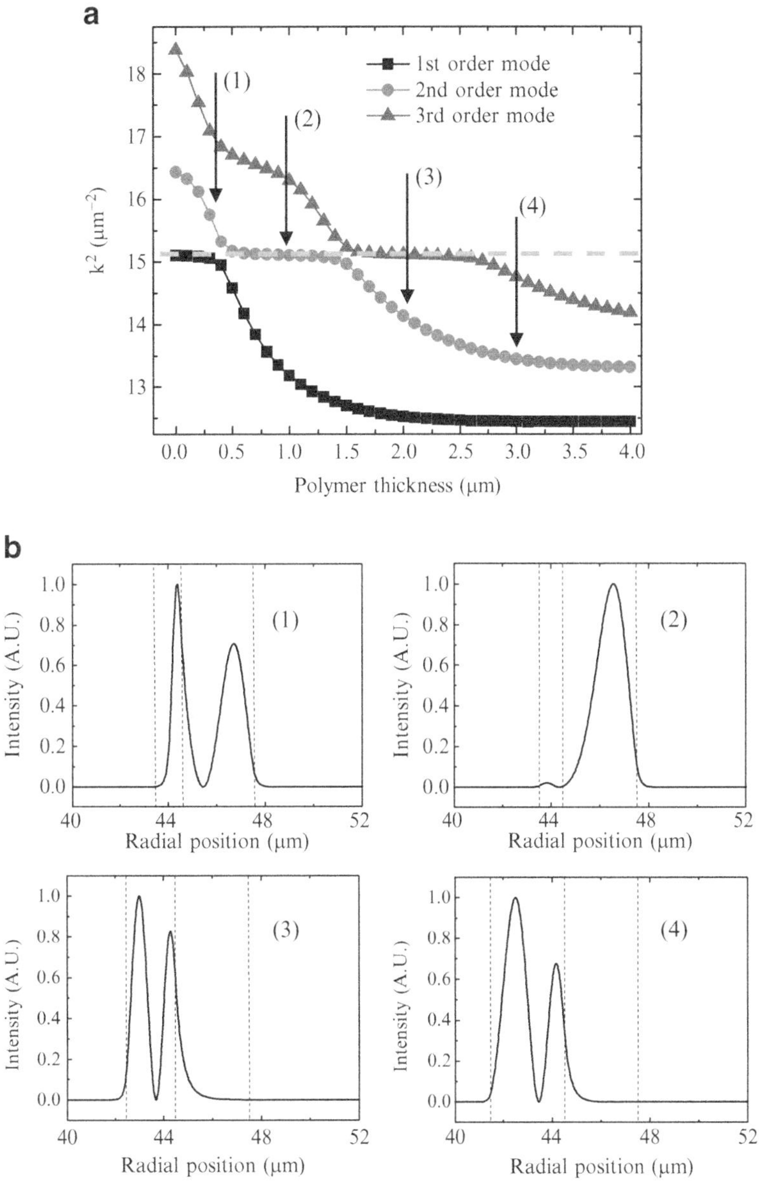

Fig. 6.6 (**a**) k^2 as a function of polymer thickness for the first three WGMs. *Dashed line* indicates the k^2 position for the first-order ring resonator wall mode in the absence of the polymer layer. The simulation parameters are the same as in Fig. 6.4, except that the polymer RI, n_2, is 1.7. (**b**) The WGM radial distribution of the second-order mode for various polymer thicknesses indicated by the *arrows* in (**a**). *Vertical lines* indicate the boundaries of the ring resonator wall and the polymer layer. Reprinted from Ref. 29 with permission. © 2008 Optical Society of America

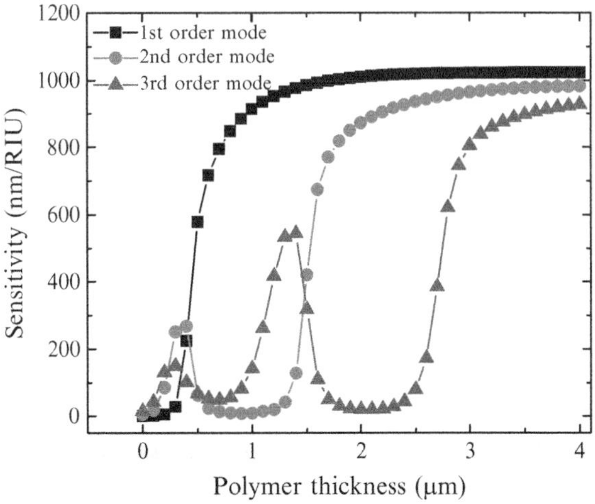

Fig. 6.7 RI sensitivity as a function of polymer thickness for the first three WGMs. The polymer is coated on the inner surface of the ring resonator. The simulation parameters are the same as in Fig. 6.5, except that the polymer RI, n_3, is 1.7. Reprinted from Ref. 29 with permission. © 2008 Optical Society of America

correspond to the plateaus in Fig. 6.6a, where the mode possesses the dominant characteristic of the wall mode (e.g., the mode in the second figure of Fig. 6.6b). After the decoupling process, all modes approach their respective maximal sensitivity. The thickness sensitivity, S_t, is generally positive. However, at the region of plateau in Fig. 6.6a, it becomes nearly zero, meaning that the OFRR sensor is insensitive to any polymer change induced by chemical vapors.

6.4 Demonstration of Using the OFRR for Chemical Vapor Sensing

6.4.1 OFRR Vapor Sensor Preparation

An OFRR of approximately 75 μm in outer diameter (OD) is produced by rapidly pulling a capillary preform (Polymicro Technologies, Phoenix, AZ) under heat, followed by HF etching to further reduce the wall thickness to around 4 μm. The OFRR pulling system is home-built and has been detailed in Chap. 14. The wall thickness is determined from measuring the bulk RI sensitivity using different concentrations of ethanol–water mixture with known RI[44].

The OFRR interior surface is then coated with various polymers, such as moderately polar methyl phenyl polysiloxane (OV-17) or highly polar polyethylene glycol (PEG), for the detection of different types of chemical vapors. The coating procedures are adapted from well-developed static coating procedures for GC microfabricated columns[32]. Briefly, the polymer dissolved in an organic solvent such as toluene, methanol, or acetone is first filled into the OFRR by capillary force. Then both ends of the OFRR are sealed with a septum and the polymer solution is left in the OFRR for 1 h. After this, low vacuum is applied at one end to gradually evaporate the coating solvent out of the OFRR, followed by high vacuum for 12 h. All the coating procedures are carried out at room temperature. The coating thickness can be controlled by varying the polymer solution concentration. In our experiment, the coating thickness is estimated to be 200 nm.

Smooth and uniform polymer surface after vacuum plays a key role to ensure good OFRR sensing performance. We have observed in experiments that toluene after vacuum is prone to leave a number of cavities of a few micrometers in diameter on the surface. These cavities will induce additional scattering loss for the WGMs in the OFRR, which greatly degrade the Q-factor, and hence the detection limit of the OFRR vapor sensor. Moreover, these small cavities have different adsorption characteristics compared to smooth polymer surface. Vapor molecules may be retained for a longer time at the cavity, which increases the response time and recovery time. Acetone and methanol are found to be better candidates for solvents because they usually leave uniform and smooth surface after vacuum.

6.4.2 *Rapid Chemical Vapor Sensing*

In this part, we demonstrate OFRR's capability as a rapid chemical vapor sensor. During experiments, ethanol and hexane vapors are used as a model system and represent polar and nonpolar analytes, respectively.

Figure 6.8 depicts the schematic of the experimental setup. Various concentrations of analyte/air vapor mixture is injected into the OFRR capillary by a syringe pump (Harvard Apparatus, MA) at a flow rate of $\sim$1 mL min^{-1}. Ultrahigh purity hydrogen is flowed into the OFRR before injection of the vapor samples to establish the sensing baseline and then after each sensing measurement to purge the analytes. A switching valve (Upchurch Scientific, WA) is used to quickly switch between the analyte channel and the purging channel. The detection part has been described in

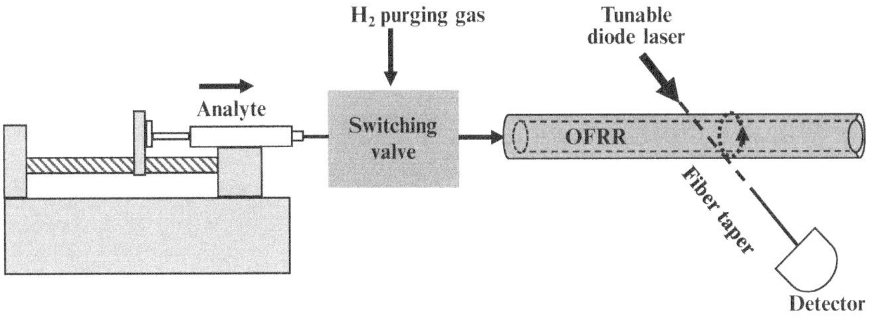

Fig. 6.8 Experimental setup for rapid chemical vapor detection

detail previously[44] (also see Chap. 14). In short, a 1,550-nm tunable diode laser scanned in wavelength (scanning rate: 5–15 Hz) is coupled into the WGM via an optical fiber taper or a waveguide in contact with the OFRR[44,45]. When the light is on resonance with the WGM, it is coupled into the OFRR, leaving an intensity dip at the fiber output end, which is used to indicate the WGM spectral position. When vapor molecules pass through the OFRR, they interact with the polymer layer, causing an RI change and thickness change in the polymer layer, which leads to a spectral shift in the WGM. The whole OFRR is placed on a plastic module covered by a glass slide to minimize temperature fluctuations.

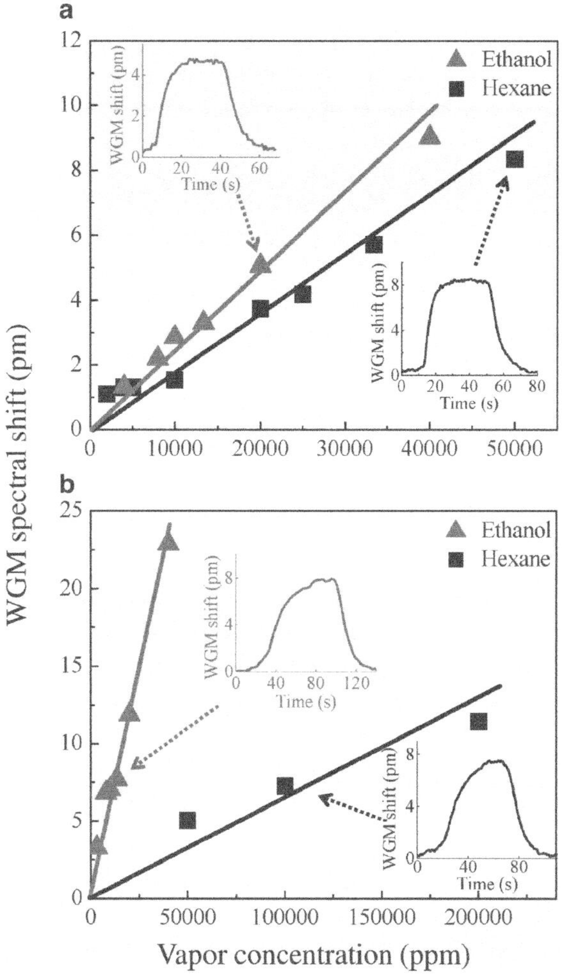

Fig. 6.9 OFRR response to various part per million (ppm) concentrations of ethanol and hexane vapors. The OFRR is coated with a 200-nm-thick OV-17 (**a**) and PEG-400 (**b**). *Insets* are the sensorgrams taken by monitoring the WGM shift in real time. Reprinted from Ref. 28 with permission. © 2008 Optical Society of America

The inset in Fig. 6.9a shows the sensorgrams when ethanol or hexane vapor is introduced to an OV-17 coated OFRR. H_2 is initially flowed through the OFRR to establish the baseline. Then the valve is switched to the analyte channel. Upon the interaction with the analyte, the WGM shifts to a longer wavelength and reaches the equilibrium value within 10 s. Then the flow is switched back to H_2 to purge the OFRR, as reflected by the decrease in the WGM spectral position. There are two phenomena worthy of pointing out, which attest to the rapid and efficient nature of the OFRR vapor sensor. First, purge is completed within 20 s, much faster than the previous ring resonator designs and many other types of optical vapor sensors. Second, thorough purge can be achieved, as evidenced by the fact that the WGM returns to the baseline every time after purge.

Figure 6.9a plots the WGM shift obtained at the equilibrium for various concentrations of ethanol or hexane vapor by the OV-17-coated OFRR. The sensitivity is nearly the same for ethanol and hexane for OV-17, which contains 50% modestly polar phenyl groups and 50% nonpolar methyl groups. For a comparison, Fig. 6.9b plots the sensitivity curve for ethanol and hexane using another OFRR coated with PEG-400. Both parts (a) and (b) of Fig. 6.9 indicate that the WGM shift linearly depends on the analyte concentration. Sensitivity for ethanol vapor with PEG is 0.5×10^{-3} pm ppm^{-1}, ten times higher than that for hexane, as PEG-400 is highly polar, hence having higher polar interactions and solubility for ethanol. To estimate the limit of detection (LOD) for ethanol vapor, we use a WGM resolution of 0.1 pm, which can be obtained without any active temperature control. Based on Fig. 6.9b, an LOD of 200 ppm for ethanol vapor can be derived.

6.4.3 OFRR-Based μGC

For gas sensors that rely on polymeric materials to achieve selectivity, it is challenging to discriminate and identify targets of interest from a mixture. Unlike antibodies used in biosensors that have high specificity toward an antigen, the polymer would respond to several analytes in a mixture that contains molecules with similar polarity, molecular structures, or functional groups. Furthermore, when samples include volatile organic background analytes, the target may be buried in the background, especially when the target has very low volatility, such as in the case of explosive detection.

Gas chromatography (GC) is a well-established and powerful analytical technique that has high selectivity in analyte detection. In GC, the selectivity is achieved by separating and recognizing analytes in the sample mixture through their unique retention times. However, traditional GC is bulky and power intensive, and requires long analysis time. Research on μGC analyzers over the past decade has shown the great potential for miniaturized, rapid, and low-power μGC systems. To date, most μGC systems are implemented using microfabrication technology, where rectangular microchannels are fabricated to replace the traditional circular GC columns. The whole system, including injector, column, and detector (e.g., flame ionization detector), are sometimes fabricated on a single chip. Although high aspect ratio rectangular cross-sectional columns have potential of greater separation efficiency and can be less affected by flow restrictions than the conventional narrow-bore circular columns, they suffer incompatibility with circular tubings that are widely adopted in GC system and other related testing equipments[46].

The OFRR offers an excellent solution to implement μGC. It has a circular cross section by design, which is well compatible with all the GC-related equipments that are currently in use. Moreover, OFRR achieves dual use as the separation column as well as the detection body, which is in sharp contrast with other μGC systems having separate column and detector. By placing an optical fiber taper or a waveguide in contact with the OFRR, we are able to achieve multipoint real-time on-column detection, which greatly simplifies the μGC system.

Here, we demonstrate the feasibility of the OFRR-based μGC, and further explore its capability in the explosive detection. In experiments, a 10-cm long OFRR is fabricated and coated with PEG-400 or PEG-1000 polymer using the same procedure described previously in the gas sensing experiments. Gas samples are extracted from the saturated vapor in the headspace of the analyte sample vials with a solid-phase microextractor (SPME) at room temperature and introduced into a GC injection system (HP 5890, injector temperature 250°C), which generates a gas pulse. The gas pulse travels along a 1.8-m, 0.25-mm inner diameter (ID) nonpolar passivated fused silica guard column (Supelco), which has minimal interaction with the analyte. The guard column is connected with the OFRR via a universal quick seal column connector (Varian). Ultrahigh purity hydrogen gas is used as the carrier gas. The whole system setup is illustrated in Fig. 6.10.

To show the separation capability of the OFRR-based μGC analyzer, we simultaneously inject mixtures of various analytes. In conventional GC and μGC systems, separation occurs as the sample travels through a separation column that is several tens of meters long for conventional GC and is on the order of a meter for μGC. In this experiment, the distance between the OFRR inlet and the location

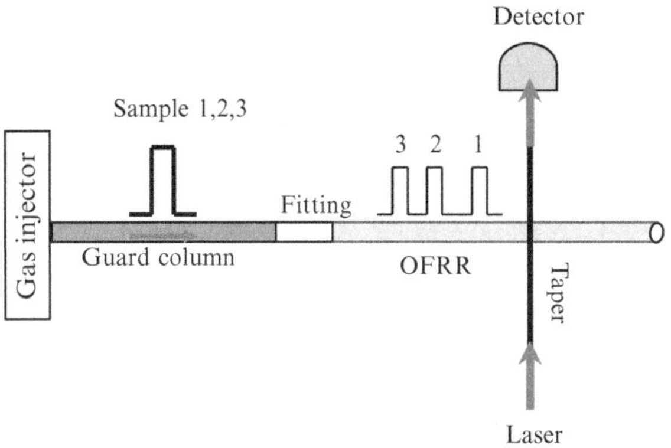

Fig. 6.10 Experimental setup for OFRR-based GC. Reprinted from Ref. 27 with permission. © 2008 American Chemical Society

where the optical taper is placed is only 1.5 cm. Figure 6.11 shows separation of ethanol/decane (a), toluene/decane (b), toluene/DMMP (Dimethyl methylphosphonate) (c), and decane/DMMP (d). The ability to separate several different samples over a short distance is enabled by the small diameter of the OFRR. Note that the sub-second peak width is achieved in ethanol detection. This peak width is limited by the laser scanning rate used in the experiment. An even sharper peak can be obtained with a higher scanning rate.

The repeatability of gas sample separation is presented in Fig. 6.12. The separation peaks of the mixed gaseous analytes match those for each individual analyte quite well, irrespective of the analyte concentration, which correlates to the peak height (Fig. 6.12a, b). Although manual sample injection may introduce a certain degree of variation in retention time measurement, using a gas marker can significantly improve such measurement, as demonstrated in Fig. 6.12c, in which decane is used as a marker.

To explore the capability of OFRR-based μGC in explosive detection, 2,4-dinitrotoluene (DNT) is chosen as a model system, due to its chemical structure

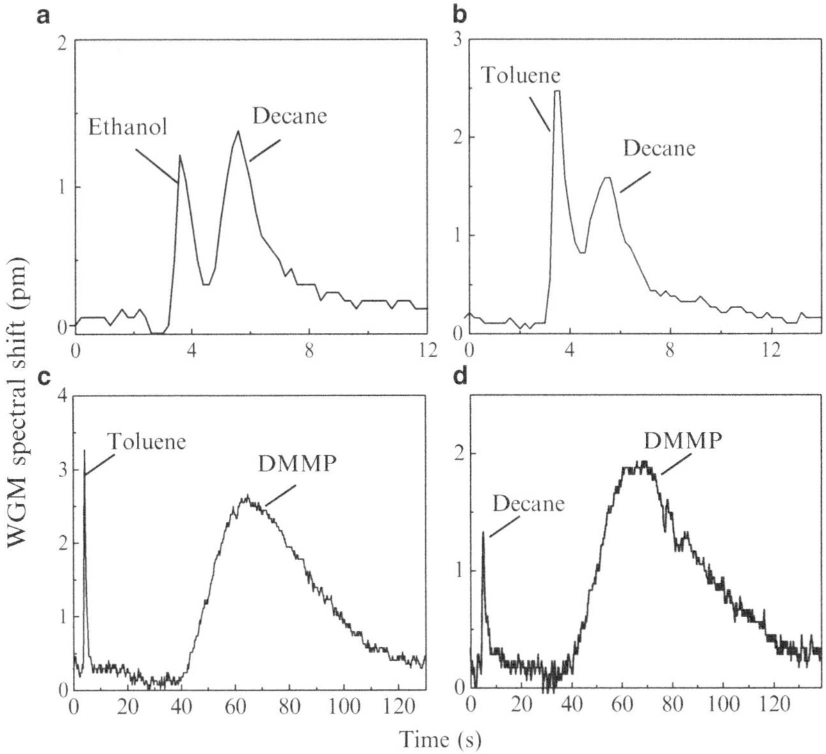

Fig. 6.11 Separation of different analytes after passing through the 1.5 cm of PEG-400 coated OFRR. Reprinted from Ref. 27 with permission. © 2008 American Chemical Society

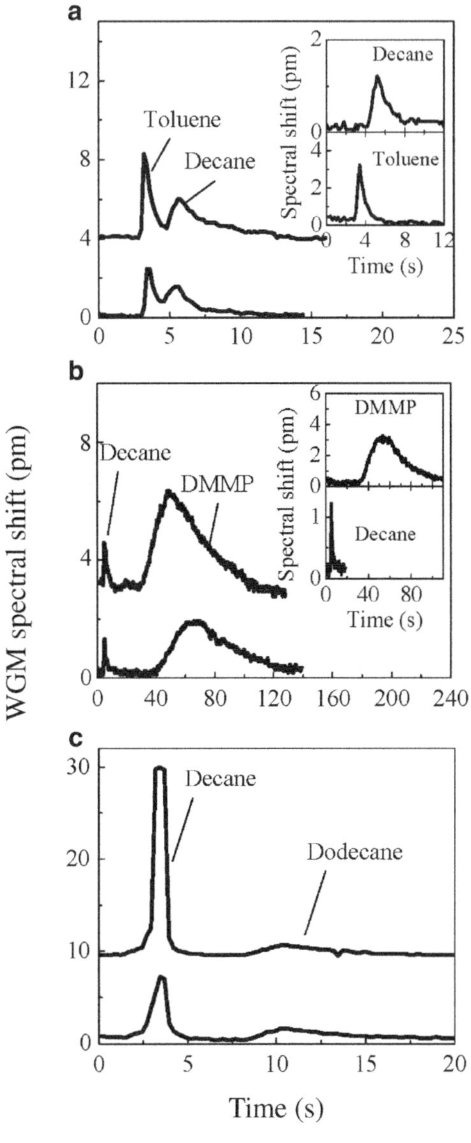

Fig. 6.12 Chromatogram repeatability test: (**a**) separation between toluene and decane; (**b**) separation between decane and DMMP; and (**c**) separation between decane and dodecane. The *insets* in parts (**a**) and (**b**) are the chromatogram for each individual gas component. Variation in peak height is due to the variations in SPME sampling time. *Curves* are sifted vertically for clarity. Reprinted from Ref. 27 with permission. © 2008 American Chemical Society

similar to trinitrotoluene (TNT). Also it is an additive or contaminant in many explosives and can be used as an indicator to predict the existence of TNT. UHP helium is used as the carrier gas with a flow rate of 10.6 mL min^{-1} measured at the

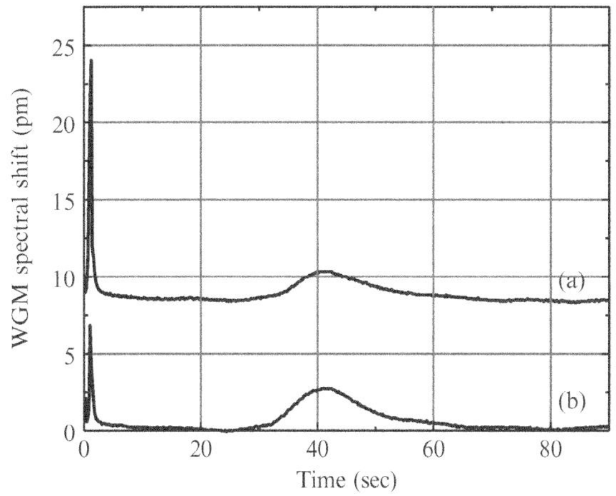

Fig. 6.13 Highly repeatability separation and detection of DNT from nitrotoluene and DNT mixtures. SPME extractions are performed at room temperature. (**a**) Nitrotoluene extraction time ~3 s, DNT extraction time ~20 s. (**b**) Nitrotoluene extraction time ~1 s, DNT extraction time ~30 s. DNT peak emerges at approximately 40 s. *Curves* are vertically shifted for clarity

outlet of the OFRR. Figure 6.13 shows the separation between DNT vapor and 2-nitrotoluene vapor. Although 2-nitrotoluene and DNT only have one nitro group difference in structure, OFRR vapor sensors achieve very efficient separation between them. We have performed numerous tests on separation between 2-nitrotoluene and DNT vapor mixtures with various mass ratios for numerous times and all the data showed consistence in retention time. In addition, the signal amplitude is proportional to the extraction time of SPME fiber in equilibrium headspace of sample vial. From the sensorgram of DNT shown in Fig. 6.13, the WGM starts to shift to a longer wavelength around 25 s after injection and reaches its peak around 40 s. The signal completely returns to baseline quickly, indicating that the polymer is completely regenerated for the following sensing activities. A single sensing event is accomplished within only 1 min at room temperature, showing great promise for detecting explosives in a rapid and simple manner, especially considering that this detection system has not yet been optimized for speed. Figure 6.14a plots the WGM spectral shift in response to various SPME extraction times of DNT sample. A good linear response range is observed. The total mass extracted by the SPME is quantified using a GC/MS system (Varian) with calibrated liquid injections and shows a linear relationship with the SPME extraction time, as shown in Fig. 6.14b. To estimate the detection limit of the OFRR vapor sensor for DNT, we use a WGM resolution of 0.1 pm, which is mainly determined by the temperature-induced WGM fluctuations. Based on the spectral shift in Fig. 6.13 and calibration curve of adsorption mass on SPME fiber in Fig. 6.14b, we estimate that the detection limit is approximately 300 pg for DNT vapor, which corresponds to about 1 s of SPME extraction at room temperature.

While the initial demonstration of this unique on-column μGC analyzer shows promise, it also highlights that there are many ways to optimize the performance of

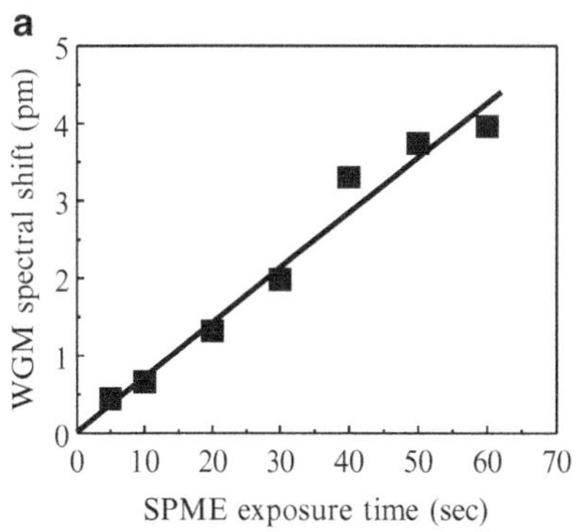

Fig. 6.14 (**a**) OFRR vapor sensor responses to DNT vapor samples extracted with various sampling time at room temperature. (**b**) Calibration curve of DNT mass extracted by on SPME fiber under various extraction times at room temperature

this technology. For example, optimization on the carrier gas flow rate, polymer-coating thickness, OFRR size, detection position along the OFRR in order to achieve better resolution, higher sample capacity, or both. Furthermore, this first version of OFRR μGC analyzer is demonstrated for isothermal operation only and the temperature for DNT detection has not been optimized. Although all RI-based optical chemical sensors are sensitive to temperature change, it is not a fundamental obstacle for OFRR-based μGC to have a temperature programming function. If the temperature ramping is controlled, it may be possible to detect sharp peaks on the moving baseline of the OFRR signal.

6.5 Conclusion

We have demonstrated rapid chemical vapor sensing based on the OFRR. The OFRR architecture is an integration of an efficient fluidic channel and a microring resonator sensing element defined by the tapered fiber in contact with the OFRR capillary. The whole sensing system is simple, cost effective, and easy to be scaled for multiplexed detection, which makes it a promising platform for the development of miniaturized, rapid, sensitive, and selective gas sensors. Based on these unique features of the OFRR, the OFRR μGC analyzer is also demonstrated with the capability of efficient separation of vapor mixtures and highly reproducible retention time for the individual analyte. Theoretical simulations on OFRR vapor sensors are also carried out. Sensitivity, detection limit, and response time are studied with respect to the sensing polymer RI, thickness, as well as ring resonator dimensions. Our future work will focus on the optimization of OFRR structure and polymer thickness to improve the detection limit that would benefit the detection of explosives and other low volatility analytes.

Acknowledgments We acknowledge support from the National Science Foundation (ECCS-0729903). We would also like to thank Mr. Aaron Thompson from ICx Nomadics for helping setup the GC system at the University of Missouri and for the GC/MS data used in Fig. 6.14b.

References

1 Stetter, J. R.; Li, J., Amperometric gas sensors-a review, *Chem. Rev.* **2008**, 108, 352–366
2 Stievater, T. H.; Rabinovich, W. S.; Ferraro, M. S.; Papanicolaou, N. A.; Bass, R.; Boos, J. B.; Stepnowski, J. L.; McGill, R. A., Photonic microharp chemical sensors, *Opt. Express* **2008**, 16, 2423–2430
3 Mah, C.; Thurbide, K. B., Acoustic methods of detection in gas chromatography, *J. Separ. Sci.* **2006**, 29, 1922–1930
4 Warken, F.; Vetsch, E.; Meschede, D.; Sokolowski, M.; Rauschenbeutel, A., Ultra-sensitive surface absorption spectroscopy using sub-wavelength diameter optical fibers, *Opt. Express* **2007**, 15, 11952–11958
5 Joel, V.; Monzón-Hernández, D., Fast detection of hydrogen with nano fiber tapers coated with ultra thin palladium layers, *Opt. Express* **2005**, 13, 5087–5092
6 Carlson, R. C.; Hayden, A. F.; Telfair, W. B., Remote observations of effluents from small building smokestacks using FTIR spectroscopy, *Appl. Opt.* **1988**, 27, 4952–4959
7 Schliesser, A.; Brehm, M.; Keilmann, F.; van der Weide, D., Frequency-comb infrared spectrometer for rapid, remote chemical sensing, *Opt. Express* **2005**, 13, 9029–9038
8 Pushkarsky, M. B., Dunayevskiy, I. G., Prasanna, M., Tsekoun, A. G., Go, R., Patel, C. K. N., High-sensitivity detection of TNT, *Proc. Natl Acad. Sci. USA* **2006**, 103, 19630–19634
9 Ramos, C.; Dagdigian, P. J., Detection of vapors of explosives and explosive-related compounds by ultraviolet cavity ringdown spectroscopy, *Appl. Opt.* **2007**, 46, 620–627
10 Stephane Content, W. C. T. M. J. S., Detection of nitrobenzene, DNT and TNT vapors by quenching of porous silicon photoluminescence, *Chem - Eur. J.* **2000**, 6, 2205–2213
11 Lipp, E. D.; Grosse, R. L., On-line monitoring of chlorosilane streams by Raman spectroscopy, *Appl. Spectrosc.* **1998**, 52, 42–46
12 Roth, E.; Kiefer, W., Surface-enhanced Raman spectroscopy as a detection method in gas chromatography, *Appl. Spectrosc.* **1994**, 48, 1193–1195
13 Skivesen, N.; Horvath, R.; Pedersen, H. C., Multimode reverse-symmetry waveguide sensor for broad-range refractometry, *Opt. Lett.* **2003**, 28, 2473–2475
14 Lowder, T. L.; Gordon, J. D.; Schultz, S. M.; Selfridge, R. H., Volatile organic compound sensing using a surface relief d-shaped fiber Bragg grating and a polydimethylsiloxane layer, *Opt. Lett.* **2007**, 32, 2523–2525
15 Butler, T. M.; Igata, E.; Sheard, S. J.; Blackie, N., Integrated optical Bragg-grating-based chemical sensor on a curved input edge waveguide structure, *Opt. Lett.* **1999**, 24, 525–527
16 Cusano, A.; Iadicicco, A.; Pilla, P.; Contessa, L.; Campopiano, S.; Cutolo, A.; Giordano, M.; Guerra, G., Coated long-period fiber gratings as high-sensitivity optochemical sensors, *J. Lightwave Technol.* **2006**, 24, 1776–1786
17 Zhang, J.; Tang, X.; Dong, J.; Wei, T.; Xiao, H., Zeolite thin film-coated long period fiber grating sensor for measuring trace chemical, *Opt. Express* **2008**, 16, 8317–8323
18 Wei, T.; Han, Y.; Li, Y.; Tsai, H.-L.; Xiao, H., Temperature-insensitive miniaturized fiber inline Fabry-Pérot interferometer for highly sensitive refractive index measurement, *Opt. Express* **2008**, 16, 5764–5769
19 Xiao, G. Z.; Adnet, A.; Zhang, Z.; Sun, F. G.; Grover, C. P., Monitoring changes in the refractive index of gases by means of a fiber optic Fabry-Pérot interferometer sensor, *Sens. Actuators A: Phys.* **2005**, 118, 177–182

20 Cross, G. H.; Ren, Y.; Swann, M. J., Refractometric discrimination of void-space filling and swelling during vapour sorption in polymer films, *Analyst* **2000**, 125, 2173–2175
21 Reichl, D.; Krage, R.; Krumme, C.; Gauglitz, G., Sensing of volatile organic compounds using a simplified reflectometric interference spectroscopy setup, *Appl. Spectrosc.* **2000**, 54, 583–586
22 Podgorsek, R. P.; Franke, H., Selective optical detection of aromatic vapors, *Appl. Opt.* **2002**, 41, 601–608
23 Ksendzov, A.; Homer, M. L.; Manfreda, A. M., Integrated optics ring-resonator chemical sensor with polymer transduction layer, *Electron. Lett.* **2004**, 40, 63–65
24 Passaro, V. M. N.; Dell'Olio, F.; Leonardis, F. D., Ammonia optical sensing by microring resonators, *Sensors* **2007**, 7, 2741–2749
25 Pang, F.; Han, X.; Chu, F.; Geng, J.; Cai, H.; Qua, R.; Fang, Z., Sensitivity to alcohols of a planar waveguide ring resonator fabricated by a sol-gel method, *Sens. Actuators B* **2007**, 120, 610–614
26 Chen, A.; Sun, H.; Pyayt, A.; Zhang, X.; Luo, J.; Jen, A.; Sullivan, P. A.; Elangovan, S.; Dalton, L. R.; Dinu, R.; Jin, D.; Huang, D., Chromophore-containing polymers for trace explosive sensors, *J. Phys. Chem. C* **2008**, 112, 8072–8078
27 Shopova, S. I.; White, I. M.; Sun, Y.; Zhu, H.; Fan, X.; Frye-Mason, G.; Thompson, A.; Ja, S.-j., On-column micro gas chromatography detection with capillary-based optical ring resonators, *Anal. Chem.* **2008**, 80, 2232–2238
28 Sun, Y.; Shopova, S. I.; Frye-Mason, G.; Fan, X., Rapid chemical vapor sensing using optofluidic ring resonators, *Opt. Lett.* **2008**, 33, 788–790
29 Sun, Y.; Fan, X., Analysis of ring resonators for chemical vapor sensor development, *Opt. Express* **2008**, 16, 10254–10268
30 Chang, R. K.; Campillo, A. J., Optical Processes in Microcavities, World Scientific, Singapore, **1996**
31 Gordon, J. D.; Lowder, T. L.; Selfridge, R. H.; Schultz, S. M., Optical d-fiber-based volatile organic compound sensor, *Appl. Opt.* **2007**, 46, 7805–7810
32 Reidy, S.; Lambertus, G.; Reece, J.; Sacks, R., High-performance, static-coated silicon microfabricated columns for gas chromatography, *Anal. Chem.* **2006**, 78, 2623–2630
33 Jaczewska, J.; Raptis, I.; Budkowski, A.; Goustouridis, D.; Raczkowska, J.; Sanopoulou, M.; Pamula, E.; Bernasik, A.; Rysz, J., Swelling of poly(3-alkylthiophene) films exposed to solvent vapors and humidity: Evaluation of solubility parameters, *Synth. Met.* **2007**, 157, 726–732
34 Chaure, S.; Yang, B.; Hassan, A. K.; Ray, A. K.; Bolognesi, A., Interaction behaviour of spun films of poly[3-(6-methoxyhexyl)thiophene] derivatives with ambient gases, *J. Phys.: Appl. Phys.* **2004**, 37, 1558–1562
35 Bohren, C. F.; Huffman, D. R., Absorption and Scattering of Light by Small Particles, Wiley, New York, NY, **1998**
36 Zhu, H.; White, I. M.; Suter, J. D.; Dale, P. S.; Fan, X., Analysis of biomolecule detection with optofluidic ring resonator sensors, *Opt. Express* **2007**, 15, 9139–9146
37 Mortensen, N. A.; Xiao, S.; Pedersen, J., Liquid-infiltrated photonic crystals: Enhanced light-matter interactions for lab-on-a-chip applications, *Microfluid. Nanofluid.* **2008**, 4, 117–127
38 White, I. M.; Fan, X., On the performance quantification of resonant refractive index sensors, *Opt. Express* **2008**, 16, 1020–1028
39 Potyrailo, R. A.; Sivavec, T. M., Boosting sensitivity of organic vapor detection with silicone block polyimide polymers, *Anal. Chem.* **2004**, 76, 7023–7027
40 Liron, Z.; Kaushansky, N.; Frishman, G.; Kaplan, D.; Greenblatt, J., The polymer-coated SAW sensor as a gravimetric sensor, *Anal. Chem.* **1997**, 69, 2848–2854
41 Fan, X.; White, I. M.; Zhu, H.; Suter, J. D.; Oveys, H., Overview of novel integrated optical ring resonator bio/chemical sensors, *Proc. SPIE* (Laser Resonators and Beam Control X) **2007**, 6452, 64520M.1–64520M.20

42 Teraoka, I.; Arnold, S., Coupled whispering gallery modes in a multilayer-coated microsphere, *Opt. Lett.* **2007**, 32, 1147–1149
43 White, I. M.; Gohring, J.; Sun, Y.; Yang, G.; Lacey, S.; Fan, X., Versatile waveguide-coupled opto-fluidic devices based on liquid core optical ring resonators, *Appl. Phys. Lett.* **2007**, 91, 241104
44 White, I. M.; Oveys, H.; Fan, X., Liquid core optical ring resonator sensors, *Opt. Lett.* **2006**, 31, 1319–1321
45 White, I. M.; Oveys, H.; Fan, X.; Smith, T. L.; Zhang, J., Integrated multiplexed biosensors based on liquid core optical ring resonators and anti-resonant reflecting optical waveguide, *Appl. Phys. Lett.* **2006**, 89, 191106
46 Dorman, F. L.; Overton, E. B.; Whiting, J. J.; Cochran, J. W.; Gardea-Torresdey, J., Gas chromatography, *Anal. Chem.* **2008**, 80, 4487–4497

Chapter 7
Miniaturized Optical Fiber Inline Interferometers for Chemical Sensing

Hai Xiao and Tao Wei

Abstract This chapter reviews the miniaturized optical fiber inline interferometers for chemical sensing based on the detection of composition variation induced refractive index changes. When used as chemical sensors, these miniaturized devices have the common advantages of small size, all-glass ruggedized structure, high sensitivity, fast response time, and large dynamic range. These advantages make them particularly attractive for real-world applications where, in situ, continuous monitoring is required. Specifically, two general types of interferometers are reviewed including the low-finesse Fabry-Perot interferometer and the core-cladding mode interferometer. The operation principles of these two types of interferometers are described. The signal processing methods are discussed. The representative structures, fabrication methods, and application examples of each interferometer type are provided with certain level of details. The advantages and disadvantages of each sensor structure are also highlighted in the discussions, with the hope that innovative researches will be stimulated to solve the technical challenges and explore future applications of these devices.

7.1 Introduction

Miniaturized and robust optical fiber sensors capable of accurate and reliable measurement of refractive index of the surrounding environment have attracted tremendous interest in recent years. One of the driving forces for the development of these fiber optic devices is their broad applications in chemical sensing. When placed in the liquid solution or gas mixture, these fiber sensors can detect the chemical composition change by monitoring its refractive index variation. These

H. Xiao (✉)
Department of Electrical and Computer Engineering, Missouri University of Science and Technology, 1870 Miner Circle, Rolla, MO 65409, USA
e-mail: xiaoha@mst.edu

X. Fan (ed.), *Advanced Photonic Structures for Biological and Chemical Detection*, Integrated Analytical Systems,
DOI 10.1007/978-0-387-98063-8_7,

devices can also be further coated with functional thin films to achieve improved detection sensitivity and selectivity. The coated thin film can selectively interact (e.g., adsorb or react) with the target analyte molecules. As a result, the film changes its refractive index that can be sensitively monitored by the fiber device to produce a signal.

Among the many fiber devices investigated for refractive index sensing, fiber inline interferometers have attracted much interest recently. So far, a number of fiber inline interferometers have been explored for refractive index sensing, including the low-finesse Fabry–Perot interferometer (FPI) with an accessible cavity, the core cladding mode Mach–Zehnder interferometer (MZI), and the core-cladding mode Michelson interferometer (MI). These inline interferometers operate based on the different principles but share the similar advantages of small size, all-glass ruggedized structure, high sensitivity, fast response time, and large dynamic range. These advantages make them attractive for various in situ monitoring applications, especially in places where other bulky sensors are hard to reach.

7.2 Fiber FPI Sensor

Fiber FPI sensors have been well demonstrated for various sensing applications in the past. The FP cavity can be either intrinsic[1] or extrinsic[2]. The two reflections at the two end surfaces of the FP cavity form an interference signal. The deflection of the cavity due to changes in environment causes a phase shift in the interference signal. As a result, a fiber FP sensor is capable for measuring various parameters. With advantages such as small size, immunity to electromagnetic interference (EMI), and corrosion resistance, these fiber FP sensors are particularly attractive for applications involving harsh environments[3,4].

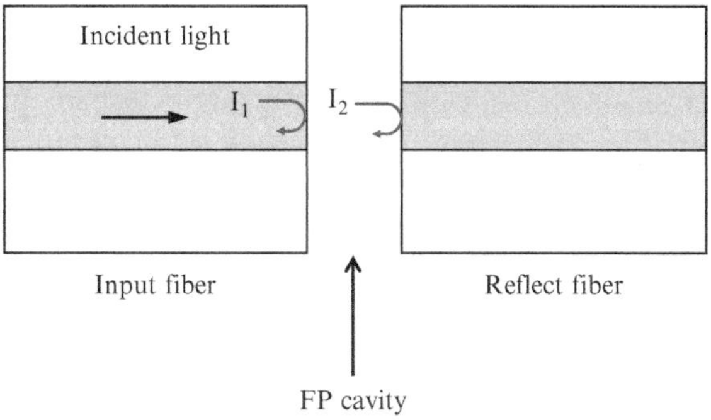

Fig. 7.1 Illustration of a fiber FPI sensor

7.2.1 FPI Principle

Although the two reflectors forming the Fabry-Perot (FP) cavity can be the surfaces of any optical components, a very simple way of constructing a fiber FPI is to directly use the cleaved endfaces of two fibers as shown in Fig. 7.1. Typically, single-mode fibers are used because of their small numerical aperture (NA). The small NA results in a small divergence angle of the exiting beam as it passes through the cavity, which is preferable to produce a high-quality interference signal. Multimode fibers have also been used for the construction of FPI sensors. The main reason of using a multimode fiber is to allow easy and efficient coupling from the light source to the fiber. As shown in Fig. 7.1, the incident light partially reflects (I_1) at the endface of the input fiber. The remainder of the light propagates across the cavity and partially reflects (I_2) at the endface of the reflect fiber. The two reflections interfere to generate a signal whose phase is linearly proportional to the optical length of the cavity, defined as the product of the cavity length and the refractive index of the medium filling the cavity.

When the reflectivity of the two end faces is low (e.g., 4% from an air-glass interface), the multiplex reflections in the cavity have negligible contribution to the optical interference. Under this circumstance, the FP cavity is commonly referred to as the low-finesse cavity and the signal can be modeled using a two-beam interference model, given by[5,6]:

$$I = I_1 + I_2 + 2\sqrt{I_1 I_2}\cos\left(\frac{4\pi Ln}{\lambda} + \varphi_0\right), \tag{7.1}$$

where I is the intensity of the interference signal; I_1 and I_2 are the reflections at the cavity surfaces, respectively; φ_0 is the initial phase of the interference; L is the length of the cavity; n is the refractive index of the medium filling the cavity; and λ is the optical wavelength in vacuum.

According to (7.1), if the FP cavity is made accessible to the external environment, the device can be used as a chemical sensor by measuring the refractive index change of a liquid solution or a gas mixture filling the open cavity. A unique advantage of such a sensor is that the phase of the interference signal is a linear function of the refractive index. Compared with other types of sensors, a fiber FPI refractive index sensor has the unique advantage of constant sensitivity over a large dynamic range.

7.2.2 Signal Processing

Equation (7.1) states that the intensity of the interference signal is a periodic function of the cavity length, the refractive index, and the wavelength. For most chemical sensing applications, the interference signal needs to be processed to obtain either the absolute value or the relative change of the refractive index that

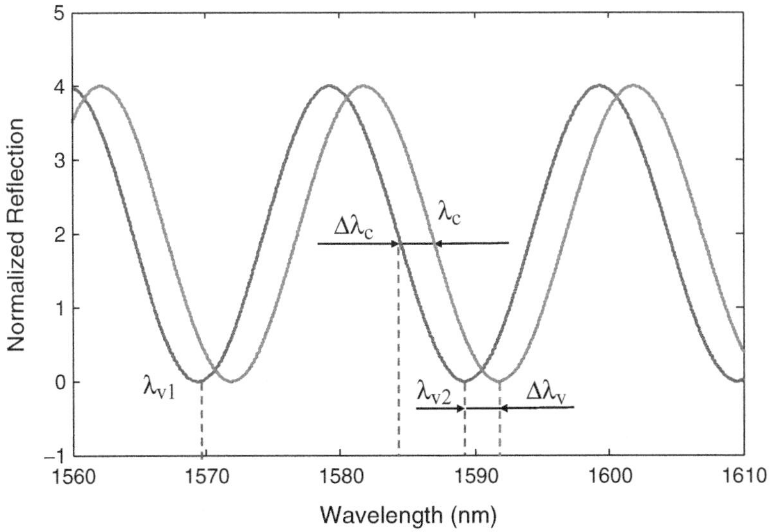

Fig. 7.2 Simulated interference spectra of a fiber FPI sensor

is related to the chemical environment of interest. Although many signal processing methods have been studied in the past, the white light interferometry method produces the highest accuracy.

White light interferometry commonly uses a broadband source to excite the interferometer and an optical spectrum analyzer (OSA) to record the interference signal in the spectral domain. Sometimes, a tunable laser is used to replace the broadband source and a regular photodetector is used to replace the OSA. The spectral signal is obtained by stepping the tunable laser over the entire available wavelength range. In either way, an interference spectrum can be obtained similar to what is given in Fig. 7.2.

According to (7.1), the two adjacent interference minimums have a phase difference of 2π. That is

$$(4\pi Ln/\lambda_{v1} + \varphi_0) - (4\pi Ln/\lambda_{v2} + \varphi_0) = 2\pi, \tag{7.2}$$

where λ_{v1} and λ_{v2} are the wavelengths of the two adjacent valleys (Fig. 7.2) in the interference spectrum. The optical length of the FP cavity, defined as the product of the cavity length and the refractive index, can thus be found as

$$L \cdot n = \frac{1}{2} \frac{\lambda_{v1}\lambda_{v2}}{\lambda_{v2} - \lambda_{v1}}. \tag{7.3}$$

In theory, (7.3) can be used to calculate either the absolute refractive index (n) or the absolute length (L) of the cavity if one of them is known. However, the measurement based on (7.3) has a poor resolution because the period is not a sensitive function of the optical path change[6,7].

In many cases, only the relative refractive index change is of interest and the range of refractive index variation is small so the phase shift is less than 2π. In this case, the phase ambiguity issue can be avoided. The relative refractive index change can be calculated based on the spectral shift of the interferogram.

According to (7.1), the interference signal reaches its minimum ($I_{\min}$) when the phase of the cosine term becomes an odd number of π. That is $I = I_{\min}$, when

$$4\pi nL/\lambda_{\mathrm{v}} + \varphi_0 = (2m+1)\pi, \tag{7.4}$$

where m is an integer and λ_{v} is the center wavelength of the specific interference valley. In (7.4), taking the derivative of n with respect to λ_{v}, one finds:

$$\mathrm{d}n/\mathrm{d}\lambda_{\mathrm{v}} = [(2m+1)\pi - \varphi_0]/4\pi L. \tag{7.5}$$

Assuming the cavity length L is maintained constant during measurement, (7.5) indicates that the refractive index is a linear function of the valley wavelength, or the sensitivity of the FPI sensor is a constant. The amount of refractive index change (n) can thus be computed based on the wavelength shift ($\Delta\lambda_{\mathrm{v}}$ as shown in Fig. 7.2) of a particular interference valley using the following equation derived based on (7.5):

$$\Delta n = n\,\Delta\lambda_{\mathrm{v}}/\lambda_{\mathrm{v}}, \tag{7.6}$$

where the relative refractive index change is directly proportional to the spectral shift of the interferogram. It is worth noting that (7.6) is also applicable to other characteristic spectral positions such as the interference peak and the center point of the interferogram (λ_{c} and $\Delta\lambda_{\mathrm{c}}$ in Fig. 7.2, where the curve is relatively linear). The advantage of using the center point in calculation is that its spectral position can be resolved with a higher resolution compared to the valley or the peak that typically has a flat bottom or top. In addition, curve fitting of the interference fringe can also improve the measurement accuracy.

Comparing (7.3) and (7.6), one finds that (7.3) can be used to calculate the absolute refractive index while (7.6) only provides the relative change of the refractive index. However, the calculation based on (7.6) has a much higher resolution than that obtained using (7.3).

7.2.3 Representative Devices and Application Examples

7.2.3.1 Holey Sleeve Hosted FPI Sensor

The fiber FPI sensor illustrated in Fig. 7.3 can be fabricated by inserting two endface-cleaved optical fibers into a ferrule. The ferrule usually has a diameter

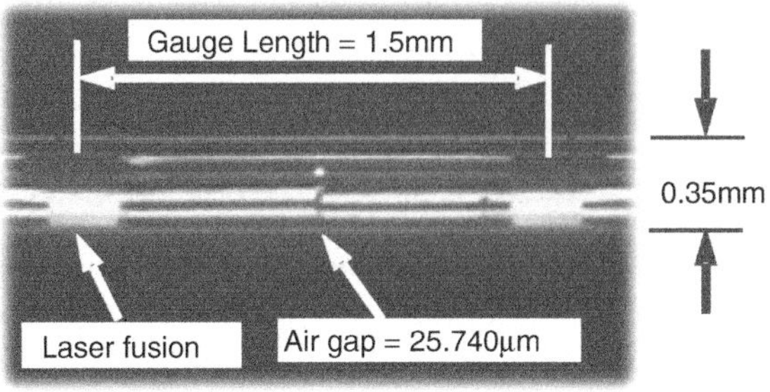

Fig. 7.3 Microscopic image of a fiber FPI sensor made by thermal fusion

slightly larger than that of an optical fiber to provide necessary alignment. After assembly, these components are bonded together to construct the FPI device. Over the years, the bonding methods have evolved from epoxy in the early days[8] to the more recent thermal fusion using a CO_2 laser[3] or an electric arc[9].

To minimize the temperature dependence and to improve the device robustness, fused glass ferrules have been used to construct the FPI cavity. Figure 7.3 shows a fiber FPI device made by thermal fusion of two single-mode fibers into a fused silica tube using a CO_2 laser. The device has been demonstrated for temperature, pressure, and strain sensing and successfully survived a high temperature (200°C) and high pressure (8,000 psi) coexisting environment[10]. Later, a short section of microfused silica tube, with an outer diameter of 126 μm and an inner diameter of 50 μm, was spliced between two cleaved single-mode fibers to construct the FPI sensor in which the tube served as a spacer[11].

In general, the microassembly methods involve complicated procedures of assembling multiple components together. The length of the cavity varies in each device, and the performance is neither predictable nor controllable. The multiple parts assembly also compromises the robustness of the device owing to the limited strength of the joints. The microassembly-based fabrication processes commonly result in a sealed FP cavity[3,9–11]. The applications of such sealed FPI sensor have been limited to the measurement of physical parameters such as displacement, temperature, strain, pressure, acoustic waves, and flow.

In order to make a FPI chemical sensor, the FP cavity needs to be made accessible by the analyte molecules. One way to achieve this is to use a holey sleeve to host the cavity. Xiao et al.[7] reported such a fiber FPI gas sensor formed by bonding two endface-polished fibers in a holey sleeve using epoxy. The holey sleeve allows gas to freely enter and leave the cavity. A resolution of 10^{-5} was estimated in monitoring the changes in the refractive index caused by varying the gas composition. However, the sensor assembly was complicated and required the use of epoxy. In addition, the various components used in sensor construction were made of different materials. As a result, the device had a strong dependence on temperature.

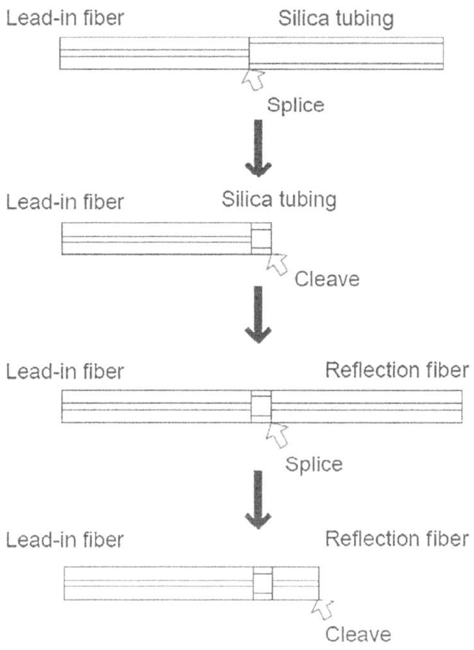

Fig. 7.4 Fabrication procedures of a multicavity fiber FPI sensor. Reprinted from Ref. 11 with permission. © 2008 Optical Society of America

7.2.3.2 Multicavity FPI Sensor

Wang et al. designed a fiber optic biosensor based on cascaded multiple FP cavities[11]. The fabrication process of the reported sensor is illustrated in Fig. 7.4. A silica capillary tube (50-μm inner diameter and 126-μm outer diameter) was spliced onto a single-mode fiber. An FP cavity was formed by cleaving the tube and splicing it to another single-mode fiber. Another fiber cavity was created by cleaving the second fiber as shown in Fig. 7.4. The fiber cavity could be as small as 10 μm, limited by cleaving and, up to the centimeter range, limited by the coherence length of the laser to interrogate the sensor. The overall fringe pattern as shown in Fig. 7.5 was due to the interference between the reflection signals from the three reflection interfaces. Multiple-beam interference was evaluated by matrix formalism.

The developed sensor was used for ultrathin-film measurement. The reflection spectrum was shifted during the deposition of thin films (e.g., self-assembly of polyelectrolyte layers) onto the sensor end. The reflection between the thin film and the fiber endface was neglected because of their similar refractive indices. As the film increased its thickness, the length of the fiber cavity changed. The amount of change was estimated by the phase shift of the interferogram. The device could also be used as an immunosensor in which the optical thickness changes were used to

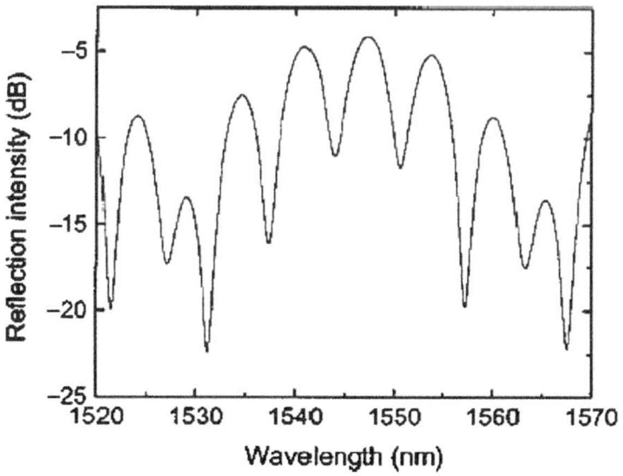

Fig. 7.5 Reflection spectrum of a multicavity FPI sensor. Reprinted from Ref. 11 with permission. © 2008 Optical Society of America

evaluate the immobilization of the IgG and the immunological activities of the immobilized layers. The binding of IgG and the corresponding antigen was observed, and the nonspecific binding characteristics were investigated.

7.2.3.3 FP Modulated tip Sensor

Recently, Ran et al. described a refractive index sensor by adding a sealed Fabry-Perot cavity near the tip of a single-mode fiber[12]. To fabricate the sensor head, a circular hole was produced using a 157-nm laser micromachining system. The cleaved endface of the PCF was exposed directly to the focused laser beam of a 157-nm pulsed laser (Coherent, LPF202). The pulse energy density, pulse width, and pulse repetition rate used were 12 J cm^{-2}, 15 ns, and 20 Hz, respectively. As a result of the focused exposure, a circular hole was formed at the center of the cross section, with a depth of ~23 μm and a diameter of ~56 μm, as shown in Fig. 7.6a. A total of 160 pulses were used to produce the hole, which took only 8 s to complete. The micromachined hole was then spliced to another fiber to form a sealed FP cavity. Finally, the spliced fiber was cleaved at a short distance from the air cavity to complete the sensor as shown in Fig. 7.6b. The length of the air cavity was 29 μm and the distance from the air cavity to the fiber end was 1,014 μm.

In this structure, the sealed FPI itself was not a sensing device but only served as a signal modulator. The external refractive index variation changed the reflectivity of the exposed fiber tip. The amount of refractive index change was determined by change in the maximum fringe contrast in the reflected spectrum. The sensor head was dipped into a glycerin solution (for $n < 1.45$) or a solution of carbon bisulfide (CS2) and alcohol (for $1.45 < n < 1.62$). The refractive index of the solution was controlled by changing the concentration of the solvent. The reflection spectrum of

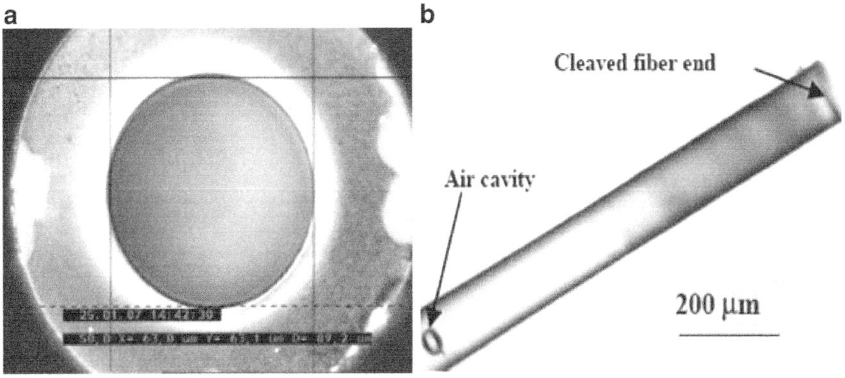

Fig. 7.6 (**a**) Microscopic image of the micromachined hole introduced on the fiber cross section. (**b**) Microscopic image of the fabricated sensor head. Reprinted from Ref. 12 with permission. © 2008 Optical Society of America

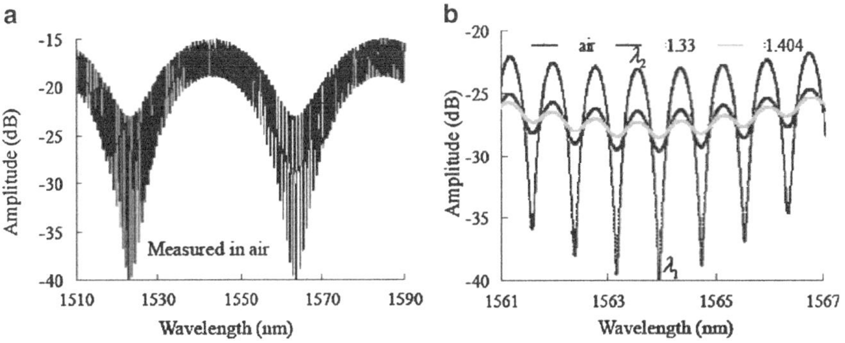

Fig. 7.7 (**a**) Reflection spectrum of the sensor measured in air. (**b**) Close-up displays of the fringes for $n_{\prime}$ = 1.0 (air), 1.33, and 1.404, respectively. Reprinted from Ref. 12 with permission. © 2008 Optical Society of America

the sensor in air is shown in Fig. 7.7a. A close-up of the fringes is shown in Fig. 7.7b for several values of RI: 1.0 (air), 1.33 (water), and 1.404 (glycerin solution). The resolution of RI measurement was estimated to be $\sim 4 \times 10^{-5}$ in its linear operating range[12].

7.2.3.4 Open Cavity FPI Sensor

The latest advancement in femtosecond (fs)-based micromachining technology has opened a new window of opportunity for fabrication of microdevices. Direct exposure of most solid materials (including fused silica glass) to high power fs laser pulses may lead to the ablation of a thin layer of materials at the laser focal point[13]. Due to the multiphoton nature of the laser-material interaction, the ablation process can be conducted on the material surface as well as within its

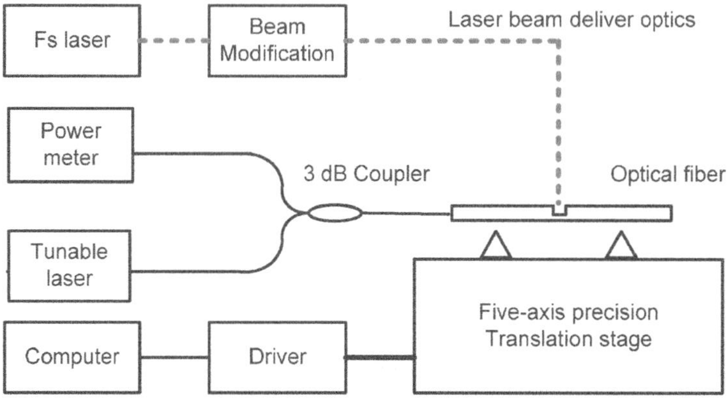

Fig. 7.8 Fiber-inline FPI device fabrication using an fs laser. Reprinted from Ref. 18 with permission. © 2008 Optical Society of America

bulk. As a result, true three-dimensional (3D) microphotonic devices can be fabricated with submicron accuracy[14,15]. Optical device examples include micro-lenses[16] microfluidic channels[17], and fiber inline Fabry-Perot interferometers[18–20].

Recently, Rao et al. reported a miniaturized fiber inline FPI device fabricated by micromachining a rectangular non-through hole into a single-mode or photonic crystal fiber using a femtosecond (fs) laser[19]. The single-mode fiber FPI device had a fringe visibility of about 2 dB. We also demonstrated an inline FPI device with a fringe visibility of 16 dB by fs laser one-step micromachining a micronotch on a single-mode fiber[18].

The device fabrication was carried out using a home-integrated fs laser 3D micromachining system as schematically shown in Fig. 7.8. The repetition rate, center wavelength, and pulse width of the fs laser (Legend-F, Coherent, Inc.) were 1 kHz, 800 nm, and 120 fs, respectively. The maximum output power of the fs laser was approximately 1 W. The actual laser power used in fabrication was reduced to about 20 mW by using the combination of waveplates, polarizers, and several neutral density (ND) filters. The attenuated laser beam was directed into an objective lens with a numerical aperture (NA) of 0.45 and focused onto the single-mode optical fiber (Corning SMF 28) mounted on a computer-controlled five-axis translation stage with a resolution of 1 μm.

During fabrication, the interference signal of the fiber FP device was continuously monitored. A tunable laser source (HP 8168E) was connected to one of the input ports of the 3-dB fiber coupler. The output port of the coupler was connected to the device under fabrication. Controlled by the computer, the tunable laser continuously scanned through its wavelength range (1,475–1,575 nm) at the rate of 1 nm per step. The signal reflected from the device at each wavelength step was recorded by an optical power meter (Agilent 8163A). The fabrication was stopped after a well-formed interference pattern was recorded.

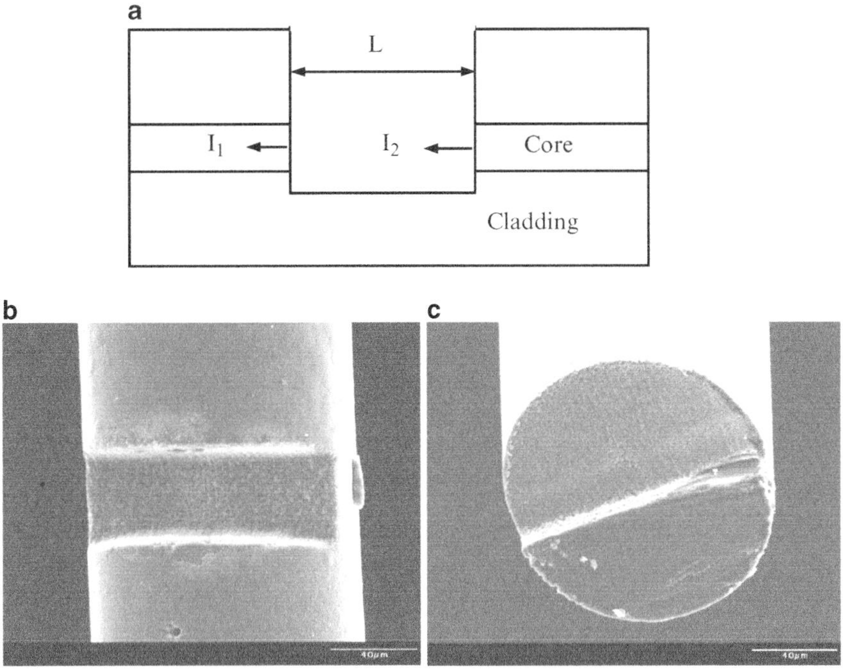

Fig. 7.9 Schematic and SEM images of a fiber-inline FPI device fabricated by fs laser. Reprinted from Ref. 18 with permission. © 2008 Optical Society of America

Figure 7.9 shows the schematic structure and scanning electron microscope (SEM) images of the fabricated fiber FP device, where Fig. 7.9b shows that a micronotch was formed on the optical fiber. Figure 7.9c shows the fs laser-ablated surface. The cavity length was about 60 μm as estimated from the SEM image. The depth of ablation was around 72 μm, just passing the fiber core. The FP cavity can be made very close (within a few hundreds of microns) to the end of the fiber. With such a short bending arm, the chance of bending induced device breakage is small.

Figure 7.10 shows the interference spectrum of an inline fiber FPI device fabricated by the fs laser. The excess loss of this particular device was about 16 dB. This relatively high loss was mainly caused by three reasons (1) the light scattering loss at the laser-ablated surface, (2) the nonperpendicular surface orientation with respect to the fiber axis, which was partially evidenced by the nonflat interference peak intensities shown in Fig. 7.10 and (3) the coupling loss as a result of recoupling the light reflected from the second endface of the FP cavity back into the fiber core. The surface roughness can be reduced by reducing the laser scanning steps, of course, at the expense of a long device fabrication time. The nonperpendicular surface orientation can also be minimized by careful adjustment of the stages. The coupling loss increases with the length of the FP cavity. As a result, it may eventually limit the practical length of the cavity. Nevertheless, the

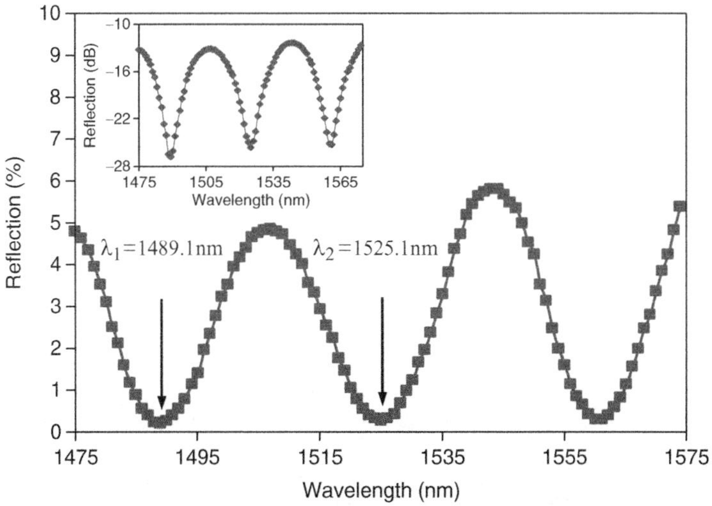

Fig. 7.10 Interference spectrum of the fs laser-fabricated fiber-inline FPI device. *Insert*: interference fringe plotted in dB scale. Reprinted from Ref. 18 with permission. © 2008 Optical Society of America

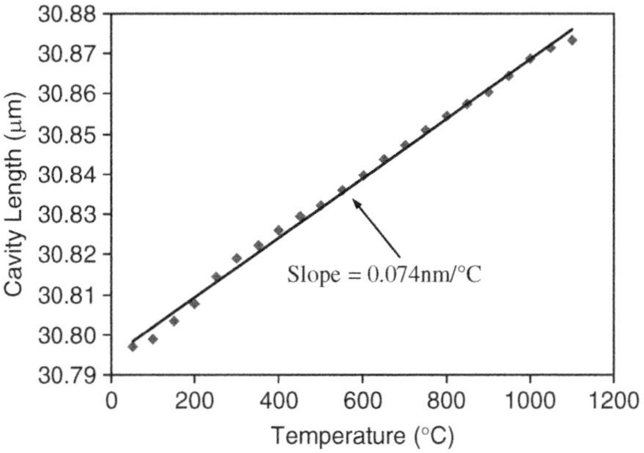

Fig. 7.11 Fiber-inline FP device in response to temperature change. Reprinted from Ref. 18 with permission. © 2008 Optical Society of America

interference spectrum indicated a high fringe visibility, exceeding 14 dB, which is sufficient for most sensing applications.

The device has an all-glass structure and does not involve assembly of multiple components. As a result, we expect that the device will have very small temperature dependence. In addition, the open micronotch FP cavity allows prompt access to gas or liquid samples for direct refractive index measurement, making it possible to be used as an ultracompact chemical sensor based on refractive index measurement.

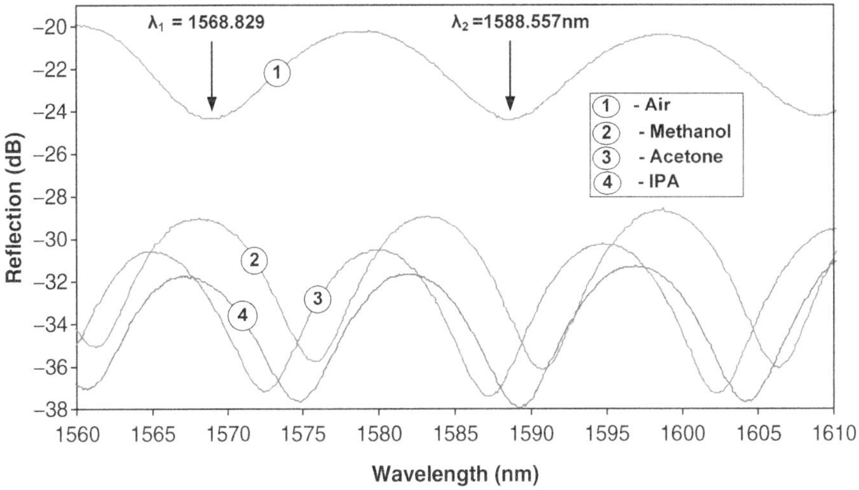

Fig. 7.12 Interference spectra of the FPI device in air, methanol, acetone, and 2-propanol (IPA). Reprinted from Ref. 20 with permission. © 2008 Optical Society of America

The temperature dependence of the fabricated open cavity FP device was evaluated experimentally. The sensor was placed in a programmable electric tubular furnace. The temperature of the furnace was increased from room temperature to 1,100°C at a step of 50°C. The cavity length as a function of the temperature is plotted in Fig. 7.11, where it increased nearly linearly following the increase of temperature. The temperature sensitivity of the particular FP device under test was estimated to be 0.074 nm $°C^{-1}$ based on the linear fit of the measurement data. The equivalent coefficient of thermal expansion (CTE) of the fiber FP device was $2.4\times10^{-6}°C^{-1}$.

To evaluate its capability for refractive index measurement, the fiber FPI device was tested using various liquids including methanol, acetone, and 2-propanol at room temperature. The interference spectra of the device immersed in various liquids are shown in Fig. 7.12 for comparison. The signal intensity dropped when the device was immersed in liquids as a result of the reduced refractive index contrast and thus lowered Fresnel reflections from the cavity endfaces. However, the interference fringes maintained a similar visibility. The spectral distance between the two adjacent valleys also decreased, indicating the increase of refractive index of the medium inside the cavity. Using (7.4), the refractive indices of the liquids were calculated to be: $n_{\text{methanol}} = 1.3283$, $n_{\text{acetone}} = 1.3577$, and $n_{\text{2-propanol}} = 1.3739$, which was close to the commonly accepted values.

The device was also evaluated for temperature-insensitive refractive index sensing by measuring the temperature-dependent refractive index of the deionized water. Figure 7.13 shows the measured refractive index of the deionized water as a function of temperature. As the temperature increases, the interference fringe shifts to a shorter wavelength indicating the decrease in its refractive index. The refractive index change was calculated using (7.6) by tracing the spectral shift of the

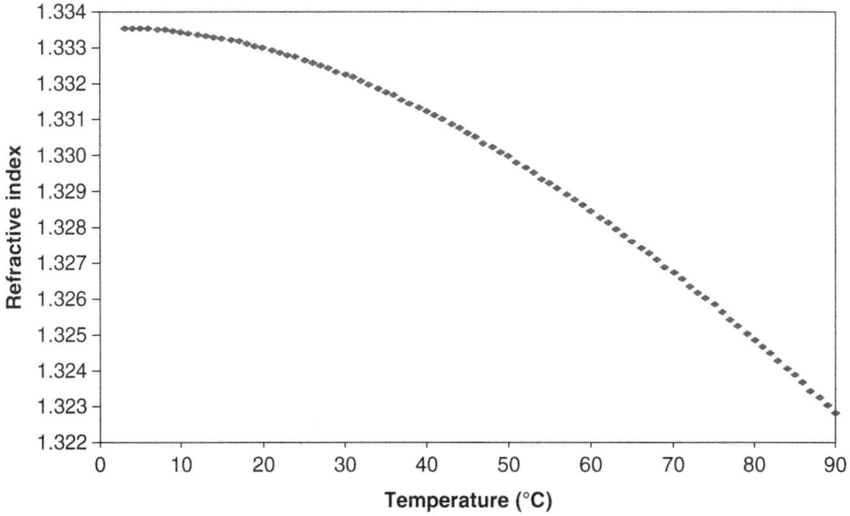

Fig. 7.13 Measured refractive index of deionized water as a function of temperature. Reprinted from Ref. 20 with permission. © 2008 Optical Society of America

interferogram, assuming a constant cavity length over the entire temperature range. To improve the accuracy of spectral shift measurement, the interference fringes were first normalized to have the same average intensities, then curve fitted using a fourth-order polynomial. The spectral shift was computed as the difference in wavelength between the two fitted curves at the center point.

The measurement results, as shown in Fig. 7.13, indicate that the refractive index of water decreases nonlinearly as its temperature increases. The amount and shape of the measured refractive index of water change as the function of temperature agreed well with the previously reported measurement data[21]. When used to monitor the relative refractive index change of water with a nominal refractive index of 1.333, the sensitivity of measurement is 1,163 nm per RIU at the wavelength of 1,550 nm according to (7.6). The actual detection limit depends on the resolution with which the spectral shift of the interferogram can be determined. If a resolution of 10 pm is achieved in the determination of interferogram shift, a detection limit of 8.6×10^{-6} RIU is attainable. With a rough resolution of 1 nm, the detection limit of 8.6×10^{-4} RIU can still be achieved.

The above measurement results also included the error contribution of the temperature cross-sensitivity of the device. From Fig. 7.11, the temperature dependence of the device was 0.074 nm $°C^{-1}$. Based on (7.6), the temperature cross-sensitivity of the device was less than 3.2×10^{-6} RIU $°C^{-1}$. Therefore, the total temperature cross-sensitivity-induced measurement error was about 2.8×10^{-4} RIU in Fig. 7.13 over the temperature variation of 87°C. The temperature dependence of the device was small and contributed only about 2.3% to the total refractive index variation over the entire temperature range.

7.2.3.5 Zeolite Thin-Film FPI Sensor

The FPI principle can also be used to develop thin-film-coating-based chemical sensors. For example, a thin layer of zeolite film has been coated to a cleaved endface of a single-mode fiber to form a low-finesse FPI sensor for chemical detection. Zeolite presents a group of crystalline aluminosilicate materials with uniform subnanometer or nanometer scale pores. Traditionally, porous zeolite materials have been used as adsorbents, catalysts, and molecular sieves for molecular or ionic separation, electrode modification, and selectivity enhancement for chemical sensors. Recently, it has been revealed that zeolites possess a unique combination of chemical and optical properties. When properly integrated with a photonic device, these unique properties may be fully utilized to develop miniaturized optical chemical sensors with high sensitivity and potentially high selectivity for various in situ monitoring applications.

Upon selective absorption of analyte molecules from the ambient environment, the zeolite thin film increases its refractive index. Correspondingly, release of adsorbed molecules from the zeolite pore results in the decrease of its refractive index. The absorption/desorption of molecules depends on the molecule concentration in the environment to be monitored. Therefore, monitoring of the refractive index change induced phase shift in the interference spectrum can detect the presence and amount of the target analyte existing in the environment.

Figure 7.14 shows the structure, microscopic images, and representative interference spectrum of the sensor. As shown in Fig. 7.14c, the two reflections

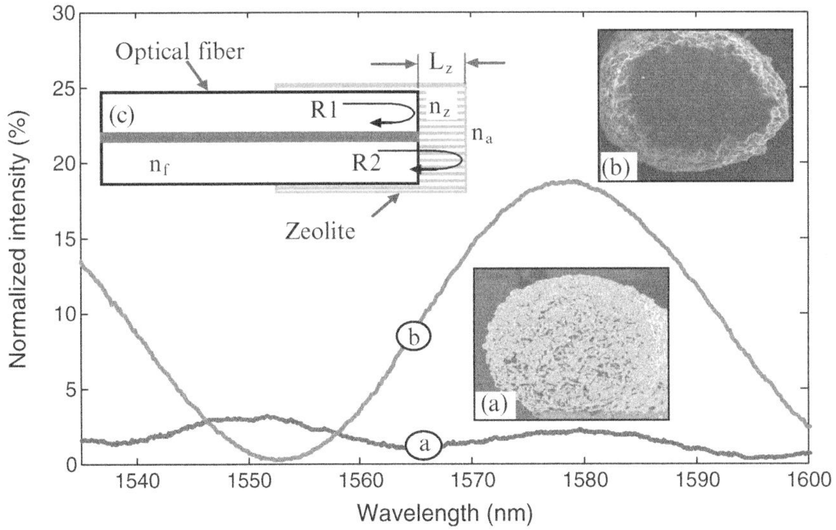

Fig. 7.14 Zeolite thin-film FPI chemical sensor. (**a**) As-synthesized outer surface and interference signal, (**b**) polished outer surface and improved interference signal, and (**c**) sensor schematic. Reprinted from Ref. 22 with permission. © 2008 Molecular Diversity Preservation International

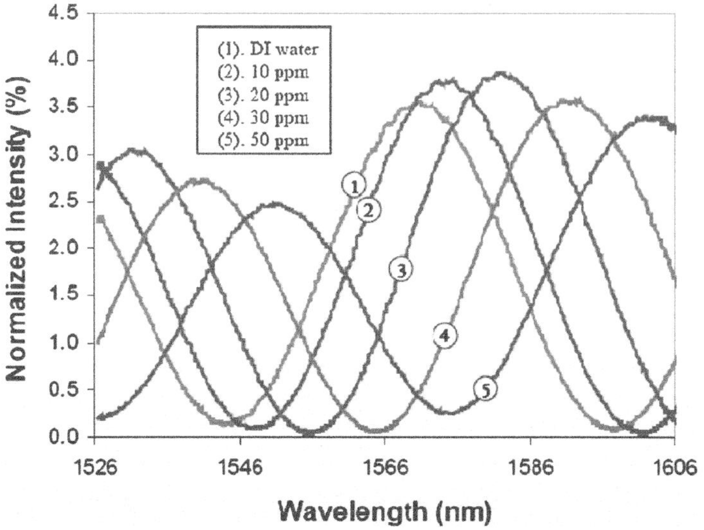

Fig. 7.15 Interferograms in correlation with toluene concentration in DI water. Reprinted from Ref. 22 with permission. © 2008 Molecular Diversity Preservation International

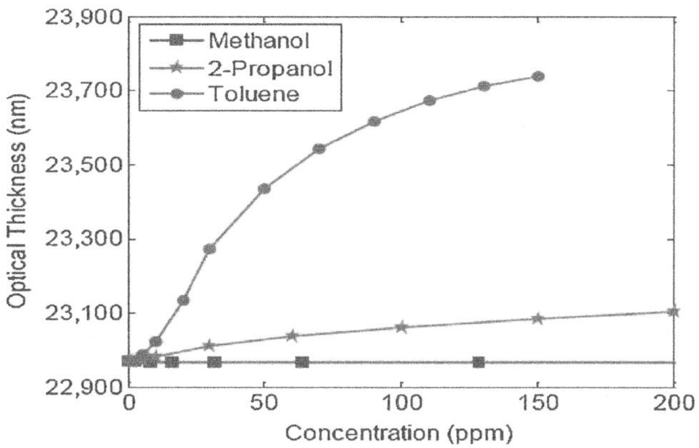

Fig. 7.16 Sensor response to the methanol, 2-propanol, and toluene concentrations in DI water. Reprinted from Ref. 22 with permission. © 2008 Molecular Diversity Preservation International

(R_1 and R_2) from the fiber-zeolite and zeolite-ambient interfaces interfere to generate the signal spectrum. Due to the polycrystalline nature of the zeolite coating, the as-synthesized zeolite film had a rough outer surface (Fig. 7.14a). As a result, the quality of the interference signal was very poor. To improve the quality of the interference signal, the outer surface of the zeolite coating was finely polished to remove the surface irregularities. As shown in Fig. 7.14b, the interference signal was significantly improved after polishing.

The polished zeolite FPI sensor was tested for solution detection of dissolved organics in water. The organics used for tests include toluene, 2-propanol, and methanol. Figure 7.15 shows the interferograms at various toluene concentrations in deionized water. It is clear that as the concentration increased, the interferogram shifted toward the short wavelength region, indicating the increase of refractive index. By tracing the wavelength shift of the interference spectrum, it was possible to calculate the optical thickness change of the zeolite thin film as a function of the organic concentration based on a low finesse FPI model. Figure 7.16 plots the optical thickness of the zeolite coating as a function of chemical concentrations dissolved in the deionized water. The sensor had a strong response to toluene, a moderate response to 2-propanol, and almost no response to methanol at low concentration level. The detection limit of the sensor was estimated to be 2 ppm for toluene, 5 ppm for 2-propanol, and 1,000 ppm for methanol. Here, the unit ppm is in volume ratio, which means, 10^{-6} L of 2-propanol with 0.999999 L deionized water gives a 2-propanol concentration of 1 ppm[22].

7.3 Fiber Core-Cladding Mode Interferometric Sensors

Fiber core-cladding mode interferometric (CCMI) sensors are relatively new. They operate based on the interference between the core mode and the cladding modes. The core mode is guided by the core-cladding interface of the fiber and well shielded by the thick cladding. Therefore its propagation is insensitive to the refractive index change of the environment. However, the cladding mode, which is guided by the cladding–ambient interface, is directly exposed to the environment. As a result, its effective propagation constant (β_{cl}) is sensitive to the ambient refractive index change. Therefore, the interference between the core mode and the cladding mode carries the ambient refractive index information, which can be used to develop compact, highly sensitive chemical sensors.

7.3.1 CCMI Principle

Based on the way the interferometer is configured, CCMI sensors can be categorized into two groups, namely the Mach-Zehnder interferometer (MZI) type and the Michelson interferometer (MI) type. The MZI sensor works in transmission mode, i.e., the transmitted interference signal is detected. The MI sensor works in reflection mode, where the light passes the interferometer twice and the reflected interference signal is detected.

The principle of operation of a typical MZI-CCMI sensor is illustrated in Fig. 7.17a, where two mode splitters/combiners (MSC) are created on a single-mode fiber. These two MSCs are separated by a distance of L, which is referred to as the length of the interferometer. The input light, originally guided inside the fiber

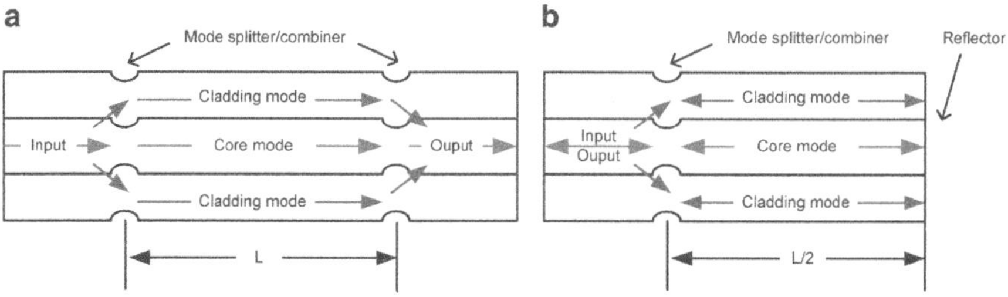

Fig. 7.17 Principle of fiber core-cladding mode interferometric (CCMI) sensors: (**a**) Mach-Zehnder type and (**b**) Michelson type

core, is split into two parts at the first MSC. Part of the light remains propagating inside the fiber core while the rest is coupled into the fiber cladding. Passing through the interferometer section, the light traveling inside cladding is coupled back to the core at the second MSC, mixing with the core-mode component to generate an interference signal.

The MI type sensor operates very similarly to as the MZI type. As shown in Fig. 7.17b, the MI type sensor is a "half" MZI. Part of the light is coupled from the core mode to the cladding mode at the MSC. They propagate through the interferometer and are reflected from the cleaved fiber endface. Sometimes, the cleaved fiber endface can be coated with a highly reflective film to improve the signal strength. The two backward propagating waves are mixed again at the MSC, generating an interference signal.

The MZI-CCMI and the MI-CCMI can be modeled using the following two-beam optical interference equation:[5, 6]

$$I = I_1 + I_2 + 2\sqrt{I_1 I_2}\cos\left[\frac{2\pi\left(n_{\mathrm{c}}^{\mathrm{eff}} - n_{\mathrm{cl}}^{\mathrm{eff}}\right)L}{\lambda} + \varphi_0\right], \tag{7.7}$$

where I is the intensity of the interference signal; I_1 and I_2 are the intensity of interfering lights propagating in the fiber core and cladding, respectively; L is the interferometer length; λ is the free space wavelength; φ_0 is the initial phase of the interference; and $n_{\mathrm{c}}^{\mathrm{eff}}$ and $n_{\mathrm{cl}}^{\mathrm{eff}}$ are the effective refractive indices of the core and cladding, respectively. The effective refractive index (n^{eff}) is related to the propagation constant (β) by $\beta = 2\pi n^{\mathrm{eff}}/\lambda$.

According to (7.7), the interference spectrum reaches its minimum ($I_{\min}$) when the phase of the cosine term becomes an odd number of π. That is

$$I = I_{\min}, \text{ when}$$

$$\frac{2\pi\left(n_{\mathrm{c}}^{\mathrm{eff}} - n_{\mathrm{cl}}^{\mathrm{eff}}\right)L}{\lambda_{\mathrm{v}}} + \varphi_0 = (2k+1)\pi, \tag{7.8}$$

where k is an integer and λ_v is the center wavelength of a specific interference valley.

It is well known that the propagation constant of the cladding mode (thus n_{cl}^{eff}) is a function of the refractive index of the surrounding medium. On the other hand, the core mode is shielded by the cladding layer so its propagation constant does not vary when the surrounding medium changes its refractive index. The interferometer can thus be used as a sensor to monitor the environmental refractive index change.

In (7.8), taking the derivative of n_{cl}^{eff} with respect to λ_v, one finds:

$$\frac{dn_{cl}^{eff}}{d\lambda_v} = -\frac{(2k+1)\pi - \varphi_0}{2\pi L} = -\frac{\left(n_c^{eff} - n_{cl}^{eff}\right)}{\lambda_v}. \tag{7.9}$$

In many cases, only the relative index change is of interest and the range of refractive index variation is small so the phase shift is less than 2π. Under this circumstance, the phase ambiguity issue can be avoided. The relative refractive index change can be computed based on the wavelength shift of a particular interference valley using the following equation derived based on (7.9):

$$\frac{\Delta n_{cl}^{eff}}{\left(n_c^{eff} - n_{cl}^{eff}\right)} = -\frac{\Delta\lambda_v}{\lambda_v}. \tag{7.10}$$

It is worth noting that (7.10) is also applicable to other characteristic spectral positions such as the interference peak and the center point of the interferogram. In addition, curve fitting of the interference fringe can also improve the measurement accuracy[20].

7.3.2 *Device and Application Examples*

The key to making the CCMI sensor is to fabricate the MSC with a high quality. Ideally, the MSC should have the function of a mode-selective, wavelength-independent 3-dB coupler that couples half of the core-mode energy into a specific cladding mode. Unfortunately, methods of fabricating such an ideal MSC is still yet to be found. Nevertheless, a number of methods have been investigated to fabricate MSCs with the basic function of core-cladding mode coupling. The reported methods include long-period fiber grating (LPFG), fiber taper, core-mismatch, and CO_2 laser irradiation. Using these MSCs, the basic operation of the CCMI sensors for refractive index monitoring has been demonstrated.

7.3.2.1 LPFG-Based CCMI Sensor

As shown in Fig. 7.18, a MZI-CCMI sensor can be constructed by an LPFG pair separated by a distance[23,24]. The first LPFG coupled part of the light energy into

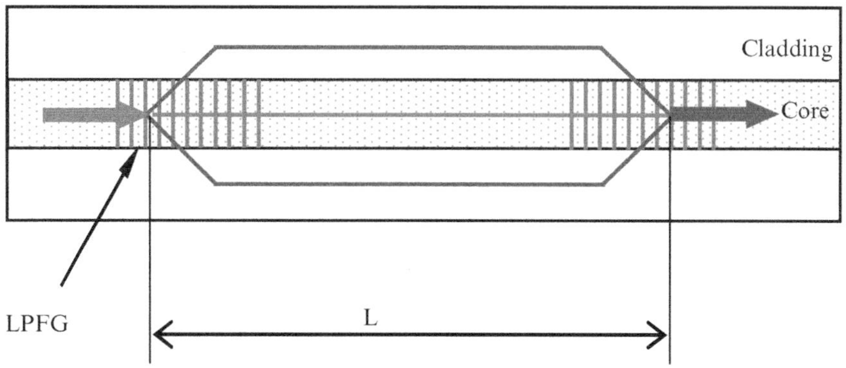

Fig. 7.18 Schematic of a LPFG-pair-based Mach-Zehnder CCMI sensor

the cladding mode while the rest remained inside the fiber core. After passing the middle section, the cladding and core modes were recoupled and mixed at the second LPFG, resulting in interference fringes in the transmission spectrum of the grating. The phase of the interference signal is proportional to the optical path difference (OPD) between the cladding and core modes propagating though the same fiber section.

A Michelson-type CCMI device can be constructed using a single LPFG and a reflective mirror[27,28]. In this configuration, the core mode light was partially coupled into a cladding mode when it propagates through the LPFG in the forward direction. After reflection at the mirror (e.g., a silver-film-coated fiber endface), the core and cladding modes are mixed at the same LPFG as they propagate in the backward direction.

Perhaps the most attractive advantage of using a LPFG pair is that the cladding mode that participates in the interference can be precisely selected by the LPFG. Therefore, the quality of the interference signal is high. This high-quality interference signal attributes to two reasons (1) the optical waves participating in the

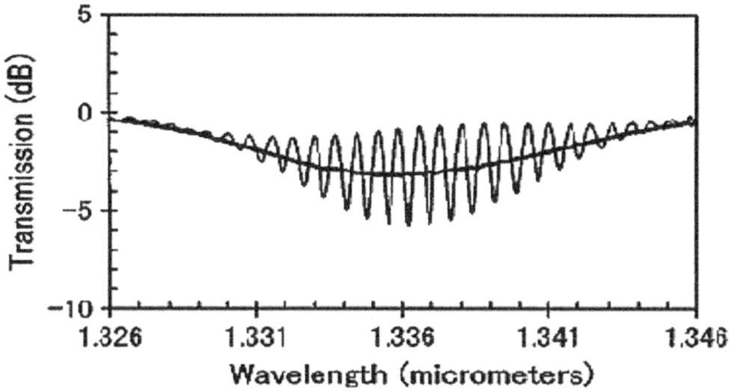

Fig. 7.19 Spectra of a weak LPG pair, separated by 30 cm, with bending in the middle (*thicker curve*, no peaks) and without bending (*peaked curve*). Each grating has 400-μm periodicity and a 20-mm length. The fringe spacing is roughly 0.72 nm. Reprinted from Ref. 25 with permission. © 2008 Optical Society of America

interference are well defined and (2) the mode splitting ratio of a LPFG can be well controlled to obtain a 3-dB coupling. In addition, the sensitivity of the device can be precisely predicted by theoretical modeling as each specific cladding mode has its own sensitivity curve. However, LPFGs are very sensitive to the environmental refractive index change. To this point, the addition of an interferometer is somewhat redundant. The interference spectrum is the superposition of both the LPFG response and the interferometer response. This is evident that the interference fringes are modulated by the LPFG transmission spectrum, as shown in Fig. 7.19.[23]

The LPFG-based CCMI sensors have been demonstrated for various applications including refractive index sensing[24]. In the reported setup, two UV inscribed LPFGs with strength of 3-dB mode splitting ratio and period of 240 μm were used to couple the core mode to the ninth cladding mode and vice versa. The interferometer was interrogated using a heterodyne technique in which a distributed feedback (DFB) laser was wavelength modulated and the first and second harmonics were extracted to compute the phase shift. The minimum detectable refractive index change was estimated about 1.8×10^{-6} RIU at the nominal value of $n = 1.43$.

7.3.2.2 Fiber-Taper-Based CCMI Sensor

Fiber tapers can also be used to couple light from core to cladding if the taper angle is large. Therefore, two cascaded fiber tapers can form a MZI-CCMI and a single-fiber taper can be used to construct a MI-CCMI[27,28]. In general, the mode coupling is dependent on the taper geometry as well as the wavelength. Fiber tapers can be made by pulling while heating the optical fiber to its softening temperature. Some commercial fusion splicers have built-in programs to make fiber tapers.

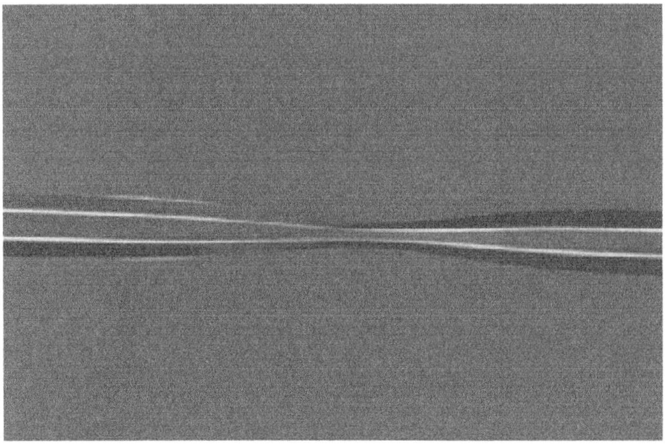

Fig. 7.20 Microscopic image of a 3-dB fiber taper with the length of 707 μm and waist diameter of 40 μm made by a fusion splicer. Reprinted from Ref. 20 with permission. © 2008 Optical Society of America

Recently, Tian et al. demonstrated both MZI- and MI-based CCMI sensors using abrupt fiber tapers[28]. The taper was made using the built-in tapering program of an Ericsson fusion splicer (FSU 995FA). The microscopic image of the taper is shown in Fig. 7.20, where the length of the taper is 707 μm and the waist diameter is 40 μm. MZI sensors of various length (24, 36, and 55 mm) were made by using two cascaded fiber tapers. A MI sensor with the length of 38 mm was also demonstrated using one fiber taper and a gold-coated fiber reflector. The maximum fringe visibility of these CCMI sensors could reach 20 dB. These sensors were tested using liquids with different refractive indices. The refractive index measurement sensitivity of these sensors was found to be similar to that of LPFGs.

The fiber taper method has the advantage of easy construction. However, the taper normally excites more than one cladding modes. As a result, the interference is the superposition of contributions from multiple waves. When the length of the interferometer is long, the interference fringe is severely distorted. Unfortunately, fabrication of a fiber taper requires a minimum lead-in length, which limits the shortest length of the device. In addition, the mode coupling of a fiber taper depends on the environmental refractive index change. In applications, contributions from the fiber taper and the interferometer itself are mixed in the same signal spectrum, resulting in a large uncertainty when a large index variation is monitored. Finally, the fiber taper has a degraded mechanical strength due to the small waist diameter. In real applications, fiber-taper-based devices are subject to large bending-induced errors.

7.3.2.3 Core-Mismatch-Based CCMI Sensor

Core mismatch between two fiber segments can also cause part of the core-mode light coupled into the cladding modes. Based on this mechanism, CCMI sensors

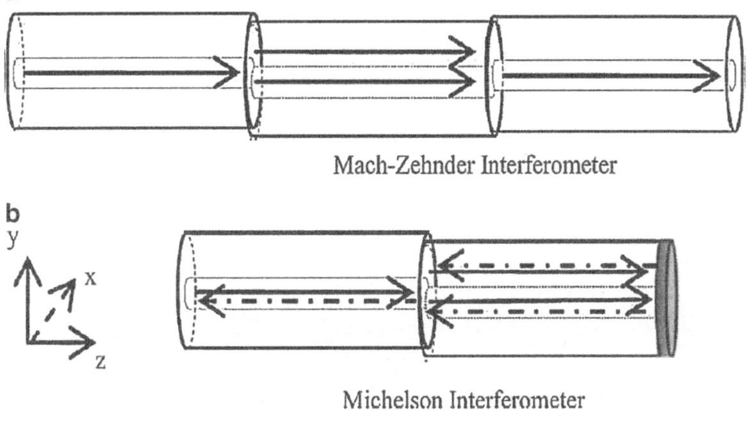

Fig. 7.21 Structures of core-offset MZI- and MI-CCMI sensors. Reprinted from Ref. 30 with permission. © 2008 Optical Society of America

have also been demonstrated using core-offset attenuators and collapse of the microstructure on a photonic crystal fiber.

As shown in Fig. 7.21, in the core-offset approach, two fibers are spliced with an intentional offset so that part of the light propagating in the first fiber leaks to the cladding portion of the second fiber through the offset[29]. The cladding modes can be either collected by another offset-spliced fiber to make a MZI sensor or the same offset joint through an endface reflector to form a MI sensor[30].

In another approach, two cascaded microstructure-collapses on a photonic crystal fiber are intentionally introduced by CO_2 laser or electric arc heating[31]. In this way, the mismatch of core size along the fiber causes light coupling between the core and cladding modes.

These interferometers have been demonstrated for refractive index sensing. The sensitivity was around 33 nm spectral shift per RIU change when they were used for refractive index measurement. The maximum fringe visibility of these interferometers was around 9 dB, which was relatively small compared with the interferometers constructed by other methods.

7.3.2.4 CCMI Sensor Fabricated by CO_2 Laser Irradiations

Very recently, we reported the fabrication MZI- and MI-type CCMI sensors by CO_2 laser irradiation. The laser irradiation induced a micronotch on the fiber structure, causing light coupling from the core mode to the cladding modes, and vice versa. Compared with other fabrication techniques, the CO_2 laser irradiation method is simple, avoids complicated assembly, and maintains the mechanical property of an optical fiber. Additionally, the fiber modification region is very small. With the help of precision transition stage, the optical length of the CCMI sensors can be very small and accurately controlled.

A CO_2 laser (SYNRAD, Inc.) with a free space wavelength of 10.6 μm and a maximum output power of 20 W was used in the system. A ZnSe cylindrical lens with a focal length of 50 mm was used to shape the CO_2 laser beam into a narrow line with a linewidth of around 220 μm. The CO_2 laser was controlled by a computer so that the output power and the exposure-time trajectory could be accurately adjusted. A 3D motorized translation stage was used to position the optical fiber to the center of the focused laser beam with the fiber axis perpendicular to the laser line. A microscope vision system was used to visualize the optical fiber. During fabrication, the CO_2 laser heated the fiber and created the first micronotch at one position. Controlled by the computer, the translation stage moved the fiber a certain distance. Then the laser was fired again to create the second micronotch. These two micronotches and the fiber in between made a fiber inline core-cladding-mode MZI.

Figure 7.22 shows the microscopic image of a typical laser-induced micronotch on a single-mode fiber (Corning SMF-28e). The laser power was set to 10 W and the exposure duration was 300 ms. Figure 7.23 shows the interference spectra of MZIs with the length of 5, 10, 20, and 40 mm, respectively. The typical interference

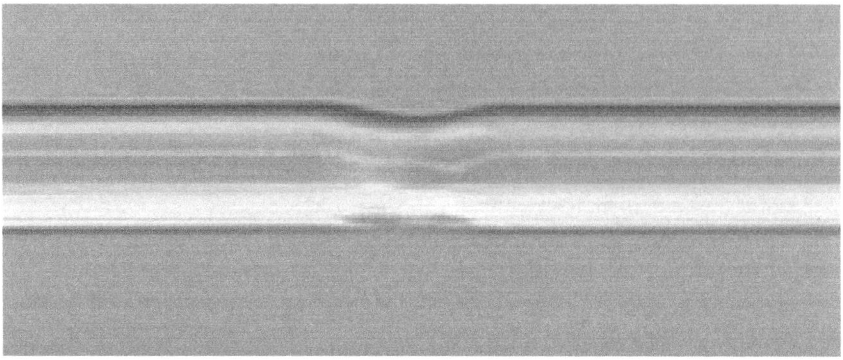

Fig. 7.22 Microscopic image of a CO_2 laser irradiation-induced core-cladding mode coupler

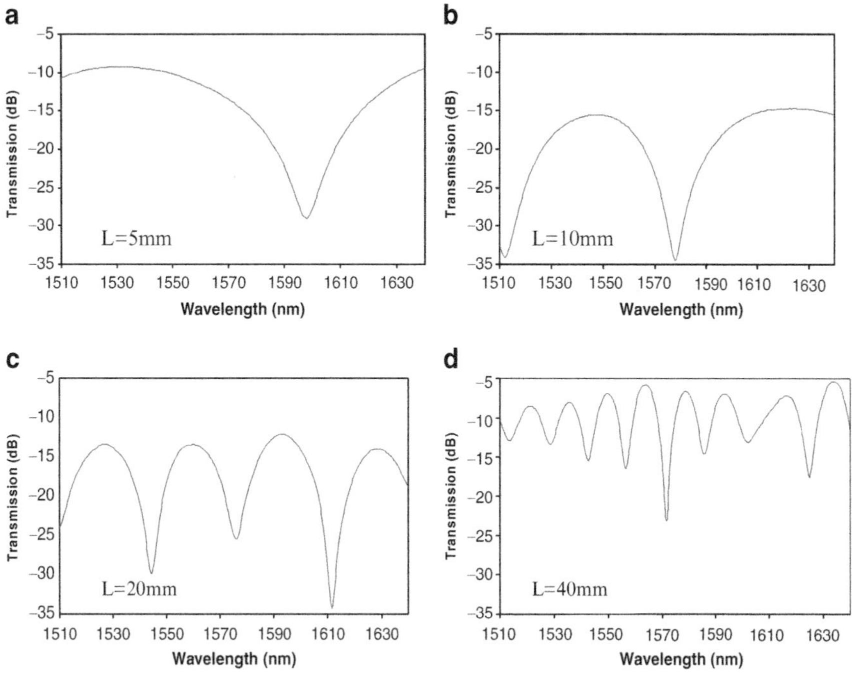

Fig. 7.23 Transmission spectra of fiber-inline core-cladding-mode MZIs with different interferometer lengths: (**a**) 5 mm; (**b**) 10 mm; (**c**) 20 mm; and (**d**) 40 mm

fringe had a visibility around 20 dB, which was sufficient for most sensing applications. The background loss varied among devices with a typical value of about 10 dB. With the increase of the interferometer length, the spectral separation between two adjacent interference fringes decreased in approximately an inversely proportional fashion.

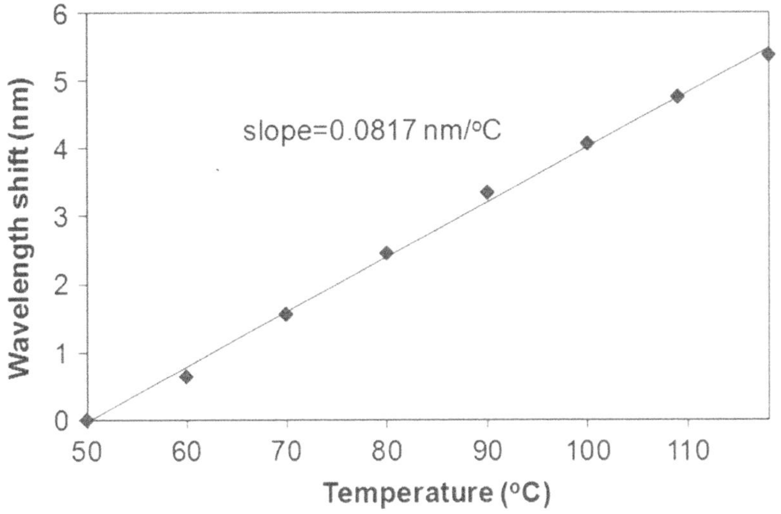

Fig. 7.24 Temperature response of a fiber-inline MZI fabricated by CO_2 laser irradiations

A fiber-inline MZI with a length of 5 mm was employed to evaluate its sensitivity toward temperature variation. The interferometer was placed in a quartz tube without stress, fixed in a tubular electric furnace (Lindberg/blue M). The device was first heated up to 120°C read from the built-in thermometer of the furnace. The furnace was then turned off to allow the temperature cool down while the interference spectra were recorded at around every 10°C of temperature dropping. The recorded interference spectra were processed to find the spectral position of the interference minima. The temperature response test ended when the temperature inside the furnace reached 50°C. Figure 7.24 shows the wavelength shift of the interference minimum as a function of the furnace temperature where the relation is roughly linear. The linear fit of the experiment data indicated that the temperature sensitivity of the device was 0.0817 nm $°C^{-1}$. The temperature sensitivity of the device can be attributed to the different thermo-optic dependence of the fiber core and cladding modes[32,33]. n_c^{eff} and n_{cl}^{eff} change differently when temperature varies, causing a phase shift of the interference fringe.

The same fiber-inline MZI used in temperature response test was evaluated for its capability for refractive index measurement at room temperature. Various liquids were used including deionized (DI) water, 2-propanol (IPA), and refractive index liquids ranging from 1.40 to 1.44. During the tests, the device was fixed on a glass plate to avoid any bending-induced signal change. Then the glass plate was immersed into different liquids and recoded the interference fringes for each test. The device was carefully cleaned by subsequent acetone and water ultrasonic baths and air dried between tests in different liquids to avoid any liquid residue from previous test.

Figure 7.25 shows the wavelength shift of the fiber-inline MZI with respect to the refractive index change. The increase of ambient refractive index caused a shift

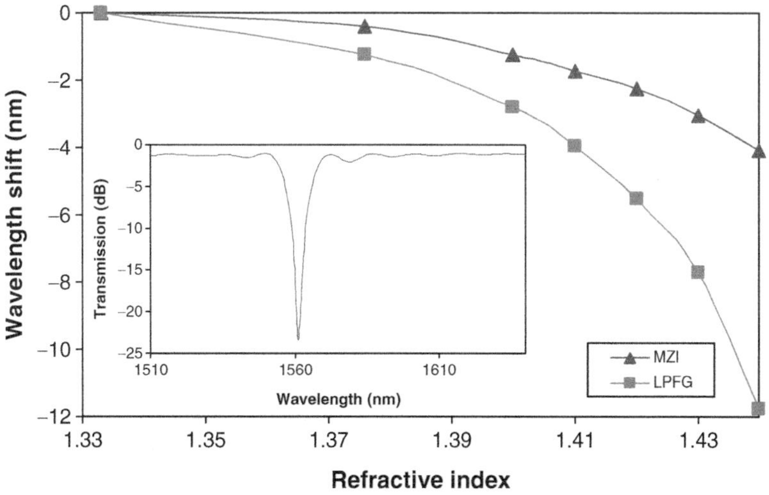

Fig. 7.25 Wavelength shift as a function of refractive index in MZI and LPFG with cladding mode of LP_{05} (*Inset*: LPFG transmission spectrum of LP_{05} mode)

of the interference spectrum toward the short wavelength. In addition, the sensitivity grew as the refractive index approached that of the cladding. This can be explained by the well-known waveguide theory[24]. The propagation constant (thus the effective index) of the cladding mode increases with the increase of refractive index of the surrounding medium. The propagation constant is also a nonlinear function of the surrounding medium. The change is more significant as the surrounding media has a refractive index approaching that of the fiber cladding.

The refractive index measurement sensitivity of the MZI-CCMI device was also compared with that of an LPFG. A LPFG fabricated by point-by-point CO_2 laser irradiations was used for comparison. The LPFG had a period of 525 μm, a length of 50 mm, and the resonance wavelength of 1,561 nm in air at room temperature. The transmission spectrum of the LPFG is shown in the insert of Fig. 7.25 and the resonance peak is resulted from light coupling from the core mode to the LP_{05} cladding mode. The shift in resonance wavelength of LPFG in response to the refractive index changes is shown in Fig. 7.25. The LPFG had a slightly higher sensitivity than the MZI.

The resonance wavelength (λ_{res}) of a LPFG can be roughly estimated by the following phase matching condition:

$$\lambda_{res} = \left(n_c^{eff} - n_{cl}^{eff}\right)\Lambda, \tag{7.11}$$

where Λ is the period of the grating. Accordingly, the relative effective refractive index change is proportional to the relative resonance wavelength, given by:

$$\frac{\Delta n_{\mathrm{cl}}^{\mathrm{eff}}}{\left(n_{\mathrm{c}}^{\mathrm{eff}} - n_{\mathrm{cl}}^{\mathrm{eff}}\right)} = -\frac{\Delta \lambda_{\mathrm{res}}}{\lambda_{\mathrm{res}}}. \tag{7.12}$$

Compare (7.11) and (7.12), one finds that a LPFG and an inline MZI operate based on the same mechanism that the propagation constant of the cladding mode is a function of the refractive index of the surrounding medium. The lower refractive index measurement sensitivity suggested that the $n_{\mathrm{cl}}^{\mathrm{eff}}$ of this MZI was higher than that of LP_{05}, indicating that the order of the cladding mode excited in this MZI was lower than that of the LP_{05}. It is worth noting that it is possible for the laser exposure to excite more than one cladding modes. However, the clean interference fringes of different interferometer lengths indicate that only one cladding mode dominates.

7.4 Conclusions

Miniaturized optical fiber interferometers have shown an increasing capability for accurate and reliable measurement of refractive index of the surrounding environment. Such a capability has made these interferometers good candidates for developing chemical and biological sensors. When placed in a liquid solution, these fiber sensors can detect the chemical composition change or interactions between biospecies by monitoring its refractive index variation. These interferometers can also be further coated with functional thin films to achieve improved detection sensitivity and selectivity. The two types of fiber-inline interferometers reviewed in this chapter, namely the FPI and CCMI, operate on different principles but share the similar advantages of small size, all-glass ruggedized structure, high sensitivity, fast response time, and large dynamic range. These advantages make the devices particularly attractive for real-world applications where, in situ, continuous monitoring is required.

The operating principles of the reviewed interferometers are well studied. However, by no means these devices are matured. For example, a mode-selective, wavelength-independent and environmental-resistant 3-dB core-cladding mode coupler is yet to be found to construct an ideal CCMI. As technology advances and research continues, we expect that more device structures will be explored and new methods will be investigated to fabricate these devices. Although the applications of these two types of sensors are yet to be explored, it is almost certain that they will find their way into real-world applications in the future.

References

1 Lee, C. E.; Taylor, H. F., Interferometric optical fiber sensors using internal mirrors, *Electron. Lett.* **1988**, 24, 193–194

2 Claus, R. O.; Gunther, M. F.; Wang, A.; Murphy, K. A., Extrinsic Fabry-Perot sensor for strain and crack opening displacement measurements for −200 to 900°C, *J. Smart Mater. Struct* **1992**, 1, 237–242

3 Xiao, H.; Deng, J.; Pickrell, G.; May, R. G.; Wang, A., Single-crystal sapphire fiber-based strain sensor for high-temperature applications, *J. Lightwave Technol.* **2003**, 21, 2276–2283
4 Wang, A.; Xiao, H.; Wang, J.; Wang, Z.; Zhao, W.; May, R. G., Self-calibrated interferometric-intensity-based optical fiber sensors, *J. Lightwave Technol.* **2001**, 19, 1495–1501
5 Hecht, E. Optics, 4th edn.; Addison Wesley, New York, NY, **2002**
6 Qi, B.; Pickrell, G. R.; Xu, J.; Zhang, P.; Duan, Y.; Peng, W.; Huang, Z.; Huo, W.; Xiao, H.; May, R. G.; Wang, A., Novel data processing techniques for dispersive white light interferometer, *Opt. Eng.* **2003**, 42, 3165–3171
7 Xiao, G. Z.; Adnet, A.; Zhang, Z. Y.; Sun, F. G.; Grover, C. P., Monitoring changes in the refractive index of gases by means of a fiber optic Fabry-Perot interferometer sensor, *Sens. Actuators A-Phys.* **2005**, 118, 177–182
8 Bhatia, V.; Murphy, K. A.; Claus, R. O.; Jones, M. E.; Grace, J. L.; Tran, T. A.; Greene, J. A., Optical fiber based absolute extrinsic Fabry-Perot interferometric sensing system, *Meas. Sci. Technol.* **1996**, 7, 58–61
9 Zhang, Y.; Chen, X.; Wang, Y.; Cooper, K. L.; Wang, A., Microgap multicavity Fabry-Pérot biosensor, *J. Lightwave Technol.* **2007**, 25, 1797–1804
10 Xiao, H.; Deng, J.; Wang, Z.; Huo, W.; Zhang, P.; Luo, M.; Pickroll, G. R.; May, R. G.; Wang, A., Fiber optic pressure sensor with self-compensation capability for harsh environment applications, *Opt. Eng.* **2005**, 44, 1–10
11 Zhang, Y.; Shibru, H.; Cooper, K. L.; Wang, A., Miniature fiber-optic multicavity Fabry-Perot interferometric biosensor, *Opt. Lett.* **2005**, 30, 1021–1023
12 Ran, Z. L.; Rao, Y. J.; Liu, W. J.; Liao, X.; Chiang, K. S., Laser-micromachined Fabry-Perot optical fiber tip sensor for high-resolution temperature-independent measurement of refractive index, *Opt. Express* **2008**, 16, 2252–2263
13 Li, M.; Menon, S.; Nibarger, J. P.; Gibson, G.N., Ultrafast electron dynamics in femtosecond optical breakdown of dielectrics, *Phys. Rev. Lett.* **1999**, 82, 2394–2397
14 Davis, K. M.; Miura, K.; Sugimoto, N; Hirao, K., Writing waveguides in glass with a femtosecond laser, *Opt. Lett.* **1996**, 21, 1729–1731
15 Szameit, A.; Bloemer, D.; Burghoff, J.; Pertsch, T.; Nolte, S.; Lederer, F.; Tuennermann, A., Hexagonal waveguide arrays written with fs-laser pulses, *Appl. Phys. B.* **2006**, 82, 507–512
16 Cheng, Y.; Tsai, H. L.; Sugioka, K.; Midorikawa, K., Fabrication of 3D microoptical lenses in photosensitive glass using femtosecond laser micromachining, *Appl. Phys. A.* **2006**, 85, 11–14
17 Sun, H.; He, F.; Zhou, Z.; Cheng, Y.; Xu, Z.; Sugioka, K.; Midorikawa, K., Fabrication of microfluidic optical waveguides on glass chips with femtosecond laser pulses, *Opt. Lett.* **2007**, 32, 1536–1538
18 Wei, T.; Han, Y.; Tsai, H. L.; Xiao, H., Miniaturized fiber inline Fabry-Perot interferometer fabricated with a femtosecond laser, *Opt. Lett.* **2008**, 33, 536–538
19 Rao, Y. J.; Deng, M.; Duan, D. W.; Yang, X. C.; Zhu, T.; Cheng, G. H., Micro Fabry-Perot interferometers in silica fibers machined by femtosecond laser, *Opt. Express* **2007**, 15, 14123–14128
20 Wei, T.; Han, Y.; Li, Y.; Tsai, H.; Xiao, H., Temperature-insensitive miniaturized fiber inline Fabry-Perot interferometer for highly sensitive refractive index measurement, *Opt. Express* **2008**, 16, 5764–5769
21 Hawkes, J. B.; Astherimer, R. W., Temperature coefficient of the refractive index of water, *J. Opt. Soc. Am.* **1948**, 38, 804–806
22 Liu, N.; Hui, J.; Sun, C.; Dong, J.; Zhang, L.; Xiao, H., Nanoporous zeolite thin film-based fiber intrinsic Fabry-Perot interferometric sensor for detection of dissolved organics in water, *Sensors* **2006**, 6, 835–847
23 Lee, B. H.; Nishii, J., Dependence of fringe spacing on the grating separation in a long-period fiber grating pair, *Appl. Opt.* **1999**, 38, 3450–3459

24 Allsop, T.; Reeves, R.; Webb, D. J.; Bennion, I.; Neal, R., A high sensitivity refractometer based upon a long period grating Mach-Zehnder interferometer, *Rev. Sci. Instrum.* **2002**, 73, 1702–1705
25 Lee, B. H.; Nishii, J., Bending sensitivity of in-series long-period fiber gratings, *Opt. Lett.* **1998**, 23, 1624–1626
26 Swart, P. L. Long-period grating Michelson refractometric sensor, *Meas. Sci. Technol.* **2004**, 15, 1576–1580
27 Tian, Z.; Yam, S. S.; Barnes, J.; Bock, W.; P. Greig; J. M. Fraser; H. P. Loock; R. D. Oleschuk, Refractive index sensing with Mach-Zehnder interferometer based on concatenating two single-mode fiber tapers, *IEEE Photon. Technol. Lett.* **2008**, 20, 626
28 Tian, Z.; Yam, S. S.; Loock, H., Refractive index sensor based on an abrupt taper Michelson interferometer in a single-mode fiber, *Opt. Lett.* **2008**, 33, 1105–1107
29 Villatoro, J.; Monzón-Hernández, D., Low-cost optical fiber refractive-index sensor based on core diameter mismatch, *J. Lightwave Technol.* **2006**, 24, 1409
30 Tian, Z.; Yam, S. S.; Loock, H. P., Single-mode fiber refractive index sensor based on core-offset attenuators, *IEEE Photon. Technol. Lett.* **2008**, 20, 1387–1389
31 Choi, H. Y.; Kim, M. J.; Lee, B. H., All-fiber Mach-Zehnder type interferometers formed in photonic crystal fiber, *Opt. Express* **2007**, 15, 5711–5720
32 Li, E.; Wang, X.; Zhang, C., Fiber-optic temperature sensor based on interference of selective higher-order modes, *Appl. Phys. Lett.* **2006**, 89, 091119
33 Kim, Y.; Paek, U.; Lee, B. H., Measurement of refractive-index variation with temperature by use of long-period fiber gratings, *Opt. Lett.* **2002**, 27, 1297–1299

Section II
Photonic Structures for Biochemical Sensing

Chapter 8
Label-Free Biochemical Sensors Based on Optical Microresonators

Chung-Yen Chao, Tao Ling, and L. Jay Guo

Abstract Biochemical sensors play a significant role in extensive applications that have tight relationship with human life. These sensors require high sensitivity and low detection limit. In this chapter, two optical sensors that meet the requirement will be discussed: polymer microring biochemical sensors and microtube resonator sensors. Both have advantages of high sensitivity, label-free detection capability, low cost, robustness, and simple fabrication process. The former devices show high sensitivity over 70 nm per RIU and low detection limit as 250 pg mm^{-2}, while the latter ones can push sensitivity to 600 nm per RIU. Moreover, polymer microring biochemical sensors are compact and integrable in an array on a substrate; while microtube-based sensors have built-in fluidic handling capability. These features facilitate the development of miniature and highly sensitive lab-on-a-chip sensors.

8.1 Introduction

Biological and chemical sensors have been widely and massively used everyday, and play a significant role in extensive applications related to human life. For example, a large amount of samples, like blood and urine, require fast and accurate testing in regular health examinations. Many patients need continuous real-time monitoring of certain critical analytes, such as blood glucose concentration for diabetic patients, in order to have timely therapeutic intervention. In addition, examinations at airports for pathogens, explosives, and lethal materials have become increasingly important to prevent terrorist attacks. Also, in situ monitoring is necessary in hospitals, manufacturing plants, and laboratories for human safety. Other applications include environmental monitoring, agricultural safety, medical diagnostics, screening of

L. Jay Guo (✉)
Department of Electrical Engineering and Computer Science, University of Michigan, Ann Arbor, MI 48109, USA
e-mail: guo@eecs.umich.edu

X. Fan (ed.), *Advanced Photonic Structures for Biological and Chemical Detection*, Integrated Analytical Systems,
DOI 10.1007/978-0-387-98063-8_8,

chemical compounds in drug discovery, and industrial process control. These aspects stimulate the demand and development of biochemical sensors. Apparently, the design of a sensor will depend upon the requirement of the particular application. Nevertheless, sensors used in various applications have similar ideal requirement. Generally speaking, an ideal biochemical sensor should confer several properties. In the aspect of device performance, it should have *high sensitivity, broad dynamic range, low detection limit, immunity to sample-matrix interferences, good reliability, and fast and reversible response*. In other aspects, *robustness, easy portability, low cost, and simple fabrication processes* are also highly desirable.

Modern biological and chemical sensors can be mainly divided into four categories based on their transduction mechanisms: electrical, thermal, mechanical, and optical transduction. Optical sensors render a number of attractive advantages over the other types[1–3]. One of the major advantages is their capability to probe surfaces in a nondestructive manner, where many important biological recognition processes take place. With a superior signal-to-noise ratio (SNR), they offer high accuracy and sensitivity. Additionally, they are robust and immune to the interference from electromagnetic fields. Optical sensors also provide fast response to permit in situ sensing and real-time measurements. Other promising features include miniaturization, inherent safety, simplicity, remote sensing capability, and multianalyte sensing capability. Optical sensors exploit various types of transducing properties such as light absorption, fluorescence, phosphorescence, bio-/chemiluminescence, reflectance, Raman scattering, effective refractive index, and phase.

Among the existing optical biochemical sensors, sensors based on evanescent wave for interrogation have been demonstrated to possess promising performance. These include surface plasmon resonance sensors[4,5], optical-waveguide sensors[6], Mach–Zehnder interferometric sensors[7,8], and grating-coupled waveguide sensors[9–11]. These devices depend upon the interaction between evanescent wave and analytes adsorbed on sensor surfaces or in the surrounding medium. Existence of analytes around a sensor changes the refractive index of the medium surrounding the device or modifies the surface structure and, in turn, its optical properties, like the effective refractive index within a waveguide. These changes can be converted into the variation of detectable parameters, such as reflection angle, optical intensity, and resonant wavelength. Such a sensing scheme eliminates the need of fluorescently labeling analyte species and facilitates label-free detection. Especially since effective index sensing does not rely on light absorption, a wide range of light sources can be chosen.

Recently, microresonators have been proposed in sensing applications[12,13]. They have a unique advantage of reducing device size (typically in the range of centimeters for straight-waveguide sensors to accumulate a detectable phase shift and to obtain a reliable SNR) by several orders of magnitude without sacrificing interaction length by virtue of high-quality factor (Q-factor) resonances. The resonance effect can greatly reduce the amount of analytes needed for detection and also provides an effective method to create equivalently long interaction length to achieve sufficient sensitivity. For example, microtoroid resonators possessing ultrahigh Q-factor over 10^8 have been demonstrated to label-freely detect single molecule of interleukin-2 in serum[14]. Microsphere resonators using whispering gallery mode (WGM) resonance have been reported to respond to minute amount of protein adsorption[15–17]. In the area of integrated optical resonators, microdisk resonators have been experimentally illustrated to render sensitivity enhancement and analyte reduction by at least one order of magnitude in comparison with straight-waveguide sensors[18]. They can also measure change in refractive index of 10^{-4} RIU (refractive index unit[19]) and detect volatile organic chemicals[20]. Microring resonators have been illustrated to distinguish minute concentration changes of biotinylated lectin with detection limit of 1.8×10^{-5} RIU and sensitivity of 150 nm per RIU[21]. In principle, large Q-factor leads to very high sensitivity and capability of detecting ultralow biochemical concentration.

In this chapter, two microresonator sensors will be discussed: nanoimprinted polymer microring sensors and prism-coupled microtube resonator sensors. Both sensing devices possess the advantages of high sensitivity, label-free detection capability, low manufacturing cost, robustness, and simple fabrication process. At first, design consideration and sensing theory using nanoimprinted polymer microring sensors will be described, followed by fabrication sequence and optimization approaches in design as well as manufacture. Subsequently, experimental demonstration will be shown to illustrate sensing performance, and two sensitivity enhancing methods in device geometry design will be discussed. Next, sensing theory with nonevanescent resonance mode and index-sensing experimental results using prism-coupled microtube resonator sensors will be presented, followed by the demonstration of detecting lipid membrane and the binding of membrane proteins. Finally, conclusions will be provided.

8.2 Nanoimprinted Polymer Microring Biochemical Sensors

Optical microresonators can be mainly created in several forms: microtoroid, microsphere, microdisk (including micropolygon, like square, hexagon, and so on), and microring (including microracetrack). They all render a number of attractive properties including high sensitivity, high specificity (label-free detection) using surface chemical modifications, compactness for use outside laboratories, and multichannel (multianalyte) sensing capability. However, each of them has its own drawbacks.

Microtoroid and microsphere resonators have the highest sensitivity due to their ultrahigh Q-factor. However, the device operation typically requires free-standing tapered fibers to in-couple and out-couple the light. Tapered fibers are very thin and brittle, which makes them lack robustness and integration capability and limits their use in practical applications. For microspheres, analytes are adsorbed over whole sphere surface, while most of them do not contribute to the WGM resonance shift. Therefore, a large amount of analytes are wasted. In addition to the obstacle in photonic integration, both of them have difficulties in microfluidic integration and mass production[22]. To solve these problems, microdisk or microring resonators can be used, albeit with reduced Q-factor and hence detection limit. These two types of resonators can be easily fabricated with current mature microfabrication technology. Very compact resonators can be made with on-chip coupled waveguides serving as light input and output. Although their Q-factor is lowered, they possess other advantages such as robustness, compatibility with microfluidic handling, and high integration capability with electronic and photonic devices. In the demonstration, microring resonators are focused.

8.2.1 Sensing Mechanisms and Transducing Methods

Integrated microring resonators typically have device size below 100 μm depending on the employed material systems. The materials could be semiconductors[23,24], silicon-on-insulators[25–27], dielectrics[28–30], and polymers[31,32]. The state-of-the-art devices show the radius as small as 1.5 μm[33]. This feature of compactness is promising for portable or even implantable sensing systems. It also enables integrated sensor arrays, and facilitates lab-on-a-chip with the aid of microfluidic integration. Furthermore, these microfabricated optical microresonators are as robust as integrated circuit chips and laser diodes made using the same technology.

A microring, consisting of a closed-loop optical waveguide, forms a resonant cavity as shown in Fig. 8.1a. Resonance occurs when the ring circumference is equal to an integer multiple of the light wavelength inside the waveguide. This leads to the expression of the resonant wavelength, λ_p, as

$$\lambda_p = n_{eff}L/m, \tag{8.1}$$

where n_{eff}, L, and m are the effective refractive index of the light in the microring waveguide, the microring circumference, and the resonance order (a positive integer), respectively. Light is coupled into and out of a microring using one or

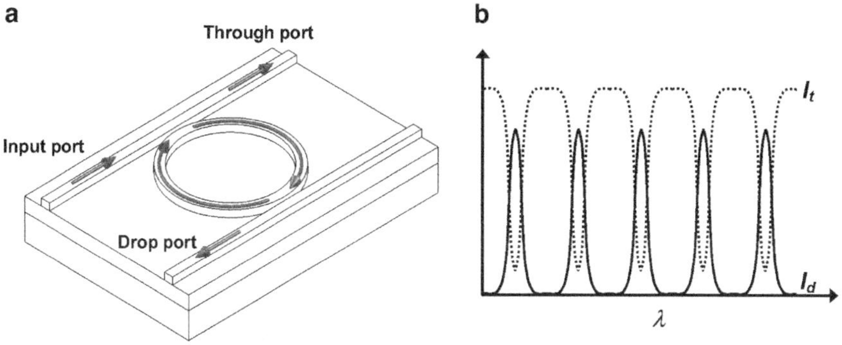

Fig. 8.1 (**a**) A microring resonator, closely coupled to two bus waveguides, has two output ports: through port and drop port. (**b**) Typical output spectra have periodic dips in the through port (*dash*) and peaks in the drop port (*solid*), respectively, resulting from resonance of the microring resonator

two bus waveguides. In the case of two bus waveguides, there are two output ports: through port and drop port as shown in Fig. 8.1a. At resonance condition, light propagates to the drop port via the microring cavity. This property gives rise to periodic dips in the spectrum of the through port and peaks in that of the drop port. Typical spectra in both output ports are plotted in Fig. 8.1b.

Microring resonators in biochemical sensing utilize evanescent field of the light guided by the ring waveguides to interrogate analytes. There are two sensing mechanisms typically employed: homogeneous sensing and surface sensing. In homogeneous sensing, a microring resonator is placed in a liquid or gaseous medium, which functions as the top cladding of the microring waveguide. When biochemical molecules are introduced in the medium as shown in Fig. 8.2a, the bulk refractive index of the medium is altered. This behavior changes the effective refractive index of the ring waveguide and in turn shifts the resonant wavelength according to (8.1). The amount of shift is closely related to the concentration of biochemical molecules. In this scheme, all biochemical molecules within the medium contribute to the shift of the resonant wavelength. Thus, this method cannot be used to distinguish different analyte species.

The other mechanism, surface sensing, requires creation of functional sites on microring surfaces. These sites allow binding of specific types of biomolecules as illustrated in Fig. 8.2b. For example, surface-immobilized antigens can be used to capture a specific kind of antibodies. After binding, other kinds of molecules that have little affinity to the binding sites are removed by a rinsing process. Addition of surface bound molecules modifies the waveguide structure and again shifts the resonant wavelength. This strategy enables specificity of biochemical sensing. In order to acquire the quantity or concentration of analytes, related parameters that can be physically measured should be obtained. As mentioned previously, existence of analytes modifies the waveguide structure of a detecting microring resonator and changes its optical properties. Two transducing methods

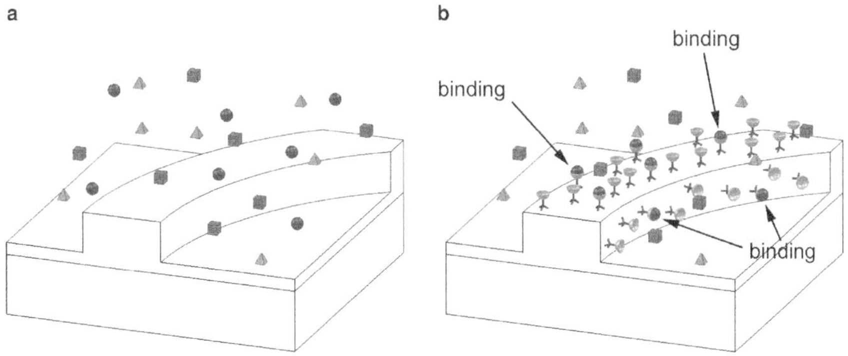

Fig. 8.2 Sensing mechanisms: (**a**) Homogeneous sensing – multiple species of biochemical molecules (three types in the figure) in the medium together alter the waveguide cladding of the microring. (**b**) Surface sensing – preimmobilized sites on the microring can bind specific analyte (*spheres* in the figure). Other types of molecules will be removed through rinsing

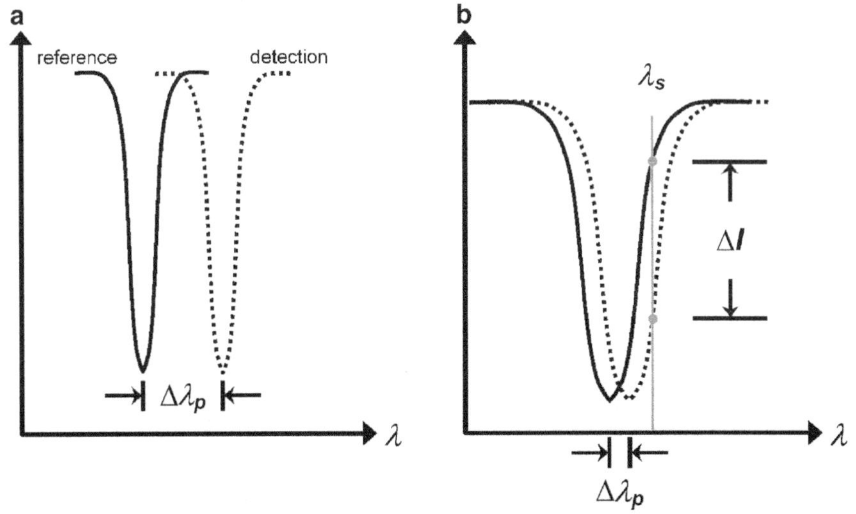

Fig. 8.3 Transducing methods: (**a**) Resonant wavelength shift scheme – monitor $\Delta\lambda_p$ between the reference resonance (*solid*) and the resonance under sensing (*dash*) and (**b**) intensity variation scheme – monitor ΔI at a selected wavelength

can be adopted in microring sensing. The first approach is to directly monitor the shift of the resonant wavelength, denoted by $\Delta\lambda_p$ in Fig. 8.3a. The magnitude of this shift depends upon the amount of analytes in either homogeneous sensing or surface sensing. The resonant wavelength shift scheme requires measurement of spectrum, and thus needs a more sophisticated system. However, this method can provide wide detection range or dynamic range as long as the same resonance is tracked. Detection limit, defined as the amount of analytes that induces the smallest detectable change of the microring property, is determined by the sharpness of resonance and the resolution of spectrum measurement. Intuitively, smaller change of the resonant wavelength can be resolved with sharper resonance (higher Q-factor).

The second transducing approach is the intensity variation scheme. It requires a single-wavelength light source with the wavelength, λ_s, at the transition of a selected resonance as shown in Fig. 8.3b. As analytes shift the resonance, output power at λ_s is also changed. This method only needs a laser with proper wavelength and a photodetector to monitor the intensity variation, which enables more compact and simpler system. In addition, it allows faster detection and smaller detection limit with sufficiently steep resonance. Detection limit of this approach is determined by the resonance slope at λ_s and the resolution of the photodetector. Steeper slope gives rise to larger intensity variation and thus higher sensitivity. However, detection fails when λ_s falls outside the linear range of the resonance, leading to nonlinear or zero-intensity variation. This feature results in narrow dynamic range. Dynamic range can be increased by widening resonant bandwidth (lowering Q-factor). But this also incurs reduction of slope. Hence, there is a trade-off between wide dynamic range and low detection limit. Accordingly, the resonant wavelength shift scheme should be chosen when wide dynamic range is desired, while the intensity variation scheme is selected to achieve low detection limit, e.g., for detecting minute amount of analytes.

8.2.2 Sensitivity Calculation

To analyze sensing capability of microring resonators, two microring waveguide structures are considered as shown in Fig. 8.4 for homogeneous sensing and surface sensing, respectively. Several assumptions are employed in both structures to simplify the analysis. Refractive index of the top cladding medium is assumed uniform in both cases, while the adsorbed analyte layer is assumed uniform in surface sensing cases. However, there might be gradient transition of concentration close to waveguide surfaces in reality, which results in nonuniform distribution of refractive index. This is also true in surface sensing cases. Additionally, an actual adsorbed layer is not a uniform film when binding sites are not fully occupied. In order to evaluate and compare device performance, sensitivity is calculated, which is defined as the ratio of the change in transducing optical parameters ($\Delta\lambda_p$ in the resonant wavelength shift scheme and ΔI in the intensity variation scheme) to the change in waveguide parameters affected by analytes (top cladding index n_c in homogeneous sensing and adsorbed film thickness t in surface sensing). According to the definition, the sensitivity S for four different combinations is listed in Table 8.1. From the mathematical expression, each sensitivity can be further divided into two parts, which are named as device sensitivity and waveguide sensitivity, respectively. Device sensitivity is defined as the ratio of the change in the transducing optical parameter to the effective index change, while waveguide

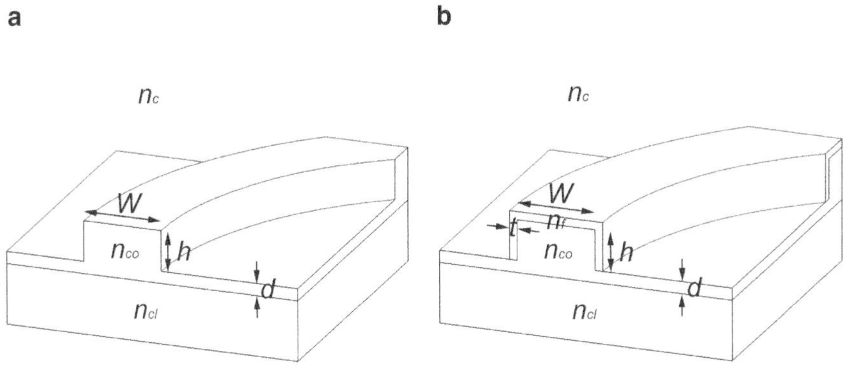

Fig. 8.4 Waveguide structure of microring sensors using (**a**) homogeneous sensing and (**b**) surface sensing: n_{co}, n_{cl}, and n_c are refractive indices of the core, the bottom cladding layer, and the medium carrying analytes for sensing (also the top cladding). W, h, and $(h + d)$ are the ridge width, the ridge height, and the height of the core layer. In surface sensing cases, an adsorbed analyte layer is formed with the refractive index of n_f and an effective thickness of t

Table 8.1 Overall sensitivity S of a microring sensor is listed based on two sensing mechanisms and two transducing schemes

Sensing mechanism	Transducing method	
	Resonant wavelength shift	Intensity variation
Homogeneous	$S \equiv \frac{\partial \lambda_p}{\partial n_c} = \frac{\partial \lambda_p}{\partial n_{eff}} \times \frac{\partial n_{eff}}{\partial n_c}$	$S \equiv \frac{\partial I}{\partial n_c} = \frac{\partial I}{\partial n_{eff}} \times \frac{\partial n_{eff}}{\partial n_c}$
Surface	$S \equiv \frac{\partial \lambda_p}{\partial t} = \frac{\partial \lambda_p}{\partial n_{eff}} \times \frac{\partial n_{eff}}{\partial t}$	$S \equiv \frac{\partial I}{\partial t} = \frac{\partial I}{\partial n_{eff}} \times \frac{\partial n_{eff}}{\partial t}$

The sensitivity in each of four cases is expressed as multiplication of device sensitivity and waveguide sensitivity

sensitivity is the ratio of the effective index change to the change in the waveguide parameter affected by analytes. Device sensitivity only depends on device properties, and waveguide sensitivity is relevant to waveguide structures regardless of the type of devices. These two parts can be enhanced separately to achieve higher overall sensitivity.

Based on the definition, device sensitivity is related to the variation of transducing optical parameters, and thus depends on the transducing method. For the resonant wavelength shift scheme, it is expressed as

$$(S_d)_{RWS} \equiv \partial \lambda_p / \partial n_{eff} = \lambda_p / n_{eff} = L/m \tag{8.2}$$

with use of (8.1), while for the intensity variation scheme, it becomes

$$(S_d)_{IV} \equiv \partial I / \partial n_{eff} = \lambda_p s / n_{eff} = Ls/m, \tag{8.3}$$

where s is the spectrum slope at λ_s. From (8.2) to (8.3), device sensitivity will be enhanced by using longer resonant wavelength or smaller effective refractive index for either case. In the intensity variation scheme, it can be further improved by increasing the spectrum slope, i.e., sharper resonance.

Waveguide sensitivity is determined by the effective index change resulting from the analyte-induced change in waveguide parameters. It varies with the sensing mechanism used, and is expressed as $\partial n_{eff}/\partial n_c$ for homogeneous sensing and $\partial n_{eff}/\partial t$ for surface sensing. Waveguide sensitivity can be calculated using the effective index method[34]. Although this approach is not as accurate as other methods such as the finite-difference time-domain (FDTD), it can provide straightforward calculation and quick analysis. To achieve more accurate results, curvature of microring waveguides should be included. It is shown that results obtained from the effective index method provide sufficient accuracy as compared with those using FDTD in most practical cases except for very small microrings.[35]

In most biochemical sensors, detection limit is a very important parameter to evaluate sensing capability of a device, which denotes the smallest detectable amount or concentration of analytes. For microring sensors, detection limit is proportional to the variation of waveguide parameters affected by occurrence of analytes: the index change of the surrounding medium, δn_c, in homogeneous sensing, or the thickness change of the adsorbed analyte film, δt, in surface sensing. This variation is called optical detection limit (in the unit of RIU or picometer) for simplicity to distinguish from detection limit. Measurable variation is determined by the resolution of optical transducing methods. According to the definitions of sensitivity in Table 8.1, optical detection limit is equal to $\delta\lambda_p/S$ or $\delta I/S$, where $\delta\lambda_p$ and δI are the resolution of spectrum measurement and photodetectors, respectively.

Theoretically, $\delta\lambda_p$ in the resonant wavelength shift scheme is independent of resonance shape or resonant bandwidth, and should be determined merely by instrument resolution, typically less than 10 pm. However, in reality, noise can perturb resonance spectra such that accurate determination of resonant wavelength shift becomes difficult for a broad resonance curve. To enhance accuracy in detecting wavelength shift, narrower resonance is required. This is equivalent to obtaining higher-Q resonance behavior. To take into account noise-included detectability of $\delta\lambda_p$, $\delta\lambda_p$ can be simply described as a fraction (ρ) of the full width at half maximum (FWHM) bandwidth of resonance, $\Delta\lambda_{FWHM}$. In this fashion, optical detection limit becomes $\rho\Delta\lambda_{FWHM}/S$ or $\rho\lambda_p/(QS)$. In practice, ρ can be chosen as a reasonable value of 0.1. In the intensity variation scheme, δI is determined by noise from environment and photodetectors. It can reach as low as several nanowatts with care.

8.2.3 Design Consideration

In sensing applications using microring resonators, one of the main objectives is to achieve low detection limit and high sensitivity. From the perspective of microring resonators, resonance properties have to be unambiguously detected. This can be achieved by pursuing high contrast ratio between ON and OFF resonance to mitigate noise disturbance, thereby increasing SNR and easily distinguishing the resonance. Higher contrast ratio can also greatly enhance sensitivity using the intensity variation scheme due to steeper spectrum slope. The highest contrast ratio is achieved by operating a microring resonator at its critically coupled point, at which the energy coupled into the resonator is balanced by the energy loss in the resonator. Another way is to reduce resonant bandwidth or to enhance Q-factor. Under the same contrast ratio, this can improve the spectrum slope used in the intensity variation scheme and lower the optical detection limit in the resonant wavelength shift scheme, which is inversely proportional to Q-factor.

The second objective is to pursue large detection range or wide dynamic range. In the intensity variation scheme, detection range is limited due to narrow resonance. It can be slightly increased at the expense of reducing spectrum slope and thus sensitivity. In the resonant wavelength shift scheme, broad detection range is achievable by tracking the specific resonance[15] if the acquisition time is faster than the rate of analyte binding. On the contrary, detection may fail when the adjacent resonance peaks confuses the resonance tracking. Therefore, it is desirable to have a wider free spectrum range (FSR), which is defined as the difference between two adjacent resonant wavelengths for a waveguide mode, and can be calculated as:

$$\mathrm{FSR} = \lambda_\mathrm{p}^2/L[n_\mathrm{eff} - \lambda_\mathrm{p} \times (\partial n_\mathrm{eff}/\partial\lambda_\mathrm{p})] = \lambda_\mathrm{p}^2/Ln_\mathrm{g}, \tag{8.4}$$

where n_g is the group velocity related to n_eff and the dispersion of the waveguide mode close to the resonance, $\partial n_\mathrm{eff}/\partial\lambda_\mathrm{p}$. From (8.4), FSR is inversely proportional to ring size. Sufficiently large FSR sets the upper limit of the radius of microrings. Furthermore, one should use single-mode microring waveguides. Otherwise the presence of higher-order modes makes detection difficult to accomplish because each mode has a different effective index and create its own periodic resonance with different FSR. This criterion sets the upper limit of waveguide width and height.

8.2.4 Why Polymer Microring Biochemical Sensors?

As previously mentioned in Sect. 8.2.1, integrated microring resonators can be built using various types of materials. The first is III–V compound semiconductors, which are widely used in optoelectronic devices. They can provide gain and modulation to optical signals propagating within the microresonators by optical or electrical injection. This feature enables versatile device functionalities. In addition, their high refractive index can help to reduce bending loss within the microresonators such that very compact devices are feasible. However, these materials are comparably expensive and require costly equipments, like molecular beam epitaxy and metal-organic chemical vapor deposition, for growth. The growth should be performed on lattice-matched substrates to avoid stress and cracking. Although their high refractive indices can make devices more compact, this feature also poses fabrication challenges to form single-mode waveguides (typically submicron size) with small sidewall roughness, which is complicated and time consuming. Additionally, large index contrast between semiconductor core and gaseous or aqueous cladding could cause more severe scattering than low index contrast waveguides. Another semiconductor system is silicon-on-insulator, which has been used to form very small silicon microresonators. This system possesses a great advantage of integration capability with electronic devices and circuits. But silicon lacks gain, and has most of the previously mentioned issues as compound semiconductors.

The other platform is dielectrics, for example, silicon dioxide, silicon nitride, silicon oxynitride, tantalum pentoxide, and titanium dioxide. They can be deposited by various methods, such as plasma-enhanced chemical vapor deposition, thermal evaporation, electron-beam evaporation, and sputtering. There are a number of dielectrics with refractive indices ranging from 1.45 to 2.4, facilitating diverse waveguide designs to satisfy different specification. Dielectrics have two other advantages: low temperature-dependent refractive index and substrate compatibility. The former property is beneficial to sensing accuracy during measurement of a minute amount of analytes, while the latter one enables device fabrication on different substrates. Furthermore, their lower refractive indices enable larger waveguide dimension. This not only relieves fabrication challenge, but also provides better waveguide modal match to optical fibers than do semiconductors, leading to higher coupling efficiency in fiber-coupled biochemical devices. However, the fabrication process is still quite involved and time consuming. Moreover, dielectrics can only form passive devices due to their high electrical insulation.

Polymers can also be employed to construct microring sensors with several advantages over the previous two material systems. First, polymers have refractive indices mostly in the range between 1.3 and 1.8. Although low index is not beneficial to ultrasmall devices, it can relieve fabrication challenge, lower surface roughness scattering, and enhance coupling efficiency to optical fibers. Secondly, polymers can be applied to any substrates by simple process, such as spin coating. This facilitates integration with other electronic and photonic devices. Also, cost can be reduced with cheaper materials and without expensive equipment. This helps development of disposable sensors. With recent rapid progress in polymer technology, a variety of polymers can be synthesized to render specific properties, including special surface functionalities, biocompatibility, electrical conduction, optical nonlinearity, and optical amplification. Diverse polymers have already been adopted in sensing applications[36]. The drawback of polymer materials is relatively high temperature-dependent refractive index coefficient, making devices more sensitive to ambient temperature variations. This can be avoided by either using active temperature control or designing sensors with a reference arm[21]. Polymers also have lower thermal stability than semiconductors and inorganic dielectrics. Elevated temperature can cause modification of physical dimension and material properties. Fortunately in most biosensing applications, detection is performed at room temperature. To apply in high temperature detection, polymers with high glass transition temperatures or cross-linked polymers should be used. With the advantages overweighing the drawbacks, polymers are selected in the development of microring sensors.

8.2.5 *Fabrication Sequence*

In the design of polymer microring resonators, polymer waveguides should have height and width in the range of 1.8–2.5 μm for single-mode operation, good optical confinement, and lower waveguide loss in the wavelength range close to 1.55 μm. In addition, a microring laterally coupled to bus waveguides requires sufficient cross-coupling of light into the cavity to produce the desired device characteristics. This can be achieved by using either a small gap separation or a longer coupling length according to the coupled-mode theory[37]. The former approach is chosen to obtain more compact devices. Based on the simulation, a suitable gap has a

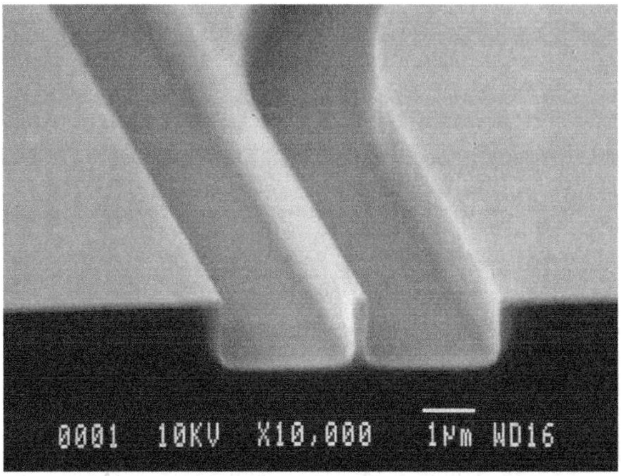

Fig. 8.5 A scanning electron micrograph of a silicon dioxide mold with reverse patterns: the left and right trenches create a bus and a microring waveguides, respectively. The high aspect ratio protruded wall between trenches is used to stamp out a narrow coupling gap. Reprinted from Ref. 42 with permission. © 2008 American Vacuum Society

dimension around 200–400 nm without sacrificing device size. This means that a coupling gap with an aspect ratio larger than five needs to be created. Such polymer microresonators can be easily fabricated by use of nanoimprint lithography[38]. It utilizes a silicon or silicon dioxide mold to directly imprint and transfer patterns onto a thin polymer layer. This approach provides several advantages, such as sub-10 nm patterning capability, precise dimension control, high fidelity, and high throughput. Moreover, the processing is simple, time saving, and cost effective[39]. The most important benefit is the ability to easily create high aspect ratio polymer structures in nanoscale.[40]

To fabricate polymer microring waveguides by nanoimprint lithography, a mold with reverse patterns of microring and bus waveguides is first fabricated using electron beam lithography and reactive ion etching (RIE) as shown in Fig. 8.5. This mold is subsequently brought into close contact to a silicon sample with thick thermally grown silicon dioxide that is covered by a spin-coated polymer on top. Silicon dioxide serves as bottom cladding. Pattern transfer from the mold to the polymer is performed under suitable heating temperature and pressure. With proper heating, the polymer becomes more fluidic and the applied pressure can help it propagate to fill rugged spacing in between. After they are cooled down, the sample is separated from the mold. It should be noted that molds need to be coated with a monolayer of surfactant to facilitate releasing[41]. After the nanoimprint process, a thin residual polymer layer is left and has to be removed with oxygen RIE. Finally, pedestal structures beneath polymer cores are created using buffered hydrofluoric acid. Figure 8.6 shows scanning electron micrographs of a nanoimprinted microring resonator. The step-by-step process schematic is shown in Fig. 8.7. Nanoimprint technique can be utilized to duplicate microresonators in a variety of polymers, such as poly(methylmethacrylate), polystyrene, and polycarbonate[42]. In the

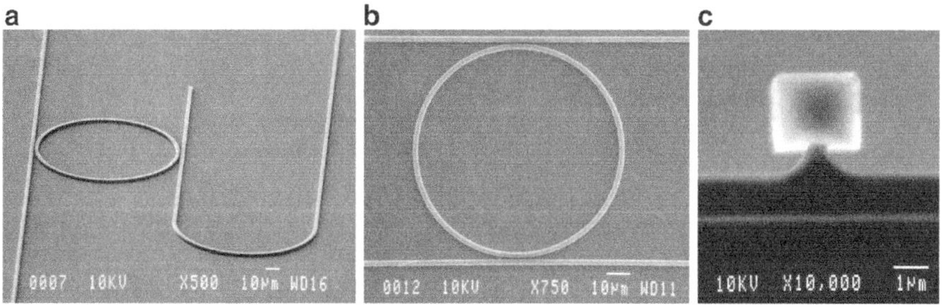

Fig. 8.6 Scanning electron micrographs of a polystyrene microring resonator closely coupled to two bus waveguides (Reproduced with permission from Ref. [43] Copyright 2006, IEEE), and its waveguide cross section showing a rectangular polymer core sitting on top of a pedestal structure. Reprinted from Ref. 42 with permission © 2008 American Vacuum Society

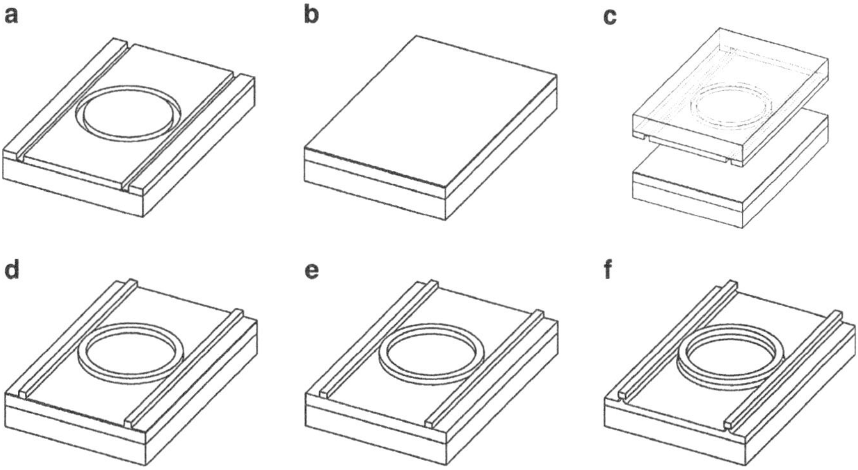

Fig. 8.7 Fabrication sequence of a polymer microring resonator: (**a**) prepare a nanoimprint mold; (**b**) spin coat a polymer thin film; (**c**) perform nanoimprinting process; (**d**) separate the sample from the mold; (**e**) dry etch the residual layer; (**f**) create pedestals by wet etch

biochemical sensing experiments, polystyrene is chosen due to its appropriate optical and mechanical properties. Figure 8.8 shows typical spectra measured at the through and drop ports of a microring resonator.

8.2.6 Optimization

In the design of a microring resonator, there are two important parameters: waveguide width (W) and microring radius (R_c). Waveguide width determines transverse modal distribution. On one hand, a narrower waveguide can lead to stronger evanescent field penetrating deeper into the surrounding medium, and provides

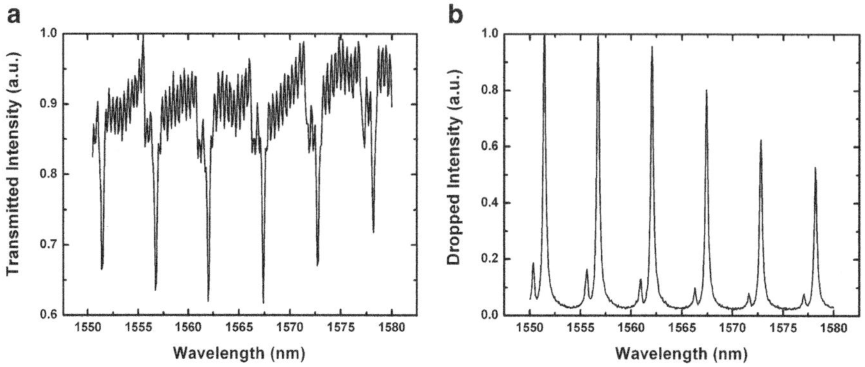

Fig. 8.8 Measured spectra of a polystyrene microring immersed in de-ionized (DI) water are plotted for (**a**) the through port and (**b**) the drop port. The drop port shows a much smoother curve. Reprinted from Ref. 43 with permission. © 2008 Institute of Electrical and Electronics Engineers

more vigorous interaction with analytes, in turn rendering higher sensitivity potentially. On the other hand, stronger field at interfaces increases sidewall roughness scattering loss, which reduces Q-factor and then lowers sensitivity. The other parameter, microring radius, determines resonator cavity length and curvature of microring waveguides. Larger radius results in less bending loss and higher Q-factor assuming fixed cavity loss. However, this also increases cavity length, and thus more loss is accumulated, leading to lower Q-factor. Hence, it is of significance to find the optimal values for the best sensing performance. Optimization can be divided into design optimization and fabrication improvement.

8.2.6.1 Design Optimization

In the optimization, a polystyrene microring sensor using the resonant wavelength shift scheme is considered as an example. This microring sits on a silicon substrate with a layer of 3.66-μm-thick thermally grown SiO_2 on top. The microring is a ridge waveguide with a polystyrene core of 1.8 μm in thickness ($h = 1.8$ μm, and $d = 0$ μm in Fig. 8.4a) operated at TE polarization. In sensing applications, detecting devices are typically immersed in aqueous environments. Thus, water is used as the medium and also as the top cladding. The indices of water, polystyrene, SiO_2, and Si near 1,550 nm are 1.3109 (n_c), 1.57 (n_{co}), 1.444 (n_{cl}), and 3.42, respectively. To simplify calculation, coupling loss in the coupling region(s) is neglected. Optimization requires calculation of coupling efficiency at the coupling region between the bus and microring waveguides, which is obtained using the coupled-mode theory[37]. The electric-field attenuation factor (due to propagation loss) per round trip of the microring (denoted by a) should also be obtained. Propagation loss within a microring includes surface roughness scattering, bending loss, leakage loss to silicon substrates, and modal mismatch loss (exclusively in microracetrack and

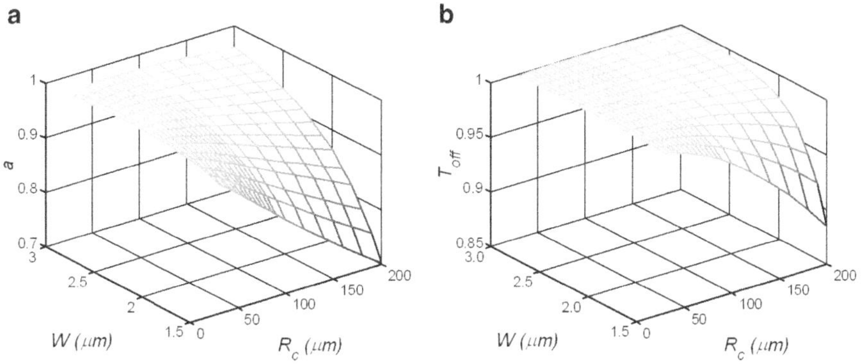

Fig. 8.9 (**a**) Round trip amplitude attenuation factor and (**b**) OFF-resonance transmission of the singly coupled microring as a function of W and R_c. Reprinted from Ref. 45 with permission. © 2008 Institute of Electrical and Electronics Engineers

ignored here for ring cavities), which are separately estimated[44]. The round trip amplitude attenuation factor is plotted in Fig. 8.9a as a function of W and R_c. As mentioned in Sect. 8.2.3, single-mode waveguide and sufficiently larger FSR are desired for better detection. To satisfy these criteria, the simulation is performed within a window of W ranging from 1.5 to 3 μm and R_c ranging from 30 to 200 μm.

A singly coupled microring operating at the critical coupling point is considered first. In this circumstance, the intensity at the through port is zero theoretically, and the highest contrast ratio is achieved. Using the critical-coupling condition, OFF-resonance transmission (T_{off}) can be calculated as shown in Fig. 8.9b. The figure reveals that narrower W and larger R_c lead to smaller OFF-resonance transmission due to increase of loss inside the microring. Even though the OFF-resonance transmission drops, the contrast ratio still can be maintained over 10 for practical applications.

In sensing applications, detection limit, which is proportional to $\Delta\lambda_{\text{FWHM}}/S$ (defined as the detection factor, P) or $1/(Q \times S)$, should be minimized, which is valid for both transducing schemes. By calculating $\Delta\lambda_{\text{FWHM}}$ from the transmission expression and S based on the aforementioned sensing theory, inverse of the detection factors, $1/P$, are plotted in Fig. 8.10 for homogeneous and surface sensing. These figures illustrate that lower detection limit, i.e., smaller P, occurs at wider waveguide width and larger radius of microrings. This behavior can be understood by separately analyzing the dependence of S and Q on W and R_c.

The overall sensitivity and the Q-factor in this case are plotted in Fig. 8.11. The figures illustrate that sensitivity has little dependence on R_c because the waveguide sensitivity is independent of R_c and the device sensitivity in (8.2) has a weak dependence on R_c. In addition, S decreases with increase of W due to shorter evanescent tail penetrating into the medium. Figure 8.11 also shows that larger Q occurs at larger W and R_c. According to another definition of Q-factor – the ratio of the energy stored in a resonator to the energy loss per cycle, higher Q implies lower energy loss and smaller $\Delta\lambda_{\text{FWHM}}$. Higher Q-factor is obtained with large R_c due to

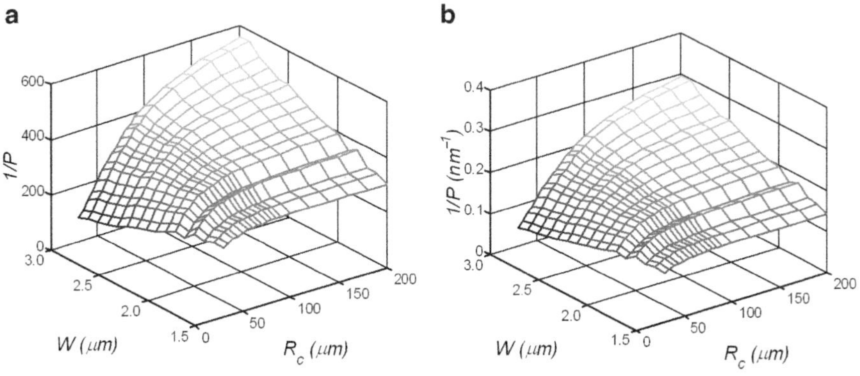

Fig. 8.10 Inverse of the detection factors are plotted as a function of W and R_c using (**a**) homogeneous and (**b**) surface sensing. The minimal P occurs at larger W and R_c in both figures. Reprinted from Ref. 45 with permission. © 2008 Institute of Electrical and Electronics Engineers

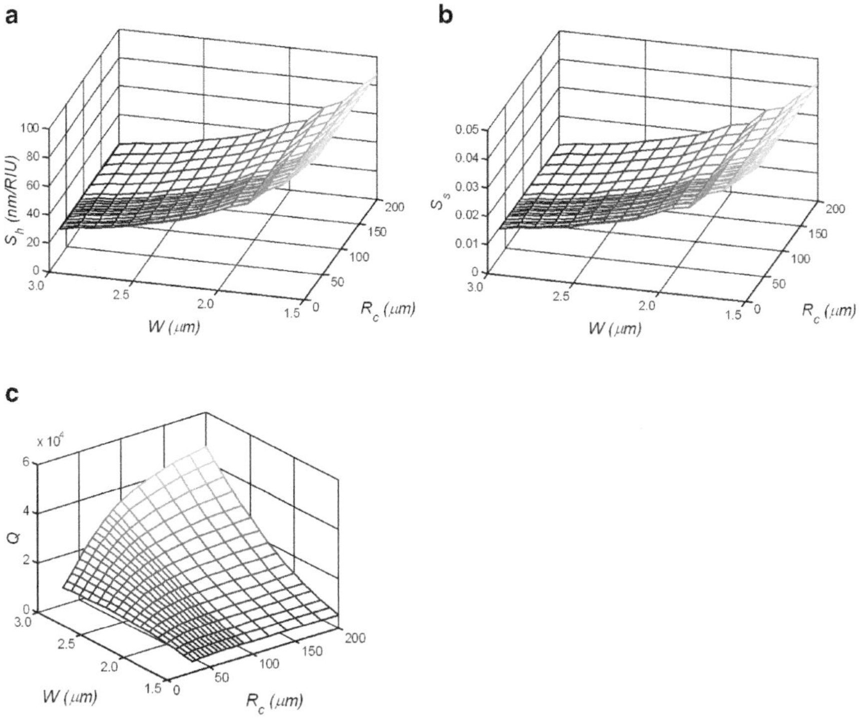

Fig. 8.11 Overall sensitivity as a function of W and R_c using (**a**) homogeneous and (**b**) surface sensing. The simulation illustrates that higher sensitivity requires smaller W but has little dependence on R_c. Q-factor is calculated and plotted in (**c**). Reprinted from Ref. 45 with permission. © 2008 Institute of Electrical and Electronics Engineers

their linear dependence[46], and a wider waveguide because of better optical confinement, which can reduce loss in the microring resonator. Since $1/P \propto Q \times S$ and the Q-factor shows similar trends to $1/P$, this implies that Q-factor has higher domination on $1/P$ than S does. Hence, higher Q should be the primary focus to achieve lower detection limit.

A doubly coupled microring sensor can be applied in sensing using its drop port. This port has less Fabry–Perot effect on the spectrum caused by end facets of the bus waveguides, and shows much smoother resonance shape experimentally (Fig. 8.8). The following analysis assumes a fixed coupling gap between the ring and bus waveguides at 250 nm. To find out the optimal W and R_c, $1/P$, contrast ratio, and ON-resonance dropping coefficient should be calculated, which are plotted in Fig. 8.12. From these figures, lower detection limit occurs at larger W and R_c for both homogeneous and surface sensing, which has a similar behavior to that of singly coupled microrings. Detection limit is still dominated by Q-factor because of the similar trend to $1/P$. However, the highest contrast ratio requires larger W and an optimal R_c. To achieve higher ON-resonance dropping coefficient, smaller W is required, providing less optical confinement but increasing overlap between the

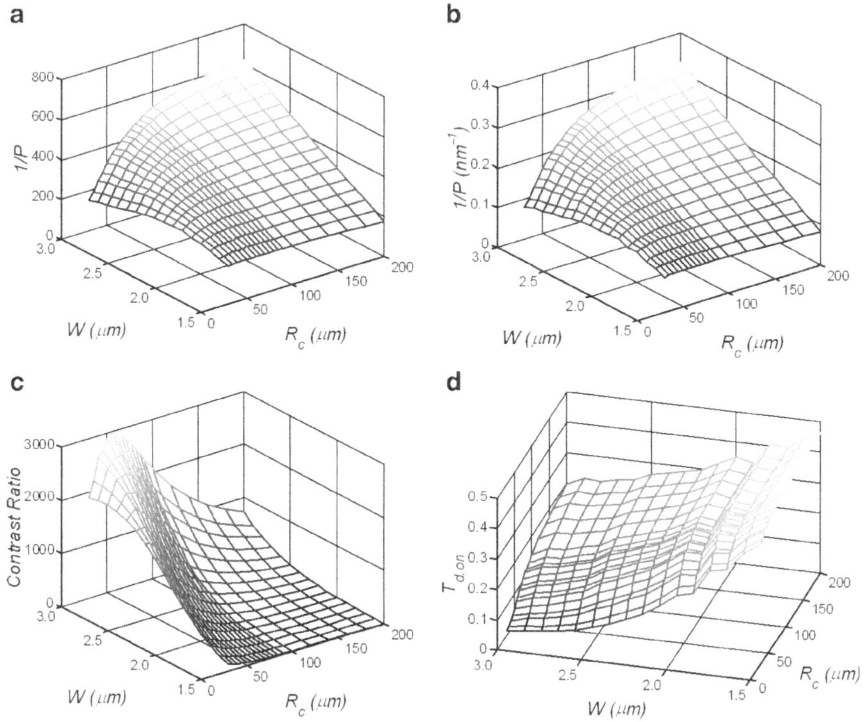

Fig. 8.12 Inverse of the detection factors using (**a**) homogeneous and (**b**) surface sensing, (**c**) contrast ratio, and (**d**) ON-resonance dropping coefficient are plotted as a function of W and R_c for a doubly coupled microring with a gap of 250 nm. Reprinted from Ref. 45 with permission. © 2008 Institute of Electrical and Electronics Engineers

modes of the straight and the curved waveguides in the coupling region. Larger mode overlap leads to higher cross coupling and, in turn, enhances ON-resonance dropping coefficient.

From the previous example, ON-resonance dropping coefficient changes substantially with W and R_c. This coefficient is very important to render a sufficiently strong output signal to maintain large SNR. Therefore, another example is considered with fixed dropping coefficient at 0.4 that can ensure good SNR. In this situation, only $1/P$ and contrast ratio should be considered, and the calculated results are illustrated in Fig. 8.13. They show that lower detection limit occurs again at larger W and R_c for both sensing mechanisms. Q-factor in Fig. 8.13c shows similar trend to $1/P$ and then still dominates the influence on detection limit. However, contrast ratio cannot reach as high as it does in the previous example. But it is still sufficient to distinguish a resonance shift. In practical uses, a contrast ratio of 10 is adequate for sensing operation.

For microring sensors, lower detection limit (i.e., higher $1/P$) can be obtained by increasing both Q-factor (i.e., smaller resonance bandwidth) and sensitivity. However, there is a tradeoff between high Q-factor and high sensitivity. The former

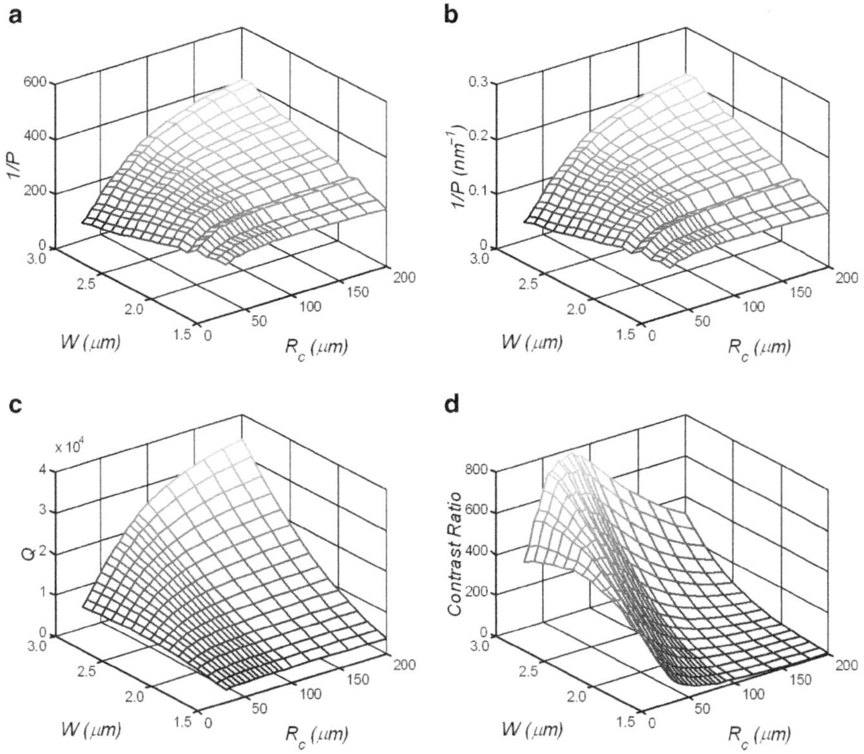

Fig. 8.13 Inverse of the detection factors using (**a**) homogeneous and (**b**) surface sensing, (**c**) Q-factor, and (**d**) contrast ratio are plotted as a function of W and R_c for a doubly coupled microring with a fixed ON-resonance dropping coefficient at 0.4. Reprinted from Ref. 45 with permission. © 2008 Institute of Electrical and Electronics Engineers

requires wider waveguides to provide better optical confinement and therefore lower loss, while the latter requires narrower waveguides to achieve stronger evanescent wave to interact with analytes. The analyses in the context of several examples conclude that detection limit is primarily determined by Q-factor more than sensitivity parameter. Thus, it is more efficient to enhance Q-factor for lowering detection limit. Q-factor can be dramatically enhanced by reducing microring loss and/or increasing the radius of microrings.

The second important consideration is to increase SNR. This requirement can be satisfied if resonance has sufficient ON-OFF contrast, which is determined by a delicate balance between coupling from a microring waveguide to a bus waveguide and loss in the microring. In practice applications, the required specification of a microring sensor should be decided beforehand, such as detection limit, sensitivity, contrast ratio, and SNR. Based on the specification, an optimal range of W and R_c can be obtained by using the optimization procedure.

8.2.6.2 Fabrication Improvement: Thermal-Reflow Technique

According to the analysis in Sect. 8.2.6.1, a direct approach to enhance sensitivity of a microring sensor is to increase its Q-factor. It can be improved by reducing its cavity loss. Loss of a microring resonator fabricated in Sect. 8.2.5 comes from surface roughness scattering, bending loss, leakage loss to silicon substrate, and modal mismatch loss. Surface roughness scattering dominates cavity loss in most cases[47] unless ring radius is too small such that bending loss increases dramatically. Surface roughness is mainly from fabrication processes, such as lithography step and metal lift-off. These steps create rough edges that are transferred to waveguide sidewalls via RIE. Typical surface roughness is in the range of several tens of nanometers, which can result in pretty large scattering loss. To greatly reduce this loss, thermal-reflow technique is developed.

Thermal-reflow technique is to apply appropriate elevated temperature to polymer devices for certain time. Each polymer has a glass transition temperature (T_g). As the applied temperature approaches the T_g, the polymer has significantly reduced viscosity and is able to reflow such that smoother surface can be achieved by surface tension. The degree of polymer's fluidity can be controlled by the applied temperature, which determines reflow rate. The degree of roughness reduction can be controlled by the reflow time. Several polystyrene microrings are thermally treated under different temperatures and heating times, illustratingdramatic improvement as shown in the scanning electron micrographs (Fig. 8.14a–c). The temperature has to be appropriately selected. Too low a temperature is not able to activate the reflow process, while too high a temperature makes it uncontrollable. For the example of polystyrene with T_g of 105°C, the heating temperature of 90°C is chosen and the typical reflow time is around 2 min.

During the thermal-reflow process, waveguides are reshaped from a rectangular shape to a mushroom one. Slight deformation does not affect the microring optical performance, but only shift its resonance position. However, too high reflow temperature and/or too long process time can dramatically deform the cross section

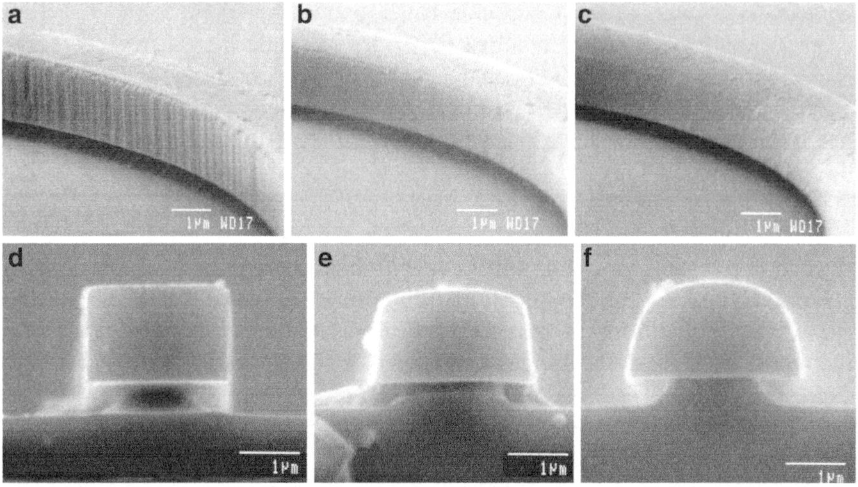

Fig. 8.14 Scanning electron micrographs shows sidewall roughness and waveguide cross section of polystyrene microring resonators (**a**, **d**) without thermal-reflow process, after reflowing at (**b**, **e**) 85°C for 120 s, and (**c**, **f**) 95°C for 60 s, respectively. Reprinted from Ref. 48 with permission. © 2008 American Institute of Physics

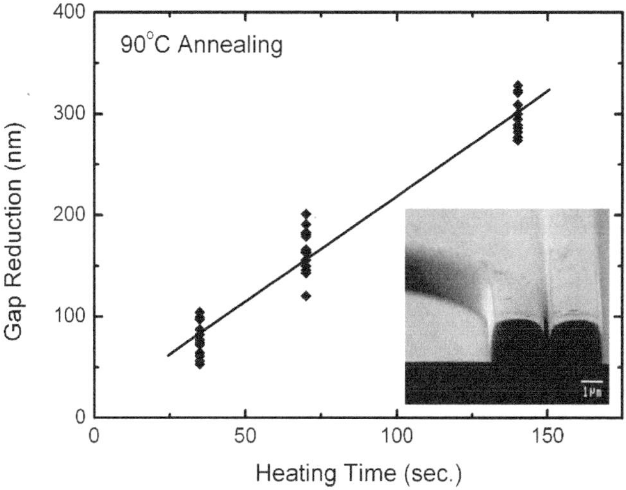

Fig. 8.15 Gap reduction has linear relationship with the reflow time at the reflow temperature of 90°C. The *inset* SEM shows a polystyrene microring with a coupling gap of ~200 nm, reduced from the initial width of 450 nm. Reprinted from Ref. 48 with permission. © 2008 American Institute of Physics

of waveguides and this should be avoided. Hence, these two parameters have to be carefully determined.

In addition to smoother sidewalls and waveguide deformation, coupling gap is reduced as shown in Fig. 8.15, which provides a way to modify coupling strength.

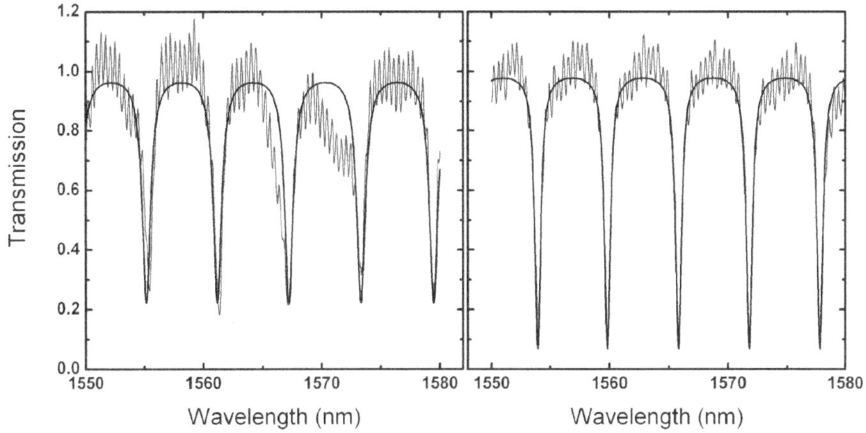

Fig. 8.16 Measured and curve-fitted (*solid*) spectra of a polystyrene microring resonator at the through port are compared before (*left*) and after (*right*) thermal-reflow process at 95°C for 60 s. Reprinted from Ref. 44 with permission. © 2008 Institute of Electrical and Electronics Engineers

In fabrication, a gap is intentionally made wider than the desired. Then it can be reduced to the desired separation together with smoothed sidewalls during the thermal-reflow. Such a process can relieve the challenge of fabricating a very small coupling gap.

To demonstrate the effectiveness of the reflow process, the spectra of a microring before and after thermal-reflow process are curve fitted to extract the coupling efficiency and the round trip amplitude attenuation factor. Measured and fitted spectra are plotted in Fig. 8.16 with reflowing at 95°C for 60 s. According to the fitting results, the amplitude cross-coupling efficiency is increased from 0.457 to 0.576 due to the shrinkage of the coupling gap. The results with the aid of calculation[44] illustrate reduction of the cavity loss by 74 dB cm^{-1}. The surface roughness is estimated to be lowered from ~45 nm to 10–15 nm. From Fig. 8.16, the resonant bandwidth is clearly narrowed, verifying the enhancement of the Q-factor. This thermal-reflow technique can be applied to a variety of polymer waveguides by modifying reflow condition to diminish surface-roughness-induced optical scattering.

8.2.7 Biosensing Experiments by High-Q Microring Biochemical Sensors

8.2.7.1 Homogeneous Sensing

To demonstrate homogeneous sensing, a polystyrene microring is immersed in a solution containing glucose as the analyte. This microring has Q-factor of ~5,000.

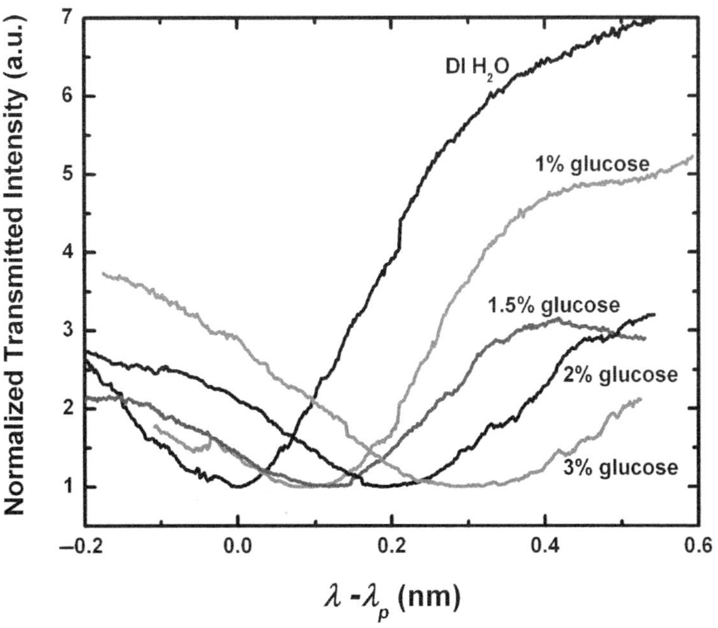

Fig. 8.17 Spectra of a polystyrene microring resonator ($Q \sim 5{,}000$) are plotted in various concentrations. The resonance red-shifts with the glucose concentration. Reprinted from Ref. 49 with permission. © 2008 American Institute of Physics

Its optical spectra are measured when the device is immersed in glucose solutions (dissolved in DI water) with different concentrations. To avoid possible experimental inaccuracy, reference spectra in DI water are measured each time before immersion into a glucose solution. Variation of the resonant wavelength of each reference measurement is within 50 pm, which might be due to temperature change, glucose adsorption on waveguide surfaces after each rinsing, and/or water penetration into polystyrene waveguide cores. Measured spectra in various glucose solutions are shown in Fig. 8.17. As expected, higher glucose concentration induces larger index change and then causes a larger shift of the resonant wavelength.

Using the resonant wavelength shift scheme for transduction, the shift of λ_p is plotted in Fig. 8.18a, showing linear relationship with the glucose concentration. Sensing by use of the same device can also be transduced with the intensity variation scheme. With fixed sensing wavelength, the output intensity is detected and normalized to the reference intensity right before each measurement sweep. Variation of the normalized intensity is plotted in Fig. 8.18b, also showing linear relationship with the glucose concentration.

In the resonant wavelength shift scheme, the microring having Q-factor of $\sim$5,000 can distinguish 50 pm shift of its resonance based on the experimental data, leading to the detection limit in terms of the effective index change as $\Delta n_{\text{eff}} \sim 5 \times 10^{-5}$ RIU by use of (8.1). On the other hand, according to the slope of the fitted curve in Fig. 8.18a, the detection limit, $\Delta\lambda_p$, of 50 pm corresponds to that in glucose

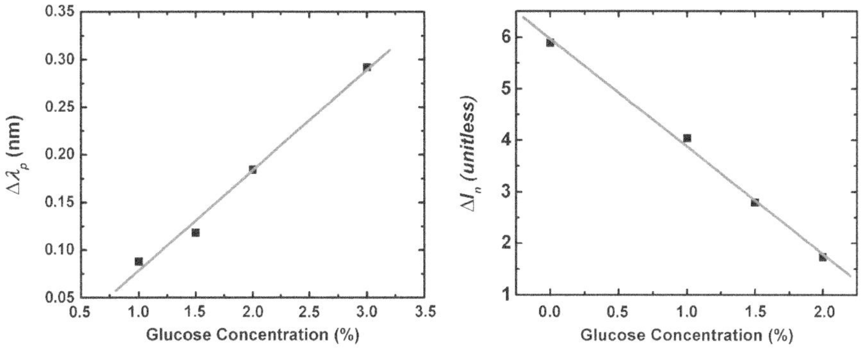

Fig. 8.18 Sensing curves of a polystyrene microring resonator ($Q \sim 5{,}000$) are plotted as a function of glucose mass concentration using (**a**) the resonant wavelength shift scheme and (**b**) the intensity variation scheme. Reprinted from Ref. 49 with permission. © 2008 American Institute of Physics

mass concentration of 0.475%. The change in mass concentration of the glucose solution is linearly related to its refractive index change, Δn_c, with the coefficient of 0.00148 RIU/% at the wavelength of 589 nm when the concentration is below 10%[50]. Due to the lack of data around 1,550 nm, this coefficient is assumed identical in this wavelength range. Hence, this microring can detect index change of glucose solution as small as 7×10^{-4} RIU. This device has the overall sensitivity around 71.4 nm per RIU.

In the intensity variation scheme, detection limit is determined by the environmental and photodetector noises. The former can be well controlled in a dark environment with noise level at 5 nW, which renders $\Delta\lambda_p \sim 1.2$ pm with the resonance slope of 4.25 μW nm^{-1}. Compared with the resonant wavelength shift scheme, the intensity variation scheme gives a 42 times higher sensitivity and lower detection limit.

According to the analyses in Sect. 8.2.6.1, detection limit can be further improved with enhanced Q-factor. With the optimization in waveguide design having less resonator loss and a larger radius of microrings (45 μm), Q-factor of polystyrene microring resonators can reach as high as 20,000. With such devices, the previous glucose experiment is repeated using the drop port because it provides larger contrast ratio and much smaller Fabry-Perot oscillation and therefore enhanced SNR and detection accuracy. Evolution of the spectra is shown in Fig. 8.19a for various concentrations of glucose solutions, and the sensing curve using the resonant wavelength shift scheme is plotted in Fig. 8.19b, again showing good linearity. The slope here is slightly smaller than that in Fig. 8.18a, because the waveguide dimension is enlarged to reduce loss. Compared with Fig. 8.17, Fig. 8.19 shows much higher resolving power due to narrower resonance. This device can potentially distinguish wavelength shift of 5 pm, giving the detection limit in terms of the effective index change as $\Delta n_{\text{eff}} \sim 5 \times 10^{-6}$ RIU, and the glucose mass concentration as 0.061%. Therefore, this high-Q microring can detect the index

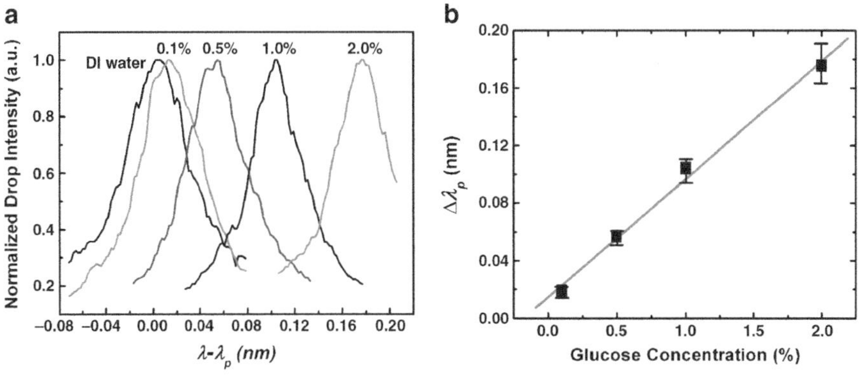

Fig. 8.19 (**a**) Spectra of a polystyrene microring resonator ($Q \sim 20{,}000$) are plotted in various concentrations, red-shifting with glucose concentration. (**b**) Sensing curve of the polystyrene microring: resonant wavelength shift is plotted as a function of glucose mass concentration using the resonant wavelength shift scheme. Reprinted from Ref. 43 with permission. © 2008 Institute of Electrical and Electronics Engineers

change of glucose solution (Δn_c) as small as 9×10^{-5} RIU, giving an overall sensitivity of 55.6 nm per RIU. Even though the sensitivity is reduced, the detection limit is improved by eight times with the enhanced Q-factor. These experiments verify that higher Q-factor provides lower detection limit as expected in Sect. 8.2.6.1.

8.2.7.2 Surface Sensing

Surface sensing requires creation of binding sites on sensor surfaces to provide specificity. The streptavidin-biotin system is chosen for demonstration. Biotin (molecular weight $M_w = 244$ Da, also known as vitamin H) is a small biomolecule, while streptavidin ($M_w = \sim 60{,}000$ Da) is a type of protein consisting of 512 amino acids, which is a relatively large biomolecule. This system has a very large adsorption coefficient ($K_a \sim 10^{15}$ M^{-1}), leading to strong affinity.[51]

In the experiment, a polystyrene microring resonator is subjected to DI water, biotinylated bovine serum albumin (BSA), streptavidin, and biotin solutions. BSA has two main functions in the process. It has strong adhesion on sensor surfaces and the biotinylation provides specific binding sites for streptavidin as shown in Fig. 8.20a. In addition, BSA can repel protein adsorption, which greatly eliminates nonspecific binding and enhances specificity. One streptavidin molecule has four binding sites for biotin. So after its binding to a biotinylated BSA, the unoccupied biotin binding sites on a streptavidin molecule can trap additional biotin molecules (Fig. 8.20b, c). The device is rinsed with DI water after each step to remove weakly bound, unbound, and nonspecific biomolecules.

Figure 8.21 shows the evolution of the resonant wavelength shift measured through the whole process. Initially, the resonant wavelength of the microring

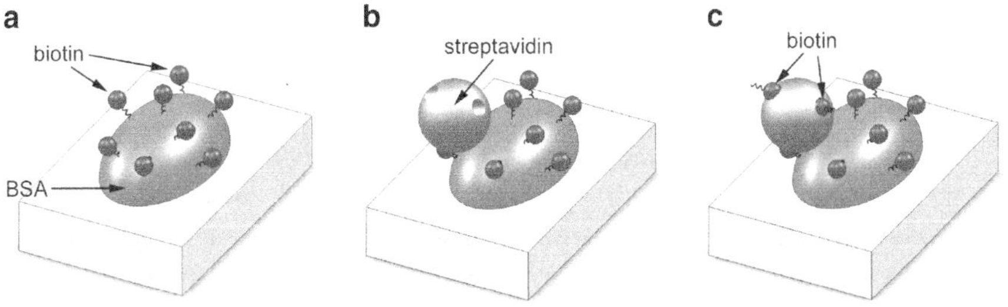

Fig. 8.20 Experimental sequence of surface sensing: (**a**) The sensor surfaces are functionalized with biotinylated BSA; (**b**) Streptavidin is specifically bound to the sensor surface with the biotin moiety; (**c**) Biotin is captured on the sensor surfaces by the streptavidin

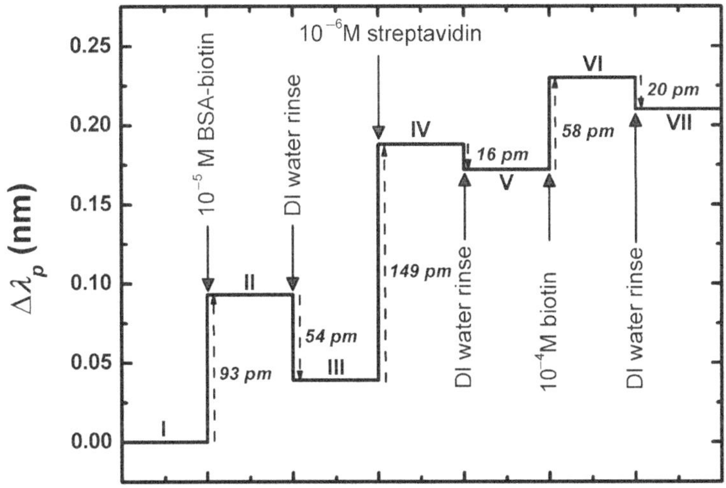

Fig. 8.21 Evolution of the resonant wavelength: I, DI water; II, 10^{-5} M biotinylated BSA; III, DI rinse; IV, 10^{-6} M streptavidin; V, DI rinse; VI, 10^{-4} M biotin; VII, DI rinse. Reprinted from Ref. 43 with permission. © 2008 Institute of Electrical and Electronics Engineers

resonator is measured in DI water as the reference. In the first step with incubation in a biotinylated BSA solution (10^{-5} M), a resonance shift of 93 pm is observed, which is caused by both adsorbed biotinylated BSA molecules on the sensor surface (specific) and freely surrounding ones in the solution (nonspecific). The effect from nonspecific binding is removed by subsequent DI water rinse, giving a net shift of 39 pm. According to the information regarding the molecular weight ($M_w = 66{,}432$ Da[15]. and size (4 nm × 4 nm × 14 nm) of BSA[52], there are sufficient amount of biotinylated BSA molecules to form a monolayer on the sample surface. In the second step, streptavidin solution (10 μL) with concentration of 10^{-6} M is introduced on the sensor followed by DI water rinse and a net shift of 133 pm is detected due to the bound streptavidin to the biotinylated BSA. Based on the size of one streptavidin molecule, 5.6 nm × 4.2 nm × 4.2 nm[53], approximate 2×10^{-12} M of streptavidin can form a monolayer on the whole sensor chip (around 28 mm^2), which implies that the free streptavidin in the solution is four times more than the captured streptavidin. Obviously, surface sensing mechanism renders a much dramatic resonance shift, facilitating accurate detection and reducing required analyte amount. With an optimized design of microfluidic system to direct solutions to only the microring area, the necessary amount of analyte can be reduced by three orders of magnitude. In the last step, biotin solution (10^{-4} M) and DI water rinse are introduced, causing a net shift of 38 pm. This experiment not only demonstrates surface sensing of specific biomolecules, but also shows that such polymer microring resonators are capable of detecting both small molecules and large biomolecules without using fluorescent labels. This approach using biotin-streptavidin system can be generally applied to detect any biotinylated protein molecules.

Based on the preliminary streptavidin experiment result, the detection limit (the smallest detectable mass coverage per unit surface area) of a microring resonator with Q-factor of 20,000 is ~250 pg mm^{-2} by assuming that resolvable wavelength shift is one-tenth of the resonant FWHM bandwidth. With further enhancement of Q-factors, optimization of microring waveguides, employment of three-dimensional matrix to create more binding sites, and control of temperature, polymer microring sensors are believed to provide superior detection limit to commercial quartz crystal microbalance biosensors typically with detection limit in the range of 50–100 pg mm^{-2} and surface plasmon resonance biosensors reaching as low as a few tens picogram per square millimeter[4].

8.2.8 *Sensitivity-Enhancing Methods*

In Sect. 8.2.6, it has been discussed that sensitivity of a polymer microring sensor can be enhanced by increasing Q-factor and reducing resonator loss. In addition, two alternative approaches will be discussed that are able to further improve the performance in parallel. The first method is to induce an asymmetric Fano-resonant line shape to narrow resonant bandwidth and to achieve steeper spectrum slope. The other method is to replace conventional channel waveguides by slot waveguides. This type of waveguides renders much larger overlap between the guided optical field and the analyte, producing much higher waveguide sensitivity in principle.

8.2.8.1 Fano-Resonant Microrings

In atomic physics, Fano-resonance originates from the interference between a discrete energy state and a coherent continuum of states, leading to an asymmetric line shape in reflection and transmission spectra[54]. Recently, Fano-resonant behavior has been illustrated in theory that it can be induced in a microcavity coupled with a single-mode waveguide by partially reflecting elements[55]. This effect does not change unloaded Q-factor of a microcavity, but modifies its resonance

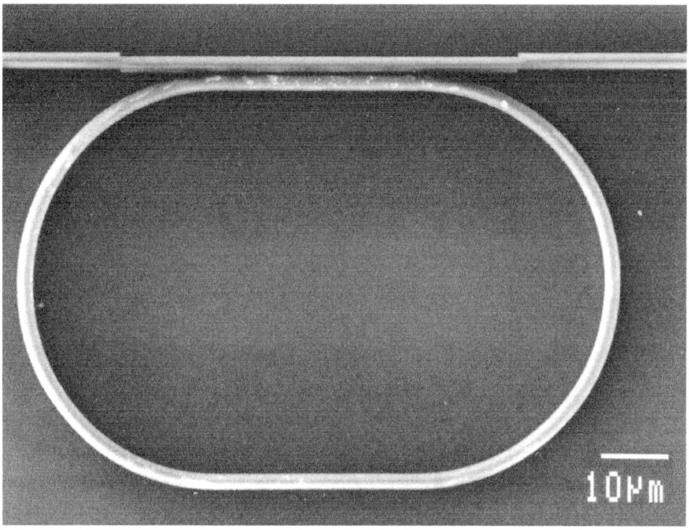

Fig. 8.22 Scanning electron micrograph shows a polystyrene microring coupled to a bus waveguide and sandwiched by two waveguide offsets. Reprinted from Ref. 49 with permission. © 2008 American Institute of Physics

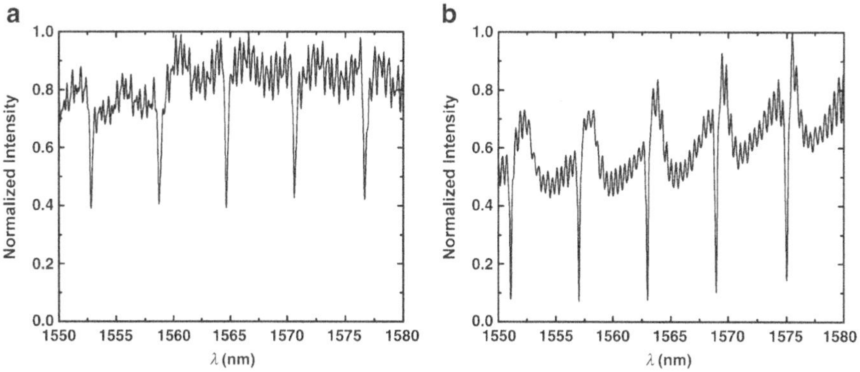

Fig. 8.23 Transmission spectra of a polystyrene microring resonator: (**a**) without waveguide offsets and (**b**) sandwiched by two waveguide offsets. Asymmetric Fano-resonant line shape is clearly observed. Reprinted from Ref. 49 with permission. © 2008 American Institute of Physics

line shape. Therefore, this approach can be adopted to further improve its performance without varying the resonator structure.

To demonstrate Fano-resonant behavior, a polystyrene microring resonator is nanoimprinted and two waveguide offsets are introduced in the bus waveguide, which provide partial reflection as shown in Fig. 8.22. Its spectrum is plotted in Fig. 8.23 together with the one without waveguide offsets. The slope is enhanced by a factor over two. Fano-resonance has also been demonstrated using a microsphere coupled with a multimode tapered fiber[56], a microtoroid resonator coupled with a Fabry-Perot etalon[57], and a Charon microresonator.[58]

8.2.8.2 Slot Waveguide Microrings

A conventional waveguide is composed of a core with high refractive index surrounded by claddings of lower indices. In such a waveguide, most of light is confined within the core. Recently, a novel waveguide structure, slot waveguide, is proposed[59], which consists of a thin low index material sandwiched between high index materials as shown in Fig. 8.24a. This slot waveguide can support stable optical modes according to the theory of electromagnetism. Based on the boundary condition that electric displacement field (*D*-field) normal to the interfaces is continuous in the absence of free surface charges, the electric field (*E*-field) within the middle material will be boosted for the TE polarized light due to lower permittivity inside the slot. The *E*-field distribution within a polystyrene slot waveguide is illustrated in Fig. 8.24b. This type of waveguide can also be applied to form slot microring resonators, which has been experimentally demonstrated using silicon-on-insulator[60,61] due to high permittivity ratio to that of air.

For an optical biosensor using evanescent field to interrogate adjacent analytes, higher percentage of evanescent wave or larger *E*-field can interact more strongly with the targeted analytes, inducing more significant shift of transducing parameters of the device. In other words, a larger overlap volume between evanescent field and environment and/or stronger *E*-field extending into an analyte medium can improve sensitivity. To prove the concept, a polystyrene slot microring is considered here and compared with an ordinary microring.

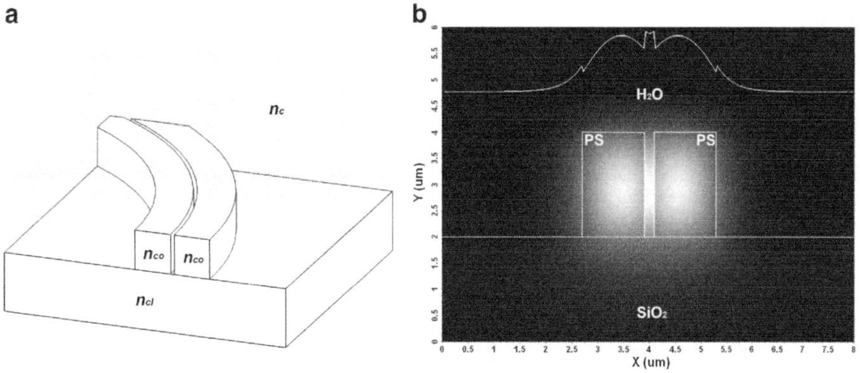

Fig. 8.24 (**a**) Schematic of a curved slot waveguide; (**b**) *E*-field distribution of a straight polystyrene slot waveguide is shown with total waveguide width of 2.6 μm including a slot of 200 nm wide (TE polarization). The *inset curve* is the field magnitude at $Y = 3$ μm, which indicates stronger *E*-field within the slot

The entire waveguide width is 2.4 μm with polystyrene as the high index material, and the height is 2 μm. Polystyrene sits on silicon dioxide bottom cladding, and is enclosed by water as top cladding. The *E*-field distributions inside each microring waveguide are calculated and plotted in Fig. 8.25 for a ring with

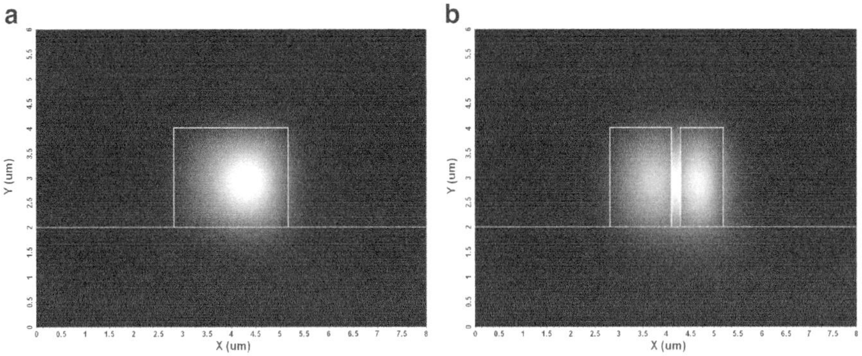

Fig. 8.25 *E*-field distribution (TE polarization) of (**a**) a conventional polystyrene microring with waveguide width of 2.4 μm and height of 2 μm, and (**b**) an asymmetric slot polystyrene microring with entire waveguide width of 2.4 μm including a slot of 200 nm. Waveguides in both cases sit on silicon dioxide and are embedded in water

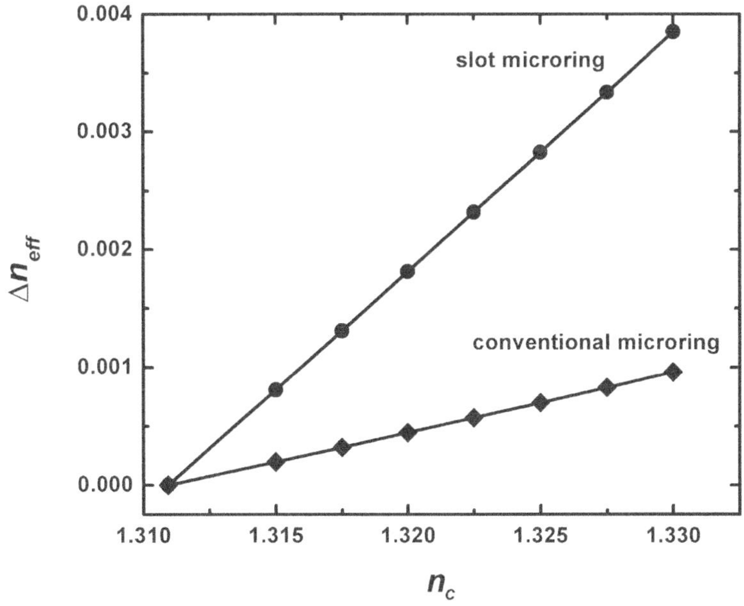

Fig. 8.26 Effective index variation is calculated as a function of the refractive index of top cladding. The result shows that slot microrings have higher waveguide sensitivity than conventional microrings

radius of 45 μm. As expected, the guided light tends to propagate closer to the outer surface (the right segment of the slot waveguide in the figure). Obviously, the *E*-field in the slot exceeds the maximum *E*-field inside the polystyrene sections. To investigate sensing performance, variation of effective refractive index is calculated using homogeneous sensing mechanism and plotted in Fig. 8.26. Waveguide sensitivity, which is the slope of the line, is enhanced by over fourfold. This sensitivity can be further optimize[62]. by varying waveguide shape, slot size, individual width of high index sections, and ring radius, or using multiple-slot microrings[63]. To analyze and optimize slot microring sensors in surface sensing, a simple method combining the effective-index method and the modified transfer-matrix method can be employed to provide quick design and evaluation of various structures[35]. A slot microring biosensor has recently been demonstrated in experiment with sensitivity of 490 nm per RIU and detection limit of 10^{-4} RIU.[64]

8.3 Prism-Coupled Microtube Resonator Sensors

Prism-coupled silica microtube resonators are another type of microresonator sensors. They have the advantages of easy integration to fluidic system, high *Q*-factor, and superior sensitivity. Two types of refractive index sensing modes were observed experimentally by tuning the incident angles in the prism-coupled configuration. A typical resonance mode interacting with fluids via evanescent wave has sensitivity below 100 nm per RIU, while a special mode with high optical field in the low index fluid region can render a record of high sensitivity of ~600 nm per RIU. In this section, the theoretical calculation of sensitivity will be discussed using the Mie scattering method and a ray optical picture will be presented to elucidate the physical nature of this special resonant mode observed in the microtube resonators. In addition, experimental results of bulk index sensing (homogeneous sensing) and surface sensing of absorbed lipid membrane on the inner wall as well as protein binding on the membrane will be demonstrated and analyzed.

8.3.1 Advantages of Prism-Coupled Microtube-Based Sensors

A silica microtube-based resonator sensor is very attractive for biochemical sensing applications due to its ability to handle aqueous analytes and its high *Q*-factor[65]. This high-*Q* characteristic (~10^6) is a result of low scattering loss due to very smooth tube surfaces. However, there are also limitations to the microtube resonators. Since only the inner surface can be used as a sensing interface, the total evanescence field interacting with analyte solutions is less than half of that in the microring resonators if these two devices have similar field distribution in the waveguide and the fluid-cladding region, thereby resulting in lower sensitivity.

Moreover, the previously demonstrated microtube resonator sensors required a tube thickness thinner than 5 μm[66] in order to obtain a refractive index sensitivity of 2.6 nm per RIU, which could cause brittle devices and handling difficulty in practical applications. These issues can be solved by using the prism-coupled microtube platform. By tuning the incident angle, such a configuration allows not only exciting the resonant modes in a microtube with much thicker tube wall, but also more importantly to excite different high order modes. These features provide the benefits of robustness, easy handling, and sensitivity tunability.

8.3.2 Experimental Setup and Results of Refractive Index Sensing

To illustrate the concept, a prism-coupled fused silica microtube is built as shown in Fig. 8.27. The fused silica microtube used in the experiment has an inner radius of $R_1 = 134$ μm and an outer radius of $R_2 = 166$ μm. The tube is positioned perpendicular to the light incident plane and connected to plastic tubes at both ends for the flow of analyte solution. Light incident onto the semicylindrical prism can be evanescently coupled into the tube through a controlled air gap between the tube and the prism surface. This gap is determined by a deposited aluminum film of 400–500 nm in thickness. The microtube is bonded to the prism and then mounted on a rotation stage. Light from a tunable diode laser around 1,550 nm in wavelength is collimated onto the prism with the incident angle that can be changed by adjusting the rotation stage. Tuning the incident angle enables incident light coupling to different resonant modes within the microtube resonator by satisfying different phase-matching conditions. The light coupled out of the resonator passes through an aperture and then through a polarizer, and finally is focused onto a photodetector. The polarizer is used to select the polarization of the output signal from the resonator. The experiment is performed for the TE polarization, due to its higher *Q-factor* than that of the TM polarization.

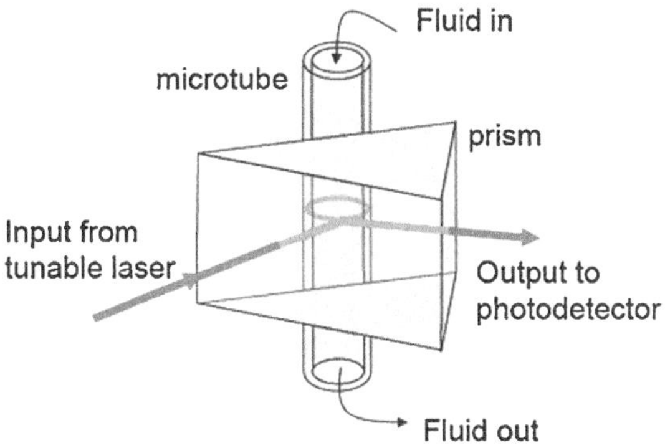

Fig. 8.27 Schematic of the experimental setup

Index-sensing experiments are performed by changing the refractive index of the liquid that flows through the microtube and simultaneously monitoring the corresponding resonant wavelength shift. The refractive indices of the liquid are controlled by premixing a very small amount of isopropyl alcohol (IPA) with deionized (DI) water. The index of the mixed solution can be estimated based on the dielectric constants of both liquids and the molar ratio of each component[67]:

$$n = \sqrt{\kappa \cdot n_{\mathrm{IPA}}^2 + (1 - \kappa) \cdot n_{\mathrm{water}}^2}, \tag{8.5}$$

where κ is the molar ratio of IPA in the solution and n_{water} and n_{IPA} are the refractive indices of water and IPA, and are taken to be 1.320 and 1.378 around the wavelength range of 1,550 nm, respectively. Figure 8.28a shows three measured resonance spectra for liquids of different refractive indices and with the incident angle to the prism at 37.5°. The refractive index difference between the adjacent spectra is 4×10^{-4}.

As seen in the figure, four resonant modes are excited within one FSR because the incident light is a Gaussian beam that covers a range of spatial wave vectors where different phase-matching conditions can be fulfilled. These four modes with different optical field distribution across the microtube respond differently to the fluidic index change: the resonant wavelength shifts for the same refractive index change $\Delta n = 4 \times 10^{-4}$ are 39 pm, 28 pm, 30 pm, and 54 pm, respectively, for the first to the fourth peaks. The resonant wavelength shift is plotted as a function of the refractive index change in Fig. 8.28b. By linear curve fitting, the sensitivity for these resonant modes are 96.7 nm per RIU, 70 nm per RIU, 74.7 nm per RIU, and 135 nm per RIU, respectively. The largest sensitivity is over 50 times higher than that reported by White et al.[66].

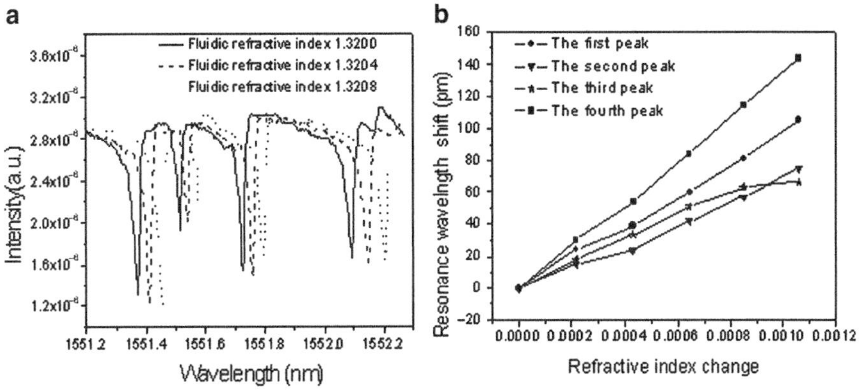

Fig. 8.28 (**a**) Resonance spectra for three different fluidic refractive indices and (**b**) resonant wavelength shift as a function of liquid refractive index change in the microtube at the incident angle of 37.5°. Reprinted from Ref. 68 with permission. © 2008 Optical Society of America

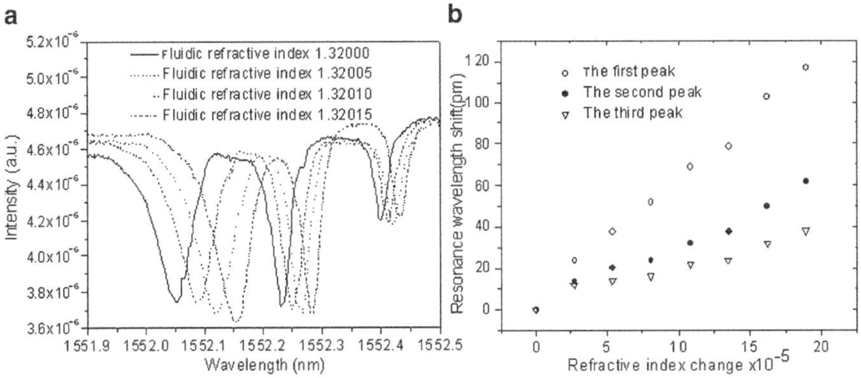

Fig. 8.29 (**a**) Resonance spectra for three different fluidic refractive indices and (**b**) resonant wavelength shift as a function of liquid refractive index change in the microtube at the incident angle of 35°. Reprinted from Ref. 68 with permission. © 2008 Optical Society of America

Further decreasing the incident angle to 35° allows the excitation of even higher order resonant modes in the microtube resonator. Figure 8.29a shows four resonance spectra measured for different liquid refractive indices with three resonant modes again showing different shifts. Responding to a refractive index change of 5×10^{-5}, the resonance shifts are 30 pm, 18 pm, and 10 pm for the three resonant peaks. The sensitivity for the three resonant modes are 600 nm per RIU, 360 nm per RIU, and 180 nm per RIU, respectively, by extracting the slope from Fig. 8.29b. The sensitivity is much higher than what can be achieved with typical evanescent sensors. Based on the highest sensitivity of 600 nm per RIU, a Q-factor of 2×10^4 and the ability of resolving a wavelength shift of 1/25th of the resonance line width, the detection limit can approach 5×10^{-6} RIU.

8.3.3 *Theory of Microtube Resonators*

8.3.3.1 Field Distribution and Sensitivity Calculation in Microtube Resonators

In order to understand the physics behind the observed super-high sensitivity, we investigated the optical field distribution in the microtube using cylindrical coordinates z, r, and ϕ. The time-independent field distribution for the resonant mode can be separated into a radial-dependent mode component and an azimuthal-dependent phase term, $\Psi_Z(r)\exp(\mathrm{i}\,m\phi)$, where Ψ_Z is the amplitude of the axial magnetic (TE) or electric (TM) modal field and m is the azimuthal quantization number. For TE mode, only three components need to be considered: H_z, E_r, and E_ϕ. The radial dependent field H_z can be expressed by Bessel functions as[69]:

$$H_Z = \begin{cases} B_1 J_m(n_1 k_0 r) & 0 < r \leq R_1 \\ [B_2 J_m(n_2 k_0 r) + B_3 N_m(n_2 k_0 r)] & R_1 < r \leq R_2 \, , \\ B_4 H_m^{(1)}(n_3 k_0 r) & R_2 \leq r < \infty \end{cases} \tag{8.6}$$

where n_1, n_2, and n_3 are the refractive indices of the liquid core region, the silica microtube region, and the outside air region; k_0 is the wave-vector in vacuum; and $J_m(nk_0 r)$, $N_m(nk_0 r)$, and $H_m^{(1)}(nk_0 r)$ are the mth order cylindrical Bessel, Neumann, and Hankel function of the first kind, respectively. By matching the boundary conditions at the liquid/silica and silica/air interfaces, we obtain the Eigen equation, which is used to calculate the resonant wavelength. The TE resonant mode with azimuthal number m is determined by:

$$\frac{n_1 J_m(k_0 n_1 R_1)}{n_2 J'_m(k_0 n_1 R_1)} = \frac{(B_2/B_3) J_m(k_0 n_2 R_1) + N_m(k_0 n_2 R_1)}{(B_2/B_3) J'_m(k_0 n_2 R_1) + N'_m(k_0 n_2 R_1)}, \tag{8.7}$$

$$\text{where } \frac{B_2}{B_3} = \frac{n_3 H_m^{(1)}(k_0 n_3 R_2) N'_m(k_0 n_2 R_2) - n_2 H_m^{(1)\prime}(k_0 n_3 R_2) N_m(k_0 n_2 R_2)}{n_2 H_m^{(1)\prime}(K_0 n_3 R_2) J_m(k_0 n_2 R_2) - n_3 H_m^{(1)}(k_0 n_3 R_2) J'_m(k_0 n_2 R_2)}.$$

For a given m, there are a series of k_0 satisfying (8.7), which are referred to as the vth ($v = 1, 2, 3, \ldots$) order resonant mode. The radial and tangential components of the electric field for TE modes can be obtained with:

$$E_r = -\frac{m}{\omega \varepsilon_0 \varepsilon_r r} H_z \tag{8.8}$$

and

$$E_\phi = \frac{1}{i\,\omega \varepsilon_0 \varepsilon_r} \frac{\partial H_z}{\partial r}. \tag{8.9}$$

To calculate the field distribution in a microtube resonator, the following parameters are chosen: $R_1 = 134$ μm, $R_2 = 166$ μm, $n_1 = 1.320$, $n_2 = 1.450$, and $n_3 = 1.000$. A TE mode has much larger E_r than E_ϕ, and thus it is appropriate to consider E_r only. Figure 8.30a shows electric field the intensity distribution of TE_{714}^{35} mode, where the subscript, 714, stands for the azimuthal mode number and the superscript, 35, stands for the radial mode number. As seen in the figure, the optical field is predominantly guided within the silica tube wall and decays exponentially in the cladding regions. Such a mode transduces refractive index change via the conventional evanescent sensing mechanism.

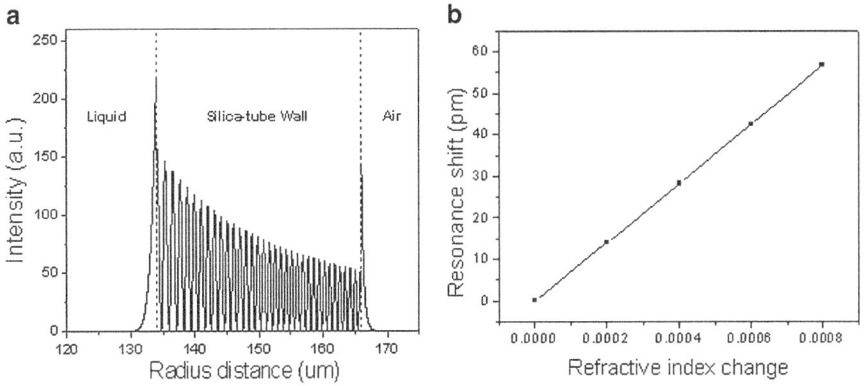

Fig. 8.30 (**a**) Radial electric field intensity distribution of E_r and (**b**) the resonance shift as a function of liquid index change for resonant mode E_{r714}^{35}. Reprinted from Ref. 68 with permission. © 2008 Optical Society of America

The sensitivity of a prism-coupled microtube resonator can be evaluated by employing the Mie theory[70]. For this calculation, the optical field in air is modified as:

$$H_z = H_z^{\text{inc.}} + H_z^{\text{scat.}}, \tag{8.10}$$

where $H_z^{\text{inc.}} = J_m(k_0 n_3 r)$ represents the incident wave and $H_z^{\text{scat.}} = B_4 H_m^{(1)}(k_0 n_3 r)$ represents the scattered wave. The coefficient of the scattered wave, B_4, is obtained by matching H_z and E_θ at $r = R_1$ and $r = R_2$. Plotting $|B_4^2|$ at a specific m as a function of wavelength gives the corresponding resonant spectrum. As expected, increasing the refractive index of the liquid gradually moves the resonance peak toward longer wavelength. The resonance shift of TE_{714}^{35} mode as a function of the refractive index change is plotted in Fig. 8.30b, giving a sensitivity of 70 nm per RIU. This result is very close to the experimental data measured at the incident angle of 37.5°. Generally speaking, such resonant modes with an evanescent wave tailing into the liquid region have a sensitivity less than 100 nm per RIU.

Using a similar approach, a higher radial mode of TE_{700}^{37} is also investigated. The electric field intensity distribution of this mode is plotted in Fig. 8.31a. The field intensity still decays exponentially in air, but is dramatically enhanced in the liquid region, which is very different from that of a typical evanescent resonant mode (e.g., TE_{714}^{35} discussed above). Not only does the field intensity of the first peak completely move into the liquid core, but it also reaches a magnitude significantly higher than the peaks in the silica tube wall. The optical field in the liquid is no longer an evanescent field, but accounts for about 60% of the total field. Such a resonant mode is very sensitive to the index change of the liquids, making it an excellent choice for sensing applications. The resonance shift as a function of the index change of liquid is calculated and shown in Fig. 8.31b. As can be seen in the figure, the sensitivity of this mode is around 560 nm per RIU, which is close to 600 nm per RIU observed in the experiment.

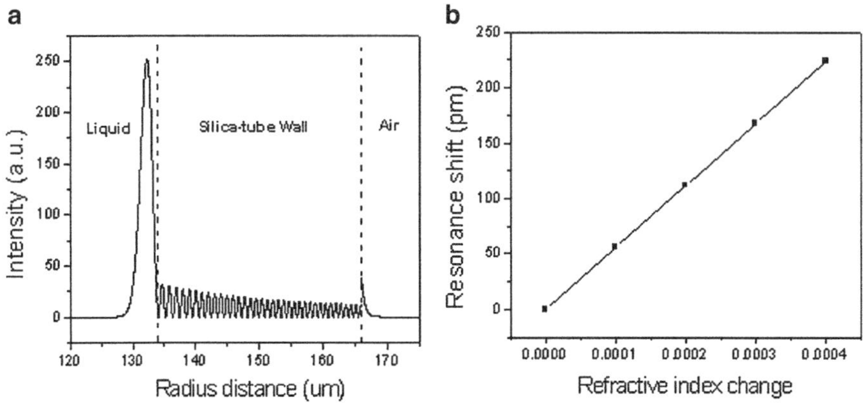

Fig. 8.31 (**a**) Radial electric field intensity distribution for resonant mode E^{37}_{r700}, and (**b**) resonance shift related as a function of refractive index change for resonant mode E^{37}_{r700}. Reprinted from Ref. 68 with permission. © 2008 Optical Society of America

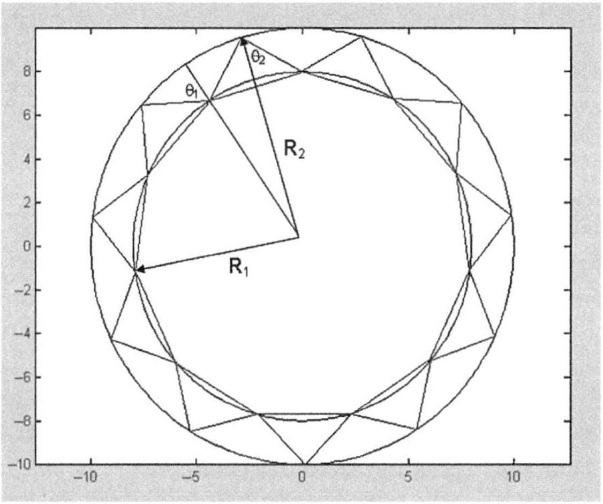

Fig. 8.32 Schematic of ray tracing picture of a resonant mode with light transmitted into the inner boundary. Reprinted from Ref. 68 with permission. © 2008 Optical Society of America

8.3.3.2 Nature of the Nonevanescent Resonant Modes

The unique field penetration into the liquid of a nonevanescent resonant mode like TE^{37}_{700} is very promising for the sensing applications. To understand the origin of this behavior and further optimize the devices, a ray optical picture[71] is presented. This type of resonant modes exist as if rays are bounced at the liquid/silica interface and confined in the liquid region as plotted in Fig. 8.32. From the viewpoint of ray optics, light is partially reflected (termed ray 1) and partially transmitted when it is incident at liquid/silica interface ($r = R_1$) from liquid side. The transmitted rays continue to propagate within the silica tube wall until they undergo total internal reflection at the silica/air interface ($r = R_2$) (termed ray 2). When these two rays meet again at the liquid/silica interface in space and in phase, they interfere constructively. Furthermore, if the two rays form a closed loop within the tube circumference, a resonant mode is created.

This type of mode can exist only under certain conditions related to the geometry of the microtube and the refractive indices of the three regions. By defining the incident and reflection angles at $r = R_1$ and $r = R_2$ as θ_1 and θ_2, the light transmitted through the inner boundary and totally internal reflected at the outer boundary should satisfy the following three criteria:

$$\frac{\sin(\pi - \theta_1)}{R_2} = \frac{\sin\theta_2}{R_1}, \quad \sin\theta_1 < \frac{n_1}{n_2}, \quad \text{and} \quad \sin\theta_2 > \frac{n_3}{n_2}, \tag{8.11}$$

These leads to an inequality: $(R_2/R_1) < (n_1/n_3)$, which sets a constraint on the ratio of the outer radius to the inner radius for the microtube. In our experiments where the inside is water and the outside is air, the requirement on the ratio of the outer radius to the inner one is $R_2/R_1 < 1.32$. This condition is satisfied for the silica microtubes used in our experiment ($R_2/R_1 = 166/134 = 1.24 < 1.32$).

The detection limit of the microresonator-based refractive index sensing device is directly related to the Q-factor of the resonator and the sensitivity of the resonant mode discussed above. The Q-factor of a microtube resonator is determined by the total loss of a resonant mode, including radiation loss, absorption loss, and surface roughness scattering loss. The overall Q-factor can be expressed as

$$\frac{1}{Q} = \frac{1}{Q_{\text{radiation}}} + \frac{1}{Q_{\text{absorption}}} + \frac{1}{Q_{\text{scattering}}}. \tag{8.12}$$

For a nonevanescent mode having high optical field penetration in the liquid, its wave is still well confined at the outer boundary due to total internal reflection, leading to low radiation loss. Scattering loss induced by the surface roughness is much lower than that in the microrings described in Sect. 8.2.6.2 due to much smoother surfaces of the microtube. Such scattering loss can be further reduced by choosing a mode with small electric field at the inner boundary. These two loss sources result in Q-factor of around 10^7 calculated by the Mie scattering method. However, if taking water absorptions into account, the Q-factor is reduced to about 5×10^4, which has the same order of magnitude as in our experiments ($Q \sim 2 \times 10^4$). To reduce the effect of water absorption and to improve sensing performance, microtube resonator sensors can be operated in the visible wavelength, enhancing intrinsic Q-factor by two orders of magnitude. This leads to detection limit of around 5×10^{-8} RIU, and elevates the microtube resonators as one of the most sensitive devices.

8.3.3.3 Sensing Property of Microtube Resonators

Sensitivity in Bulk Index Sensing

Using the method described in Sect. 8.3.3.1 to obtain the electric field in a microtube resonator, the device sensitivity can be calculated with the perturbation theory for different modes. This provides more systematic insight into the operation of microtube resonators. First, bulk index sensing is considered. The sensitivity is proportional to the integrated optical field inside the liquid core region over that of the entire space, and can be expressed as[72]:

$$S_{\text{bulk}} = \frac{\delta\lambda}{\delta n} = \lambda \frac{\int_{V_1} |E_z|^2 n \mathrm{d}\vec{r}}{\int_V n^2 |E_z|^2 \mathrm{d}\vec{r}}, \tag{8.13}$$

where λ is the resonant wavelength, V is the entire space, and V_1 is the space within the inner boundary of the microtube.

Sensitivity for different radial mode numbers (N) is calculated and plotted in Fig. 8.33. One noticeable feature is that the sensitivity dramatically increases with the radial mode number from $N = 32$ to $N = 37$. This is because the ratio of the electric field intensity of the resonant mode in the liquid region to the total electric field intensity is increased with the resonant mode radial number. By calculating the electric field distribution in each case, we find that the microtube is operated in the evanescent sensing regime for $N < 35$, while it starts to approach the nonevanescent sensing regime for $N > 35$. Sensitivity can be enhanced roughly six times comparing the maximal sensitivity in the nonevanescent regime to that in the evanescent regime. Also it is interesting that the sensitivity in the nonevanescence region with different radial order modes (i.e., different field penetration in

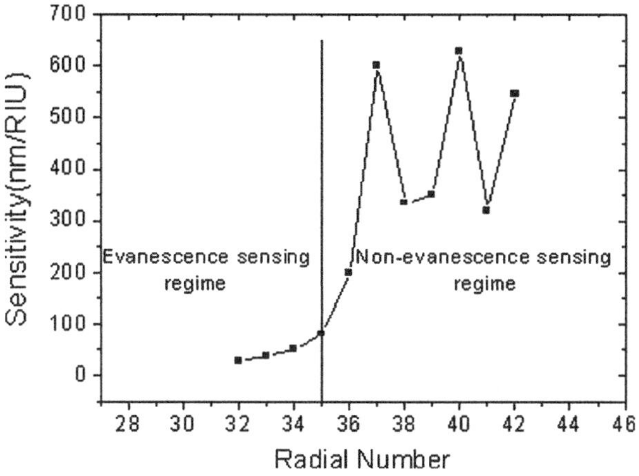

Fig. 8.33 Calculated sensitivity in bulk index sensing for different radial mode numbers with the same azimuthal number $m = 700$

the liquid region) oscillates between 300 nm per RIU and 600 nm per RIU, which is somewhat contradictory to our expectation that as the optical field penetrates deeper into the fluid region the sensitivity would continue to increase rather than to oscillate. We believe that this behavior is due to an intricate interference effect between the two optical rays depicted in Fig. 8.32.

Sensitivity in Surface Sensing

In surface sensing, the adsorbed biomolecules onto the inner wall of a microtube will change the wave vector of a supported mode from k_0 to k_1 and the electric field from E_0 to E_1. These unperturbed (E_0) and perturbed (E_1) electric fields still satisfied wave equation:

$$\nabla \times \nabla \times E_0 = k_0^2 \varepsilon E_0, \tag{8.14}$$

$$\nabla \times \nabla \times E_1 = k_1^2 \varepsilon_1 E_1, \tag{8.15}$$

where ε and ε_1 are the dielectric constant distributions before and after the biomolecule adsorption. Multiplying (8.15) by E_0^* and integrating over the entire space gives

$$\int_v E_0^* \cdot \nabla \times \nabla \times E_1 \mathrm{d}\vec{r} = k_1^2 \int_v E_0^* \cdot \varepsilon E_1 \mathrm{d}\vec{r} + k_1^2 \int_{v_1} E_0^* \cdot (\varepsilon_{\text{film}} - \varepsilon_{\text{liquid}}) E_1 \mathrm{d}\vec{r}, \tag{8.16}$$

where $\varepsilon_{\text{film}}$ and $\varepsilon_{\text{liquid}}$ are the dielectric constant of the biomolecular film and the liquid, v is the total space of the microtube, and v_1 is the space that biofilm absorbed on the microtube. By using[73]:

$$\int_v E_0^* \cdot \nabla \times \nabla \times E_1 \mathrm{d}\vec{r} = \int_v E_1 \cdot \nabla \times \nabla \times E_0^* \mathrm{d}\vec{r} = k_0^2 \int_v E_1 \cdot \varepsilon E_0^* \mathrm{d}\vec{r}.$$

The following equation can be derived

$$\left(k_0^2 - k_1^2\right) \int_v E_1 \cdot \varepsilon E_0^* \mathrm{d}\vec{r} = k_1^2 \int_{v_1} E_0^* \cdot (\varepsilon_{\text{film}} - \varepsilon_{\text{liquid}}) E_1 \mathrm{d}\vec{r}. \tag{8.17}$$

Assuming that the adsorbed biomolecular layer thickness T is very small, the electric field within the film can be taken to be a constant and the field distribution in the resonator change is very small, so E_1 can be approximated by E_0, which lead to

$$\frac{\delta\lambda}{\lambda} = \frac{\int_{v_1} E_0^* \cdot (\varepsilon_{\text{film}} - \varepsilon_{\text{liquid}}) E_1 \mathrm{d}\vec{r}}{2\int_v E_1 \cdot \varepsilon E_0^* \mathrm{d}\vec{r}} \approx \frac{|E_{R_1}|^2 (\varepsilon_{\text{film}} - \varepsilon_{\text{liquid}}) \pi R_1 T}{\int_v E_0 \cdot \varepsilon E_0^* \mathrm{d}\vec{r}}. \tag{8.18}$$

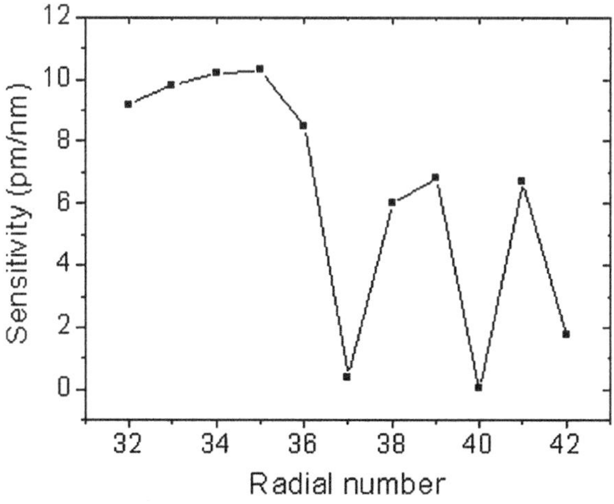

Fig. 8.34 Surface sensing sensitivity of different radial order modes are simulated by using the perturbation method with the same azimuthal number $m = 700$.The adsorbed polymer layer is assumed to have a refractive index of 1.46

Using the same definition in Table 8.1, the sensitivity in surface sensing can be then expressed as

$$S_{\text{surface}} = \frac{\delta\lambda}{T} = \lambda \frac{|E_{R_1}|^2(\varepsilon_{\text{film}} - \varepsilon_{\text{liquid}})\pi R_1}{\int_v E_0 \cdot \varepsilon E_0^* \mathrm{d}\vec{r}}. \tag{8.19}$$

From (8.19), we can see that the sensitivity can be enhanced by increasing the ratio of the field intensity at the boundary to the total field, reducing the tube thickness, or enlarging the inner radius of the microtube. The first condition can be achieved by exciting higher-order evanescent modes. Reducing the tube thickness also decreases the optical field inside the tube wall, which has a similar effect to the first condition. Increasing the radius of the microtube will increase the evanescent field at the inner surface, which leads to improved sensitivity. Calculated sensitivity for different radial mode number is plotted in Fig. 8.34. In the evanescent sensing regime, the sensitivity increases with radial number because of the increase of the electric field at the inner boundary. In the nonevanescent sensing regime, the oscillatory behavior as a function of the mode number, similar to the case of bulk index sensing, is observed but with an opposite trend.

Sensing by Light Absorption in Microtube-Based Resonator Sensor

Detection of the presence of certain analytes in the fluid can also be accomplished if the analytes have light absorption in the wavelength range of interest. The sensitivity can be enhanced by using the microresonator structure due to the increased interaction length between the light propagating in the resonator and the analytes. For simplicity, we consider the case of analytes dissolved in the bulk liquid flowing through the microtube. Light absorption can be included in our model by taking into account the imaginary part of the refractive indices. The presence of the light absorbing analytes changes the imaginary part of refractive index of the liquid from κ_1 to κ_2, which will cause the imaginary part of the wave vector to change from k_1 to k_2 and the electric field of the resonant mode changes from E_1 to E_2. The unperturbed and the perturbed electric field still satisfy the wave vector equation:

$$\nabla \times \nabla \times E_1 = (k_0 - \mathrm{i}k_1)^2 \varepsilon_1 E_1, \tag{8.20}$$

$$\nabla \times \nabla \times E_2 = (k_0 - \mathrm{i}k_2)^2 \varepsilon_2 E_2, \tag{8.21}$$

where ε_1 and ε_2 are the dielectric constant distributions for microtube filled with liquid with optical absorption coefficients κ_1 and κ_2. In the liquid region, the dielectric constant are $\varepsilon_1 = n_0^2 - \kappa_1^2 + 2\mathrm{i}\, n_0\kappa_1$ and $\varepsilon_2 = n_0^2 - \kappa_2^2 + 2\mathrm{i}\, n_0\kappa_2$. In typical sensing experiment, both κ_1 and κ_2 are much smaller than n_0, so to the first-order approximation the resonant wavelength and the field distribution will remain unchanged. However, the resonant mode's Q-factor will change with the optical absorption coefficient. From (8.20) and (8.21), we can obtain the following equation

$$(k_0 - \mathrm{i}k_1)^2 \int \varepsilon_1 E_1 \cdot E_1^* \mathrm{d}\vec{r} = (k_0 - \mathrm{i}k_2)^2 \int \varepsilon_2 E_2 \cdot E_2^* \mathrm{d}\vec{r}. \tag{8.22}$$

By comparing the real part and imaginary part of the (8.22) we can obtain the following two equations

$$\frac{k_2 - k_1}{k_0} = \frac{n_0 \int_{v_1} E_1 \cdot E_1^* \mathrm{d}\vec{r}}{\int_v \varepsilon E_1 \cdot E_1^* \mathrm{d}\vec{r}} (\kappa_2 - \kappa_1), \tag{8.23}$$

$$k_1 \kappa_2 = k_2 \kappa_1. \tag{8.24}$$

The quality factor of the microtube resonator can be defined as flowing: $Q_i = k_0/2k_i$, where the $i = 1{,}2$. If we define

$$\Gamma = \frac{2n_0 \int_{v_1} E_1 \cdot E_1^* \mathrm{d}\vec{r}}{\int_v \varepsilon E_1 \cdot E_1^* \mathrm{d}\vec{r}},$$

which is interaction factor of the resonant mode, we can get

$$Q = \frac{1}{\kappa \Gamma} = \frac{\lambda}{4\pi\alpha\Gamma}. \tag{8.25}$$

Then we define the sensitivity as the change in Q-factor with the change of the imaginary part of the refractive index of the liquid:

$$S_{\text{abs}} = \left|\frac{\delta Q}{\delta \alpha}\right| = \frac{\lambda}{4\pi\alpha^2 \Gamma} = \frac{Q}{\alpha}. \tag{8.26}$$

Clearly high Q-factor resonance also benefits the detection of light-absorbing analytes in the fluid region.

8.3.4 *Surface Sensing Experiment: Detection of Lipid Membrane and Membrane Proteins*

Surface sensing relies on the interaction between the electric field and the adsorbed analytes at the sensing surface. According to (8.22), a microtube sensor is more sensitive with a larger field at the inner boundary. Unlike the nonevanescent modes with the highest sensitivity exploited in the bulk index sensing but having a very small field at the inner boundary, a different type of mode should be chosen that has sufficiently high optical field magnitude at the boundary. Figure 8.36 shows the electric field distribution of such a preferred resonant mode. The selected mode, corresponding to TE_{707}^{36}, is predominantly guided within the silica tube wall and decays exponentially in the low-index cladding regions. This mode has large electric field at the sensing surface (i.e., the liquid/silica interface) to allow maximal interaction with the bound biomolecules. Note that typically such a mode will only render moderate bulk index sensitivity due to the reduction in the ratio of the optical field residing in the liquid region.

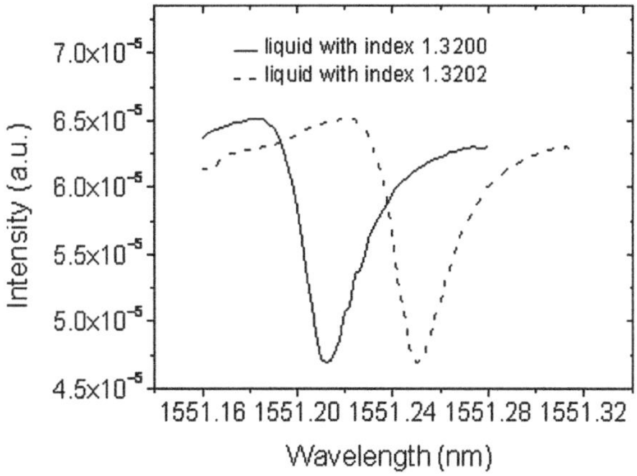

Fig. 8.35 Resonance spectra for two different liquid indices inside the micro-tube: 1.3200 and 1.3202

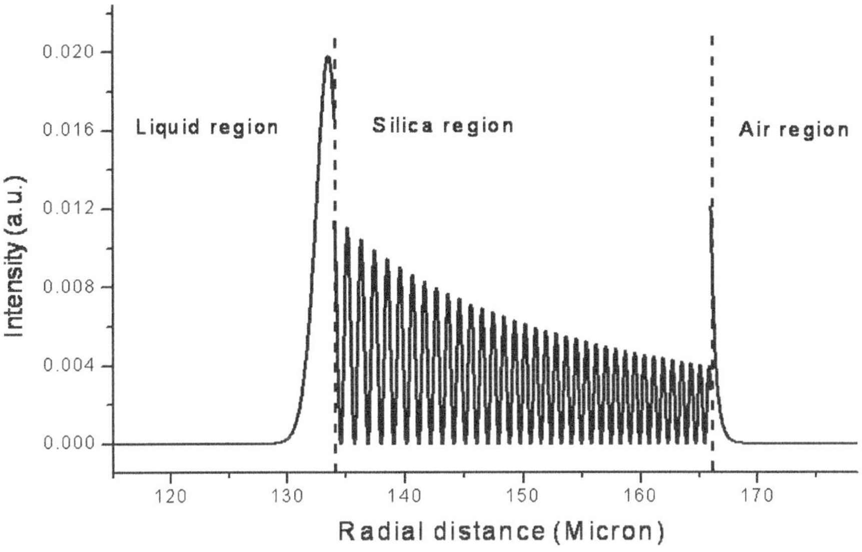

Fig. 8.36 Radial electric field intensity distribution for the resonant mode E_{r707}^{36}

To experimentally select a proper resonant mode for high surface sensitivity, the bulk index sensing experiment as described in Sect. 8.3.2 is first performed by varying the incident angles. Figure 8.35 shows the output resonance spectra for two fluids with index difference of 2×10^{-4}. The resonant wavelength shift is around 38 pm, resulting in a sensitivity of 190 nm per RIU. Based on the previous discussion, such a resonant mode should have a reasonably high optical field at the liquid/silica boundary.

The surface sensing experiment is designed to detect the formation of lipid bilayer membranes on the silica tube's inner wall. The lipid membrane is formed by introducing solutions containing the liposome into the tube. The liposomes rupture upon contacting the silica wall and subsequently fuse together and form a lipid bilayer membrane onto the silica tube wall. In the experiment, a microtube is first flowed with Tris buffer solution to establish a baseline. Then 1-palmitoyl-2-oleoyl-sn-glycero-3-phosphocholine (POPC) liposome solution is injected into the tube. After 30 min, the resonant spectrum was measured and showed a red shift of ~75 pm, which is due to the formation of lipid membrane on the microtube wall and the unbound liposome remained in the solution. Next a rinsing step with Tris buffer washes away any unbound liposome and the resonance is observed to shift back by ~21 pm. The resonance evolution is shown in Fig. 8.37. The net shift due to the adsorbed lipid membrane on the tube wall is around 44 pm.

Next we want to use the microresonator to detect the binding of a membrane protein to the lipid bilayer formed on the tube wall. The protein we choose is Annexin V, since its binding to the lipid membrane has been well studied in the

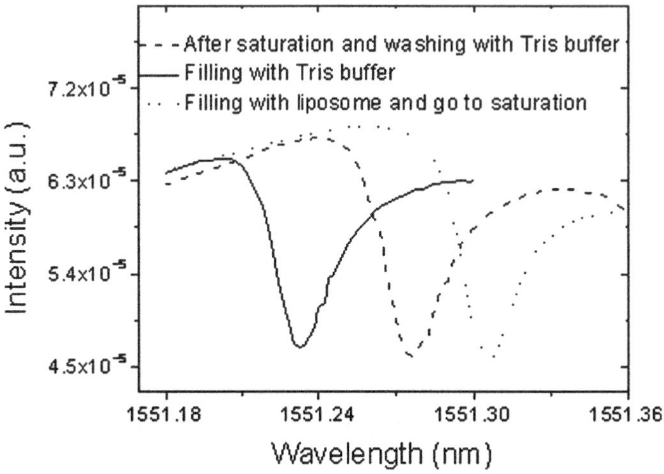

Fig. 8.37 Resonance spectra of the microtube filled with Tris buffer, liposome solution in saturation, and Tris buffer after rinsing, respectively

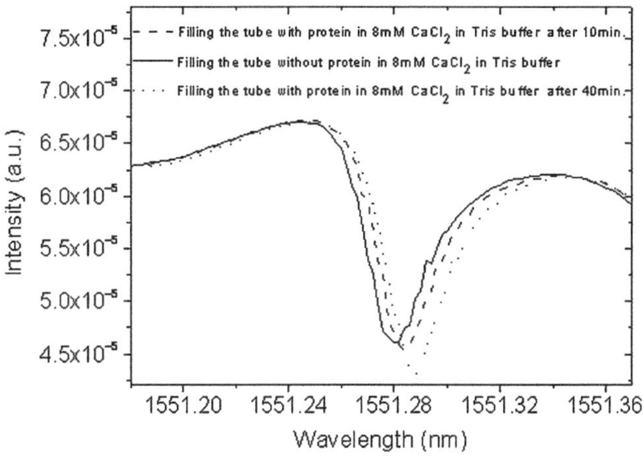

Fig. 8.38 Resonance curve shift due to Annex V proteins binding to the adsorbed lipid membrane on the silica wall

past[74]. After coated with lipid bilayer, the microtube is flowed with Annexin V proteins dissolved in 8 mM $CaCl_2$ at a concentration of ~26.67 ng mL^{-1}. The annexin V protein will bind to the lipid membrane in the presence of Ca^{2+}. Figure 8.38 shows a series of resonance curves when the microtube is filled with 8 mM $CaCl_2$ buffer solution to establish the baseline (left); after 10-min continuous flow of Annexin V solution followed by $CaCl_2$ rinsing (middle); and 40-min continuous flow of Annexin V solution followed by $CaCl_2$ rinsing (right). The resonance shift is relatively small as compared with that observed in the formation of lipid bilyer. This is due to the partial coverage of lipid membrane by the Annexin V protein.[74]

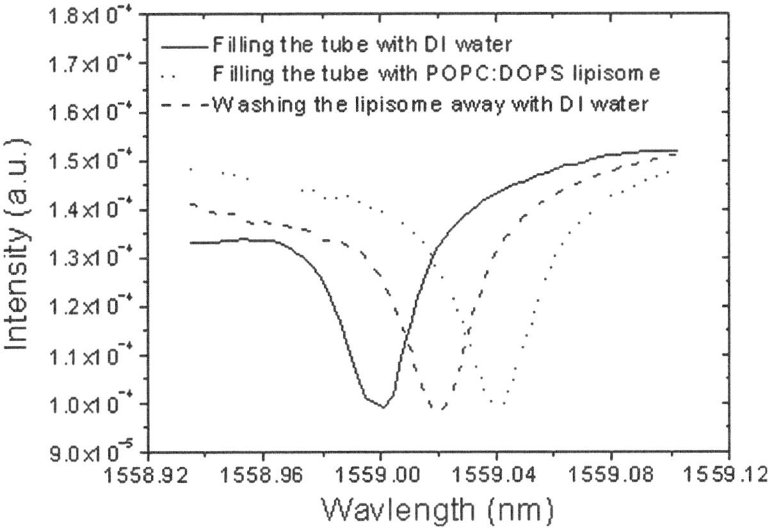

Fig. 8.39 Evolution of resonance curves due to the formation of lipid monolayer onto the inner wall of a microtube

In order to verify that the adsorbed lipid membrane indeed forms a bilayer film, another experiment is conducted with an aim to detect the formation of a monolayer lipid. It starts with a piranha-cleaned micro-tube treated with silane to render its inner surface hydrophobic. POPC liposome is then injected into the microtube. It is known that POPC lipid will form a monolayer to such a surface by orienting their hydrophobic tails toward the hydrophobic wall. The experimental results using a mode with similar sensitivity as the previous experiment are shown in Fig. 8.39. The resonance shift in this case is ~22 pm, which is about half of that observed for the adsorption of a lipid bilayer. These two experiments suggest that the microtube resonator is capable of accurately determining an adsorbed biomolecular layer down to a few nm thicknesses.

To further investigate the capability and the repeatability of the microtube system in detecting the adsorption of a ultrathin film, we performed another experiment by investigating a layer-by-layer polymer film formation on the silica tube wall. This experiment is done by using the electrostatic self-assembly of nanometer-thick polyelectrolyte films. Electrostatic self-assembly relies on the electrostatic interactions between an adjacent polycation and polyanion layers. The substrate is preprocessed to have charges (e.g., the oxide surface has a negative charge in dilute acid solution). In our experiment, we used poly-diallyldimethylammonium chloride (PDDA) as the polycation and polydye s-119 as the polyanion. To start, the microtube is cleaned with piranha solution for 30 min, which helps to form -O^-H^+ groups on tube surface where H^+ ions can be readily exchanged with other cations. Then cationic solution PDDA solution is flown through the tube to allow the negatively charged silica wall surface to be completely

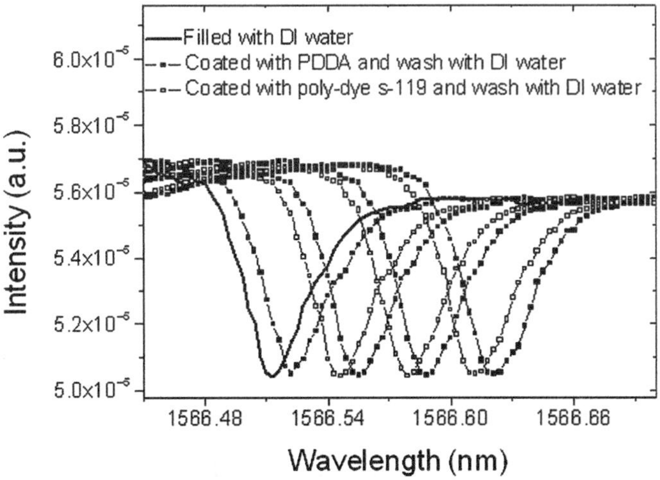

Fig. 8.40 Evolution of the resonance curves during the self-assembly of seven alternating electrostatic monolayers

covered with a layer of positively charged PDDA molecules. After a DI water rinse, anionic solution poly dye S-119 flows through the tube and forms a negatively charged monolayer that bonds to the pervious cationic layer by electrostatic interaction[75]. This sequence can be repeated to form a multilayer coating at the inner tube wall. The experiment results of coating seven alternating mono-layers are shown in Fig. 8.40. The resonance shift induced by PDDA monolayer is ~9 pm, while that of poly dye s-119 monolayer is ~24 pm. The reason for this difference is that the refractive index of poly dye s-119 is much large than that of PDDA.

8.3.5 *Quantification: Lipid Membrane and Binding of Membrane Proteins*

Next we will calculate the resonant wavelength shift due to the adsorption of an organic layer on the inner tube wall by using the Mie theory. We consider the resonant mode TE_{707}^{36} as shown in Fig. 8.36. First, we calculate the bulk index sensitivity by using the procedure described previously. The sensitivity for TE_{707}^{36} mode is found to be ~170 nm per RIU. This value is very close to the experimental result obtained in the index sensing experiment, 190 nm per RIU. This agreement also supports our interpretation of the optical field distribution shown in Fig. 8.36 for such a resonant mode.

Using the Mie scattering method, the resonant wavelength shift due to the adsorbed lipid membrane onto the tube wall is simulated with the microtube dimensions identical to the previous simulations. The refractive index of lipid bilayer is assumed to be 1.46. When the inner surface of a tube is coated with lipid membrane, the magnetic field for TE mode can be described in the following form:

$$H_z = \begin{cases} B_1 J_m(k_0 n_1 \rho) & 0 \le \rho < d_1 \\ [B_2 J_m(k_0 n_2 \rho) + B_3 N_m(k_0 n_2 \rho)] & d_1 \le \rho < d_2 \\ [B_4 J_m(k_0 n_3 \rho) + B_5 N_m(k_0 n_3 \rho)] & d_2 \le \rho < d_3 \\ B_6 H_m^{(1)}(k_0 n_4 \rho) & d_2 \le \rho < +\infty \end{cases} \tag{8.27}$$

where n_1, n_2, n_3, and n_4 are the refractive indices of the liquid, the lipid membrane, the silica microtube, and the air, respectively. After matching the boundary conditions, the resonance condition can be given by:

$$\frac{(B_2/B_3)J_m(k_0 n_2 d_2) + N_m(k_0 n_2 d_2)}{(B_2/B_3)n_2^{-1}J'_m(k_0 n_2 d_2) + n_2^{-1}N'_m(k_0 n_2 d_2)} = \frac{(B_4/B_5)J_m(k_0 n_3 d_2) + N_m(k_0 n_3 d_2)}{(B_4/B_5)n_3^{-1}J'_m(k_0 n_3 d_2) + n_3^{-1}N'_m(k_0 n_3 d_2)}, \tag{8.28}$$

where

$$\frac{B_2}{B_3} = \frac{n_2 J'_m(k_0 n_1 d_1)N_m(k_0 n_2 d_1) - n_1 J_m(k_0 n_1 d_1)N'_m(k_0 n_2 d_1)}{n_1 J_m(k_0 n_1 d_1)J'_m(k_0 n_2 d_1) - n_2 J'_m(k_0 n_1 d_1)J_m(k_0 n_2 d_1)}$$

and

$$\frac{B_4}{B_5} = \frac{n_3 H_m^{(1)'}(k_0 n_4 d_3)N_m(k_0 n_3 d_3) - n_4 H_m^{(1)}(k_0 n_4 d_3)N'_m(k_0 n_3 d_3)}{n_4 H_m^{(1)}(k_0 n_4 d_3)J'_m(k_0 n_3 d_3) - n_3 H_m^{(1)'}(k_0 n_4 d_3)J_m(k_0 n_3 d_3)}$$

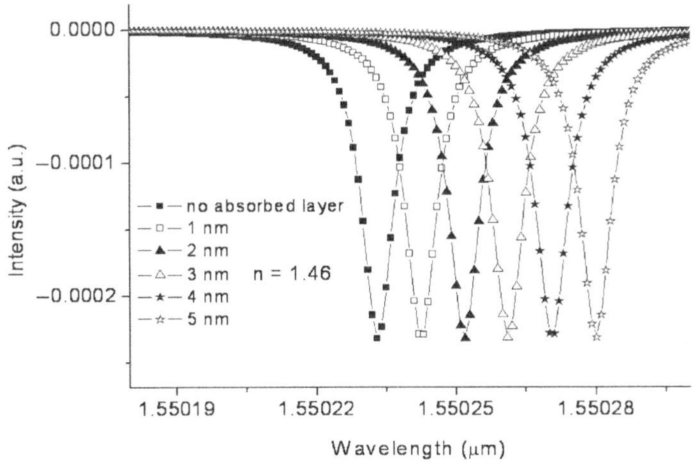

Fig. 8.41 Simulated resonance curve shift with the thickness change of the lipid membrane

Figure 8.41 shows the evolution of resonance curves when the thickness of the absorbed film increases incrementally by 1.0 nm. The curve shifts to longer wavelength with increased film thickness due to the increase of effective refractive index in the cavity. The resonant wavelength shift is calculated with respect to the thickness change of lipid membrane, and we find that the experimentally observed shift of 44 pm corresponds to a uniform thickness of 4–5 nm using the typical index of lipid membrane (1.46). This result is consistent with the thickness of a bilayer lipid membrane reported in the literature[76]. Performing the same calculation for the layer-by-layer adsorption experiment (PDDA and poly dye s-119 with the refractive indices of 1.46 and 1.6, respectively), we find the resonance shift would correspond to a monolayer thickness of 1 nm, which is also consistent to that reported in previous papers.[77]

8.4 Conclusions

Two types of optical microresonator sensors, polymer microring biochemical sensors and microtube resonator sensors, have been demonstrated in homogeneous and surface sensing. The former devices show high sensitivity over 70 nm per RIU and low detection limit as 250 pg mm^{-2}, while the latter ones enhance sensitivity up to 600 nm per RIU. Both illustrate high sensitivity, label-free detection capability, low cost, robustness, and simple fabrication process. Moreover, the theoretical calculation to analyze their sensing process has been derived, which well explained the experimental results. This not only facilitates the understanding of the device working principles, but also provides the directions to optimize sensing performance. In addition to the aforementioned advantages, polymer microring biochemical sensors are compact and highly integrable in an array format, while microtube-based sensors have built-in fluidic handling capability. These features could benefit the development of miniature and highly sensitive sensors with potential applications in many areas such as disease diagnostics, drug discovery, chemical analysis, and environmental monitoring.

Acknowledgment The authors would like to thank the support by the AFOSR and a University of Michigan Life Science Institute and Fisher-Thermo Pilot grant. We would also like to thank Wayne Fung for his assistance in the microring biosensor experiment, and Dr. Michael Mayer and Sheereen Majdzarringhalamaraghy for providing the liposome and membrane protein samples used in the microtube sensor experiment.

References

1 Collins, A. F.; Caruso, F., Biosensors: Recent advances, *Rep. Prog. Phys.* **1997**, 60, 1397–1445
2 Ramsden, J. J., Optical biosensors, *J. Mol. Recognit.* **1997**, 10, 109–120

3 Potyrailo, R. A.; Hobbs, S. E.; Hieftje, G. M., Optical waveguide sensors in analytical chemistry: Today's instrumentation, applications and trends for future development, *Fresenius J. Anal. Chem.* **1998**, 362, 349–373
4 Homola, J.; Yee, S. S.; Gauglitz, G., Surface plasmon resonance sensors: Review, *Sens. Actuators B.* **1999**, 54, 3–15
5 Homola, J., Surface Plasmon Resonance Based Sensors, Springer, Berlin, **2006**
6 Yin, D.-L.; Deamer, D. W.; Schmidt, H.; Barber, J. P.; Hawkins, A. R., Single-molecule detection sensitivity using planar integrated optics on a chip, *Opt. Lett.* **2006**, 31, 2136–2138
7 Luff, B. J.; Wilkinson, J. S.; Piehler, J.; Hollenbach, U.; Ingenhoff, J.; Fabricius, N., Integrated optical Mach-Zehnder Biosensor, *J. Lightwave Technol.* **1998**, 16, 583–592
8 Densmore, A.; Xu, D.-X.; Waldron, P.; Janz, S.; Cheben, P.; Lapointe, J.; Delage, A.; Lamontagne, B.; Schmid, J. H.; Post, E., A silicon-on-insulator photonic wire based evanescent field sensor, *IEEE Photon. Technol. Lett.* **2006**, 18, 2520–2522
9 Voros, J.; Ramsden, J. J.; Csucs, G.; Szendro, I.; De Paul, S. M.; Textor, M.; Spencer, N. D., Optical grating coupler biosensors, *Biomaterials* **2002**, 23, 3699–3710
10 Horvath, R.; Pedersen, H. C.; Skivesen, N.; Selmeczi, D.; Larsen, N. B., Optical waveguide sensor for on-line monitoring of bacteria, *Opt. Lett.* **2003**, 28, 1233–1235
11 Horvath, R.; Pedersen, H. C.; Skivesen, N.; Selmeczi, D.; Larsen, N. B., Monitoring of living cell attachment and spreading using reverse symmetry waveguide sensing, *Appl. Phys. Lett.* **2005**, 86, 071101
12 Blair, S.; Chen, Y., Resonant-enhanced evanescent-wave fluorescence biosensing with cylindrical optical cavities, *Appl. Opt.* **2001**, 40, 570–582
13 Boyd, R. W.; Heebner, J. E., Sensitive disk resonator photonic biosensor, *Appl. Opt.* **2001**, 40, 5742–5747
14 Armani, A. M.; Kulkarni, R. P.; Fraser, S. E.; Flagan, R. C.; Vahala, K. J., Label-free, single-molecule detection with optical microcavities, *Science* **2007**, 317, 783–787
15 Vollmer, F.; Braun, D.; Libchaber, A.; Khoshsima, M.; Teraoka, I.; Arnold, S., Protein detection by optical shift of a resonant microcavity, *Appl. Phys. Lett.* **2002**, 80, 4057–4059
16 Nadeau, J. L.; Iltchenko, V. S.; Kossakovski, D.; Bearman, G. H.; Maleki, L., High-Q whispering-gallery mode sensor in liquids, *Proc. SPIE* **2002**, 4629, 172–180
17 Noto, M.; Vollmer, F.; Keng, D.; Teraoka, I.; Arnold, S., Nanolayer characterization through wavelength multiplexing of microsphere resonator, *Opt. Lett.* **2005**, 30, 510–512
18 Krioukov, E.; Klunder, D. J. W.; Driessen, A.; Greve, J.; Otto, C., Integrated optical microcavities for enhanced evanescent-wave spectroscopy, *Opt. Lett.* **2002**, 27, 1504–1506
19 Krioukov, E.; Klunder, D. J. W.; Driessen, A.; Greve, J.; Otto, C., Sensor based on an integrated optical microcavity, *Opt. Lett.* **2002**, 27, 512–514
20 Fang, W.; Buchholz, D. B.; Bailey, R. C.; Hupp, J. T.; Chang, R. P. H.; Cao, H., Detection of chemical species using ultraviolet microdisk lasers, *Appl. Phys. Lett.* **2004**, 85, 3666–3668
21 Yalcin, A.; Popat, K. C.; Aldridge, J. C.; Desai, T. A.; Hryniewicz, J.; Chbouki, N.; Little, B. E.; King, O.; Van, V.; Chu, S.; Gill, D.; Anthes-Washburn, M.; Unlu, M. S.; Goldberg, B. B., Optical sensing of biomolecules using microring resonators, *IEEE J. Sel. Top. Quantum Electron.* **2006**, 12, 148–155
22 Fan, X.; White, I. M.; Zhu, H.; Suter, J. D.; Oveys, H., Overview of novel integrated optical ring resonator bio/chemical sensors, *Proc. SPIE* **2007**, 6452, 64520M
23 Rafizadeh, D.; Zhang, J. P.; Hagness, S. C.; Taflove, A.; Stair, K. A.; Ho, S. T.; Tiberio, R. C., Waveguide-coupled AlGaAs/GaAs microcavity ring and disk resonators with high finesse and 21.6-nm free spectral range, *Opt. Lett.* **1997**, 22, 1244–1246
24 Choi, S. J.; Djordjev, K.; Peng, Z.; Yang, Q.; Choi, S. J.; Dapkus, P. D., Laterally coupled buried heterostructure high-Q ring resonators, *IEEE Photon. Technol. Lett.* **2004**, 16, 2266–2268
25 Niehusmann, J.; Vorckel, A.; Bolivar, P. H.; Wahlbrink, T.; Henschel, W.; Kurz, H., Ultra-high-quality-factor silicon-on-insulator microring resonator, *Opt. Lett.* **2004**, 29, 2861–2863
26 Baehr-Jones, T.; Hochberg, M.; Walker, C.; Scherer, A., High-Q ring resonators in thin silicon-on-insulator, *Appl. Phys. Lett.* **2004**, 85, 3346–3348

27 Almeida, V. R.; Barrios, C. A.; Panepucci, R. R.; Lipson, M., All-optical control of light on a silicon chip, *Nature* **2004**, 431, 1081–1084
28 Klunder, D. J. W.; Tan, F. S.; van der Veen, T.; Bulthuis, H. F.; Sengo, G.; Docter, B.; Hoekstra, H. J. W. M.; Driessen, A., Experimental and numerical study of SiON microresonators with air and polymer cladding, *J. Lightwave Technol.* **2003**, 21, 1099–1110
29 Barwicz, T.; Popovic, M. A.; Rakich, P. T.; Watts, M. R.; Haus, H. A.; Ippen, E. P.; Smith, H. I., Microring-resonator-based add-drop filters in SiN: Fabrication and analysis, *Opt. Exp.* **2004**, 12, 1437–1442
30 Chu, S. T.; Little, B. E.; Pan, W.; Kaneko, T.; Kokubun, Y., Cascaded microring resonators for crosstalk reduction and spectrum cleanup in add-drop filters, *IEEE Photon. Technol. Lett.* **1999**, 11, 1423–1425
31 Rabiei, P.; Steier, W. H.; Zhang, C.; Dalton, L. R., Polymer micro-ring filters and modulators, *J. Lightwave Technol.* **2002**, 20, 1968–1975
32 Huang, Y.; Paloczi, G. T.; Scheuer, J.; Yariv, A., Soft lithography replication of polymeric microring optical resonators, *Opt. Exp.* **2003**, 11, 2452–2458
33 Xu, Q.; Fattal, D.; Beausoleil, R. G., Silicon microring resonators with 1.5-μm radius, *Opt. Exp.* **2008**, 16, 4309–4315
34 Nishihara, H.; Haruna, M.; Suhara, T., Optical Integrated Circuits, McGraw-Hill, New York, NY, **1989**
35 Chao, C. Y., Simple and effective calculation of modal properties of bent slot waveguides, *J. Opt. Soc. Am. B* **2007**, 24, 2373–2377
36 Adhikari, B.; Majumdar, S., Polymers in sensor applications, *Prog. Polym. Sci.* **2004**, 29, 699–766
37 Okamoto, K., Fundamentals of Optical Waveguides, Academic, San Diego, CA, 2000
38 Chou, S. Y.; Krauss, P. R.; Zhang, W.; Guo, L. J.; Zhuang, L., Sub-10 nm imprint lithography and applications, *J. Vac. Sci. Technol. B* **1997**, 15, 2897–2904
39 Guo, L. J., Topical review: Recent progress in nanoimprint technology and its applications, *J. Phys. D: Appl. Phys.* **2004**, 37, R123–R141
40 Hirai, Y.; Konishi, T.; Yoshikawa, T.; Yoshida, S., Simulation and experimental study of polymer deformation in nanoimprint lithography, *J. Vac. Sci. Technol. B* **2004**, 22, 3288–3293
41 Guo, L. J., Nanoimprint lithography: Methods and material requirement, *Adv. Mater.* **2007**, 19, 495–513
42 Chao, C. Y.; Guo, L. J., Polymer micro-ring resonators fabricated by nanoimprint technique, *J. Vac. Sci. Technol. B* **2002**, 20, 2862–2866
43 Chao, C. Y.; Fung, W.; Guo, L. J., Polymer microring resonators for biochemical sensing applications, *IEEE J. Sel. Top. Quantum Electron.* **2006**, 12, 134–142
44 Chao, C. Y.; Guo, L. J., Reduction of surface scattering loss in polymer microrings using thermal-reflow technique, *IEEE Photon. Technol. Lett.* **2004**, 16, 1498–1500
45 Chao, C. Y.; Guo, L. J., Design and optimization of microring resonators in biochemical sensing applications, *J. Lightwave Technol.* **2006**, 24, 1395–1401
46 Little, B. E.; Chu, S. T.; Haus, H. A.; Foresi, J.; Laine, J.-P., Microring resonator channel dropping filters, *J. Lightwave Technol.* **1997**, 15, 998–1005
47 Van, V.; Absil, P. P.; Hryniewicz, J. V.; Ho, P.-T., Propagation loss in single-mode GaAs-AlGaAs microring resonators: Measurement and model, *J. Lightwave Technol.* **2001**, 19, 1734–1739
48 Chao, C. Y.; Guo, L. J., Thermal-flow technique for reducing surface roughness and controlling gap size in polymer microring resonators, *Appl. Phys. Lett.* **2004**, 84, 2479–2481
49 Chao, C. Y.; Guo, L. J., Biochemical sensors based on polymer microrings with sharp asymmetrical resonances, *Appl. Phys. Lett.* **2003**, 83, 1527–1529
50 Lide, D. R., CRC Handbook of Chemistry and Physics, 88th edn.; CRC, Cleveland, OH, 2007
51 Diamandis, E. P.; Christopoulos, T. K., The biotin-(strept)avidin system: Principles and applications in biotechnology, *Clin. Chem.* **1991**, 37, 625–636
52 Peters, T., Jr., All About Albumin: Biochemistry, Genetics, and Medical Applications, Academic, San Diego, CA, **1996**

53 Yang, V. C.; Ngo, T. T., Biosensors and Their Applications, Kluwer/Plenum, New York, NY, **2000**
54 Fano, U., Effects of configuration interaction on intensities and phase shifts, *Phys. Rev.* **1961**, 124, 1866–1878
55 Fan, S., Sharp asymmetric lineshapes in side-coupled waveguide-cavity systems, *Appl. Phys. Lett.* **2002**, 80, 908–910
56 Chiba, A.; Fujiwara, H.; Hotta, J.; Takeuchi, S.; Sasaki, K., Fano resonance in a multimode tapered fiber coupled with a microspherical cavity, *Appl. Phys. Lett.* **2005**, 86, 261106
57 Liang, W.; Yang, L.; Poon, J. K.; Huang, Y.; Vahala, K. J.; Yariv, A., Transmission characteristics of a Fabry-Perot etalon-microtoroid resonator coupled systems, *Opt. Lett.* **2006**, 31, 510–512
58 Rezzonico, D.; Jazbinsek, M.; Guarino, A.; Kwon, O.; Gunter, P., Electro-optic Charon polymeric microring modulators, *Opt. Exp.* **2008**, 16, 613–627
59 Almeida, V. R.; Xu, Q.; Barrios, C. A.; Lipson, M., Guiding and confining light in void nanostructure, *Opt. Lett.* **2004**, 29, 1209–1211
60 Xu, Q.; Almeida, V. R.; Panepucci, R. R.; Lipson, M., Experimental demonstration of guiding and confining light in nanometer-size low-refractive-index material, *Opt. Lett.* **2004**, 29, 1626–1628
61 Baehr-Jones, T.; Hochberg, M.; Walker, C.; Scherer, A., High-Q optical resonators in silicon-on-insulator-based slot waveguides, *Appl. Phys. Lett.* **2005**, 86, 081101
62 Dell'Olio, F.; Passaro, V. M. N., Optical sensing by optimized silicon slot waveguide, *Opt. Exp.* **2007**, 15, 4977–4993
63 Feng, N.-N., Michel, J.; Kimerling, L. C., Optical field concentration in low-index waveguides, *IEEE J. Quantum Electron.* **2006**, 42, 885–890
64 Robinson, J. T.; Chen, L.; Lipson, M., On-chip gas detection in silicon optical microcavities, *Opt. Exp.* **2008**, 16, 4296–4301
65 White, I. M.; Suter, J. D.; Oveys, H.; Fan, X., Universal coupling between metal-clad waveguide and optical ring resonators, *Opt. Express* **2007**, 15, 646–651
66 White, I. M.; Oveys, H.; Fan, X., Liquid-core optical ring-resonator sensors, *Opt. Lett.* **2006**, 31, 1319–1321
67 Liang, W.; Huang, Y. Y.; Xu, Y.; Lee, R. K.; Yariv, A., High sensitive fiber Bragg grating refractive index sensors, *Appl. Phys. Lett.* **2005**, 86, 151122
68 Ling, T.; Guo, L. J., A unique resonance mode observed in a prism coupled micro-tube resonator sensor with superior index sensitivity, *Opt. Exp.* **2007**, 15, 17427–17432
69 Bohren, C. F.; Huffman, D. R., Absorption and Scattering of Light by Small Particles, Wiley, New York, NY, **1998**
70 Barber, P. W.; Hill, S. C., Light Scattering by Particles: Computational Methods, World Scientific, Teaneck, NJ, **1990**
71 Moon, H. J.; An, K., Interferential coupling effect on the whispering-gallery mode lasing in a double-layered microcylinder, *Appl. Phys. Lett.* **2002**, 80, 3250–3252
72 Teraoka, I.; Arnold, S., Enhancing the sensitivity of a whispering gallery mode microsphere sensor by a high-refractive-index surface layer, *J. Opt. Soc. Am. B* **2006**, 23, 1434–1441
73 Teraoka, I.; Arnold, S., Theory of resonance shifts in TE and TM whispering gallery modes by nonradial perturbations for sensing applications, *J. Opt. Soc. Am. B* **2006**, 23, 1381–1389
74 Voges, D.; Beredes, R.; Burger, A.; Demange, P.; Baumeister, W.; Huber, R., Three-dimensional structure of membrane-bound annexin V. A correlative electron microscopy-X-ray crystallography study, *J. Mol. Biol.* **1994**, 238, 199–213
75 Cheng, X.; Guo, L. J., Electrostatic self assembly of nano-composite polymer in grating structure, *J. Vac. Sci. Technol. B* **2001**, 19, 2736–2740
76 Howland, M. C.; Szmodis, A. W.; Sanii, B.; Parikh, A. N., Characterization of physical properties of supported phospholipid membranes using imaging ellipsometry at optical wavelengths, *Biophys. J.* **2007**, 92, 1306–1317
77 Miyashita, T.; Mizuta, Y.; Matsuda, M., Studies on Langmuir-Blodgett multilayer formation from preformed poly(*N*-alkylacrylamides), *Br. Polym. J.* **1990**, 22, 327–331

Chapter 9
Silicon Photonic Wire Waveguide Sensors

S. Janz, A. Densmore, D.-X. Xu, P. Waldron, J. Lapointe, J. H. Schmid, T. Mischki, G. Lopinski, A. Delâge, R. McKinnon, P. Cheben, and B. Lamontagne

Abstract Silicon photonic wire evanescent field (PWEF) waveguide sensors have a very high intrinsic response to molecular adsorption. This response arises from the high refractive index contrast and small size of these waveguides, which in combination cause much of the electric field to be localized to a thin layer adjacent to the waveguide surface. We describe the basic theory of PWEF waveguide sensors and compare the predicted PWEF sensor performance with that of surface plasmon resonance (SPR) and other waveguide systems. Finally, we present experimental results for PWEF sensors incorporated into integrated optical circuits employing Mach–Zehnder interferometers (MZIs), ring resonators, and folded waveguide structures that both amplify and facilitate the interrogation of the PWEF sensor response.

9.1 Introduction

Technologies for the detection of trace amounts of molecules are the basis for chemical and biological sensors, diagnostic methods relying on the presence or absence of molecular biomarkers, and also enable basic research in proteomics, genomics, and pharmacology. In most of these technologies, specificity, or the ability to recognize a specific target among a random assortment of molecules, is achieved through the use of affinity-binding reactions involving complementary pairs of molecules. Such chemical reactions are essential and integral to all living organisms, with examples including the binding of complementary DNA strands, enzymatic processes, and a wide range of antibody–antigen reactions. Similar selective receptor–target reactions can also be found for various inorganic elements

S. Janz (✉)
Institute for Microstructural Sciences, National Research Council Canada, Ottawa, ONK1A 0R6, Canada
e-mail: Siegfried.Janz@nrc-cnrc.gc.ca

X. Fan (ed.), *Advanced Photonic Structures for Biological and Chemical Detection*, Integrated Analytical Systems,
DOI 10.1007/978-0-387-98063-8_9,

and compounds, although these reactions are usually less specific than those between complex biological molecules. Given suitable binding partners, a sensor can be made by attaching one molecule (the receptor) to the surface, and detecting the subsequent binding of the second molecule (the analyte, or target) to the receptor via optical, mechanical, or electrochemical means.

In an optical sensor that detects the binding of molecules to a surface, the fundamental challenge is to obtain an optical transduction signal that is specific to the surface, yet is measurable over the background contribution of light reflected and scattered from the bulk substrate and surrounding media. In the case of simple reflection and transmission, the optical response is determined almost entirely by bulk material properties, since a surface molecular monolayer is at most only a few nanometers thick. The change in transmission through a monolayer film will be negligible. Similarly, a thin film causes a noticeable reflectivity change only once the film thickness approaches a significant fraction of the light's wavelength. At wavelengths in the range from $\lambda \sim 400$ to 2,000 nm for visible and near infrared light, the reflectivity change caused by the presence of a layer of bound molecules is typically less than about 0.1% that of the substrate reflectivity[1–6]. Although precise measurements of complex reflection coefficients by ellipsometry are routinely used to characterize thin films, the technique is near the limit of capabilities for single and fractional monolayer films[1,2]. Real time monitoring of reactions in liquid environments is also difficult with such reflectance-based techniques. Techniques with a higher surface specificity must be used to probe surface processes in any useful detail.

There are different paths to achieving surface specificity. One can exploit optical susceptibilities and resonances that are nonzero only at the surface or only for the molecular species of interest adsorbed on the surface. Examples include the use of second-order nonlinear mixing processes such as second harmonic generation[7–9] for which the nonlinear susceptibility tensor is nonzero only where inversion symmetry is broken. Spectroscopic techniques with very high selectivity for molecular resonances such as surface-enhanced infrared or Raman spectroscopy[10–12] may also be used.

Another strategy is to use guided optical modes that are localized to a narrow region near the surface and propagate along the surface, thereby enhancing the optical interaction cross-section with surface bound species. Surface plasmon resonance (SPR) sensing[13–15] and evanescent waveguide sensing[16–22] are two examples. In these guided wave methods the linear optical properties are measured, i.e., absorption or refractive index, and the wavelength dependence of the signal is not critical. Interpretation of the data and the necessary optical apparatus are therefore simpler than for nonlinear and spectroscopic techniques. On the other hand, specialized optical structures must be used to confine light to the region of interest; multilayer dielectric waveguides in the case of evanescent field waveguide sensors, and prism-coupled metal films in the case of SPR sensing[13–15]. However, waveguide and SPR substrates are available and usable for many types of biological experiments, as long as a suitable receptor molecule can be bound to the sensor surface – a process sometimes referred to as functionalization. SPR systems currently dominate the market for label-free optical monitoring of molecular-binding reactions. However, evanescent field waveguide technology is undergoing a major transformation in capabilities with the adoption of silicon photonics[23,24] and in particular the submicrometer silicon photonic wire platform[25–27].

In this chapter we will review the theory of silicon photonic wire evanescent field (PWEF) waveguide sensors, and compare PWEF sensor performance with other waveguide and guided-wave optical sensing techniques. In particular, Sect. 9.3 is devoted to a comparison of the physical mechanism and attainable performance of waveguide sensors and SPR. SPR systems for molecular interaction studies have been commercially available for many years[28], and set the present standard to which other optical sensing platforms must be compared. Finally, we will present recent experimental results for different silicon PWEF sensor circuits.

9.2 Optical Waveguides and Silicon Photonics

The simple three layer optical waveguide in Fig. 9.1 consists of a core layer surrounded by upper and lower cladding layers. When the core index of refraction, n_1, is sufficiently larger than the cladding layer indices n_2 and n_3, light coupled into the end facet of the waveguide structure undergoes total internal reflection at each core-cladding interface and thereby remains confined within the waveguide core as it propagates. In a more precise description based on electromagnetic theory, light propagates in the waveguide as a superposition of modes, the eigenvalue solutions of Maxwell's equations for the electromagnetic field in the waveguide structure[29,30]. In general, a waveguide may support one or more optical modes depending on the waveguide dimensions and materials[29,30] and each mode will have a different intensity profile and propagation velocity. However, in the remainder of the chapter we assume that the sensor waveguides are designed so that only one fundamental mode can propagate, since this is an essential condition for accurate sensor measurements to be made. In principle, the light can travel along the

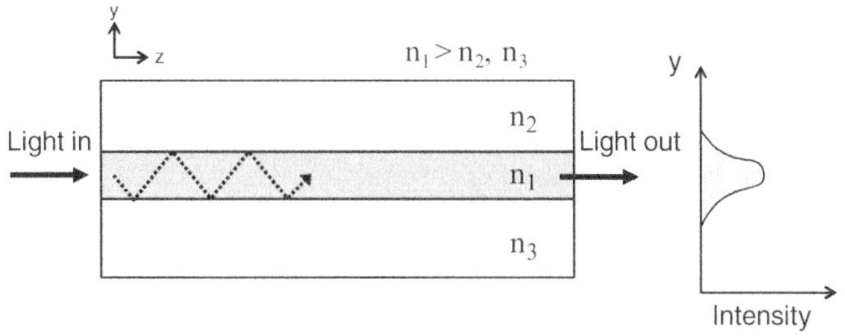

Fig. 9.1 Cross-section of a simple three-layer slab waveguide, and a plot of the fundamental mode intensity profile. Light rays (*dashed line*) in the waveguide are confined by total internal reflection at the core-cladding interfaces

waveguide forever with no loss, but in practice the propagation distance is limited since losses are always present to some degree. For example, in optical telecommunication fiber, light will travel for many tens of kilometers in the fiber core with no need for regeneration. In semiconductor waveguides, however, losses are usually on the order of 0.1–1 dB cm^{-1}, because of microscopic roughness at the core-cladding boundary, and intrinsic optical absorption in the waveguide materials.

Waveguides can take different forms and be built with many different materials, but the operating principle is the same as the simple three-layer waveguide of Fig. 9.1. Optical fibers have a cylindrical glass core of higher refractive index surrounded by a lower index glass cladding, with core diameters ranging from less than 10 μm for single mode fiber to 50 μm or more for multimode fibers. The index step between fiber core and cladding is usually on the order of 1%, depending on the fiber type.

The waveguides used in semiconductor lasers and integrated optical components are formed from layer structures similar to that of Fig. 9.1, and usually grown by molecular beam epitaxy (MBE) or chemical vapor deposition (CVD) on a semiconductor substrate wafer. The basic layer structure as in Fig. 9.1 provides only vertical confinement of light, but narrow ridges or channels, as in Fig. 9.2, can be etched into the waveguide layers to provide lateral optical confinement. The buried channel waveguide in Fig. 9.2b is formed by first etching the core layer, and then depositing a final dielectric layer. These ridge or channel waveguides are the optical equivalent of metal lines in electronic chips and circuit boards, and can be arranged in a similar way to form optical circuits that guide light to various points on a semiconductor chip. Integrated optical circuits perform essential functions, with examples ranging from the semiconductor lasers used in telecommunications, to complex devices that perform high-speed optical modulation and wavelength division multiplexing in long distance and local area optical communication networks.

At the smallest scale, the silicon photonic wire waveguides that are the subject of this chapter have a silicon core only a few hundred nanometers across and refractive index of $n = 3.47$ at a wavelength of $\lambda = 1550$ nm[31]. Figure 9.3 shows an example of such a photonic wire waveguide. The silicon core is bounded by an underlying SiO_2 cladding with an index of $n = 1.44$[31], and an upper cladding that may be

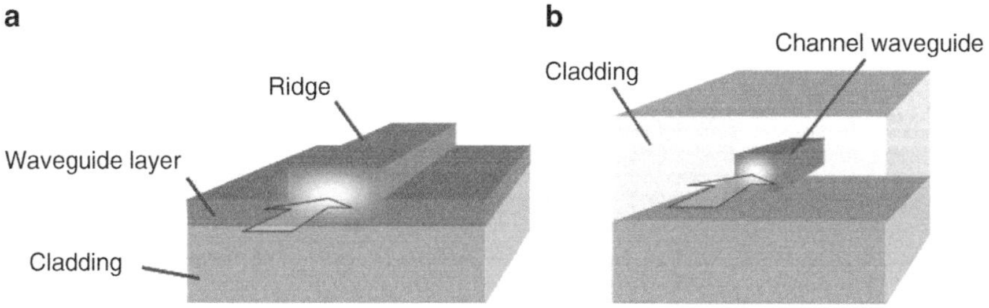

Fig. 9.2 Views of (**a**) a simple ridge waveguide and (**b**) a channel waveguide on a semiconductor substrate. The *arrow* indicates the direction of light propagation

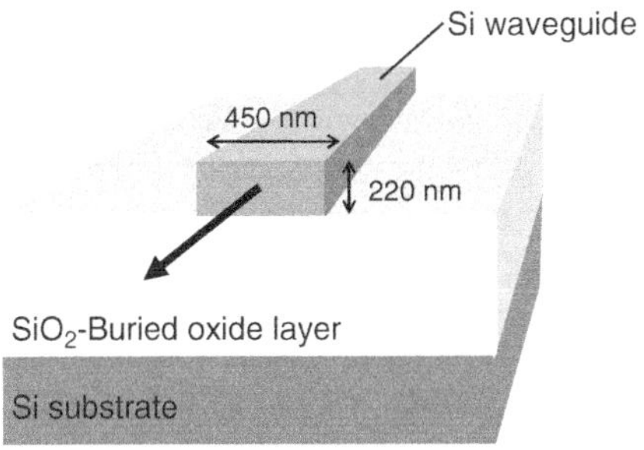

Fig. 9.3 Oblique cross-section view of the structure of a silicon photonic wire waveguide with a 220 nm × 450 nm core

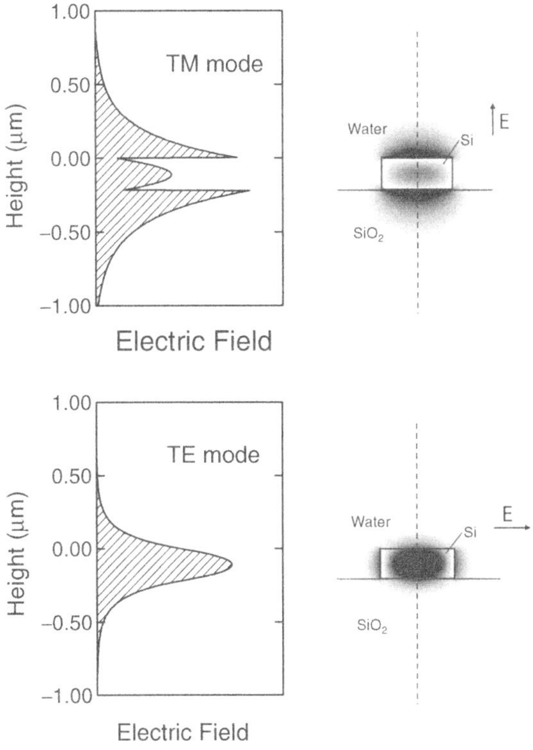

Fig. 9.4 The calculated TM and TE modes for $\lambda = 1550$ nm light propagating in the 220 nm × 450 nm photonic wire waveguide shown in Fig.9.3. The mode electric field profiles are shown as *line plots* along the center axis of the waveguide (indicated by the *dashed lines*), as well as in full cross-section by *gray* scale contour plots

air ($n = 1.0$), SiO_2, polymer ($n \sim 1.4$–1.6), or water ($n = 1.32$) as in the case of the exposed molecular sensor element. For all of these, the index difference between the Si core and cladding is more than a factor of two, and hence silicon waveguides are often described as a high index contrast system.

The importance of index contrast is evident in Fig. 9.4, which shows the cross-section of a photonic wire waveguide and the calculated transverse electric (TE) and transverse magnetic (TM) mode profiles. For the TE mode the electric field of the light is mainly polarized parallel to the wafer plane, while the TM mode the field is mainly polarized normal to the wafer plane. The differences in the two mode field distributions are a direct result of the high core-cladding index contrast combined with electromagnetic boundary conditions, and have a direct bearing on the application of these waveguides as molecular sensors. The high refractive index of the silicon core along with the large index contrast allows waveguides to have cross-sectional dimensions as small as 200 nm, and bend radii of a few micrometers. The chip area occupied by an integrated optical waveguide circuit depends on how sharply waveguides can be curved before bend losses become too high. Silicon photonic wire optical waveguide circuits can occupy areas less than 1 mm^2 on a silicon chip. This is much smaller than for low index contrast glass waveguides, for which the minimum waveguide core size is about 5 μm and the bend radius is usually larger than 5 mm to avoid high bend losses.

Most silicon waveguide devices are fabricated on silicon-on-insulator (SOI) wafers. An SOI wafer consists of a single crystal silicon substrate, typically about 600-μm thick. The uppermost layer is also single crystal silicon, but separated from the substrate by a buried oxide (BOx) SiO_2 layer. The top silicon and BOx layers can both range in thickness from a few hundred nanometers to several micrometers, depending on the ultimate application. As an example, the photonic wire waveguide shown in Fig. 9.3 is etched into the top Si layer, which is usually between 200 and 300-nm thick. The thickness of the underlying BOx layer is about 2 μm or more to ensure that guided light will not leak into the substrate.

One of the advantages of SOI-based silicon photonics is that high quality SOI wafers are readily available in bulk quantities from a number of manufacturers, since SOI is used extensively in the electronics industry. SOI wafers are manufactured either by implantation of oxygen ions into the silicon surface followed by annealing to form a homogeneous SiO_2 BOx layer (a process known as SIMOX – separation by implantation of oxygen), or alternatively by wafer bonding two Si wafers that have been covered with a surface SiO_2 layer. The bonded oxide layer becomes the BOx, while most of one of the two bonded Si wafers is removed by etching and polishing (bond and etch back process) or by cleaving along a previously defined defect layer created by ion implantation into one of the starting wafers. The latter technique is the one most often used to create the thin and extremely uniform Si layers required for photonic wire devices.

9.3 Silicon Photonic Wire Evanescent Field Sensors

The evanescent field waveguide sensor shown in Fig. 9.5 has the same basic structure as the three layer waveguide of Fig. 9.1, but the upper cladding is the fluid or gas that is being probed. During sensor operation, the fundamental

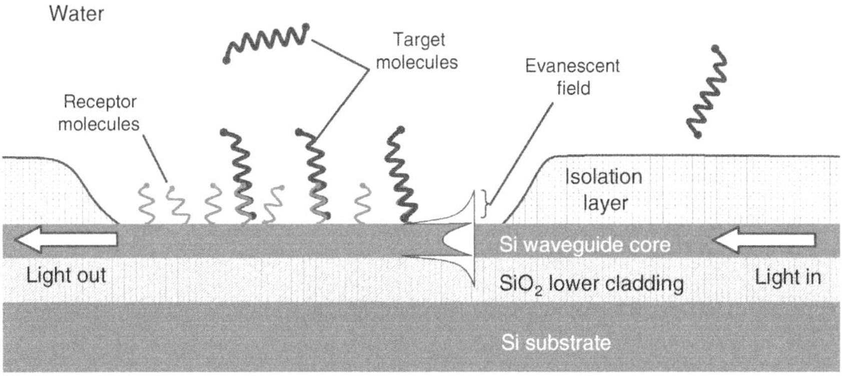

Fig. 9.5 Cross-section of an SOI evanescent field (EF) waveguide sensor

propagating waveguide mode is excited by launching light into the waveguide core. For any waveguide, the modal field distribution will always have an evanescent tail that extends a short distance into the cladding layers. This evanescent light effectively travels in the cladding, and any changes in the optical properties of the cladding material will change the propagation velocity or optical absorption of the waveguide mode as a whole. Sensor transduction is a result of the interaction of the guided mode evanescent tail with the ambient liquid and molecules adjacent to the waveguide surface.

The structure and mode profile in Figs. 9.3 and 9.4 are in fact designed for an SOI waveguide evanescent field sensor optimized for molecular affinity binding. Figure 9.5 shows a longitudinal cross-section of such a waveguide incorporated into an evanescent field sensor. Most of the waveguide circuit is covered with an impermeable dielectric layer, which may be SiO_2 or a polymer such as SU-8. The only restriction is that the material has negligible absorption and a lower index than the core, and is compatible with the biological samples and chemicals used in sensor operation. A window in this isolation layer exposes the waveguide surface to the sample fluid. For an affinity-binding sensor, the sensing window area is functionalized by attaching receptor molecules to the exposed surface. The receptor species is chosen for its ability to bind specifically and exclusively to the analyte molecule.

The two important properties of the waveguide mode are its cross-sectional electric field amplitude profile $A(x,y)$ and the effective index N_{eff}. Here x and y are the coordinates in the plane perpendicular to z, the waveguide propagation direction. The electric field at any point along the waveguide is given by

$$E(x, y, z) = A(x, y)\mathrm{e}^{\mathrm{i}(\beta z - \omega t)}, \tag{9.1}$$

where $\beta = N_{\mathrm{eff}}\,(2\pi\,/\lambda)$ is the propagation wave vector, and the corresponding phase velocity of the guided optical mode is $v = c/N_{\mathrm{eff}}$. Here c and λ are, respectively, the speed and the wavelength of light in vacuum. The value of N_{eff} is determined by the index of refraction of the core layer and both the upper and lower claddings[29,30]. In the sensor window in Fig. 9.5, the index of refraction immediately above the waveguide surface will increase as the analyte molecules bind to the receptor molecules, since most molecules of interest have a larger index of refraction than the water they displace. This will shift the effective index of the waveguide mode by an amount δN_{eff} that varies with both the size and density of the adsorbed molecules. When molecules bind to the sensor surface the phase, $\phi = (\beta z - \omega t)$, in (9.1) for light propagating through the sensor window of length L, will shift by

$$\Delta\phi = (2\pi/\lambda)\ \delta N_{\text{eff}} L. \tag{9.2}$$

The optical phase change $\Delta\phi$ provides a transduction signal proportional to the molecular density on the surface, and can be measured using interferometric techniques that will be described in Sect. 9.5. The change in effective index δN_{eff} depends on the interaction of the modal field distribution in (9.1) with the molecular layer. Using a perturbation theory approach[30], δN_{eff} can be expressed as an overlap integral of the waveguide mode field distribution with the local dielectric function perturbation $\Delta\varepsilon$ or the equivalent material index perturbation Δn.

$$\delta N_{\text{eff}} = c \int \Delta\varepsilon\, EE^{*} \mathrm{d}y \mathrm{d}x = 2c\varepsilon_0 \int n\Delta n\, EE^{*} \mathrm{d}y\, \mathrm{d}x. \tag{9.3}$$

As before, x and y are the coordinates perpendicular to the waveguide propagation direction.

Equations (9.2) and (9.3) suggest two independent strategies to increase waveguide sensor response. First, (9.2) states that the total phase shift increases in proportion the waveguide sensor length L. This interaction length is determined by details of the optical waveguide circuit layout, as will be discussed in Sect. 9.4. On the other hand, the effective index change given by (9.3) depends on the overlap of the waveguide mode with the surface region where molecular binding takes place. This overlap can be optimized by engineering the properties of the waveguide itself. While increasing the waveguide length, L, increases the response to the molecular density adsorbed on the waveguide, the phase response per adsorbed molecule is unchanged. By optimizing electric field overlap with the near surface region, however, the absolute response per adsorbed molecule can be increased.

Reducing either the index step between the waveguide core and upper cladding or the waveguide core thickness will both act to increase the fraction of the mode outside the core in the evanescent tail. For example, in glass waveguides with an index step near 1%, reducing the waveguide core thickness below about 3 μm causes the optical mode width to expand into the cladding. In SOI waveguides, the high index contrast is much more effective in confining the field to the waveguide core, and the core thickness must be reduced to less than 0.5 μm before a significant fraction of the mode electric field propagates in the cladding region.

Increasing the relative optical power contained in the evanescent field can increase the sensor response δN_{eff} to changes in the cladding material, but it does not necessarily lead to an effective surface-binding sensor. On the contrary, if the evanescent field extends too far into the cladding, the mode overlap with the adsorbed molecule layer at the surface will become smaller. The evanescent field must also be tightly localized in the vicinity of the adsorbed layer, which typically extends only a few nanometers from the waveguide surface. The waveguide evanescent field decays exponentially into the cladding as $\exp(-y/\delta)$, where the penetration depth δ depends on the difference between the mode effective index and the index of refraction of the cladding layer[29]:

$$\frac{1}{\delta} = \left(\frac{2\pi}{\lambda}\right)\sqrt{N_{\text{eff}}^2 - n_{\text{clad}}^2}. \tag{9.4}$$

Electric field localization to the near surface region therefore requires a large difference between N_{eff} and the cladding refractive index, and hence for a given waveguide core thickness the core material should have the highest possible refractive index.

Given these considerations, an evanescent field sensor waveguide should (a) be thin to force most of the mode distribution to propagate in the cladding and (b) have a much higher core refractive index than the lower and upper claddings, to effectively concentrate the evanescent field at the waveguide surface where the molecular-binding interactions take place. A survey of the commonly available waveguide systems suggests that the thin SOI-based photonic wire waveguides similar to the ones shown in Figs. 9.3 and 9.4 are the optimal choice for fabricating waveguide sensor devices.

Finally, the surface response of high index contrast waveguides such as SOI can be enhanced further by the electromagnetic boundary conditions at the waveguide surface. The tangential component of the electric field vector $\boldsymbol{E}$ must be continuous across the core-cladding interface, while the perpendicular component of the displacement $\boldsymbol{D} = \varepsilon\,\boldsymbol{E}$ must be continuous. The latter condition demands that the perpendicular electric field have a discontinuous step across the core-cladding boundary to accommodate the step in the dielectric function, ε, such that the ratio of the field just outside the core, $E(y^+)$, and just inside the surface, $E(y^-)$, is inversely proportional to the ratio of the core and cladding dielectric constants:

$$\frac{E(y^+)}{E(y^-)} = \frac{\varepsilon_{\text{core}}}{\varepsilon_{\text{clad}}} = \left(\frac{n_{\text{core}}}{n_{\text{clad}}}\right)^2. \tag{9.5}$$

Equation (9.5) suggests that once again, choosing a waveguide system with a high core refractive index will enhance modal electric field $E(y^+)$ in the near surface region of the sensor waveguide. In this case, the polarization of the waveguide mode must be the transverse magnetic (TM) mode, which has the dominant electric field component directed normal to the waveguide surface. Figure 9.4 shows the TM waveguide mode for a 220 nm × 450 nm SOI photonic wire waveguide, calculated using a mode expansion method. For the TM mode, more than half of the electric field intensity is in the cladding, yet the evanescent field remains localized within an attenuation length of only $\delta = 164$ nm from the surface; this is precisely the requirement for an optimized evanescent field waveguide sensor[32].

9.3.1 Comparison of Waveguide Sensors

There are many competing optical technologies addressing the need for a label-free sensing tool, which is reliable, inexpensive, and has the potential for high-throughput multivariate analysis. These include various evanescent field waveguide sensors[16–22] and SPR sensors[13–15]. Among the new approaches are sensors based on grating formation by molecular adsorption[33] and porous silicon-based devices[34,35]. There are also several refinements of reflectometry or ellipsometry[3–6], which are used to detect and monitor the extremely small changes in the reflectance or transmittance of a surface upon which binding reactions take place. While the physical considerations presented in the previous section suggest that silicon photonic wire waveguides are very well suited for sensing applications, a more quantitative assessment of PWEF and competing optical sensor platforms may be useful. In this section, the intrinsic surface response of silicon PWEF sensors will be compared only against other waveguide material platforms and SPR. Since SPR is the only fully established and widely deployed label-free affinity-binding sensor, it provides the natural benchmark against which emerging sensor platforms can be assessed. The physical assumptions used in this comparison can be easily adapted to give a relevant comparison of other sensor types.

Waveguide mode calculations can be used to make quantitative predictions of the effective index response of an SOI PWEF sensor to adsorption of a molecular monolayer, to optimize the sensor waveguide design, and also to compare the response for several different waveguide types. Sensor performance and practicality are complex issues that involve intrinsic sensitivity, ease of use, surface chemistry, and also the complexity, cost, and size of measurement apparatus. However, in most early stage development work, the initial point of comparison for these technologies is the intrinsic response of the sensor, and there are a number of detailed discussions on the sensitivity of optical sensor devices[16,36,37]. Nevertheless, the comparative merits of various sensor platforms are sometimes obscured by the use of different or inappropriate figures of merit to describe and compare surface affinity-binding sensors. In particular, the ability of a sensor to respond to small refractive index changes in the overlying sample fluid is often quoted (i.e., response per refractive index unit (RIU)), despite the fact that the response to bulk material properties has little relevance for monitoring surface reactions at the molecular scale. As discussed in the previous section, in a waveguide sensor the response to a surface monolayer will decline if the evanescent field extends too far into the cladding. However, the mode overlap with the bulk cladding material increases, and therefore so does the response to small index changes in a cladding liquid or gas. Clearly optimizing a sensor to measure bulk material properties is independent and incompatible with designing a sensor to monitor surface chemical reactions.

The fundamental response of both waveguide sensors and SPR sensors is the change in effective index, δN_{eff}, of the guided waveguide mode or surface plasmon mode. The physical perturbation is the attachment of molecules on the waveguide surface. The optical constants of such molecular layers are not well defined, particularly for submonolayer coverage, since the index of refraction, n, and dielectric function ε are concepts that have a well-defined meaning only for homogeneous bulk materials. It is not obvious that at 1% monolayer coverage, the index of refraction of the partial layer of molecules should be 1% that of a homogeneous layer of the same molecule. However it is generally accepted that this is the case, and work to date suggests that this is at least a reasonable working approximation[2,38,39]. In our numerical modeling, we therefore take the perturbation as being a continuous layer of refractive index n and thickness d. It has been suggested that at least for roughly spheroidal molecules, the measured n is proportional to surface density, while d is related to the size of the molecule on the surface[39]. Ambiguity still arises because many combinations of d and n can induce the same δN_{eff} in the waveguide mode or surface plasmon.

In many optical problems, the integrated optical path length $P_{opt} = \int n \, dl$ provides a convenient measure when evaluating the phase change of light after propagating a distance $L = \int dl$ through a medium of varying refractive index n. Here the integrals are line integrals along the light path. The optical thickness, D_{opt}, of a thin film can be similarly defined as the optical path length taken normal to the film plane, which in the case of a uniform film is the product of the index of refraction and film thickness, $D_{opt} = nd$. Equation (9.3) suggests that the effective index changes for both waveguide modes and surface plasmon modes are proportional to and uniquely determined by the optical thickness of the surface layer, provided Δn (or $\Delta\varepsilon$) is small and $d << \delta$, so that the electric field amplitude does not vary significantly over the thickness of the surface film.

Calculations of the effective index response to deposition of a thin film have been carried out using transfer matrix methods[30,40], for both silicon waveguide modes and surface plasmons under typical operating conditions. The surface plasmon mode was calculated assuming excitation by light at a vacuum wavelength of $\lambda = 800$ nm in a 50-nm thick gold film deposited on a glass substrate, with the metal surface bounded by water. The silicon waveguide was a 220-nm thick Si slab waveguide clad with SiO_2 underneath and water above, and an incident wavelength of $\lambda = 1550$ nm. In both cases, light is polarized perpendicular to the surface. The results in Fig. 9.6 show that for both PWEF and surface plasmon-guided modes, the film-induced δN_{eff} remains approximately constant as film index n and thickness d are varied over a wide range with the constraint that their product, the optical thickness, remains constant at $D_{opt} = 0.4$ nm and $D_{opt} = 0.1$ nm. For very thin films δN_{eff} in Fig. 9.6 does begin to decrease noticeably; a result of the large film index required to keep D_{opt} constant. In this regime, the assumption of small Δn (or $\Delta\varepsilon$)

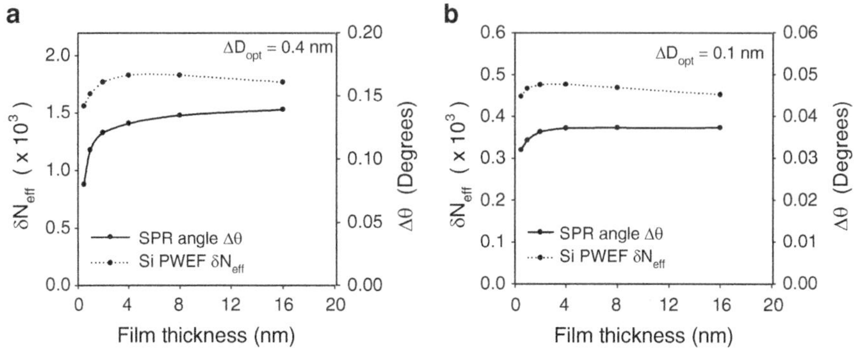

Fig. 9.6 The calculated effective index change in the silicon PWEF waveguide of Fig. 9.3 induced by adsorbed films of constant optical thickness (**a**) $D_{opt} = 0.4$ nm and (**b**) $D_{opt} = 0.1$ nm for film thickness between 0.5 and 16 nm. As the film thickness changes, the refractive index is adjusted so that the optical thickness remains constant. For comparison, the graphs also show δN_{eff} and the corresponding SPR angle shift $\Delta\theta$ for an SPR experiment

leading to (9.3) is no longer valid. As the film index becomes large, electric field is excluded from the film layer leading to a decrease in the induced δN_{eff}. For thicker films (9.3) remains valid, and δN_{eff} changes very little as long as the adsorbed film thickness is much less than the evanescent field penetration depth, δ.

Therefore D_{opt} provides a convenient single parameter that can be used to estimate the response of guided modes to very thin adsorbed layers. To assess and compare surface sensitivity, in this chapter, we will use the differential change of mode effective index with the optical thickness:

$$S_{mod} = \partial N_{eff} / \partial D_{opt}. \tag{9.6}$$

Unlike the dependence of N_{eff} on film thickness alone ($\partial N_{eff}/\partial d$) that is sometimes used as a figure of merit for guided mode molecular sensors, S_{mod} captures both the index and thickness dependence of the sensor response in a single parameter. While D_{opt} does not uniquely determine the film response for other optical techniques such as ellipsometry and reflectance difference, once d and n of the film are known, the optical thickness can be evaluated and comparisons are made between guided mode sensors and other techniques.

Guided mode calculations were also carried out to compare the sensor response of several waveguide systems. In these simulations a model molecular monolayer is represented by a 2-nm thick layer with a refractive index of $n = 1.5$. The optical properties of this model layer are typical of a dense layer of organic molecules on a substrate[1,41], and are a reasonable approximation for a streptavidin protein layer bound to a biotinylated surface, the experimental model system we use to characterize our sensors. The ambient upper cladding was assumed to be water with a refractive index of $n = 1.32$. For all examples, the lower cladding was assumed to be SiO_2 with an index of $n = 1.44$. In the simulations, the effective index of

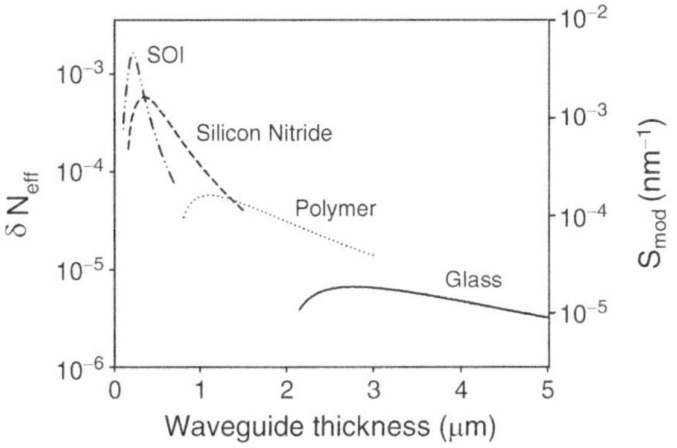

Fig. 9.7 The effective index change induced by the adsorption of a 2-nm thick molecular layer with index $n = 1.5$, calculated as a function of core thickness for the TM mode propagating in glass, polymer, silicon nitride, and silicon waveguides. The *right vertical axis* shows the equivalent modal sensitivity

a waveguide was calculated before and after the addition of the 2-nm molecular adlayer to the surface, and the difference in effective index δN_{eff} was found. The wavelength in all cases was assumed to be $\lambda = 1550$ nm.

The results of these simulations are shown in Fig. 9.7 for glass ($n_{core} = 1.454$), polymer ($n_{core} = 1.60$), silicon nitride ($n_{core} = 2.03$), and silicon ($n_{core} = 3.47$) waveguide cores. In all four cases, the lower cladding was taken to be SiO_2 ($n = 1.444$) and the upper cladding was water ($n = 1.32$). The minimum extent of each curve is the point at which the fundamental waveguide mode is cut-off. The results shown in Fig. 9.7 clearly confirm that sensor response to molecular adsorption increases with increasing waveguide core refractive index. The response also improves rapidly as the core thickness is decreased, but does reaches a maximum below which the sensitivity drops. As discussed above, once the core becomes too thin the mode expands so far into the upper and lower cladding that the relative overlap of the mode with the adsorbed layer (see 9.3) begins to decrease. The optimal thickness for a silicon waveguide sensor operating at this wavelength is 0.22 μm, with an induced index change of approximately $\delta N_{eff} = 1.7 \times 10^{-3}$ for the model adlayer film, corresponding to a modal sensitivity of $S_{mod} = 4.6 \times 10^{-3}\ \text{nm}^{-1}$.

Calculations for TE polarized light give similar results[42], but the maximum induced δN_{eff} in the PWEF waveguide is about three times smaller than for the TM mode. This difference is largely due to the absence of the surface field enhancement factor of (9.5), since the electric field of the TE mode is parallel to the waveguide surface and hence there is no electric field discontinuity.

Figure 9.8 shows a variant of the silicon photonic wire waveguide known as a slot waveguide, which has attracted considerable recent interest[43–45]. The structure consists of two parallel Si channel waveguides separated by a narrow gap. Here almost the entire mode field is concentrated in the gap between the two

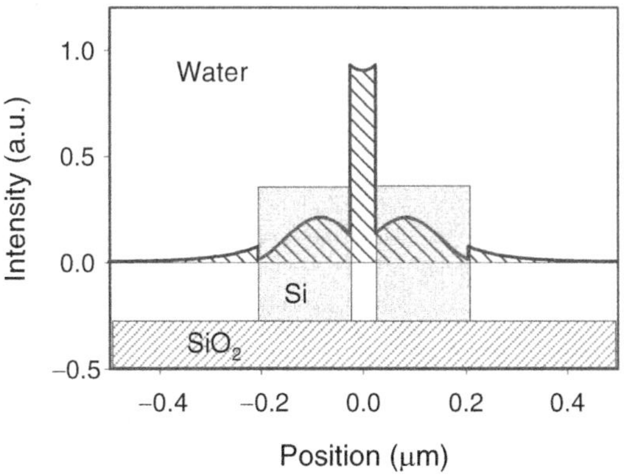

Fig. 9.8 Cross-section of a silicon slot waveguide consisting of two 180 nm × 250 nm silicon channels, separated by a 50-nm gap. The *solid line* represents a line plot of the electric field amplitude of the horizontally polarized TE mode, taken along the *horizontal midline* of the waveguide

channels[43]. This is again a consequence of small waveguide size and the electromagnetic boundary conditions that produce the strong surface field above a simple photonic wire, but here the electric field is polarized in the horizontal plane so that the main field discontinuity is at the vertical sidewalls. The slot waveguide is attractive for molecular affinity-binding sensing[44] and homogeneous material sensing[45], because of the enhanced overlap of the waveguide mode with the surface layer on the sidewalls of the gap and the bulk material filling the gap. This overlap is large because the extent of the external field is limited by the gap width (e.g., ~50 nm) rather than the penetration depth δ given by (9.4), and for narrow gap widths the field is largely excluded from the silicon channels. Waveguide calculations predict that the modal sensitivity may be five or more times larger than a simple photonic wire, depending on exact channel and gap dimensions. It is too early to say whether this advantage will be offset by practical difficulties associated with the narrow gap and high aspect ratio of a 50-nm wide slot. These challenges include reproducible slot fabrication, filling the slot with conformal cladding materials, as well as effective sample fluid flow through such a gap. In the remainder of the chapter our examples are based on the simpler photonic wire waveguide of Figs. 9.3 and 9.4, but the essential principles of sensor operation are the same for slot and PWEF waveguide sensors.

In practice, target molecules do not attach directly onto the waveguide core surface. To create a working sensor, the waveguide surface must be functionalized with receptor molecules that selectively bind the targeted analyte molecule. The creation of this functional layer usually involves the introduction of one or more intervening layers between the waveguide core surface and final receptor molecule. For example, for silicon PWEF sensors it is useful to form a thin layer of silicon dioxide on the surface to form a stable layer for attachment of subsequent molecular layers. The presence of such intervening layers displaces the receptor-target interaction interface away from the waveguide core and thereby may reduce the optical sensor response.

Mode calculations show that the insertion of up to 10-nm layer of SiO_2 on the waveguide surface reduces the sensor response by less than 10%[32,42]. Therefore Si sensor waveguides can be coated with a thin layer of SiO_2 with little loss in sensitivity, and the oxide surface can be functionalized to activate the sensor. This is an important point when considering the prospects for practical device implementation, since most commercial biosensor microarray devices use glass substrates. The chemistries already established for these devices can therefore be directly adapted to the Si PWEF sensor platform. More generally, the index of refraction for most organic molecules is comparable to that of SiO_2 ($n = 1.44$)[31], and the molecule size is a few nanometers, so the insertion of multiple molecular layers on the waveguide surface will have negligible effect on sensor response. For the Si photonic wire waveguide of Fig. 9.3, the presence of the functionalization layers should not seriously degrade sensor response as long as the total intervening thickness is much less than the extent of the evanescent field into the cladding (i.e., $\ll$ 100 nm).

9.3.2 Comparison of PWEF and SPR Sensors

In SPR the guided optical mode is an electron density wave propagating along a metal surface. The differences between SPR and waveguide sensors are best understood by considering the mode electric field distributions in the two cases. The mode profiles for the SPR and PWEF waveguide were calculated for a multilayer slab models as described in the previous section, using simple but exact transfer matrix methods[30,40]. As in Sect. 9.3.1, the SPR model consisted of a 50-nm thick Au (gold) film placed on a glass substrate and the excitation wavelength was $\lambda = 800$ nm. This corresponds to a standard SPR configuration at a near optimal operating wavelength[13,14]. The Au film is covered in water with index $n = 1.32$, and the dielectric function of gold is taken as $\varepsilon = -26.15 + i1.85$[31]. The silicon waveguide mode profile response is calculated using a slab waveguide model consisting of a buried oxide layer, a 220-nm thick Si core layer and a water upper cladding. The operating wavelength here was $\lambda = 1550$ nm, since this is the standard operating wavelength for most optoelectronic devices intended for telecommunications. The multilayer slab model is an accurate representation of a typical SPR system since there is no lateral optical confinement in the gold film plane. In the case of the silicon PWEF sensor, the slab layer model results will differ slightly from those determined for the channel waveguide of Fig. 9.3, but the small differences are negligible for the purpose of this comparison. The calculated plasmon mode profile for this system is shown in Fig. 9.9a and the silicon PWEF waveguide mode profile is shown in Fig. 9.9b. In the case of the plasmon mode, the electric field is confined to within $\delta = 362$ nm ($1/e$ penetration depth), more than double the evanescent field extent of $\delta \sim 164$ nm for the silicon waveguide. Since

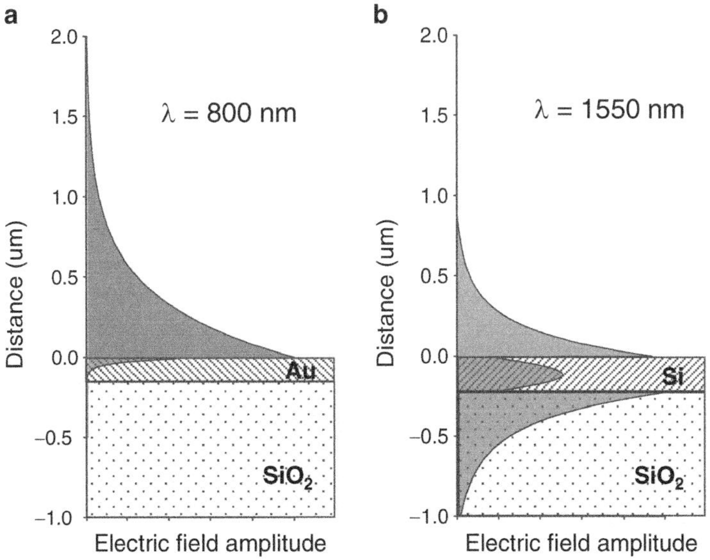

Fig. 9.9 The calculated mode profiles for (**a**) a surface plasmon mode propagating along a 50-nm thick Au film, with an excitation wavelength of $\lambda = 800$ nm (note: the plasmon field amplitude within the Au film is multiplied by 10 for clarity), and (**b**) a silicon PWEF waveguide with silicon core thickness of 220 nm, for an input wavelength of $\lambda = 1550$ nm

Table 9.1 Response of SPR and PWEF sensors to a film with $n = 1.5$ and $d = 2$ nm

	δN_{eff}	$\delta N_{eff}/N_{eff}$	S_{mod} (nm^{-1})
SPR (Au, $\lambda = 800$ nm)	1.85×10^{-3}	0.136	5.14×10^{-3}
PWEF (Si, $\lambda = 1550$ nm)	1.59×10^{-3}	0.08	4.42×10^{-3}

the sensor depends on the overlap of the mode profile with the molecular layer, a qualitative comparison of Fig. 9.9a and b suggests that both SPR and PWEF will have a comparable response to surface adsorption. The PWEF evanescent tail is better localized to the surface region, but a significant fraction of the mode is confined to the Si core and lower cladding. In the Au film of Fig. 9.9a, almost the entire plasmon mode field propagates in the overlying water, but it extends so far above the surface that the overlap with a surface monolayer is not substantially larger than in the PWEF case.

The effective index response δN_{eff} of the SPR and PWEF sensors can be evaluated as in the case of the waveguide comparison of Fig. 9.7, by comparing the modal effective index with and without the presence of a thin film on the waveguide or Au surface. Once the molecule induced index change has been calculated, the modal sensitivity $S_{mod} = \partial N_{eff}/\partial D_{opt}$ can also be determined. These S_{mod} values, given in Table 9.1, are approximately the same for both optimized PWEF waveguide and SPR sensors, confirming the qualitative conclusion drawn from consideration of the mode profiles of Fig. 9.9.

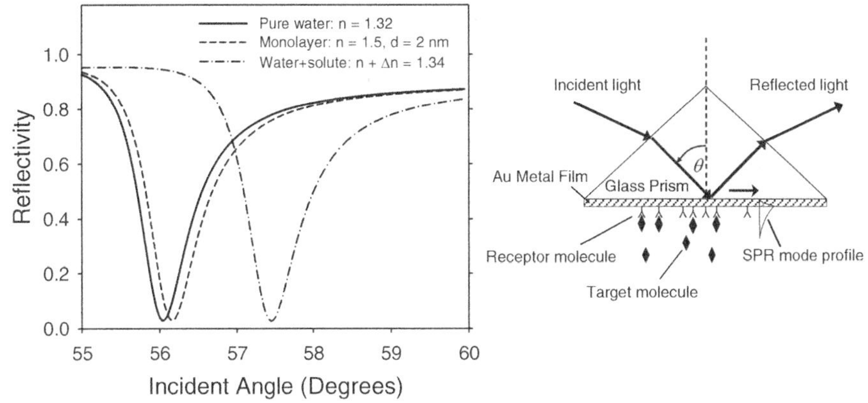

Fig. 9.10 The calculated angular reflection spectra for an SPR sensor with and without a 2-nm thick film on the gold surface. The assumed film index is $n = 1.5$. The diagram at *right* shows the schematic prism coupling geometry for measuring the SPR reflectivity angle dependence

SPR and PWEF sensors have comparable intrinsic modal sensitivity to molecular binding, but there does remain a fundamental physical difference between an optical mode in a waveguide and a plasmon mode propagating along the metal surface. A waveguide mode in a medium with low optical absorption or scattering losses will propagate long distances with little attenuation. A surface plasmon has a very short attenuation length, since the refractive index for metals always has a large imaginary part and hence a very large absorption coefficient. Since this absorption is an intrinsic property of all metals, plasmon attenuation is unavoidable in working with plasmon-based devices. In the case of a surface plasmon at a gold surface, the attenuation length is only about 18 μm for light at $\lambda = 800$ nm. Therefore, even though the molecule-induced δN_{eff} may be similar in both SPR and PWEF sensors, the short absorption length of a surface plasmon limits the accuracy with which δN_{eff} can be measured. Figure 9.10 shows the calculated angular reflection spectrum for a prism-coupled SPR measurement geometry. A beam at a wavelength of $\lambda = 800$ nm is incident from the prism side of a 50-nm Au film deposited on glass. The angular reflection spectrum is shown for an Au film with only water ($n = 1.32$) above the surface and with a 2-nm molecular layer on the surface. As the angle of incidence is varied, the reflected power has a minimum when the parallel component of the incident wave vector matches that of the surface plasmon in the Au film. When this condition is satisfied optical power from the incident beam is coupled to the plasmon, resulting in a strongly attenuated reflected beam. Since the wave vector is proportional to the effective index, the SPR angle gives a direct measure of N_{eff}. Monitoring the resonant angle is the most common method used to measure plasmon effective index in SPR sensors. The important point is that the width of the SPR resonance is determined by the attenuation length of the surface plasmon; the shorter the attenuation length, the broader the line width. This is a consequence of the uncertainty principle occurring in optics and quantum mechanics that relates the wave vector/momentum and spatial localization of a wave excitation. In Fig. 9.10, the shift in resonant angle is a very small fraction of the line width of the SPR resonance, so that the intensity change at any given angle is only a few percent. As a result, extracting accurate δN_{eff} measurements from SPR data requires very precise opto-mechanics and data fitting.

In the case of a PWEF waveguide sensor, the transduction signal is the cumulative phase shift $\Delta\phi$ of the light as it passes through the sensor waveguide. Equation (9.2) predicts that the sensor response can be increased without limit simply by increasing the length of the sensing waveguide. In practice, waveguides always have some loss due to interface roughness and material defects, but the typical propagation lengths of a centimeter or longer are hundreds of times larger than for surface plasmons. Therefore, even though SPR and PWEF waveguide sensors have comparable intrinsic response (i.e., δN_{eff}) to molecular binding, the ability to measure δN_{eff} in SPR is fundamentally limited by the absorption in metal. The cumulative phase response of PWEF waveguide sensors is in principle unlimited and only constrained by the waveguide circuit design, practical sensor size, and quality of waveguide fabrication. To illustrate this point, Fig. 9.11 shows the calculated response of an $L = 1.5$-mm long PWEF waveguide sensor as our model molecular film grows in thickness from 0 to 2 nm, again assuming a molecular layer index of $n = 1.5$. The transduction signal shown here is the output intensity of a Mach-Zehnder interferometer (MZI), which will be described in detail in Sect. 9.4.2. A comparison of Figs. 9.10 and 9.11 suggests that an appropriately designed waveguide sensor can have a much more robust response than SPR, particularly at very small fractions of a surface layer.

The previous sections show that silicon PWEF waveguide sensors have a higher response to surface molecular binding than all other commonly available

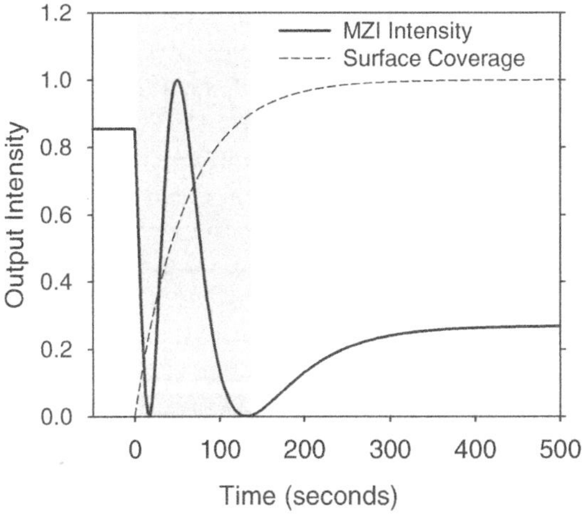

Fig. 9.11 The calculated response of a 1.5-mm long silicon PWEF sensor in a Mach–Zehnder interferometer configuration, as a film of final thickness $d = 2$ nm and index $n = 1.5$ grows on the waveguide surface. The normalized fractional surface coverage is also shown

waveguide platforms, and that the intrinsic modal sensitivity is comparable to standard SPR systems. However, as evidenced by Fig. 9.10 and Fig. 9.11, the potential of PWEF sensor technology lies in the ability to design sensor waveguide circuits to enhance device response far beyond SPR-based sensors, and to add additional capabilities to the sensor chip such as multisensor arrays, all on a single silicon chip. Some of the preliminary work in combining integrated optics with sensor technology will be discussed in the next section.

9.4 Photonic Wire Sensor Circuits

Most optical sensors rely on free space beam propagation and bulk optics. An assemblage of lenses, mirrors, and various opto-mechanical components directs light to and collects light from the sensor, and ultimately extracts a useful transduction signal from the light. The prism coupling geometry of Fig. 9.10 used to interrogate surface plasmon resonance sensors is a typical example. Waveguide sensors, on the other hand, can be monolithically integrated on a semiconductor chip with input and output waveguides and waveguide circuits that convert the molecule-induced optical phase change into a measurable transduction signal. Monolithic integration means that the alignment of optical elements is fixed and immune to disturbance. Since no bulk opto-mechanical components are required, an entire sensor array can be incorporated into very small chip areas. A waveguide also allows one to manipulate the propagation direction, phase, and intensity of light in much the same way that metal lines on an electronic integrated circuit deliver electrical signals to any point within the chip. This ability to manipulate light in complex ways, without the need for discrete optical components, is the second important advantage of waveguide-based sensors. In the following, we review and demonstrate a few basic waveguide circuit configurations that perform the optical signal functions necessary for a working sensor chip.

9.4.1 Folded Waveguide Path Sensors

The binding of molecules to the waveguide sensor surface and resulting change in effective index, ΔN_{eff}, create a phase change in the light passing through the sensor. According to (9.2) the cumulative phase response to a given surface density of molecules can be amplified, in principle without limit, by extending the length, L, of the active sensor waveguide. In practice, there are physical constraints on sensor area and chip size. In particular, long straight sensor waveguide geometries are not compatible with the microfluidics systems used to functionalize the sensor and also deliver sample liquids to the sensor. Multisensor array biochips are functionalized using automated spotting machines, which produce fluid droplets a few hundred

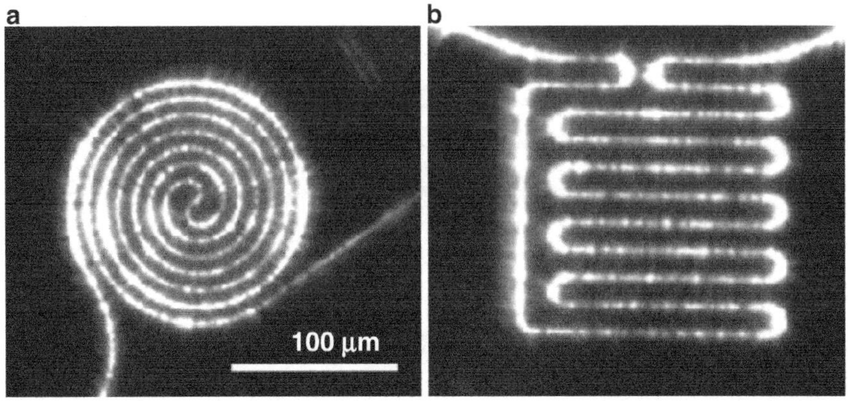

Fig. 9.12 *Top-view* of folded path waveguide layouts using (**a**) a double spiral and (**b**) a grid configuration. Both images are taken using an InGaAs infra-red camera while $\lambda = 1550$ nm light is coupled into the silicon waveguide chip

micrometers or smaller in diameter. The cross-section dimensions of microfluidic channels integrated on sensor chips are also a few hundred micrometers, and are determined by requirements of flow rate and fabrication constraints.

The tightly folded spiral and grid waveguide paths shown in Fig. 9.12 are used to optimize sensor response within the geometrical constraints. In these configurations, sensor elements with total waveguide lengths up to $L = 5$ mm have been demonstrated, with sensor diameters of only 150 μm. Although using long waveguides seems obvious, the practical optimization of the waveguide sensor geometry in this way is only possible because of the high refractive index contrast of the silicon photonic wire waveguides, which allows waveguide bend radii to be as small as 5 μm with negligible bend loss. For example, such compact structures would be impossible to make in conventional glass planar waveguide platforms, where the minimum bend radius is approximately 5 mm. In the cases of the folded waveguide paths of Fig. 9.12, the minimum bend radius used is 7 μm. The excess loss incurred by introducing a folded path waveguide is less than -1 dB for both the grid and spiral structures, when measured relative to a straight waveguide with the same end point separation.

Folded path waveguide structures can also enhance the cross-section of the active sensor surface with the microfluidic channel that delivers sample fluid to the sensor. A single straight photonic wire sensor embedded in a 100-μm wide microfluidic channel occupies less than 0.5% of the channel floor. Since the entire channel floor is functionalized with receptor molecules, only a small fraction of the analyte molecules will attach to the sensor waveguide surface and contribute to the measured response. By folding the sensing waveguide so that it weaves back and forth across the channel floor, the molecule collection efficiency can be increased many times for the same volume of sample fluid. Again, the high index contrast ratio and tight bend radii possible with recently developed silicon photonic wire platform make this strategy practical for the first time.

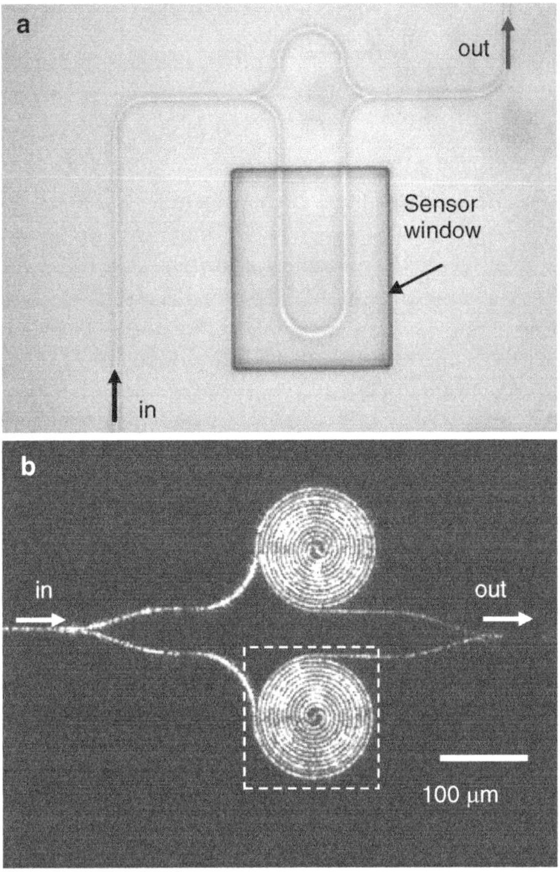

Fig. 9.13 Mach–Zehnder interferometer (MZI) sensor interrogation circuit using (**a**) conventional linear waveguide geometry and (**b**) folded spiral waveguide geometry. The *dashed line* in (**b**) indicates the sensor window boundaries in this infra-red camera image

9.4.2 Mach–Zehnder Interferometers

The first signal-processing task of a sensor waveguide circuit is to translate the phase shift caused by molecular binding into a measurable signal. This requires the use of some form of interferometer or an optical resonator. The MZI is the most common waveguide interferometer. Figure 9.13 shows two examples of MZI sensor circuits, one using a simple straight waveguide layout and one with folded waveguide segments. In either MZI geometry, the input waveguide is split into two arms and then recombined. The light that is split and passed through the two arms is recombined and will interfere to give a phase-dependent MZI output intensity. The active sensing section of the waveguide sensor is incorporated into one arm of the MZI, while the other arm serves as a reference. If the phase, $\Delta\phi$, of the light passing through the sensor arm changes according to (9.2), the output intensity of the MZI undergoes a sinusoidal modulation of form

$$I = \cos^2(\delta N_{\text{eff}}(2\pi/\lambda)L_s + \phi_0), \tag{9.7}$$

which can be measured with a photodetector. Here L_s is the active sensor length, and ϕ_0 is the baseline phase difference between light traveling through the two arms of the interferometer, prior to molecular adsorption. This phase difference will depend on the total optical path length of the arm. Changes in δN_{eff} can be detected either by operating at one fixed wavelength and monitoring the MZI intensity or by measuring the wavelength spectrum of the MZI output and extracting the wavelength shift of the MZI output fringes. Assuming the MZI is biased at its maximum output intensity (e.g., $\phi_0 = 0$), a 3 dB (i.e., a factor of 2) change in MZI output intensity requires an effective index perturbation

$$\delta N_{\text{eff}} = \frac{\lambda}{8L_s}. \tag{9.8}$$

Using the modal sensitivity results in Table 9.1, this corresponds to the binding of a film with an optical thickness change of $\delta D_{\text{opt}} = 0.04$ nm (relative to a layer of water of the same thickness), if we assume a $L_s =$ 1-mm long sensor and an operating wavelength of $\lambda = 1550$ nm. This is estimated to be equivalent to a surface coverage of 0.2 monolayer of streptavidin or an organic molecule of similar size.

Intensity measurements at a fixed wavelength are simple and require little equipment beyond an appropriate light source and photodetector, but the sensitivity of the MZI sensor can be zero if the device is biased at the extrema of the cosine function in (9.7). Although this is not a problem if the sensor output is expected to sweep through many fringe cycles in response to a target, detecting very small signals may be problematic unless the MZI is biased a priori near the operating point of maximum sensitivity. Wavelength interrogation can circumvent this problem by monitoring a spectral bandwidth encompassing at least one full fringe spacing, at the cost of requiring a tunable laser source or a spectrometer as part of the sensor interrogation system.

Once the modal sensitivity and sensor length are given, signal-to-noise and sensor drift are the remaining factors that determine performance. In the case of sensors based on optical phase measurements, even small fluctuations in local temperature can be an ongoing source of spurious sensor drift. The parameters in (9.7) that determine the background phase, ϕ_0, all vary with temperature. These include the refractive index of the waveguide core, cladding materials, and sample liquids. For silicon waveguides, in particular, the thermo-optic coefficient of 1.8×10^{-4} is relatively large. A temperature change of a few degrees C can produce an effective index shift in a sensor waveguide that is comparable to that induced by the binding of a monolayer of proteins.

The MZI interrogation circuit of Fig. 9.13b offers a simple method to eliminate sensor drift due to changes in temperature or other common mode variables.

By introducing a second-folded waveguide spiral into the reference arm of the MZI such that the optical path length matches the sensor arm, the base line phase difference ϕ_0 between light traveling through the two arms will be zero. Phase changes due to variables such as temperature that affect both arms in the same way will also cancel when the light propagating in the two arms is recombined. For the design shown, the temperature dependence of the measured signal is reduced by a factor of 20 compared with that for an unbalanced sensor. The remaining temperature dependence arises from a residual optical path length difference and differences in thermo-optic coefficients of the two MZI waveguide arms. Even though the geometrical waveguide lengths and waveguide core are identical, the material index and thermo-optic coefficients of the SU-8 and water claddings over the sensor and reference arms are different. However given the sufficient knowledge of the cladding optical constants, the MZI path lengths can easily be adjusted to compensate for these differences.

Assessment of PWEF sensor performance has been carried out using the biotin-streptavidin-binding reaction. This system is well suited to calibrating surface response since the biotin-streptavidin reaction is effectively irreversible, with a dissociation constant of $K_d = 10^{-15}$ mol L^{-1}, one of the strongest noncovalent bonds known. As a result, the surface-binding reaction proceeds in one direction only, so interpretation of sensor data is not complicated by details of equilibrium kinetics. The biotin-streptavidin affinity forms the basis of many processes used in molecular research, and hence this work also provides a basis for future work on related systems. Streptavidin is a protein with a molecular weight of approximately 60 kDa, and an estimated surface footprint of 4×5 nm[46]. Hence its optical properties should be representative of many medium and large size proteins.

The PWEF waveguides were all 260 nm × 450 nm silicon photonic wires. These were patterned using e-beam lithography, and etched using inductively coupled plasma (ICP) etching with an SF_6-C_4F_8 plasma chemistry. The entire chip was coated with SU-8 polymer, except for the folded sensor section of the MZI sensor arm where the SU-8 was removed to expose the waveguide to the sample fluid. Finally, a polydimethylsiloxane (PDMS) block with molded microfluidic channels was bonded to the MZI chip surface. Solution containing target molecules is delivered to the exposed sensor waveguide through these channels. In this and similar experiments, the sensor waveguide surface is functionalized as described in Ref. 41. A silane molecule (3-aminopropyltriethoxysile, APTES) is attached to the SiO_2-covered surface, and then a biotin molecule is attached to the end of the APTES chain. The underlying sensor surface has a thin 3-nm SiO_2 layer over the silicon, so the APTES is attached to the SiO_2 rather than bare Si. After functionalization, the biotinylated sensors are dried and stored until ready for use.

In these experiments, a MZI with a 2-mm long double spiral configuration, as in Fig. 9.13b, was used. TM polarized light at a wavelength of $\lambda = 1550$ nm was coupled into the waveguide sensor from a tapered fiber. The output light was collected and projected onto a photodetector with a microscope objective. Figure Figure 9.14a shows an example of a MZI sensor output collected in real time as the sensor is exposed to a 10 μg/ml solution of streptavidin molecules, during which time streptavidin monolayer builds up on a biotin functionalized surface.

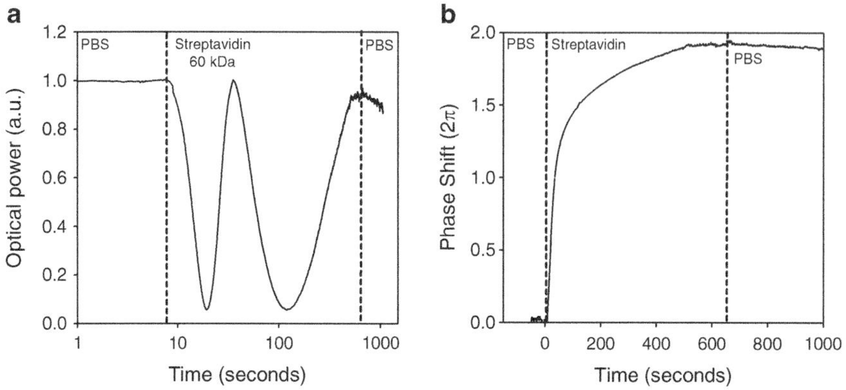

Fig. 9.14 (**a**) The measured variation of the Mach–Zehnder interferometer (MZI) output intensity as a monolayer of streptavidin is bound to the surface, and (**b**) the optical phase change calculated from this intensity data

The measured intensity modulation can then be used to recover the original optical phase change $\Delta\phi$ in the sensor arm. This phase shift, shown in Fig. 9.14b, is directly proportional to the density of molecules on the surface, as long as the film thickness is much less than the evanescent field penetration depth of $\delta \sim 162$ nm.

In the data of Fig. 9.14, the sensor phase response initially increases rapidly as the streptavidin solution is introduced, and then approaches a saturation level. Qualitatively, this behavior is typical for adsorption of molecules on a surface, although the curve in Fig. 9.14b does not follow the classical Langmuir adsorption model likely because of transport and surface site blocking effects[47,48]. We interpret this response saturation as the completion of a monolayer of streptavidin on the sensor surface. The precise surface density of such a monolayer is somewhat uncertain, because the biotin molecule (molecular weight 244 Da) is much smaller than streptavidin. Therefore the streptavidin monolayer completion is determined by the spatial blocking of remaining biotin-binding sites, rather than the filling of all biotin sites. Blocking also modifies the observed binding rate as the monolayer approaches completion[47,48].

Depending on the dynamics and deformability of the bound streptavidin, the density of a completed monolayer may approach the density of a full close packed arrangement of streptavidin, or be as low as 54% of the close packed density predicted by rigid sphere blocking models[47,48]. Keeping these uncertainties in mind, we have estimated a level of detection of about 0.3% of a monolayer or an equivalent mass density of 7 pg mm^{-2}, using both ellipsometry and waveguide measurements[41].

In the present experiment, the optical fiber is aligned to the sensor chip using a separate micropositioner, and no antireflection coatings are used on the chip facets. As a result, the observed noise floor is limited by vibrations and interference effects of the present experimental setup. With the adoption of well-established packaging

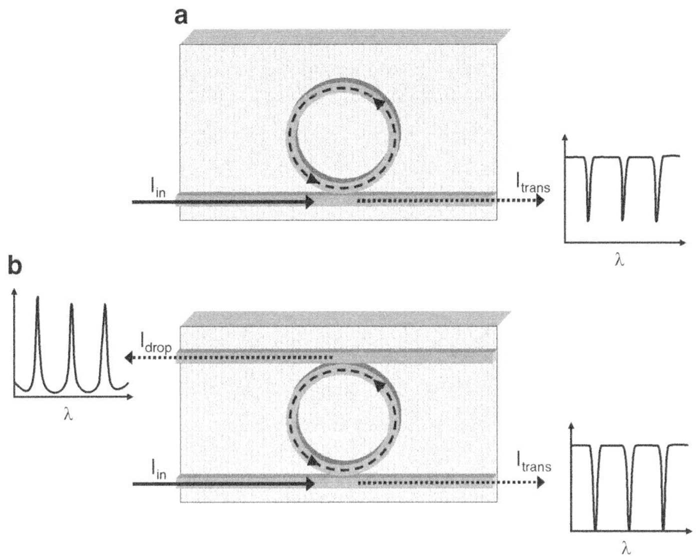

Fig. 9.15 A ring resonator sensor interrogation circuits with (**a**) one bus waveguide, and (**b**) two bus waveguides. The plots give a qualitative indication of the typical intensity spectrum that would be observed at the respective output ports

technologies used in the optical telecommunications industry, we expect a significant improvement of the signal-to-noise ratio.

9.4.3 Ring Resonator Sensors

The second approach for obtaining accurate optical phase measurements is the use of resonators. In waveguide optics, the ring resonator configurations of Fig. 9.15 have been explored by many groups, with Refs. 49–53 representing only a very small sample of the available literature. Rings have been used for wavelength selective filtering, optical modulation, and wavelength routing, as well as optical sensing. The ring resonator consists of a circular or oval waveguide resonator cavity that is coupled to a straight bus waveguide. A second separate bus waveguide as in Fig. 9.15b may also be present, if the ring is used as a wavelength selective filter, but this is not necessary for simple sensor applications. Hence, in the following discussion we focus on the single bus waveguide geometry of Fig. 9.15a.

In operation, a fraction of light propagating along the bus waveguide in Fig. 9.15a will couple into the ring waveguide with an electric field amplitude coupling coefficient, κ, while the remainder is transmitted through the coupler to the waveguide output with a transmitted amplitude t. After one trip around the ring, a fraction of this light will be coupled back to the bus waveguide again with coupling coefficient κ, while the rest is transmitted back into the ring with

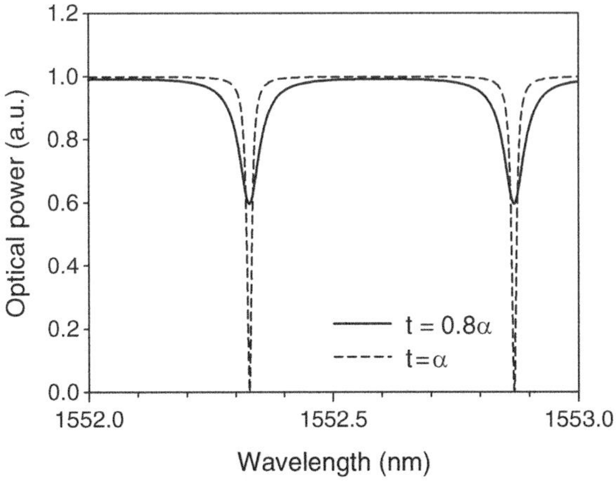

Fig. 9.16 The calculated output spectrum of a silicon waveguide ring resonator sensor at critical coupling ($t = \alpha$) and for $t = 0.8\alpha$. The assumed ring radius is $R = 150$ μm

transmission coefficient t. The ring resonators in Fig. 9.15 behave much the same as a classical Fabry-Perot resonator where the waveguide couplers take the place of mirrors. The coupler transmission coefficient, t, plays the role of Fabry-Perot mirror reflectivity while the coupling coefficient, κ, corresponds to mirror transmission. For the ring with only one bus waveguide as in Fig. 9.15a, the classical analog is a Fabry-Perot cavity with one mirror having 100% reflectivity. As in a Fabry-Perot resonator, the light transmitted past the ring resonator will display strong resonances when the incident wavelength is resonant with the ring cavity, i.e., the total ring length is equal to an integer number, m, of wavelengths. Two examples are given in Fig. 9.16. Since the physical wavelength in the waveguide is $\lambda = \lambda_0/N_{\mathrm{eff}}$, the resonance wavelength will vary with the waveguide effective index. If the ring is used as the waveguide sensor element, the binding of molecules to the ring resonator waveguide surface will change N_{eff} and cause a corresponding shift of the ring resonance.

Before considering the use of rings as sensor elements, it is useful to review a few basic ring properties and definitions. The output intensity of a ring with a single bus waveguide as in Fig. 9.15a is described by the equation

$$I_{\mathrm{out}} = I_0 \cdot \frac{\alpha^2 - 2\alpha|t|\cos\phi + |t|^2}{1 - 2\alpha|t|\cos\phi + \alpha^2|t|^2}, \tag{9.9}$$

where $|t|^2 = tt^*$ and $|\kappa|^2 = \kappa\kappa^*$ represent the power coupling from the input bus waveguide to the output waveguide, and from the input waveguide to the ring, respectively. The term α^2 describes the total power transmission for one pass around the ring[54] (i.e., $\alpha = 1$ corresponds to no loss). Since the phase in (9.9) is given by $\phi = N_{\mathrm{eff}}\, L(2\pi/\lambda)$, a shift in N_{eff} caused by molecular attachment will cause a resonance wavelength shift, and change the output power at any fixed wavelength.

The phase shift induced by an effective index perturbation, δN_{eff}, caused by molecular binding is the same as in (9.2), but the sensor length L is now the ring circumference.

The calculated ring resonator output spectrum in Fig. 9.16 is based on (9.9) and shows a series of resonance minima spaced at intervals known as the free spectral range $\Delta\lambda_{\mathrm{FSR}}$:

$$\Delta\lambda_{\mathrm{FSR}} = \frac{\lambda^2}{N_g L}. \tag{9.10}$$

Here $N_g = N_{\mathrm{eff}} - \lambda(\partial N_{\mathrm{eff}}/\partial\lambda)$ is the group effective index of the ring waveguide, a parameter that takes into account the wavelength dependence of the effective index. In the absence of dispersion $N_g = N_{\mathrm{eff}}$, but in these silicon photonic wire waveguides, $N_g = 4.8$ is more than twice as large as $N_{\mathrm{eff}} = 2.2$.

The ability to measure small index perturbations, δN_{eff}, depends on the depth (i.e., the extinction ratio at resonance) and width of these resonances. The depth and width in turn depend on the relative values of the coupling coefficient t and the resonator loss α. Figure 9.16 shows typical ring resonator spectra calculated for two different loss values. The maximum extinction ratio is obtained at critical coupling when $t = \alpha$, and the output intensity at resonance is exactly zero. The width of the resonances depends on the total round trip resonator loss, αt, which is the product of waveguide losses within the ring (α) and coupling loss (t) at the coupler.

The ratio of the 3 dB resonance line width, $\Delta\lambda_{3\mathrm{dB}}$, to the free spectral range is known as the finesse:

$$F = \frac{\Delta\lambda_{\mathrm{FSR}}}{\Delta\lambda_{3\,\mathrm{dB}}} = \pi\left[\arccos\left(\frac{2\alpha|t|}{1+\alpha^2|t|^2}\right)\right]^{-1}. \tag{9.11}$$

The finesse is related to the resonator quality factor, Q,

$$Q = \frac{\lambda}{\Delta\lambda_{3\,\mathrm{dB}}} = \frac{\lambda}{\Delta\lambda_{\mathrm{FSR}}}F \tag{9.12}$$

that is often used to describe resonators in optics and electronics. In optics, the quality factor is the ratio of the energy stored in the resonator to the energy dissipated per optical cycle. However, taking Q as the ratio between the wavelength and resonance line width, as implicit in (9.11) and (9.12), is a more practical working definition especially when dealing with very lossy resonators. Since the resolution with which shifts in ring resonator wavelength can be measured depends on the resonance width and depth, the sensor must have both a reasonably high finesse (or Q) in addition to a large extinction ratio.

As in the previous case of a MZI, a ring resonator sensor can be interrogated by collecting an output spectrum and following the wavelength shift of a resonance, or by monitoring the change in output intensity at a fixed wavelength. The limitation of intensity monitoring alone is that for higher values of finesse, there are large dead zones between the resonances where the intensity change in response to a small perturbation is vanishingly small (e.g., from $\lambda \sim 1552.5$ to 1552.7 nm in Fig. 9.16). Therefore, the interrogation wavelength needs to be selected to be at or near resonance.

The ring resonator is obviously a more complicated circuit element than the simple MZI. The great advantage that a resonator can offer is that light circulating repeatedly around the resonator cavity interacts with the same molecule more than once, amplifying the phase shift induced by adsorption of a molecule. To facilitate a comparison between ring resonators and MZI interrogation circuits, it is useful to consider the effective index change required to cause a 3-dB change in intensity for each. Using (9.9)–(9.11), one can show that the effective index shift needed to move the ring resonance wavelength by half the resonance linewidth, $\Delta\lambda_{3\text{dB}}$, is

$$\delta N_{\text{eff}} = \frac{\lambda}{2LF}. \tag{9.13}$$

This relation provides a convenient criterion to compare Mach-Zehnder and ring resonator interrogation circuits.

Comparing (9.8) and (9.13), the ring resonator response is comparable to a MZI interferometer with a sensor arm of length $L' = FL/4$, where L is the geometrical ring length. In a MZI, the output signal depends only on sensor waveguide length alone, while the response of a ring resonator of any given length fixed can be enhanced by increasing the cavity finesse F, thus providing an additional degree of design freedom. For example, a ring resonator with equivalent sensitivity to a Mach–Zehnder with sensor length L can be made much smaller provided F is large enough to compensate the decreased length L. Therefore a ring resonator sensor can potentially have both a higher sensitivity and a smaller size than a Mach-Zehnder configuration.

The use of ring resonators for evanescent field sensing has been demonstrated by a number of groups[55–57]. Nevertheless exploiting ring resonators to full advantage requires careful attention to the interplay between resonance extinction ratio, resonator losses, coupling coefficients, and resonator length. Since achieving the enhanced response predicted by (9.13) requires that the resonator have a high extinction ratio and therefore operates near the critical coupling condition, resonator loss must therefore be quantified and controlled. Waveguide loss is largely due to optical scattering by surface roughness or internal material defects and absorption. Minimizing waveguide loss is mainly a problem of optimizing fabrication processes. To date losses as low as -2 dB cm^{-1} have been reported in silicon photonic wires[58,59].

The overall resonator loss per pass, αt, and therefore the finesse, is ultimately limited by the coupling coefficient t. To achieve the highest possible finesse one could use a coupler with t approaching $t = 1$. While this can in principle be done by using a directional coupler (DC) formed by two closely spaced waveguides as in Fig. 9.17a. Here optical coupling from bus waveguide to ring and back occurs

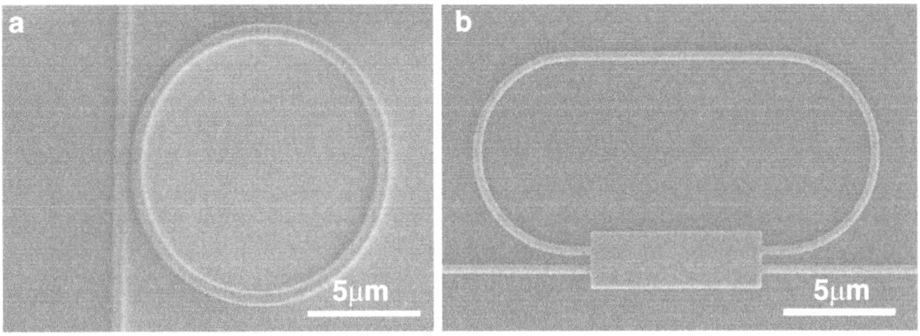

Fig. 9.17 Silicon waveguide ring resonators employing (**a**) a directional coupler (DC), and (**b**) a multimode interference (MMI) coupler, as viewed from above using a scanning electron microscope

through light leaking across the narrow gap into the adjacent waveguide. Fabricating such DC couplers to exact specifications is difficult due to the narrow gap required.

Furthermore, as in the case of the slot waveguide of Fig. 9.8, the behavior of narrow gaps between waveguides can be unpredictable in the presence of fluids or after the deposition of cladding materials. An alternative that is particularly promising for sensor applications is to replace DC couplers by multimode interference (MMI) couplers[59,60]. An example of an MMI coupler is shown in Fig. 9.17b. The coupler is simply a wide section of waveguide with two input and two output waveguides. Efficient coupling is the result of the evolving interference pattern of light as it propagates along the MMI section. Since a narrow gap is not required, MMI couplers are much easier to fabricate and combine with various overcladding materials. The disadvantage is that simple rectangular MMI couplers can only be designed with the specific coupling ratios: $|t|^2 = 0.5$, $|\kappa|^2 = 0.5$, and $|t|^2 = 0.85$, $|\kappa|^2 = 0.15$[59,60]. In the absence of waveguide loss, these coupling ratios limit the maximum finesse to F = 9.2 and F = 38.7, respectively. However, the restricted range of coupling ratios in MMI couplers can be compensated by the second free parameter in (9.13) – the resonator length.

Equation (9.13) predicts that the minimum level of detection should scale with resonator length. In particular, when waveguide loss is negligible extending ring length allows one to improve the ring sensor response independently of the coupling coefficients: this is a particularly useful degree of freedom if MMI couplers with a restricted set of coupling ratios are used. Of course, the cavity finesse (9.11) itself depends on length through the loss term α, which depends exponentially on L through Beer's law relation for waveguide loss ($\alpha \sim e^{-\gamma L}$). Once the cavity length L is comparable to the attenuation length $1/\gamma$ of light in the ring waveguide, further length increases are counterproductive since most of the light is lost during one trip around the ring, and the finesse drops.

While the ring length can be increased simply by increasing the ring radius, this again leads to large structures that have a poor cross-section with the microfluidic channels. As in the case of the MZI-based devices, incorporating a folded

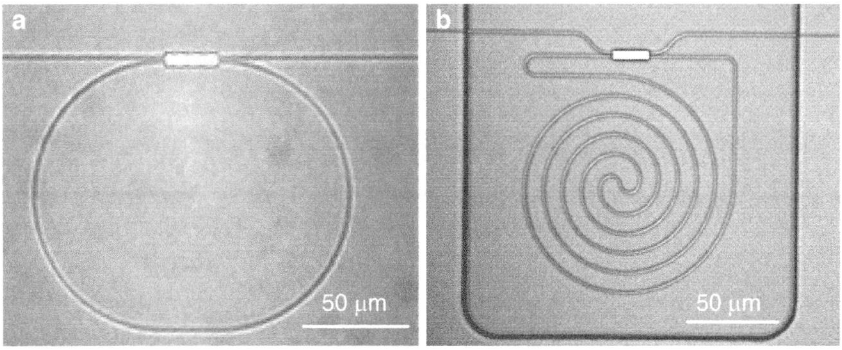

Fig. 9.18 Examples of ring resonators with (**a**) conventional race track and (**b**) folded cavity geometries

waveguide design into the ring allows us to significantly increase the ring cavity length without increasing overall ring diameter. Examples of folded cavity ring resonator sensors are shown in Fig. 9.18. Using these and similar designs it is possible to fabricate ring sensors with long cavity lengths and hence larger Q, and also with improved sensor surface area cross-section with available fluid volume in the microfluidic channel. The folded ring design also facilitates the arraying of sensors on a chip with a format that facilitates independently functionalizing each sensor element using spotting technology.

Unlike the balanced MZI interrogation circuit as in Fig. 9.13b, a single ring resonator cannot be self-referenced to eliminate response to temperature fluctuations. However, rings of different diameters can be placed sequentially along a single input waveguide. Since the resonance wavelengths will be different, two or more rings can be monitored simultaneously using essentially the same wavelength scanning procedure as before. One ring can therefore be isolated from the analyte solution and used as a standard reference. By comparing the wavelength shifts of reference rings and sensor rings when observed simultaneously, the effects of temperature can be removed from the sensor data in real time[61].

Folded cavity spiral ring resonator sensors have been demonstrated, again using the biotin-streptavidin model system[63]. The ring and input waveguides were formed from 450 nm × 260 nm silicon photonic wire channels using the same process as for the MZI devices. MMI coupler with a 3-dB power splitting ratio was used to couple the ring to the bus waveguide. The length of the spiral ring used in this experiment is $L = 1.27$ mm, and the measured finesse was $F = 5.0$ (corresponding to a quality factor of $Q = 19{,}600$). The chips were covered in SU-8 polymer with windows opened over the active sensor rings, and the rings were functionalized for streptavidin binding by biotin attachment as described previously. Light from a tunable laser was coupled into the input waveguide. By rapidly scanning the tunable laser, a 3-nm wide ring spectrum was generated every few seconds as the sensor was exposed to the analyte liquid. A peak-fitting algorithm was then used to recover the position of the ring resonances. Figure 9.19 displays

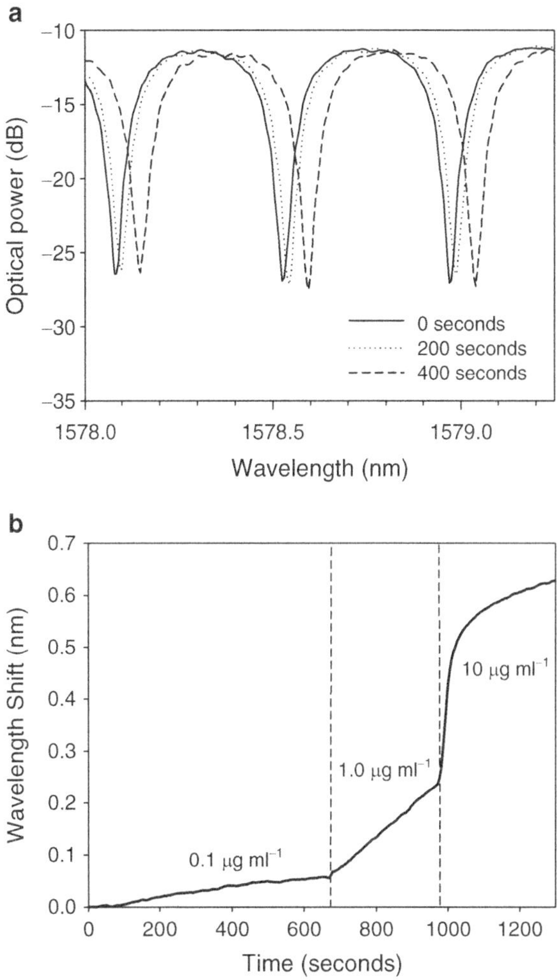

Fig. 9.19 Resonace wavelength shift during streptavidin binding to a ring resonator sensor surface. The concentration of strepatavidin is increased in one decade steps during the experiment

the resonance wavelength shift as the ring sensor was exposed to streptavidin solutions at three different concentrations, increasing in tenfold steps. The induced resonance wavelength shift is related to the effective index change of the sensor waveguide through the relation

$$\Delta\lambda = \lambda\left(\frac{\delta N_{\text{eff}}}{N_{\text{g}}}\right), \tag{9.14}$$

and hence is directly proportional to surface density of bound molecules. As the streptavidin concentration increases, the film growth rate changes consistently with the increase in streptavidin concentration. Overall, the measured binding curve shows the same rapid initial growth followed by saturation as the streptavidin monolayer forms. Based on this and similar data, we can estimate a minimum level of detection of 0.2% of a monolayer ($\sim$4 pg mm^{-2}). This level of detection is comparable to the estimated 0.3% monolayer detection limit achieved by the MZI sensor used to obtain the data of Fig. 9.14, under similar experimental conditions. This is as may be expected from (9.7) and (9.12), when the appropriate parameters for the ring and MZI sensors are used. In future work based on the same ring layout, considerably higher sensitivity can be achieved if the finesse is increased by using an MMI with a higher power splitting ratio and reducing ring waveguide loss.

9.5 Summary: from Waveguide Sensor to Biochip Array

In this chapter, we have reviewed the recent work in our laboratory toward creating molecular biosensors using the silicon photonic wire platform. Although the evanescent field waveguide sensor concept has been known and used for more than two decades, the adoption of silicon photonics for this application brings some major improvements on the existing prior art in waveguide sensing. By using very small photonic wire waveguides, the intrinsic surface sensitivity is comparable to that of SPR systems. However, the most important enhancement silicon photonics brings to sensor technology is the ability to combine silicon PWEF sensor elements with complex integrated optical waveguide circuits. This is of course immediately attractive because most optical functions can then be carried out by integrated components fabricated on a single monolithic chip. Apart from the external light source and photodetector, no assembly and optical alignment is involved.

Even though silicon PWEF molecular sensor devices provide intrinsic sensitivities comparable to existing label-free technologies such as SPR, the measured signal can be easily amplified using long sensor waveguides densely folded to form compact sensor elements with diameters of a few hundred micrometers or less. Given their small size, tens or hundreds of sensor elements can be combined to form waveguide-based sensor arrays on 1-cm^2 silicon chips. The small size of each sensor element allows simple waveguide comparators to be designed that use two or more sensors to calibrate out environmental fluctuations, and thereby lessen the demands on sensor packaging.

The simple MZI, ring resonator circuits, and folded waveguide sensor configurations described in this chapter only begin to suggest the new potential silicon-based integrated optics can bring to sensor technology. The examples given in this chapter include simple MZI comparators that can eliminate the temperature dependence of the sensor output and multiple ring resonator circuits. The brief review of Mach–Zehnder and ring sensor integration circuits in this chapter suggests that MZI-based configurations lend themselves well to building robust self-referencing devices that require little in the way of external environmental control, and which can be operated using only a simple fixed wavelength light source and a photodetector. Ring resonators require more complex wavelength tunable interrogation optics, but have additional degrees of design freedom. By optimizing both finesse and resonator length, the sensor size can be further reduced, while at the same time maintaining or even decreasing the minimum level of detection.

With further work on sensor arraying, more effective means of sensor referencing may be developed to eliminate the effects not only of temperature but also nonspecific binding and test sample contamination. The overall objective is to create label-free optical sensor arrays that require minimal if any temperature stabilization, and as little a priori sample purification as possible.

Despite the promise of silicon photonics, much work remains to be done to arrive at a practical sensor technology. As is always the case in silicon microphotonics, coupling light into and off the chip is a critical issue. While easily addressed in an optics research lab using tapered fibers, microscopes, and opto-mechanical components, a viable sensor technology requires that biochips can be exchanged, aligned, and operational within a few minutes. At the same time, microfluidics must be incorporated on chip to allow simple and effective sample liquid delivery to the test chip. It may be important to remember that the success of commercial SPR tools[28] is as much due to developments in SPR chip reading technology and sample handling capability, as to the fundamental physical capabilities of SPR sensors themselves.

Thus, future work on silicon-based waveguide sensors must address the need for easily reconfigurable optical input/output coupling arrangements and microfluidic integration with the silicon chip. Finally, the promise of a waveguide-based biosensor array requires a means of simultaneously addressing multiple sensors on the same chip. Although these challenges are difficult, related technologies already exist in the optical telecommunications arena, where multiplexing, routing, and demultiplexing many different data streams are essential operations in electronic and optical communication links. It seems likely that many of the strategies already developed in telecommunications can be reshaped and applied to the problem of integrated optical sensor arrays.

Acknowledgments This work has been supported by the National Research Council Canada Genomic and Health Initiative.

References

1 Elwing, H., Protein adsorption and ellipsometry in biomaterial research, *Biomaterials* **1998**, 19, 397–406

2 Smith, T., Ellipsometry for measurements at and below monolayer coverage, *J. Opt. Soc. Am.* **1968**, 58, 1069–1079

3 Patskovsky, S.; Meunier, M.; Kabashin, A. V., Phase-sensitive silicon-based total internal reflection sensor, *Opt. Express* **2007**, 15, 12523–12528

4 Gao, T.; Rothberg, L. J., Label-free sensing of binding to microarrays using Brewster angle straddle interferometry, *Anal. Chem.* **2007**, 79, 7589–7595
5 Zhu, X.; Landry, J. P.; Sun, Y. S.; Gregg, J. P.; Lam, K. S.; Guo, X., Oblique incidence reflectivity difference microscope for label-free high-throughput biochemical reactions in a microarray format, *Appl. Opt.* **2007**, 46, 1890–1895
6 Joo, C.; de Boer, J. F., Spectral-domain optical coherence reflectometric sensor for highly sensitive molecular detection, *Opt. Lett.* **2007**, 32, 2426–2428
7 Shen, Y. R., Optical second-harmonic generation at interfaces, *Annu. Rev. Phys. Chem.* **1989**, 40, 327–350
8 Shen, Y. R., Surface properties probed by second harmonic and sum-frequency generation, *Nature* **1989**, 337, 519–525
9 Corn, R. M.; Higgins, D. A., Optical second-harmonic generation as a probe of surface chemistry, *Chem. Rev.* **1994**, 94, 107–125
10 Vo-Dinh, T.; Stokes, D. L., Surface enhanced Raman scattering for biomedical diagnostics, *Biomedical Photonics Handbook*, Vo-Dinh, T., Ed. CRC Press, Boca Raton, FL, **2003**, 64.1–64.39
11 Moskovits, M., Surface-enhanced Raman spectroscopy: a brief retrospective, *J. Raman Spectrosc.* **2005**, 36, 485–496
12 Osawa, M., Surface enhanced infrared absorption, In Near Field Optics and Plasmon Polaritons; Kawata, S., Ed.; Springer, Berlin, **2001**, 163–187
13 Homola, J., Ed., Surface Plasmon Based Resonance Sensors, Springer, Berlin, 2006
14 Homola, J.; Yee, S. S.; Gauglitz, G., Surface plasmon resonance sensors: review, *Sens. Actuators B* **1999**, 54, 3–15
15 Karlsson, R., SPR for molecular interaction analysis: a review of emerging application areas, *J. Molec. Recognit.* **2004**, 17, 151–161
16 Lukosz, W., Principles and sensitivities of integrated optical and surface plasmon sensors for direct affinity sensing and immunosensing, *Biosens. Bioelectron.* **1991**, 6, 215–225
17 Luff, B. J.; Wilkinson, J. S.; Piehler, J.; Hollenback, U.; Ingenhoff, J.; Fabricius, N.; Integrated optical Mach-Zehnder biosensor, *J. Lightwave Technol.*, **1998**, 16, 583–591
18 Prieto, F.; Sepulveda, B.; Calle, A.; Llobera, A.; Dominguez, C.; Abad, A.; Montoya, A.; Lechuga, L. M., An integrated optical interferometric nanodevice based on silicon technology for biosensor applications, *Nanotechnology* **2003**, 14, 907–912
19 Boyd, R. W.; Heebner, J. E., Sensitive disk resonator photonic biosensor, *Appl. Opt.* **2001**, 40, 5742–5747
20 Weisser, M.; Tovar, G.; Mittler-Neher, S.; Knoll, W.; Brosinger, F.; Freimuth, H.; Lacher, M.; Ehrfeld, W., Specific biorecognition reactions observed with and integrated Mach-Zehnder interferometer, *Biosens. Bioelectron.* **1999**, 14, 405–411
21 Luff, B. J.; Harris, R. D.; Wilkinson, J. S.; Wilson, R.; Schiffrin, D. J., Integrated-optical directional coupler biosensor, *Opt. Lett.* **1996**, 21, 618–620
22 Horvath, R.; Pedersen, H. C.; Skivesen, N.; Selmeczi, D.; Larsen, N. B., Monitoring living cell attachment and spreading using reverse symmetry waveguide sensing, *Appl. Phys. Lett.* **2005**, 86, 071101–071103
23 Reed, G. T.; Knights, A. P., Silicon Photonics: An Introduction, Wiley, Chichester 2004
24 Pavesi, L.; Lockwood, D. J., Silicon Photonics, Springer, Berlin, 2004
25 Bogaerts, W.; Tailaert, D.; Luyssaert, B.; Dumon, P.; van Campenhout, J.; Bientsman, P.; van Thourhout, D.; Baets, R., Basic structures for photonic integrated circuits in silicon-on-insulator, *Opt. Express* **2004**, 12, 1583–1591
26 Ahmad, R. U.; Pizzuto, F.; Camarda, G. S.; Espinola, R. L.; Rao, H.; Osgoode, R. M., Ultracompact corner-mirrors and T-branches in silicon-on-insulator, *IEEE Phot. Technol. Lett.* **2002**, 14, 65–67
27 Dulkeith, E.; Xia, F.; Schares, L.; Green, W. M. J.; Vlasov, Y. A., Group index and group velocity dispersion in silicon-on-insulator photonic wires, *Opt. Express* **2006**, 14, 3853–3863

28 Rich, R. L.; Myszka, D. G., BIACORE J: a new platform for routine biomolecular interaction analysis, *J. Mol. Recognit.* **2001**, 14, 223–228
29 Hunsperger, R. G., Integrated Optics: Theory and Technology, 3rd edn.; Springer, Berlin, 1991
30 Kogelnik, H., Theory of optical waveguides, In Guided Wave Optoelectronics; 2nd edn, Tamir, T., Ed.; Springer, Berlin, **1990**, 7–87
31 Palik, E. D., Handbook of Optical Constants of Solids, Academic Press, Orlando, FL, 1985
32 Densmore, A.; Xu, D.-X.; Waldron, P.; Janz, S.; Cheben, P.; Lapointe, J.; Delâge, A.; Lamontagne, B.; Schmid, J. H.; Post, E., A silicon-on-insulator photonic wire based evanescent field sensor, *IEEE Photon. Technol. Lett.* **2006**, 18, 2520–2522
33 Goh, J. B.; Loo, R. W.; Goh, M. C., Label free monitoring of multiple biomolecular binding interactions in real-time with diffraction-based sensing, *Sens. Actuators B: Chem.* **2005**, 106, 243–248
34 Ouyang, H.; Striemer, C. C.; Fauchet, P. M., Quantitative analysis of the sensitivity of porous silicon optical biosensors, *Appl. Phys. Lett.* **2006**, 88, 163108–163110
35 Saarinen, J. J.; Weiss, S. M.; Fauchet, P. M.; Sipe, J. E., Optical sensor based on resonant porous silicon structures, *Opt. Express* **2005**, 13, 3754–3764
36 Tiefenthaler, K.; Lukosz, W., Sensitivity of grating couplers as integrated-optical chemical sensors, *J. Opt. Soc. Am. B* **1989**, 6, 209–220
37 Parriaux, O.; Veldhuis, G. J., Normalized analysis for the sensitivity optimization of integrated optical evanescent-wave sensors, *J. Lightwave Technol.* **1998**, 16, 573–582
38 Mann, E. K.; Heinrich, L.; Schaaf, P.,Validity of the uniform thin-film approximation for the optical analysis of particulate films, *Langmuir* **1997**, 13, 4906–4909
39 Guemouri, L.; Olgier, J.; Ramsden, J. J., Optical properties of protein monolayers during assembly, *J. Chem Phys.* **1998**, 109, 3265–3268
40 Sipe, J. E., New Green-function formalism for surface optics, *J. Opt. Soc. Am. B* **1987**, 4, 481–489
41 Densmore, A.; Xu, D.-X.; Janz, S.; Waldron, P.; Mischki, T.; Lopinski, G.; Delâge, A.; Lapointe, J.; Cheben, P.; Lamontagne, B.; Schmid, J. H, Spiral-path high-sensitivity silicon photonic wire molecular sensor with temperature-independent response, *Opt. Lett.* **2008**, 6, 596–598
42 Densmore, A.; Xu, D.-X.; Waldron, P.; Janz, S.; Delâge, A.; Cheben, P.; Lapointe, J., Thin silicon waveguides for biological and chemical sensing, In Proceedings of the SPIE Photonics West, paper 6477–43, SPIE, Bellingham, WA, 2007
43 Almeida, V. R.; Xu, Q.; Barrios, C. A.; Lipson, M., Guiding and confining light in void nanostructure, *Opt. Lett.* **2004**, 29, 1209–1211
44 Barrios, C. A.; Banuls, M. J.; Gonzalez-Pedro, V.; Gylfasson, K. B.; Sanchez, B.; Griol, A.; Maquieira, A.; Sohlstom, H.; Holgado, M.; Casquel, R.; Label-free optical biosensing with slot-waveguides, *Opt. Lett.* **2008**, 33, 708–710
45 Robinson, J. T.; Chen, L.; Lipson, M., On-chip gas detection in silicon optical microcavities, In Proceedings of the Conference on Lasers and Electro-optics, paper CMJJ6, Optical Society of America, Washington, DC, 2008
46 Cooper, J. M.; Shen, J.; Young, F. M.; Connolly, P.; Barker, J. R.; Moores, G., The imaging of streptavidin and avidin using scanning tunneling microscopy, *J. Mater. Sci.: Mater. Electron.* **1994**, 5, 106–110
47 Schaaf, P.; Talbot, J., Surface exclusion effects in adsorption processes, *J. Chem. Phys.* **1989**, 91, 4401–4409
48 Ramsden, J. J., Review of new experimental techniques for investigating random sequential adsorption, *J. Statist. Phys.* **1993**, 73, 853–877
49 Little, B. E.; Chu, S. T.; Haus, H. A.; Foresi, J.; Laine, J.-P., Microring resonator channel dropping filters, *J. Lightwave Technol.* **1997**, 15, 998–1005
50 Rabiei, P.; Steier, W.; Zhang, C.; Dalton, L. R., Polymer micro-ring filters and modulators, *J. Lightwave Technol.* **2002**, 20, 1968–1975
51 Xu Q.; Lipson, M., All-optical logic based on silicon micro-ring resonators, *Opt. Express*, **2007**, 15, 924–929

52 Blair, S.; Heebner, J. E.; Boyd, R. W., Beyond the absorption-limited nonlinear phase shift with microring resonators, *Opt. Lett.* **2002**, 27, 357–359
53 Xu, Q.; Schmidt, B.; Pradhan, S.; Lipson, M., Micrometre-scale silicon electro-optic modulator, *Nature* **2005**, 435, 325–327
54 Yariv, A., Universal relations for coupling of optical power between microresonators and dielectric waveguides, *Electron. Lett.* **2000**, 36, 321–323
55 Ksendzov, A.; Homer, M. L.; Manfreda, A. M., Integrated optics ring-resonator chemical sensor with polymer transduction layer, *Electron. Lett.* **2004**, 40, 63–65
56 Yalcin, A.; Popat, K. C.; Aldridge, J. C.; Desai, T. A.; Hryniewicz, J.; Chbouki, N.; Little, B. E.; King, O.; Van, V.; Chu, S.; Gill, D.; Anthes-Washburn, M.; Unlu, M. S.; Goldberg, B. B., Optical sensing of biomolecules using microring resonators, *IEEE J. Selected Topics Quant. Electron.* **2006**, 12, 148–155
57 De Vos, K.; Bartolozzi, I.; Schacht, E.; Bienstman, P.; Baets, R., Silicon-on-insulator microring resonator for sensitive and label-free biosensing, *Optics Exp.* **2007**, 15, 7610–7615
58 Dumon, P.; Bogaerts, W.; Wiaux, V.; Wouters, J.; Beckx, S.; van Campenhout, J.; Taillaert, D.; Luyssaert, B.; Bienstman, P.; van Thourhout, D.; Baets, R.; Low-loss SOI photonic wires and ring resonators fabricated with deep UV lithography, *IEEE Photonic Technol. Lett.* **2004**, 16, 1328–1330
59 Niehhusmann, J.; Vorckel, A.; Bolivar, P. H.; Wahlbrink, T.; Henschel, W.; Kurz, H., Ultrahigh qulaity factor silicon-on-insulator microring resonator, *Opt. Lett.* **2004**, 29, 2861–2863
60 Xu, D.-X.; Densmore, A.; Waldron, P.; Lapointe, J.; Post, E.; Delâge, A.; Janz, S.; Cheben, P.; Schmid, J. H.; Lamontagne, B., High bandwidth SOI photonic wire ring resonators using MMI couplers, *Opt. Express* **2007**, 15, 3149–3155
61 Xu, D.-X.; Janz, S.; Cheben, P.; Design of polarization-insensitive ring resonators in silicon-on-insulator using MMI couplers and cladding stress engineering, *IEEE Photonic Technol. Lett.* **2006**, 18, 343–345
62 Xu, D.-X.; Post, E.; Densmore, A.; Waldron, P.; Janz, S.; Lapointe, J.; Delâge, A.; Cheben, P.; Schmid, J. H., Cancellation of the temperature dependence in SOI photonic wire waveguide ring resonator biological sensors, accepted, In Proceedings of the IEEE/LEOS Conference on Group IV Photonics, IEEE, Piscataway, NJ, 2008
63 Xu, D.-X.; Densmore, A.; Delâge, A.; Waldron, P.; Janz, S.; Lapointe, J.; Lopinski, G.; Mischki, T.; McKinnon, R.; Cheben, P.; Schmid, J. H., Folded cavity SOI microring resonator sensors for high sensitivity and real time measurement of biomolecular binding, *Opt. Express* **2008**, 16, 15137–15148

Chapter 10
A Fast and Sensitive Integrated Young Interferometer Biosensor

Johannes S. Kanger, Vinod Subramaniam, Paul H. J. Nederkoorn, and Aurel Ymeti

Abstract We have developed an ultrasensitive biosensor based on an integrated optical Young interferometer. Key features of this sensor are that it is very compact, extremely sensitive, label free, and very fast. Therefore the Young interferometer has significant potential to be developed into a handheld, point-of-care device. In this chapter we review the progress that has been made on the development of integrated Young interferometer sensors. The sensor developed in our lab is discussed in detail. We demonstrate various applications of the current sensor. Special attention is paid to the detection of viruses. Finally a discussion on future prospects of this sensor for diagnostics is given.

10.1 Introduction

The integrated Young interferometer (YI) "immuno sensor" that is discussed in this chapter is one example of a whole class of so-called integrated optical (IO)-based chemical and immuno-sensors. In the past several types of IO sensors have been developed, including waveguide-based sensors such as interferometric sensors[1–6], grating couplers[7], resonant mirrors[8], integrated microcavity-based sensor[9–12], surface plasmon resonance (SPR)-based sensor[13], photonic wire sensors[14], and reflectance-based sensors like reflectance interference device[15].

One of the main advantages of the optical-based sensors is their high resolution. Integrated optical sensors[16], especially the interferometric ones that have been developed in recent years, such as the Mach-Zehnder interferometer (MZI)[17] and the Young interferometer [18–21], show an extremely high refractive index resolution in the range of $10^{-7}-10^{-8}$ RIU (refractive index units) [17,21], which is equivalent to

J.S. Kanger (✉)
Faculty of Science and Technology, University of Twente, Enschede, The Netherlands
e-mail: J.S.Kanger@utwente.nl

X. Fan (ed.), *Advanced Photonic Structures for Biological and Chemical Detection*, Integrated Analytical Systems,
DOI 10.1007/978-0-387-98063-8_10,

detection of a protein mass coverage of ~30 fg/mm^2. Furthermore, the IO readout systems are robust and small, allowing for miniaturization where many elements of the device can be integrated in a single chip occupying a relatively small area. These features offer the prospect for development of multisensing systems. Furthermore, this implies an increase of the analysis throughput, a reduction in the consumption of biomaterials, and cost reduction. Moreover, the optical-based sensors are label free sensing techniques, which require less complex sample preparation, do not need special laboratories and skilled personnel, and result in lower test costs. An additional advantage is that usually a simpler and faster signal analysis can be achieved.

Integrated optical waveguide-based sensors can monitor antibody-antigen interactions by making use of the evanescent field sensitivity of guided modes propagating through the waveguide structure of the sensor. The principle of evanescent field sensing has the advantage that it matches well with the availability of many chemical interfaces. Figure 10.1 shows a scheme of a binding event taking place at the core-cover interface of a three-layer waveguide structure. Here, the core-cover interface of the optical waveguide structure is coated with a chemo-optical transducer receptor layer, which can selectively bind to specific analyte molecules present in the cover medium.

When the binding of an analyte occurs, the substitution of solvent by the analyte molecules will result in an increase of the refractive index n at the core-cover interface. Therefore, the effective refractive index N_{eff} of a guided mode, which probes the core-cover interface during its propagation through the waveguide, will be changed within the interaction length, resulting in a phase change of the mode at the exit of the interaction region.

Here, two different situations can be distinguished: in the first situation, the evanescent region is completely occupied with one material, e.g., a solution of one given analyte (homogeneous sensing) without the benefit of a receptor transduction

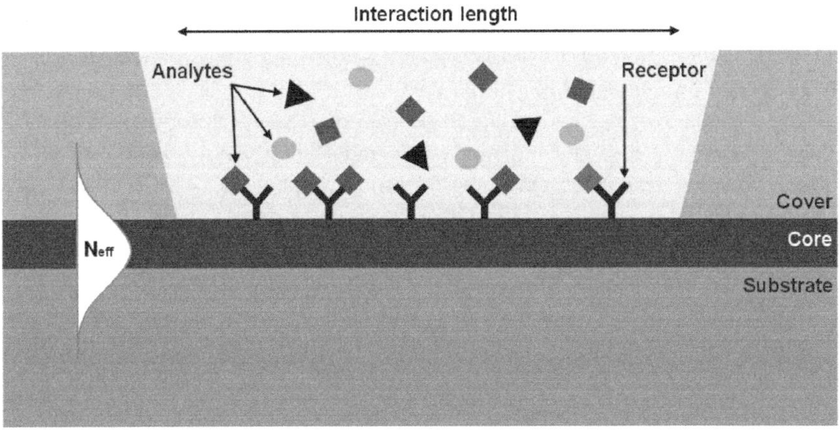

Fig. 10.1 Binding event between analyte and receptor molecules occurring at the core-cover interface of a three-layer waveguide structure within the evanescent region of a guided mode. N_{eff} indicates the effective refractive index of the fundamental guided mode

layer. In this case, the refractive index change is representative of the analyte concentration C in the solution. In the second situation, a chemo-optical transduction layer is immobilized at the core-cover interface of the waveguide structure. For example, in an immunoreaction, this transducer can be an antibody layer (surface sensing).

The phase change that results from a binding event can be measured with a high sensitivity by using an interferometric sensor, e.g., the MZI[17], the YI[19,21], or differential interferometer[22], due to the relatively long interaction lengths of these devices (up to several centimeters). In the group of interferometric devices, those that have a reference branch to compensate for common mode effects, such as temperature changes, pressure changes, fluctuations of light source power, etc., such as the MZI and the YI, are preferable and yield more robust and reproducible results.

Here, we focus on the development of a YI sensor due to its simpler chip design (no optical phase modulation is required to achieve a high sensitivity as in the case of an MZI) and flexibility for development of a multichannel device. In Fig. 10.2, a schematic layout of a two-channel IO YI is shown. Light is coupled into an IO channel waveguide and split by a Y-junction to two parallel output channels positioned at a certain distance from each other, as shown in Fig. 10.2. An interference pattern is generated as a result of the overlap of both outgoing divergent beams. The interference pattern is recorded by a CCD camera, which is placed at a given distance from the endface of the integrated chip. A sensing window is realized on the top of the core layer of the measuring channel where the samples to be analyzed are applied. When a binding event takes place on the sensing window of the measuring channel a phase change between reference and

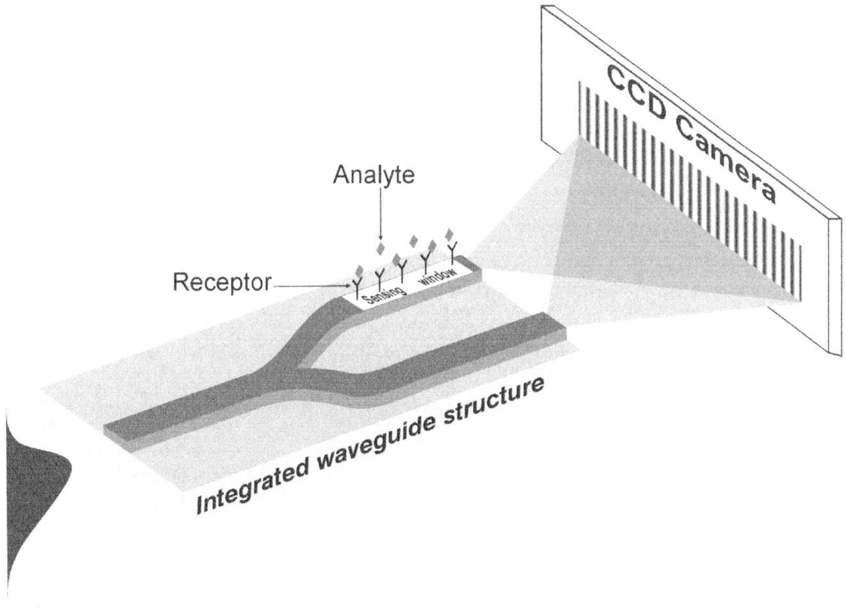

Fig. 10.2 Two-channel integrated optical Young interferometer sensor

measuring channels occurs and the interference pattern will be shifted with respect to the camera surface. By measuring the spatial shift of the interference pattern, the phase change can be calculated, yielding an accurate determination of the refractive index change occurring on the sensing window of the measuring channel.

The interferometric sensors have been mainly used for detection of proteins, e.g., binding of antigen molecules to a specific antibody layer that is immobilized on the sensor surface[19,21,23–26]. It is scientifically interesting and relevant from the application point of view to explore the use of these sensors not only for monitoring proteins, but also for other types of analytes, e.g., viruses, which so far have been detected with time-consuming methods[27]. The principle of the sensor makes possible the detection of relevant viruses such as HIV, SARS, Hepatitis B and C provided that the appropriate specific cognate receptor or antibody is available. Recently the YI sensor has been successfully used to measure low concentrations of HSV-1 virus[28] (see also result section for a detailed description) and Avian Influenza virus[29].

In this chapter we first discuss the fundamentals and the design aspects of an integrated optical YI sensor (Sect. 10.2), followed by a description of the experimental setup (Sect. 10.3). In the result section (Sect. 10.4) both protein and virus detection experiments are discussed. Section 10.5 demonstrates the use of microfluidic chips for efficient sample handling in combination with the YI sensor. This chapter concludes with a discussion on the prospects of the sensor for point-of-care diagnostics.

10.2 Fundamentals

10.2.1 Optical Waveguides

In a three-layer waveguide structure, guiding of the light is performed by engineering appropriate refractive index contrast between the core layer and the cladding. In Fig. 10.1, such a structure is presented, consisting of a substrate, core, and cover layer, which have, respectively, a refractive index of n_S, n_F, and n_C. A higher refractive index of the core layer allows total internal reflection of the light at the core-cladding interface, as shown schematically in Fig. 10.3, making possible propagation of the light through the slab waveguide.

Under conditions of total internal reflection, guided modes are obtained due to the constructive interference between the light rays propagating through the waveguide[30]

$$\frac{2\pi}{\lambda}2d_F n_F \cos\theta - \Phi_C - \Phi_S = m2\pi, \qquad (10.1)$$

where λ is the vacuum wavelength of the guided light, d_F is the thickness of the core layer, θ is the angle of incidence at the core-cladding boundary, Φ_C and Φ_S are the

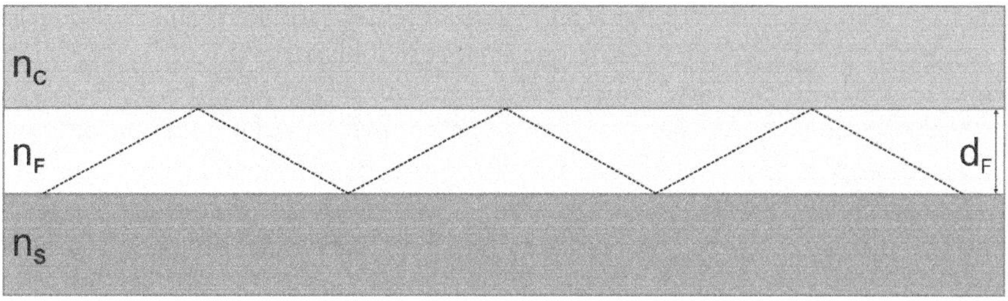

Fig. 10.3 Total internal reflection in a three-layer waveguide structure. n_i (i = C, F, S) indicates the refractive index of layer i of the waveguide structure, $n_F > n_{C,S}$; d_F is the thickness of the core layer

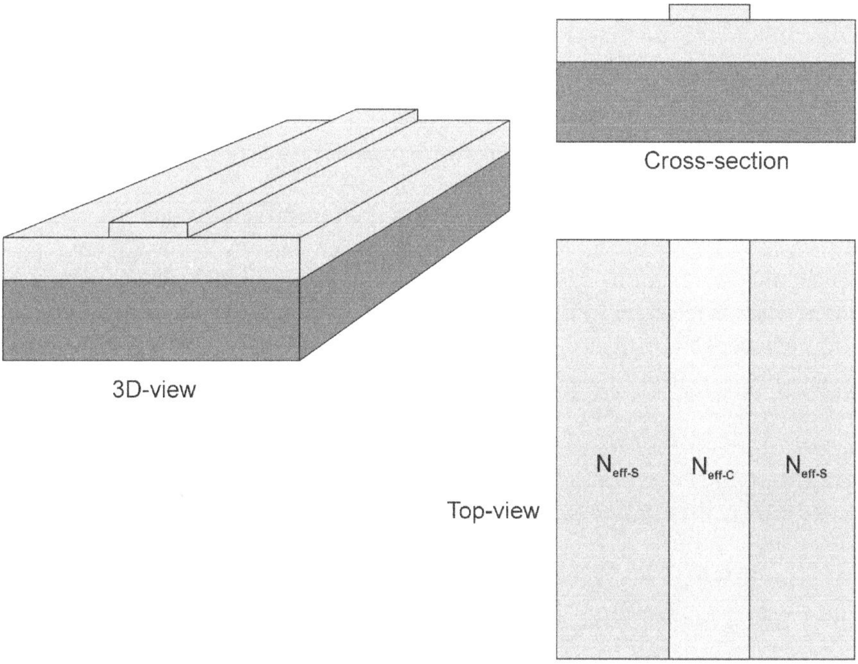

Fig. 10.4 3D-view, cross-section, and *top-view* of a channel waveguide structure; N_{eff-C} and N_{eff-S} indicate the effective refractive indices of the channel and slab region, respectively

phase changes arising from the reflection at the core-cover and core-substrate boundary, respectively, and m is the order of the guided modes.

In a channel waveguide structure, schematically shown in Fig. 10.4, the light is confined not only in the transverse direction, but also in the lateral one. This confinement is caused by the lateral contrast of the (effective) refractive indices between the channel region, N_{eff-C}, and the slab, N_{eff-S}, i.e., $N_{eff-C} > N_{eff-S}$. This contrast can be established by the presence of a ridge in the core section, where the effective refractive index is defined as $N_{\text{eff}} = n_{\text{F}} \sin\theta$. Generally, for a given channel structure the parameters expressing quantitatively the propagation of guided light beams (modes) can be calculated using Maxwell theory-based software.

Finally, channel waveguide structures can be used to guide the light not only along a straight section, but also in a bent one, e.g., in an S-bend, etc.

10.2.2 The YI Sensor

A YI can be designed to have more than two channels. As the two-channel YI sensor is a special case of the more general *N*-channel or multichannel YI, the theory of the YI is treated for the general *N*-channel YI. As shown in this section, in the multichannel YI all pairs of parallel output channels act as a two-channel YI, each with a unique distance between its two arms. In Fig. 10.5 the layout of the 4-channel YI sensor described in this chapter is shown.

In a multichannel YI, monochromatic laser light with a vacuum wavelength λ is coupled into an input channel waveguide and split to N output parallel channels, as shown in Fig. 10.5 (in this figure a network of three Y-junctions is used to split the light to four output channels).

The output divergent beams overlap with each other and the final interference pattern is a superposition of individual interference patterns, each of them representing the overlap of the divergent beams of a specific channel pair. The interference pattern is recorded by a CCD camera, which is placed at a distance L from the chip endface. The irradiance distribution of the sum interference pattern at the input face of the camera can be derived to be[32]

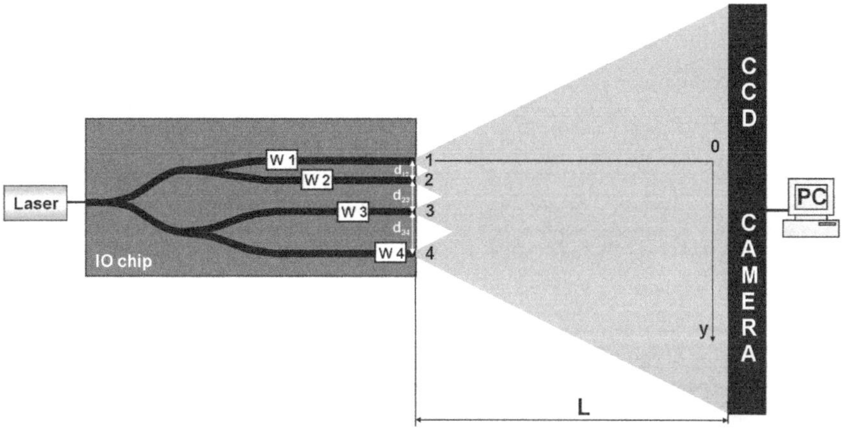

Fig. 10.5 Layout of the four-channel IO YI (not to scale); W_i indicates the sensing window on channel i, d_{ij} is the distance between channels i and j, L indicates the distance of the CCD camera from the chip end-face. Reprinted from Ref. 31 with permission. © 2008 Optical Society of America

$$I(y) \approx \sum_{i=1}^{N} I_i + 2 \sum_{i,j=1;i<j}^{N} (I_i I_j)^{\frac{1}{2}} \cos(\Delta\Phi_{ij}(y) - \Delta\phi_{ij}), \tag{10.2}$$

where I_i and I_j are the output powers from channels i and j, respectively. A channel output is approximated as a point source. Equation (10.2) is only valid if L is chosen such that the output divergent beams fully overlap at the CCD camera. In (10.2), $\Delta\Phi_{ij}(y)$ is the phase difference between the interfering beams coupled out from channels i and j as a result of their different optical path lengths at the interfering position y on the CCD camera. The phase change $\Delta\phi_{ij}$ represents the sum of the initial phase difference at the channel outputs in absence of the analyte, $\Delta\phi_{ij}{}^0$, and the phase change $\Delta\varphi_{ij} = \Delta\varphi_i - \Delta\varphi_j$ induced by refractive index changes Δn_i and Δn_j in channels i and j, respectively, i.e., $\Delta\phi_{ij} = \Delta\phi_{ij}{}^0 + \Delta\varphi_{ij}$. The phase difference $\Delta\Phi_{ij}(y)$ is given by[32]

$$\Delta\Phi_{ij}(y) = \frac{2\pi}{\lambda}\frac{d_{ij}}{L}y - \frac{2\pi}{\lambda}\frac{d_{ij}}{L}\left(d_{1i} + \frac{1}{2}d_{ij}\right) \tag{10.3}$$

provided that the distance between channels i and j $d_{ij} << L$ for all i, j values, a condition that is well satisfied in most cases. Hence, an individual interference pattern shows a spatial frequency k_{ij} given as

$$k_{ij} = \frac{1}{\lambda}\frac{d_{ij}}{L}. \tag{10.4}$$

The second term in (10.3) corrects for the shift of the center between two channels i and j with respect to the origin of the y-axis along the CCD camera (see Fig. 10.5).

The refractive index changes in channels i and j can be induced by binding between analyte and receptor molecules in the sensing window of each channel, or alternatively by a change of analyte concentration in the bulk solution. For both cases the refractive index in (part of) the sensing area of channel i and channel j that is probed by the evanescent field of the guided mode will be affected. As a result, the effective refractive index of guided modes in channels i and j ($N_{\mathrm{eff},i}$ and $N_{\mathrm{eff},j}$ respectively) will be changed, and consequently the phase changes $\Delta\varphi_i$ and $\Delta\varphi_j$ will result. The phase change $\Delta\varphi_{ij} = \Delta\varphi_i - \Delta\varphi_j$ can then be derived as:

$$\Delta\varphi_{ij} = \frac{2\pi}{\lambda} l(\Delta N_{\mathrm{eff},i} - \Delta N_{\mathrm{eff},j}) = \frac{2\pi}{\lambda} l\,\Delta N_{\mathrm{eff},ij}, \tag{10.5}$$

where l is the length of the sensing window realized on top of each output channel and $\Delta N_{\mathrm{eff},ij} = \Delta N_{\mathrm{eff},i} - \Delta N_{\mathrm{eff},j}$. The resulting refractive index change between channels i and j can cause a spatial shift of Δy_{ij} of the individual interference pattern for this channel pair along the camera surface. This shift can be calculated based on (10.3) and (10.5) as

$$\Delta y_{ij} = \frac{lL}{d_{ij}}\Delta N_{\mathrm{eff},ij}. \tag{10.6}$$

If one chooses channel i, $i = 1, 2, \ldots, N - 1$, to be the measuring channels, and channel N as a reference channel, where N is the total number of the channels, the refractive index changes $\Delta N_{\mathrm{eff},iN}$ may be determined using (10.6) by measuring the spatial shifts Δy_{iN}. Section 10.2.3 demonstrates that this approach allows determining either the amount of adsorbed analytes on the sensor surface, or alternatively the change in concentration of analytes in the bulk solution.

If $N = 2$, there will be one measuring channel, and this is the case of a two-channel YI device[34]. In case of $N = 4$, the refractive index changes $\Delta N_{\mathrm{eff},14}$, $\Delta N_{\mathrm{eff},24}$, and $\Delta N_{\mathrm{eff},34}$ can be calculated based on the measurement of the spatial shifts Δy_{14}, Δy_{24}, and Δy_{34}.[31] An essential condition is that these spatial shifts have to be measured simultaneously and independently of each other, requiring a proper configuration of the four-channel YI device, which is described next.

The output parallel channels of the multichannel device are positioned at different distances d_{ij} from each other in such a way that channels of each pair are separated by a unique distance, as shown in Fig. 10.5. In other words, $d_{12} \neq d_{23} \neq \ldots \neq d_{ij}$. In this way all spatial frequencies k_{ij}, each of them corresponding to the specific distance d_{ij} as shown from relation (2.4), are different, which makes it possible to well-separate different peaks in the amplitude spectrum of the Fourier-transformed interference pattern and select the spatial frequency for each peak. As a result, the phase change $\Delta\phi_{ij}$ between both channels of one pair, corresponding to the selected spatial frequency k_{ij}, can be monitored independently of the other channel pairs in the phase spectrum of the Fourier-transformed interference pattern.

A more detailed description of the working principle of the multichannel YI is given for a four-channel device ($N = 4$). The distances between the channels have been chosen such that $d_{12} \neq d_{23} \neq d_{34} \neq d_{13} \neq d_{24} \neq d_{14}$. There are six possible different channel pairs corresponding to six different distances of $d_{12} = 60$ µm, $d_{23} = 80$ µm, $d_{34} = 100$ µm, $d_{13} = 140$ µm, $d_{24} = 180$ µm, and $d_{14} = 240$ µm. These distances match the realized YI sensor structure described in Sect. 10.3. The final interference pattern will thus be a superposition of six two-channel interference patterns. The calculated interference pattern for the four-channel YI is shown in Fig. 10.6a. The amplitude spectrum (lower graph) and the phase spectrum (upper graph) of the Fourier-transformed interference pattern are presented in Fig. 10.6b.

As one might expect, there are six different peaks in the amplitude spectrum of the Fourier-transformed interference pattern located at six different spatial frequencies k_{ij}, each of them corresponding to the interference pattern obtained as a result of overlap of the channels that are separated by the specific distance d_{ij}. Looking at the selected spatial frequencies at the phase spectrum of the Fourier-transformed interference pattern, the phase of each "two-channel sensor" can be monitored simultaneously and independently from each other. The difference between distances d_{ij} should be designed such that they allow a good separation of the six different peaks.

If an extra phase change occur in one of the channels, e.g., in channel 1 as a result of the binding of an analyte, the same phase change is expected between channels of those pairs that involve channel 1. In Fig. 10.6a, the dotted line represents the calculated interference pattern when an extra phase change is

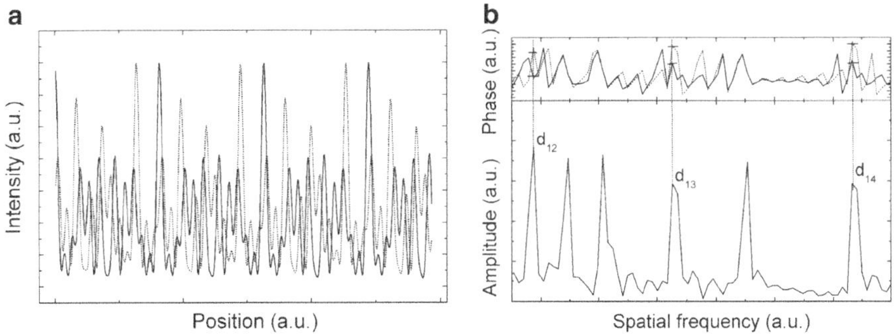

Fig. 10.6 (**a**) The interference pattern in the four-channel YI calculated for $L = 6$ cm, $d_{12} = 60$ μm, $d_{23} = 80$ μm, $d_{34} = 100$ μm (*solid line*). The *dotted line* represents the interference pattern calculated when an extra phase change is introduced on channel 1. (**b**) Amplitude spectrum (lower graph) and the phase spectrum (upper graph) of the Fourier-transformed interference patterns shown in (**a**) (*solid* and *dotted line*, respectively). Reprinted from Ref. 34 with permission. © 2008 Elsevier

introduced in channel 1. Calculation of the Fast Fourier Transform (FFT) of this interference pattern results in six different peaks in the amplitude spectrum of the Fourier-transformed interference pattern, located at the same spatial frequencies when no extra phase change is introduced in one of the channels. In the phase spectrum of the Fourier Transform, one can clearly observe a phase change corresponding to those channel pairs where channel 1 is involved (see dotted line in Fig. 10.6b).

If channels 1, 2, and 3 are chosen as measuring channels and channel 4 as a reference channel, it will be possible in principle to monitor independently and simultaneously three different binding events, each of them taking place in one of the measuring channels.

10.2.3 Analysis of Sensitivity

The sensitivity of the YI describes how the signal (shift of the interference pattern) changes in response to a change of concentration ΔC of the analyte (either adsorbed to the surface of the sensing area or in the bulk solution). Here, the sensitivity chain $\Delta C \Rightarrow \Delta N_{\text{eff}} \Rightarrow \Delta\varphi \Rightarrow \Delta$ (interference pattern) is discussed in detail.

Two different modes of operation are distinguished. The first mode is the detection of changes in bulk concentrations. In this case a change in analyte concentration causes a change of the refractive index of the bulk solution. For the relation between the refractive index change Δn_i and the change in concentration ΔC_i we have:

$$\Delta n_i = \frac{\partial n}{\partial C} \Delta C_i. \tag{10.7}$$

The value of $\partial n/\partial C$ depends on the properties of the material, e.g., in case of proteins this coefficient is ~0.188 ml/g[35] and in case of glucose dissolved in water (used for calibration measurements as discussed later in this chapter) this coefficient is ~0.069 ml/g[36]. In the second mode of operation, analytes bind to the sensor surface (e.g., mediated by a receptor layer). In this case a thin layer with thickness w and refractive index n_w is formed by the adsorbed analytes. Because the value of n_w (e.g., 1.45 for proteins) is usually different than the refractive index of the solution (e.g., 1.33 for water) that contains the analyte molecules, a phase change is induced. The average layer growth (Δw) on the sensor surface can be related to the mass change (Δm) per surface area (A):

$$\Delta w_i = \rho_w^{-1} \frac{\Delta m_i}{A}, \tag{10.8}$$

where ρ_w denotes the mass density of the adsorbed molecules. Here it is assumed that all channels are identical i.e., $A_i = A$ for all channels i. A value of $\rho_w = 0.6$ mg/mm^3 is estimated for proteins[35]. The changes in either refractive index or adsorbed layer thickness affect the effective refractive index of the guided mode:

$$\Delta N_{\text{eff},i} = \frac{\partial N_{\text{eff}}}{\partial n} \Delta n_i, \tag{10.9a}$$

$$\Delta N_{\text{eff},i} = \frac{\partial N_{\text{eff}}}{\partial w} \Delta w_i. \tag{10.9b}$$

$\partial N_{\text{eff}}/\partial n$ is the sensitivity coefficient of N_{eff} with respect to the change of the refractive index of the bulk, whereas $\partial N_{\text{eff}}/\partial w$ is the sensitivity coefficient of N_{eff} with respect to the change of the adsorbed layer thickness. Again it is assumed that all channels are identical, i.e., $\partial N_{\text{eff},i}/\partial n_i = \partial N_{\text{eff}}/\partial n$ and $\partial N_{\text{eff},i}/\partial w_i = \partial N_{\text{eff}}/\partial w$ for all values of i. In order to have a high sensitivity, $\partial N_{\text{eff}}/\partial n$ and $\partial N_{\text{eff}}/\partial w$ should be maximized. These sensitivity coefficients depends on the refractive indices of the substrate, core, and analyte solution (n_S, n_F, and n_{sol}, respectively), the thickness of the core layer, d_F, the wavelength of the light source, λ, and polarization used either Transverse Electric (TE) or Transverse Magnetic (TM).

Using the SiO_xN_y technology, highest sensitivity is obtained[37] by using Si_3N_4 as material for the core layer of the structure, having the highest refractive index of the SiON family, in combination with the choice of SiO_2, a low refractive index material, for the substrate and cover layers.

In addition, $\partial N_{\text{eff}}/\partial n$ and $\partial N_{\text{eff}}/\partial w$ depend on the thickness of the core layer, which at the same time should be chosen such that given the wavelength and polarization, the channel waveguide should be monomode (interference of one mode only is required) in the transverse direction. The width and ridge height of the ridge type channel waveguide should be chosen such that the channel is additionally monomode in the lateral direction, and in such a way that modal field should be maximally confined in the lateral direction. The lateral field profile is preferred to be as small as possible, because this gives the best agreement with the point source approximation used to describe the sensing system (see Sect. 10.2.3).

For the waveguide design as described in Sect. 10.3, the calculated values for both sensitivities are $\partial N_{\text{eff}}/\partial n = 0.21$; and $\partial N_{\text{eff}}/\partial w = 3.1 \times 10^5\ \text{m}^{-1}$ (for $n_w = 1.45$).

The change in N_{eff} causes a change in the phase $\Delta\varphi_{ij}$ of the guided light at the end of the sensing window. Using (10.5) and (10.7–10.9) allows deriving the complete sensitivity chain as (where the subscript "*bulk*" and "*layer*" are used to distinguish between both modes of operation):

$$\Delta\varphi_{ij}^{\text{bulk}} = \frac{2\pi}{\lambda} l \frac{\partial N_{\text{eff}}}{\partial n} \frac{\partial n}{\partial C} \Delta C_{ij}, \tag{10.10a}$$

$$\Delta\varphi_{ij}^{\text{layer}} = \frac{2\pi}{\lambda} l \frac{\partial N_{\text{eff}}}{\partial w} \rho_w^{-1} \frac{\Delta m_{ij}}{A}. \tag{10.10b}$$

Increasing the interaction length, l, or choosing a smaller wavelength λ results in a higher sensitivity of $\Delta\varphi_{ij}$ with respect to N_{eff}. However, l is limited by the size of the chip and for choosing λ, other factors such as scattering losses, which increase when λ decreases, should be considered.

In a final stage, the change of the phase difference causes a change of the interference pattern from which, using Fourier analysis, the $\Delta\varphi_{ij}$ can be derived. This has been treated in Sect. 10.2.2 and has consequences for the distance between the output channels only.

To give an example both sensitivity coefficients are evaluated for the current sensor (see Sect. 10.3 for details). For the bulk detection of glucose this results in $\Delta\varphi^{\text{bulk}}$ (rad) $= 5.6 \times 10^2\ \Delta C$ (g/ml), whereas for the adsorption of proteins on the sensor surface the overall sensitivity of the sensor is evaluated as $\Delta\varphi^{\text{layer}}$ (rad) $=$ $2.0 \times 10^{-5}\ \Delta m/\text{A}$ (fg/mm^2). Measuring the phase change $\Delta\varphi_{ij}$ between any of the two channels i and j can thus give an estimation on the change in analyte concentrations between those two channels. If one channel (e.g., channel N) is used as a reference channel, then $\Delta C_N = 0$ and $\Delta m_N = 0$ and absolute analyte concentrations can be determined.

10.3 Fabrication and Characterization

10.3.1 The Sensor Chip

Based on the analysis presented in the previous section, here the design of the four-channel YI sensor is discussed.

In the four-channel YI sensor single mode channel waveguides are used to split the incoupled light into four parallel channels using a network of three Y-junctions. The single mode character is important as different modes are affected differently by refractive index changes due to, e.g., analyte concentration changes. This will strongly complicate the signal analysis. Coupling of the light to the channel waveguide is achieved by very efficient fiber-to-chip coupling[17], which results in a robust device. Alternative coupling schemes such as end-fire coupling[38] or grating coupling[39] may also be applied. For optimal fiber-to-chip coupling adiabatic transition from the channel waveguide end-face cross-section to the cross-section of the common waveguide requires transverse adiabatic tapers, which have a small taper angle (<0.030) to have functional tapering losses <0.1 dB[40]. The distances between the four channels are chosen such that all six possible distances between a set of two channels are distinct in order to allow independent measurements of phase differences (see discussion in Sect. 10.2.2). Furthermore, the distances are chosen such that (a) a reasonable number (>2) of fringes are visible at the CCD interface (the channels should not be too close) and (b) there is a good overlap of the light from the different channels at the surface of the CCD (the channels should not be too far apart). Although the distances are not device-critical, it is important to take these considerations into account for an optimal device design.

The final structure design of the four-channel YI sensor is shown in Fig. 10.7: top-view (a), cross-section perpendicular to output channels (b), and side-view along the optical path (c). The distance between the mutually parallel output branches of the first Y-junction is chosen as $D = 160$ μm and the length of the splitting section is $L_{\mathrm{split}} \sim 10$ mm.

Besides the limited size of the wafer (4 in.) and the required minimum length of other subfunctions in the optical chip, the need for integrating the four sensing windows with a microfluidic system or a flow-through cuvette further reduces the interaction length. In addition, an interdistance between sensing windows along the channel direction is needed to allow a good alignment of the flow-through cuvette. The length of the four sensing windows is chosen to be $l = 4$ mm (width $w_{\mathrm{win}} = 100$ μm) with an interdistance $l_{\mathrm{int}} = 3$ mm.

The last two restrictions can be overcome by using a microfluidic system where the sensing windows can be implemented parallel to each other and no interdistance is necessary. In that case, each sensing window can be as long as $\sim$28 mm, resulting in an improvement of the sensitivity by almost 1 order of magnitude.

Table 10.1 summarizes the parameters of the final four-channel YI structure. The IO chips are fabricated using Silicon-oxynitride (cleanroom) technology. Details of the fabrication process have been described extensively elsewhere[41].

10.3.2 Applying the Analytes

Application of sample solutions to the sensing windows of the four-channel sensor is achieved by integrating either a flow-through cuvette or a microfluidic system[42]. These systems are required to simultaneously apply different sample solutions to all four sensing windows. The flow-through cuvette is fixed onto the optical chip by slightly pressing it on the top, whereas the microfluidic system can be permanently bonded to the optical chip.

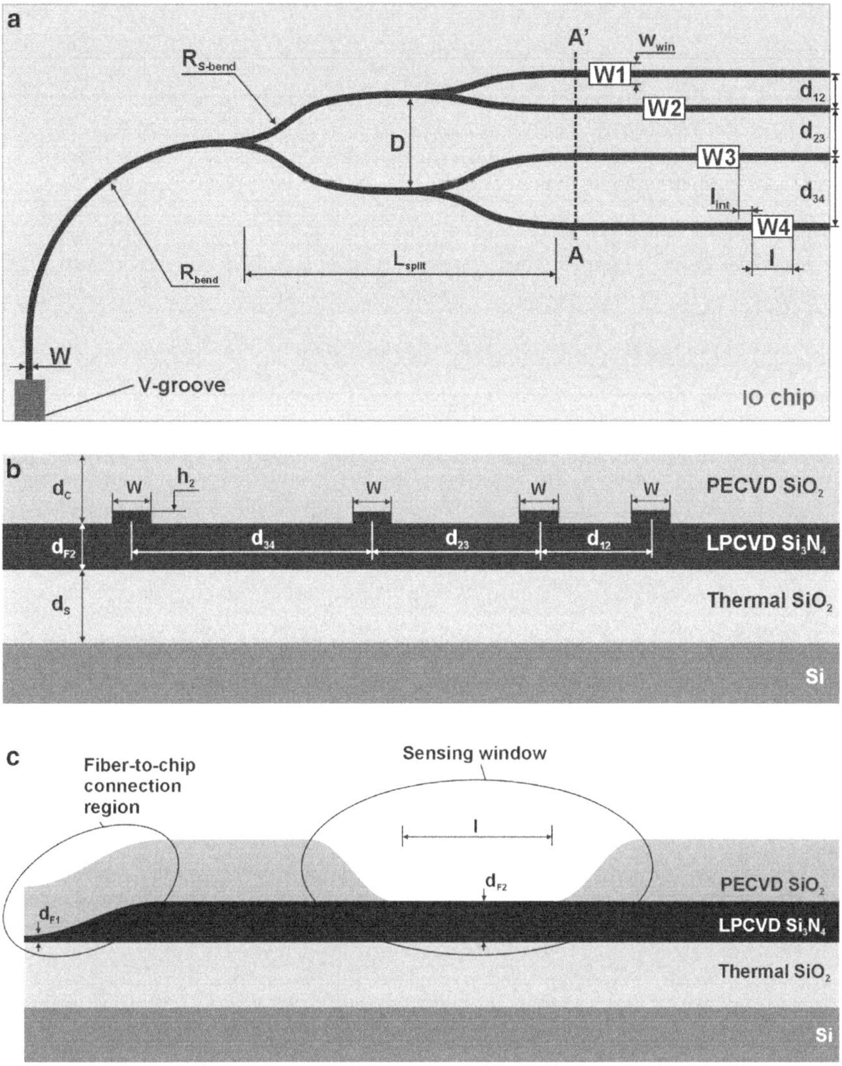

Fig. 10.7 *Top-view* (**a**), cross-section along AA' (**b**), and *side-view* (**c**) of the four-channel IO YI sensor; W_i indicates the sensing window on channel i, d_{ij} indicates the distance between channels i and j, R_{bend} and $R_{\text{S-bend}}$ are the radius of 90°-bend and S-bends, respectively, W is the width of the channel waveguide. The dimensions of the chip are 63 mm × 24 mm

The flow-through cuvette that is used for the four-channel sensor is made from Perspex and is 31-mm long and 7-mm wide. It has four flow chambers each with a volume of 1.2 μl (6 mm long and 3 mm wide), see Fig. 10.8. Each chamber of the cuvette has an inlet and outlet that are connected via a tubing system with the sampling reservoirs, which contain solutions to be monitored, and to the waste, respectively. Samples are flowed by means of a peristaltic pump (Ismatec

Table 10.1 Parameters of the four-channel IO YI sensor

Parameter	Symbol	Value	Dimension
Wavelength of used light	λ	647	nm
Waveguide mode	–	TE_{00}	–
Thickness of the core layer	d_{F1}	15	nm
	d_{F2}	70	nm
Height of the channel ridge	h_1	2	nm
	h_2	0.7	nm
Thickness of SiO_2 subs. layer	d_S	1.25	μm
Thickness of SiO_2 cover layer	d_C	1.5	μm
Channel width	W	4	μm
Radius of 90°-bend	R_{bend}	10	mm
Radius of S-bend	$R_{S\text{-}bend}$	50	mm
Length of sensing windows	l	4	mm
Width of sensing windows	w_{win}	100	μm
Windows interdistance	l_{int}	3	mm
Distance between channels	d_{12}	60	μm
	d_{23}	80	μm
	d_{34}	100	μm

See text and Fig. 10.7 for a detailed description of each parameter

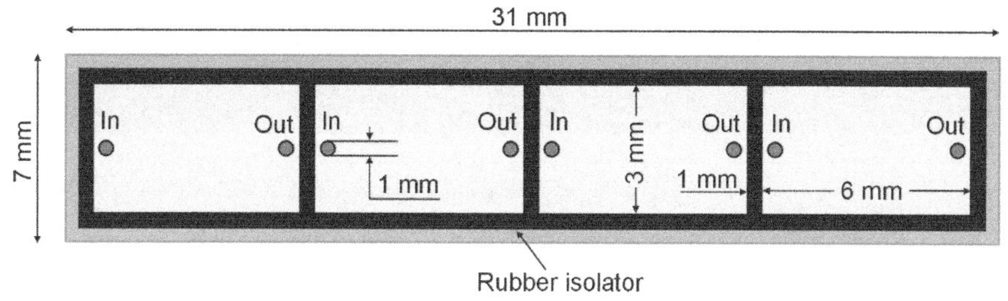

Fig. 10.8 *Bottom-view* of the four-chamber flow-through cuvette; each chamber has an In(let) and an Out(let), through which the sampling liquid flows to/from the sensing window. Reprinted from Ref. 31 with permission. © 2008 Optical Society of America

MS-REGLO; Ismatec SA., Glattbrugg, Switzerland) that has four separate pumping channels, each of them functioning independently from each other. In Fig. 10.9 a photograph of a flow-through cuvette assembled on the YI chip is shown.

The hybrid integration of the IO chip with a microfluidic system is treated in detail in Sect. 10.5.

10.3.3 Experimental Setup

Monochromatic laser light of an Ar^+ laser (Innova 70 Spectrum; Coherent, Inc., Santa Clara, CA) with a wavelength of 647 nm is end-fire coupled into the input fiber connected to the YI chip. The four output divergent beams overlap one another on the front side of a CCD camera (Teli CS-3440; 756 × 575 pixels, $w_{pixel} = 11.6$

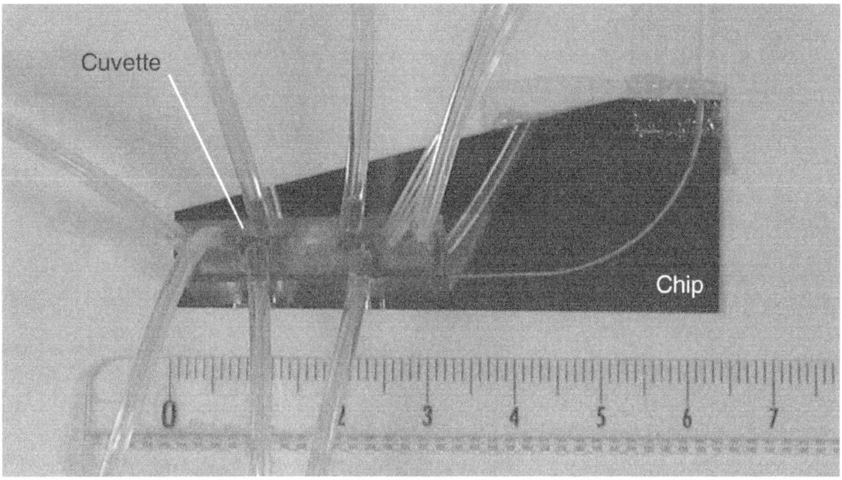

Fig. 10.9 Photograph of the sensor chip and the flow-through cuvette. The chip contains 11 four-channel YI sensors. The light is coupled into the chip by means of an optical fiber at the top-right-hand side. The light is transported to the sensing area by means of a curved channel waveguide, visible as a *white curved line*. Ruler indicates the dimensions in centimeters. Reprinted from Ref. 28 with permission. © 2008 American Chemical Society

μm; Tokyo Electronic Industry Co., Ltd., Tokyo, Japan) placed at a distance of ~60 mm from the chip endface and parallel to the endface plane for recording the interference pattern. The distance between the CCD camera and chip endface is chosen such that the recording area of the CCD camera can cover the total sum interference pattern, making use of all available pixels. A cylindrical lens is positioned between the chip endface and the CCD array for collimating the output divergent beams in the direction perpendicular to the sensor chip upper surface to collect all light emitted from the output channels. The recorded sum interference pattern is digitized to 12-bit information and analyzed by custom-written software in which a two-dimensional FFT algorithm is implemented. The algorithm recognizes and selects six different peaks that correspond to the six pairs of channels in the amplitude spectrum of the fast-Fourier-transformed interference pattern. The phase value corresponding to each pair of channels is extracted in the phase spectrum of the FFT at the given spatial frequencies. The maximum bandwidth of the measurement is 25 Hz, which is limited by the frame rate of the CCD camera.

10.3.4 Characterization

The performance of the realized YI sensor was first tested by applying solutions of different concentrations of glucose in water, thus inducing well-defined refractive index changes between measuring and reference channels of the device. In

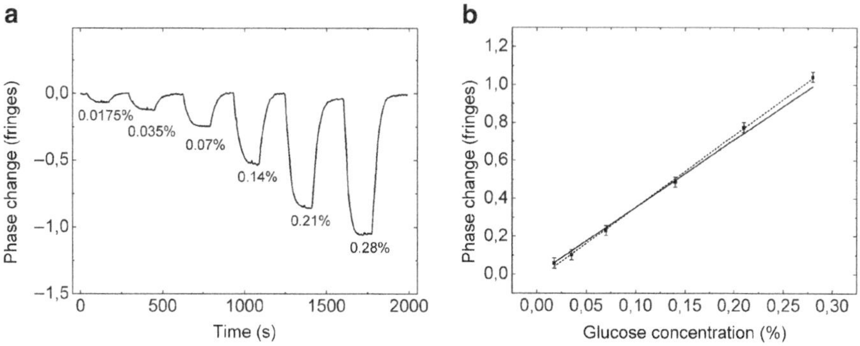

Fig. 10.10 (**a**) Response of the two-channel YI sensor to refractive index changes caused by applying six different glucose concentrations of 0.0175%, 0.035%, 0.07%, 0.14%, 0.21%, and 0.28% (by weight) intermitted by pure water to the measuring channel of the device. (**b**) Phase change vs. the glucose concentration: calculated (*line*), measured (*filled square*)), linear fit (*dotted line*)

Fig. 10.10a, the response of the sensor to different glucose concentrations applied in the sensing window of the measuring arm is shown. First pure water was flowed simultaneously over both sensing windows, and after several minutes, the flow in the measuring window was changed from water to 0.0175% (by weight) solution of glucose in pure water. When the signal stabilized, water was introduced and flowed over the measuring window. This procedure was repeated for five other glucose concentrations, respectively, 0.035%, 0.07%, 0.14%, 0.21%, and 0.28% (by weight). The induced refractive index changes, which can be calculated from the literature[36], vary from $\sim 2.4 \times 10^{-5}$ to $\sim 3.8 \times 10^{-4}$, whereas the corresponding phase changes vary from $\sim 0.062 \times 2\pi$ to $\sim 0.991 \times 2\pi$.

In Fig. 10.10b, the measured and expected phase changes as a function of the glucose concentration are plotted, including a linear fit of the measured data. The experimental plot (slope $\sim 3.8 \times 10^{-2} \times 2\pi$ ml/g) is in good accordance with the expected values (slope $\sim 3.5 \times 10^{-2} \times 2\pi$ ml/g).

The multichannel character of the YI sensor was tested by applying different glucose solutions to the different channels. In Fig. 10.11a, the response of the sensor is shown when a phase change of 2π was introduced in channel 1 and 3 and a phase change of $2 \times 2\pi$ was introduced in channel 2 simultaneously, using 0.616% and 1.232% (by weight) glucose solutions. Pure water was continuously flowed in channel 4. After a stable signal was reached, the solutions were exchanged for pure water, at first in channel 2, and next in channels 3 and 1, as shown in Fig. 10.11a. From the measured values, $\Delta\varphi_{12}$, $\Delta\varphi_{23}$, and $\Delta\varphi_{14}$, the phase change introduced in the channels, $\Delta\phi_1$, $\Delta\phi_2$, and $\Delta\phi_3$ can be determined, e.g., $\Delta\phi_1 = \Delta\varphi_{14} + \Delta\phi_4$, etc. Note that $\Delta\phi_4 = 0$, because no phase change was introduced in channel 4. The time-dependence of the phase changes are shown in Fig. 10.11b.

The obtained phase changes correspond well to the introduced changes, demonstrating one of the multipurpose features of the multichannel device with which

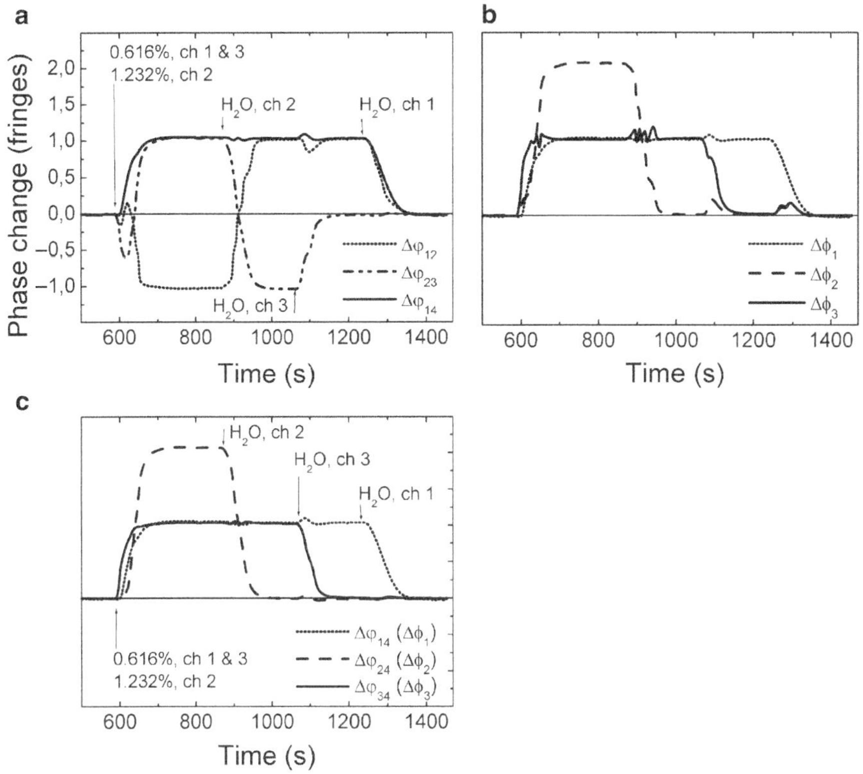

Fig. 10.11 (**a**) Measured phase changes $\Delta\varphi_{12}$, $\Delta\varphi_{23}$, and $\Delta\varphi_{14}$ as a function of time in the four-channel YI when a phase change of 2π was introduced in channel 1 and 3 and a phase change of $2 \times 2\pi$ in channel 2 simultaneously. (**b**) Phase changes that occur in channel 1, 2, and 3 ($\Delta\phi_1$, $\Delta\phi_2$, and $\Delta\phi_3$), calculated from measured phase changes $\Delta\varphi_{12}$, $\Delta\varphi_{23}$, and $\Delta\varphi_{14}$. (**c**) Measured phase changes $\Delta\varphi_{14}$, $\Delta\varphi_{24}$, and $\Delta\varphi_{34}$, which correspond to phase changes occurring in channel 1, 2, and 3, respectively. Reprinted from Ref. 31 with permission. © 2008 Optical Society of America

three different glucose concentrations were measured simultaneously and independently of each other. Alternatively, these channel phase changes can also be determined by using the measured phase changes $\Delta\varphi_{14}$, $\Delta\varphi_{24}$, and $\Delta\varphi_{34}$, as shown in Fig. 10.11c. Results obtained by both approaches correspond well, highlighting the consistency of the realized device. Fluctuations in the phase signal observed in Fig. 10.11b, c are caused by cross-talk[31] that is present when phase changes introduced in the channels are different from an integer number of fringes.

Figure 10.12 shows the phase difference measured between channels of each pair when no phase change was introduced in any of the channels. The phase resolution ($S/N = 1$) for each pair of channels, as determined from these measurements, is $\sim 1 \times 10^{-4} \times 2\pi$, which corresponds to a refractive index resolution of $\sim 8.5 \times 10^{-8}$ RIU (refractive index units) at a bandwidth of 1 Hz.

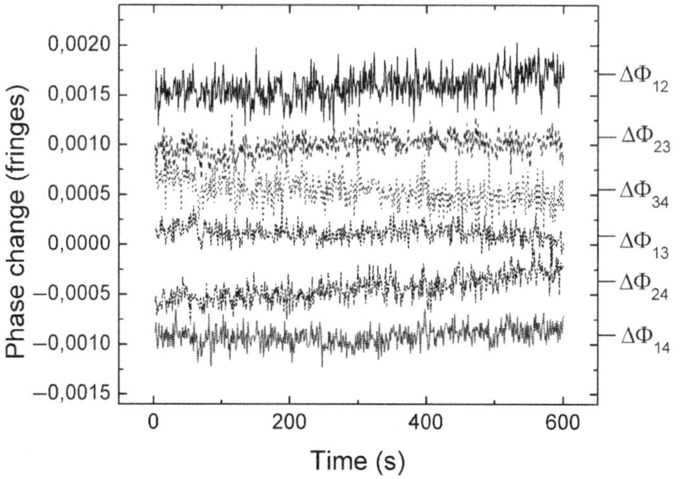

Fig. 10.12 Phase changes measured between two channels of each channel pair in the four-channel YI when no phase change was introduced at the channels. A shift of phase plots along vertical axis is applied to distinguish them from each other

10.4 Results

10.4.1 Immunosensing Using the YI Sensor

In an immunosensor the core-cover interface of an optical waveguide structure is coated with a chemo-optical transducer receptor layer, which can selectively bind to specific analyte molecules present in the cover medium. The receptor-analyte reaction obeys the law of mass action, which states that the rate of a reaction is proportional to the concentration of the reactants. At equilibrium, the rate of formation of the receptor-analyte complex is equal to the rate of breaking, and the equilibrium constant, K, can be written as

$$K = k_f/k_b = [\mathrm{Re\,An}]/[\mathrm{Re}]\cdot[\mathrm{An}], \tag{10.11}$$

where k_f and k_b are the formation and breaking rate constants and [Re], [An], and [ReAn] are, respectively, the concentration of receptor, analyte, and receptor-analyte complex. The thickness of the layer grown on the sensor surface due to the binding of analyte molecules to the receptor, Δw, depends on the receptor density, Γ_0, and the volume of the analyte molecules, V, and can be written as

$$\Delta w \approx \Delta F \Gamma_0 V, \tag{10.12}$$

where ΔF is the fraction of receptor molecules that have bound an analyte molecule. The maximum fraction of receptor molecules bound to an analyte molecule as a function of concentration of the analyte molecules can be described by the Langmuir isotherm[43]

$$\Delta F_{\max} = \frac{K[\mathrm{An}]}{1 + K[\mathrm{An}]}, \tag{10.13}$$

where K is the equilibrium constant for binding of the analyte molecules to the receptor molecules. In an interferometric sensor, such as the YI, the thickness of the protein layer grown on the sensor surface can be calculated as:

$$\Delta w = \frac{\lambda}{2\pi l}\left(\frac{\partial N_{\mathrm{eff}}}{\partial w}\right)^{-1} \Delta\varphi. \tag{10.14}$$

Equations (10.14) and (10.8) can be used for estimation of the thickness and mass coverage of the receptor adsorbed to the sensor surface and of the analyte layer bound to the receptor layer.

The ratio between the number of bound analytes, N_{An}, and the number of receptors adsorbed, N_{Re}, can be derived as

$$\frac{N_{\mathrm{An}}}{N_{Re}} \approx \frac{M_{Re}}{M_{\mathrm{An}}} \cdot \frac{\Delta\varphi_{\mathrm{An}}}{\Delta\varphi_{Re}}, \tag{10.15}$$

where M_{An} and M_{Re} are the molecular weight of analyte and receptor, respectively, and $\Delta\varphi_{\mathrm{An}}$ and $\Delta\varphi_{\mathrm{Re}}$ are the phase changes that result from binding of analyte and receptor, respectively. Evaluation of this quantity yields information on the quality of the adhered receptor layer, e.g., maximum $N_{\mathrm{An}}/N_{\mathrm{Re}} = 2$ (receptor can bind maximum two analytes).

10.4.2 Materials

The protein A (pA), antihuman serum albumin (α-HSA, $M \sim 150$ kD), and human serum albumin (HSA, $M \sim 65$ kD) were provided by Paradocs BV (Tiel, The Netherlands). The Herpes Simplex Virus type 1 (HSV-1) and anti-HSV-1 gG glycoprotein G monoclonal antibody (α-HSV-1 gG) were purchased from Virusys Corporation (Marriottsville, MD, USA). Bovine serum albumin (BSA, $M \sim 50$ kD) was purchased from Sigma-Aldrich Chemie BV (Zwijndrecht, The Netherlands). Synthetic surface protein of Hepatitis-B virus generated in Hep-G2 cell-line (HEP G2, $M \sim 25$ kD) was provided by BioMerieux BV (Boxtel, The Netherlands). Phosphate buffered saline (PBS) was used for all experiments.

10.4.3 Protein Detection

In this section, use of the IO YI as an immunosensor is demonstrated, being relevant, e.g., for identification of biomarkers for early disease detection. The sensor is used to monitor the anti-human serum albumin - human serum albumin (α-HSA-HSA) immunoreaction.

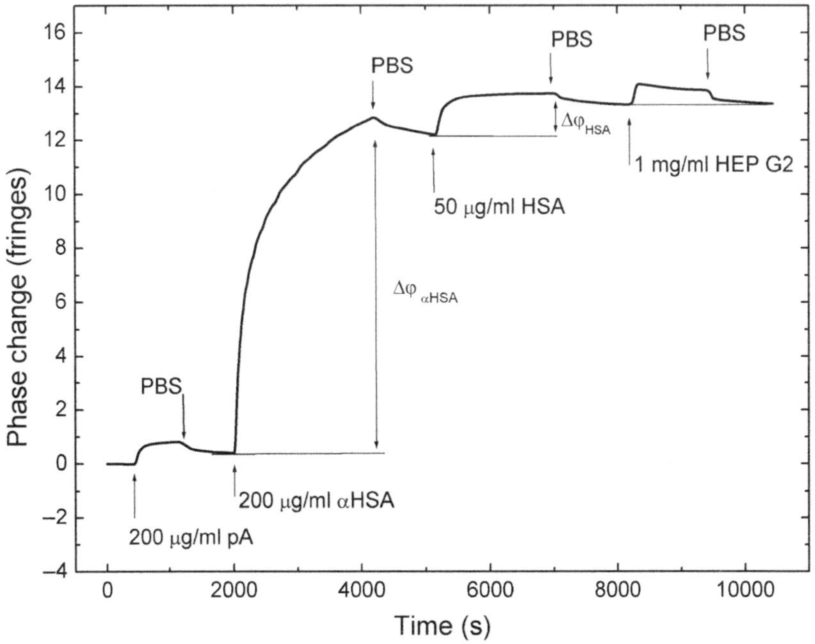

Fig. 10.13 Time-response between channels 1 and 4 in the four-channel YI sensor when the α-HSA-HSA immunoreaction was taking place in channel 1 and PBS buffer solution was continuously flowed in channel 4

Figure 10.13 shows the phase change measured between channels 1 and 4 of the four-channel YI when the PBS buffer solution has been continuously flowed in channels 2, 3, and 4, while in channel 1 the following steps were applied: first, a PBS buffer solution was flowed for ~10 min. Then the flow was changed to a solution of protein A (pA) prepared in PBS with a concentration of 200 μg/ml. A pA-modified sensing surface is implemented in order to promote the binding and to ensure proper orientation of antibodies for further antigen binding[45,46]. After an equilibrium in the signal was achieved in ~10 min, indicating the maximum amount of the pA adsorbed, the PBS buffer solution was flowed in channel 1 in order to get rid of the bulk refractive index effect. Physical adsorption of pA resulted in a phase change of $\sim 0.5 \times 2\pi$, which is equivalent to a mass coverage of ~0.12 mg/m^2, see (10.8).

Next, when a stable base line was achieved in ~13 min, a concentration of 200 μg/ml solution of α-HSA in PBS was flowed in channel 1 for ~30 min, and the flow was changed back to PBS buffer solution to correct for the bulk effect of α-HSA. Binding of α-HSA resulted in a phase change of $\sim 12 \times 2\pi$, which corresponds to a mass coverage of ~3.1 mg/m^2. Next, after the buffer solution was flowed for ~10 min until a stable signal was achieved, a solution of 50 μg/ml HSA prepared in PBS was applied. Finally, the flow in channel 1 was changed back to PBS solution and the phase change due to antigen layer formation was extracted. Binding of HSA caused a phase change of $\sim 1.3 \times 2\pi$. The thickness of the bound HSA layer to α-HSA molecules is found to be ~0.6 nm, see (10.14), assuming that there is only specific binding and a homogeneous layer was formed. The average number of HSA antigen molecules bound per one α-HSA antibody molecule is low (~0.25), see (10.15), indicating that there may be still antibody molecules available for binding. A larger binding in that case can be achieved by increasing the concentration of HSA antigen solutions, until the saturation level is reached, meaning that nearly all the available binding sites in the α-HSA layer are occupied by antigens. If the low number of 0.25 is due to improper orientation of α-HSA molecules, then little can be done.

Specificity of the α-HSA-HSA immunoreaction was checked by flowing in channel 1 a high concentration (1 mg/ml) of HEP G2 antigen solution prepared in PBS. After the flow was changed back to buffer solution, no response was observed, indicating that nonspecific binding of HEP G2 is negligible. This clearly demonstrates the specificity of α-HSA-HSA interaction.

Another possibility of the four-channel YI sensor is to measure different concentrations of a given analyte simultaneously. This feature was demonstrated experimentally by measuring simultaneously different concentrations of HSA, based on which the Langmuir binding curve (see (10.13)) was generated. For that purpose, three measuring channels, i.e., channel 1, 2, and 3, were simultaneously coated using a concentration of 200 μg/ml α-HSA prepared in PBS (PBS buffer solution was continuously flowed in channel 4). The next step consists of simultaneously applying three different concentrations of HSA, i.e., 500 ng/ml, 1 μg/ml, and 5 μg/ml in measuring channel 1, 2, and 3, respectively. This step was repeated by applying concentrations of 25 μg/ml, 50 μg/ml, and 100 μg/ml, in measuring channel 1, 2, and 3 simultaneously, followed by another set of concentrations (250 μg/ml, 500 μg/ml, and 1 mg/ml). After each set of concentrations was applied, PBS buffer solution was simultaneously flowed in the measuring channels to get rid of the bulk refractive index of HSA solutions applied. Finally, after coating channel 4 using 200 μg/ml α-HSA solution, concentrations of 100 ng/ml and 250 ng/ml HSA were applied successively (PBS was flowed in all other channels). The phase changes measured for the different HSA concentrations are plotted in Fig. 10.14.

The experimentally measured data are fitted with the Langmuir isotherm[43]. According to these measurements, when the concentration of HSA approaches ~1 mg/ml (1.5×10^{-5} M), a saturation level is reached, meaning that the antigen molecules already occupy all available binding sites. The total phase change measured due to HSA binding was $\sim 1.13 \times 2\pi$, corresponding to a layer thickness of ~0.5 nm. The equilibrium constant is estimated to be $\sim 3 \times 10^{7}\ \mathrm{M}^{-1}$, as extracted from a fitted Langmuir curve shown in Fig. 10.14, and is consistent with previous results[24].

From these results it is concluded that a pA-modified sensing surface is efficiently used to immobilize antibodies at the Si_3N_4 surface with a surface coverage of $\sim 3\ \mathrm{mg/m^2}$ corresponding to values found in literature[45,46]. However, the ratio

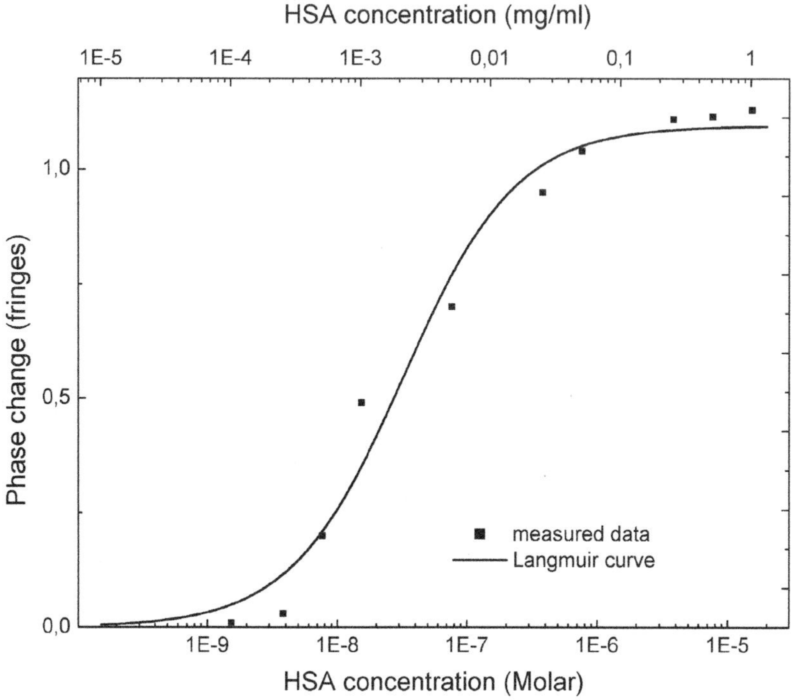

Fig. 10.14 Phase change measured in the four-channel YI for different concentrations of HSA solution. *Solid line* represents the Langmuir curve fitted to the experimental data

between the number of bound antigens and the number of antibodies (~0.25) was lower than expected since theoretically two antigens can bind to one antibody. It should be note, however, that the used antigen concentrations were below the saturation value. The use of higher concentrations of HSA yielded slightly better results (ratio ~0.3) indicating that there is still room for improvement. In addition, physical adsorption was used for immobilization of antibodies as an alternative to a pA-modified surface. In that case, postcoating with BSA after antibody adsorption was applied to prevent nonspecific binding of antigen. The results achieved with this method were less effective concerning the antibody immobilization (~1.6 mg/m^2) and specificity. In addition, a signal drift after BSA adsorption was observed, which most likely is due to desorption of BSA from the surface.

In general, it is observed that after the binding step of proteins to the interface a gradual change of the phase is present (see, e.g., Fig. 10.13 after the α-HSA adsorption step). This effect can be explained by desorption of bound molecules not firmly attached. Given the time span used in these experiments, desorption did not seriously hamper the detection of analytes.

10.4.4 Virus Detection

One of the major challenges in the field of biosensors is the ability to detect and identify viruses. Viral outbreaks are a continuous threat to human populations and economies. Examples of recent viral outbreaks are H5N1 avian influenza and SARS. The extent of a viral outbreak depends, among other things, on currently available diagnostic capabilities. Existing methods like polymerase chain reaction (PCR)[47], branched-chain DNA (bDNA) tests[48], and enzyme-linked immunosorbent assay (ELISA)[49] are very sensitive, but have some major drawbacks. These techniques are very labor-intensive and require time-consuming sample preparation procedures. Therefore these techniques may have limited use for rapid responses in case of serious viral outbreaks. This drawback has motivated the development of many alternative techniques that may overcome the limitations of the existing methods. Among these, label-free detection schemes such as Raman[50,51], rupture event scanning[52], nanowires[53], microring resonators[10], and cantilever based techniques[54,55] are among the most promising. The YI sensor described here also is a good candidate for rapid viral detection.

The possibility to use the YI sensor for virus detection was explored by monitoring the interaction between α-HSV-1 gG antibody and HSV-1 virus particles. To this end, channel 1 was coated with protein pA as described in Sect. 10.4.2 followed by the immobilization of a α-HSV-1 gG layer on the sensing surface of channel 1. Channel 4 was used as a reference channel. Finally a solution with HSV-1 virus particles at a concentration of 10^5 particles/ml was added to channel 1. Figure 10.2 shows the phase change measured between channel 1 and reference channel 4, clearly demonstrating the detection of virus particles by the YI sensor (Fig. 10.15).

As a next step, the dynamic range and the sensitivity of the sensor was explored. Solutions with varying viral concentrations, ranging from 8.5×10^2 to 8.5×10^6 particles/ml were analyzed by measuring the corresponding phase changes when applied to the measuring channel of the sensor. The concentration range used corresponds to the classification "very low" to "very high" in terms of the viral load[56]. The result of this experiment is shown in Fig. 10.16a and demonstrates that a large dynamic range of at least 4 orders of magnitude is achieved, covering the entire clinically relevant concentration range.

Finally, the specificity of the virus sensor was addressed by detecting virus particles suspended in complex samples such as human blood serum. To do so, HSV-1 virus particles were spiked into human blood serum at a concentration of 10^5 particles/ml. The response of the sensor to the application of the HSV-1/serum sample is shown in Fig. 4.4b. Here we first added serum (with no virus particles present) to the measuring channel, which was followed by adding virus-containing serum. The results as shown in Fig. 4.4b, clearly demonstrates the feasibility to detect the binding of HSV-1 in the presence of serum. The observed phase change is in good agreement with the measurement of HSV-1 in buffer as indicated in Fig. 10.16a. To further confirm the specificity of HSV-1–α-HSV-1 gG interaction, a

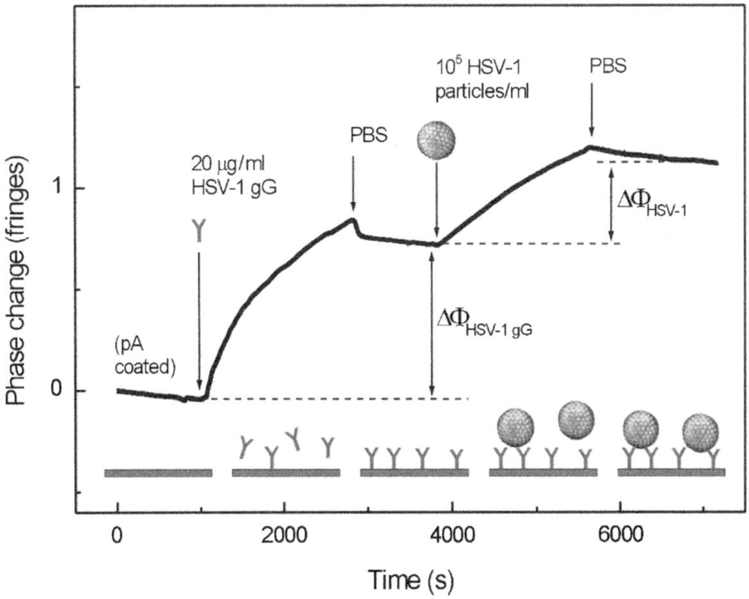

Fig. 10.15 Virus detection test. Sensor signal (phase change) measured between channel 1 and the reference channel for the immobilization of anti-HSV-1 glycoprotein G monoclonal antibody layer on the sensing surface of channel 1 ($\Delta\Phi_{\text{HSV-1 gG}}$) and the binding of HSV-1 particles to this layer ($\Delta\Phi_{\text{HSV-1}}$). Reprinted from Ref. 28 with permission. © 2008 American Chemical Society

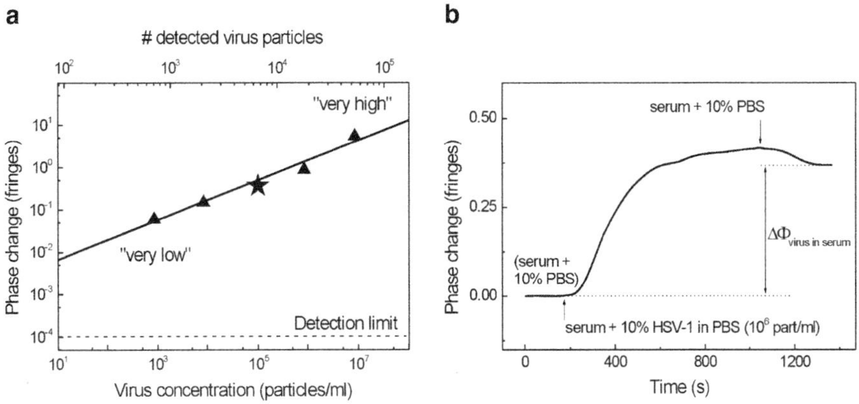

Fig. 10.16 Measurement of different HSV-1 concentrations and detection in serum. (**a**) Phase change measured for different concentrations of HSV-1 sample solutions in PBS applied in the measuring channel of the YI sensor (*filled triangle*). *Solid line* is a linear-fit of the experimental data, φ represents the phase change measured for HSV-1 diluted in serum (see Fig. 16b), *dashed line* indicates the phase detection limit of the sensor. (**b**) Sensor response due to the binding of HSV-1 diluted in serum. Final concentration of HSV-1 was 10^5 particles/ml. The total signal is estimated to be $\Delta\Phi_{\text{virus in serum}} = 0.37$ fringes, consistent with results obtained in PBS (see φ in Fig. 16a). Reprinted from Ref. 28 with permission. © 2008 American Chemical Society

control experiment in which no antibody was immobilized in the measuring channel was performed. No response was measured when the virus solution in serum was applied (graph not shown), indicating the specificity of HSV-1–α-HSV-1 gG interaction. These results show the possibility to specifically detect capture of HSV-1 particles, even at very low particle concentrations and at a low number of virus particles.

In the clinically relevant concentration range[56] the relation between the response of the sensor and the viral concentration is linear (a linear fit through the data points in Fig. 10.16a gives a correlation coefficient of 0.98) facilitating easy virus concentration predictions with a calibrated sensor. Furthermore, even at the lowest measured virus concentration (850 particles/ml) a high signal-to-noise ratio of 2×10^2 (at a bandwidth of 0.1 Hz) is achieved. Therefore it is predicted that the current YI sensor is able to detect even much lower concentrations. An estimation of the number of captured HSV-1 particles can be made given the size (150–200 nm)[57] and the refractive index (~1.41)[58]. This results in a phase change of $\sim 1.1 \times 10^{-4}$ fringes for the binding of a single virus particle. From this estimation it is derived that for the lowest measured concentration ~700 virus particles were detected with an average binding rate of 1 virus every 4 s. Given the detection limit of the sensor (10^{-4} fringes) it can be argued that the capture of a single virus particle at the sensor surface might yield a detectable signal.

The YI sensor also enables specific detection of virus particles in human blood serum. Even though there is a background signal caused by nonspecific binding of serum proteins, the signal due to binding of virus particles is readily detected. Although the background remains a drawback of this approach, alternative detection schemes with the YI sensor such as the use of differential phase change information from different measuring channels may be applied to reduce the influence of the background considerably. Moreover, attention should be paid to improvement of the measuring channel coatings, which should result in further reduction of the background.

10.4.5 Multianalyte Detection

Having demonstrated the possibility to use the IO YI as an immunosensor, in this section the multipurpose feature of the sensor is shown by monitoring more than one analyte simultaneously. These tests are performed by immobilizing different receptor layers in adjacent measuring channels and monitoring the sensor response to different analyte solutions. To do so, antihuman serum albumin (α-HSA; Sigma-Aldrich, St. Louis, MO, United States) was immobilized in channel 1 by flowing a 200 μg/ml α-HSA solution prepared in PBS. In channel 2, α-HSV-1 gG was immobilized. Next, a solution of 50 μg/ml human serum albumin (HSA; Sigma-Aldrich, St. Louis, MO, United States) in PBS was simultaneously applied in channels 1 and 2 and after approximately 30 min, flow was changed back to PBS for both channels. The observed binding curves are shown in Fig. 10.17 (graphs A1

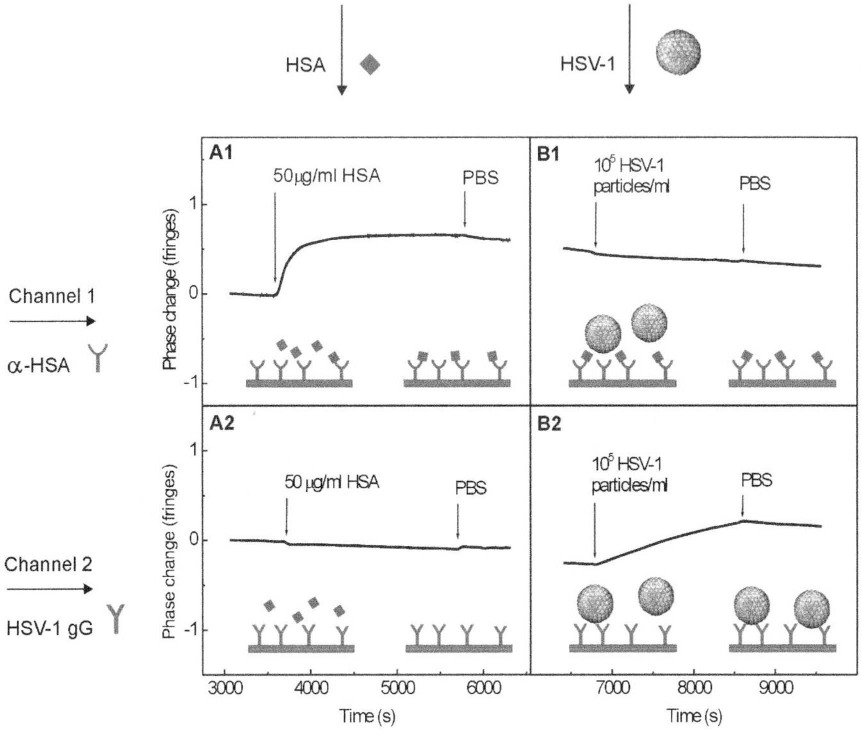

Fig. 10.17 Specific detection of HSV-1. Phase changes $\Delta\Phi 14$ and $\Delta\Phi 24$ in the four-channel YI sensor as a function of time during several processes. HSA solution was first flowed through channels 1 and 2 simultaneously (A1 and A2). Next, after washing with PBS, HSV-1 solution was flowed in channels 1 and 2 simultaneously (B1 and B2); PBS was continuously flowed in reference channel 4. Thus, the four graphs show the following interactions: (A1) α-HSA-HSA, (A2) α-HSV-1 gG-HSA, (B1) α-HSA-HSV-1, (B2) α-HSV-1 gG-HSV-1. Note that initial phases in A1 and A2 were shifted to 0 for clarity. Reprinted from Ref. 28 with permission. © 2008 American Chemical Society

and A2, respectively). After achievement of a stable baseline, a 10^5 particles/ml HSV-1 solution was flowed in both channels (see B1 and B2 in Fig. 10.17, respectively). The observation that a response is measured only for the α-HSA-HSA and α-HSV-1 gG-HSV-1 interactions is a clear indication that the signals are caused by specific interactions and that crossreactivity between coatings is negligible.

10.5 A Hybrid YI-Sensor Microfluidic Chip

The main advantages of the current YI sensor are that it is extremely sensitive, is a label-free technique, has multiplexing capabilities, and allows for a compact design. As such it is a good candidate for the development of a reliable handheld device that can be used in the field to quickly and easily detect, e.g., viruses from either human or animal body fluid samples. To this end the integration of the chip with microfluidic sample handling would strongly enhance the robustness and compactness of the sensor.

Figure 10.18 shows a four-channel YI sensor chip integrated with a microfluidic system[42]. The microfluidic system was designed such that each measuring channel can be addressed individually. The channels are fabricated in a transparent glass plate with cross-sectional dimensions of 200 μm × 15 μm. The glass plate is glued onto the IO-chip using a UV curable resin. A major advantage of the microfluidic system over the cuvette system is, apart from the improved robustness of the sensor, the strongly enhanced time-response of the sensor. Experiments on monitoring the antihuman serum albumin/human serum albumin immunoreaction show that the response time of the sensor of 100 s achieved with a bulky cuvette, is reduced to 4 s (at similar sample flow). This improvement is due to the much smaller dimensions of the microfluidic channels compared to those of the cuvette. Reduced dimensions, especially the height of the flow channel, imply higher flow velocities at the sample-sensor surface interface and shorter diffusion times of molecules/analytes.

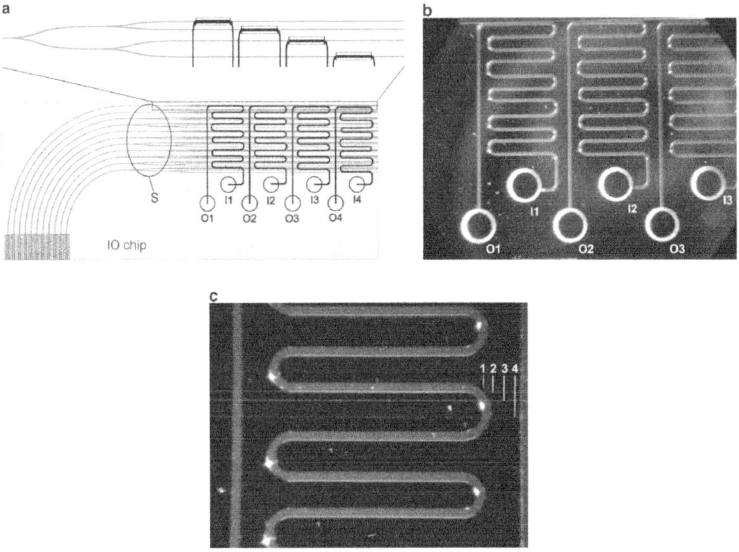

Fig. 10.18 Hybrid sensing platform obtained by bonding a glass microfluidic system to an IO four-channel YI chip. (**a**) *Top-view* of the microfluidic system. The optical chip contains configurations of YI sensor structure, which have different distances between their output channels, resulting in a different length of splitting function, manifested by different starting positions of the first Y-junction (see S in **a**). (**b**) *Close-view* micrograph of microfluidic channel 1, 2, and 3. (**c**) *Close-view* micrograph showing the light propagating through waveguide channel 1, 2, 3, and 4, and the microfluidic channel 2 addressing the sensing window of the waveguide channel 2. Ii and Oi indicate the input and output, respectively, of microfluidic channel *i*, addressing the *i*th bundle of sensing windows. Dimensions of the microfluidic sensing system are 63 mm × 24 mm. (**a**) is reprinted from Ref. 42 with permission. © 2008 Elsevier. (**b**) and (**c**) are reprinted from Ref. 59 with permission. © 2008 Expert Reviews Ltd

Both aspects improve the time-response of the sensor. In addition the total sample volume required is also strongly reduced from the sub-ml to the μl range.

10.6 Discussion and Conclusions

As demonstrated, the YI sensor can be used to detect analytes such as proteins and viruses with a very high sensitivity. The sensitivity obtained with, the 4-channel YI sensor is ~20 fg/mm^2 with, ~2 orders of magnitude higher than other label-free sensor techniques such as the well established SPR sensor[60]. Additional experiments with a two-channel YI sensor with longer interaction lengths (20 mm rather that 4 mm) show an even higher sensitivity of ~1 fg/mm^2 [41]. Moreover, the YI sensor is simple, easy to use, and compact, providing excellent possibilities for development of portable point-of-care devices. The detectable concentrations of analytes in a sample approach those required in clinical diagnostics assays, which could be an attractive market for these types of sensors. This is further motivated by the ability to integrate the sensor with efficient microfluidic-based fluid delivery systems[42], resulting in a much faster time response of the sensor. In addition, the sensor manufacturing technology is suitable for mass production, allowing important scale-based costs reduction.

The high sensitivity of the sensor that is demonstrated is obviously very promising. However, to be suitable for application in the field of clinical diagnostics assays the sensor also has to specific. The sensor signal should only be sensitive to the binding of the analyte of interest and not to other molecules or viruses in the sample. Therefore it is extremely important that these types of sensors are tested with clinically relevant samples. Specificity of the sensor thus strongly depends on the "chemical" coating of the sensor surface. The use of a pA-based coating was used to successfully detect HSV-1 viruses in human blood serum (this work) at clinically relevant concentrations. More recently it was shown that tuberculosis-specific antibodies can be detected with a YI sensor in undiluted human blood serum[39]. The availability of antibodies with high-specificity and low crossreactivity for relevant antigens, combined with these results suggest that YI-based sensors can combine very high sensitivity with good specificity and therefore provide a solid foundation for further development of YI sensors for clinical diagnostic assays.

References

1 Hong, J. G.; Choi, J. S.; Han, G. Y.; Kang, J. K.; Kim, C. M.; Kim, T. S.; Yoon, D. S., A Mach-Zehnder interferometer based on silicon oxides for biosensor applications, *Anal. Chim. Acta* **2006**, 573, 97–103

2 Heideman, R. G.; Kooyman, R. P. H.; Greve, J., Performance of a highly sensitive optical wave-guide Mach-Zehnder interferometer immunosensor, *Sens. Actuators B-Chem.* **1993**, 10, 209–217

3 Schipper, E. F.; Brugman, A. M.; Dominguez, C.; Lechuga, L. M.; Kooyman, R. P. H.; Greve, J., The realization of an integrated Mach-Zehnder waveguide immunosensor in silicon technology, *Sens. Actuators B-Chem.* **1997**, 40, 147–1534.
4 Duport, I. S.; Benech, P.; Rimet, R., New integrated-optics interferometer in planar technology, *Appl. Opt.* **1994**, 33, 5954–5958
5 Stamm, C.; Lukosz, W., Integrated optical difference interferometer as immunosensor, *Sens. Actuators B-Chem.* **1996**, 31, 203–207
6 Cox, E. R.; Jones, B. E. Fiber optic color sensors based on Fabry-Perot interferometry, In First International Conference on Optical Fiber Sensors, London, 1993
7 Clerc, D.; Lukosz, W., Integrated optical output grating coupler as biochemical sensor, *Sens. Actuators B-Chem.* **1994**, 19, 581–586
8 Cush, R.; Cronin, J. M.; Stewart, W. J.; Maule, C. H.; Molloy, J.; Goddard, N. J., The resonant mirror - a novel optical biosensor for direct sensing of biomolecular interactions. 1. Principle of operation and associated instrumentation, *Biosens. Bioelectron.* **1993**, 8, 347–353
9 Krioukov, E.; Klunder, D. J. W.; Driessen, A.; Greve, J.; Otto, C., Sensor based on an integrated optical microcavity, *Opt. Lett.* **2002**, 27, 512–514
10 Zhu, H. Y.; White, I. M.; Suter, J. D.; Zourob, M.; Fan, X. D., Opto-fluidic micro-ring resonator for sensitive label-free viral detection, *Analyst* **2008**, 133, 356–360
11 Armani, A. M.; Kulkarni, R. P.; Fraser, S. E.; Flagan, R. C.; Vahala, K. J., Label-free, single-molecule detection with optical microcavities, *Science* **2007**, 317, 783–787
12 Sun, Y. Z.; Shopova, S. I.; Frye-Mason, G.; Fan, X. D., Rapid chemical-vapor sensing using optofluidic ring resonators, *Opt. Lett.* **2008**, 33, 788–790
13 Kooyman, R. P. H.; Lenferink, A. T. M.; Eenink, R. G.; Greve, J., Vibrating mirror surface-plasmon resonance immunosensor, *Anal. Chem.* **1991**, 63, 83–85
14 Densmore, A.; Xu, D. X.; Janz, S.; Waldron, P.; Mischki, T.; Lopinski, G.; Delage, A.; Lapointe, J.; Cheben, P.; Lamontagne, B.; Schmid, J. H., Spiral-path high-sensitivity silicon photonic wire molecular sensor with temperature-independent response, *Opt. Lett.* **2008**, 33, 596–598
15 Brecht, A.; Piehler, J.; Lang, G.; Gauglitz, G., A direct optical immunosensor for atrazine detection, *Anal. Chim. Acta* **1995**, 311, 289–299
16 Lambeck, P. V., Integrated opto-chemical sensors, *Sens. Actuators B-Chem.* **1992**, 8, 103–116
17 Heideman, R. G.; Lambeck, P. V., Remote opto-chemical sensing with extreme sensitivity: design, fabrication and performance of a pigtailed integrated optical phase-modulated Mach-Zehnder interferometer system, *Sens. Actuators B-Chem.* **1999**, 61, 100–127
18 Brandenburg, A., Differential refractometry by an integrated-optical Young interferometer, *Sens. Actuators B-Chem.* **1997**, 39, 266–271
19 Brandenburg, A.; Krauter, R.; Kunzel, C.; Stefan, M.; Schulte, H., Interferometric sensor for detection of surface-bound bioreactions, *Appl. Opt.* **2000**, 39, 6396–6405
20 Cross, G. H.; Ren, Y. T.; Freeman, N. J., Young's fringes from vertically integrated slab waveguides: Applications to humidity sensing, *J. Appl. Phys.* **1999**, 86, 6483–6488
21 Cross, G. H.; Reeves, A. A.; Brand, S.; Popplewell, J. F.; Peel, L. L.; Swann, M. J.; Freeman, N. J., A new quantitative optical biosensor for protein characterisation, *Biosens. Bioelectron.* **2003**, 19, 383–390
22 Koster, T.; Lambeck, P., Fully integrated optical polarimeter, *Sens. Actuators B-Chem.* **2002**, 82, 213–226
23 Heideman, R. G. Optical Wabeguide Based Evanescent Field Immunosensors, University of Twente, Enschede, The Netherlands, 1993
24 Heideman, R. G.; Kooyman, R. P. H.; Greve, J., Immunoreactivity of adsorbed antihuman chorionic-gonadotropin studied with an optical wave-guide interferometric sensor, *Biosens. Bioelectron.* **1994**, 9, 33–43
25 Schipper, E. F.; Bergevoet, A. J. H.; Kooyman, R. P. H.; Greve, J., New detection method for atrazine pesticides with the optical waveguide Mach-Zehnder immunosensor, *Anal. Chim. Acta* **1997**, 341, 171–176

26 Schipper, E. F. Waveguide Immunosensing of Small Molecules, University of Twente, Enschede, The Netherlands, 1997
27 Minson, A. C., Alphaherpesviruses: herpes simplex and varicella zoster virus infection, In Topley and Wilson's Microbiology And Microbial Infections, 9th edn.; Mahy, B.; Collier, L., Eds.; Arnold, London, 1998, Vol. 1, 325–339
28 Ymeti, A.; Greve, J.; Lambeck, P. V.; Wink, T.; van Hovell, S.; Beumer, T. A. M.; Wijn, R. R.; Heideman, R. G.; Subramaniam, V.; Kanger, J. S., Fast, ultrasensitive virus detection using a young interferometer sensor, *Nano Lett.* **2007**, 7, 394–397
29 Xu, J.; Suarez, D.; Gottfried, D. S., Detection of avian influenza virus using an interferometric biosensor, *Anal. Bioanal. Chem.* **2007**, 389, 1193–1199
30 Wilson, J.; Hawkes, J. F. B., *Optoelectronics, An Introduction.* Prentice Hall, London, 1989
31 Ymeti, A.; Kanger, J. S.; Greve, J.; Lambeck, P. V.; Wijn, R.; Heideman, R. G., Realization of a multichannel integrated Young interferometer chemical sensor, *Appl. Opt.* **2003**, 42, 5649–5660
32 Hecht, E., *Optics*, Addison Wesley, Reading, MA, 1998
33 Ymeti, A.; Kanger, J. S.; Wijn, R.; Lambeck, P. V.; Greve, J., Development of a multichannel integrated interferometer immunosensor, *Sens. Actuators B-Chem.* **2002**, 83, 1–7
34 Defeijter, J. A.; Benjamins, J.; Veer, F. A., Ellipsometry as a tool to study adsorption behavior of synthetic and biopolymers at air-water-interface, *Biopolymers* **1978**, 17, 1759–1772
35 Weast, R. C., Handbook of Chemistry and Physics, 65th edn.; CRC Press, Boca Raton, FL, 1984
36 Parriaux, O.; Veldhuis, G. J., Normalized analysis for the sensitivity optimization of integrated optical evanescent-wave sensors, *J. Lightwave Technol.* **1998**, 16, 573–582
37 Haruna, M.; Segawa, Y.; Nishihara, H., Nondestructive and simple method of optical-waveguide loss measurement with optimization of end-fire coupling, *Electron. Lett.* **1992**, 28, 1612–1613
38 Nagel, T.; Ehrentreich-Forster, E.; Singh, M.; Schmitt, K.; Brandenburg, A.; Berka, A.; Bier, F. F., Direct detection of tuberculosis infection in blood serum using three optical label-free approaches, *Sens. Actuators B-Chem.* **2008**, 129, 934–940
39 Parishi, G. I.; Haszko, S. E.; Rozgonyi, G. A., Tapered windows in SiO_2:the effect of NH4F: HF dilution and etching temperature, *J. Electrochem. Soc. Solid State Sci. Technol.* **1997**, 124, 917–921
40 Ymeti, A. Development of a Multichannel Integrated Young Interferometer Immunosensor, University of Twente, Enschede, The Netherlands, 2004
41 Ymeti, A.; Kanger, J. S.; Greve, J.; Besselink, G. A. J.; Lambeck, P. V.; Wijn, R.; Heideman, R. G., Integration of microfluidics with a four-channel integrated optical Young interferometer immunosensor, *Biosens. Bioelectron.* **2005**, 20, 1417–1421
42 Antrade, J. D., Principles of protein adsorption, In Surface and Interfacial Aspects of Biomedical Polymers; Antrade, J. D., Ed.; Plenum Press, New York, 1985, Vol. 2
43 Owaku, K.; Goto, M.; Ikariyama, Y.; Aizawa, M., Optical immunosensing for IgG, *Sens. Actuators B-Chem.* **1993**, 14, 723–724
44 Lu, B.; Smyth, M. R.; Okennedy, R., Oriented immobilization of antibodies and its applications in immunoassays and immunosensors, *Analyst* **1996**, 121, R29–R32
45 Piatak, M.; Saag, M. S.; Yang, L. C.; Clark, S. J.; Kappes, J. C.; Luk, K. C.; Hahn, B. H.; Shaw, G. M.; Lifson, J. D., High-levels of HIV-1 in plasma during all stages of infection determined by competitive PCR, *Science* **1993**, 259, 1749–1754
46 Pachl, C.; Todd, J. A.; Kern, D. G.; Sheridan, P. J.; Fong, S. J.; Stempien, M.; Hoo, B.; Besemer, D.; Yeghiazarian, T.; Irvine, B.; Kolberg, J.; Kokka, R.; Neuwald, P.; Urdea, M. S., Rapid and precise quantification of HIV-1 RNA in plasma using a branched DNA signal amplification assay, *J. Acquir. Immune Defic. Syndr. Hum. Retrovirol.* **1995**, 8, 446–454
47 Kumarasamy, V.; Wahab, A. H. A.; Chua, S. K.; Hassan, Z.; Chem, Y. K.; Mohamad, M.; Chua, K. B., Evaluation of a commercial dengue NS1 antigen-capture ELISA for laboratory diagnosis of acute dengue virus infection, *J. Virolog. Methods* **2007**, 140, 75–79

48 Shanmukh, S.; Jones, L.; Driskell, J.; Zhao, Y. P.; Dluhy, R.; Tripp, R. A., Rapid and sensitive detection of respiratory virus molecular signatures using a silver nanorod array SERS substrate, *Nano Lett.* **2006**, 6, 2630–2636

49 Shanmukh, S.; Jones, L.; Zhao, Y. P.; Driskell, J. D.; Tripp, R. A.; Dluhy, R. A., Identification and classification of respiratory syncytial virus (RSV) strains by surface-enhanced Raman spectroscopy and multivariate statistical techniques, *Anal. Bioanal. Chem.* **2008**, 390, 1551–1555

50 Cooper, M. A.; Dultsev, F. N.; Minson, T.; Ostanin, V. P.; Abell, C.; Klenerman, D., Direct and sensitive detection of a human virus by rupture event scanning, *Nat. Biotechnol.* **2001**, 19, 33–837

51 Patolsky, F.; Zheng, G. F.; Hayden, O.; Lakadamyali, M.; Zhuang, X. W.; Lieber, C. M., Electrical detection of single viruses, *Proc. Natl. Acad. Sci. USA* **2004**, 101, 14017–14022

52 Gupta, A.; Akin, D.; Bashir, R., Single virus particle mass detection using microresonators with nanoscale thickness, *Appl. Phys. Lett.* **2004**, 84, 1976–1978

53 Owen, T. W.; Al-Kaysi, R. O.; Bardeen, C. J.; Cheng, Q., Microgravimetric immunosensor for direct detection of aerosolized influenza A virus particles, *Sens. Actuators B-Chem.* **2007**, 126, 691–699

54 Mellors, J. W.; Rinaldo, C. R.; Gupta, P.; White, R. M.; Todd, J. A.; Kingsley, L. A., Prognosis in HIV-1 infection predicted by the quantity of virus in plasma, *Science* **1996**, 272, 1167–1170

55 Levine, A. J., Viruses, Scientific American Library, New York, 1992

56 Balch, W. M.; Vaughn, J.; Novotny, J.; Drapeau, D. T.; Vaillancourt, R.; Lapierre, J.; Ashe, A., Light scattering by viral suspensions, *Limnol. Oceanogr.* **2000**, 45, 492–498

57 Ymeti, A.; Subramaniam, V.; Beumer, T. A. M.; Kanger, J. S., An ultrasensitive Young interferometer handheld sensor for rapid virus detection, *Expert Rev. Med. Devices* **2007**, 4, 447–454

58 Berger, C. E. H.; Beumer, T. A. M.; Kooyman, R. P. H.; Greve, J., Surface plasmon resonance multisensing, *Anal. Chem.* **1998**, 70, 703–706

Chapter 11
The BioCD: High-Speed Interferometric Optical Biosensor

David D. Nolte, Ming Zhao, and Xuefeng Wang

Abstract The bio-optical compact disc (BioCD) is an optical biochip that uses common-path interferometry on a disc spinning at high speed to detect captured proteins. High-speed scanning moves the detection frequency far from $1/f$ noise, providing high sensitivity and enabling rapid measurement of high-throughput multiplexed assays. The common-path configuration makes it ultra stable with surface height precision down to 20 pm within the focused probe area. This chapter reviews the state of the art in interferometric detection of proteins using spinning-disc interferometry. There are several common-path configurations that achieve phase quadrature for sensitive detection of surface-immobilized proteins. We have implemented differential phase contrast, in-line, microdiffraction, and adaptive optical approaches. Protein patterning provides spatial frequencies for Fourier-domain detection and spatial multiplexing on the BioCD surface. The detection limits of protein are set by a scaling surface mass density, with a metrology limit below 1 pg/mm. Specific immunoassay applications are described for prostate-specific antigen and haptoglobin. A highly multiplexed platform like the BioCD may enable a Moore's Law of protein detection as the scaling capabilities of protein patterning coevolve with proteomics to explore increasingly complex protein interaction networks.

11.1 Scalable Biosensor Technology

Our body's machinery is built of proteins. They are the functional elements that make everything work, keeping us alive and healthy. But when we are not healthy, when disease strikes or when our inner mechanisms go astray, then the surest way to track illness is through the concentrations and relationships of all of our proteins. This is truly a daunting task. Our blood alone carries over 10,000 proteins.

D.D. Nolte (✉)
Department of Physics, Purdue University, Lafayette, IN 47907, USA
e-mail: nolte@purdue.edu

X. Fan (ed.), *Advanced Photonic Structures for Biological and Chemical Detection*, Integrated Analytical Systems,
DOI 10.1007/978-0-387-98063-8_11,

And contained within the sum of all of our bodily fluids are over a hundred thousand proteins. These proteins do not act in isolation, but form extended protein interaction networks[1]. This expands the protein interaction space into millions of measurements that could be needed to specify our state of health.

The technology that will be needed to match the scale of the proteomic challenge is one that has scalability from hundreds to tens of thousands of measurements. Scalability for a technology means that the resource cost rises sublinearly with the capacity of the technology. When a technology is scalable, then doubled capacity costs less than a factor of 2. Scalability lies behind the success of the electronic industry with its famous Moore's Law[2]. The challenge of proteomics is to find an analytical technology that is scalable to sufficient capacity to match the proteomic measurement size.

One class of analytical techniques that has many of the pieces of the scalability puzzle is label-free optical biosensors[3,4]. Biosensors measure biological molecules in biological fluids. Being label-free means that the sensors detect the target molecules directly, without requiring them to be tagged or labeled with a secondary chemical reagent, such as a fluorescent molecule. Freedom from secondary reagents is a primary requirement for a scalable biosensor technology. Biosensors that use optical transduction to read out the molecular signals have the additional advantages of the parallelism of light in which multiple channels can be acquired simultaneously providing a multiplex advantage. Therefore, optical information transduction and transfer has the hallmarks of scalability, and label-free optical biosensors may provide the scalable technology that will match the scale of the proteomic measurement problem.

11.2 Spinning-Disc Interferometry on the BioCD

The bio-optical compact disc (BioCD) is an optical biochip that uses common-path interferometry on a disk spinning at high speed to detect captured proteins[5]. High-speed scanning moves the detection frequency far from $1/f$ noise, providing high sensitivity and enabling rapid measurement of high-throughput multiplexed assays. The common-path configuration makes it ultra stable with surface height precision down to 20 pm within the focused probe area. The basic optical configuration of the BioCD system is shown in Fig. 11.1. The optical schematic uses a single probe laser that can measure interferometric signals as well as fluorescence[6]. The oblique incidence in Fig. 11.1 is used to separate the fluorescence channel from the interferometric channel, but can be surface-normal for interferometry alone. Spinning-disc interferometry (SDI) has several advantages that make it attractive for label-free direct detection applications relative to conventional approaches. The first advantage is the high-speed scanning that allows the detection of tens of thousands of antibody spots in several minutes. The high-speed data acquisition reduces the noise floor as the sampling frequency moves far off of $1/f$ noise that dominates many other approaches. The circular scan geometry also enables fast data averaging over the same antibody spots to further reduce noise. The second

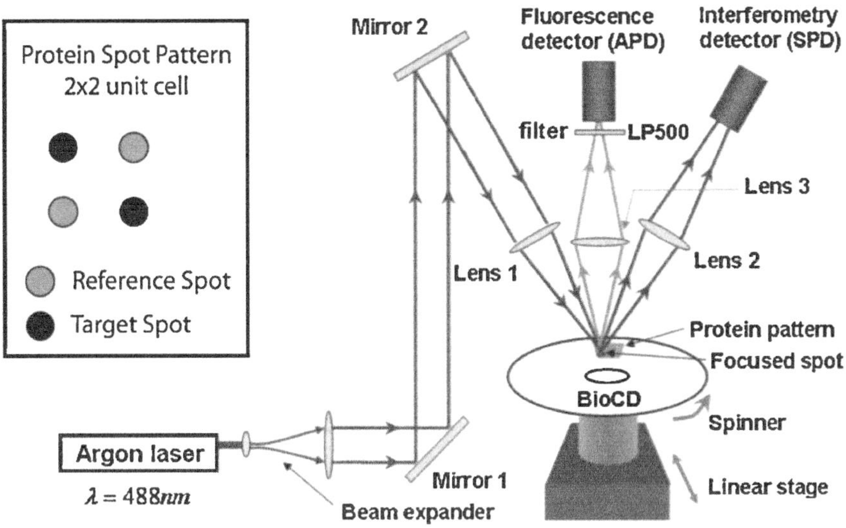

Fig. 11.1 Optical setup of the BioCD system with 2-channel detection for fluorescence and interferometry. The focused laser spot is approximately 30 μm in diameter with 15 μm pitch between tracks. The protein spots are approximately 150 μm in diameter

advantage is the interferometric sensitivity associated with phase quadrature. Interferometry is one of the most sensitive detection techniques, with subpicometer height sensitivity at the scale of protein spots, enabling high surface sensitivity for immunoassay applications.

11.2.1 Phase Quadrature

All two-port interferometers have the universal intensity output given by

$$I(\phi) = I_1 + I_2 + 2\sqrt{I_1 I_2}\cos\phi, \tag{11.1}$$

where $I_{1,2}$ are the input intensities, and ϕ is the fixed relative phase difference between the waves. When a phase modulation $\Delta\phi(t)$ is present on the signal wave, the intensity dependence is

$$\begin{aligned} I(t) &= I_1 + I_2 + 2\sqrt{I_1 I_2}\cos(\phi + \Delta\phi(t)), \\ &= I(\phi) - 2\sqrt{I_1 I_2}\sin\phi\sin\Delta\phi(t). \end{aligned} \tag{11.2}$$

In the case when the phase modulation is small, the intensity modulation is

$$\begin{aligned} \Delta I(t) &= \left[2\sqrt{I_1 I_2}\sin\phi\right]\Delta\phi(t) \\ &= \frac{\mathrm{d}I}{\mathrm{d}\phi}\Delta\phi(t), \end{aligned} \tag{11.3}$$

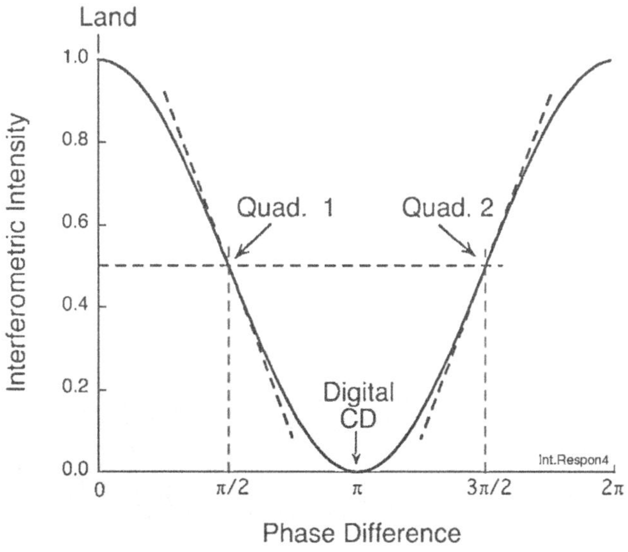

Fig. 11.2 Universal response curve for a two-port interferometer as a function of phase offset. The positions of quadrature occur at the points of largest slopes where phase-to-intensity transduction is largest

which is a differential response that has a maximum magnitude when the intensity curve has a maximum slope.

The universal interferometric response of a balanced two-port interferometer is shown in Fig. 11.2 as a function of the fixed phase offset between the two waves. The maximum slope of the intensity curve occurs when the fixed phase offset between the waves is an odd integer of $\phi = \pi/2$. These conditions of maximum slope are called the conditions of phase quadrature. There are two quadrature conditions per cycle, with opposite slopes and hence opposite signed responses to modulated phase. These are the positions of maximum phase-to-intensity transduction and are the operating points for interferometric detection of protein or DNA on spinning discs.

The detection geometries described below for the BioCD represent different optical means for establishing and maintaining phase quadrature. The key to achieving robust performance on a spinning disc, which experiences wobble and other mechanism effects, is to establish and lock the phase quadrature so that it is insensitive to mechanical vibrations or thermal drifts. This is accomplished using the concept of common-path interferometry. In common-path configurations, both the signal and the reference waves share identical ray paths from the signal source to the detector. In this way, any phase or other optical aberration is shared in common by both waves, and will not appear in the interferometric response. This common-mode rejection is what makes it possible to resolve picometer surface heights (at high detection frequency) in the presence of micron-scale (low frequency) surface movement.

11.2.2 Spinning and 1/f Noise

The BioCD laser scanning takes place on circular "tracks" of fixed radius r on the disc that is spinning with angular frequency ω. The linear velocity is

$$v_{\text{linear}} = \omega r \tag{11.4}$$

and the transit time for the laser spot to traverse its own radius ρ_0 is

$$t_{\text{transit}} = \frac{\rho_0}{v_{\text{linear}}} = \frac{\rho_0}{\omega r}. \tag{11.5}$$

The natural sampling rate for data acquisition is then

$$f_{\text{samp}} = \frac{1}{t_{\text{transit}}} = \frac{\omega r}{\rho_0}. \tag{11.6}$$

For a disc spinning at 80 Hz probed at a fixed radius of 30 mm with a laser spot diameter of 30 μm, this corresponds to a sampling frequency of 0.5 MHz and a sampling time of 2 μs. In contrast, the disc wobble frequency is 80 Hz, and most mechanical vibrations occur below 100 Hz. Therefore, the sampling frequency is over a thousand times higher than mechanical perturbations in the system. Furthermore, laser intensity drifts typically have $1/f$ noise below several kilohertz, with flat Gaussian noise at higher frequency. Therefore, at the sampling frequency of the data acquisition system, there is low noise and low drift.

In practical implementations of spinning disc interferometry, the noise rejection is approximately 50 dB. This represents a significant advantage for the BioCD over many label-free detection approaches. The key figure for all detection systems is not signal alone, but is the signal-to-noise ratio, S/N. There are two opposite ways to make S/N as large as possible. One is to boost signal. This is the approach taken by many evanescent wave and resonant detection systems. However, with resonant enhancement of signal often there comes resonant enhancement of noise, and resonant gain in a system may not necessarily improve S/N. In addition, the ability to engineer 50 dB of gain is often difficult, requiring extremely tight fabrication tolerances that prevent large-area devices from being easily manufactured.

Our approach to maximizing S/N takes the opposite approach by reducing the noise. It is relatively straightforward to reduce noise by 50 dB. In addition, by requiring no resonant enhancement of the signal, the tolerances for fabrication become extremely easy to match over large areas, such as the size of a 100-mm diameter BioCD. For these reasons, the high-frequency nonresonant approach we take with the BioCD leads to high signal-to-noise and high surface mass sensitivities.

11.3 BioCD Common-Path Quadrature Classes

Phase quadrature is an over-arching concept for all interferometric detection. However, achieving a common-path configuration that locks in a stable quadrature condition puts constraints on possible system designs. This section reviews the several configurations of common-path quadrature that have been demonstrated so far in spinning-disc systems. These are the micro-diffraction class, the adaptive optical class, the phase contrast and the in-line class. At the end of this section, we show that the phase-contrast and in-line classes are conjugate quadratures of each other.

11.3.1 Microdiffraction (MD) Quadrature

The first BioCD took its inspiration from the compact disc. The compact disc was invented in 1970 by Claus Campaan of Phillips Laboratory. The concept is purely digital and uses null interferometers that are far from quadrature, as appropriate for the readout of two binary intensity states. The interferometers were common-path and stable, as required for the mechanical environment of portable compact disc readers. The original BioCD used the same physics as the compact disc, but modified the on-disc microstructures to change from the digital readout to an analog readout that operated in quadrature for sensitive detection of surface-bound proteins[7,8]. Because the quadrature condition is established by diffraction off of microstructures on the disc, this is called the microdiffraction-class (MD-Class) of BioCD.

The MD-Class disc is composed of high-reflectance rectangular ridges on a high-reflectance substrate called the land. A stationary Gaussian laser beam is focused onto the disc surface that spins underneath. When the Gaussian beam is incident on the high-reflectance land, the far-field intensity distribution is unity. As the disc spins and the ridge sweeps through the focused beam, the configuration reaches a symmetric arrangement when the beam straddles the ridge in an intensity-balanced condition in which half of the intensity falls on the land and half of the intensity falls on the ridge. If the ridge height is a quarter wave (digital case), then the far-field diffraction distribution has an intensity null on the optical axis. This null is caused by the destructive interference between the partial wave coming from the ridge relative to the partial wave coming from the land. The quarter-wave height locks the relative phase difference between these waves at π. This is the null-interferometer condition of the digital compact disc.

Half-way between the null and the land conditions is the position of quadrature, where the intensity curve has the steepest slope. This occurs for a ridge height of an eighth-wave. The far-field diffraction for this case has half-intensity along the optic axis. At the quadrature condition, a small increment in surface height of the ridge or of the land leads to the largest intensity change in the far-field, whereas for the digital null condition, the response is extremely small. The intensity change for the MD-class BioCD is

$$\begin{aligned}\Delta I(t) &= \left[2\sqrt{I_1 I_2}\sin\phi\right]\Delta\phi(t) \\ &= 2\sqrt{I_1 I_2}\sin\left(\frac{4\pi h}{\lambda}\right)\frac{4\pi\Delta h}{\lambda},\end{aligned} \tag{11.7}$$

where h is the ridge height, and Δh is the change in the relative optical path length of the ridge and the land. The maximum response to Δh occurs when h is an eighth wave.

When the disc structure consists of numerous spokes and is spun, these data are acquired at a high sampling rate that moves the acquisition frequency far from $1/f$ noise. A time trace of a spinning BioCD with immobilized protein alternating on half of the spokes is shown in Fig. 11.3a. The presence of the alternating protein is clearly seen as a spoke-to-spoke oscillation. The half-harmonic is included in the figure to guide the eye. The power spectrum of the time series is shown in Fig. 11.3b. The carrier peak arises from the spoke periodicity, while the immobilized protein appears as a half-harmonic. Note that the noise floor around the protein half-harmonic is -45 dB below the peak of the $1/f$ noise of the system, and the protein signal-to-noise is nearly 30 dB.

The BioCD structure for the data in Fig. 11.3 was a dielectric Bragg stack designed to have a high reflectance at 633 nm (the HeNe laser source wavelength) and an antinode condition on the disc surface. The antinode condition is important to maximize the electric field at the disc surface for maximum interaction with the protein dipole moments. The BioCD was fabricated using photolithography to pattern 1,024 gold spokes 50 μm-wide deposited by gold evaporation on the disc surface. The protein immobilization used a second step of photolithography to immobilize protein on the land around alternating spokes. The MD-class BioCD was used to perform basic immunoassays by immobilizing antibody on the discs

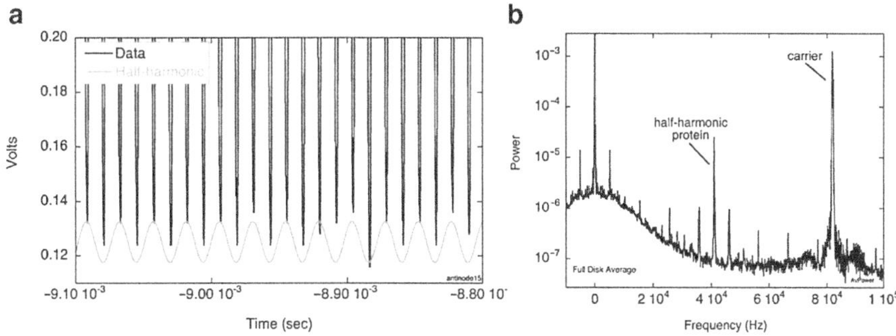

Fig. 11.3 Data from an MD-Class BioCD. (**a**) Time trace of gold spokes on an antinode dielectric disk with alternating immobilized antibody. A half harmonic sine-wave is shown for comparision. (**b**) Power spectrum of the signal, showing the carrier frequency and the half-harmonic protein pattern

and exposing them to antigen with a detection sensitivity down to 1 ng/mL[9]. In addition, multiple quadratures were implemented to show the sign-change in the response to protein[10].

11.3.2 Adaptive Optical (AO) Quadrature

A drawback of the MD-class BioCD is the microfabrication required to pattern gold spokes on the disc to set the quadrature condition. To remove the microfabrication, an alternative means to establish a quadrature condition uses adaptive optical mixers in the far field to establish and lock phase quadrature. In this case, the disc can be a high-reflectance antinode surface with protein patterned directly on the disc surface without any need for surface structuring. As the disc spins, the immobilized protein causes phase modulation that is detected in the quadrature condition set up by the adaptive mixer.

Our adaptive mixer uses photorefractive quantum wells[11–13] that are dynamic holographic thin films. They perform as an adaptive beam combiner in an adaptive interferometer[14] that is insensitive to mechanical vibrations and has been used for remote laser-based detection of ultrasound[15]. Because the beam combiner operates on principles of adaptive holography, the quadrature condition is maintained in the presence of mechanical vibrations by moving interference fringes in the photorefractive quantum wells. We modified our adaptive interferometer to work with spinning discs carrying protein patterns in both a transmission configuration[16] and in a reflection configuration on an antinode disc[17].

11.3.3 Differential Phase Contrast (DPC) Quadrature

Although the AO-class BioCD removed the need for disc microfabrication, it shifted the technical complexity to the far field in the form of the sophisticated adaptive mixer. An approach that retains the simplicity of direct patterning of proteins on the disc, while also simplifying the far-field detection is based on phase-contrast laser scanning. Phase-contrast laser scanning has been used for surface metrology applications[18,19] primarily in the semiconductor industry[20]. We modified the approach to detect proteins and protein binding on phase contrast BioCDs[21].

As the disc spins, a protein edge passes through a focused Gaussian beam that has a beam radius ρ_0. At the moment when the optic axis coincides with the protein edge in the near field, there is a quadrature condition in the far field set by

$$\begin{aligned} \rho_0 \sin\theta_Q &= \pm\lambda/4 \\ \theta_Q &= \pm\sin^{-1}\left(\frac{\lambda}{4\rho_0}\right), \end{aligned} \tag{11.8}$$

where θ_Q is the quadrature angle. The spatial intensity changes are extrema at the quadrature angles. These angles also coincide with the steepest change in the far-field Gaussian intensity distribution. The asymmetry of the intensity change is detected using a split detector with an inverting and summing circuit. This type of phase-sensitive detection uses differential phase contrast that detects the slope of surface topology rather than surface topology directly. The detector current is given by

$$i_d = \tfrac{1}{2}\phi_1 \left[\frac{I(x)}{I(0)} \otimes \frac{\mathrm{d}h}{\mathrm{d}x}\right], \tag{11.9}$$

where h is the surface height, $I(x)$ is the near-field focused Gaussian beam, and

$$\phi_1 = \frac{4\pi(n_\mathrm{p}^2 - 1)}{\lambda} \tag{11.10}$$

is the phase shift for an antinodal surface[21]. The detector current is proportional to the convolution of the near-field focused beam profile and the derivative of the surface height.

An example of the differential intensity for a series of 2 nm high protein spokes immobilized on an antinodal high-reflectance Bragg stack is shown in Fig. 11.4a. The spokes were 110-μm wide. The signal is negative for a rising protein edge, and positive for a falling protein edge. The "noisy" features in Fig. 11.4a are not noise, but are actually heterogeneous protein structures that are fully repeated on each cycle around the disc. The laser and electronic noise are not visible in this graph. The signal in Fig. 11.4a is integrated to yield the surface protein height profile in Fig. 11.4b. The ridges are approximately 2 nm high. The profile resolution is set by the diameter of the focused Gaussian beam, which was approximately 20 μm for these data.

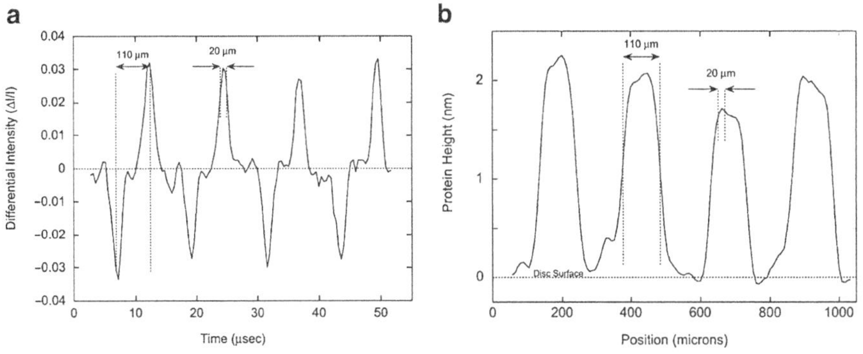

Fig. 11.4 Results from a PC-Class BioCD containing printed protein stripes. (**a**) The raw signal as differential change in intensity sweeping across 4 stripes. (**b**) The integrated protein profile with 2 nm high and 110 μm wide protein ridges

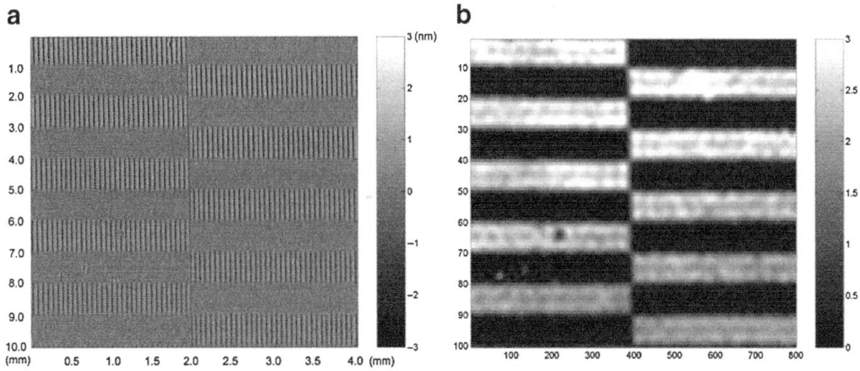

Fig. 11.5 Differential phase contrast detection of patterned protein in a 10 mm by 30 mm region. (**a**) Protein height signal showing ridges of protein in a checker-board pattern. (**b**) Side-band demodulated signal image in which the carrier frequency of the ridges is removed to show only the protein envelope. Reprinted from Ref. 21. with permission. © 2008 Optical Society of America

By scanning successive annular tracks with increasing radius, the differential surface profile is mapped out in two dimensions, as shown in Fig. 11.5a[21]. The surface pattern for these data is a series of protein spokes approximately 3-nm high in a "chess-board" pattern. The protein is immunoglobulin IgG that was immobilized using a polysuccinimide surface chemistry, and was patterned using conventional photolithography. For periodic patterns, the average height of the pattern is extracted using Fourier techniques that demodulate the periodicity as a side-band that shifts it to DC. The resulting Fourier amplitude is plotted in Fig. 11.8b.

11.3.4 In-line (IL) Quadrature

Although the differential phase-contrast quadrature has maximum sensitivity to rising and falling protein edges, it is insensitive to broad expanses of uniform protein height. A way of establishing quadrature that enables the full protein profile to be extracted is called in-line quadrature. To establish in-line quadrature, a transparent spacer layer that is approximately an eighth-wave in thickness is placed on top of a high-reflectance substrate, in this case silicon. Protein is patterned on top of the spacer layer. Light that is incident on the layer has partial reflections from the upper and lower surfaces of the spacer. Because of the choice of the layer thickness, these partial reflections differ in phase by $\pi/2$ and hence are in a quadrature condition. Therefore, when the surface height changes, as when protein is immobilized on the top surface, the extra phase that is acquired upon light propagation through the protein is converted to intensity by interfering with the reference wave (in phase quadrature) that originates from the high-reflectance substrate. This configuration is called in-line because the signal and the reference are colinear.

The reflectance is shown in Fig. 11.6a as a function of wavelength for three oxide thicknesses of 80, 100, and 120 nm on silicon. The corresponding intensity

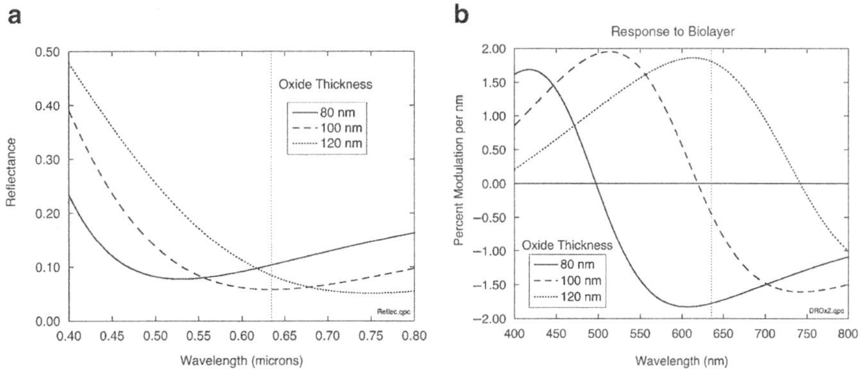

Fig. 11.6 Optical properties of thermal oxide on silicon for three oxide thickness of 80, 100, and 120 nm. (**a**) The reflectance as a function of wavelength. (**b**) The relative change in reflectance upon immobilization of 1 nm protein

response to a biolayer of 1 nm thickness is shown in Fig. 11.6b. The maximum modulation is approximately 1.8% per nanometer of protein. The maximum response does not occur exactly when the oxide is an eighth-wave thick. This is because the total response to protein is a combination of the quadrature condition and also of the field strength at the surface of the oxide layer, which is set by the electromagnetic boundary conditions. This pulls the operating wavelengths toward the minimum reflectance wavelength.

11.3.5 Common-Path Interferometry and Conjugate Quadratures

The configurations of differential phase contrast and in-line quadrature share much in common. In each case the signal response is a convolution of a surface height property over the focused beam profile. For phase contrast, the property is the derivative of surface height, while for in-line it is directly the height. For phase contrast, the signal in the far field is asymmetric, while for in-line it is symmetric. These two quadrature detection classes represent conjugate quadratures. The two responses can be expressed as[22]

$$\begin{aligned} i_{\mathrm{IL}} &= -2\phi_{Im}|r|^2[g^2 \otimes h] \\ i_{\mathrm{DPC}} &= -2\phi_{Re}|r|^2\frac{\rho_0}{2}\left[g^2 \otimes \frac{\mathrm{d}h}{\mathrm{d}x}\right], \end{aligned} \tag{11.11}$$

where

$$\phi_{Re} + i\phi_{Im} = \frac{4\pi n_{\mathrm{p}}}{\lambda}\frac{(r_{\mathrm{p}} - r)(1 - rr_{\mathrm{p}})}{r(1 - r_{\mathrm{p}}^2)}, \tag{11.12}$$

where ρ_0 is the Gaussian beam radius, r is the complex reflection coefficient of the bare surface, and r_p is the reflection coefficient of bulk protein. In (11.11), the normalized Gaussian field profile is given by g, and hence g^2 represents the intensity profile.

11.4 Sensitivity Limits

There are two detection sensitivities that are of interest for characterizing the BioCD performance. The first is a metrology sensitivity that includes experimental noise sources such as the laser intensity fluctuations and detector noise, but does not include variability of the protein spots on the disc. The second detection sensitivity is under actual assay conditions in which antibody spot variability plays a dominant role. For both of these sensitivities, it is important to define a scale-free sensitivity that is an intrinsic property of the detection platform.

If the height fluctuations for a specified laser beam area a are uncorrelated and characterized by a standard deviation Δh_a, then the standard error on N measurements of spot height is simply $h_{\min} = \Delta h_a/\sqrt{N}$. To express this in terms of the total scanned area A of a protein layer, it becomes

$$\Delta h_{\min} = \Delta h_a \sqrt{\frac{a}{A}} = h' \frac{1}{\sqrt{A}}. \tag{11.13}$$

The parameter h' is

$$h' = \Delta h_a \sqrt{a} \tag{11.14}$$

and is scale free, meaning that it is an intrinsic property of the detection platform. It has units of length times root area.

The minimum detectable surface mass density is likewise

$$\Delta S_{\min} = \rho \Delta h_a \sqrt{\frac{a}{A}} = S' \frac{1}{\sqrt{A}}, \tag{11.15}$$

where

$$S' = \rho \Delta h_a \sqrt{a} \tag{11.16}$$

is scale free and has units of mass per root area, or mass per length.

In experiments[23], we have observed a metrology-limited height of 20 pm per pixel, and an assay-limited height of 75 pm per protein spot. Based on these values, the scale-free minimum detectable height

$$\begin{aligned} h'_{\text{met}} &= 20\,\text{pm}\sqrt{20\,\mu\text{m} \cdot 20\,\mu\text{m}} \\ &= 4 \times 10^{-10}\text{mm}^2 \end{aligned}$$

and

$$\begin{aligned} h'_{\text{assay}} &= 75\,\text{pm}\sqrt{\pi(75\,\mu\text{m})^2} \\ &= 10 \times 10^{-9}\text{mm}^2, \end{aligned}$$

where 75 pm is the detection limit based on one spot, and 75 μm is the spot radius. The scale-free sensitivities are

$$\begin{aligned} S'_{\text{met}} &= 1.3\,\text{g/cm}^3 20\,\text{pm}\sqrt{20\mu\text{m} \cdot 20\mu\text{m}} \\ &= 0.5\,\text{pg/mm} \end{aligned}$$

and

$$\begin{aligned} S'_{\text{assay}} &= 1.3\,\text{g/cm}^3 75\,\text{pm}\sqrt{\pi \cdot (75\,\mu\text{m})^2} \\ &= 13\,\text{pg/mm}. \end{aligned}$$

Note that the assay sensitivity under assay conditions is over an order of magnitude larger than the metrology limit of the in-line BioCD. The assay is performed in a protein background on a hydrophobic butylaldehyde surface chemistry that concentrates droplets during dry-down on the antibody spots, leading to randomly distributed residues at the level of 75 pm per spot. Improved surface treatments should decrease the current value of S' toward the metrology limit.

11.5 Immunoassays

The performance of the BioCD under assay conditions has been tested using several "gold standard" systems. These are assays of anti-rabbit and anti-mouse IgG systems, prostate specific antigen (PSA), and haptoglobin. Incubations have been performed under equilibrium conditions without transport limitation, and also under transient conditions as ambient assays that are diffusion limited. Ambient assays are performed in practice, while equilibrium assays provide more information about the performance of the antibodies and provides a quantitative estimate for equilibrium dissociation constants.

11.5.1 Equilibrium Assays

We perform equilibrium assays using an orbital shaker and incubation times of 20 h. The active flow of the fluid over the disc surface and the long incubation times ensure

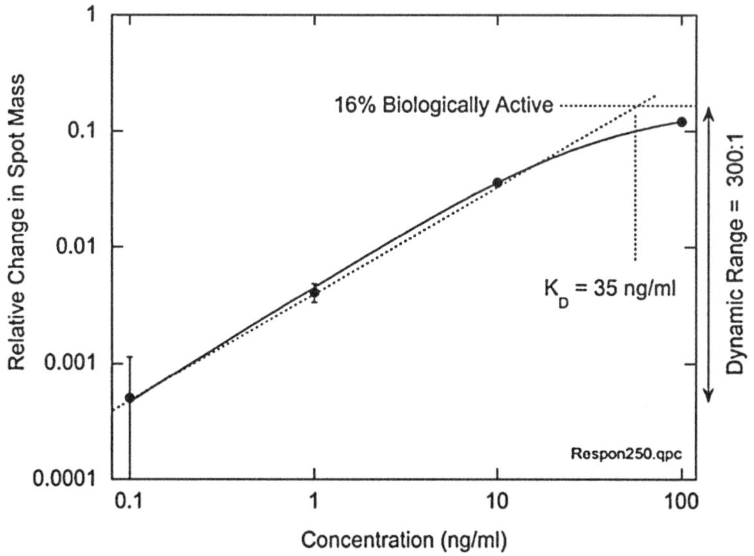

Fig. 11.7 Equilibrium assay using immbolized mouse IgG antigen capturing anti-mouse antibody from solution. The dynamic range is 300:1, showing 16% biological activity and an affinity of 35 ng/mL

that we achieve an equilibrium surface binding. The results for an in-line BioCD printed with 25,000 IgG spots are shown in Fig. 11.7. The spots are approximately 100 μm in diameter and several nanometers high. The IgG spots for this assay were the target antigens, and the sample solution contained the associated polyclonal affinity-purified antibodies. This reverse-phase capture is known as a reverse-phase assay and is more robust than forward-phase assays because the immobilization of antigen retains more activity than immobilization of antibody. The entire disc was incubated with increasing concentrations of antibody at concentrations of 0, 0.1, 1, 10, and 100 ng/mL, with each incubation lasting 20 h in the orbital shaker. The surface immobilization chemistry was a di-isocyanate that has extremely high stability to withstand the accumulated 100 h of incubation.

The results of the successive incubations with increasing concentrations in Fig. 11.7 show the accumulated change in spot mass as a function of increasing antibody concentration. The minimum detectable spot height change was approximately one part per thousand. For a 1 nm printed protein spot height, this corresponds to a spot height of approximately 1 pm. These data are fit with a simple Langmuir function with an equilibrium dissociation constant $k_D = 35$ ng/mL $= 300$ pM, which represents a high-affinity reaction. The saturated spot height change was approximately 0.16, suggesting that the biological activity of the printed antigen was only 16%. This low biological activity for a printed antigen is likely due to the isocyanate immobilization chemistry with strong covalent bonds that may cause conformation changes in the printed protein. The vertical dynamic range in this experiment was 300:1 with a comparable horizontal dynamic range in concentration.

11.5.2 Prostate-Specific Antigen (PSA)

Forward-phase assays are the most common form of practical assays as printed antibodies bind their antigens from sample fluids. We have performed equilibrium forward-phase assays of PSA with antibodies printed directly to the disc. To maintain the biological activity of the antibody, the surface immobilization chemistry was changed to an aldehyde chemistry, which is more benign than (but also not as durable as) the isocyanate chemistry. Therefore, equilibrium assays were performed only at a single concentration, and detection limits were extrapolated based on the signal-to-noise ratio for the detected concentration. Half of the spots were anti-PSA, while the other half were reference IgY molecules. The spots were approximately 100 μm in diameter and several nanometers thick. The disc was incubated against 10 ng/mL PSA in PBST buffer with 2 mg/mL BSA protein background[24].

The histograms of the spot height changes for both target and reference spots in the forward-phase assay are shown in Fig. 11.8a. The average spot height change between the target and reference spots was 80 pm with a standard error of 1.5 pm (using 11,418 pairs of spots in the average). The extrapolated detection limit is 300 pg/mL averaged over an area of 1 cm^2. The histograms after a sandwich assay are shown in Fig. 11.8b for 9,712 pairs of spots. For the sandwich assay, a secondary anti-PSA antibody was incubated with the captured antigen at a concentration of 10 μg/mL. The average height change was 512 pm with a standard error of 0.6 pm. The extrapolated limit of detection in this case is 20 pg/mL.

11.5.3 Haptoglobin Standard Curve

To recover quantitative concentrations from an assay, it is necessary to establish standard response curves for a particular antibody-antigen reaction. As an example,

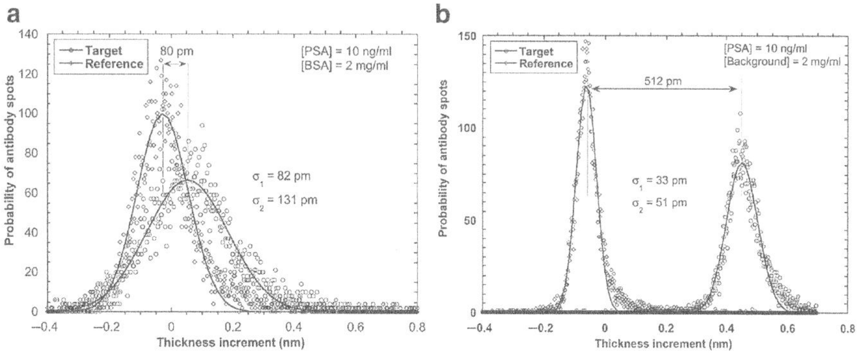

Fig. 11.8 Assays for 10 ng/mL PSA on IL-Class BioCDs. (**a**) Forward label-free assay at 10 ng/mL showing the histrogram of approximately 10,000 spots pairs. (**b**) Sandwich assay shows clear separation

we performed assays against haptoglobin, a blood-related protein with an average molecular weight of 150 kDa. The disc format for the standard curve and concentration recovery was a 96-well disc with 8 × 8 arrays of 32 target and 32 reference spots in each well. The anti-haptoglobin antibodies were attached to printed spots of the fusion protein A/G. The A/G immobilization captures the Fc region of the antibody, orienting the antibody for maximum activity. The reference spots were chicken IgY. The applied concentrations across the wells were 0, 0.3, 1, 3, 10, 30, 300, 1,000, 3,000, 10,000 ng/mL, applied to 9 wells each. The incubation was performed under ambient diffusion (hand-spotted into the wells) for 1 h.

The assay results are shown in Fig. 11.9, which plots the height change of the target spots relative to the reference spots as a function of haptoglobin concentration. The detection floor was approximately 50 pm with a limit-of-detection of 1 ng/mL. The effective dissociation constant was $k_D = 180$ ng/mL = 1 nM. The higher value of the dissociation constant for this reaction compared with equilibrium assays is likely because the transport of antigen to the surface is diffusion limited. The biological activity for this assay was nearly 100%, with a surface height change over 1 nm. These data are fit with a stretched Langmuir function given by

$$\Delta h = h_{\text{sat}} \frac{[C]^e}{[C]^e + k_D^e}, \tag{11.17}$$

where h_{sat} is the saturation height change. The exponent e describes a "stretch" to the usual Langmuir curve (for which $e = 1$). Smaller exponents produce shallower response curves for lower concentrations. Stretched response curves are commonly

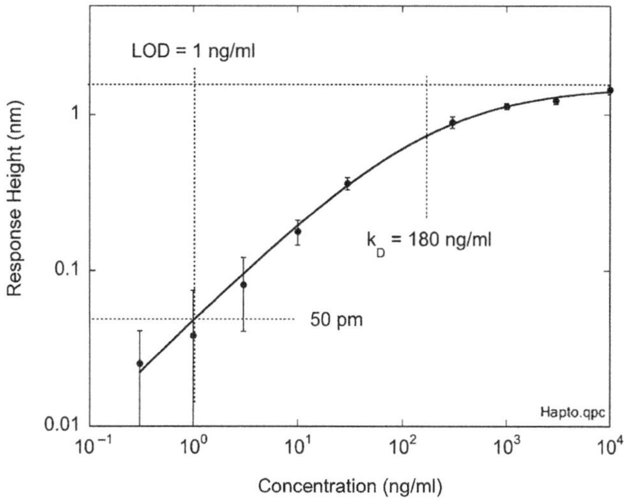

Fig. 11.9 Response curve for haptoglobin on a 96-well in-line BioCD with protein A/G immobilization. Error bars are standard deviation of 9 wells

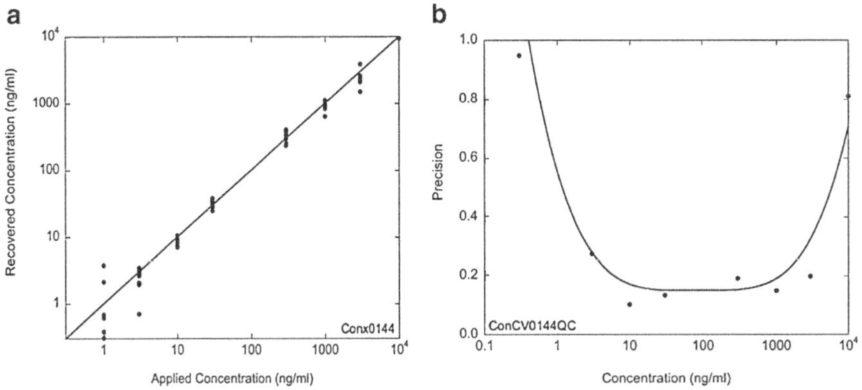

Fig. 11.10 Concentration recovery on the haptoglobin assay. (**a**) The recovered concentration against the applied concentration. (**b**) The precision of recovery, achieving 80% recovery over a dynamic range of 1,000:1

experienced in nonequilibrium assays for which there is a disproportionately higher response at lower concentrations. The best fit stretch exponent in Fig. 11.9 is approximately $e = 0.6$, which is nearly a square-root.

The stretched Langmuir fit is used to perform concentration recovery on the well-by-well data. The response of a well is compared against the fit curve and then projected to the concentration axis. The recovered concentrations are plotted against the applied concentrations in Fig. 11.10a. The straight line is the one-to-one correspondence. The smallest variances occur where the response curve is steepest, around the mid-range of the response, while the variances become large at both low and high concentrations when the response curve flattens out. The coefficient of variance of the recovered concentrations, known as the assay precision, is plotted in Fig. 11.10b against the applied concentration. The recovery precision for this assay is about 20% (known as 80% recovery) over a concentration dynamic range of about 1,000:1.

11.6 Future Prospects

The BioCD draws from the sensitivity and speed of SDI, combined with the surface-normal configuration analogous to optical compact discs, to produce element densities that have a capacity of over 50,000 protein measurements per disc. The high-speed, high-sensitivity, and high spatial multiplexing capabilities of the BioCD give this approach considerable promise for applications in molecular diagnostics and high-throughput screening.

Perhaps the most important aspect of the detection of protein binding on the BioCD is the scalability of the technology. The principle of scalability requires that the resource cost for doubling capacity is sublinear. The best sublinear scalability occurs when the cost scales logarithmically with the capacity. In this case, each doubling of capacity has a fixed incremental cost. Scalability is at the heart of Moore's Law of integrated circuits, which has been a self-fulfilling economic driving force for over 40 years.

The surface-normal character of the interferometer elements on the BioCD, combined with the high-speed laser scanning, makes the BioCD platform intrinsically scalable. The ultimate capacity for an optical biosensor is one measurement per optical mode (or square wavelength). This is a density of about 10,000 measurements per square centimeter. On a 100 mm diameter BioCD the total disc capacity is over a billion measurements per disc. Indeed, a 650 MB compact disc has about this many digital pits, as the optical engineers use the ultimate mode density for digital storage.

The ability of protein sensors to access the ultimate mode density is complicated by several challenges that digital storage did not encounter. The principle challenge is the stockpiling of libraries of up to a million different molecules, and then dispensing them to the disc while maintaining their biological activity. Clearly, serial printing approaches, such as ink-jet or piezoelectric printing, would fail at this scale. Parallel printing methods are possible that use up to 1,536 individual pins that dip into a 1,536-well plate. The pin pattern from multiple plates can be interleaved to produce protein spot densities that fill the disc. This approach would enable the assaying for thousands of different analytes per disc, with further improvements possible pending new developments in parallel protein printing.

Perhaps a more serious challenge to the scalability of the BioCD is the limit of understanding of what multiple measurements mean biologically. Today, when a doctor orders a panel of clinical tests, perhaps only a dozen or so protein measurements are made. The reason for this small number, in the face of 10,000 available blood proteins, is that doctors would not know what to do with all the measurements that could be made available. Therefore, the prospect of making thousands of analyte measurements per patient today is largely meaningless.

However, the understanding of the molecular signatures of health and disease are steadily progressing. Although only a few biomarkers are known and tested for today, that number will double in a few years...and then double again as we move into the future. The field of proteomics is advancing steadily, and the challenge for a scalable technology is to coevolve with proteomics. Therefore, as the panel of biomarkers doubles, the sensor technology, if it is scalable, would keep pace, so that when doctors eventually *do* know what to do with a thousand measurements, the sensor technology would be in place to provide it.

An important part of scalability is not just in capacity, but also in cost and in size. As electronic chips have seen the numbers of transistors rise to over a billion per chip, the cost per chip has remained low and the chip sizes similarly have stayed relatively constant. This aspect of scalability is what makes it possible for increasingly powerful computers to arrive at our desktops at reasonable cost.

The equivalent scalability for molecular sensors applies to the point of care. Scalable biosensors will enable high-capacity diagnostic capabilities to remain sufficiently affordable and small in size to stay in doctors offices or in hospitals close to the patient. The proximity of fast diagnostics to the patient shortens diagnosis time, while the increasing capacity and multiplexing can allow highly personalized analysis of a patient's biochemical traits, with its idiosyncratic response to therapies. The availability of a scalable technology could usher in the new era of personalized medicine, when the individual response to a medicine can replace the mean-based response that is relied upon today.

Acknowledgments The authors gratefully appreciate all the help and hard work provided by the scientists and engineers at Quadraspec, Inc., and in particular the inspiration and support of Chad Barden.

References

1 Barabasi, A. L.; Oltvai, Z. N., Network biology: Understanding the cell's functional organization, *Nat. Rev. Genet.* **2004**, 5, 101–115
2 Lundstrom, M., Moore's law forever? *Science* **2003**, 299, 210–211
3 Gauglitz, G., Direct optical sensors: principles and selected applications, *Anal. Bioanal. Chem.* **2005**, 381, 141–155
4 Fan, X. D.; White, I. M.; Shopoua, S. I.; Zhu, H. Y.; Suter, J. D.; Sun, Y. Z., Sensitive optical biosensors for unlabeled targets: A review, *Anal. Chim. Acta* **2008**, 620, 8–26
5 Nolte, D. D.; Regnier, F. E., Spinning-disk interferometry: The BioCD, *Opt. Photon. News* **2004**, 48–53
6 Wang, X.; Zhao, M.; Nolte, D. D., Combined fluorescent and interferometric detection of protein on a BioCD, *Appl. Opt.* **2008**, 47, 2779–2789
7 Varma, M.; Nolte, D. D.; Inerowicz, H. D.; Regnier, F. E., Multi-analyte array micro-diffraction interferometry, In Microarrays: Design, fabrication and reading; Bornhop, B. J., et al., Ed.; *SPIE* **2002**, 4626, 69–77.
8 Varma, M.; Nolte, D. D.; Inerowicz, H. D.; Regnier, F. E., High-speed label-free multi-analyte detection through micro-interferometry, In Microarrays and combinatorial technologies for biomedical applications: Design, fabrication, and analysis; Nicolau, D. V.; Raghavachari, R., Eds.; *SPIE* **2003**, 4966, 58–64
9 Varma, M. M., Inerowicz, H. D., Regnier, F. E., Nolte, D. D., High-speed label-free detection by spinning-disk micro-interferometry, *Biosens. Bioelectron.* **2004**, 19, 1371–1376
10 Varma, M. M.; Nolte, D. D.; Inerowicz, H. D.; Regnier, F. E., Spinning-disk self-referencing interferometry of antigen-antibody recognition, *Opt. Lett.* **2004**, 29, 950–952
11 Nolte, D. D.; Olson, D. H.; Doran, G. E.; Knox, W. H.; Glass, A. M., Resonant photodiffractive effect in semi-insulating multiple quantum wells, *J. Opt. Soc. Am.* **1990**, B7, 2217
12 Nolte, D. D.; Melloch, M. R., Photorefractive quantum wells and thin films, In Photorefractive Effects and Materials; Nolte, D. D., Ed.; Kluwer Academic Publishers, Dordrecht, 1995
13 Nolte, D. D., Semi-insulating semiconductor heterostructures: Optoelectronic properties and applications, *J. Appl. Phys.* **1999**, 85, 6259
14 Nolte, D. D.; Cubel, T.; Pyrak-Nolte, L. J.; Melloch, M. R., Adaptive beam combining and interferometry using photorefractive quantum wells, *J. Opt. Soc. Am. B* **2001**, 18, 195–205
15 Lahiri, I.; Pyrak-Nolte, L. J.; Nolte, D. D.; Melloch, M. R.; Kruger, R. A.; Bacher, G. D.; Klein, M. B., Laser-based ultrasound detection using photorefractive quantum wells, *Appl. Phys. Lett.* **1998**, 73, 1041–43
16 Peng, L.; Varma, M.; Inerowicz, H. D.; Regnier, F. E.; Nolte, D. D., Adaptive optical biocd for biosensing, *Appl. Phys. Lett.* **2005**, 86, 183902-1-3

17 Peng, L.; Varma, M. M.; Cho, W.; Regnier, F. E.; Nolte, D. D., Adaptive interferometry of protein on a BioCD, *Appl. Opt.* **2007**, 46, 5384–5395
18 See, C. W.; Iravani, M. V.; Wickramasinghe, H. K., Scanning differential phase-contrast optical microscope - application to surface studies, *Appl. Opt.* **1985**, 24, 2373–2379
19 Suddendorf, M. B.; Somekh, M. G.; See, C. W., Single-probe-beam differential amplitude and phase-scanning interferometer, *Appl. Opt.* **1997**, 36, 6202–6210
20 Takao, A.; Steigmeier, E. F.; Hagleitner, W.; Pidduck, A. J., Microroughness measurements on polished silicon wafers, *Jpn. J. Appl. Phys.* **1992**, 31, 721–728
21 Zhao, M.; Cho, W.; Regnier, F.; Nolte, D., Differential phase-contrast BioCD biosensor, *Appl. Opt.* **2007**, 46, 6196–6209
22 Wang, X.; Zhao, M.; Nolte, D. D., Common-path interferometric detection of protein on the BioCD, *Appl. Opt.* **2007**, 46, 7836–7849
23 Wang, X.; Zhao, M.; Nolte, D. D., Area-scaling of interferometric and fluoresecnt detection of protein on antibody microarrays, *Biosen. Bioelectron.* 2008
24 Wang, X.; Zhao, M.; Nolte, D. D., Prostate-specific antigen immunoassays on the BioCD, *Anal. Bioanal. Chem.* **2009**, 393, 1151–1156

Chapter 12
Ultra-Sensitive Biochemical Optical Detection Using Distributed Feedback Nanolasers

Jacob Scheuer

Abstract Circular resonators are promising candidates for a wide range of applications, ranging from research involving highly confined fields and strong photon-atom interactions such as cavity QED to optical communication systems and biochemical sensing. For sensing applications, circular cavities exhibit a great potential for achieving ultra-high sensitivity while retaining compact dimensions. The main characteristics of circular resonators are the Q-factor, the free spectral range (FSR), and the modal volume, where the last two are primarily determined by the resonator radius. The total-internal-reflection mechanism employed in "conventional" resonators couples between these characteristics and limits the ability to realize compact devices exhibiting large FSR, small modal volume, and high Q. Recently, a new class of annular resonator, based on a single defect surrounded by radial Bragg reflectors, has been proposed and analyzed. The radial Bragg confinement decouples the modal volume and the Q and paves a new way for the realization of compact and low loss resonators. Such properties as well as the unique mode profile of these circular Bragg nanolasers make this class of devices an excellent tool for ultra-sensitive biochemical detection as well as for studies in nonlinear optics.

12.1 Introduction

Circular resonators, ring and disk shaped, are key ingredients in the realization of many basic components needed for advanced optical communication systems. During the last decade, numerous circular-resonator based applications such as lasers, filters[1], add/drop multiplexers[2], modulators[3], and delay lines[4] have been suggested and demonstrated. In addition, the applicability of circular resonators was

J. Scheuer
Department of Physical Electronics, School of Electrical Engineering, Tel-Aviv University, Ramat-Aviv, Israel
e-mail: kobys@eng.tau.ac.il

X. Fan (ed.), *Advanced Photonic Structures for Biological and Chemical Detection*, Integrated Analytical Systems,
DOI 10.1007/978-0-387-98063-8_12,

shown to extend beyond telecommunication to the fields of sensing[5], spectroscopy, and standardization[6], as well as to basic research in QED, nonlinear optics, and other related fields[7].

The realization of highly miniaturized sensors capable of detecting selective binding of ultra-small quantities of chemical or biological agents as well as motion (in particular rotation) is a significant step toward the development of compact and versatile "lab-on-chip" devices. One of the most attractive approaches for the realization of compact integrated optical biochemical sensors is detecting the shifts of the resonance frequency of a micro-resonator caused by changes in the refractive index of the device surroundings[5,8–10]. To achieve high sensitivity, large interaction volume between the modal field profile and the analyte is desired. In conventional resonators, employing total internal reflection (TIR) as the radial confinement mechanism, only the small evanescent tail of the field interacts with the surroundings, which consequently limits the sensitivity.

Recently, we have proposed utilizing Bragg reflection instead of TIR as the radial confinement mechanism in order to break the link between the free spectral range (FSR) and the loss and facilitate low-loss, large FSR circular resonators[11–13]. This concept is illustrated in Fig. 12.1. A pillar (Fig. 12.1a) or an annular defect (Fig. 12.1b) is located within a medium, which consists of annular Bragg layers. The confinement of the modal field within the defect (Fig. 12.1b) or in the disk (Fig. 12.1a) is accomplished by Bragg reflection. Unlike conventional resonators, the reflectivity of the Bragg mirrors can be increased simply by adding more layers. As a result, the radius of the defect (or the disk) can be reduced almost arbitrarily without increasing the bending losses. Such microcavities, employing distributed Bragg reflection as the transverse confinement mechanism, e.g., photonic crystal (PC) defect cavities[9,10] and annular Bragg resonators[11–13], exhibit unique mode profile, which can offer enhanced sensitivity because of the larger interaction volume with the surrounding (see also Fig. 12.9).

Here, we describe the advantages o CBNLs for biochemical sensing applications. The biochemical sensor is shown to exhibit high spectral resolution and excellent sensitivity to changes in the absorption or refractive index of its

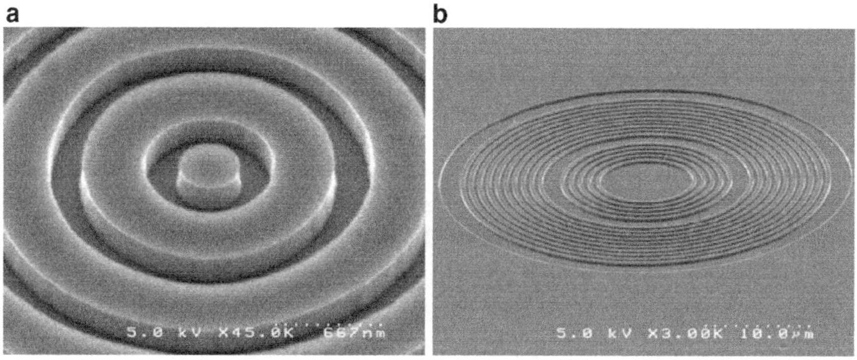

Fig. 12.1 SEM images of (**a**) a disk CBNL; (**b**) a ring CBNL

surrounding. We compare the sensitivity of the CBNL sensor to that of a conventional, TIR based, resonator of similar material and dimensions and find that the measurement resolution of the CBNL sensor is ten times larger than that of the conventional device.

12.2 Basic Theory

We consider a radially symmetric structure as illustrated in Fig. 12.2. The guiding defect, consisting of a material of refractive index n_{defect}, is surrounded by distributed Bragg reflectors on both sides, where the reflectors' layers are of refractive indices n_1 and n_2. All the electromagnetic field components can be expressed in terms of the z-component of the electric and magnetic fields[14]. These components satisfy the scalar Helmholtz equation, which in cylindrical coordinates is given by:

$$\left[\rho^2 \frac{\partial^2}{\partial \rho^2} + \rho \frac{\partial}{\partial \rho} + \left(k^2(\rho) - \beta^2\right)\rho^2 - m^2\right] \begin{pmatrix} E_z \\ H_z \end{pmatrix} = 0, \qquad (12.1)$$

where ρ and θ are the radial and azimuthal coordinates respectively, m is the azimuthal wavenumber, β is the z-component propagation coefficient, $k(\rho) = k_0.n(\rho)$, k_0 is the wavenumber in vacuum, and the refractive index $n(\rho)$ equals either n_{defect}, n_1, or n_2 according to the radius ρ.

The general solution of (12.1) can be expressed by a superposition of the Bessel functions of the first and second kind.

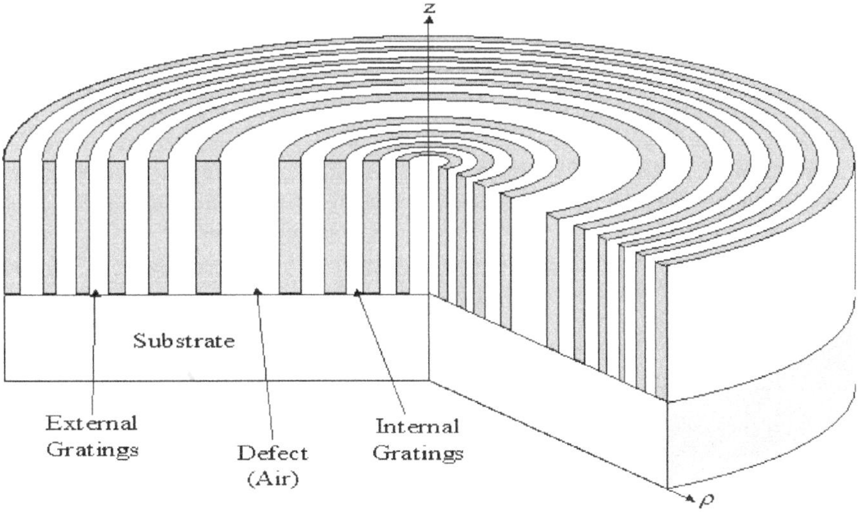

Fig. 12.2 An illustration of a ring shaped CBNL structure. Reprinted from Ref. 21 with permission. © 2008 Institute of Electrical and Electronics Engineers

$$E_z = \left[A \cdot J_m\left(\sqrt{k_j^2 - \beta^2}\rho\right) + B \cdot Y_m\left(\sqrt{k_j^2 - \beta^2}\rho\right)\right] \cdot \cos(\beta \cdot z + \varphi) \cdot \exp(im\theta),$$
$$H_z = \left[C \cdot J_m\left(\sqrt{k_j^2 - \beta^2}\rho\right) + D \cdot Y_m\left(\sqrt{k_j^2 - \beta^2}\rho\right)\right] \cdot \sin(\beta \cdot z + \varphi) \cdot \exp(im\theta), \tag{12.2}$$

where k_j is the material wavenumber in the jth layer.

The other field components can be easily derived from E_z and H_z[14]. Introducing (12.2) into (12.3) yields all the field components in the jth layer. The parallel component of the fields – E_z, H_z, E_θ, H_θ must be continuous at the interfaces separating successive layers. This requirement can be written in from of a transfer matrix, connecting the amplitude vector $[A\ B\ C\ D]$ in the jth and $j + 1$ layers:

$$[A \quad B \quad C \quad D]_{j+1}^{\mathrm{T}} = \tilde{M}_{j+1}^{-1}\left(\rho_{j+1}\right) \cdot \tilde{M}_j\left(\rho_j\right) \cdot [A \quad B \quad C \quad D]_j^{\mathrm{T}} \tag{12.3}$$

and $\tilde{M}_j$ is given by:

$$\tilde{M}_j = \begin{pmatrix} J(\gamma_j\rho) & Y(\gamma_j\rho) & 0 & 0 \\ \frac{n_j^2}{\gamma_j} J'(\gamma_j\rho) & \frac{n_j^2}{\gamma_j} Y'(\gamma_j\rho) & \frac{m\beta}{\rho\omega\varepsilon_0\gamma_j^2} J(\gamma_j\rho) & \frac{m\beta}{\rho\omega\varepsilon_0\gamma_j^2} Y(\gamma_j\rho) \\ 0 & 0 & J(\gamma_j\rho) & Y(\gamma_j\rho) \\ \frac{m\beta}{\rho\omega\mu\gamma_j^2} J(\gamma_j\rho) & \frac{m\beta}{\rho\omega\mu\gamma_j^2} Y(\gamma_j\rho) & \frac{1}{\gamma_j} J'(\gamma_j\rho) & \frac{1}{\gamma_j} Y'(\gamma_j\rho) \end{pmatrix}, \tag{12.4}$$

where ε and μ are the dielectric and magnetic susceptibilities, ω is the optical angular frequency, and the primes indicate derivative with respect to the function argument.

Assuming strong vertical confinement (i.e., $\beta < < k_j$), the modal field solutions can be separated into two distinct polarizations: TE consisting of H_z, E_p, and E_θ and TM consisting of E_z, H_r, and H_θ. The z component of the relevant field for each polarization can be described by two coefficients in each layer: A_j and B_j for TM and C_j and D_j for TE. For each polarization, the boundary conditions at the interfaces between successive layers can be represented by a simpler 2×2 transfer matrix:

$$\tilde{M}_j^{\mathrm{TM}} = \begin{pmatrix} J(\gamma_j\rho) & Y(\gamma_j\rho) \\ \frac{n_j^2}{\gamma_j} J'(\gamma_j\rho) & \frac{n_j^2}{\gamma_j} Y'(\gamma_j\rho) \end{pmatrix} \quad \tilde{M}_j^{\mathrm{TE}} = \begin{pmatrix} J(\gamma_j\rho) & Y(\gamma_j\rho) \\ \frac{1}{\gamma_j} J'(\gamma_j\rho) & \frac{1}{\gamma_j} Y'(\gamma_j\rho) \end{pmatrix}. \tag{12.5}$$

The relation (12.3) and the transfer matrices (12.5) can be used to "propagate" the field components from the inner layers to the external layers. The boundary condition at $\rho = 0$ is the finiteness of the field so that $B_1 = D_1 = 0$. The second boundary condition is that there are no inward propagating field beyond the last layer, so that $B_{N+1} = -iA_{N+1}$ for TM and $C_{N+1} = -iD_{N+1}$ for TE, where N is the number of layers.

12.3 Design Rules

The transfer matrix formalism enables us to find the modal field profile in the case of an arbitrary arrangement of annular concentric dielectric rings. However, we are especially interested in structures that can confine the modal energy near a predetermined radial distance, i.e. within the defect.

It has been shown that the best strategy to attain an exponential decrease (or increase) in the field intensity in the grating region is to position the interfaces of the layers at the zeros and extrema of the z component of the field[14]. The index profile and the field are calculated simultaneously, using the index to find the field and the field to determine the position of the interfaces. The resulting optimal index profile exhibits an inherent resemblance to the conventional (Cartesian) Bragg reflector. The optimal Cartesian Bragg reflector can be designed in a similar way, leading to layers that are quarter-wavelength thick[15]. Here the resulting layers are also "quarter-wavelength" thick but in the sense of the quasi-periodicity of the mth order Bessel function. The defect (again, as in the Cartesian case) should be "half-wavelength" wide, i.e., its interfaces should be located at successive zeros of the field.

To attain a transverse field profile that is confined within the defect, the profile must exponentially increase for $\rho < \rho_{\text{def}}$, and exponentially decrease for $\rho > \rho_{\text{def}}$. This requirement determines which index-interfaces (lowhigh or highlow) should be positioned at zeros of the field and which at the extrema of the field. The constraints on the index profile are similar to the Cartesian case and differ for the TE and TM polarizations. For the TE polarization, the interfaces for decreasing (increasing) field should be at the zeros (extrema) of H_z if $n(\rho^-) > n(\rho^+)$ at the interface and at the extrema (zeros) of H_z if $n(\rho^-) < n(\rho^+)$ at the interface. For the TM polarization, the interfaces for decreasing (increasing) field should be at the extrema (zeros) of E_z if $n(\rho^-) > n(\rho^+)$ at the interface and at the zeros (extrema) of E_z if $n(\rho^-) < n(\rho^+)$ at the interface. The interfaces of the defect must be located at zeros of H_z for TE and of E_z for TM.

Figure 12.3 depicts the refractive index (Fig. 12.3a) and the TE modal field (Fig. 12.3b) profiles of a CBNL designed for a 0.55-μm thick InGaAsP layer suspended in air. The device is designed to have a mode with an angular propagation coefficient of $m = 7$ at $\lambda_{\text{res}} = 0.852$ μm. The effective index approximation in the vertical dimension is used to reduce the 3D problem to a 2D equivalent one. As can be seen in the figure, the field is primarily confined in the defect and it decays while oscillating in the Bragg reflectors.

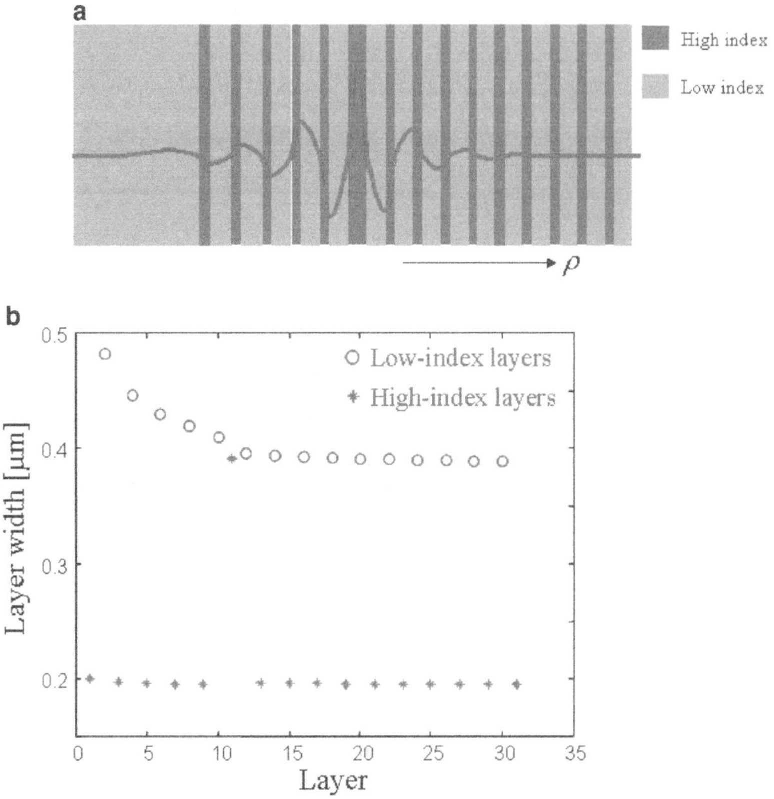

Fig. 12.3 Refractive index (**a**) and TE-polarized modal field profiles of a CBNL designed for $m = 7$, $\lambda_{\text{res}} = 0.852$ μm, with 5 internal and 10 external Bragg layers. (**b**) A nonperiodic distribution of high and low-index material is required in an annular Bragg resonator

Figure 12.3b shows the widths of the Bragg layers as a function of the layer number of the CBNL depicted in Fig. 12.3a. There are two notable properties of the Bragg layers: (1) the width of the high-index layers is smaller than the width of the low-index layers, and (2) the width of the layers decreases exponentially as a function of the radius, converging asymptotically to a constant value. The first property exists in conventional DBRs as well and stems from the dependence of the spatial oscillation period, or the "wavelength," on the index of refraction. The second property is unique to the cylindrical geometry and arises from the nonperiodic nature of the solutions of the wave equation (Bessel or Hankel functions) in this geometry.

At large radii the widths of the Bragg layers converges asymptotically to the conventional (Cartesian) quarter-wavelength condition. Mathematically, this can be explained by noting that for large radii the Bessel function can be approximated by a sinusoidal function divided by square root of the radius. From the physical point of view, at large radii the curvature of the Bragg layers is small and, therefore, the structure of a radial reflector structure at large radii is similar to the structure of a Cartesian reflector. Figure 12.3b shows the radial mode profile of the CBNL structure of Fig. 12.3a. The field is confined within the defect and it decays while oscillating in the Bragg region. Although the overall envelope of the radial modal profile (averaging over the oscillations) is quite similar to that of a conventional ring resonator, the oscillatory behavior of the field in the reflectors regions is the fingerprint of the distributed feedback mechanism.

12.4 Key Features and Advantages

The Bragg confinement mechanism allows great flexibility in engineering the radial mode profile. For example, it is possible to design a resonator in which the light is confined within a defect composed of low refractive index material or even air. By contrast, this would be impossible in the case of conventional TIR-based resonators. Such configuration is useful for sensing applications as discussed later.

The Bragg reflection concept allows one to tailor the reflector structure to a desired radial field profile and decouple between the modal volume (or cavity dimensions) and the radiation losses. In fact, the radiation losses can be reduced almost arbitrarily, without significantly changing the modal volume, simply by adding more layers to the (external) Bragg reflector. Therefore, resonators exhibiting both high Q and small modal volume can be engineered and their modal filed profile can be tailored for specific applications.

Another interesting and important property of the CBNL structure is that it supports slow light propagation along the circumference of the defect. Intuitively, this property can be understood as follows: Because the Bragg reflector does not require the incidence angle of the EM wave to be larger than some "critical" angle (as in TIR based devices), it is possible to design an CBNL in which the incidence angle of the wave is almost perpendicular to the Bragg layers. The angular component of the wave-vector in this case is very small, and thus, the angular propagation velocity is very low. Slow whispering gallery modes have several advantages, especially for applications requiring strong atom-field interactions. For a given power level in the guiding channel, the corresponding intensity of the electric field is inversely proportional to the group velocity – $|E|^2 \sim P/v_g$. Therefore, such modes can be exploited to attain large gain (if the device is a laser), large and controllable dispersion, and enhanced nonlinear response.

More specifically, the unique modal field profile and characteristics of the CBNL structure are advantageous for biochemical sensing applications, but also for surface emitting lasers and for studies involving strong atom-field interactions such as nonlinear optics and cavity QED.

12.5 Biochemical Sensing

Among the most straightforward approaches for optoelectronic (bio)chemical sensing is to detect the change in the refractive index or the absorption caused by the presence of a (bio)chemical agent. Several schemes have been suggested to detect these types of changes, employing directional couplers[16], Mach-Zehnder interferometers (MZI)[17], or high-*Q* optical resonators[18].

The two later concepts are illustrated in Fig. 12.4. The detection mechanism underlying these sensors is the modification of the phase accumulation rate (i.e., the propagation coefficient) of the field due to the interaction of the evanescent tail of the field with the analyte. In the MZI-based senor, one of the channels is exposed to the surrounding while the other is not. The existence of the biochemical agent modifies the phase shift accumulated by the signal traveling through the exposed channel, thus modulating the intensity of the signal at the MZI output port. In the micro-ring-based sensor, the existence of the biochemical agent modifies the phase accumulated by the electromagnetic wave circulating in the cavity, thus modifying the resonance wavelength of the micro-ring. The detection of this shift (e.g., by monitoring the output intensity of light launched in the original resonance wavelength) indicates the existence of the biochemical reagent.

The primary disadvantage of these detection methods is that the interaction of the field with the surrounding environment is weak, and therefore, the influence of the analyte on the propagation coefficient is small. As a result, achieving high

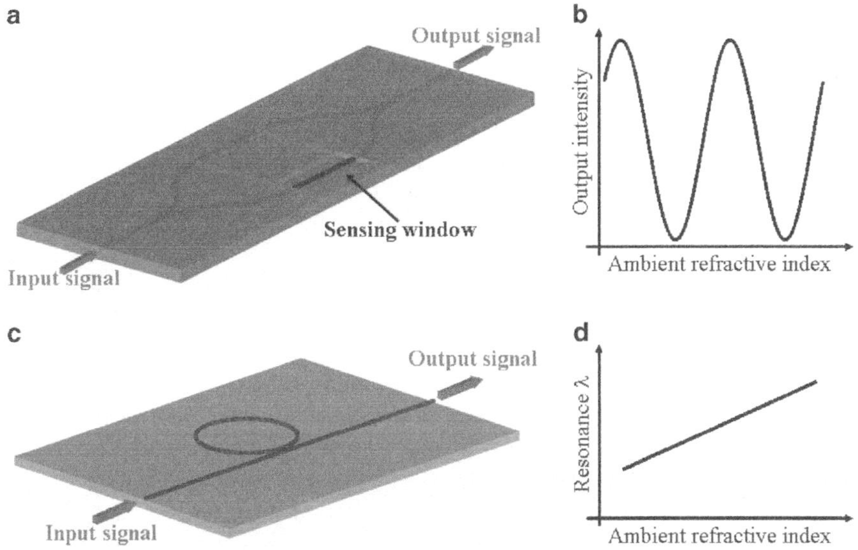

Fig. 12.4 (**a**) MZI-based biochemical sensor; (**b**) Output signal dependence on the ambient refractive index for the MZI sensor; (**c**) Micro-ring based sensor; (**d**) Resonance wavelength dependence on the ambient refractive index for the micro-ring sensor

sensitivity requires large interaction length leading to the requirement of long interferometers and very high-Q resonators. In addition, the MZI-type sensors might have difficulties detecting small numbers (or single) molecules, regardless of their length. The CBNL structure and mode profile (see Fig. 12.3) allow for the interaction of the *nonevanescent* part of the field, especially when the device is designed to include an air defect. As a result, CBNL-based sensors are expected to offer significantly enhanced sensitivity compared with sensors that are based on conventional resonators of similar dimensions and materials.

Figure 12.5 shows a comparison between the shifts of the resonance frequency of a CBNL and a conventional ring resonator due to changes in the refractive index of the surroundings. The CBNL structure consists of alternating layers with refractive indices of 1.545 and 1.0 and an air defect. The conventional resonator consists of $n = 1.545$ core surrounded by air cladding. Both resonators are approximately 16 μm in diameter and designed to resonate in the visible wavelength regime. The sensitivity of each device is indicated by the slope of the curves shown in Fig. 12.5. The resonance wavelength of the conventional ring resonator shifts by approximately 0.007 nm for an increase of 10^{-3} in the refractive index. For the same index change, the Bragg resonator's resonance wavelength shifts by 0.4 nm, i.e., the CBNL exhibits higher sensitivity by a factor of 60.

In addition to sensitivity, an important feature for any biochemical sensor is selectivity, i.e., the ability to response to a specific chemical or agent. In additional to the obvious significance (reducing the "false alarm" rate, etc.), this ability is

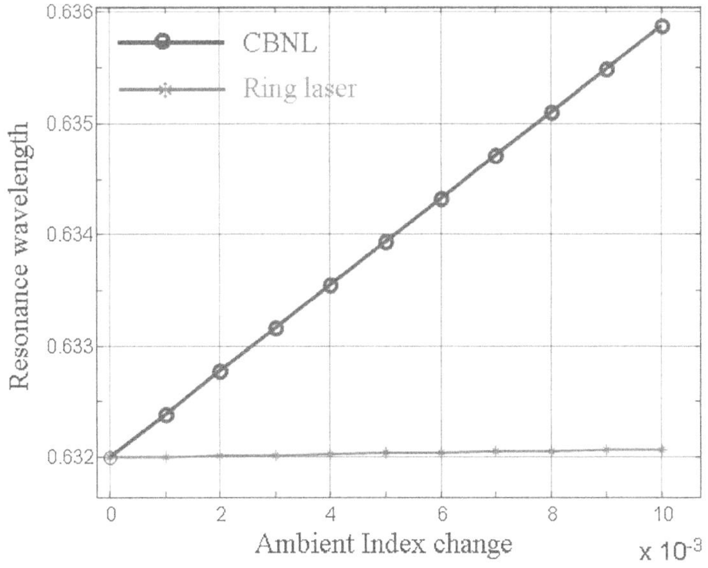

Fig. 12.5 Comparison of the sensitivity of a CBNL and a conventional ring resonator to changes in the refractive index of the surroundings. Reprinted from Ref. 21 with permission. © 2008 Institute of Electrical and Electronics Engineers

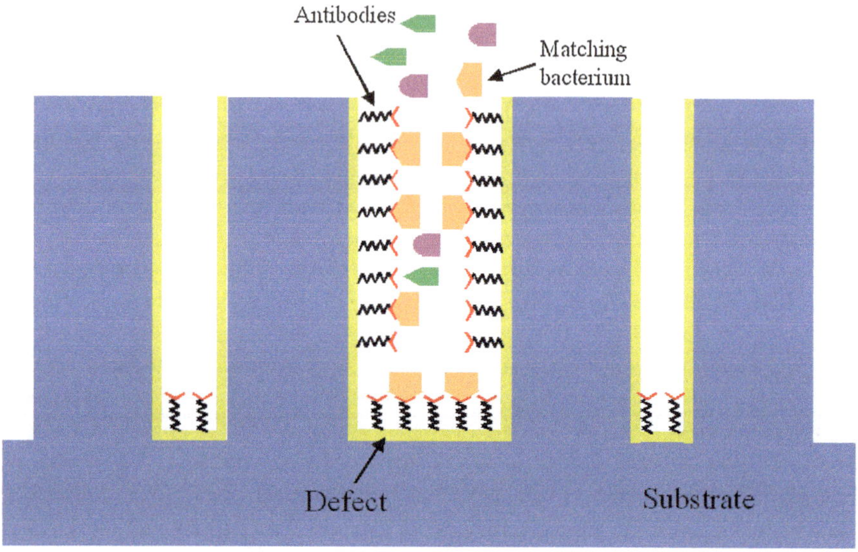

Fig. 12.6 Antibodies coated Bragg resonator for enhanced selectivity

important because it provides a mean of identifying biochemical materials and not only detecting their presence. The combination of compact yet sensitive sensors with biochemical selectivity allows constructing an array of sensors where each one is dedicated to a specific agent. Such an array is literally a complete biochemical lab implemented on a single chip, in particular, if combined with an efficient delivery system such as a micro-fluidic channels network[19].

To achieve selectivity, the resonator surface must be coated with specific binder with high affinity to the analyte[20]. This is the most commonly used method for realizing selectivity in optical sensors. For bacterial sensing, for instance, the sensor's surface is usually coated with a layer of specific antibodies, which are capable of binding specific types of bacteria to the sensor surface (see Fig. 12.6). This coating ensures that only specific kind of bacteria, which matches the antibodies, can bind and influence the resonator optical properties. Unlike conventional resonators, for which the evanescent part probes only a small part of the cell (~200 nm), the Bragg resonator modal field can interact with the whole cell – attaining higher sensitivity without sacrificing the selectivity properties

12.6 Device Fabrication

Circular Bragg nanolasers (CBNLs) of several geometries and Bragg reflector orders were fabricated within a thin membrane of InGaAsP semiconductor material[21]. A cross-section of the semiconductor epitaxial structure used is illustrated

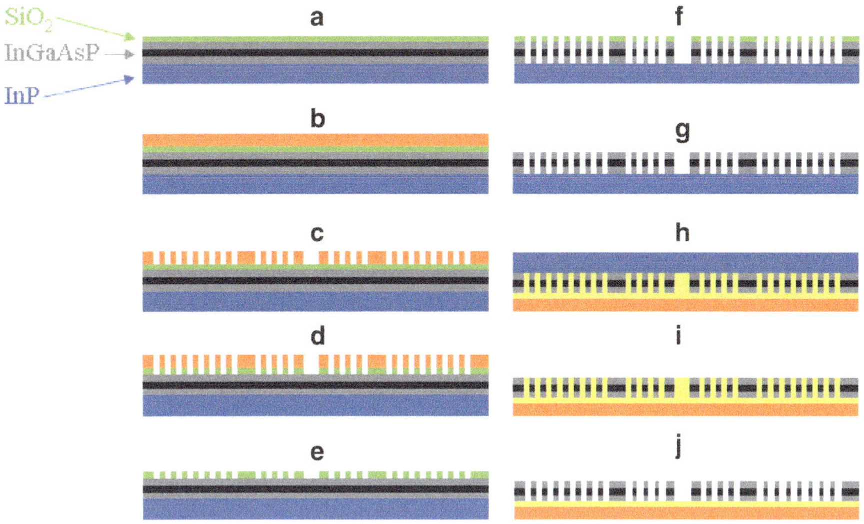

Fig. 12.7 InGaAsP/InP multi-quantum-well semiconductor structure process (**a**) SiO_2 etch mask deposition; (**b**) PMMA spin coating; (**c**) E-beam lithography and develop; (**d**) SiO_2 etch; (**e**) PMMA stripping; (**f**) InGaAsP membrane etch; (**g**) SiO_2 stripping; (**h**) Chip flipping and bonding to sapphire; (**i**) InP substrate etch; (**j**) Adhesive etch

in Fig. 12.7. The layers were grown by MOCVD on a (100) oriented InP substrate. This structure consists of a 500-nm thick InP buffer layer, followed by a 50 nm InGaAsP ($\lambda_g = 1.3$ μm) stop etch, and a 250 nm InP sacrificial layer. A 250 nm membrane composed of 605 Å InGaAsP ($\lambda_g = 1.1$ μm) layers sandwiching six 75 Å quantum wells (1% compressive strain) separated by 120 Å InGaAsP barriers ($\lambda_g = 1.2$ μm, 0.5% tensile strain) completes the structure. Photoluminescence from the quantum wells peaks at a wavelength $\lambda = 1{,}559$ nm. Annular Bragg resonators were designed to have resonant wavelengths between 1.5 and 1.6 μm, for large overlap with the gain spectrum of the multi-quantum-well material.

The optical gain of the semiconductor epitaxial structure used favors the TE polarization (H_z), because of the optical properties of compressively strained quantum wells[22]. Therefore, the design of the fabricated devices is optimized for this polarization. To simplify the design calculations, we employ the effective index approximation in the vertical dimension. An effective index $n_{eff} = 2.8$ is found by solving for the TE-polarized mode of the transferred InGaAsP slab. To facilitate the fabrication of the device, a mixed Bragg order scheme is used, with second-order ($3\lambda/4 \sim 430$ nm) high-index layers and first-order ($\lambda/4 \sim 400$ nm) low-index layers.

In addition to relaxing the fabrication tolerances, the mixed Bragg order implementation induces a coherent diffraction component in the vertical direction[23]. Although this mechanism reduces the overall Q of the cavity, it facilitates the observation and measurement of the resonator emission.

The procedure followed to fabricate the annular Bragg resonators is as follows. First, an etch mask consisting of 120 nm of PECVD SiO_2 was deposited on the semiconductor substrate. A 550-nm thick layer of PMMA electron-sensitive resist was spin-coated onto the substrate, and electron beam lithography was used to define the resonator patterns in the PMMA. The patterns were then transferred to the SiO_2 layer with a reactive ion etching (RIE) step in after which the remaining PMMA was stripped using O_2 plasma. Then, a second RIE step was used to transfer the SiO_2 pattern through the 250-nm thick InGaAsP membrane. Finally, the remaining SiO_2 etch mask was removed with a wet chemical etch.

The high refractive index of the InP substrate ($n \approx 3.17$ at $\lambda = 1.55$ μm) reduces the vertical confinement of the guided optical mode within the membrane. Therefore, to achieve strong vertical optical confinement and improve the quality factor Q of the resonators, the patterned InGaAsP membrane must be surrounded by low index material. One method for achieving this is to generate an air-suspended membrane by selective removal of the InP beneath the device, as is commonly done with photonic crystal devices[23–25]. However, this technique is not applicable to the resonators studied here, because the concentric ring structure would collapse if the solid substrate were removed. Therefore, the resonators were flip-bonded to a double-side polished sapphire substrate with a thin layer of UV curable optical adhesive (Norland NOA 73, $n \approx 1.54$ at $\lambda = 1.55$ μm). After fully curing the adhesive under a UV lamp, the InP substrate was removed using a combination of mechanical lapping and selective chemical etching. Note that the excellent thermal properties of sapphire can improve the conduction of heat generated during optical pumping away from the membrane. In addition, the large bandgap energy of sapphire, and associated transparency at visible and infrared wavelengths, permits optical pumping and imaging of the resonator luminescence from either the top or bottom of the device. Transfer of the semiconductor membrane to a transparent substrate also facilitates use of annular Bragg resonators for sensing applications. For instance, a solution containing an analyte whose concentration is to be monitored by the resonator could be introduced via the exposed topside of the device, while optical pumping of the resonator and imaging of the sensor signal could be accomplished through the transparent substrate.

12.7 Experimental Results

We measured and analyzed the vertical emission from the resonators under pulsed optical pumping. The experimental setup is illustrated in Fig. 12.8a: A Ti/sapphire mode-locked laser was used to optically pump the devices at a center wavelength of 980 nm, repetition rate of 76.6 MHz, and pulse duration of approximately 150 fs. A variable attenuator was used to control the pump power. The average pump power and center wavelength were monitored by a wavemeter, through a 50/50 beamsplitter. The pump beam is focused on the back side of the sample with a 50× objective lens. A 20× objective lens is used to collect the vertical emission from the sample and to focus it on an IR camera to obtain the NF intensity pattern and to

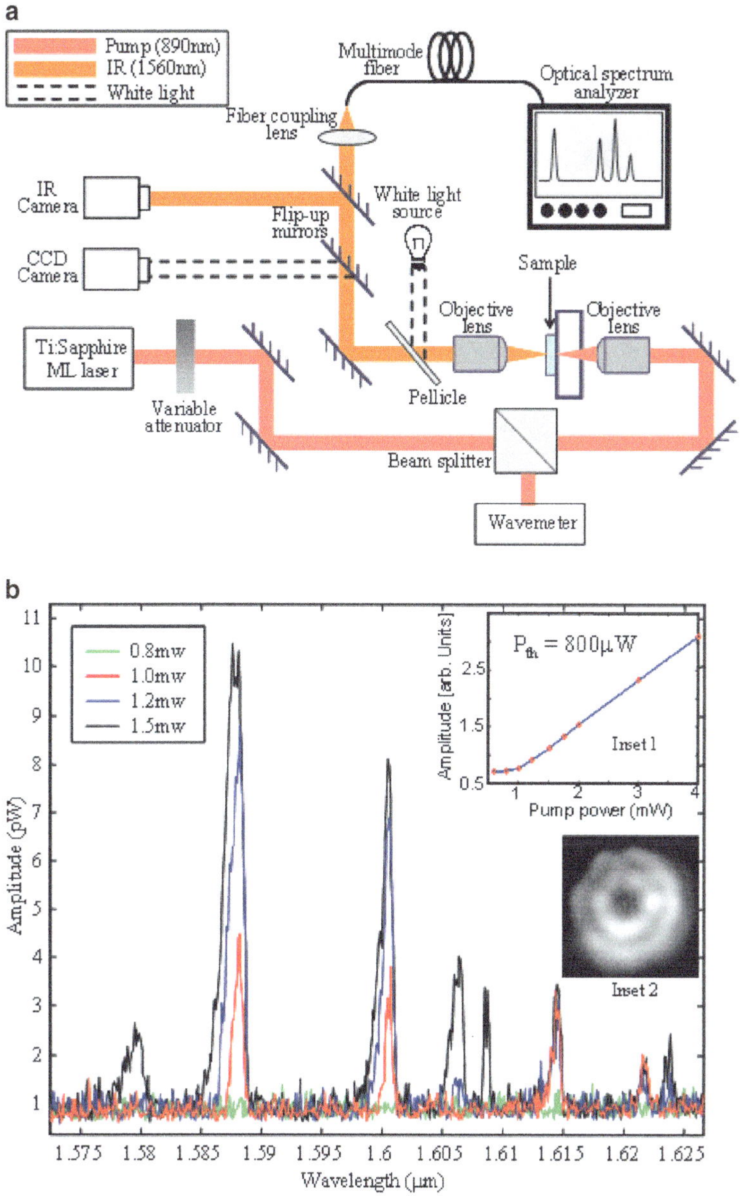

Fig. 12.8 (**a**) The experimental setup. (**b**) Optical spectra emitted from a CBNL under different pump levels. *Inset*: Integrated emitted power vs. pump power, showing laser threshold at ~800 mW. Inset 2 – Lasing pattern. Reprinted from Ref. 21 with permission. Copyright 2005 IEEE

couple the light into a multimode fiber to obtain the emitted spectrum. A CCD camera is used to image the resonators and the pump spot, for alignment purposes, using a white light source.

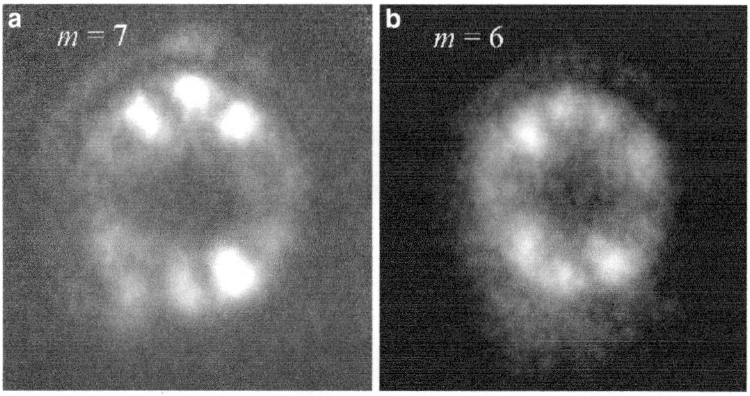

Fig. 12.9 IR images of emitted patterns with small angular modal number (**a**) $m = 7$ and (**b**) $m = 6$

When the unpatterned QW layer structure is pumped, the emitted spectrum consists of a wide peak centered at 1,559 nm. As the pumping power is increased from 1 mw to 20 mW, the FWHM of the luminescence broadens from approximately 70–110 nm, and the peak of the photoluminescence shifts toward longer wavelength due to heating. No significant shift is observed when the pump power is below 5 mW, indicating that heating is of less significance at these pump levels. When a CBNL is pumped, the emission characteristics change significantly. While the specific details (threshold levels, emitted wavelengths, etc.) vary from device to device, the overall behavior is similar. Once a certain pump intensity threshold is exceeded, clear and narrow (~0.5 nm FWHM) emission lines appear in the spectrum (see Fig. 12.8b). As the pump intensity is increased, the intensity of the emission lines increase as well and they broaden toward shorter wavelengths. Increasing the pump power further results in the appearance of additional emission lines.

Depending on the defect width and design, the CBNL can emit patterns with very small angular modal number (see Fig. 12.9). These modes correspond to effective index, which is lower than 1 and, thus, cannot evolve in conventional, TIR based, micro-ring lasers. The ability to support such lasing modes provides substantial flexibility and control over the properties of the modal field, allowing for the engineering of field profiles that are highly susceptible to changes in the ambient refractive index.

To demonstrate the unique advantages of the CBNL structure, we focus on two applications – surface emitting lasers (SEL) and biochemical sensing.

12.7.1 Surface Emitting Nanolasers

Figure 12.10b depicts the emitted spectra from the $m = 0$ laser for various pumping levels above the lasing threshold. An SEM image of the resonator is shown in Fig. 12.10a. The emitted light consists primarily of a single wavelength

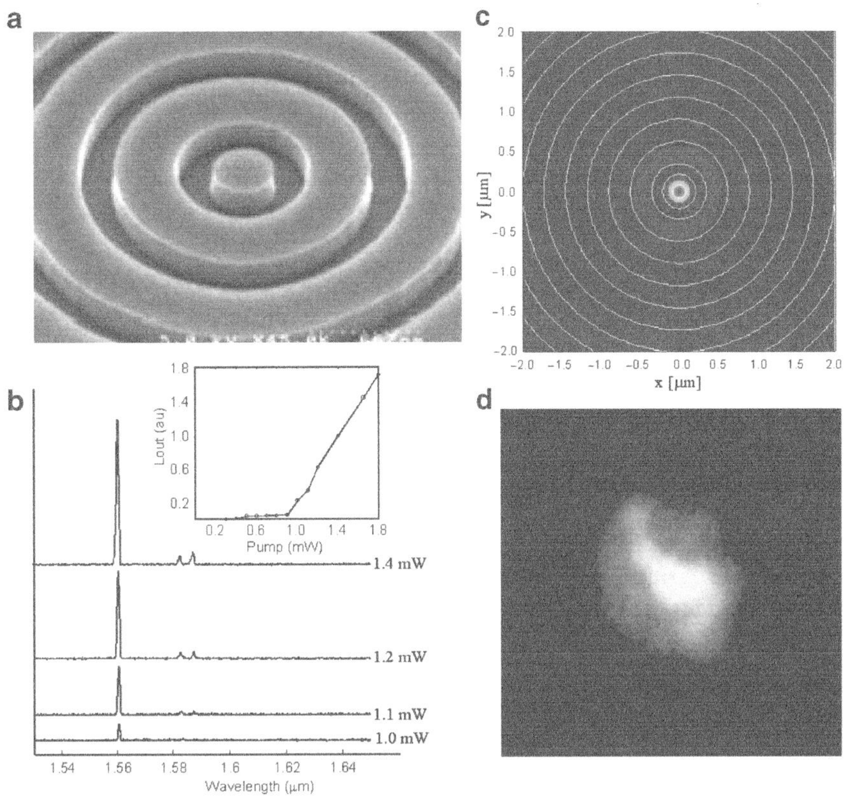

Fig. 12.10 (**a**) SEM image of the circular Bragg nanocavity designed to support the $m = 0$ mode in the 300-nm wide central pillar. (**b**) The evolution of the emitted spectrum from the device shown in Fig. 12.9a as a function of the pump intensity. Inset – L-L curve, indicating a lasing threshold of $P_{th} = 900\ \mu W$. (**c**) Calculated modal intensity profile of the nanocavity. (**d**) IR image of the emitted beam profile

at $\lambda = 1.56\ \mu m$, very close to the target design wavelength of 1.55 μm. The inset of Fig. 12.10b shows an L-L curve of the same device, indicating a threshold at $P_{th} = 900\ \lambda W$. It should be noted that the pump powers quoted indicate the overall power carried by the pump beam while the actual power absorbed by the QWs is significantly lower. At high pump levels ($\sim 1.5 \times P_{th}$), two additional low-intensity emission lines appear at longer wavelengths ($\sim$1.59 μm), although the main emission line remained the dominant one. We attribute these modes to emission from the external Bragg grating region.

Figure 12.10c shows a contour plot of the nanolaser index profile superimposed on a cross-section of the modal field intensity profile in the center of the active medium. As shown in the figure, the modal profile of the nanocavity is confined almost completely in the 300-nm wide central pillar with a modal volume of 0.213 $(\lambda/n)^3$ (0.024 μm^3) – only 1.75 times the theoretically possible limit of a cubic half wavelength. This modal volume is a 30% lower than the modal volume demonstrated in PC defect cavities[24,25]. This nonnegligible improvement stems from the optimized match between the cavity dimensions and the (quasi) periodicity of the grating, made possible because of the cylindrically symmetric geometry.

The employment of second order radial Bragg gratings in the nanolaser provides, simultaneously, the radial feedback necessary for lasing and the vertical output coupling. Compared with conventional vertical cavity surface emitting lasers, the circular Bragg grating based SEL is simpler to fabricate and expected to provide higher output powers because of the larger field-gain overlap in the active material.

12.7.2 Ultra-Sensitive Sensors

Figure 12.11a, b shows an SEM micrograph of a CBNL and conventional ring-based sensors realized within a thin membrane of InGaAsP active material, respectively. Figure 12.11c, d depicts a cross-section of the mode profile of the CBNL and the conventional ring. The index structure is superimposed on the calculated mode profiles. The field profile of the CBNL peaks in the defect and decays while oscillating in the reflector regions. Figure 12.11c, d also illustrates the advantages

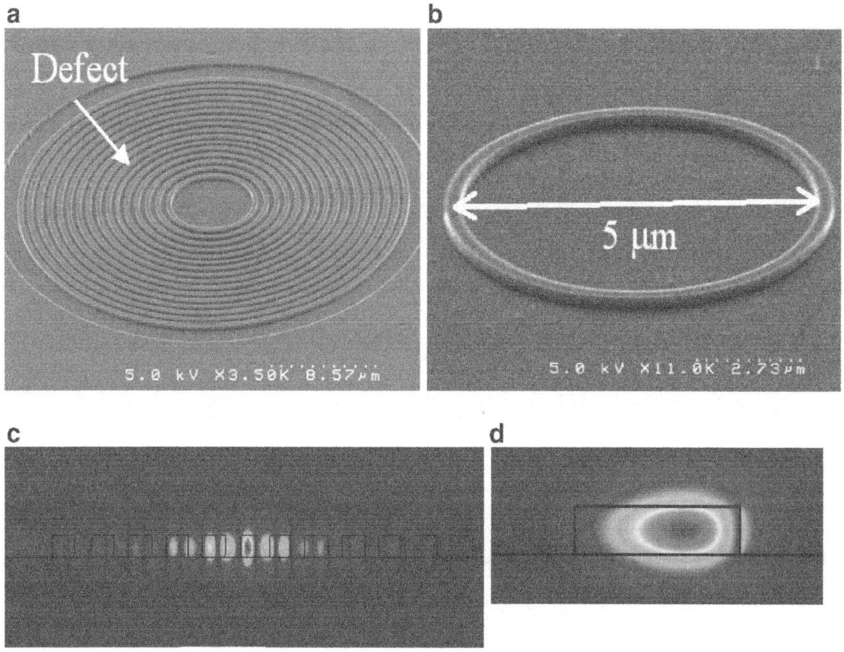

Fig. 12.11 SEM images of the CBNL (**a**) and the conventional ring (**b**) sensors, and their corresponding modal field profiles (**c**, **d**)

of the CBNL structure for sensing applications. Even though the main lobe of the mode profile does not interact with the device surroundings, significant part of the field within the trenches between the semiconductor rings are exposed to the surroundings, thus yielding large interaction volume (compared with a conventional resonator). Thus, a CBNL-based sensor is expected to exhibit enhanced sensitivity compared with a conventional resonator of similar dimensions.

To test the influence of a change in the ambient refractive index on the laser spectrum, the CBNL was immersed in a specially designed beaker containing index-matching fluids with different index of refraction. For each fluid, the emitted spectrum is measured ($P_{pump} \sim 2.5$ mW).

To demonstrate the advantages of CBNL-sensors in terms of sensitivity and resolution, we compare the performance of our device to that of a conventional ring resonator with 5-μm radius consisting of the same material[26]. The conventional ring was tested under the same conditions (pump power, spot-size, etc.) that were used to characterize the CBNL sensor. Figure 12.12 depicts the relative shifts in the resonance wavelengths of the CBNL and the conventional micro-rings as a function of the ambient refractive index. The sensitivity of the conventional resonator is approximately $d\lambda/dn \sim 33$ nm (dashed) and the resonance FWHM is ~1.4-nm wide, while the sensitivity of the CBNL is $d\lambda/dn \sim 130$ nm (solid) and the

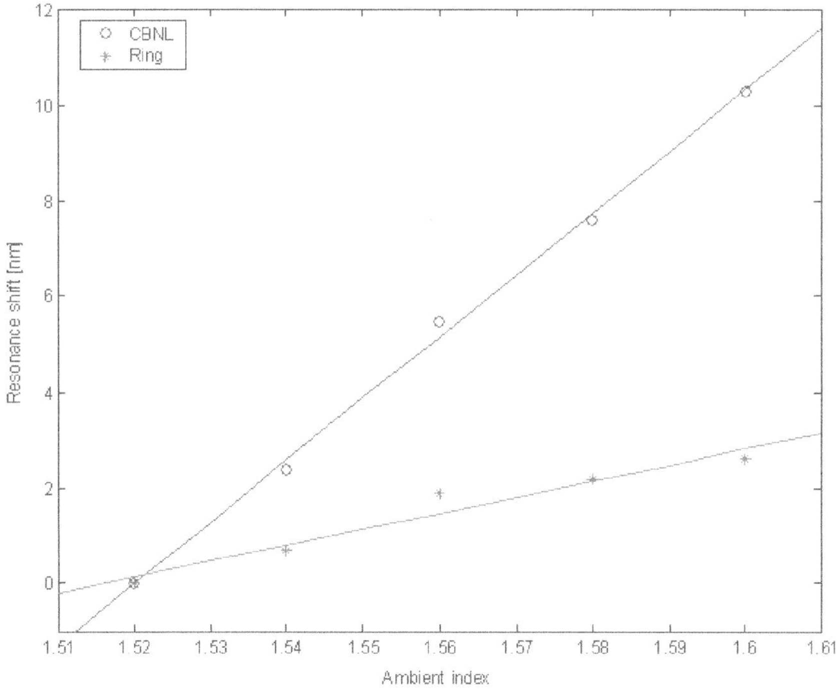

Fig. 12.12 Comparison between the shifts in the emitted wavelength of a CBNL and a conventional ring laser for changes in the ambient refractive index

resonance FWHM is ~1 nm wide. As a result, the advantage of the CBNL as a sensor is twofold: (1) The sensitivity of the CBNL is four times larger than that of the conventional device because of the larger interaction volume, and (2) the conventional resonator has wider linewidth because of its lower Q, which reduces the ability to resolve the resonance wavelength.

For the 1 nm linewidth of the CBNL, shifts in the resonance wavelength of 0.1 nm can be easily resolved, thus allowing the detection of ambient index changes of $\Delta n \sim 5 \times 10^{-4}$, which to our knowledge is the one of the best resolutions ever demonstrated using integrated optical nanosensors[27]. As a comparison, for 1.4 nm FWHM of the conventional ring, the ability to resolve the resonance wavelength is limited to $\Delta \sim 0.14$ nm, which when combined with the lower sensitivity of the conventional device yields a minimum detectable index change of $\Delta n \sim 4 \times 10^{-3}$ – approximately an order of magnitude less than that of the CBNL sensor.

12.8 Conclusions

We have studied, theoretically and experimentally, the characteristics of a novel class of lasers that are based on radial Bragg reflectors. Lasing action with low threshold levels are demonstrated at room temperature under pulsed optical pumping. The observed Q factors are in the order of several thousands. The unique characteristics on these lasers make them promising candidates for numerous applications in telecommunications, sensing, and basic research.

Circular Bragg nano-cavities offer significant enhancement of the ability to detect small changes in the ambient refractive index, compared with conventional micro-rings, because of their lager interaction volume and Q. We demonstrated, experimentally, sensitivity enhancement of an order of magnitude in a CBNL compared with conventional resonators. Further enhancement of the CBNL sensor sensitivity can be achieved by employing an air-defect design. The lasers' highly compact dimensions and enhanced resolution make them excellent candidates for becoming key components for the realization of compact, integrated biochemical sensor arrays.

Acknowledgments This research was supported by the Israeli Ministry of Science and Technology.

References

1 Madsen, C. K.; Zhao, J. H., *Optical Filter Design and Analysis: A Signal Processing Approach*, Wiley-Interscience Publications, New York, NY, **1999**, and references therein.
2 Little, B. E.; Chu, T.; Haus, H. A., Second-order filtering and sensing with partially coupled traveling waves in a single resonator, *Opt. Lett.* **1998**, 23, 1570–1572
3 Yariv, A., Critical coupling and its control in optical waveguide-ring resonator systems, *IEEE Photonics Technol. Lett.* **2002**, 14, 483–485
4 Heebner, J. E.; Boyd, R. W., Slow and fast light in resonator-coupled waveguides, *J. Mod. Opt.* **2002**, 49, 2629–2636

5 Chao, C. Y.; Gao, L. J., Biochemical sensors based on polymer microrings with sharp asymmetrical resonance, *Appl. Phys. Lett.* **2003**, 83, 1527–1529
6 Matsko, A. B.; Maleki, L.; Savchenkov, A. A.; Ilchenko, V. S., Whispering gallery mode based optoelectronic microwave oscillator, *J. Mod. Opt.* **2003**, 50, 2523–2542
7 Vahala, K.J., Optical microcavities, *Nature* **2003**, 424, 839–846
8 Boyd, R. W., et al, Nanofabrication of optical structures and devices for photonics and biophotonics, *J. Mod. Opt.* **2003**, 50, 2543
9 Loncar, M.; Scherer, A.; Qiu, Y., Photonic crystal laser sources for chemical detection, *Appl. Phys. Lett.* **2003**, 82, 4648
10 Chow, E.; Grot, A.; Mirkarimi, L. W.; Sigalas, M.; Girolami, G., Ultracompact biochemical sensor built with two-dimensional photonic crystal microcavity, *Opt. Lett.* **2004**, 29, 1093
11 Scheuer, J.; Yariv, A., Annular Bragg defect mode resonators, *J. Opt. Soc. Am. B.* **2003**, 20, 2285–2291
12 Scheuer, J.; Green, W. M. J.; DeRose, G.; Yariv, A., Low threshold two-dimensional annular Bragg lasers, *Opt. Lett.* **2004**, 29, 2641–2643
13 Scheuer, J.; Green,W. M. J.; DeRose, G.; Yariv, A., Lasing from a circular Bragg nanocavity with an ultra-small modal volume, *Appl. Phys. Lett.* **2005**, 86, 251101
14 Yeh, P.; Yariv, A.; Marom, E., Theory of Bragg fiber, *J. Opt. Soc. Am.* **1978**, 68, 1196–1201
15 Yariv, A., Optical Electronics in Modern Communications, 5th edn., Oxford University Press, New York, NY, **1997**
16 Luff, B. J.; Harris, R. D.; Wilkinson, J. S.; Wilson, R.; Schiffrin, D. J., Integrated-optical directional coupler biosensor, *Opt. Lett.* **1996**, 21, 618–620
17 Luff, B. J.; Wilkinson, J. S.; Piehler, J.; Hollenbach, U.; Igenhoff, J.; Fabricius, N., Integrated optical Mach-Zehnder biosensor, *IEEE J. Lightwave Technol.* **1998**, 16, 583–592
18 Boyd, R. W.; Heebner, J., Sensitive disk resonator photonic biosensor, *Appl. Opt.* **2001**, 40, 5742–5747
19 Levy, U.; Shamai, R., Tunable optofluidic devices, *Microfluidics and Nanofluidics* **2008**, 4, 97–105
20 Luff, B. J.; Harris, R. D.; Wilkinson, J. S.; Wilson, R.; Schiffrin, D. J., Integrated-optical directional coupler biosensor, *Opt. Lett.* **1996**, 21, 618
21 Scheuer, J.; Green, W. M. J.; DeRose, G.; Yariv, A., InGaAsP annular Bragg lasers: Theory, applications and modal properties, *IEEE J. Sel. Top. Quantum Electron.* **2005**, 11, 476–484
22 Coldren, L. A.; Corzine, S. W., Diode Lasers and Photonic Integrated Circuits, Wiley-Interscience Publications, New York, NY, **1995**.
23 Kim, S.; Ryu, H.; Park, H.; Kim, G.; Choi, Y.; Lee, Y.; Kim, J., Two-dimensional photonic crystal hexagonal waveguide ring laser, *Appl. Phys. Lett.* **2002**, 81, 2499–2501
24 Yoshie, T.; Shchekin, O. B.; Chen, H.; Deppe, D. G.; Scherer, A., Planar photonic crystal nanolasers (II): Low-threshold quantum dot lasers, *IEICE Trans. Electron.* **2004, E87-C**, 300–307
25 Painter, O.; Lee, R. K.; Scherer, A.; Yariv, A.; O'Brien, J. D.; Dapkus, P. D.; Kim, I., Two-dimensional photonic band-gap defect mode laser, *Science* **1999**, 284, 1819–1821
26 Green, W.; Scheuer, J.; DeRose, G.; Yariv, A., Ultra-sensitive biochemical sensor based on circular bragg micro-cavities, CLEO/QELS **2005**, Baltimore, Maryland, paper CPDA7
27 Chow, E.; Grot, A.; Mirkarimi, L. W.; Sigalas, M.; Girolami, G., Ultracompact biochemical sensor built with two-dimensional photonic crystal microcavity, *Opt. Lett.* **2004**, 29, 1093–1095

Chapter 13
Optical Micro/Nanofibers for Sensing Applications

M. Sumetsky

Abstract This chapter reviews the applications of optical micro/nanofibers (MNFs) and MNF-based photonic devices for sensing of the ambient medium. An optical MNF is usually fabricated by drawing a glass or a polymer optical fiber and has a diameter between ~100 nm and several μm. The MNF sensors can be either based or not based on the interaction of the ambient medium with the evanescent part of the mode propagating along the MNF. The simplest MNF evanescent sensor is an MNF, which detects changes caused by proximity of chemical/biological species and microparticles at the MNF surface. The advantage of the MNF evanescent sensor compared with other evanescent sensors is that the MNF is more open to the environment and therefore more sensitive to ambient changes. This chapter starts with a brief review of the theory of uniform MNFs, MNF tapers, and MNF-based or MNF-supported photonic resonators (MNF loop resonator, MNF/microsphere resonator, MNF/microdisk resonator, MNF coil resonator, and MNF/microcylinder resonator), which are used as optical sensors. The description of the theory is followed by the review of applications of these devices for sensing. First, the simplest MNF sensors based on straight MNF are considered. They include microfluidic sensor, hydrogen sensor, molecular absorption sensor, humidity and gas sensor, optical fiber surface sensor, and atomic fluorescence sensor. Then the applications of looped and coiled MNFs as a direct-contact gas temperature sensor, an infrared radiation sensor, and a microfluidic sensor are reviewed. Finally, more complex sensors, which consist of MNFs coupled to microspheres, microdisks, microcylinders, and microcapillaries, are considered as refractive index sensors, nanolayer sensors, surface sensors, and also as individual atom and molecule sensors.

M. Sumetsky
OFS Laboratories, 19 Schoolhouse Road, Somerset, NJ 08873, USA
e-mail: sumetski@ofsoptics.com

X. Fan (ed.), *Advanced Photonic Structures for Biological and Chemical Detection*, Integrated Analytical Systems,
DOI 10.1007/978-0-387-98063-8_13, © Springer Science+Business Media, LLC 2009

13.1 Introduction

An optical micro/nanofiber (MNF) is usually fabricated by drawing a glass or a polymer optical fiber and has a diameter between ~100 nm and several μm. Optical MNFs have been fabricated and explored for different applications since the late 1950s of the previous century. Early application of MNFs included using MNF bundles for transmission of images in late 1950s[1,2], Later, in 1990, an MNF was implemented for optical amplification[3]. The large fraction of the evanescent field, which was propagated in a dye solution, allowed a significant reduction in the pump power[3,4]. In 1997, an MNF was applied for low-loss excitation of the whispering gallery mode (WGM) resonances in a microsphere[5] and, in 2000, for supercontinuum generation[6]. In 2003, fabrication of very thin and uniform MNFs with diameters as small as 50 nm was demonstrated[7]. An MNF-based technology for the fabrication of photonic circuits by wrapping MNFs onto central optical rods has been suggested in Ref. 8. Extremely low transmission losses of an MNF had been demonstrated experimentally.[9,10] In Refs.[11–13], a theory of MNF tapers was developed. It was shown that very thin MNFs cannot transmit light due to fundamental limitations.

This chapter considers the transmission properties of MNFs and photonic devices that can be fabricated from or using MNFs and used as sensors that detect changes in the ambient medium by monitoring changes in the transmission power of light. These changes may be caused by variation in temperature, radiation, concentration of chemical or biological species, microparticles, etc. The considered MNFs represent the waists of fiber tapers, rather than the sharply tapered fibers conventionally used in scanning near-field optical microscopes[14]. Everywhere throughout this chapter it is assumed (unless otherwise specified) that an MNF is connected to a light source at one end and to an optical spectrum analyzer or a power meter at the other end. An electromagnetic field mode supported by an MNF has an evanescent part, which is distributed outside the MNF. The evanescent part may be very sensitive to changes which take place at or near the surface of the MNF. Figure 13.1 illustrates principle of performance of an optical MNF sensor (Fig. 13.1a–c) and specific realizations of photonic sensors where an MNF either performs sensing (Fig. 13.1a–f) or serves as a waveguide, which connects a photonic sensor to the source and the detector of light (Fig. 13.1c, g–k). Figure 13.1a shows a simplest uncoated MNF sensor in which the thickness is usually smaller than the radiation wavelength. For this sensor, it is assumed that a significant (evanescent) part of the fundamental mode of radiation propagates in the ambient medium. Transmission power of this sensor changes with its complex effective refractive index, which depends on the properties of the ambient medium. The sensor illustrated in Fig. 13.1b is an MNF coated with chemical or biological reagents, which are sensitive to selective species. Figure 13.1c illustrates a generic photonic sensor having the MNF input and output. Figure 13.1d–j depicts the straight MNF sensor[15–22], the MNF loop sensor[23–26], the MNF coil sensor[8,26–33], the MNF/microsphere sensor[5,34–45], the MNF/microdisk sensor[46–49], the MNF/microcylinder sensor[18] and the MNF/microcapillary sensor,[50–54], respectively. Figure 13.1k, which shows a multiple MNF/microcylinder sensor, illustrates possible assembly of several MNF-based sensors. The sensors shown in Fig. 13.1 can be either not

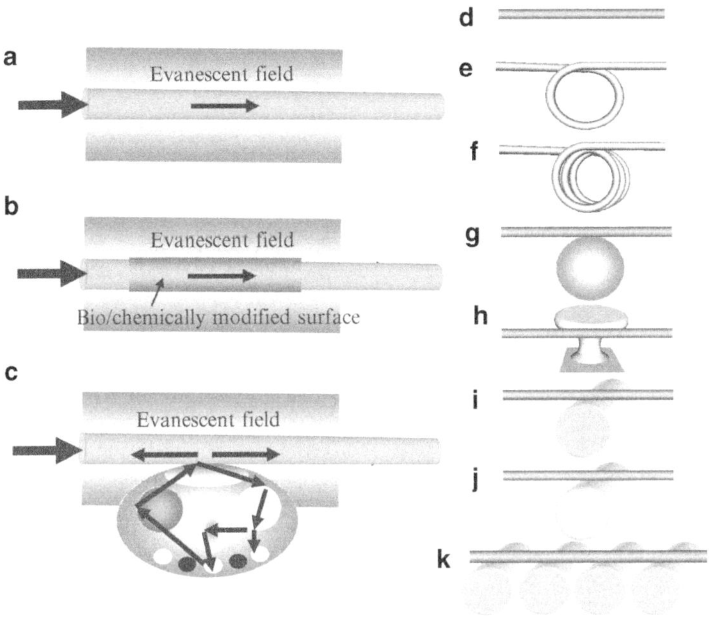

Fig. 13.1 Photonic sensors that are either based on MNF sensing ((**a**),(**b**), (**d**), (**e**), (**f**)) or use MNF for the input and output connection (**c**), (**g**)–(**j**). (**a**) – straight MNF sensor with surrounding evanescent field; (**b**) – straight MNF sensor coated with bio or chemical layer and surrounding evanescent field; (**c**) – generic structure of MNF-based optical sensor with surrounding evanescent field; (**d**) – straight MNF sensor; (**e**) – MNF loop resonator (MLR) sensor; (**f**) – MNF coil resonator (MCR) sensor; (**g**) – MNF/microsphere sensor; (**h**) – MNF/microdisk sensor; (**i**) – MNF/microcylinder sensor (**j**) – MNF/microcapillary sensor; (**k**) – a sensor composed of an MNF coupled to a series of microcylinders (optical fibers)

based or based on the interaction of the ambient medium with the evanescent field. The sensors of the first group are MNF devices, whose transmission characteristics vary due to changes in optical and/or geometrical properties of the MNF itself and/or of the propagating mode inside the MNF. Changes of this kind can be caused by outside radiation, ambient temperature, mechanical vibration, etc. For example, an MNF loop resonator (MLR) illustrated in Fig. 13.1e has been demonstrated as a fast direct-contact temperature sensor and also as an infrared radiation sensor[23]. The simplest of the sensors of the second group is a single MNF shown in Fig. 13.1b, which detects the changes caused by proximity of chemical/biological species and microparticles at the MNF surface. This sensor was demonstrated both for silica MNFs[15–21] and polymer MNFs.[22]. The MNF evanescent sensor belongs to a family of optical waveguide evanescent sensors illustrated in Fig. 13.2. They are a planar waveguide sensor (Fig. 13.2a), a polished optical fiber sensor (Fig. 13.2b), and an MNF sensor (Fig. 13.2c). The advantage of the MNF evanescent sensor, compared with other sensors shown in Fig. 13.1[55–58], is that the MNF is not immersed into a substrate, and therefore it is more open and more sensitive to ambient changes. In addition, the more complex sensors based on MNF, which are illustrated in Fig. 13.1,

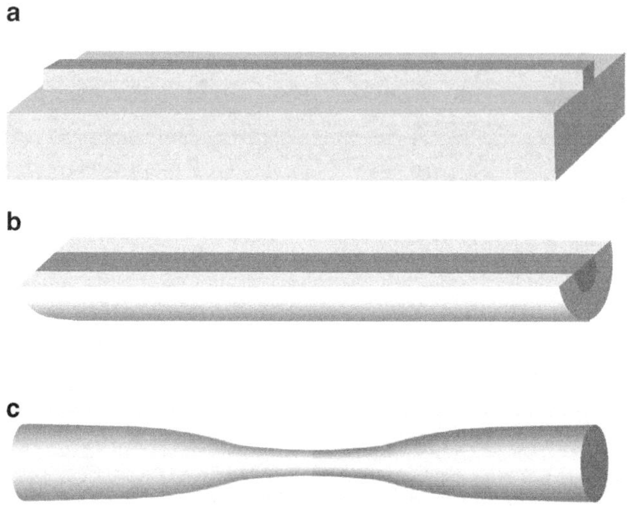

Fig. 13.2 Different types of evanescent sensors. (**a**) – a planar waveguide; (**b**) – a polished optical fiber; (**c**) an MNF taper

can be often more compact than similar optical sensors fabricated with planar lithographic technology.

Section 13.2 of this chapter describes the theory of uniform MNFs and MNF tapers, which is important for sensing applications of MNFs. Section 13.3 briefly introduces the theory of MNF-based or MNF-supported photonic resonators, which are used as optical sensors: MLR, MNF/microsphere resonator, MNF/microdisk resonator, MNF coil resonator (MCR), and MNF/microcylinder resonators. In Sect. 13.4, the simplest MNF sensors are considered. These sensors probe the environment by monitoring changes of the evanescent field and also of the refractive index and geometric parameters of a straight MNF. They include microfluidic sensor, hydrogen sensor, molecular absorption sensor, humidity and gas sensor, optical fiber surface sensor, and atomic fluorescence sensor. Section 13.5 describes application of looped and coiled MNFs for sensing. This section considers a direct-contact gas temperature sensor, an infrared radiation sensor, and a microfluidic sensor. Section 13.6 considers more complex sensors, which consist of MNFs coupled to microspheres, microdisks, microcylinders, and microcapillaries. These photonic devices were demonstrated as refractive index sensors, nanolayer sensors, surface sensors, and also as individual atom and molecule sensors. Section 13.7 summarizes this chapter.

13.2 Guiding Properties of MNFs

An optical MNF is a waveguide that transmits light. Generally, an MNF is a multimode waveguide. In this chapter, we primarily consider single mode MNFs, which support the fundamental mode only. The diameter of a single mode MNF is less than any of the cutoff diameters of the higher order modes. It is usually around a micron or less. Formally, there is no cutoff diameter for the fundamental mode. The most interesting for sensing applications are very thin MNFs, in which the diameter is significantly less than the radiation wavelength. For these MNFs, the evanescent component of fundamental mode, which propagates outside the MNFs, is significant and can probe changes of the ambient medium. The thinner the MNF the larger is the effective width of the evanescent component of its fundamental mode. This section reviews the basic transmission properties of the uniform optical MNF, MNF tapers, and discusses how thin the MNF can be in practice.

13.2.1 A Uniform MNF

The theory of a straight and uniform MNF with a circular cross-section of radius a and a step-index profile is well developed[59–61]. A section of an MNF is shown in Fig. 13.3 in Cartesian coordinates (x,y,z). The distribution of refractive index along the radius $r = \sqrt{x^2 + y^2}$ of such MNF is

$$n(r) = \begin{cases} n_1 & r < a \\ n_2 & r \geq a \end{cases}. \tag{13.1}$$

Generally, the propagation modes of an optical MNF can be written in the form.[59]:

$$F(x, y, z) = G_\beta(x, y) \exp(i\beta z), \tag{13.2}$$

where β is the propagation constant that is determined by the eigenvalue equation:

$$\left[\frac{J_\nu'(U)}{UJ_\nu(U)} + \frac{K_\nu'(W)}{WK_\nu(W)}\right]\left[\frac{J_\nu'(U)}{UJ_\nu(U)} + \frac{n_2^2}{n_1^2}\frac{K_\nu'(W)}{WK_\nu(W)}\right] = \left(\frac{\nu\beta}{k_0 n_1}\right)^2\left(\frac{V}{UW}\right)^4, \tag{13.3}$$

In this equation, J_ν is the Bessel function of the first kind, and K_ν is the modified Bessel function of the second kind, $U = a({k_0}^2{n_1}^2 - \beta^2)^{1/2}$, $W = a(\beta^2 - {k_0}^2{n_2}^2)^{1/2}$,

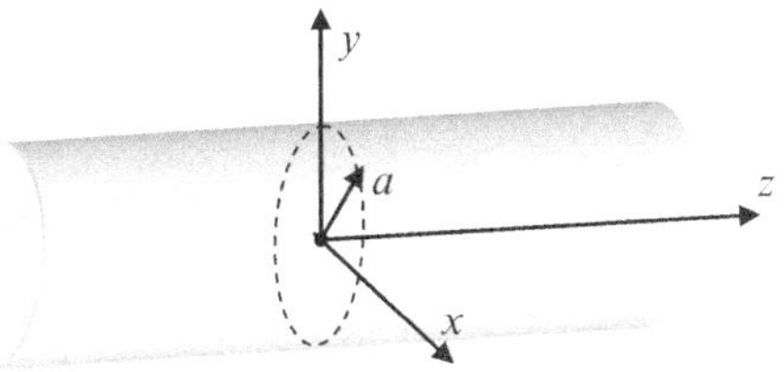

Fig. 13.3 Straight MNF

$V = k_0 a(n_1{}^2 - n_2{}^2)^{1/2}$, $k_0 = 2\pi/\lambda$, λ is the radiation wavelength, and "$'$" means derivative. The analytical expressions for $G_\beta(x, y)$ can be found in Refs. [59–61]. The propagation constant β is expressed through the propagation constant of the ambient medium, $\beta_0 = 2\pi n_2/\lambda$, as follows:

$$\beta = \sqrt{\beta_0^2 + \gamma^2}, \tag{13.4}$$

where γ is the transversal component of the propagation constant. For a given β_0, the constant γ is uniquely determined through the MNF taper diameter from (13.2). If the MNF diameter is noticeably smaller than the radiation wavelength, then the fundamental mode is mainly represented by its evanescent component, which is propagating outside the MNF. As an example, Fig. 13.4 shows the distribution of electromagnetic field components for the silica fiber with parameters $a = 0.2$ μm, $n_1 = 1.4469$, $n_2 = 1$ and radiation wavelength $\lambda = 1.3$ μm (from Ref. 61). It is seen

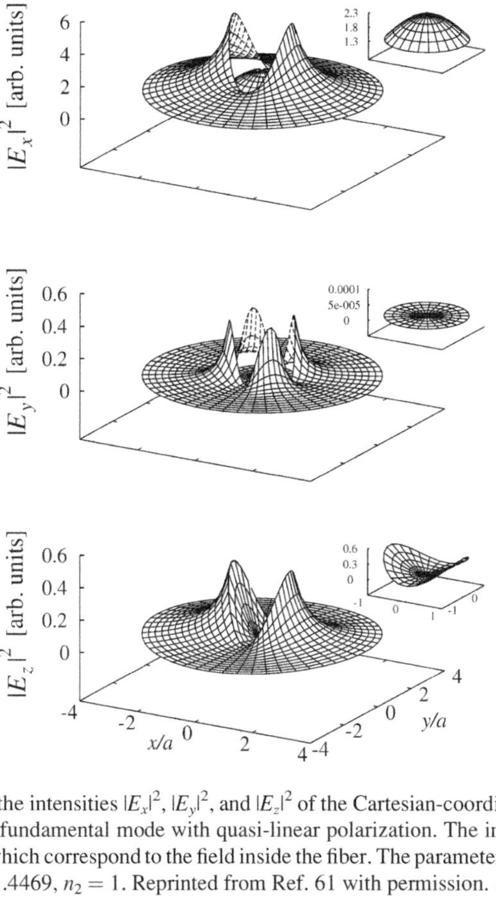

Fig. 13.4 Profiles of the intensities $|E_x|^2$, $|E_y|^2$, and $|E_z|^2$ of the Cartesian-coordinate components of the electric field in a fundamental mode with quasi-linear polarization. The insets show the inner parts of the profiles, which correspond to the field inside the fiber. The parameters used: $a = 200$ nm, $\lambda = 1{,}300$ nm, $n_1 = 1.4469$, $n_2 = 1$. Reprinted from Ref. 61 with permission. © 2008 Elsevier

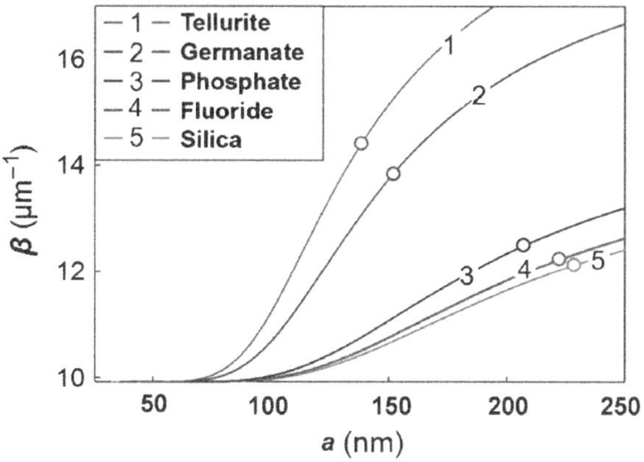

Fig. 13.5 Calculated propagation constants (β) for the fundamental modes of glass MNFs with refractive indices of 1.46 (silica), 1.48 (fluoride), 1.54 (phosphate), 1.89 (germinate), and 2.02 (tellurite), respectively. A circle marked on each curve corresponds to the maximum radius of the single-mode MF. Radiation wavelength is $\lambda = 633$ nm. Reprinted from Ref. 62 with permission.

that the mode distribution is axially asymmetric and has a break at the MNF surface. Figure 13.5 shows the dependence of the propagation constant, β, on the MNF radius a for glasses with different refractive indices[62].

For small values of γ, the characteristic dimension of the evanescent field, ρ_{ch}, becomes exponentially large. For example, for the radiation wavelength λ=1.5μm and the MNF diameter d=0.4 μm, we have $\rho_{ch} = 6\ \mu m$, 0.1 mm, and 0.5 m, respectively. It is not a problem to fabricate an MNF with a diameter as thin as 0.2 μm and less[7,62,63]. However, the characteristic dimension of an evanescent field for the diameter of 0.2 μm, which is equal to 0.5 m, is unrealistic due to the dramatic growth of radiation losses. The problem of losses in very thin MNF tapers is discussed in Sect. 13.2.3[11,64].

13.2.2 Adiabatic MNF Tapers

Usually, an MNF sensor represents the waist of a fiber taper. For this reason, to understand the performance of an MNF evanescent sensor, it is important to understand the structure of the evanescent field of an MNF taper. This subsection briefly describes the theory of MNF tapers.

In an adiabatically tapered axially symmetric MNF, the propagation constant of the fundamental mode, β(z), is a slow function of the coordinate z along the axis of the MNF and (13.4) is modified as follows:

$$\beta(z) = \sqrt{\beta_0^2 + \gamma(z)^2}. \tag{13.5}$$

In the adiabatic limit and for very thin MNF, when $\gamma(z) << \beta(z)$, the evanescent part of the fundamental mode is axially symmetric[59,13]:

$$U(r,z)\tilde{\gamma}(z)K_0(|\gamma(z)|r)\exp\left[i\int^{z} dz\beta(z)\right]. \tag{13.6}$$

However, using (13.6), it is not possible to determine the radiation loss caused by tapering. The problem of radiation loss can be solved with the asymptotic solution, which can be constructed with the semiclassical theory[13]. In Ref. 13, it was shown that the radiation part is split off from the fundamental mode locally, near focal circumferences. The interference between the guiding and radiating parts give rise to a complex structure of the evanescent electromagnetic field propagating along the MNF taper illustrated in Fig. 13.6. The optics of light propagation along the tapered MNF can be visualized with this figure as follows. The input mode is lunched from the left hand side of Fig. 13.6. In the neighborhood of each focal circumference, the mode is split into two components: the component, which is guided by the MNF and the component that contributes to the exponentially small radiation loss. As an example, consider a biconical MNF taper, which is described by the following Lorentzian variation of the transversal propagation constant:

$$\gamma^{(0)}(z) = \gamma_\infty + \frac{\gamma_0 - \gamma_\infty}{1+(z/L)^2} \tag{13.7}$$

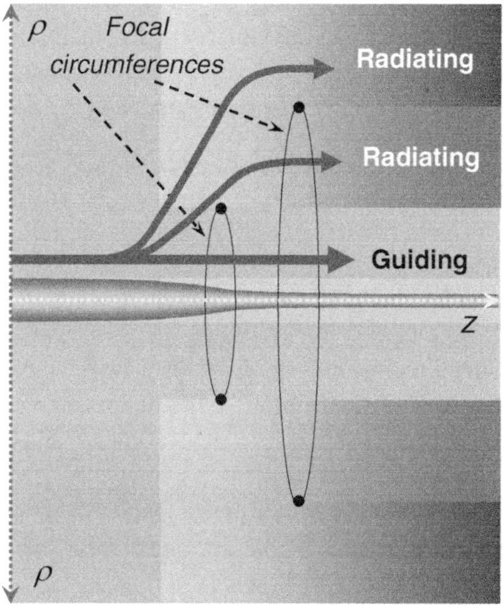

Fig. 13.6 Optics of radiation losses in a MF taper, which shows the splitting of the guiding and radiating components of the propagating electromagnetic field near focal circumferences

Here L is the characteristic length of the taper and γ_0 and γ_w are the values of transversal propagation constants in the center of the taper waist and on the infinity, respectively. For this taper, the analytical expression for the radiation loss was found in Ref. 13. In particular, for a strong taper, when $|\gamma_0| << |\gamma_\infty|$, the radiation loss has the form:

$$P = \frac{\pi^{3/2}}{2(30)^{1/2}S^{1/2}} \exp(-S), \quad S = \frac{15L\gamma_0^{5/2}}{8\beta_0\gamma_\infty^{1/2}}. \tag{13.8}$$

This expression is useful for estimating of the radiation loss of an adiabatic taper with characteristic length L. Figure 13.7 shows the distribution of the evanescent electromagnetic field intensity in the vicinity of the taper for $\gamma_0 = 0.2\ \mu\text{m}^{-1}$, $\gamma_\infty = 0.4\ \mu\text{m}^{-1}$, $L = 500\ \mu\text{m}$, and $\beta_0 = 4\ \mu\text{m}^{-1}$. In agreement with illustration in Fig. 13.6, the radiating and guiding components are split off near the focal point shown as a small circle. The interference between these components gives rise to the oscillations of the evanescent field and to the appearance of quasiperiodic dips, which are clearly seen in Fig. 13.7.

13.2.3 The Thinnest MNF Optical Waveguide

The fundamental mode of a uniform lossless dielectric waveguide and, in particular, an MNF exists independently of its thickness. However, in practice, waveguiding is limited by losses due to material absorption and geometric nonuniformities. For a very thin MNF, the transmission loss is primarily determined by input and output losses, which, in practice, cannot be reduced significantly[11,64]. As an example, Refs. 11 and 13 theoretically explored an MNF with adiabatically

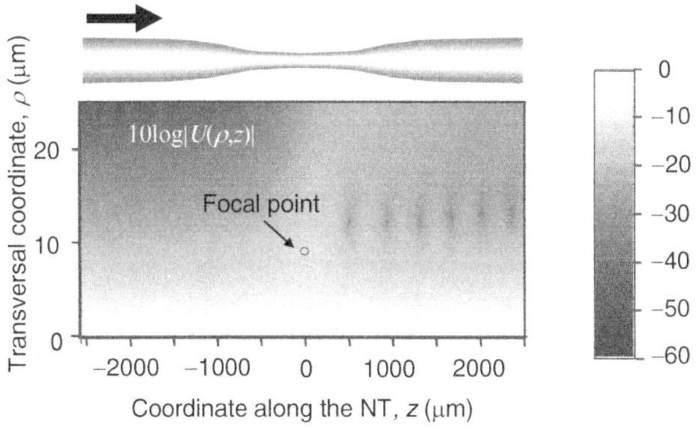

Fig. 13.7 Distribution of the electromagnetic field intensity near the MF taper defined by (13.6) with parameters $\gamma_0 = 0.2\ \mu\text{m}^{-1}$, $\gamma_\infty = 0.4\ \mu\text{m}^{-1}$, $L = 500\ \mu\text{m}$, and $\beta_0 = 4\ \mu\text{m}^{-1}$. Reprinted from Ref. 13 with permission. © 2008 Optical Society of America

tapered connections to a regular fiber. The radiation loss of an MNF taper can be estimated with exponential accuracy for the taper having the Lorentzian radius variation,

$$a(z) = a_\infty - \frac{(a_\infty - a_0)}{1 - (z - z_0)^2/L^2}, \tag{13.9}$$

where α_w and α_0 are the taper radii at its ends and at the center, respectively. For a silica MNF taper (refractive index 1.45) with a small waist diameter 2a << λ, it can be shown that the radiation loss depends on the waist radius a_0 double-exponentially[11,12]:

$$P \sim \exp\left[-\frac{0.51L}{a_0^{1/2}(a_\infty - a_0)^{1/2}} \exp\left(-\frac{0.143\lambda^2}{a_0^2}\right)\right] \tag{13.10}$$

The MNF is waveguiding only if $P \ll 1$, while the condition $p \sim 1$ corresponds to the threshold where waveguiding disappears. Figure 13.8 shows the transmission loss P as a function of the microfiber diameter $2a_0$ for radiation wavelength $\lambda =$ 1.55 μm and a_w μm. It demonstrates the dramatic threshold behavior of the transmission loss predicted by (13.10) and shows no significant difference between the threshold values of the MNF diameter for a typical $L = 10$ mm and for a gigantic L 10 km. These diameters are of the same order of magnitude ~0.2–0.4 μm and only a few times less than the radiation wavelength. These predictions have been confirmed experimentally in Ref. 64. It was shown that, while a silica MNF with the radius smaller than the threshold radius determined by (13.10) is a low loss

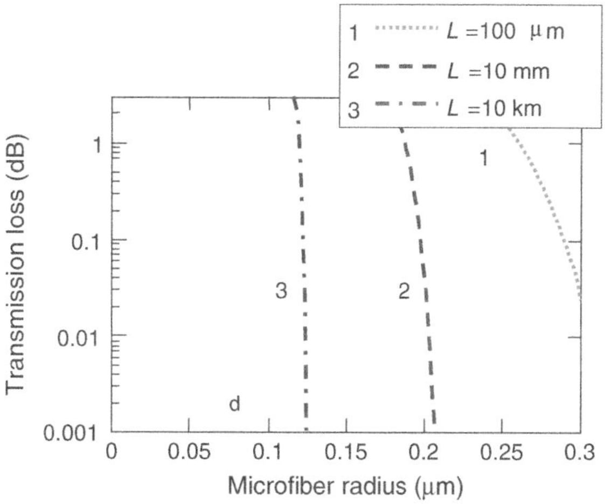

Fig. 13.8 Transmission loss as a function of the MNF diameter calculated with (13.10) for different characteristic lengths L of the MF taper (curves 1, 2, and 3) and $a_\infty = 0.5$ μm. Reprinted from Ref. 12 with permission. © 2008 Optical Society of America

waveguide, the MNF with larger radius is very lossy and, practically, does not transmit light. The fundamental character of losses of very thin MNFs predicted by (13.10) was experimentally confirmed in Ref. 65 for terahertz fibers.

13.3 Photonic Devices Employing MNFs

The principle of an MNF performance as an optical sensor is based on the variation of the MNF transmission spectrum in response to changes in the ambient medium. More advanced photonic sensors exploiting MNFs allow to enhance these variations and/or to make them more selective. This section briefly reviews the basic theory of simplest resonant photonic sensors. In these devices, an MNF either form an optical resonator or serve as an input and output connection to an optical resonator. These devices include the MLR[23–26], the MCR[8,26–33], the MNF/microsphere resonator[34–45], the MNF/microdisk resonators[46–49], and the MNF/microcylinder resonator[18,50–54].

13.3.1 MNF Loop Resonator, MNF/Microsphere Resonator, and MNF/Microdisk Resonator

An MLR, which is illustrated in Fig. 13.1e, is a miniature version of a fiber loop resonator created from an MNF. An MNF/microsphere resonator and MNF/microdisk resonator, which are illustrated in Fig. 13.1g, h, consist of an MNF coupled to a microsphere or a microdisk. The excited WGMs are localized in the neighborhood of the microsphere (microdisk) circumference situated in the plane of symmetry of these devices. The transmission power of an MLR, a microsphere, or a microdisk sensor, $P(\lambda)$, near the resonance wavelength λ_0 can be found in the form similar to the transmission power of a ring, a disk, and a microsphere resonator[66,67]:

$$P(\lambda) = P_0(\lambda)\frac{(\lambda-\lambda_0)^2 + (\gamma_a-\gamma_c)^2}{(\lambda-\lambda_0)^2 + (\gamma_a+\gamma_c)^2}. \tag{13.11}$$

Here, P_0 is the input transmission power, and the parameters γ_c and γ_a are the coupling and attenuation parameters, respectively. They are expressed through the attenuation constant, α, the circumference, S, the effective refractive index, n, and the coupling coefficient, κ, as follows:

$$\gamma_a = \frac{\alpha\lambda_0^2}{2\pi n}, \quad \gamma_c = \frac{(1-\kappa)\lambda_0^2}{2\pi n S}. \tag{13.12}$$

The Q-factor of an optical resonator is determined as a ratio of the wavelength of the resonance and its FWHM, $\Delta\lambda$: $Q = \lambda_0/\Delta\lambda$. From (13.11), the Q-factor of the resonance is:

$$Q = \frac{\lambda_0}{2(\gamma_a + \gamma_c)}. \tag{13.13}$$

From (13.11), the extinction ratio of the resonance is:

$$R = 10 \cdot \log\left[\left(\frac{\gamma_c + \gamma_a}{\gamma_c - \gamma_a}\right)^2\right]. \tag{13.14}$$

The total power transmitted through the considered resonators, P, is a sum of partial transmission powers for TE and TM polarizations:

$$P = P_1 + P_2, \tag{13.15}$$

where each of the transmission powers, P_1 and P_2 corresponds to one of the polarization states, TM or TE. Each of the transmission powers P_j in (13.15) are defined by (13.11). For a given loss of an optical resonator, its performance as a sensor can be optimized[67,68].

Generally, the resonances in transmission spectrum can overlap. Then, the transmission spectrum of these devices depends on the coupling of resonant mode to the MNF/MNF or MNF/microsphere (microdisk) as well as on the coupling of different WGMs to each other in the MNF/microsphere contact region.

13.3.2 MNF Coil Resonator

An MCR, which is illustrated in Fig. 13. 1f, is an MNF wrapped on a central rod, forming a coil. If the distance between the turns of the coil is significantly greater than the radiation wavelength, the MNF turns are not coupling electromagnetically, and the coil behaves as a delay line with a smooth broadband transmission spectrum. Coupling between turns leads to interference phenomena and to the appearance of resonances in the transmission spectrum of an MNF coil, which turns into an MCR. Propagation of light along an MCR is described by the coupling mode equations[8,26]. The structure of transmission resonances of an MCR can be very complex[8,26,28]. In the simplest case of a single turn, an MCR coincides with an MLR. An important case of an MCR is a higher-order MLR, determined as a coiled MNF, which consists of several independent MLRs. The transmission coefficient of a higher-order MLR is simply a product of individual transmission coefficients of MLRs that compose this MCR. In particular, if the resonances of individual MLRs are close to each other then

$$P(\lambda) = P_0(\lambda)\prod_{n=1}^{N} P_n(\lambda),$$
$$P_n(\lambda) = \frac{(\lambda - \lambda_n)^2 + (\gamma_a^{(n)} - \gamma_c^{(n)})^2}{(\lambda - \lambda_n)^2 + (\gamma_a^{(n)} + \gamma_c^{(n)})^2}, \tag{13.16}$$

where λ_n, $\gamma_c^{(n)}$, and $\gamma_a^{(n)}$ are the neighboring resonant wavelengths, coupling parameters, and attenuation parameters of individual loops. The transmission spectrum of MCRs with more complex configurations and interturn coupling can be determined using the coupling wave equations[8].

13.3.3 MNF/Microcylinder Resonator

An MNF/microcylinder sensor exploits WGMs resonances in a cylinder (optical fiber), which are excited by an MNF. The arrangement of an MNF and a cylinder is shown in Fig. 13.1i. As opposed to the WGM in a microsphere and microdisk considered in Sect. 13.3.1, the beam launched from the MNF into the cylinder spreads along the cylinder surface and eventually vanishes, even if there is no loss. The theory of resonant transmission of the MNF/microcylinder sensor was developed in Ref. 18. The resonant transmission power of this device can be modeled by a self-interference of a Gaussian beam that made n turns along the cylinder circumference:

$$P \left|\sigma \sum_{n=0}^{\infty} (1-\sigma)^{n/2}\left(\beta_t(\lambda)^2 w^4 + 4S^2 n^2\right)^{-1/2} \exp(in\beta_t(\lambda)S)\right|^2, \tag{13.17}$$

where S is the circumference length, σ, is the roundtrip power loss (which takes place primarily near the microfiber-cylinder contact point), w is the initial beam width, and $\beta_t(\lambda)$ is the effective transversal propagation constant. At no loss ($\sigma = 0$) and at the resonance condition of the constructive interference ($\beta_t S = 2\pi \times \text{integer}$), the sum in (13.17) diverges. For small $\sigma << 1$ and close to the resonance, i.e. for small $\Delta\beta_t = (\beta_t - 2\pi \times \text{integer})S^{-1} = (\mathrm{d}\beta_t/\mathrm{d}\lambda)(\lambda - \lambda_0)$, the sum in (13.17) exhibits a negative peak:

$$P \sim \frac{\sigma^2}{\beta_t^2 w^4}\left\{1 + \frac{\beta_t w^2}{S} \ln\left[(\sigma/2)^2 + \left(S\frac{\mathrm{d}\beta_t}{\mathrm{d}\lambda}\right)^2 (\lambda - \lambda_0)^2\right]\right\}. \tag{13.18}$$

This logarithmic peak is much more gradual than the Lorentzian peak of the conventional transmission resonance of (13.11). It can, however, be sharp as well.

13.4 Application of a Straight MNF for Sensing

To probe specific physical, chemical, and biological elements in the ambient medium, an MNF should change its transmission properties in response to variation of the sample properties. For this purpose, an MNF must be coated with reagents. These reagents change the refractive index/light absorption in the process of interaction with the chemical/biological elements under test. An evanescent MNF hydrogen sensor was demonstrated in Ref. 15. The MNF was coated with a palladium film with a thickness of 4 nm. The exposure of the coated MNF to hydrogen gave rise to the creation of the palladium hydride layer and changed the MNF transmission loss. Reference [16] demonstrated another type of an MNF sensor, which has been immersed into a lower-index cured polymer having a channel in proximity to the MNF. This MNF sensor was used for measuring the refractive index of liquid in the channel. In Ref. 19, an MNF was demonstrated as a molecular absorption sensor. In Refs. 17 and 18, an MNF was used as a sensor of the optical fiber radius variation and surface contamination. In Ref. 22, polymer MNFs were used for humidity and gas sensing. The authors of Refs. 20 and 21 suggested and demonstrated an MNF for atom probing and trapping. The principle of this sensor is based on the detection of light scattered by atoms into the fundamental mode of an MNF. This section reviews the above-mentioned applications of straight MNFs for optical sensing.

13.4.1 Microfluidic Refractive Index MNF Sensor

The authors of Ref. 16 demonstrated an MNF-based optical sensing device capable of measuring the refractive index of liquids propagating in a microfluidic channel. The sensor is based on a MNF fabricated from a conventional single-mode optical fiber and immersed in a transparent curable soft polymer. The sensing device, which is illustrated in Fig. 13.9a, was fabricated as follows. A rectangular bath was filled with the transparent curable soft polymer (PDMS). A slit was machined through the bottom of the bath to fit a 3-mm-thick glass rod sticking out of the bottom of the bath by approximately one half of its diameter. The rod position defined the channel for the liquid analyte that had a semicircular cross section. After the bath is filled with liquid PDMS, the fiber taper oriented orthogonally to the glass rod is slowly lowered into the uncured polymer from the top while the optical transmission through the taper is monitored. After the PDMS is cured, the device is flipped over and the glass rod is removed, exposing the channel for the sample liquid.

The principle of operation of this sensor is based on the fact that, as the fundamental optical mode travels through the MNF, its shape is modified depending on the index contrast between the solution in the channel and the polymer. Consequently, the change of the fundamental mode results in variation of the MNF

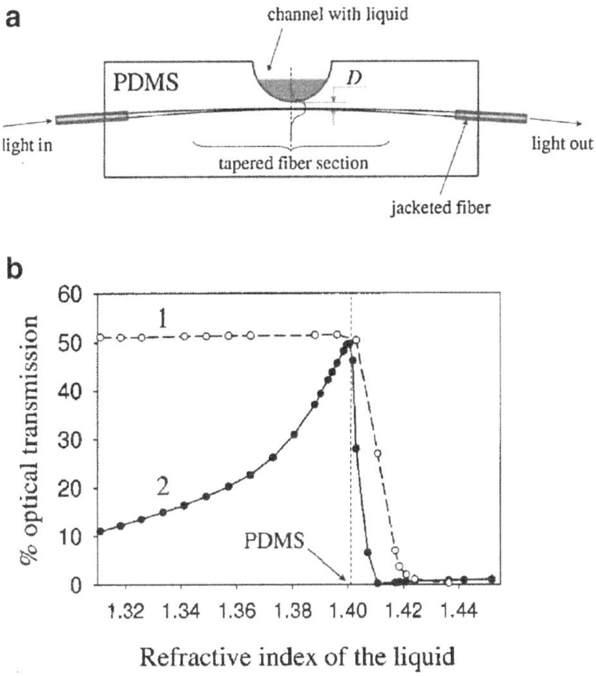

Fig. 13.9 (**a**) -The MNF sensor with microfluidic channel. (**b**) - Measured optical transmission versus refractive index of the liquid in the channel for two sensors with different MNF diameter at the waist: *curve 1* – 1.6 μm; *curve 2* – 700 nm. *Circles* are measurement data points. The *arrow* shows the refractive index of the surrounding polymer. Reprinted from Ref. 16 with permission. © 2008 Optical Society of America

transmission. The sensitivity of the device was measured by monitoring the optical transmission through the MNF with the channel filled with solutions of glycerol in water. The refractive index of the solution was changed by using different concentrations of glycerol. The resulting calibration curves are shown in Fig. 13.9b for two different sensors, fabricated with (1.6 ± 0.2) μm diameter and with (0.7 ± 0.2) μm diameter MNF sections. It was found that, as expected, in both cases the transmission achieved maximum when the refractive index of the liquid matches the index of the surrounding polymer, and the thinner MNF is more sensitive. The results of measurements plotted in Fig. 13.9b show the potential measurement accuracy of the refractive index $\sim 10^{-4}$.

13.4.2 Hydrogen MNF Sensor

In Ref. 15, an optical MNF hydrogen sensor was demonstrated. Similar to the MNF sensor described in the Sect. 13.4.1, this sensor monitors variation of transmission power of an MNF representing a waist of a biconical taper. The sensor is shown in

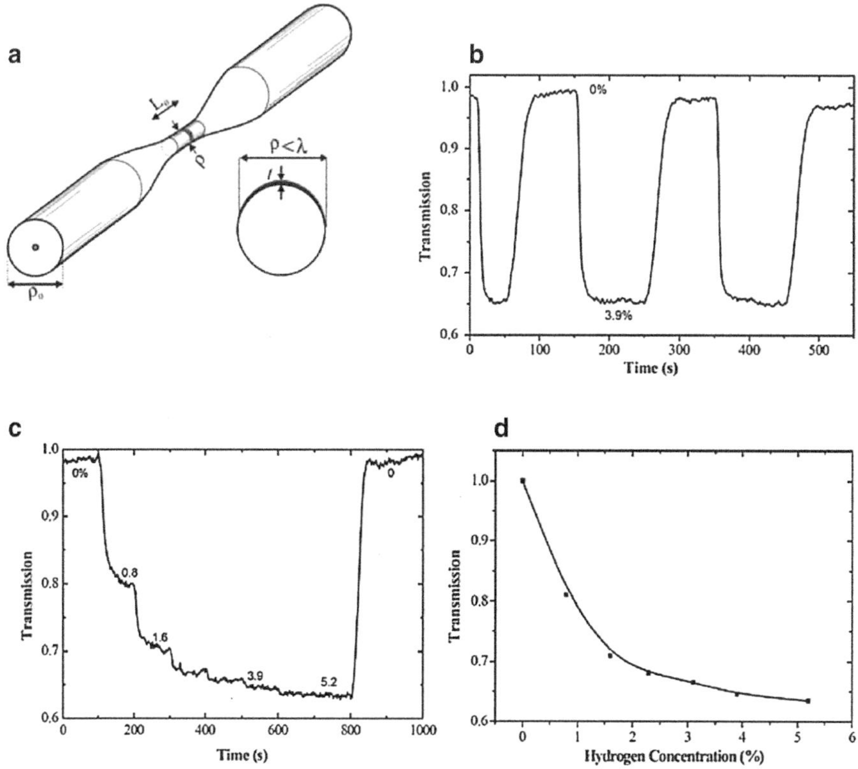

Fig. 13.10 (**a**) – Tapered optical fiber. ρ_0 is the initial diameter, inset – schematic cross section of the device; ρ is the waist diameter, L_0 is the length of the waist, t is the maximum thickness of the palladium film (*shadowed area*) and λ is radiation wavelength. (**b**) – Time response of the sensor to periodic cycles from a pure nitrogen atmosphere to a mixture of 3.9% hydrogen in nitrogen. (**c**) – Time response of a sensor when it was exposed to different hydrogen concentrations. (**d**) – Transmission versus hydrogen concentration; sensor parameters: $\rho = 1{,}300$ nm, $L = 2$ mm, and $t = 4$ nm. Reprinted from Ref. 15 with permission. © 2008 Optical Society of America

Fig. 13.10a. An MNF was coated with an ultra thin palladium film. The operational principle of this sensor was based on the fact that a thin palladium film has the ability to selectively absorb hydrogen. If a palladium film is exposed to hydrogen, its refractive index, and, in particular, absorbance, changes. The change in refractive index causes a change in transmission power of an MNF. The MNF fabricated in Ref. 15 had a palladium film of 4 nm in thickness and 2 mm in length. In Fig. 13.10b, the transmission power of the MNF is shown as a function of time when the sensor was exposed successively to a 3.9% concentration of hydrogen. The response time calculated from the plot was ~10 s. This response time is 3–5 times faster than that of other optical hydrogen sensors and about 15 times faster than that of some electrical nano hydrogen sensors. The fast response of the sensor is, presumably, due to the ultra small thickness of the palladium film that is rapidly filled with hydrogen. Figure 13.10c shows the transmission of this sensor as a function of time for different concentrations of hydrogen. A calibration curve of the sensor is shown in Fig. 13.10d. The authors of Ref. 15 suggested that with thinner tapers the detected changes in transmission power could be larger means that the length of the palladium film can be made even shorter, probably of a few hundred microns. It was also mentioned that that the demonstrated sensor has preserved its characteristics, sensitivity, and fast response time, after some months of its fabrication.

13.4.3 Molecular Absorption MNF Sensor

The authors of Ref. 19 presented an optical MNF sensor of submonolayers of perylene tetracarboxylic dianhydride molecules (PTCDA, see inset in Fig. 13.11a). A 500-nm diameter 3-mm long MNF waist of a tapered single mode fiber was used for measurements. Figure 13.11a shows the experimental setup using a conventional absorption spectrometer configuration with a tungsten light source and a CCD spectrograph. The molecules are deposited on the MNF by placing a crucible with PTCDA crystals below the fiber and by heating it up to 250°C. The air carries sublimated molecules to the MNF where they are adsorbed. The transmission spectrum of the MNF was recorded during deposition with an integration time of 1 s. Figure 13.11b displays a series of absorption spectra not saturating the molecules. Qualitatively, they agree well with spectra of submonolayers of PTCDA on mica. The absorbance spectrum can be measured very rapidly and with an excellent signal to noise ratio. Figure 13.11c displays a series of spectra that monitor the post-deposition evolution, i.e., the ripening of the film after stopping the deposition of the molecules. It demonstrates that a PTCDA monolayer on a glass surface is metastable at ambient conditions and transforms into islands with a thickness of at least two monolayers within minutes. The sensitivity of these measurements exceeded the sensitivity of previous studies by two orders of magnitude.

13.4.4 Humidity and Gas Polymer MNF Sensor

The authors of Ref. 22 demonstrated polymer MNFs that were used for humidity sensing with a response time of 30 ms, and for NO_2 and NH_3 detection down to subparts-per-million level. The polymer MNFs were fabricated by direct drawing of solvated polymers. As illustrated in Fig. 13.12a, two fiber tapers drawn from a single mode fiber with the end diameter about 500 nm were placed in parallel and close contact with the ends of a polymer MNF supported by a low-index substrate. Because of the strong evanescent coupling between the nanowire and the fiber taper, light can be efficiently launched into and picked up from the nanowire within a few micrometers overlap. The authors employ a PAM MNF for relative humidity sensing. The response of the MNF was tested by alternately cycling 75% and 88% relative humidity air inside the chamber, with an excellent reversibility shown in

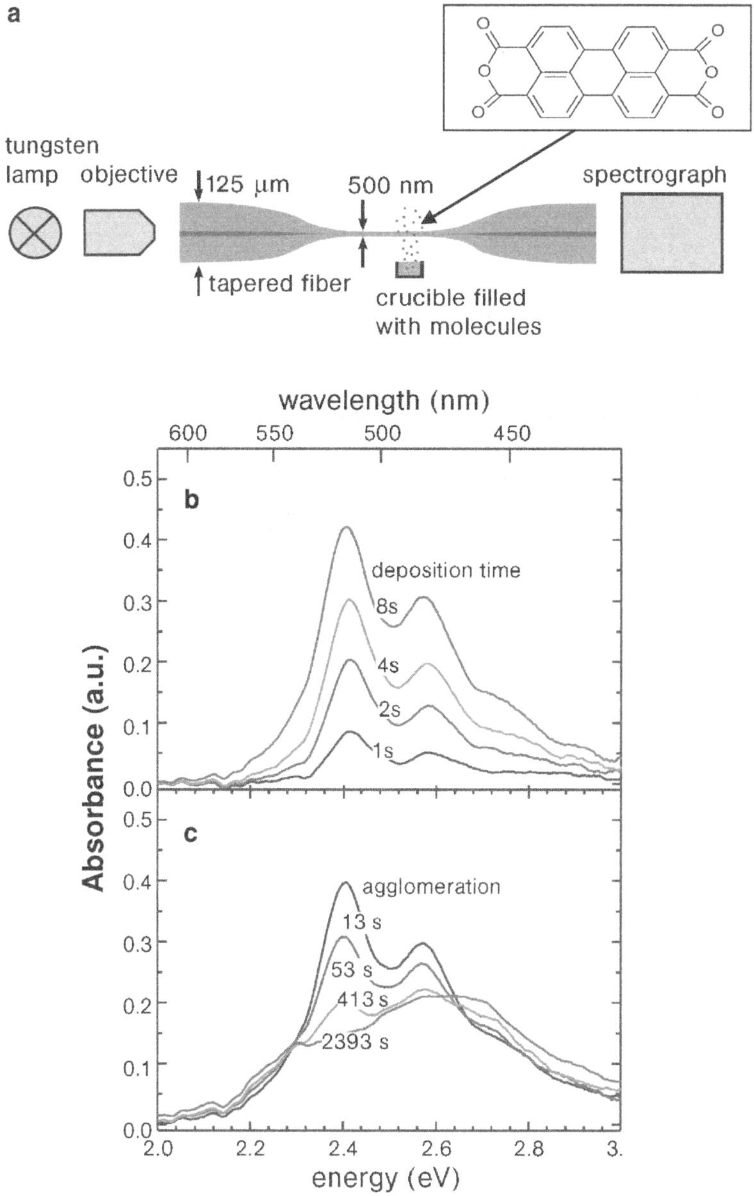

Fig. 13.11 (**a**) –The molecular absorbance MNF sensor (inset – PTCDA molecule); (**b**) – Evolution of spectrum of submonolayer deposition at 1, 2, 4, and 8 s after beginning of molecule exposure; (**c**) – Subsequent evolution of the spectral absorption at constant molecule number at 13, 53, 413, and 2,393 s after beginning of molecule exposure. Reprinted from Ref. 19 with permission. © 2008 Optical Society of America

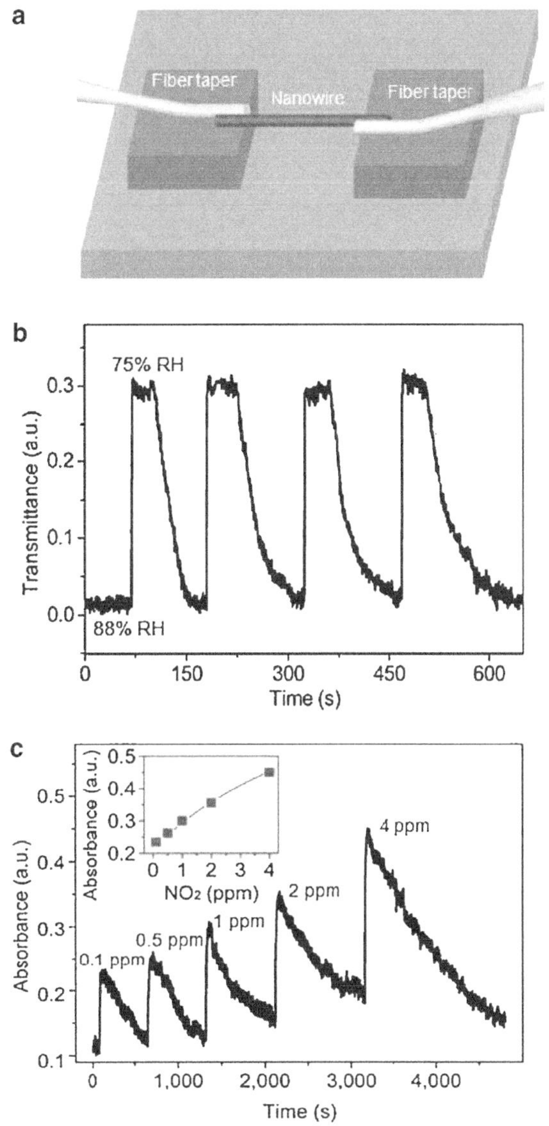

Fig. 13.12 (**a**) – Supported polymer MNF with two ends coupled to fiber tapers. (**b**) – Time dependence of the MNF transmittance to alternately cycling air with 75% and 88% relative humidity. (**c**) – Time-dependence of the MNF absorbance to cyclic NO_2/nitrogen exposure with NO_2 concentration from 0.1 to 4 ppm. Inset, dependence of the absorbance over the NO_2 concentration ranging from 0.1 to 4 ppm. Reprinted from Ref. 22 with permission. © 2008 American Chemical Society

Fig. 13.12c. The estimated response time (baseline to 90% signal saturation) of this sensor was about 25–30 ms, which is one or two orders of magnitude faster than those of existing relative humidity sensors. The remarkably fast response of this sensor is due to the small diameter and large surface-to-volume ratio of the nanowire that enable rapid diffusion or evaporation of the water molecules.

When blended or doped with other functional materials, polymer MNFs can be used for highly fast and efficient optical sensing. As an example, in Ref. 22 a PANI/PS MNF was employed for gas sensing. When exposed to NO_2, the increase of the oxidation degree of PANI results in change of the absorption of the propagating light. The time-dependent absorbance of the MNF at room temperature to periodic NO_2/nitrogen exposure with NO_2 concentration changing between 0.1 and 4 ppm is given in Fig. 13.12c. This figure demonstrates good reversibility of the MNF response.

Since polymers can be doped and coated with a wide range of functionalized materials, polymer MNF optical sensors can be realized for detecting various other chemical and biological species.

13.4.5 Optical Fiber Surface MNF Sensor

References 17 and 18 developed a very accurate experimental method for sensing the optical fiber surface and, in particular, the fiber radius variation, which is important for understanding the range of diameter variation of the optical fiber, the surface contamination and adhesion. In addition, the optical fiber can be used as a substrate for sensing the absorbates disposed on its surface, e.g., thin films, microparticles, and biological species. The developed method is based on detecting of the WGM resonances in the optical fiber using an MNF moving along its surface. The experimental setup used in Refs. 17 and 18 is illustrated in Fig. 13.13a. It consists of a fiber under test and an MNF touching this fiber and aligned normally to it. The MNF was connected to a broadband light source at one end and to the OSA at the other. The measured WGM transmission spectrum is shown in Fig. 13.13b. From this spectrum, a well-pronounced resonance (circled in the inset of Fig. 13.13b) was chosen. The position of this resonance was monitored during the translation of the MNF along the fiber and used to determine the variation of the optical fiber effective diameter. In addition, the change of the broadband spectrum was used for more advanced sensoring of the optical fiber surface. The optical fiber effective diameter variation, Δd, was calculated from the displacement of the chosen resonance, $\Delta\lambda$, with a simple equation, $\Delta d = d\Delta\lambda/\lambda$, where d = 125 μm is the diameter of the fiber and λ =1.55 μm is the wavelength of the resonance. Figure 13.14a–c show the results of measurement of the fiber diameter variation with 250, 50, and 5 μm steps, respectively[18]. Each measurement was made twice in order to estimate the accuracy of the method. Small local spatial shifts between the two similar measurements are caused by the stick-slip problem. Nevertheless, Fig. 13.14 clearly demonstrates the achieved angstrom accuracy of measurement of the optical fiber diameter variations. More detailed information about the properties of the optical fiber surface can be obtained by treating of the broadband spectrum

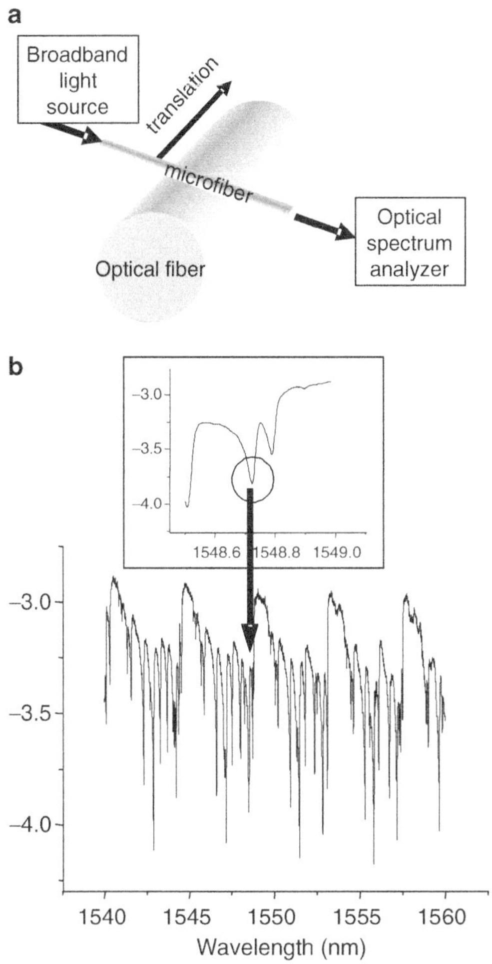

Fig. 13.13 (**a**) – Experimental setup. (**b**) – Transmission spectrum of an MNF touching the optical fiber; inset shows the spectral band and transmission dip (*circled*) used for measurement of the fiber diameter variation. Reprinted from Ref. 18 with permission. © 2008 Optical Society of America

of light transmitted through the microfiber. As an example, Fig. 13.15a shows the results of measurements of a segment of a contaminated optical fiber with 5 μm steps in the wavelength interval from 1,552.3 to 1,552.8 nm. In Fig. 13.15b, the same measurement is repeated. For comparison, Fig. 13.15c shows the spectrum variation at the 250 μm segment of a clean fiber in the wavelength interval from 1,948.5 to 1,549 nm. Figure 13.15a, b demonstrates that the developed method has a very good spatial resolution and clearly distinguishes features at the 5 μm scale. The developed method can be useful for the highly accurate measurement of the optical fiber diameter variation as well as for sensing the absorbates at the fiber surface.

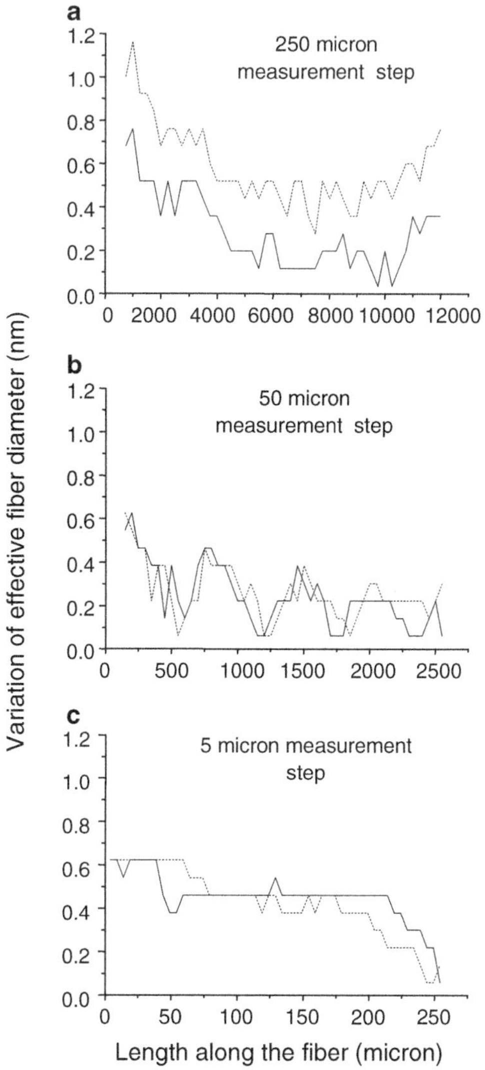

Fig. 13.14 Demonstration of angstrom accuracy and reproducibility of the measurement of the effective fiber diameter variation. (**a**) – measurement along the 12 mm segment with 250 μm steps. (**b**) – measurement along the 2.5 mm segment with 50 μm steps. (**b**) – measurement along the 250 μm segment with 5 μm steps. *Solid curve* – first measurement, *dashed curve* – second measurement. Reprinted from Ref. 18 with permission. © 2008 Optical Society of America

13.4.6 Atomic Fluorescence MNF Sensor

In Refs. 20 and 21, an optical MNF was suggested and demonstrated as a sensor of atomic fluorescence. The experimental setup used in Ref. 21 is illustrated in

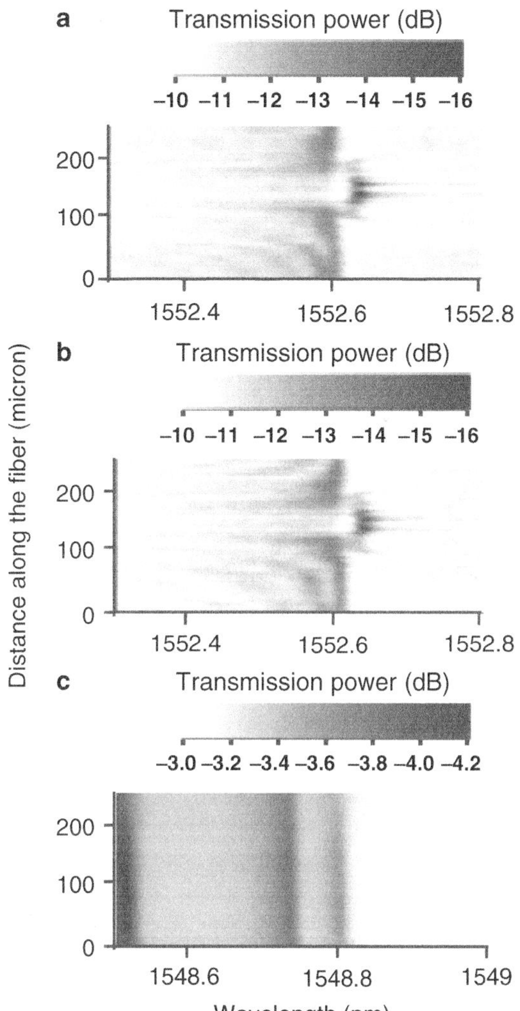

Fig. 13.15 Variation of the transmission power spectrum along the 250 μm segments of a dirty (**a**, **b**) and clean (**c**) optical fibers. (**a**) and (**b**) are two measurements of the same segment, which demonstrate the reproducibility of our measurements. Reprinted from Ref. 18 with permission. © 2008 Optical Society of America

Fig. 13.16a. As an atom source, a magneto-optical trap (MOT) for cold Cs-atoms was used. The fluorescence of MOT atoms around the MNF was detected by the measurement of fluorescence photons with an avalanche photodiode connected to one end of the fiber. Signals are accumulated and recorded on a PC using a photon-counting.

PC-board: It was shown that a very small number of atoms can be detected by monitoring through single-mode optical fiber under strong resonant laser

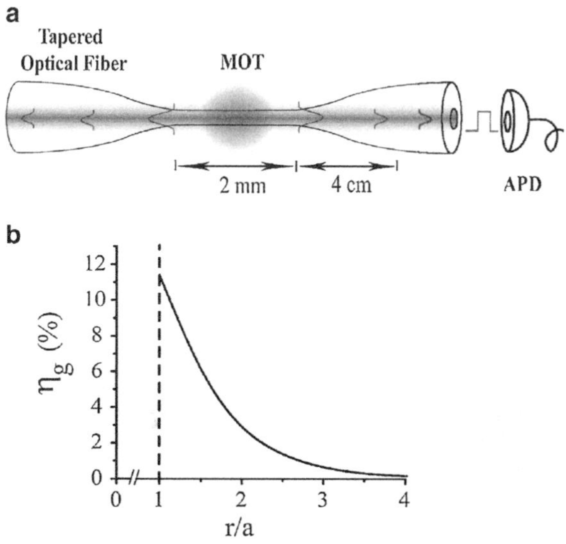

Fig. 13.16 (**a**) – Experiment on detection of atomic clouds with an MNF. APD is the avalanche photodiode. MOT is the magneto-optical trap. (**b**) – Coupling efficiency of spontaneous emission into each direction of nanofiber propagation mode, η_g, vs. atom position r/a, where r and a are distance from nanofiber axis and radius of nanofiber, respectively. Reprinted from Ref. 21 with permission. © 2008 Optical Society of America

irradiation. These results show that the optical MNF may serve as an efficient tool for quantum optics. The approach developed in Ref. 21 can naturally be extended to other systems than atoms, such as molecules or quantum dots.

13.5 Application of Looped and Coiled MNF for Sensing

The sensitivity of an MNF sensor can be significantly enhanced by employing interferometric measurement. Figure 13.1e shows an MLR sensor representing a Sagnac interferometer. Fabrication of the MLR consists of drawing a MF and bending it into a self-coupling loop. Two types of MLR have been demonstrated experimentally: a regular MLR illustrated in Fig. 13.1d[23] and a knot MLR, which has a twisted self-coupling region[24,25]. The MCR can be fabricated by wrapping an MNF on a central optical rod[27,30–33]. This section reviews application of an MLR for gas temperature sensing and an MCR for microfluidic sensing.

13.5.1 Ultra-Fast Direct Contact Gas Temperature Sensor

Narrow spectral resonances of an MLR are very sensitive to variations in the optical properties of the ambient medium. In addition, these parameters are changing with

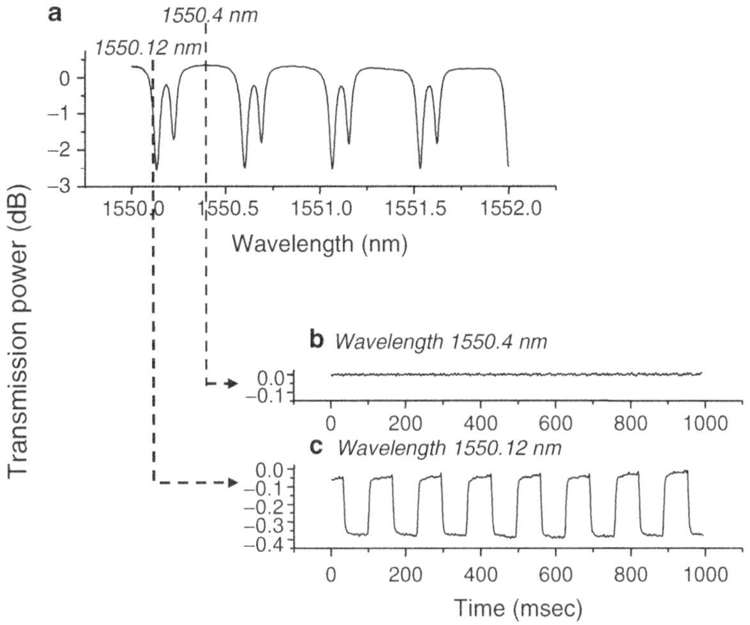

Fig. 13.17 (**a**) – transmission spectrum of MLR used as a temperature sensor. Transient behavior of MRL periodically heated by CO_2 laser at wavelength of (**b**) – 1550.4 nm (flat spectrum) and (**c**) – 1550.12 nm (steep spectrum). Reprinted from Ref. 23 with permission. © 2008 Institute of Electrical and Electronics Engineers

temperature, pressure, and applied radiation. An MLR positioned in free space has a much larger interfacial contact area with the environment than a microsphere and a microring resonator mounted on a substrate. Therefore, the MLR can perform as a more sensitive detector of environmental changes. In this subsection, the performance of an MLR as an ambient gas local temperature sensor[23] is briefly reviewed.

The sensitivity of an MLR is enhanced compared with the sensitivity of a straight MNF considered in Sect. 13.4 because the MNF loop performs as a resonator having very narrow dips in transmission spectrum. The position of these dips is very sensitive to changes in the ambient medium. As an example, Fig. 13.17a shows the transmission spectrum of an MLR, which was demonstrated in Ref. 23 as a fast direct gas temperature sensor and infrared radiation sensor. The resonator was periodically heated by a CO_2 laser beam transmitted through a beam chopper as illustrated in Fig. 13.18. Figure 13.17b, c shows the transmission power variation of this resonator at wavelengths corresponding to a flat region of spectrum (no effect) and to a resonance region of spectrum (strong oscillations). From the transmission power oscillations shown in Fig. 13.17c, the magnitude of temperature variation was estimated as 0.4 K, and the thermal equilibration time constant as 3 ms, in agreement with the theoretical prediction. For a regular optical fiber, which has a diameter 100 times greater, this constant is, proportionally, 0.3 s. The measured temperature range in this experiment was $\sim$1 K. The temperature

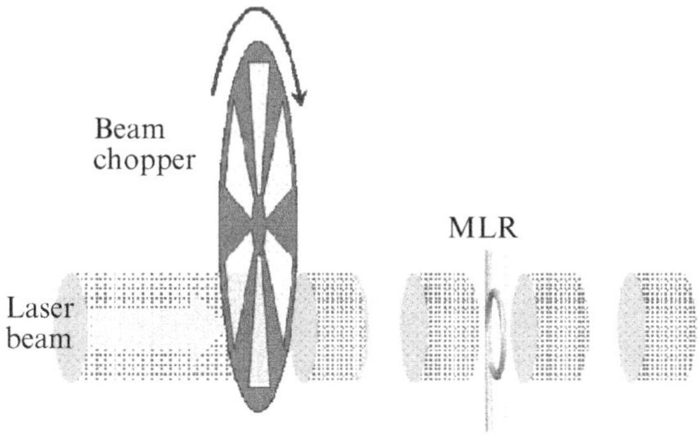

Fig. 13.18 Periodic on/off heating of a MLR using a CO_2 laser beam and a beam chopper. Reprinted from Ref. 23 with permission. © 2008 Institute of Electrical and Electronics Engineers

range can be increased with decreasing Q or choosing a region of the transmission spectrum having smaller steepness.

13.5.2 MCR Microfluidic Sensor

The fragileness of free MNF devices as, e.g., an MLR sensor considered in previous subsection restricts its applications. The design of an optical sensor by embedding an MNF into the curable liquid can increase the robustness of the device as well as broaden its functionality. Fabrication of a knot MLR embedded into a polymer matrix was considered[25]. The MCR wrapped on a supporting optical rod imbedded in low-index polymer was demonstrated in Refs. 27, 31, 33. The authors of Refs. 32 and 33 suggested and demonstrated the MCR microfluidic sensor that was created from an embedded MCR by removing its supporting rod. The MCR was fabricated by wrapping a 2.5-μm diameter MNF on a 1-mm diameter polymethylmethacrylate (PMMA) rod. The PMMA refractive index was in the range of 1.49–1.51. Then, the MCR was coated by the low-index Teflon solution 601S1–100–6 (DuPont, USA), which was dried in air and solidified. Next, the MCL was left soaking into acetone to dissolve the PMMA rod. Finally, the microfluidic MCR sensor with a ~1 mm diameter microchannel was fabricated. The sensor consisted of a MCR with five turns with a channel inside as illustrated in Fig. 13.19. The thickness of the Teflon film separating the MCR from the open channel was very small. The sensitivity of this device was measured by inserting the sensor in a beaker containing mixtures of isopropyl and methanol with ratios 60%, 61.5%, 63%, 64.3%, 65.5%, 66.7%, and 67.7%. Figure 13.20 shows the transmission spectra of the MCR recorded near 1,530 nm. It is seen that the resonance peak shifted toward longer wavelengths with growth of the refractive index. The calculated sensitivity of this MCR sensor was ~40 nm/RIU.

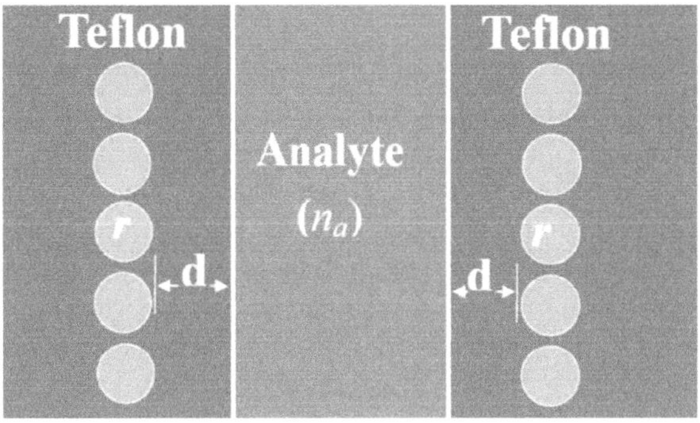

Fig. 13.19 MCR microfluidic sensor. Reprinted from Ref. 33 with permission. © 2008 American Institute of Physics

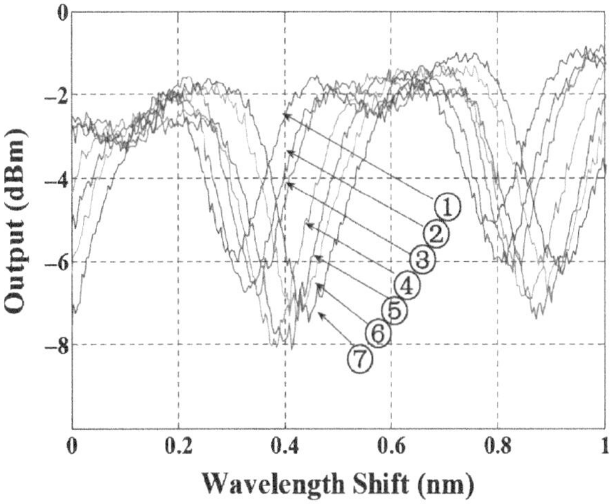

Fig. 13.20 Output spectrum of the MCR microfluidic sensor in mixtures of isopropyl of methanol. The isopropyl fraction is (1) – 60%, (2) – 61.5%, (3) – 63%, (4) – 64.3%, (5) – 65.5%, (6) – 66.7%, and (7) – 67.7%, respectively. Reprinted from Ref. 33 with permission. © 2008 American Institute of Physics

13.6 Resonant Photonic Sensors Using MNFs for Input and Output Connections

Assembling MNFs with other photonic elements can broaden their functionality and make them more robust. The simplest elements that can be joined with MNFs are microspheres, microdisks, microcylinder, and microcapillary resonators. This section briefly considers several designs of optical sensors where an MNF performs coupling of light into a microresonator, which probe the environment with excited WGMs. The principle of operation of these devices is similar. It is based on controlling the shift and width (generally, the shape) of transmission resonances, which respond to changes in the ambient medium.

13.6.1 MNF/Microsphere and MNF/Microdisk Sensor

An adiabatic optical fiber taper with MNF waist is convenient for coupling light into a microsphere or a microdisk and exciting the WGMs[34–49]. The MNF/microsphere and MNF/microdisk photonic devices, which are illustrated in Fig. 13.1g, h, have been suggested and demonstrated for evanescent sensing of various physical, chemical, and biological elements situated near or at the microsphere (microdisk) surface[34–49]. This subsection reviews application of microsphere and microdisk resonators for sensing of complex-valued refractive index of the ambient medium, of deposited nanolayer films, and of absorbed microparticles, molecules, and atoms.

13.6.1.1 Refractive Index Sensor

The performance of the microsphere and microdisk photonic sensors is similar to that of the MNF resonant sensors described in Sect. 13.5. The WGMs excited by an MNF are partly propagating in the ambient medium. For this reason, the resonant peaks in the transmission spectrum change in response to variation of the refractive index of the environment. In the simplest situation, the sample distribution is uniform in space surrounding the resonator. Then, the only parameter, which affects the resonant spectrum, is the refractive index of the sample. The narrow transmission resonance of a microsphere (microdisk), which is described by (13.11), has three parameters: λ_0, γ_a, and γ_c. These parameters are simply determined from (13.13) and (13.14) using the experimentally measured values of resonance peak position, Q-factor, and extinction ratio. Variation of these parameters with refractive index of the ambient medium can be calibrated with a known sample and then used for determination of the real and imaginary parts of the refractive index of unknown samples. For a small change of refractive index, Δn, variations of λ_0, γ_a and γ_c are proportional to Δ_n. As an example, Ref. 47 demonstrated very accurate determination of concentration of heavy water D_2O in H_2O with the MNF/microdisk resonator. It was noticed that the refractive indices of H_2O and D_2O are the same. For this reason, the resonant wavelength λ_0 does not depend on the D_2O concentration ($\Delta\lambda_0 = 0$). However, the absorption of H_2O is larger than that of D_2O and, therefore, the Q-factor of the resonator immersed in H_2O is smaller than when it is immersed in D_2O. The corresponding variation of parameter γ_a, and γ_c can be calibrated and used for determination of the D_2O concentration. The results of the D_2O concentration measurement are shown in Fig. 13.21. By monitoring Q, concentrations of 0.0001% of D_2O in H_2O have been detected. Reversibility of detection was demonstrated by periodic introduction and flushing of D_2O.

13.6.1.2 Nanolayer Sensor

The refractive index analysis considered in the previous subsection was based on monitoring of a single resonance in the transmission spectrum. The information

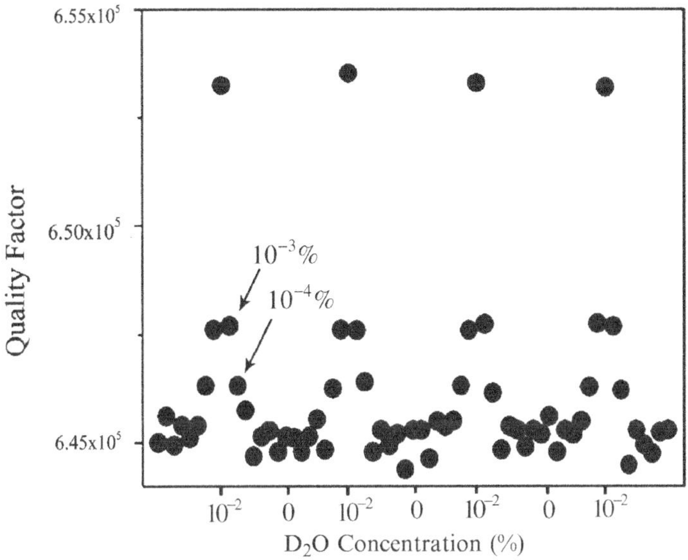

Fig. 13.21 Variation of the Q-factor in response to periodic change of the concentration of D_2O in H_2O. Reprinted from Ref. 47 with permission. © 2008 Optical Society of America

collected from a larger region of transmission spectrum allows to determine further details of the medium adjacent to the surface of the photonic microresonator. Generally, the transmission spectrum depends on wavelength λ and a set of parameters of the ambient medium. The dependence of transmission spectrum on these parameters can be determined by calibration measurements and theoretically. As an example, Ref. 36 suggested a method that allows simultaneous detection of the refractive index and thickness of a nanolayer adsorbed to the surface of a silica microsphere. As opposed to the case considered in the previous subsection, only the shifts of the resonant levels were taken into account. It was found that the shift $\Delta\lambda_0^{(j)}$ of the resonant level $\lambda_0^{(j)}$ due to the change Δ_n in the refractive index of the ambient medium, which has thickness t, can be estimated by the equation:

$$\frac{\Delta\lambda_0^{(j)}}{\lambda_0^{(j)}} = \frac{\lambda_0^{(j)} n_m}{2\pi R(n_s^2 - n_m^2)^{3/2}} \left[1 - \exp\left(-4\pi(n_s^2 - n_m^2)^{1/2} \frac{t}{\lambda_0^{(j)}}\right)\right] \Delta n. \tag{13.19}$$

Here n_5 and n_{m} are the refractive indices of the microsphere and ambient medium, respectively, and R is the microsphere radius. To determine Δ_{n} and t from this equation, it is sufficient to measure $\Delta\lambda_0^{(j)}$ for two wavelength, $\lambda_0^{(1)}$ and $\lambda_0^{(2)}$. In Ref. 36, this was done for $\lambda_0^{(1)} = 760\,\mathrm{nm}$ and $\lambda_0^{(1)} = 1,310\,\mathrm{nm}$. As the result the authors optically characterized a hydrogel nanolayer with 110-nm thickness and an extremely small excess refractive index of 0.0012, which was formed in situ in an aqueous environment.

13.6.1.3 Surface Sensor

The MNF/microsphere resonator can be applied for sensing of flat surfaces. The corresponding setup, which has been suggested and demonstrated in Ref. [42], is illustrated in Fig. 13.22a. It consists of an MNF positioned in direct contact with a microsphere contacting a sample surface. To enhance the response of the device to nonuniformities of the sample surface, the transmission power was recorded at the wavelength corresponding to the highly gradient slope of a resonant peak. The sample surface, which was a quartz diffraction grating, was scanned by the MNF/microsphere along the direction normal to the grooves. Variation of the transmission power, which is shown in Fig. 13.22b, exhibits oscillations, which characterizes the grooves. In this particular measurement, the segment with defected grooves is followed by the segment with good grooves. The micron resolution, which was achieved in Ref. [42], can be further enhanced by attaching a metal nanoparticle to the bottom of the microsphere. In doing so, an optical surface probe with subwavelength resolution, which is much more robust and efficient than the conventional ultra-small probes used in near-field microscopy, can be fabricated.

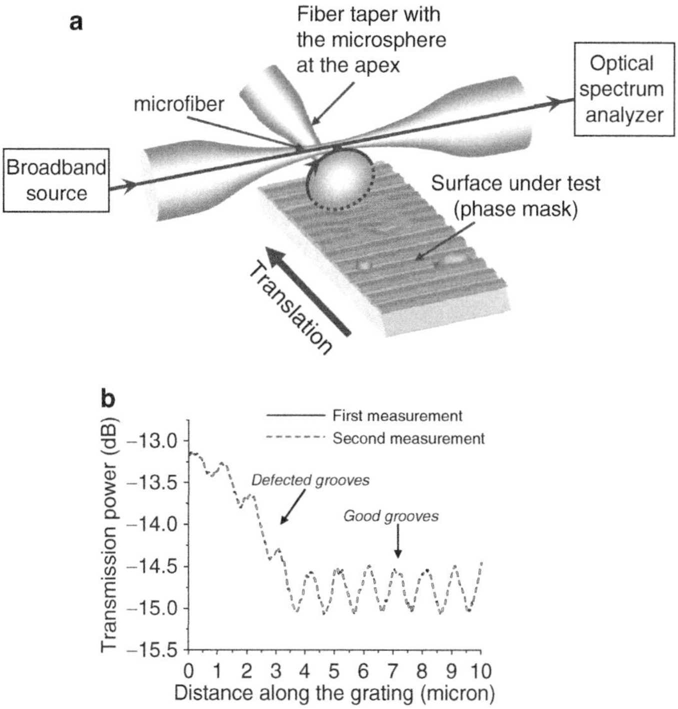

Fig. 13.22 (**a**) – MNF/microsphere tool for surface sensing. (**b**) – Transmission power dependence on the position of the microsphere at the diffraction grating surface. Comparison of two measurements of the same grating region shows the reproducibility of measurements. Reprinted from Ref. 42 with permission. © 2008 Optical Society of America

13.6.1.4 Individual Atom and Molecule Sensor

Very high Q-factor optical microresonators are extremely sensitive to changes in ambient medium. In Ref. 48, it was shown that coupling between individual caesium atoms and the fields of a high-quality toroidal microresonator, illustrated in Fig. 13.1h and in the inset in Fig. 13.23a, can be detected by monitoring the transmission spectrum of this resonator. As the result, the authors of Ref. 48 were able to detect the transit events for single atoms falling through the resonator's evanescent field. The MNF/microtoroid resonator was tuned to the condition of critical coupling, whose transmission spectrum is shown in Fig. 13.23a. Cold atoms were delivered to the vicinity of the resonator from a small cloud of caesium atoms cooled to $T < 10$ mK. The interaction of an individual atom with the evanescent field destroys the condition of critical coupling, leading to an increase in transmission power. Figure 13.23b shows records $C(t)$ for the number of single photon detection events within time bins of 2 ms as functions of time t. This figure

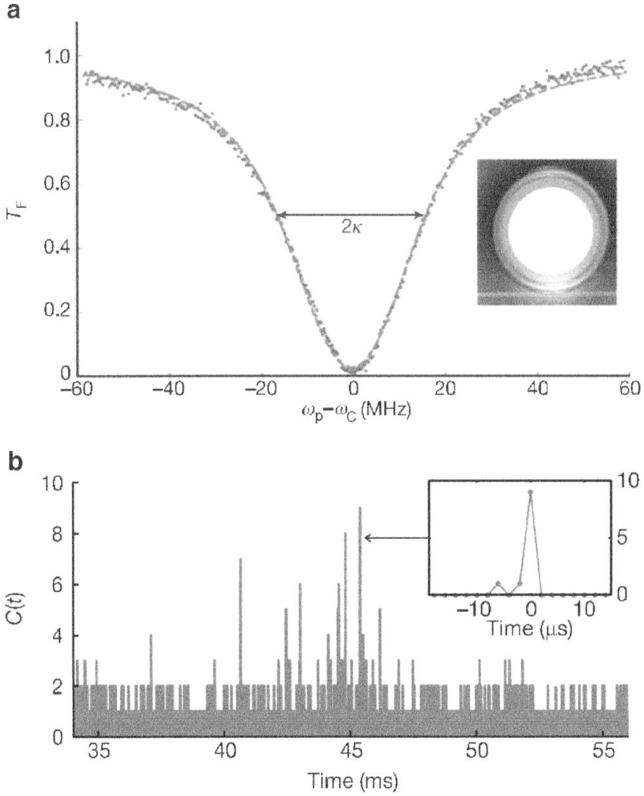

Fig. 13.23 (**a**) – Transmission power of microtoroid resonator near the condition of critical coupling (inset shows the MNF/microtoroid sensor). (**b**) – Single-photon counting events $C(t)$ as a function of time t after the release of the cold atom cloud at $t = 0$. Reprinted from Ref. 48 with permission. © 2008 Nature Publishing Group

clearly shows sharp peaks of duration < 2 ms that were attributed to the transit of a single atom through the resonant mode of the microtoroid.

In Ref. 49, a similar MNF/microtoroid sensor was used to demonstrate the single molecule adsorption at the surface of the toroidal microresonator. The effect was demonstrated with the interleukin-2 (IL-2) molecules that are instrumental in the body's natural response to microbial infection. By optimizing concentration and injection rate of solution of these molecules as well as acquisition rates, shifts caused by individual molecules binding to the surface of the microtoroid were resolved. A new mechanism of the effect of absorbed molecules on the transmission spectrum of the microtoroid was suggested. According to this mechanism, the high circulating field intensity within the resonator locally heats molecules. The corresponding temperature increase results in a red-shift of the resonant wavelength through the thermo-optic effect, when the microresonator material is heated by the molecule.

13.6.2 MNF/Microcylinder Sensor

A particular case of a MNF/microcylinder sensor has already been considered in Sect. 13.4.5, where an MNF was used for probing nonuniformities of an optical microcylinder (fiber). The same MNF/microcylinder configuration can be used for sensing of changes, which take place in the medium adjacent to the circumference of this cylinder contacting the MNF. The corresponding theory of transmission resonances was briefly reviewed in Sect. 13.3.3.

13.6.3 MNF/Microcapillary Sensor

An interesting approach for evanescent field sensing was suggested in Refs. 50 and 51. The authors suggested a liquid core WGM optical sensor, which consists of an optical capillary coupled to an MNF or another optical waveguide called liquid-core optical ring-resonator sensor (LCORRS). This device is illustrated in Fig. 13.24a. If the walls of the capillary are thin enough, the WGM transmission spectrum of the MNF can detect changes happening along the internal surface of the capillary. The LCORRS has been demonstrated as a very efficient tool for chemical and biological sensing[50–53]. The complete review of LCORRS is given in chapter … [the chapter of Fan et. al]. Here, we briefly review a specific realization of LCORRS suggested in Ref. 54, which exhibits the highest sensitivity and robustness. To increase the sensitivity, the wall of the capillary has to be decreased, which increases the fragility of this device. In addition, the waist of the optical fiber is prone to contamination and corrosion, which reduces the transmission power of the detected light. These problems are solved in Ref. 54 by immersing of the LCORRS into a curable low index polymer with refractive index ~1.38. After curing of the polymer, the taper and the capillary surfaces of the LCORRS, which is illustrated in Fig. 13.24b, were fully protected from corrosion and contamination. Furthermore,

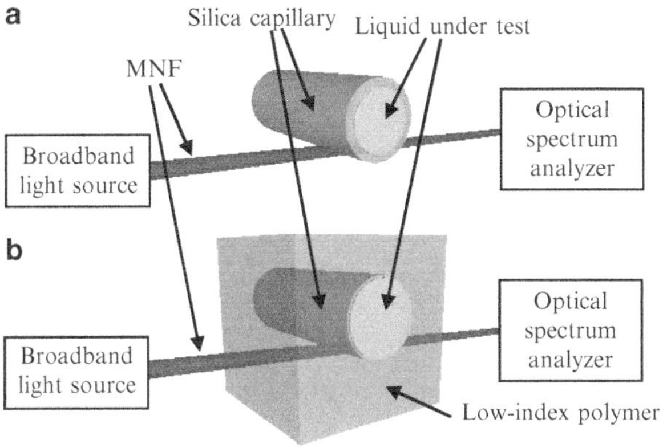

Fig. 13.24 (**a**) – LCORRS. (**b**) – Illustration of an LCORRS imbedded into a low-index polymer matrix

because the capillary was imbedded into the solid polymer matrix, the authors of Ref. 54 were able to etch the capillary wall down to submicron values and, eventually, fully eliminate it. In this sensor, a significant part of resonant WGMs is located inside the tested liquid (rather than propagates primarily along the capillary wall, as in the free-standing LCORRS), which considerably increases its sensitivity. The sensitivity of the device was measured with optical refractive index matching liquids. Figure 13.25a shows an example of measurement of the sensitivity. The spectrum of the LCORRS was recorded for close refractive indices $n_1 = 1.444$ and $n_2 = 1.444$, and the sensitivity found from the shift of the dips in these spectra, $\Delta\theta$, as $S = \Delta\lambda/\Delta n$. For the spectra depicted in Fig. 13.25a, $\Delta\lambda \approx 1$ nm and $\Delta n = n_2 - n_1 = 0.002$ resulting in $S \approx 500$ nm/RIU. The plot of sensitivity of the LCORRS is shown in Fig. 13.25b, curve 1. For the test liquids with larger refractive indices, $n_c > 1.5$, the sensitivity was as high as ~700 nm/RIU. For smaller indices, the sensitivity decreased and was ~25 nm/RIU at $n_c = 1.296$. To further increase the sensitivity of the LRROS, the microcapillary was completely removed by etching. The corresponding sensitivity of the LRROS is given by the curve 2 in Fig. 13.25b. This curve demonstrates the increase of the sensitivity at larger indices up to ~800 nm/RIU and also a significant growth of sensitivity at lower indices. The authors of Ref. 54 suggested that with replacement of the polymer matrix used in their experiment with lower index Teflon AF, which has the refractive index ~1.29, the sensitivity of the LCORRS can be further increased to the sensitivity given by the curve 3 in Fig. 13.25b.

13.6.4 Multiple-Cavity Sensors Supported by MNFs

MNFs, optical cylinders, disks, and microspheres can be assembled in more complex multifunctional sensing structures. One of the suggested structures is

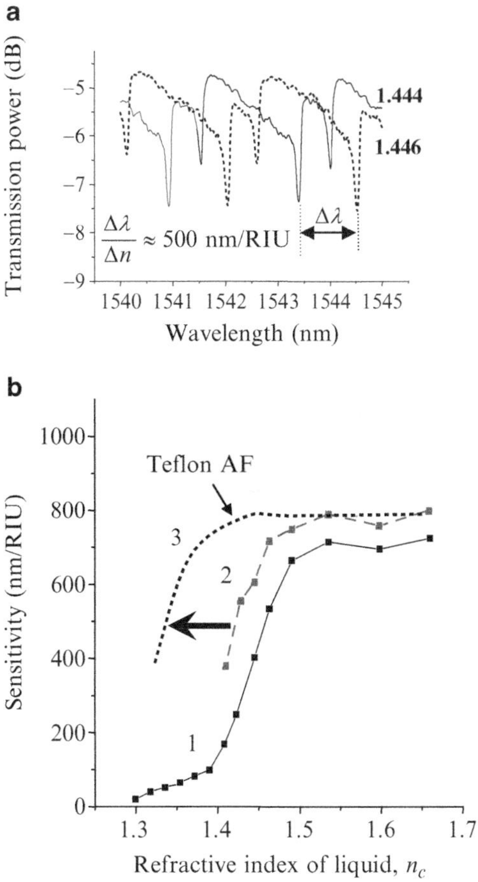

Fig. 13.25 (**a**) – Measurement of the LCORRS sensitivity. (**b**) – The sensitivity of the LCORRS measured when the silica microcapillary was not removed completely (*curve 1*) and for the removed microcapillary (*curve 2*). *Curve 3* is the sensitivity predicted for the Teflon AF matrix. Reprinted from Ref. 54 with permission. © 2008 Optical Society of America

illustrated in Fig. 13.26a. This is a matrix of parallel MNFs positioned in touch with parallel optical microcylinders. Each of the MNFs excites WGMs in each of the cylinders. The transmission spectrum of each MNF contains resonances from WGMs excited at the intersection of this MNF with each of the cylinders. If the cylinders are uniform in diameter, it is not possible to identify the changes in particular cylinders from the MNF transmission spectrum. Therefore, it is assumed that the diameters of the cylinders, d_i, are different from each other (which always happens in practice). Then the contributions of individual cylinders to the transmission spectrum are linear independent. A small change in the refractive index localized near an intersection of a certain cylinder with a certain MNF will result in a complex linear shift of the transmission spectrum of this cylinder. Because of

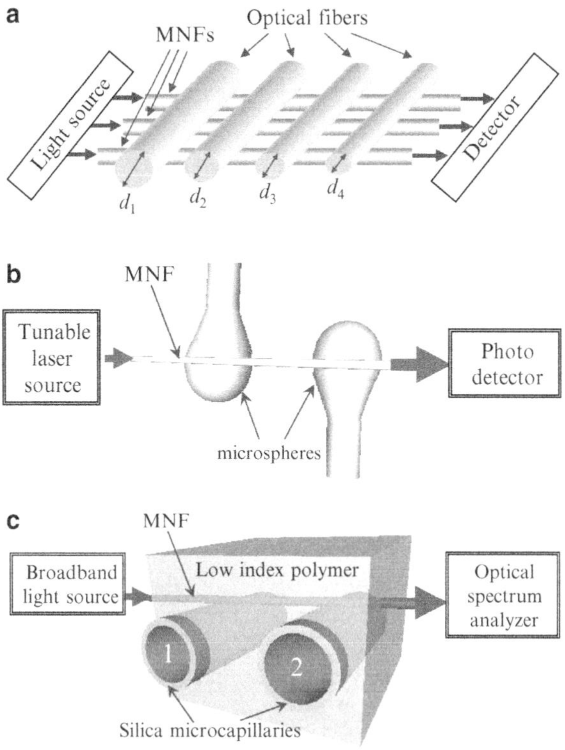

Fig. 13.26 (**a**) – Set of parallel MNFs and parallel microcylinders coupled to each other. (**b**) – Two microspheres coupled to an MNF (from Ref. 38). (**c**) – Two microcapillaries coupled to an MNF imbedded in a low index polymer. Reprinted from Ref. 69 with permission. © 2008 Optical Society of America

linear independency, the positions of local changes and their values can be separated and determined (see later). Obviously, the cylinders in Fig. 13.26a can be replaced by microcapillaries, microspheres, or microdisks.

If the resonances in the transmission spectrum, which are introduced by separate microcylinders (microspheres, microdisks, etc.), coupled to the same MNF are narrow and not overlapping, then the resonances of individual microcylinders can be separated. In this case, variations in transmission spectrum recorded from an individual MNF gives the information about changes happen locally at each of the microcylinders coupled to this MNF. A sensor based on this principle was demonstrated in Ref. [38] where the DNA detection was performed by using two microspheres connected to a single MNF illustrated in Fig. 13.26b. The multiplexed signal from two microspheres allowed the authors to discriminate a single nucleotide mismatch in an 11-mer oligonucleotide with a high signal-to-noise ratio of 54.

Generally, the transmission resonances of individual microresonators connected to a single input/output MNF can overlap and their separation may not be simple. However, in the case of relatively small changes, the separation of individual transmission spectra can be performed by the method suggested in Ref. 69. The transmission power of an optical microresonator sensor, $P(\lambda, n_1, n_2, ...n_K, T, p)$ depends on the wavelength, λ, refractive indices of components, n_k, temperature, T, and other parameters, e.g., the pressure inside the channels, p. If refractive index variations, $\Delta n_k = n_k - n_{0k}$, temperature variation, $\Delta T = T - T_0$, and pressure variations, $\Delta p = p - p_0$ are relatively small then the corresponding variation of $P(\lambda, n_1, n_2, ...n_K, T, p)$ is a linear function of Δn_{k}, ΔT, and Δp:

$$\begin{aligned} P(\lambda, n_1(T), n_2(T), ...n_K(T), T, p) - P(\lambda, n_{01}, n_{02}, ...n_{0K}, T_0, p_0) \\ = \sum_{k=1}^{K} Q_{n_k}(\lambda)\Delta n_k + Q_{\mathrm{T}}(\lambda)\Delta T + Q_{\mathrm{p}}(\lambda)\Delta p. \end{aligned} \tag{13.20}$$

Here functions $Q_{n_k}(\lambda)$, $Q_{\mathrm{T}}(\lambda)$, and $Q_{\mathrm{p}}(\lambda)$ can be determined experimentally using calibration samples. If these functions are linear independent then the parameters Δn_{k}, Δ_T, and Δ_p can be uniquely determined from the variation of $P(\lambda, n_1, n_2, ...n_K, T, p)$considered as a function of λ. In particular, the side effects, i.e., the temperature and pressure dependences, can be eliminated from the transmission spectrum. The sensing method based on this simple idea was applied in Ref. [69] for determination of microfluidic refractive index changes in two microcapillaries coupled to a single MNF illustrated in Fig. 13.26c. The developed approach allowed to compensate the side temperature and pressure variation effects.

13.7 Summary

This chapter reviewed applications of MNFs for optical sensing. It is shown that the MNF can be the basic element as well as a light input/output waveguide of miniature photonic sensors. One of the advantages of MNFs as optical sensors is their free-standing position enabling much larger contact with the environment compared with other types of evanescent waveguide sensors. In addition, large evanescent field, extremely small cross-sections and weight, very small losses, and availability of miniature MNF-based optical resonators, including MNF supported optical microresonators, allow the MNFs to perform detection of extremely small variations of physical, chemical, and biological properties of the ambient medium.

References

1 Kapany, N. S., High-resolution fibre optics using sub-micron multiple fibres, *Nature* **1959**, 184, 881–883

2 Kapany, N. S.; Burke, J. J., Fiber optics. IX. Waveguide effects, *J. Opt. Soc. Am.* **1961**, 51, 1067–1078

3 Mackenzie, H. S.; Payne, F. P., Evanescent field amplification in a tapered single-mode fibre, *Electron. Lett.* **1990**, 26, 130–132

4 Pendock, G. J.; MacKenzie, H. S.; Payne, F. P., Dye lasers using tapered optical fibers, *Appl. Opt.* **1993**, 32, 5236–5242

5 Knight, J. C.; Cheung, G.; Jacques, F.; Birks, T. A., Phase-matched excitation of whispering-gallery mode resonances by a fiber taper, *Opt. Lett.* **1997**, 22, 1129–1131

6 Birks, T. A.; Wadsworth, W. J.; Russell, P. St. J., Supercontinuum generation in tapered fibers, *Opt. Lett.* **2000**, 25, 1415–1417

7 Tong, L.; Gattass, R. R.; Ashcom, J. B.; He, S.; Lou, J.; Shen, M.; Maxwell, I.; Mazur, E., Subwavelength-diameter silica wires for low-loss optical wave guiding, *Nature* **2003**, 426, 816–819

8 Sumetsky, M., Optical fiber microcoil resonators, *Opt. Express* **2004**, 12, 2303–2316

9 Leon-Saval, S. G.; Birks, T. A.; Wadsworth, W. J.; Russell, P. S. J.; Mason, M. W., Supercontinuum generation in submicron fibre waveguides, *Opt. Express* 2004, 12, 2864–2869

10 Brambilla, G.; Finazzi, V.; Richardson, D. J.; Ultra-low-loss optical fiber nanotapers, *Opt. Express* 2004, 12, 2258–2263

11 Sumetsky, M., How thin can a microfiber be and still guide light?, *Opt. Lett.* **2006**, 31, 870–872

12 Sumetsky, M., How thin can a microfiber be and still guide light? Errata, *Opt. Lett.* **2006**, 31, 3577–3578

13 Sumetsky, M., Optics of tunneling from adiabatic nanotapers, *Opt. Lett.* **2006**, 31, 3420–3422

14 Novotny, L.; Hecht, B., Principles of Nano-Optics, Cambridge University Press, Cambridge, 2006.

15 Villatoro, J.; Monzón-Hernández, D., Fast detection of hydrogen with nano fiber tapers coated with ultra thin palladium layers, *Opt. Express* **2005**, 13, 5087–5092

16 Polynkin, P.; Polynkin, A.; Peyghambarian, N.; Mansuripur, M., Evanescent field-based optical fiber sensing device for measuring the refractive index of liquids in microfluidic channels, *Opt. Lett.* **2005**, 30, 1273–1275

17 Birks, T. A.; Knight, J. C.; Dimmick, T. E., High-resolution measurement of the fiber diameter variations using whispering gallery modes and no optical alignment, *IEEE Photon. Technol. Lett.* **2000**, 12, 182–184

18 Sumetsky, M.; Dulashko, Y., Sensing an optical fiber surface by a microfiber with angstrom accuracy, In Optical Fiber Communication conference, Anaheim, 2006

19 Warken, F.; Vetsch, E.; Meschede, D.; Sokolowski, M.; Rauschenbeutel, A., Ultra-sensitive surface absorption spectroscopy using sub-wavelength diameter optical fibers, *Opt. Express* **2007**, 15, 11952–11958

20 Balykin, V. I.; Hakuta, K.; Kien, F. L.; Liang, J. Q.; Morinagal, M., Atom trapping and guiding with a subwavelength-diameter optical fiber, *Phys. Rev.* **2004**, A 70, 011401

21 Nayak, K. P.; Melentiev, P. N.; Morinaga, M.; Kien, F. L.; Balykin, V. I.; Hakuta, K., Optical nanofiber as an efficient tool for manipulating and probing atomic Fluorescence, *Opt. Express* **2007**, 15, 5431–5438

22 Gu, F.; Zhang, L.; Yin, X.; Tong, L., Polymer single-nanowire optical sensors, *Nano Lett.* **2008**, 8, 2757–2761

23 Sumetsky, M.; Dulashko, Y.; Fini, J. M.; Hale, A.; DiGiovanni, D. J., The microfiber loop resonator: Theory, experiment, and application, *IEEE J. Lightwave Technol.* **2006**, 24, 242–250

24 Jiang, X.; Tong, L.; Vienne, G.; Guo, X.; Tsao, A.; Yang, Q.; Yang, D., Demonstration of optical microfiber knot resonators, *Appl. Phys. Lett.* **2006**, 88, 223501

25 Vienne, G.; Li, Y.; Tong, L., Microfiber resonator in polymer matrix, *IEICE Trans. Electron.* **2007**, E90, 415–421

26 Sumetsky, M., Uniform coil optical resonator and waveguide: transmission spectrum, eigenmodes, and dispersion relation, *Opt. Express* **2005**, 13, 4331–4340

27 Sumetsky, M., Basic elements for microfiber photonics: micro/nanofibers and microfiber coil resonators, *IEEE J. Lightwave Technol.* **2008**, 26, 21–27

28 Xu, F.; Horak, P.; Brambilla, G., Conical and biconical ultra-high-Q optical-fiber nanowire microcoil resonator, *Appl. Opt.* **2007**, 46, 570–573
29 Xu, F.; Horak, P.; Brambilla, G., Optimized design of microcoil resonators, *J. Lightwave Technol.* **2007**, 25, 1561–1567
30 Fei, X.; Brambilla, G., Manufacture of 3-D microfiber coil resonators, *IEEE Photon. Technol. Lett.*, **2007**, 19, 1481–1483
31 Xu, F.; Brambilla, G., Embedding optical microfiber coil resonators in Teflon, *Opt. Lett.* **2007**, 32, 2164–2166
32 Xu, F.; Horak, P.; Brambilla, G., Optical microfiber coil resonator refractometric sensor, *Opt. Express* **2007**, 15, 7888–7893
33 Xu, F.; Brambilla, G., Demonstration of a refractometric sensor based on optical microfiber coil resonator, *Appl. Phys. Lett.* **2008**, 92, 101126
34 Vahala, K. J., Optical microcavities, *Nature* **2003**, 424, 839–846
35 Vollmer, F.; Braun, D.; Libchaber, A., Protein detection by optical shift of a resonant microcavity, *Appl. Phys. Lett.* **2002**, 80, 4057–4059
36 Noto, M.; Vollmer, F.; Keng, D.; Teraoka, I.; Arnold, S., Nanolayer characterization through wavelength multiplexing of a microsphere resonator, *Opt. Lett.* **2005**, 30, 510–512
37 Arnold, S.; Khoshsima, M.; Teraoka, I.; Holler, S.; Vollmer, F., Shift of whispering-gallery modes in microspheres by protein adsorption, *Opt. Lett.* **2003**, 28, 272–274
38 Vollmer, F.; Arnold, S.; Braun, D.; Teraoka, I.; Libchaber, A., Multiplexed DNA quantification by spectroscopic shift of two microsphere cavities, *Biophys. J.* **2003**, 85 1974–1979
39 Teraoka, I.; Arnold, S.; Vollmer, F., Perturbation approach to resonance shifts of whispering-gallery modes in a dielectric microsphere as a probe of a surrounding medium, *J. Opt. Soc. Am. B* **2003**, 20, 1937–1946
40 Hanumegowda, N. M.; Stica, C. J.; Patel, B. C.; White, I. M.; Fan, X., Refractometric sensors based on microsphere resonators, *Appl. Phys. Lett.* **2005**, 87, 201107
41 White, I. M.; Hanumegowda, N. M.; Fan, X., Subfemtomole detection of small molecules with microsphere sensors, *Opt. Lett.* **2005**, 30, 3189–3191
42 Sumetsky, M.; Dulashko, Y.; DiGiovanni, D. J., Optical surface microscopy with a moving microsphere, In Nanophotonics Topical Meeting, OSA, Uncasville, 2006
43 Ren, H.; Vollmer, F.; Arnold, S.; Libchaber, A., High-Q microsphere biosensor - analysis for adsorption of rodlike bacteria, *Opt. Express* **2007**, 15, 17410–17423
44 Matsko, A. B.; Ilchenko, V. S., Optical resonators with whispering-gallery modes-part I: basics, *IEEE J. Sel. Top. Quantum Electron.* **2006**, 12, 3–14
45 Ilchenko, V.S.; Matsko, A.B., Optical resonators with whispering-gallery modes-part II: applications, *IEEE J. Sel. Top. Quantum Electron.* **2006**, 12, 15–32
46 Armani, D. K.; Kippenberg, T. J.; Spillane, S. M.; Vahala, K. J., Ultra-high-Q toroid microcavity on a chip, *Nature* **2003**, 421, 925–928
47 Armani, A. M.; Vahala, K. J., Heavy water detection using ultra-high-Q microcavities, *Opt. Lett.* **2006**, 31, 1896–1898
48 Aoki, T.; Dayan, B.; Wilcut, E.; Bowen, W. P.; Parkins, A. S.; Kippenberg, T. J.; Vahala, K. J.; Kimble, H. J., Observation of strong coupling between one atom and a monolithic microresonator, *Nature* **2006**, 443, 671–674
49 Armani, A. M.; Kulkarni, R. P.; Fraser, S. E.; Flagan, R. C.; Vahala, K. J., Label-free, single-molecule detection with optical microcavities, *Science* **2007**, 317, 783–787
50 White, I. M.; Oveys, H.; Fan, X., Liquid core optical ring resonator sensors, *Opt. Lett.* **2006**, 31, 1319–1321
51 White, I. M.; Oveys, H.; Fan, X.; Smith, T. L.; Zhang, J., Integrated multiplexed biosensors based on liquid core optical ring resonators and antiresonant reflecting optical waveguides, *Appl. Phys. Lett.* **2006**, 89, 191106
52 Fan, X.; White, I. M.; Zhu, H.; Suter, J. D.; Oveys, H., Overview of novel integrated optical ring resonator bio/chemical sensors, *Proc. SPIE* **2007**, 6452, 1–20

53 Zhu, H.; White, I. M.; Suter, J. D.; Dale, P. S.; Fan, X., Analysis of biomolecule detection with optofluidic ring resonator sensors, *Opt. Express* **2007**, 15, 9139–9146
54 Sumetsky, M.; Windeler, R. S.; Dulashko, Y.; Fan, X., Optical liquid ring resonator sensor, *Opt. Express* **2007**, 15, 14376–14381
55 Paul, P. H.; Kychakoff, G., Fiber-optic evanescent field absorption sensor, *Appl. Phys. Lett.* **1987**, 51, 12–14
56 Piraud, C.; Mwarania, E. K.; Yao, J.; O'Dwyer, K.; Schiffrin, D. J.; Wilkinson, J. S., Optoelectrochemical transduction on planar optical waveguides, *J. Lightwave Technol.* **1992**, 10, 693–699
57 Ruddy, V.; MacCraith, B. D.; Murphy, J. A., Evanescent wave absorption spectroscopy using multimode fibers, *J. Appl. Phys.* **1990**, 67, 6070–6074
58 James, S. W.; Tatam, R. P., Fibre optic sensors with nano-structured coatings, *J. Opt. A: Pure Appl. Opt.* **2006**, 8, S430–S444
59 Snyder, A. W.; Love, J. D., *Optical Waveguide Theory*, Kluwer Academic Publishers, Norwell, 2000
60 Tong, L.; Lou, J.; Mazur, E.; Single-mode guiding properties of subwavelength-diameter silica and silicon wire waveguides, *Opt. Express* **2004**, 12, 1025–1035
61 Le Kien, F.; Liang, J. Q.; Hakuta, K.; Balykin, V. I., Field intensity distributions and polarization orientations in a vacuum-clad subwavelength-diameter optical fiber, *Opt. Commun.* **2004**, 242, 445–456
62 Tong, L. M.; Hu, L. L.; Zhang, J. J.; Qiu, J. R.; Yang, Q.; Lou, J. Y.; Shen, Y. H.; He, J. L.; Ye, Z. Z., Photonic nanowires directly drawn from bulk glasses, *Opt. Express* **2006**, 14, 82–87
63 Sumetsky, M.; Dulashko, Y.; Hale, A., Fabrication and study of bent and coiled free silica nanowires: Self-coupling microloop optical interferometer, *Opt. Express* **2004**, 12, 3521–3531
64 Sumetsky, M.; Dulashko, Y.; Domachuk, P.; Eggleton, B. J., Thinnest optical waveguide: Experimental test, *Opt. Lett.* **2007**, 32, 754–756
65 Chen, H. W.; Li, Y. T.; Pan, C. L.; Kuo, J. L.; Lu, J. Y.; Chen, L. J.; Sun, C. K., Investigation on spectral loss characteristics of subwavelength terahertz fibers, *Opt. Lett.* **2007**, 32, 1017–1019
66 Heebner, J.; Grover, R.; Ibrahim, T., Optical Micro-Resonators: Theory, Fabrication, and Applications, Springer-Verlag, New York, 2007.
67 Sumetsky, M., Optimization of optical ring resonator devices for sensing applications, *Opt. Lett.* **2007**, 32, 2577–2579
68 Sumetsky, M., Optimization of resonant optical sensors, *Opt. Express* **2007**, 15, 17449–17457
69 Sumetsky, M.; Dulashko, Y.; Windeler, R. S., Temperature and pressure compensated microfluidic optical Sensor, In *Conference on lasers and electro-optics/quantum electronics and laser science conference and photonic applications systems technologies*, OSA Technical Digest (CD), Optical Society of America, 2008, paper CMJ6

Chapter 14
Label-Free Biosensing with the Optofluidic Ring Resonator

Ian M. White, Hongying Zhu, Jonathan D. Suter, and Xudong Fan

Abstract The optofluidic ring resonator (OFRR) is a label-free refractive index biosensing platform that uses a glass capillary as a microfluidic sample delivery mechanism with an inherently integrated optical ring resonator. The device delivers molecular detection limits on par with the top-performing commercial label-free sensing technologies while reducing cost and complexity compared with fluorescence-based biosensing techniques as well as other label-free devices. The OFRR has been shown to quantify target molecules in a sample through specific biomolecular recognition events at the surface of the capillary. Examples presented here include protein detection, DNA sequence detection, and pathogen detection. The design of the OFRR makes it well-suited for integration into a lab-on-a-chip platform that will have significant implications for the diagnosis and understanding of disease.

14.1 OFRR Label-Free Refractive Index Biosensor

Modern science and medicine have made great strides in the previous decade in the diagnosis and study of disease. Nonetheless, early disease detection capabilities must be improved. Also, new tools are needed for the study of diseases such as cancer, Alzheimer's, heart disease, and many others to enable greater progress toward cure and prevention. Today, the field of optical biosensing is poised to develop tools that will enable earlier diagnosis and that will allow scientists to better study diseases at the molecular level, leading to the development of cures and prevention methods.

Photonics technologies are used widely today for applications in disease diagnosis at the molecular level, as well as in molecular biology studies of disease. Often, the sensing technique involves the detection of photons emitted by

X. Fan (✉)
Department of Biological Engineering, University of Missouri, Columbia, MO 65211, USA
e-mail: fanxud@missouri.edu

X. Fan (ed.), *Advanced Photonic Structures for Biological and Chemical Detection*, Integrated Analytical Systems,
DOI 10.1007/978-0-387-98063-8_14,

luminescent molecules indicating the presence of the target analyte. Techniques utilizing luminescence include enzyme-linked immunosorbent assay (ELISA), immuno-histochemistry, flow cytometry, immunoblots, and DNA microarrays. By now, luminescence-based detection is a well-understood and reliable technique that offers incredible potential for sensitivity. In fact, single-molecule studies often use fluorescence-based detection mechanisms.

However, luminescence-based detection techniques often require a high number of steps. Consider ELISA as an example. As a first step, the sample is introduced into a 96-well plate; an antibody targeting the antigen of interest has been immobilized to the wells of the plate. After a rinse, the wells contain the antibody and any bound antigen. However, although the antigen has been isolated, the protocol is nowhere near completion. The remaining steps include another antibody (different from the first) to form a sandwich assay, a secondary antibody with an enzymatic label, and a substrate that is luminescent when activated by the enzyme. Finally, the sample is analyzed by relatively expensive detection optics to determine the amount of analyte that was captured in the assay. The steps are illustrated in Fig. 14.1a.

The elimination of these extra steps and of the costly optics may result in low-cost, high-throughput photonics-based detection. A "label-free" photonics-based technique involves refractive index (RI) sensing as the molecular detection technique. A number of RI sensing techniques have been developed over the last 15 years[1]. Surface plasmon resonance (SPR) is a RI sensing technique that has already achieved some degree of commercial success. However, although commercialized SPR has set the benchmark for biomolecule detection limits, it has not fulfilled the

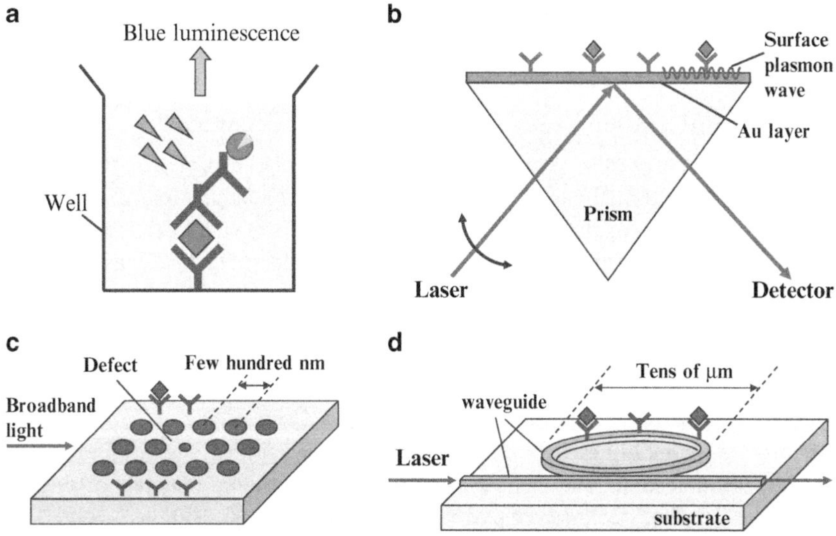

Fig. 14.1 (**a**) ELISA configuration. An antibody coated to the well captures the analyte; then another antibody labels the analyte and an enzyme-labeled secondary antibody labels the system. (**b**) SPR prism-coupling configuration. Molecules captured at the gold surface change the resonant coupling angle. (**c**) Photonic crystal sensing chip. (**d**) Waveguide-based ring resonator sensing chip

promise of low-cost, high-throughput sensing. As a result, further research and development into SPR and alternative RI sensing mechanisms is necessary.

In SPR biosensing[2–4], a thin gold layer acts as the sensing surface. In one popular configuration, shown in Fig. 14.1b, a thin gold layer on top of a prism is functionalized with biomolecular layers, including biorecognition molecules. When light enters the prism at a certain angle, it couples into a surface plasmon wave at the top of the gold surface. An evanescent field extends beyond the surface and interacts with the molecules at the surface. When analytes are captured by the biorecognition molecules, the RI at the surface changes, and as a result, the critical angle of coupling changes. The change in angle can be monitored in real time and is used as the sensing signal. Commercial SPR units generally have a RI detection limit of 10^{-7} refractive index units (RIU). One hot area of research and development is the technique of SPR imaging (SPRI), in which a large area of the sensor surface is probed by an expanded beam, and a CCD array is used to analyze discretized sensing regions simultaneously[5]. This may enable SPR technology to be applied to high-throughput applications.

Integrated interferometry techniques may offer a lower cost biosensing tool because they generally do not require the mechanically rotating optical source (as commercial SPR tools do) or expensive spectral analysis techniques. Interferometry configurations used for biosensing include the Mach-Zehnder interferometer[6] and the Young's interferometer[7]. RI detection limits of 10^{-7} RIU have been demonstrated for both of these configurations. One drawback of these configurations when compared with other types of sensors is that high sensitivity requires a long interaction length between the light in the interferometer arms and the sample. Thus, the size of each sensing device can be measured in centimeters. More recently, backscattering interferometry configurations have shown excellent RI detection limits without the need for a large sensor area[8,9].

Photonic crystal structures are an emerging technology for RI-based biosensing[10]. These structures are periodic crystal structures with a single defect manufactured into a substrate, as illustrated in Fig. 14.1c. The structures are designed to resonate at a particular wavelength; the transmission or reflection spectrum is measured to determine the exact resonant wavelength. Just as in SPR, when the RI at the surface changes, such as when biomolecules are captured at the surface, the resonant condition changes, and the shift in the measured resonant wavelength acts as the sensing signal. Although the detection limit of these devices has not yet caught up with other technologies, photonic crystal sensors are today an interesting approach that may in the future offer many advantages for integration into lab-on-a-chip devices.

Meanwhile, optical ring resonators are currently aggressively studied for biosensing applications. Figure 14.1d shows an example of a ring resonator structure. Light is coupled into the ring, where it is trapped and guided around the ring repeatedly. A resonant condition exists for frequencies of light that exactly match the optical circumference around the ring. The resonating wavelength can be observed and measured quite easily in real time. As with the other RI sensors described earlier, an evanescent field exists just beyond the surface of the ring resonator. When the RI changes near the surface of the resonator, such as when biomolecules bind at the surface, the resonant frequency changes due to an effective change in the optical circumference. This change of resonant wavelength can be measured in real time to generate a sensing signal. Using this ring resonator sensing concept, a number of biosensing demonstrations have been achieved[11–20].

Ring resonator sensors have been implemented in a number of different forms. The device illustrated in Fig. 14.1d is a planar waveguide ring resonator. These sensors can be produced with conventional planar waveguide manufacturing techniques[18] or with soft lithography[15]. This provides an advantage in terms of sensor integration and mass production. However, the resonant quality factor (Q-factor) is much lower (10^4–10^5) when compared with other techniques. Low Q-factor has a detrimental effect on the detection limit because detection resolution is compromised[21]. Also, a lower Q-factor results in the need for higher-cost photonics to measure the broad linewidth, as explained later in Sect. 14.3. Microsphere[12,14] and microtoroid[22] ring resonators have also been demonstrated for sensing. Q-factors range from 10^6–10^8, which enables excellent detection resolution. However, microsphere resonators are difficult to mass-produce and difficult to integrate into a sensing platform with an excitation waveguide and microfluidics. Microtoroids can be mass-produced with semiconductor processing techniques, but the device requires an externally mounted fiber taper, which is a drawback in terms of integration and robustness. Also, integrating external microfluidics with the sensing device may pose challenges.

The optofluidic ring resonator (OFRR) was designed to achieve the high resonating performance of microsphere resonators but to offer a more practical solution in terms of microfluidics[23–25]. The OFRR uses a silica capillary as the sample delivery mechanism and as a ring-resonator-based RI sensor. The sensing concept is illustrated in Fig. 14.2. Laser light is coupled into the wall of the capillary perpendicular to the longitudinal axis where it continually circulates due to the guiding capability of the glass wall. This creates a resonant condition for certain optical frequencies based on the length of the optical path around the capillary. If the wall of the capillary is approximately as thin as a few wavelengths, then an evanescent field will exist at the inner surface of the capillary. Changes in RI within

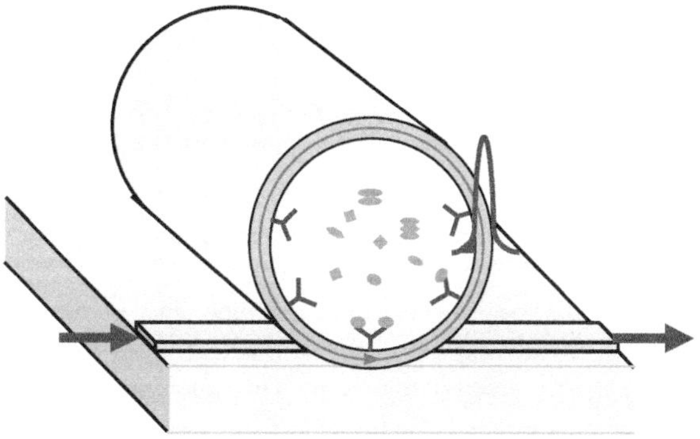

Fig. 14.2 OFRR concept: The capillary cross section is the ring resonator; the biological sample flows through the center. The capture of biomolecules at the inner surface causes a change in resonance frequency

about 100 nm of the inner surface of the capillary cause a change in the effective optical path length around the capillary circumference, and thus the resonant wavelength changes. This change of resonant wavelength can be measured in real time to generate a sensing signal.

To perform biosensing, biorecognition molecules (e.g., antibodies, single-stranded oligonucleotides, etc.) are immobilized onto the inner surface of the capillary. As the sample is passed through the capillary, target analytes bind to the biorecognition molecules. Any molecules located within the evanescent field will change the RI when compared with the liquid buffer, and thus will cause the resonant optical mode to shift in frequency. After rinsing the inside of the capillary, specifically captured analytes remain bound at the surface, causing a net frequency shift of the resonant mode. The magnitude of the shift is related to the number of biomolecules captured within the evanescent field of the resonant mode.

The OFRR has a number of advantages compared with other sensing technologies. First, the microfluidics channel is the sensor, which results in a robust, low-cost design. Second, the silica capillary maintains a high Q-factor compared with planar ring resonators; the high Q-factor enables the use of low-cost photonics because of the low bandwidth requirement. Also, the small-diameter capillary design has a high sensor-surface-area to sample-volume ratio, which increases capture efficiency and decreases assay time. For example, compared with a 200 μL well in a 96-well plate, an OFRR with a 50 μm diameter has a surface-area to volume ratio 100 times larger. This also implies that only a very small sample/reagent volume is consumed in an assay, which saves costs.

14.2 Theoretical Analysis of Biosensing with the OFRR

The sensitivity of the resonant mode frequency to biomolecule binding events at the inner surface of the capillary can be evaluated theoretically. As described earlier, the spectral position of the resonant mode in the OFRR is dependent upon the effective RI experienced by the resonant mode. If the RI of the fluid inside the core of the capillary changes, the resonant mode will shift spectrally. The sensitivity of the resonant mode's position with respect to the RI of the fluid in the capillary is referred to as the bulk refractive index sensitivity (BRIS). A BRIS of 800 nm/RIU has been experimentally demonstrated using a capillary-based ring resonator[26]. Intuitively, the biomolecule detection limit of the OFRR is directly related to the BRIS, while the BRIS is proportional to the fraction of the resonant mode contained in the evanescent field inside the surface of the capillary. The relationship between the BRIS and the biomolecule detection limit has been developed rigorously in Zhu et al.[27] Under the assumption of a longitudinally and radially confined optical mode, the resonant mode wavelength, λ, can be written as:

$$2\pi n_{\mathrm{eff}} R = l \cdot \lambda, \tag{14.1}$$

where n_{eff} is the effective RI experienced by the optical mode, R is the resonant mode radius, and l is the resonant mode angular momentum, which must be an integer. In plain terms, if the wave travels an integer number of wavelengths along the optical path before returning to where the path begins, then the optical mode resonates. As an approximation, the effective RI can be written as a weighted function of the refractive indices of the sample in the core, the capillary wall, and the surrounding medium, where the fraction of the optical mode in the respective medium is the weighting factor. The resonant wavelength can then be written as:

$$2\pi R \cdot [n_1\eta_1 + n_3\eta_3 + n_2(1 - \eta_1 - \eta_3)] = l \cdot \lambda, \tag{14.2}$$

where η_1 and η_3 are the fraction of the light energy beyond the outer surface and beyond the inner surface, respectively. The BRIS is defined as the change in the resonant mode's spectral position per change in RI in the core, and can be found by taking the derivative of (14.2):

$$S = \left(\frac{\delta\lambda}{\delta n_3}\right) = \frac{2\pi R}{l}\eta_3 \approx \frac{\lambda}{n_2}\eta_3. \tag{14.3}$$

Equation (14.3) uses the approximation $\lambda = 2\pi n_2 R/l$, as most of the light is contained within the capillary wall. This result verifies that the BRIS is directly proportional to the fraction of the resonant mode's intensity that exists beyond the inner surface of the capillary. In general, a thinner capillary wall results in a higher evanescent fraction at the inner surface of the capillary. Therefore, as a general rule, a thinner capillary wall yields a more highly sensitive device. A model has been developed using Mie theory to calculate the mode fraction for a given capillary thickness[23]. The mode fraction interacting with the sample is also dependent on the radial order of the resonant mode. Similar to multimode fiber optic cable, a number of propagating modes exist along the radial axis. Generally, higher order radial modes have a higher fraction of evanescent intensity at the inner surface of the capillary, and thus a higher sensitivity[23].

The OFRR's sensitivity to biomolecules can be analyzed as a function of the BRIS. The evanescent field in the core decays exponentially and thus η_3 can be calculated as:

$$\eta_3 = \frac{2\pi h \times \int_0^R \varepsilon_0\varepsilon_3 |E_0|^2 \cdot e^{-(R-r)/L} r\,dr}{\int \varepsilon_0\varepsilon(r) \cdot |E(r)|^2 dV} \approx 2\pi R h \varepsilon_0 n_3^2 |E_0|^2 L, \tag{14.4}$$

where E_0 is the electric field at the OFRR inner surface and h is an arbitrary length along the OFRR longitudinal direction. L is the light intensity decay constant. Meanwhile, the resonant mode's frequency shift in response to the attachment of molecules to the OFRR surface, $\delta\lambda$, is calculated by using the equation developed by Arnold et al.[28] and by considering (14.2)–(14.4):

$$\frac{\delta\lambda}{\lambda} = \frac{\sigma_p \alpha_{ex} \cdot 2\pi R h \cdot |E_0|^2}{2\int \varepsilon_0 \varepsilon(r) \cdot |E(r)|^2 \mathrm{d}V} = \sigma_p \alpha_{ex} \frac{2\pi\sqrt{n_2^2 - n_3^2}}{\varepsilon_0 \lambda^2} \frac{n_2}{n_3^2} S, \tag{14.5}$$

where σ_p is the biomolecule surface density and α_{ex} is the excess polarizability of the molecule. The excess polarizability refers to the additional energy required to polarize the captured molecule when compared with the buffer molecules (e.g., H_2O) that were replaced by the captured molecule at the surface.

Equation (14.5) establishes the linear relationship between the RI sensitivity and the resonant mode's frequency shift for a given number of captured molecules at the OFRR surface. Also, note that the magnitude of the shift is directly proportional to the surface density and the excess polarizability of the bound molecules. Although (14.5) is developed for the OFRR, it is generally applicable to other types of ring resonator sensors, as is shown by Zhu, et al.[27]

The biomolecule sensitivity is defined as the magnitude of the resonant mode shift for a given biomolecule surface density, i.e. $\delta\lambda/\sigma_p$. Upon determining the biomolecule sensitivity, the detection limit (DL) can be calculated. The DL is the minimum surface density of biomolecules that can possibly be measured by the sensing device:

$$\mathrm{DL} = \sigma_{\min} = \frac{\delta\lambda_{\min}}{\delta\lambda/\sigma_p}. \tag{14.6}$$

Therefore, the smallest measurable frequency shift ($\delta\lambda_{\min}$) must be determined. This value depends on the spectral width of the resonant mode, amplitude noise (e.g., thermal and shot noise), spectral noise (e.g., thermo-optic variations in the sensing device), and the spectral resolution of the measurement technique[21]. Minimizing these parameters will improve the DL.

14.3 Experimental Configuration for Biosensing

The capillaries used for OFRR biosensing experiments are generally between 100 and 150 μm in diameter and have a wall thickness of around 3 μm to have high sensitivity. At this time, glass capillaries of these dimensions are not commercially available and thus must be produced in a prototyping environment. For the experiments described here, the OFRRs were pulled under heat from a silica preform. The preform is generally a glass tube with a high ratio of diameter to wall thickness (aspect ratio). Two CO_2 lasers on opposing sides of the glass tube heat a small spot on the tube while a motorized stage draws one end of the tube away from the heating zone. The other end of the tube is fed slowly into the heating zone with a motorized stage to maintain a constant mass in the zone. Generally, the aspect ratio of the preform is nearly preserved during the drawing process. However, because glass tubes with the necessary aspect ratio are not readily available, further wall thickness reduction is necessary after the drawing step. Hydrofluoric acid (HF) is pumped through the capillary to trim the wall thickness gradually. This step can be monitored in real time so that it can be well controlled. Alternative methods for pulling thin-walled glass capillaries have been presented in Refs 26 and 29. As an alternative to using thin-walled capillaries, Ling and Guo have demonstrated a procedure that enables the use of capillaries with a wall thickness of tens of microns, although the excitation of the resonant mode for sensing is more difficult[30]. (also see Chap. 8)

To use the OFRR as a biosensing device, the optical resonant mode is excited and the resonant frequency is measured continuously in real time. The conceptual measurement setup is illustrated in Fig. 14.3. Laser light from a distributed feedback (DFB) laser is delivered to the OFRR using fiber optic cable. One method that has been used to excite the resonant modes is to place a tapered fiber optic cable with a diameter less than 4 μm in contact with the OFRR. The evanescent field of the tapered fiber overlaps with the evanescent field outside of the capillary wall, which enables mode coupling between the two media[24].

Alternatively, a pedestal waveguide device can replace the fiber taper. This may offer a more robust design and lead to a multiplexed lab-on-a-chip architecture. This has been demonstrated with two different waveguiding structures, shown in Fig. 14.4. The first uses an antiresonant reflecting optical waveguide (ARROW) pedestal structure in which alternating layers of silicon nitride and silica are used as a cladding below the waveguide[24]. The OFRR is placed in contact with the top of the waveguide, where the evanescent field at the top of the ARROW excites the resonant modes in the OFRR. In the second design, a gold layer is used as a

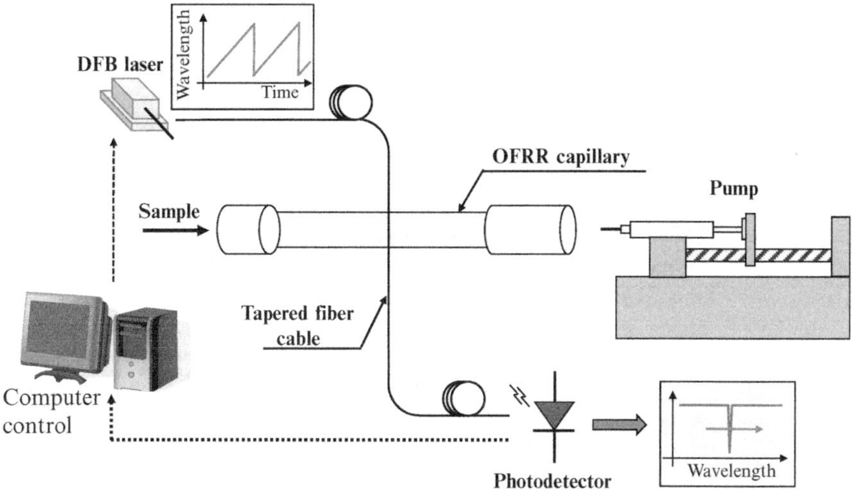

Fig. 14.3 Experimental configuration: A DFB periodically scans across a small wavelength region. The relative wavelength of the resonance mode is identified by intensity dips on a photodetector at the output of the excitation waveguide

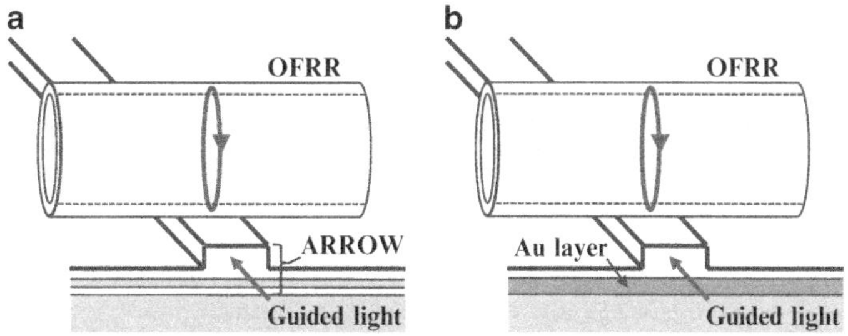

Fig. 14.4 Waveguide excitation of the OFRR resonant modes using (**a**) an ARROW-clad waveguide and (**b**) a gold-clad waveguide

reflective cladding below the silica waveguide[25]. In contrast with the ARROW design, the gold-clad structure offers light-guiding and coupling capabilities throughout much of the visible and near-infrared spectrum.

In the fiber taper and waveguide mode-excitation configurations, light not only couples from the fiber or waveguide into the OFRR, but it also couples back from the OFRR into the waveguide. This allows the real-time measurement of the resonant mode. In the condition of resonance, the light in the resonator continues to add in phase with light coupling into the device. By mode coupling theory, this implies that light coupling from the resonator back into the waveguide is exactly out of phase with the light propagating down the waveguide. Therefore, at the resonant frequency, the intensity of the light at the output of the waveguide is reduced when compared with all nonresonant frequencies.

As shown in the experimental configuration in Fig. 14.3, the resonant mode can be observed and tracked in real time using a DFB laser at the waveguide input and a photodetector at the waveguide output. The output wavelength of the DFB laser is periodically scanned simply by modulating the gain current into the laser. The laser scans across a wavelength range of several tens of picometers; as the laser output wavelength scans through the resonant wavelengths, the intensity dip is observed at the output of the photodetector. The spectral position of the intensity dip is recorded as the sensing is performed, and thus the sensing signal can be observed in real time.

The use of a 1550 nm DFB laser and photodetector offers a very low-cost solution compared with many other photonics-based technologies, including fluorescence-based technologies. However, the OFRR is one of the few RI-based resonant sensing technologies that can take advantage of this low-cost design. It is the OFRR's high Q-factor (around 10^6 for biosensing) and thus the narrow mode linewidth that enables the use of the low-cost tunable laser. For other designs, such as SPR and planar ring resonators, the larger linewidth requires the use of a broadband optical source and a spectral analysis device for observation of the mode. Thus, the OFRR is one of the first devices to achieve the promise of RI-based sensing with low-cost photonics when compared with fluorescence-based sensing technologies.

14.4 Surface Chemistry

To produce a specific sensing response (i.e., resonant frequency shift), the target molecules must be captured at the inner surface of the OFRR. The chemistry on the inner surface of the OFRR is critically important to the capture efficiency of the device as well as to the specificity. Thus, good surface chemistry has an impact on the detection limit of the device. Generally, the surface should present a molecule with high affinity to the target and low affinity to any other molecules. For most protein targets, this implies that a well-designed antibody should be immobilized onto the surface. For specific DNA sequences, a single-stranded oligonucleotide with the complementary sequence makes an excellent biorecognition probe at the surface.

Immobilizing the biorecognition probes with high surface density and proper orientation requires a few surface preparation steps. The surface chemistry for immobilizing antibodies onto the OFRR surface is depicted in Fig. 14.5. The first step is to functionalize the silica surface for the capture of biomolecules. A common technique is to use 3-aminopropyltrimethoxysilane (3APS). This molecule binds to glass and presents an amine (NH_2) functional group, making the surface capable of binding biomolecules. The next surface chemistry step is to immobilize protein A or protein G onto the surface. These proteins have a high affinity for the Fc region of the antibody, and thus they serve to capture and orient the antibody onto the sensor surface[31]. Protein A/G can be attached to the surface of the OFRR using a homo-biofunctional cross-linker that binds to the amine group of the 3APS and to an amine group of the protein A/G. Crosslinkers used with the OFRR include bis (sulfosuccinimidyl) substrate (BS3), dimethyl adipimidate (DMA), dimethyl pimelimidate (DMP), and glutaraldehyde. After the protein A/G solution is passed through the capillary and all unbound molecules are rinsed away, the antibody solution is passed into the OFRR. The final surface preparation step is to block the

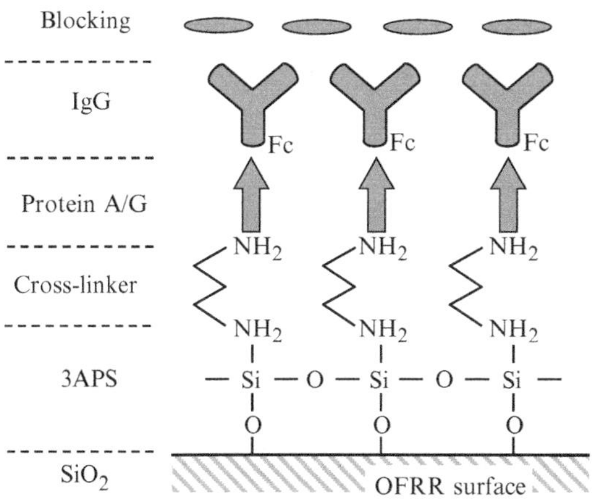

Fig. 14.5 Surface chemistry: The OFRR glass surface is biofunctionalized with 3APS. Then biorecognition molecules are immobilized using a cross-linker

surface to prevent molecules in the sample other than the target from binding to available binding sites within the surface chemistry, such as amine or carboxyl groups on the antibodies or protein A/G molecules. Typical blocking proteins include casein and bovine serum albumin (BSA). It also may be desirable to use poly(ethylene glycol) (PEG) with an amine terminal to block amine-binding regions. After the blocking step is complete, the OFRR is prepared to analyze the sample.

14.5 Biosensing Demonstrations with the OFRR

14.5.1 Quantitative Biomolecule Detection

As an instructional demonstration of the OFRR's capability to detect biomolecules in a sample, BSA was used as a target analyte, and the concentration-dependent sensing response was shown (for more details, see Ref. 27). A number of OFRRs were produced with incrementally increasing sensitivity. The RI sensitivity of each OFRR was characterized by passing solutions of ethanol in water with known RI through the capillary and measuring the resulting resonant mode spectral shift. After RI characterization, a solution of 1% 3APS was passed through the OFRR to functionalize the inner surface for biomolecule capture. Then, a 5% solution of glutaraldehyde was passed through the OFRR to provide a cross-linker for BSA capture. Finally, a solution of BSA in phosphate buffered saline (PBS) at a concentration of 100 μg/mL was pumped through the OFRR capillary. This concentration is sufficient to saturate the sensing response (i.e., adding more BSA will not produce a larger response because the sensor surface has captured the maximum possible number of BSA molecules).

The saturated sensing response for each of the OFRRs of increasing sensitivity is plotted in Fig. 14.6. Also plotted is the calculated sensing signal magnitude assuming a perfect monolayer of BSA molecules assembled on the inner surface of the OFRRs of increasing sensitivity. The calculation is performed as described above and uses the parameters of BSA given in Ref. 27. The excellent correspondence between the measured and calculated data suggests two conclusions. First, the mathematical model outlined above appears to be accurate. Second, it appears that in fact the OFRR captures the BSA molecules with a high surface density. This is in agreement with previous observations[11,14,28].

14.5.2 DNA Sequence Detection

In these previous experiments, BSA is captured at the interior surface of the OFRR capillary, but these instructional experiments do not use a specific biorecognition

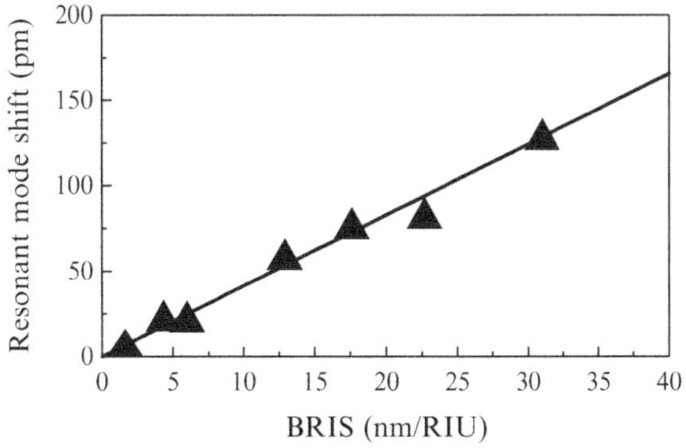

Fig. 14.6 Resonant mode spectral shift for a saturated surface of BSA. *Triangles* represent measured data points, while the *line* is calculated from the theory presented in Sect. 14.2. Reprinted from Ref. 27 with permission. © 2008 Optical Society of America

capture technique. Experimental detection of specific DNA sequences using the OFRR demonstrates the capability to capture target molecules specifically at the sensor surface. This method offers a simpler alternative to fluorescence-based DNA microarray detection systems. In the assay, oligonucleotides containing the complementary sequence to the target sequence are immobilized onto the interior wall of the OFRR. As the biological sample is subsequently passed through the OFRR, oligonucleotides containing the target sequence hybridize with the capture probes at the OFRR inner surface, producing a visible sensing signal.

Preliminary results of DNA sequence detection using the OFRR are thoroughly described by Suter, et al.[32] The experimental setup described above for tracking the resonant mode shift is used. An OFRR with RI sensitivity of about 7 nm/RIU was produced and used for these experiments. The OFRR surface was functionalized with 3APS and then DMA was used as an amine-amine crosslinker. The single-stranded oligonucleotide capture probe was synthesized with an amine functional group connected to the $5'$ end by a 6-carbon linker, and has a 25 base-pair sequence[32].

The target oligonucleotide sequence was a single-stranded 25 base-pair sequence exactly complementary to the capture probe. Concentrations were prepared from 500 pM to 500 nM and then sequentially passed through the OFRR. Figure 14.7a shows the resonant mode spectral shift for each target oligo concentration. Additionally, a Michaelis–Menton curve fit is plotted using a dissociation constant K_d of 2.9 nM and a maximum wavelength shift of 5.1 pm[32].

All DNA hybridization assays are subject to cross-hybridization, in which an oligonucleotide that is not a perfect sequence match hybridizes with the capture probe. The cross-hybridization in the OFRR was investigated using samples of oligonucleotides with either 0-, 1-, 2-, 5-, or 25-base mismatches when compared with the 25 base-pair biorecognition capture probe. The resulting resonant mode spectral shifts for the respective mismatch are plotted in Fig. 14.7b. The measurements show a difference of 1.3 pm and 2.8 pm for one and two base-pair

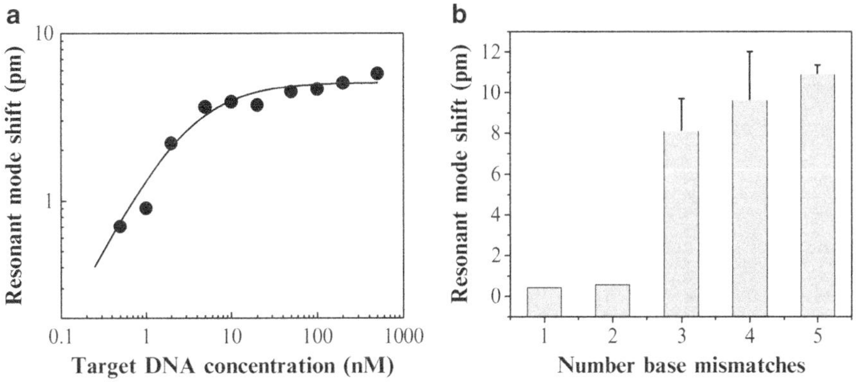

Fig. 14.7 (**a**) Measured resonant mode spectral shift for increasing concentrations of target oligonucleotides. The line fit is a Michaelis-Menton curve with a K_d of 2.9 nM and a maximum resonant wavelength shift of 5.1 pm. (**b**) Measured resonant mode shift for oligonucleotides of increasing number of base-pair-mismatches with the capture probe. Reprinted from Ref. 32 with permission. © 2008 Elsevier

mismatches, respectively, when compared with the perfect complement. A sample with 5-base-pair mismatches results in essentially no signal. These signal contrasts between the target oligonucleotide and oligonucleotides of one or two nucleotide polymorphisms likely can be made somewhat larger by conducting a rinse just below the melting temperature.

14.5.3 Pathogen Detection

In food safety and health care services, the traditional method for detection of pathogens is to take a sample, place it in a growth medium, and wait for cultures to grow. However, a low-cost real-time assay for pathogens could save costs and improve safety and public health. The OFRR can be designed for this assay in a similar way that it is designed to capture and quantify proteins or DNA sequences. Many bacteria and viruses have characteristic surface proteins that can be targeted by antibody capture probes. These probes can be immobilized on the OFRR surface, where they can capture pathogens from a sample that is passed through the OFRR. Pathogens captured at the OFRR surface cause the resonant mode to shift, generating the corresponding sensing signal.

Highly sensitive detection of the M13 filamentous bacteriophage has been achieved with the OFRR[33]. M13 is about 900 nm long and 10 nm wide and is composed of circular single-stranded DNA encapsulated in approximately 2700 copies of the major coat protein g8p. M13 has similar molecular weight to hazardous human pathogens, and thus is an excellent model to study the OFRR's capability for highly sensitive pathogen detection. An OFRR with a RI sensitivity of 22 nm/RIU was produced and then functionalized with 3APS. Glutaraldehyde was used to

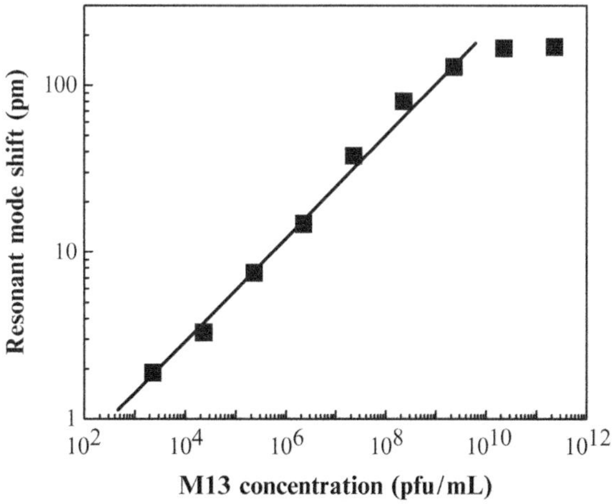

Fig. 14.8 Measured resonant mode spectral shift for increasing concentration of M13 phage. Reprinted from Ref. 27 with permission.© 2008 The Royal Society of Chemistry

cross-link protein A to the OFRR surface, and then anti-M13 antibodies were immobilized at the surface. After this, the M13 phage suspended in PBS was passed through the OFRR. The resonant mode frequency was monitored in real time as the phage was captured at the OFRR surface by the antibody.

M13 suspensions in concentrations of 2.3×10^3 to 2.3×10^{11} plaque forming units per mL (pfu/mL) were sequentially passed through the OFRR. Figure 14.8 plots the magnitude of the sensor response vs. the M13 sample concentration. An easily observable resonant mode wavelength shift of about 2 pm was recorded for the low concentration of 2.3×10^3 pfu/mL. The sensing response saturates at around 2.3×10^{10} pfu/mL because the sensor surface had become densely occupied with M13 phage. This ultra-low detection limit for the M13 bacteriophage and the operation range of seven orders of magnitude is highly encouraging in terms of utilizing the OFRR as a pathogen sensor. Furthermore, the high surface area to volume ratio of the OFRR is shown to perform pathogen sensing faster than conventional waveguide sensing configurations, as demonstrated by Zhu et al.[33]

14.6 Summary and Conclusions

These preliminary experimental results illustrate the power of the OFRR to detect specifically biomolecules while using low-cost photonics and simple protocols. The OFRR is a label-free RI-based sensor, and thus it avoids the extra steps and costly photonics of fluorescence-based techniques. Even compared with other RI-based techniques, such as SPR, the OFRR is able to use low-cost simple photonics because of its high Q-factor and thus minimal bandwidth requirements. Additionally, the sensor is inherently integrated into the microfluidic sample delivery channel.

Furthermore, the micro-capillary-shaped sensor has an excellent sensor-surface-area to sample-volume ratio, which enables quick and efficient capture of biomolecules.

The experimental results presented in this chapter demonstrate the biosensing performance achieved by the first-generation OFRR. Quantification of BSA protein binding to the surface illustrates the capability of the OFRR to provide real-time quantitative analysis of biomolecule concentrations in a sample. These experiments also highlight the competitive detection limit of the OFRR. A measurable signal was recorded for concentrations of BSA as low as 10 pM, and a predicted detection limit of less than 1 pg/mm^2 was determined based on the results[27]. Specific analyte capture using a probe at the OFRR surface was illustrated by the DNA sequence detection experiments. Detection of 10 pM target analyte was also demonstrated in these experiments[32]. Finally, real-time biomolecular-based recognition of pathogens was demonstrated using the M13 phage as a representative model. Detection of 2,000 pfu/mL was performed.

This diverse set of biosensing experimental demonstrations illustrates the flexibility of the OFRR device. Nearly any biomolecular recognition event can be detected. The examples illustrated with the previously described experiments include DNA sequence detection and virus detection through surface proteins. Additional biosensing examples for which the OFRR is well-suited include site-specific cleavage, protein-protein interactions, and cell genotype/phenotype identification through receptors. Furthermore, as shown by the theory outlined above, the OFRR can be accurately and precisely quantitative.

Looking to the future, it is anticipated that the OFRR technology will be integrated into a lab-on-a-chip platform that would perform sample preparation and processing, and then sample analysis using the OFRR. This integrated device has tremendous potential as a solution for early disease detection and analysis. An example application is the detection of glycoprotein biomarkers in serum, such as the CA15–3 breast cancer biomarker. In addition, the OFRR integrated device may find a number of applications in molecular biology laboratories for tasks such as quantitative protein expression in cell lysates, protein-protein interaction studies, cell signaling studies, and many others. In fact, it is possible that the device could replace existing fluorescence-based techniques while saving time and costs. Given these highly relevant clinical and biomedical research applications, it is possible that the OFRR technology may be one key to better understanding, diagnosing, and treating disease.

Acknowledgment This work is supported by the Wallace H. Coulter Foundation Early Career Award and the NIH (1K25 EB006011).

References

1 Fan, X.; White, I. M.; Shopova, S. I.; Zhu, H.; Suter, J. D.; Sun, Y., Sensitive optical biosensors for unlabeled targets: a review, *Anal. Chim. Acta* **2008**, 620, 8–26

2 Homola, J.; Yee, S. S.; Gauglitz, G., Surface plasmon resonance sensors: Review, *Sens. Actuators B* **1999**, 54, 3–15

3 Hoa, X. D.; Kirk, A. G.; Tabrizian, M., Towards sensitive and integrated surface plasmon resonance biosensors: A review of recent progress, *Biosens. Bioelectron.* **2007**, 23, 151–160

4 Homola, J., Surface plasmon resonance sensors for detection of chemical and biological species, *Chem. Rev.* **2008**, 108, 462–493

5 Shumaker-Parry, J. S.; Campbell, C. T., Quantitative methods for spatially resolved adsorption/desorption measurements in real time by surface plasmon resonance microscopy, *Anal. Chem.* **2004**, 76, 907–917

6 Heideman, R. G.; Lambeck, P. V., Remote opto-chemical sensing with extreme sensitivity: Design, fabrication and performance of a pigtailed integrated optical phase-modulated Mach-Zehnder interferometer system, *Sens. Actuators B* **1999**, 61, 100–127

7 Ymeti, A.; Kanger, J. S.; Greve, J.; Lambeck, P. V.; Wijn, R.; Heideman, R. G., Realization of a multi-channel integrated Young interferometer chemical sensor, *Appl. Opt.* **2003**, 42, 5649–5660

8 Markov, D. A.; Swinney, K.; Bornhop, D. J., Label-free molecular interaction determinations with nanoscale interferometry, *J. Am. Chem. Soc.* **2004**, 126, 16659–16664

9 Zhao, M.; Nolte, D.; Cho, W.; Regnier, F.; Varma, M.; Lawrence, G.; Pasqua, J., High-speed interferometric detection of label-free immunoassays on the biological compact disc, *Clin. Chem.* **2006**, 52, 2135–2140

10 Lee, M. R.; Fauchet, P. M., Two-dimensional silicon photonic crystal based biosensing platform for protein detection, *Opt. Express* **2007**, 15, 4530–4535

11 Vollmer, F.; Braun, D.; Libchaber, A.; Koshsima, M.; Teraoka, I.; Arnold, S., Protein detection by optical shift of a resonant microcavity, *Appl. Phys. Lett.* **2002**, 80, 4057–4059

12 Vollmer, F.; Arnold, S.; Braun, D.; Teraoka, I.; Libchaber, A., Multiplexed DNA quantification by spectroscopic shift of two microsphere cavities, *Biophys. J.* **2003**, 85, 1974–1979

13 Ksendzov, A.; Lin, Y., Integrated optics ring resonator for protein detection, *Opt. Lett.* **2005**, 30, 3344–3346

14 Hanumegowda, N. M.; White, I. M.; Oveys, H.; Fan, X., Label-free protease sensors based on optical microsphere resonators, *Sens. Lett.* **2005**, 3, 315–319

15 Chao, C-Y.; Fung, W.; Guo, L. J., Polymer microring resonators for biochemical sensing applications, *IEEE J. Sel. Top. Quantum Electron.* **2006**, 12, 134–142

16 Zhu, H.; Suter, J. D.; White, I. M.; Fan, X., Aptamer based microsphere biosensor for thrombin detection, *Sensors* **2006**, 6, 785–795

17 Yalcin, A.; Popat, K. C.; Aldridge, O. C.; Desai, T. A.; Hryniewicz, J.; Chbouki, N.; Little, B. E.; King, O.; Van, V.; Chu, S.; Gill, D.; Anthes-Washburn, M.; Unlu, M. S.; Goldberg, B. B., Optical sensing of biomolecules using microring resonators, *IEEE J. Sel. Top. Quantum Electron.* **2006**, 12, 148–155

18 De Vos, K.; Bartolozzi, I.; Schacht, E.; Bienstman, P.; Baets, R., Silicon on insulator microring resonator for sensitive and label-free biosensing, *Opt. Express* **2007**, 15, 7610–7615

19 Ren, H-C.; Vollmer, F.; Arnold, S.; Libchaber, A., High-Q microsphere biosensor – analysis for adsorption of rod-like bacteria, *Opt. Express* **2007**, 15, 17410–17423

20 Ramachandran, A.; Wang, S.; Clarke, J.; Ja, S. J.; Goad, D.; Wald, L.; Flood, E. M.; Knobbe, E.; Hryniewicz, J. V.; Chu, S. T.; Gill, D.; Chen, W.; King, O.; Little, B. E., A universal biosensing platform based on optical micro-ring resonators, *Biosens. Bioelectron.* **2008**, 23, 939–944

21 White, I. M.; Fan, X., On the performance quantification of resonant refractive index sensors, *Opt. Express* **2008**, 16, 1020–1028

22 Armani, A. M.; Vahala, K. J., Heavy water detection using ultra-high-Q microcavities, *Opt. Lett.* **2006**, 31, 1896–1898

23 White, I. M.; Oveys, H.; Fan, X., Liquid core optical ring resonator sensors, *Opt. Lett.* **2006**, 31, 1319–1321

24 White, I. M.; Oveys, H.; Fan, X.; Smith, T. L.; Zhang, J., Integrated multiplexed biosensors based on liquid core optical ring resonators and antiresonant reflecting optical waveguides, *Appl. Phys. Lett.* **2006**, 89, 191106

25 White, I. M.; Suter, J. D.; Oveys, H.; Fan, X., Universal coupling between metal-clad waveguides and optical ring resonators, *Opt. Express* **2007**, 15, 646–651
26 Sumetsky, M.; Windeler, R. S.; Dulashko, Y.; Fan, X., Optical liquid ring resonator sensor, *Opt. Express* **2007**, 15, 14376–14381
27 Zhu, H.; White, I. M.; Suter, J. D.; Dale, P. S.; Fan, X., Analysis of biomolecule detection with optofluidic ring resonator sensors, *Opt. Express* **2007**, 15, 9139–9146
28 Arnold, S.; Khoshsima, M.; Teraoka, I.; Holler, S.; Vollmer, F., Shift of whispering-gallery modes in microspheres by protein adsorption, *Opt. Lett.* **2003**, 28, 272–274
29 Zamora, V.; Diez, A.; Andres, M. V.; Gimeno, B., Refractometric sensor based on whispering gallery modes of thin capillaries, *Opt. Express* **2007**, 15, 12011–12016
30 Ling, T.; Guo, J. L., A unique resonance mode observed in a prism-coupled micro-tube resonator sensor with superior index sensitivity, *Opt. Express* **2007**, 15, 17424–17432
31 Jung, Y.; Jeong, J. Y.; Chung, B. H., Recent advances in immobilization methods of antibodies on solid supports, *Analyst* **2008**, 133, 697–701
32 Suter, J. D.; White, I. M.; Zhu, H.; Shi, H.; Caldwell, C. W.; Fan, X., Label-free quantitative DNA detection using the liquid core optical ring resonator, *Biosens. Bioelectron.* **2008**, 23, 1003–1009
33 Zhu, H.; White, I. M.; Suter, J. D.; Zourob, M.; Fan, X., Opto-fluidic micro-ring resonator for sensitive label-free viral detection, *Analyst* **2008**, 133, 356–360

Chapter 15
Deep-Probe Optical Waveguides for Chemical and Biosensors

Mohammed Zourob, Nina Skivesen, Robert Horvath, Stephan Mohr, and Nicholas J. Goddard

Abstract Typical evanescent wave biosensors generate an electromagnetic wave at the sensor surface that penetrates 100–200 nm into the analysed medium. This has proven to be a highly sensitive tool to monitor refractive index changes in the close vicinity of the sensor surface. The sensitivity of such sensors can be enhanced significantly to monitor interactions caused by large micron scale objects such as bacterial and mammalian cells by increasing the penetration depth of the evanescent field. Recently, different formats of deep-probe optical waveguides including reverse waveguides (RW) based on low refractive index substrates (below 1.33) and metal-clad leaky waveguides (MCLW) have been developed for various sensing applications. These sensors are designed to maximize the overlap between the optical mode and the adlayer (superstrate layer) to be sensed. Increasing the penetration depth of the evanescent field opens up new perspectives for the detection of larger biological objects as it accommodates the majority of their body within the evanescent field. RWs use substrate materials with lower refractive index than that of the monitored superstrate layer (aqueous solution). In MCLWs, a thin metal layer is inserted between the substrate and the thicker waveguide layer. These sensor designs facilitate both increasing and tuning the penetration depth of the modes into the monitored aqueous solution and thereby significantly extend the range of possible application areas of optical waveguide sensors. The developed devices have been used for a range of biosensing applications, including the detection of bacteria, mammalian cells, organophosphorous pesticides and glucose using refractive index changes, absorbance and fluorescence monitoring. Integrating deep-probe sensors with an external electrical field or ultrasonic standing waves shortens analysis time significantly and reduces non-specific binding due to enhanced diffusion of analytes to the immobilized recognition receptors, and thus improves the detection limit by a few orders of magnitude.

M. Zourob (✉)
GROUPE GDG ENVIRONNEMENT LTÉE, 430 rue Saint-Laurent, Trois-Rivières (QC), Canada G8T 6H3
e-mail: mohammed.zourob@gdg.ca

X. Fan (ed.), *Advanced Photonic Structures for Biological and Chemical Detection*, Integrated Analytical Systems,
DOI 10.1007/978-0-387-98063-8_15,

15.1 Introduction

Evanescent wave techniques have been widely applied for sensing at surfaces[1]. In many cases, optical waveguides have been used to provide high sensitivity, examples of such devices include the resonant mirror (RM)[2,3], differential interferometry sensor[4,5], and the grating coupler[6–8]. To ensure high sensitivity, the waveguides are usually made thin enough to support only a single mode and are fabricated from very high refractive index (RI) materials such as silicon nitride, hafnium dioxide or tantalum pentaoxide, much higher RIs than the sample to be measured. For each polarization there is an optimum waveguide thickness[8,9], which maximizes the evanescent field that interacts with the biochemical species immobilized on the waveguide and thus maximizes the sensitivity of the waveguide modes. The mode of conventional waveguide devices decays exponentially away from the surface of the actual waveguiding layer into the low RI medium where sensing occurs[10,11]. Because the intensity of the mode is higher at the waveguide surface than in the bulk of the sample, these devices are very sensitive to imperfections on the surface, to the obvious detriment of performance. Typically, defects in the waveguiding layer cause scattering, giving relatively high light losses[11].

A widely used strategy to increase the detection sensitivity is to deposit recognition molecules, such as antibodies or other chemically selective species on top of the waveguide layer for example entrapped in a 3D-porous polymer or hydrogel film or coupled to a thin chemically active polymer. This provides a higher density of binding sites compared to a monolayer of molecules directly immobilized on the sensor surface. This adlayer will have, in general, a slightly higher refractive index than the aqueous solution, as a result of the higher polarizability of the polymer strands. Typically, the RI of the polymer will be of the order of 0.005–0.01 greater than that of the aqueous solution. This is enough to permit total internal reflection (TIR) at the adlayer/solution interface. In contrast to evanescent wave sensing, this configuration is called guided wave sensing, due to guiding of the waveguide mode (not evanescent) in the adlayer. This configuration has in addition to biosensing also been successfully applied for monitoring polymer films[12].

To increase the sensitivity to RI changes, a larger proportion of the optical mode must interact with the chemically sensitive layer. In the case of evanescent wave sensors, this can be done by increasing the RI of the waveguiding layer and simultaneously reducing its thickness and reducing the RI of the waveguide film support[9,11]. In the case of true waveguiding in a gel adlayer with aqueous cover media, a low RI substrate material is required to obtain conventional waveguiding by TIR. The RI of a hydrogel layer, for example, is only slightly higher than water, and thus, only special materials like fluorinated polymers such as Teflon AF or highly nanoporous silica have a sufficiently low refractive index to act as substrates[9,11]. Hence, material considerations are especially important to increase sensitivity for both the cases.

Previously, Goddard and co-workers used anti–resonant Fabry–Perot reflectors to form antiresonant reflecting optical waveguides (ARROWs) in a gel layer[13,14].

More recently, Horvath and co-workers[9,15–18] reported on reverse symmetry waveguides also called reverse waveguides (RWs) based on ultra-low refractive index waveguide supports, and Zourob et al.[19–24] and Skivesen et al.[25–27] reported on metal-clad leaky waveguides (MCLWs), for which evanescent field penetration depths of around 1 μm can be obtained. These novel designs are appropriate for the detection of micron scale biological objects, such as bacterial and mammalian cells.

In this chapter, various deep-probe optical waveguide configurations for chemical and biosensing applications, including the sensitive detection of particles and mammalian cells, will be reviewed. The design of the MCLW sensors can be exploited to integrate an electric field perpendicular to the sensor surface. The integration of external fields can be used to overcome some of the problems in the actual use of optical waveguide sensors, such as long analysis times and nonspecific binding.

15.2 Evanescent Waves of Guided and Leaky Waveguide Modes

The sensor systems outlined in the present chapter use evanescent electromagnetic radiation to monitor various analytes in aqueous solutions. Therefore, as a beginning, the basic properties of evanescent electromagnetic waves and the so-called TIR phenomena are summarized. Afterwards, two types of waveguide modes will be briefly discussed; guided and leaky modes, which both generate evanescent waves at a solid/liquid boundary.

Let us consider a smooth and planar interface between two different non-absorbing and uniform materials, having refractive indices n_1 and n_2 ($n_1 > n_2$). Depending on the incident angle, θ_I, in the material with RI, n_1, the angle of the refracted light, θ_R, is given by the well-known Snell's law: $n_1\sin(\theta_I) = n_2\sin(\theta_R)$. At a specific angle of incidence, the angle of refracted light reaches $\theta_R = 90°$. This incident angle is called the critical angle: $\theta_C = \sin^{-1}(n_2/n_1)$. If the incident angle is small and below the critical angle a fraction of light is reflected back into medium n_1 at an angle equal to the incident angle and the rest is transmitted into medium n_2 at the angle θ_R. Increasing θ_I also results in an increase of θ_R, and at θ_C the refracted wave propagates parallel to the interface. When the angle of incidence exceeds the critical angle, no refraction occurs; instead, the light is reflected back into the medium of incidence and TIR is obtained. We emphasize that the critical angle only exists when the light is incident on the boundary from within a high RI medium.

A more thorough analysis shows that during TIR the light penetrates into the low index (cover) media before "turning back" into the media with higher RI[10,28]. The "short journey" in the low index media also means that the light experiences a phase shift during TIR. However, it is important to emphasize that under TIR condition no net flow of energy occurs across the interface; hence, all of the light intensities reflect back and the interface in many senses acts like a perfect mirror.

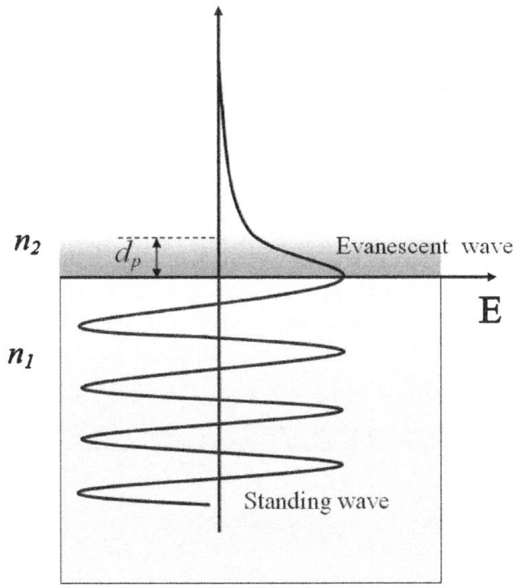

Fig. 15.1 Schematic of the field structure when TIR occurs at a smooth planar interface

When considering light as optical waves instead of rays, the light that penetrates and propagates in the low index media is called the *evanescent wave*. Evanescent waves have an exponentially decaying amplitude away from the interface, described by $A = A_0\exp(-z/d_P)$, where $z > 0$ is the distance from the interface, see Fig. 15.1 for illustration. The word "evanescent" refers exactly to this decaying light amplitude perpendicular to the surface. The exponentially decaying amplitude results in an intensity decrease away from the interface, and thus the most sensitive area is that closest to the sensor surface. It is important to note that the penetration depth can be controlled by an appropriate choice of the ratio of refractive indices of the two materials, the angle of incidence and the wavelength of the incident light.

In many practical situations, the waveguiding layer consists of an optically dense media formed as a thin film with a thickness d_F and refractive index n_F. This film is usually applied on a solid substrate (typically glass) with refractive index n_S less than the RI of the film. The macroscopically thick substrate creates a mechanically stable foundation for the above film, which has a thickness typically in the submicron range. In this case, the light can travel by TIR inside the film, in a zigzag manner (see Fig. 15.2a), creating the evanescent wave both in the aqueous cover solution with refractive index n_C and in the substrate[10].

When the travelling light experiences TIR both at the film/cover and film/substrate boundaries, the "light propagation form" is called a *guided mode* of the waveguide (see Fig. 15.2a). In the literature, these modes are often simply just called waveguide modes. When light propagates in the waveguide film, it experiences a phase shift due to both the distance travelled in the film and due to the TIR at the interfaces. To achieve guided modes, incident light and twice-reflected light at any position in the film must be in phase, which requires the magnitude of the

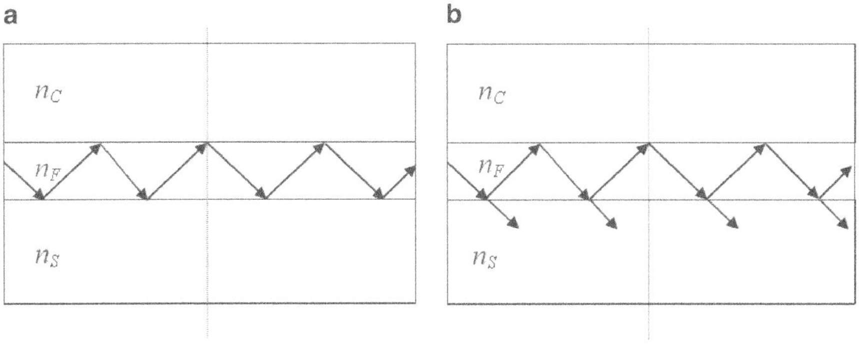

Fig. 15.2 Illustration of (**a**) guided waveguide modes and (**b**) leaky waveguide modes in a three-layered waveguide structure

total phase change after a complete cycle in the film to be an integer multiple of 2π[10]. Thus, not all the arbitrary beams at incident angles beyond the critical angle can achieve a guided mode. Only waves with specific discrete incident angles can propagate as guided modes, when considering the phase change at every reflection. These angles are called the resonant angles. For a more comprehensive review on the above-mentioned interference phenomena and waveguide resonances, the reader can consult a review paper by Tien[10].

When the RI of the cover medium is changed, the incident angle, for which waveguide modes are excited, shifts; thus, the resonant angle corresponds to a given RI of the cover medium for the specific waveguide design. The resonant angle of the waveguide mode is also influenced by the RI of the film and the surrounding media and the modes can therefore be characterized by a so-called effective refractive index, N, given by $n_F \cos(\theta_I)$, representing the refractive index of an equivalent bulk medium having identical values of phase velocity and propagation constant.

Planar optical waveguides can be designed to support many modes propagating along the guide axis, each with a different mode order, m (0, 1, 2, 3,. . . .) and effective index, N_m. The effective index of all the guided modes varies between n_F and $\max\{n_S, n_C\}$[10]. Generally, the number of waveguide modes that can be excited in the film increases with increasing film thickness d_F and increasing n_F. The lowest thickness of the film at which the mode with the mth mode order can no longer exist is called the cut-off thickness for the given mode[10].

In some cases the light experiences TIR at the film/cover boundary only and the light energy leaks out into the substrate (see Fig. 15.2b). These types of modes are called *leaky modes*. This happens if the RI of the substrate is larger than the RI of the film (TIR is not possible at the film/substrate boundary); however, in these situations usually the light energy leakage into the substrate is large, and the propagating mode damps significantly during propagation.

One possibility to obtain a relatively small leakage into the substrate is to introduce a thin film of metal or absorbing layer such as a polymer or a dye with a complex dielectric function, or a thin layer of low refractive index material between the substrate and the film. In this situation, the so-called frustrated TIR (FTIR) arises at the films' lower boundary reflecting significant portion of the light energy back into the film.

Leaky modes can also be characterized by a mode order, m, and an effective index, N. As described for guided waveguide modes the effective refractive index of the leaky modes, N_m, also shifts due to RI changes in the sensing layer. Thus the sensor principle using leaky modes is similar to that for sensors exciting guided waveguide modes: The shift in the angular position of the resonance is observed to determine the change in RI, similar to the sensors using guided waveguide modes. However, it has to be noted that for leaky modes the effective refractive index is usually complex and its imaginary part reflects the damping of the mode due to the energy leakage into the substrate. The modes can also differ in their polarizations, TE and TM, which refer to transverse electric and transverse magnetic polarization. In the first case the electric, while in the second case the magnetic field vector is perpendicular to both the propagation direction of the mode and the normal vector of the waveguide film interfaces[10].

15.3 Principle of Operation of Optical Waveguide Sensors

In Sect. 15.2 we concluded that refractive index variation in the aqueous cover media shifts the effective refractive index of the waveguide modes. Thus, by monitoring on-line the effective refractive index, the refractive index change can be followed. This is the basic principle of waveguide sensing.

15.3.1 Mode Equations to Determine the Effective Refractive Indices

Various sensor arrangements depend on the actual way how the shift in the mode's effective refractive index is followed. In order to design a sensitive waveguide sensor and to quantify the refractive index changes, the effective refractive index has to be connected with the physical parameters of the waveguide. To this end, one needs to solve the Maxwell's equations using the appropriate boundary conditions[10].

First, we consider the three-layer (substrate/waveguide film/cover) structure, assuming that the waveguide structure is left alone without any field incident from the outside. The solutions are obtained by using the solution ansatz ψ comprising plane waves present in the individual layers:

$$\psi = \begin{cases} A_C \exp[i(k_x x + k_{z,C} z)] & \text{(in the cover)} \\ A_F \exp[i(k_x x + k_{z,F} z)] + B_F \exp[i(k_x x - k_{z,F} z)] & \text{(in the film)} \\ B_S \exp[i(k_x x - k_{z,S} z)] & \text{(in the substrate)} \end{cases}, \quad (15.1)$$

where ψ represents the E-field for TE-polarized light and the H-field for TM-polarized light, k_x is the x-component of the wave vector and $k_{z,I} = k\sqrt{n_I^2 - N^2}$ is the wave vector component along z in medium I. From the boundary conditions, (I) ψ and $\partial_z^2\psi$ are continuous across a boundary for TE-polarized light and (II) ψ and $n^{-2}\partial_z^2\psi$ are continuous across a boundary for TM-polarized light. For each polarization, four equations can be obtained, relating the four amplitudes in (15.1). By introducing a polarization index $\rho = 0, 1$ for the TE and TM polarized case, respectively, the two systems of equations can be written in the form:[18,27]

$$\overline{\overline{A_\rho}}\vec{\psi} = \vec{0}, \tag{15.2}$$

where $\vec{\psi} = \{A_\mathrm{C}, A_\mathrm{F}, B_\mathrm{F}, B_\mathrm{S}\}$

In order to get non-trivial solutions the determinant of $\overline{\overline{A_\rho}}$ needs to be zero: $\mathrm{Det}\left[\overline{\overline{A_\rho}}\right] = 0$, which leads to the so-called three-layer mode equation:

$$\begin{aligned} 2m\pi = 2d_\mathrm{F}k\left(n_\mathrm{F}^2 - N_\mathrm{m}^2\right)^{0.5} &- 2\arctan\left[\left(\frac{n_\mathrm{F}}{n_\mathrm{S}}\right)^{2\rho}\left(\frac{N_\mathrm{m}^2 - n_\mathrm{S}^2}{n_\mathrm{F}^2 - N_\mathrm{m}^2}\right)^{0.5}\right] \\ &- 2\arctan\left[\left(\frac{n_\mathrm{C}}{n_\mathrm{S}}\right)^{2\rho}\left(\frac{N_\mathrm{m}^2 - n_\mathrm{C}^2}{n_\mathrm{C}^2 - N_\mathrm{m}^2}\right)^{0.5}\right]. \end{aligned} \tag{15.3}$$

The physical meaning of the above equation is that reflected and twice reflected waves in the waveguide film has to be in phase hence, the difference between their phase has to be $2m\pi$. The first term on the right-hand side of (15.3) represents the phase shift due to the light travelling inside the film while the second and third terms give the phase change due to TIRs at the film/substrate and film/cover boundaries[10,18,29]. The mode equation includes the value N_m, which is referred to as the effective refractive index of the mode and corresponds to the resonance angle of the propagating beam in the film, $N_m = n_\mathrm{F}\sin(\theta_m)$.

If $N_m > n_\mathrm{F}$, all three values of $k_{z,I}$, $(I = S, F, C)$ are imaginary and the solution is non-physical.

For $n_\mathrm{F} > N_m > \max\{n_\mathrm{S}, n_\mathrm{C}\}$, exponential decaying fields are formed in the cover and the substrate, and a discrete set of sinusoidal fields inside the film are generated; this is the range of the guided modes. If $\max\{n_\mathrm{S}, n_\mathrm{C}\} > N_m > \min\{n_\mathrm{S}, n_\mathrm{C}\}$, the field will be exponentially decreasing in the cover and sinusoidal in the film and substrate layers, if $n_\mathrm{S} > n_\mathrm{C}$. These modes, known as substrate radiation modes, propagate into the substrate and are damped out over a short distance in the z-direction. The case where $\min\{n_\mathrm{S}, n_\mathrm{C}\} > N_m$ results in an oscillatory field in all the three regions. These non-guided modes lose energy through both substrate and cover regions and are called air radiation modes[10].

For waveguide modes in a pure dielectric structure, N_m is $n_{\max} = \mathrm{Max}\{n_\mathrm{C}, n_\mathrm{S}\}$ at $d_\mathrm{F} = d_\mathrm{CUTOFF}$ and n_F at $d_\mathrm{F} \to \infty$. The so-called cut off thickness, d_CUTOFF, is defined as the film thickness at which the field's propagation angle in the film gets

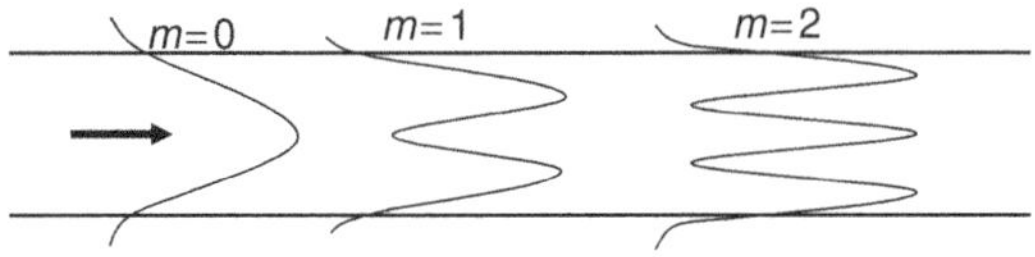

Fig. 15.3 The three lowest-order mode-power profiles in a three-layer waveguide

below the critical angle at either the film/cover or film/substrate interface, in which case the light escapes from the waveguide. The fact that $n_{\max} < N_m < n_F$ implies that $k_{z,C} = k\left(n_C^2 - N_m^2\right)^{0.5}$ and $k_{z,S} = k\left(n_S^2 - N_m^2\right)^{0.5}$ are always purely imaginary and the electromagnetic fields in the substrate and cover media are therefore evanescent. However, in the film, the result of the real-valued $k_{z,F} = k\left(n_F^2 - N_m^2\right)^{0.5}$ is that the electromagnetic field consists of two propagating waves. Hence, the solutions assume characteristic mode power profiles as shown in Fig. 15.3.

From the three-layer mode equation, (15.3), the values d_{CUTOFF} and the penetration depth of the evanescent wave in the cover media, $d_{P,C}$, can be found for the three-layer structure as:

$$d_{\mathrm{CUTOFF}} = \frac{1}{k\sqrt{n_F^2 - n_{\max}^2}}\left[\arctan\left(\frac{n_F^{2\rho}}{n_{\min}^{2\rho}}\frac{\sqrt{n_{\max}^2 - n_{\min}^2}}{n_F^2 - n_{\max}^2}\right) + 2\pi m\right], \qquad (15.4)$$

$$d_{P,C} = k^{-1}\sqrt{N_m^2 - n_C^2}\left[(N_m/n_F)^2 + (N_m/n_C)^2 - 1\right]^{\rho}. \qquad (15.5)$$

As mentioned in Sect. 15.2, sometimes a 4th thin layer (M) of metal or die is incorporated between the substrate and waveguide film to decrease the radiation loss into the substrate of the substrate radiation modes. These modes are referred to as leaky modes and the obtained structure is the MCLW. This configuration is also broadly used in evanescent wave sensor systems.

For an exhaustive analysis of the MCLW structure, the field distribution in the layers should be considered, similar to the analysis of the three-layer structure in the previous section. The field description of the three-layered structure in (15.1) can be expanded and applied for the four-layer structure to derive a mode equation derived in the same manner as the three-layer mode equation (15.1 and 15.2), e.g. by adding the electromagnetic fields in layer M to the solution ansatz ψ in (15.1):

$$A_M \exp[i(k_x x + k_{z,M} z)] + B_M \exp[i(k_x x - k_{z,M} z)].$$

This results in a mode equation quite similar to the three-layer mode equation, where only the term for the phase shift at the film/substrate boundary is interchanged with a new phase shift for the film/metal/substrate boundary:

$$\phi_{F,M,S} = 2\arctan\left[i\frac{(1 - r_{F,M})(1 - r_{M,S}\exp[i2k_{z,M}d_M])}{(1 + r_{F,M})(1 + r_{M,S}\exp[i2k_{z,M}d_M])}\right],$$

where

$$r_{I,J} = \frac{n_J^{2\rho} k_{z,I} - n_I^{2\rho} k_{z,J}}{n_J^{2\rho} k_{z,I} + n_I^{2\rho} k_{z,J}}$$

and

$$k_{z,I} = \pm k\sqrt{n_I^2 - N^2}.$$

The sign of $k_{z,I}$ should be shifted to "−" if N is complex and $\mathrm{Re}[N] > \mathrm{Re}[n_I]$. From the mode equation, the allowed values of $N = N_{\mathrm{RE}} + \mathrm{i}N_{\mathrm{IM}}$ is derived for which the *ansatz* ψ for the electromagnetic field is a solution to the Maxwell's equations. Generally, N is complex due to the complex RI of the cladding.

The phase shift at the film/cover boundary for the four-layer mode equation is given in exactly the same way as for the three-layer mode equation and thus, again a change in n_c results in a change in the effective mode RI, N_m of the MCLW and thus in the resonance angle of the waveguide mode. However, here the waveguide modes are leaky in the cladding and substrate due to the complex RI of the cladding layer and hence, the intensity of the waveguide mode is attenuated along the waveguide film.

15.3.2 Light Coupling Techniques and Mode Excitation

Several techniques can be employed to couple an optical beam propagating in free space into a thin-film optical waveguide. Grating coupling and prism coupling (Fig. 15.4) are briefly outlined here, since these methods are applied in actual reverse waveguide and MCLW configurations, respectively.

In general, the coupling elements make "communication" of the outside world with the waveguide possible and are used to monitor the above-mentioned effective refractive indices.

When a grating is used to excite the modes, usually the waveguide film is periodically modulated at the film/cover or film/substrate (or at both) interface (Fig. 15.4b). The incident beam illuminates this grating and one of the diffraction orders will excite the modes. The mode's effective refractive index can be calculated from the so-called grating equation[10,30].

$$N_m = n_{\mathrm{air}} \sin(\theta_m) + l\frac{\lambda}{\Lambda}, \tag{15.6}$$

where n_{air} is the RI of air, λ is the vacuum wavelength of the incident light, Λ is the periodicity of the grating modulation and l is the diffraction order. This technique is

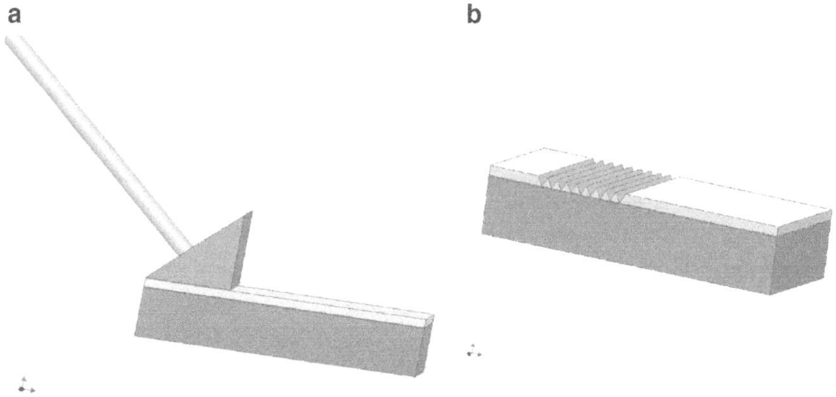

Fig. 15.4 Two common light coupling elements used for waveguides (**a**) a prism and (**b**) a grating

broadly used to excite the modes of a three-layer dielectric planar optical waveguide sensor. Here, typically, the in-coupled light intensity is measured by a photodiode at the end of the waveguide and the incident angle on the grating is varied, resulting in sharp resonance peaks corresponding to the modes of the waveguide[7,18]. This method is usually termed as angular interrogation.

The mode coupling condition for prism coupling (Fig. 15.4a) is similar to the grating condition. For prism coupling the prism often constitute the substrate of the waveguide structure, or the RI of the prism and that of the substrate is matched, and thus the mode's effective index can be calculated from

$$N_m = n_{\text{prism}} \sin(\theta_{\text{prism}}), \tag{15.7}$$

where n_{prism} is the RI of the substrate and θ_{prism} is the incident angle inside the prism at the interface of the prism and the remaining waveguide structure. In prism coupling, the incident light from the prism usually tunnels through a thin coupling layer with low refractive index or a metal layer with complex dielectric constant and excites the modes of the waveguide film on top of the coupling layer[10,20,21,25–27].

In both cases, the incident angle is monitored using a goniometer or a CCD and converted to effective refractive index values using the above equations.

Prism coupling is often used for reflectance measurements from waveguide structures rather than measurements of the guided light in the waveguide structure. This is advantageous when the excited mode damps during propagation, which is the case for MCLW modes. Therefore, for prism coupling the Fresnel reflection laws can be applied for calculating the intensity ratio between the reflected light and the incoupled light[25–27].

In actual arrangements typically the reflected light intensity is measured while the incident light angle is varied, resulting in peaks or dips in the reflected light intensity[19–21,23–25]. These peaks and dips can be connected to the leaky modes of the system and their position and shape is used to monitor the effective refractive index[19–21,24–27].

15.4 Waveguide Sensors Using Low Index Substrates: Reverse Waveguide Configuration

A three-layer, planar optical waveguide typically consists of a high-refractive-index film with RI n_F and thickness d_F sandwiched between two lower-refractive-index materials referred to as the substrate and the cover media with RIs, n_S and n_C, respectively. The cover media to be monitored is usually an aqueous solution with refractive index around 1.33. In conventional waveguides, the substrate is usually glass or polymer, which has refractive indices larger than that of the (aqueous) cover sample being investigated, typically 1.53 for glass. This geometry causes the evanescent field to extend deeply into the substrate and shortly into the cover[9,18]. However, by making the refractive index of the waveguide support less than the refractive index of the aqueous cover solution (for example using nanoporous silica with RI around 1.2 as the substrate) the waveguide mode will penetrate deeper into the cover media than into the substrate. The mode profile is "reversed". In addition, it is found that the mode penetration depth into the aqueous solution can be tuned, in principle, up to infinity. This is in contrast with the conventional geometry where the penetration depth into the cover solution is limited, typically to around 180–200 nm[9,18]. This type of waveguide geometry – using low RI substrates – is often termed "reverse symmetry waveguide" or simply "reverse waveguide" and was introduced by Horvath and co-workers[9,12,15] Waveguide sensors with this design has subsequently been analysed in several studies[16–18,31].

To understand the extended mode penetration depth using reverse waveguides, (15.5) needs to be revisited for the mode's penetration depth into the cover media. From (15.5) it is seen that the penetration depth depends on the wavelength, cover media index and the effective refractive index of the mode (N_m). For waveguide modes, the effective refractive index is less than the refractive index of the waveguiding film and larger than the maximum value of the refractive index of the substrate and cover media: $n_F > N_m > n_{max}$. The effective refractive index can be tuned in this range by changing for example the thickness of the waveguiding film (see mode equation (15.3)).

Figure 15.5 shows the cover penetration depth as a function of effective refractive index. It is immediately seen that for $n_S > n_C$ the penetration depth is limited and reaching its maximum value at $N = n_s$. For $n_s = 1.53$ (glass substrate) the maximum penetration depth is around 180 nm, using a light wavelength of 633 nm. However, by choosing a substrate with a refractive index *less* than the refractive index of the cover medium $n_C > n_S$ the allowed range of effective refractive indices is between n_C and n_F. Since, at $N = n_C$ the cover penetration depth goes to

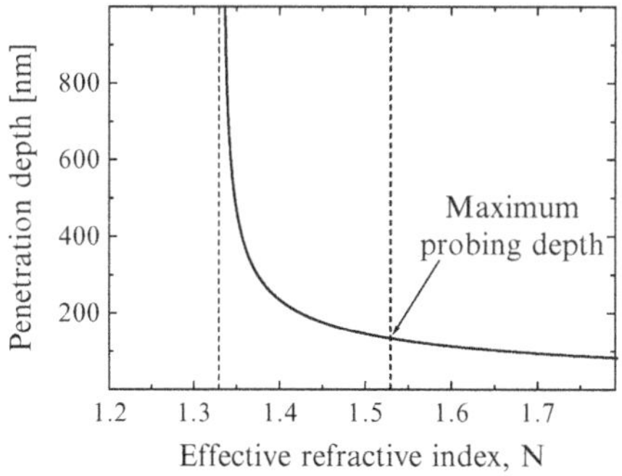

Fig. 15.5 Cover penetration depth as a function of the effective refractive index. Using a substrate with RI less than the RI of the aqueous cover solution, the penetration depth into the aqueous cover can be tuned up to infinity. While using glass as a substrate with RI of 1.53 the probing depth has a maximum value around 180 nm, using a light wavelength of 633 nm

infinity (referred to as cover cutoff) now, the penetration depth can be tuned in principle up to infinity.

Reverse waveguides can be realized by using freestanding waveguiding films (air as the "substrate")[9,32,33] or using nanoporous silica substrates with very low refractive index values, around 1.2[17,18]. Nanoporous silica contains nanometer-sized air pores embedded in glass, lowering the averaged (or effective) refractive index of the material without introducing significant loss effects. Using nanoporous silica, a reverse waveguide with a substrate RI around 1.2 has been realized and applied for various sensor applications by Horvath and co-workers[12,16,17,31].

More detailed analysis shows that the application of the low refractive index substrates not only increases the penetration depth into the cover media, but also – since the mode profile is reversed – places a larger portion of the electromagnetic power flowing in the waveguide structure into the cover media, thus increasing the sensor sensitivity for refractive index variations in the cover solution[9,15].

In the following, we compare the sensor performances of the conventional waveguide (CW) and the reverse waveguide (RW) configuration through the investigation of concrete waveguide structures[34]. In Fig. 15.6 the cover penetration depths Δz_C and cover (assumed to be homogeneous material) RI sensitivities $S\{n_C\} \equiv \partial N/\partial n_C$, are plotted for the two different waveguide configurations (CW and RW) as a function of waveguide film thickness d_F for the TE_0, TM_0, TE_1, and TM_1 modes[9,30] using a light wavelength of 633 nm. The modes of the CW ($n_S = 1.53$, $n_F = 1.8$) has a quite narrow possible range of penetration depths (around 50 nm) and the maximum possible penetration depth is only 140 nm.

The situation is dramatically different for the RW configuration ($n_S = 1.2$, $n_F = 1.575$), for which the cover penetration depth can be varied between ~100 nm and infinity for all the modes, with significantly improved cover index sensitivities (see Fig. 15.6b)[9,15]. For the CW configuration, the maximum obtainable sensitivity decreases with increasing mode orders compared to the zeroth-order modes. Contrarily, the RW configuration offers the possibility to use higher order modes for

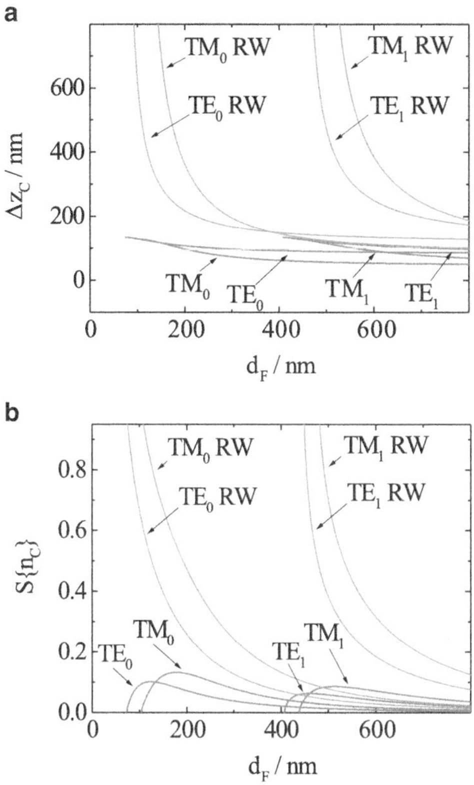

Fig. 15.6 Comparison of the cover penetration depths (**a**) and cover RI sensitivities (**b**) of conventional and reverse waveguides (marked with RW) for the zeroth- and first-order TE and TM modes. Reprinted from Ref. 34 with permission. © 2008 Elsevier

sensing since the maximum obtainable sensitivity for higher order modes is 1, just like for the zeroth-order modes (see Fig. 15.6).

15.4.1 Refractometry Using Nanoporous-Silica-Supported and Freestanding Waveguides

To fabricate nanoporous-silica-supported reverse waveguides, a glass substrate is first spin coated with a 1-μm thick layer of nanoporous silica (Nanopore, Inc., New Mexico, USA) and then coated with a 160-nm thick layer of polystyrene. In the polystyrene film, a surface grating with a period of approximately 480 nm and an amplitude of 10–15 nm is imprinted using a polydimethylsiloxane (PDMS) stamp with the grating structure[17]. It should be noted that the nanoporous silica layer is so thick that the evanescent field is practically vanishing in the glass support.

Hence, the mode only experiences the nanoporous silica, the film, and the cover medium resulting in a pure monomode operation.

To apply the waveguide for refractometry, the sensor chip is inserted into a goniometer setup. Here, the waveguide chip is rotated by a high precision rotation stage while a fixed He–Ne laser with wavelength $\lambda = 632.8$ nm illuminates the coupling grating. The grating is positioned at the centre of rotation[9,18]. During rotation the in-coupled light intensity is measured at the edge of the waveguide and monitored online as a function of illumination angle, α. In order to shift between TE and TM illumination, the polarization of the incoming laser light is controlled by a polarizer and a half-wave plate. A cuvette with tubes connected to a liquid pump is positioned on top of the waveguide, which makes it possible to flow various liquids across the surface of the waveguide.

In Fig. 15.7, four sensorgrams are shown for air and water cover media and for TE/TM polarized light. The peaks are correctly centred at the angles given by the grating equation $\sin(\alpha_0) = N_0 - \lambda/\Lambda$ and the half-widths are approximately 0.05°. It is seen that the shift in peak position from air to water is largest for the TM mode, which is due to the fact that the TM mode is closest to cutoff and therefore exhibit the largest sensitivity.

By altering the cover medium between pure water and a water/glycerol solution (corresponding to RIs of 1.331 and 1.355, respectively) the TE and TM peaks shift. This way 3.5 times larger sensitivities than that of a typical conventional waveguide was demonstrated in good agreement with theoretical predictions[15].

The main advantage of using the reverse waveguide configuration for extending the penetration depth into the cover medium is based on operating the waveguide close to the cover cutoff, i.e. close to the point at which the evanescent wave in the cover transforms into a propagating wave. Obviously, in that transition another phenomenon takes place, namely that the guided mode vanishes. This phenomenon was demonstrated experimentally[35] and showed that the incoupling resonance peaks decrease significantly in height, as the modes get closer to cutoff (as the RI

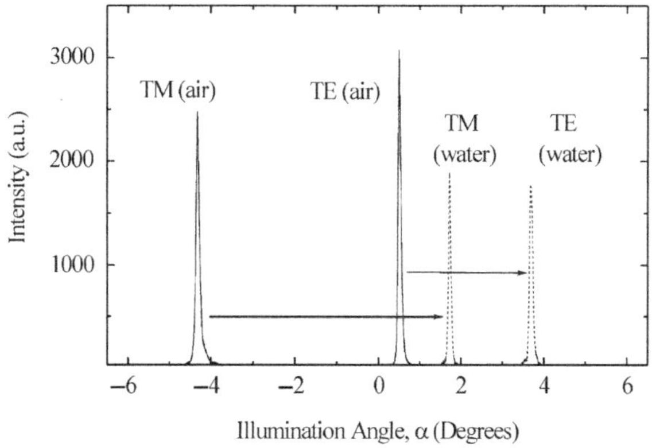

Fig. 15.7 Incoupling resonant peaks of nanoporous silica supported polystyrene waveguide. Reprinted from Ref. 13 with permission. © 2008 American Institute of Physics

of cover media is increased). This height change was suggested as an alternative transduction parameter to the typically used peak position transduction[35]. In Fig. 15.8, the two transduction mechanisms are compared by alternately changing the refractive index of the cover medium. The main difference in performance between the two mechanisms is seen to be the noise levels. Hence, TM detection limits of 5×10^{-6} and 6×10^{-5} refractive index units are obtained for the peak-angle and peak-height measurements, respectively. However, one has to emphasize that measuring the intensity of a signal is generally much easier than measuring an angle, so to keep the instrumentation price low the peak-height detection method may be a practical alternative.

The RW configuration supporting several modes was also demonstrated for refractometry. Here a 50 μm-thick freestanding glass plate supporting several modes was employed as a waveguide film. In this case, a simple refractometer was constructed using the disappearance–appearance of the higher order modes due to RI changes[33].

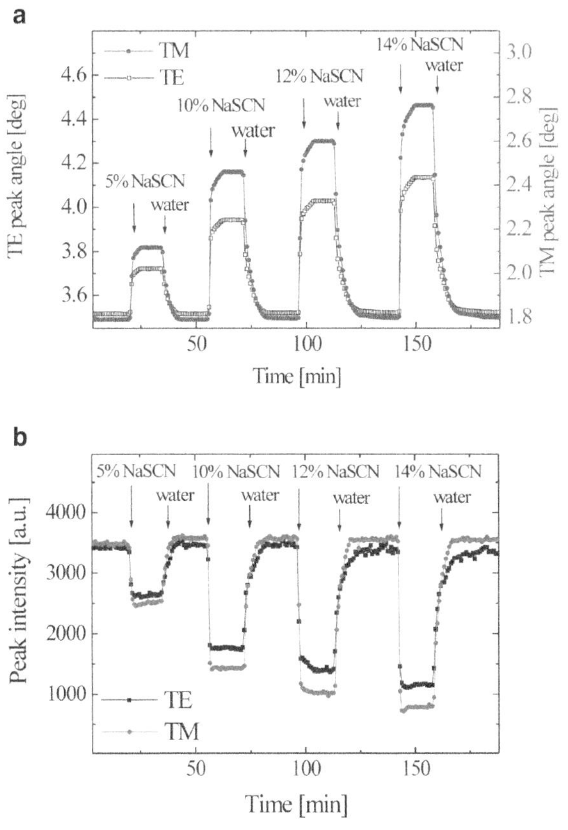

Fig. 15.8 Comparison of resonant peak position (**a**) and peak amplitude (**b**) transduction for a grating coupled reverse waveguide. Reprinted from Ref. 36 with permission. © 2008 American Institute of Physics

15.4.2 Bacterial Cell Detection Using Nanoporous-Silica-Supported Reverse Waveguides

The reverse waveguide sensor has been demonstrated for bacterial detection. In order to attach bacterial cells to the waveguide surface, the reverse waveguide chip was first plasma treated and coated with a thin layer of poly-L-lysine, a positively charged polymer. After this, a bacterial solution consisting of phosphate-buffered saline (PBS) with a concentration of 3×10^7 *Escherichia coli* K12 cells per mL was pumped through the cuvette. The resulting changes in TE and TM peak positions were monitored using the above-mentioned goniometer setup. It was seen that after exposing the waveguide surface to the bacterial solution, the peak positions shifted gradually to higher values, as bacterial cells attach to the surface and locally increasing the refractive index of the cover solution. After rinsing the cuvette system with PBS the peak positions levelled off, indicating that the bacterial cells that are already attached to the surface stayed immobilized by electrostatic forces. The waveguide chip was afterwards investigated under microscope, where clear images of the bacterial cells, 1×3 μm in size, were observed. The surface density was estimated to be 12,400 cells per mm^2 leading to a surface cell sensitivity of 1.3×10^{-6} and $1.65 \times 10^{-6^\circ}$ mm^2 per cell for the TE and TM mode operation, respectively. In view of the noise level of 10^{-4°, these sensitivities correspond to detection limits of 78 and 60 cells per mm2 for TE and TM mode detection[31].

15.4.3 Living Cell Detection Using Nanoporous-Silica-Supported Reverse Waveguides

The grating-coupled nanoporous-silica-supported reverse waveguide chip was also applied for monitoring the attachment and spreading of Human Dermal Fibroblast cells to the surface[16]. As in the bacteria experiments, the waveguide surface was coated with a thin layer of poly-L-lysine layer to improve cell attachment and spreading.

It was demonstrated that, as the fibroblast cells started to attach to the surface, the resonant peak shapes changed considerably. The peaks became broadened and appeared to split into two peaks[16]. We believe that the main reason for not observing a simple peak shift, as is the case for protein attachment and bacterial attachment, is due to the size of the cells, which are 10–100 times larger than the wavelength of the light. This implies that rather than averaging the refractive index of the cover layer being exposed to the evanescent field, the electromagnetic field considers the location of the cells as local waveguide areas with their own distinct cover refractive index. This type of peak splitting due to inhomogeneous coverage was predicted theoretically[36] and investigated by numerical simulations using the Local Interference Method suggesting a novel imageless microscopy to monitor surface cell patterns by analysing the resonance peak shapes[37].

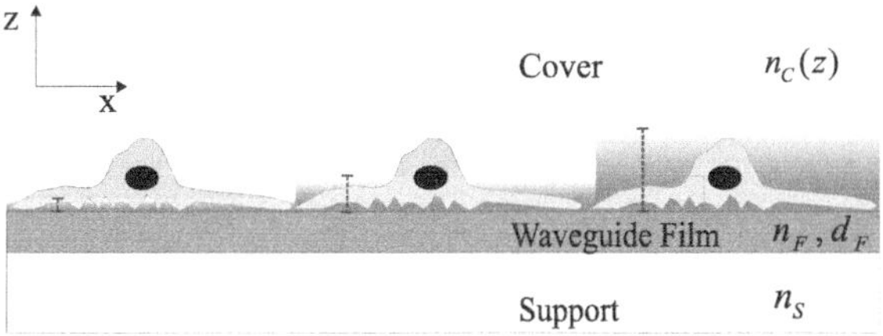

Fig. 15.9 Schematic illustration of multidepth waveguide sensing of living cells. Reprinted from Ref. 34 with permission. © 2008 Elsevier

The RW configuration has clear advantages in cell screening as different modes with significantly different penetration depths can be used to monitor RI variations at different depths (see Fig. 15.9)[34]. This is not possible by employing conventional waveguide configurations, since higher order modes have poor sensitivity and the possible range of penetration depths is limited, typically, to a range around 50 nm[34]. This multidepth label-free sensing approach was demonstrated by Horvath and co-workers using two- and four-mode waveguides supported by low refractive index nanoporous silica[34]. This type of approach can be advantageous – for example – when RI variations of cell organelles (like nucleus) or the movement of these organelles have to be monitored, or when these contributions have to be separated from – for example – secreted protein adsorption on the surface.

Combining the above-mentioned results suggests that quantitative information from grating-coupled RWs can be obtained for sample inhomogeneities parallel to the waveguide surface by analysing the resonant peak shapes[36,37]. The present technique can be extended to analyse 3D sample inhomogeneities.

15.5 Guided Wave Sensing of Polyelectrolytes

In recent years, the measurement of thick adlayers has received an increasing interest in the field of so-called polyelectrolyte multilayer films. These films sometimes have a thickness of several micrometers. Such large thicknesses are obviously not easily monitored using conventional waveguide sensors because of the limited penetration depth into the adlayer. Simply, waveguide sensors loses their sensitivity when the adsorbed layer thickness exceeds 2–3 times the penetration depth of the evanescent field and cannot be used to monitor films thicker than ~350 nm[12].

Horvath and co-workers extended the powerful waveguide sensing technique for thick film detection (>350 nm) by introducing an alternative detection strategy. Instead of evanescent wave sensing, they proposed to monitor the films by using waves that are guided in the adsorbed layer[12].

In order to achieve guiding in the deposited polyelectrolyte adlayer, the effective refractive index of the guided mode N must be lower than the RI of the adlayer, n_A, which is typically 1.36–1.5, while still being larger than the RI of the substrate, n_S.[10] This is not possible with commonly used substrate materials like glass (RI ~ 1.53). However, low refractive index substrate supported waveguides to overcome this limitation and the build up of thick adlayers can be monitored by reverse waveguide sensors[12].

15.6 Metal-Clad Leaky Waveguides

Deep-probe waveguide sensors can also be designed using more prevalent materials than presented for the reverse symmetry waveguide, namely by introducing a fourth layer to the waveguide structure. The extra layer should comprise a material with a complex RI value. This can be for example a metal layer, a polymer film or a dye layer. In this section, bacterial cell detection will be presented using waveguide designs including a metal layer or a dye layer. In the literature, these structures are referred to in various ways for example the metal-clad leaky waveguide (MCLW), metal-clad waveguide (MCWG), coupled plasmon resonance waveguides (CPRW) and, for the dye layer, dye clad leaky waveguide (DCLW). MCLW will be used *as a general term* and for a specific waveguide design, the material used for the cladding will be stated as part of the abbreviation by changing the M with the specific material used, for example Ag-CLW for a MCWG where silver is used for the cladding.

The deep-probe four-layered waveguides comprises a planar structure of a substrate S, a thin cladding of metal or dye or other complex-valued materials M, a dielectric waveguide film F and a cover medium C, with refractive indices and layers thickness n_l and d_l, respectively. The structure of these waveguides are thus similar to the conventional dielectric waveguide sensor but with an extra layer introduced between the substrate and the waveguide film. The MCLW can also be compared to the structure of an SPR-sensor. The difference between the two structures is that a dielectric waveguide film is added on top of the metal-coated layer for the MCLW structure.

The MCLW sensor can be operated in a number of operating setups. Here, two different setups will be discussed: reflection mode and scattering mode. In both situations, light is incident on the structure through the substrate. But, for the reflection mode, the reflected intensity is measured, while for the scattering mode the intensity of the light scattered from the captured biological particles on the sensor surface through the cover medium is measured. The light is coupled into the waveguide structure through a prism or a grating coupler. Similar to the dielectric waveguide, a waveguide mode can be excited in the waveguide film; however, for the MCLW it will be a leaky waveguide mode.

In reflection mode, the angle of incidence is scanned and thus the propagation angle of light in the waveguide film is varied, while monitoring the intensity of the reflected light by a detector. The measured sensorgram shows the reflectance vs. angle of incidence, and a dip in reflectance is identified at the resonance angle, θ_m, of the waveguide mode. The operation of the MCLW sensor is similar to the operation of the dielectric waveguide sensor; the measured change in position of the resonance angle corresponds to a given change in n_c (cover medium). Similar to the dielectric waveguide, waveguide modes are excited in the MCLW and are guided in the waveguide film. However, contrary to the dielectric waveguides, where the guided light experiences TIR at both the film/cover and the film/substrate boundary, the guided light in the MCLW is only totally internally reflected at the film/cover boundary; while at the film/metal boundary, a normal reflectance of the light occurs, where the light is partly transmitted into the metal cladding and the substrate and partly reflected back into the waveguide film at the film/ metal interface, which gives rise to an attenuation of the guided light in the waveguide film.

In scattering mode, the incident light is fixed at the resonance angle, θ_m, and scattered light from the sensor surface due to the presence of biological objects within the evanescent field is imaged in real time using a CCD camera. At the resonance angle, the light is reflected by TIR from the film/cover interface and an evanescent wave exists in the cover medium. The presence of bacterial cells on the surface will cause scattering of light. Evanescent wave light illumination is a favourable method of generating scattering and fluorescence excitation of objects near the sensor surface. This technique gives a dark background and provides a significant improvement in spatial resolution for objects near surface structures, in comparison with other microscopy techniques[38]. Scattering and fluorescence are very sensitive techniques, which are capable of detecting low concentrations of bacteria where most refractive index sensors give a negligible response[39].

15.6.1 MCLW as Detectors for μ-TAS Applications

It has become apparent that the limiting factor in reducing the size of a micro-total analytical system (μ-TAS) device is primarily set by the analyte detector. To date, small-volume detection has generally focused on optical measurements[40–42]. Optical detection in μ-TAS systems usually takes place through the depth of the microfabricated channels, which has resulted in comprehensive work on the fabrication of high aspect ratio channel structures. Recently, the use of optical waveguides as detectors in μ-TAS systems has been reported[19,40–42]. In these systems, a high-index waveguide forms the top or bottom wall of the channel, permitting the optical detection via the evanescent field present at the channel-waveguide boundary. While this sensing scheme has many advantages, it has the major disadvantage of only sensing a thin layer (about half a wavelength) at the channel wall. If the channel is much deeper than half a wavelength, much of the material in the channel cannot be sensed. This is a particular problem when using pressure-driven flow in microchannels, as the parabolic flow profile results in sensing of the stagnant layer at the channel walls. Using other transport methods such as electroosmosis limits, but not eliminates, this problem. To overcome this disadvantage requires a structure is required that permits waveguiding in a low index medium (the normally aqueous medium in the channel) bounded by higher index materials (the μ-TAS substrate). Ideally, the optical mode should reach its highest intensity in the centre of the channel, but if this cannot be achieved, an increase of the penetration depth of the evanescent field at the channel walls will increase the sensitivity to the material to be sensed.

Here, the MCLW devices are demonstrated as detectors for μ-TAS applications[19]. The chips were fabricated using silica as the leaky waveguiding layer in combination with diffraction grating coupling elements and an injection-moulded flow-channel structure. These devices are used designed in order to increase the overlap between the extension of the evanescent field at the channel wall and the materials within the channel, the "cover" solution. These devices are used to detect the channel contents by monitoring the resonance angle at the liquid–waveguide interface; these variations are evidenced through changes of the decoupling angle measured using a linear CCD. Figure 15.10 shows the results of MCLW chips irrigating with glycerol/water solutions [0–10% (w/v)]. Glycerol was used as model, for simplicity, since this compound has a low molecular weight and it is notoriously difficult to analyse using more conventional optical detection techniques. This is due to the fact that this compound does not contain chromophores and hence direct quantitative analysis is not possible by using UV, absorption or fluorescence techniques. Figure 15.11 shows the linear response for the lowest order mode of the MCLW and the resonant mirror (RM) sensor to glycerol concentration. This clearly demonstrates the increased sensitivity of the MCLW TE_0 mode compared with the RM TM_0 mode. Theoretical predictions indicated that

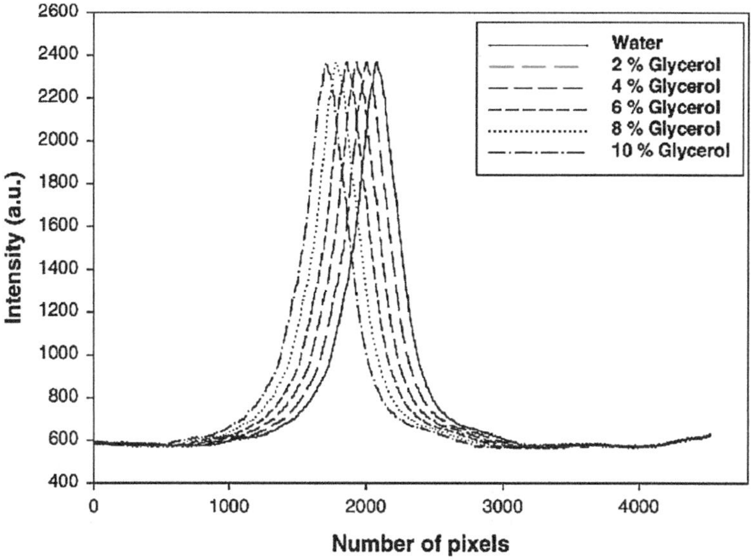

Fig. 15.10 The sensitivity of the MCLW device for refractive index monitoring of various concentrations of glycerol in aqueous solution. Reprinted from Ref. 19 with permission. © 2008 Elsevier

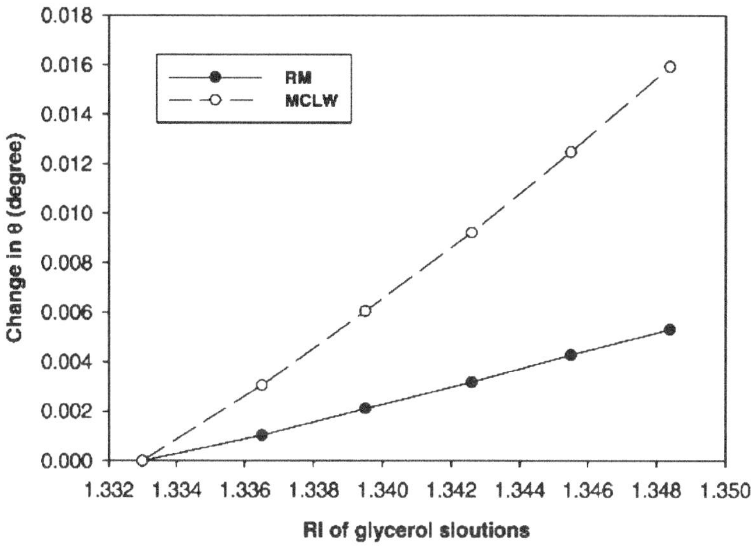

Fig. 15.11 Comparison of the mode movement of the RM and MCLW devices for different concentrations of glycerol in aqueous solution. Reprinted from Ref. 19 with permission. © 2008 Elsevier

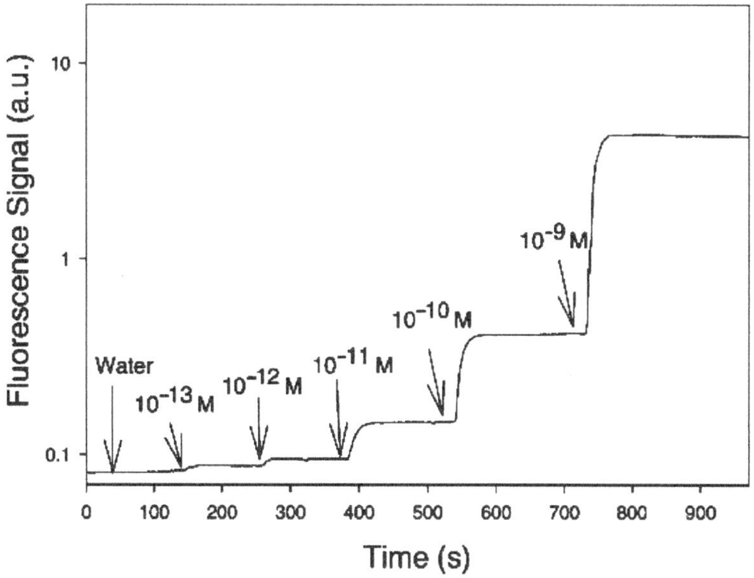

Fig. 15.12 The detected emission of various concentrations of fluorescein solutions excited by a 473-nm blue laser using the MCLW chip (*arrows* indicate the times at which various solutions were added). Reprinted from Ref. 19 with permission. © 2008 Elsevier

the sensitivity of the MCLW TE_0 mode should be 3.33 times greater than that of the RM TM_0, while the experimental results show a factor of 3.10. This value is in good agreement with the predicted value, and the small discrepancy can be attributed to differences in the thickness of actual and modelled waveguide layers.

The MCLW chip with moulded grating and flow cell is also demonstrated for fluorescence detection. Figure 15.12 shows the MCLW for fluorescence detection using different concentrations of fluorescein solutions (10^{-13}–10^{-7} M). The limit of detection was determined to be 10^{-13} M for fluorescein solution.

15.6.2 Chemical and Biochemical Applications

The MCLW sensor has been exploited by several groups for a wide range of experiments. However, both the design of the sensor and the detection scheme used in connection with the monitoring vary. This section reports on experiments using angular interrogation, absorbance detection and fluorescence detection.

15.6.2.1 Angular Interrogation

Salamon et al. have used the MCLW for a wide range of experiments and uses the term Coupled Plasmon-Waveguide Resonator (CPWR) for the configuration[43–45]. The group has employed different configurations, but the most used configuration comprises a three-layer structure with a BK-7 glass prism ($n = 1.515$) as substrate, a silver cladding with thickness ~50 nm ($n = 0.065 + i4$; $\times = n2 = -16 + i0.52$) and a dielectric waveguide film of 460-nm SiO_2 ($n = 1.47$), which constitutes the surface to the sample to be analysed – in this case the solid-supported lipid bilayer. As the light source, the group has most often used a He–Ne laser with wavelength 632.8 nm.

Coupled plasmon-waveguide resonance (CPWR) spectroscopy was used to evaluate the birefringence and linear dichroism of anisotropic thin films such as proteolipid membranes. These optical properties are related to molecular properties such as polarizability, shape and orientation. The proteolipid membrane consists of 1–10 mol% of an acyl chain chromophore-labelled phosphatidylcholine (PC) incorporated into a solid-supported PC bilayer deposited onto a hydrated silica surface. The CPWR measurements were made of refractive index and extinction coefficient anisotropies with two exciting light wavelengths, one of which is absorbed and one of which is not absorbed by the chromophore. The generated results were used to calculate longitudinal and transverse molecular polarizabilities, the orientational order parameter and average angle between the longitudinal axis of the lipid molecule and the membrane normal and the molecular shape factors of the lipid molecules. The values thereby obtained are in excellent agreement with parameters determined by other techniques, and provide a powerful tool for analysing lipid–protein, protein–protein, and protein–ligand interactions in proteolipid films[45]. Salamon et al. also used the CPWR and impedance spectroscopy for studying the interaction between the cell-penetrating peptide, penetratin and solid-supported lipid bilayer membranes consisting of either egg phosphatidylcholine (PC) or a 75/25 mol% mixture of egg PC and palmitoyloleylphosphatidylglycerol[46].

Signal transduction in biological membrane systems has also been studied using the CPWR. In this case, the sensor was applied to characterize the kinetics, affinities and conformational changes of cloned δ-opioid receptor from human brain incorporated into a lipid bilayer involved in the ligand binding to G-protein coupled receptors. The authors concluded that the receptors exist in at least four different conformational states depending upon whether it is unoccupied by a ligand or occupied by an agonist, an antagonist or an inverse agonist and each of these states interact with G-proteins quite differently[47–51].

Zourob et al. demonstrated the MCLW sensors for protein A binding to immobilized anti-protein A and organophosphorous pesticides using refractive index changes[52,53]. The anti-protein A was immobilized on agarose hydrogel as waveguiding layer[52]. The used MCLW consists of glass slides ($n = 1.51$) coated with 8.5 nm Ti ($n = 2.11–2.88i$ at 2 eV per 620 nm) followed by a 50 nm thick silica layer. The silanized chip was then coated with a hybrid of 4% agarose IIA and IIB hydrogel. In this system, the antibody for protein A was immobilized on the agarose hydrogel using glutaraldehyde as cross-linker. The sensor showed improved limits of detection compared with the conventional evanescent field sensors[54]. This is due to the fact that the MCLW sensor entrapped more anti-protein A as it has a thicker sensing layer than the evanescent sensors (see Fig. 15.13).

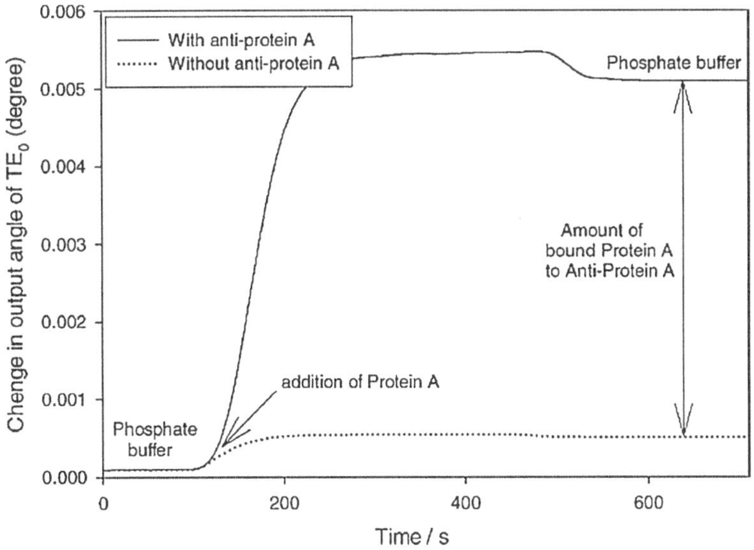

Fig. 15.13 Change in the output angle of the TE_0 MCLW mode as protein A binds to anti-protein A. Reprinted from Ref. 52 with permission. © 2008 Elsevier

Absorbing materials solvent blue-clad leaky waveguide chips were employed for the detection of OP pesticides by monitoring the change in the refractive index of a pH-sensitive hydrogel[53]. The chip consists of a polymethylmethacrylate (PMMA) polymer substrate integrated with injection-moulded gratings coated with 30-nm thick solvent blue absorbing layer followed by a 3-μm thick waveguiding layer where the enzyme was entrapped. The advantage of inserting the absorbing material layer is that it absorbs a large proportion of any scattered light or autofluorescent background from the polymer substrate. This is particularly useful when the injection-moulded chips with the integrated grating are being used. Hence, optical anisotropy is introduced into the finished device. The waveguiding layer of the devices was designed to support one guided mode (TE_0) and to allow the maximum interaction of the light with the sensing layer, where the enzyme was entrapped along with the suitable indicator. This thick sensing layer increased the amount of entrapped enzyme within the sensing layer, and resulted in an enhanced sensitivity through increasing the change in the pH around the enzyme.

The OPH enzyme was covalently immobilized on a UV curable pH-sensitive hydrogel waveguiding layer. The pH-sensitive hydrogel consists of poly-HEMA gel containing 5 mol% ethylene glycol dimethyacrylate (EDMA) and 6 mol% of the functional monomers methacrylic acid coated onto silanized chip. Control waveguiding layers were also constructed by immobilizing bovine serum albumin using the same procedure.

It is known that OPH catalyses the hydrolysis of a broad range of OP pesticide compounds, which generates up to two protons as a result of each hydrolysis reaction, thus making it possible to direct neurotoxin detection by measurement of the pH change associated with the enzyme activity[55,56]. Using one or more ionic or ionizable monomers in a cross-linked polymeric backbone makes it possible to detect the change in the pH around the enzyme as a result of influencing the charge of the ionic monomers[57–59]. This causes the hydrogel to swell or to shrink depending on the degree of influence on the charges of the ionic monomers, which can be detected easily by the optical waveguide sensors. This is a favourable approach, as unlabelled detection for OP can be achieved without the need for chromophore/fluorophore labelling, which suffers from major problems such as being time consuming, moderately expensive and problems with dye bleaching in the case of entrapment or photobleaching in the case of using fluorophores. In addition, to avoid these problems the hydrogel has a high percentage of water (aqueous environment), which is a good environment for preserving the enzyme activity. It was found that using 5 mol% of EDMA and 6 mol% of MAA provided a flow-stable hydrogel (able to shrink/swell many times without collapsing) with a stable and sensitive response. Operating under these conditions, the biosensor was able to detect as low as 1.5×10^{-7} M of paraoxon with very good accuracy and selectivity. Other non-organophosphate pesticides, such as simazine and triazine, did not give any signal when they were incubated with the sensor for more than 5 min. The control waveguide with immobilized bovine serum albumin (BSA) did not give any response when incubated with paraoxon. The biosensor was completely stable for at least 30 days when it was stored at 4°C in 20 mM phosphate buffer

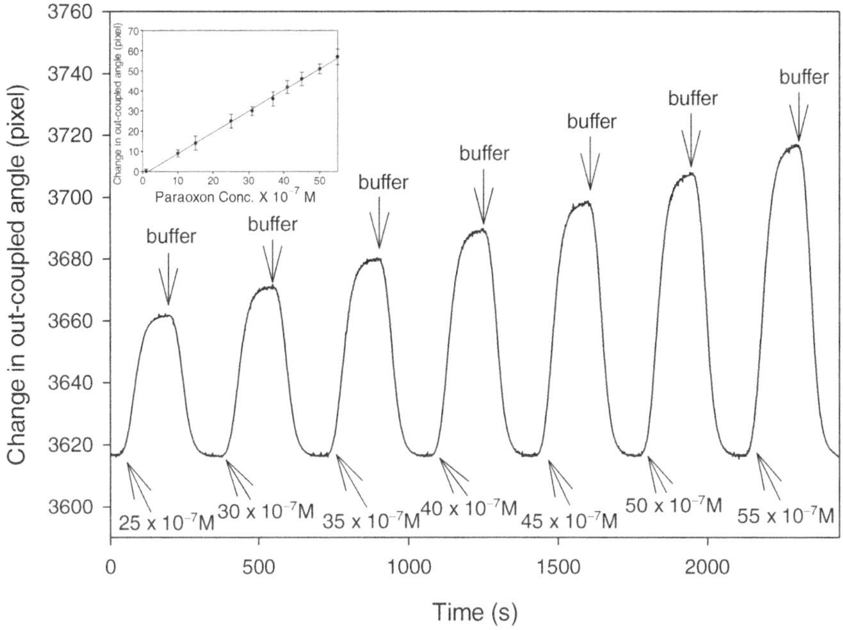

Fig. 15.14 The response of the TE_0 mode to different concentrations of paraoxon and the calibration curve. Reprinted from Ref. 53 with permission. © 2008 The Royal Society of Chemistry

(pH = 8.3) in 100 mM NaCl, 2.7 mM KCl and 30 mM $CoCl_2$. Figure 15.14 shows the OPH enzyme biosensor response and the calibration curve when incubated with different concentrations of OP samples. These results revealed that after the sensor had been incubated with paraoxon, the hydrogel swelling and the resulting shift in the out-coupled angle is proportional to paraoxon concentration. There was no response when the sensor was incubated with triazine as a control analyte. The OPH sensor response showed the high reproducibility to a periodic exposure to 27×10^{-7} M paraoxon [relative standard deviation (R.S.D) = 4.8%, $n = 7$].

15.6.2.2 Absorbance Detection

The same group demonstrated the MCLW for the detection of organophosphorous pesticides using change in absorbance[52]. In this case, acetylcholinesterase enzyme was entrapped in a sol–gel waveguiding layer with bromocresol green acid=base indicator. Acetylcholine was used as the test substrate for the characterization of the biosensor enzyme assay and to quantify the activity of the entrapped enzyme. When acetylcholine is introduced into the enzyme sensor, it is biocatalysed into acetic acid and choline. The liberated acetic acid causes a pH decrease in the local environment of the enzyme sensor. This pH change is apparent as a colour change (yellow to blue) of bromocresol green, leading to a decrease in the reflectivity of mode intensity when measured at 610 nm. An indirect method was used for the detection of organophosphorous pesticides using a three-step assay. First step requires incubating the sensor surface with acetylcholine to find the enzyme activity followed by incubating the sensor with the organophosphorous pesticide, which inhibits the enzyme. Finally, the activity of the enzyme sensors will be determined by incubating the sensor again with acetylcholine to find the activity of the enzyme.

The good reproducibility of the sensor to repetitive injections of acetycholine is very important to the measurements accuracy due to the fact that the AchE-based sensor to the determination of inhibitors depends on its reproducible response to the substrate solution. This requirement arises from the fact that the detection process involves the measurement of the enzyme activity before and after the exposure to the inhibitor. The change in sensor response must be attributable to the inhibitor and not to poor or non-reproducible sensor performance. It was documented that the response of the AchE-based optical sensor to a periodic exposure of 50×10^{-4} M acetylcholine chloride was reproducible (R.S.D. = 1.83%, $n = 7$). Depending on the acetylcholine chloride substrate concentration, it takes approximately 1 min to reach the maximum change in the mode intensity and approximately 2.4 min to recover the original mode intensity. The response of the sensor is linear to acetylcholine concentration in the range 2.5×10^{-5}–40×10^{-5} M ($R^2 = 0.9994$). Figure 15.15 shows example of the response of the AchE-based sensor with a 1-μM

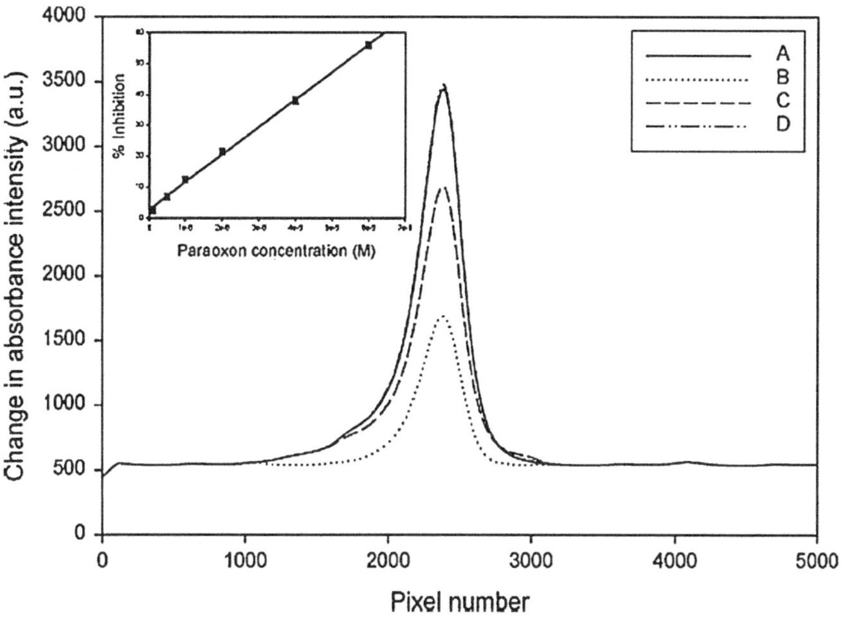

Fig. 15.15 Response of the AchE TE_0 mode of the MCLW sensor to 1 μM paraoxon, regeneration by 2-PAM and the calibration curve. (A) Tris buffer, (B) AchCl before paraoxon inhibition, (C) AchCl after paraoxon inhibition and (D) regeneration by 2-PAM. Reprinted from Ref. 52 with permission. © 2008 Elsevier

paraoxon sample and its regeneration by 2-pyrimdine aldoxime (2-PAM). It is seen that a reduction in absorption of BCG is obtained for the AchE enzyme after the sensor has been incubated with paraoxon, compared with that obtained prior to the incubation, indicating that the enzyme remains inhibited. The difference in the pH shown in Fig. 15.15 before (b) and after (c) pesticide exposure can be used for the analysis of pesticides. A range of concentrations of freshly prepared paraoxon solutions were analysed at a flow rate of 500 μL min^{-1} for 5 min. Figure 15.15 shows the corresponding calibration curve for the paraoxon using the porous sol–gel and demonstrates good reproducibility ($R^2 = 0.9995$) for three consecutive analyses. The limit of detection of paraoxon was calculated to be 6×10^{-9} M. This concentration corresponds to the response that has been taken as the value of the intercept plus three times the standard deviation of this value.

The same method was used in the solvent blue-clad leaky waveguide chips for the direct detection of paraoxon using OPH enzyme[53]. In this part of the work, bromocresol green (BCG) acid-base indicator was entrapped within a sol–gel matrix along with the OPH enzyme. The liberated two protons from the hydrolysis process of the OP decreased the pH in the local environment of the enzyme sensor. This pH change was apparent as a colour change of bromocresol green, leading to an increase in reflectivity of the TE_0 mode intensity when measured at 610 nm. Figure 15.16 shows examples of the response of the OPH-based sensor and the

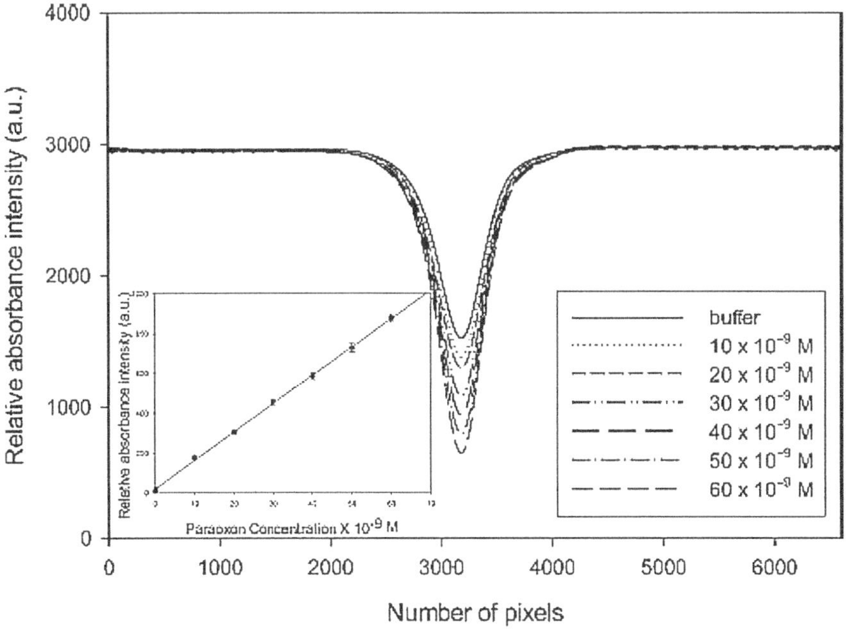

Fig. 15.16 The change in the TE_0 solvent blue-CLW mode intensity and the calibration curve for different concentrations of paraoxon. *Error bars* represent ±1 standard deviation, $n = 4$. Reprinted from Ref. 53 with permission. © 2008 The Royal Society of Chemistry

calibration curve for different concentrations of paraoxon. The limit of detection was calculated to be 4×10^{-9} M for paraoxon and parathion and 8.1×10^{-8} M for diazinon. In this work, the detection limit was set to the concentration giving a signal, which was three times higher than the background noise. The response of the OPH-based optical sensor to a periodic exposure of 30×10^{-9} M paraoxon was reproducible (R.S.D. = 1.7%, $n = 6$).

15.6.2.3 Fluorescence Detection

Thick planar sol–gel film waveguide-based MCLW biosensor was used for the detection of glucose[52]. The detection in this system was based on a fluorescent, pH-sensitive, ruthenium-ligand complex entrapped in a sol–gel, along with glucose oxidase enzyme. The sensing scheme is based on the measurement of the quenching of the fluorescence of the ruthenium complex due to protonation of the excited state of the complex when gluconic acid is released by the biooxidation of glucose[60]. The sensor was not tested directly after preparation, but tested after storage at 4°C for 20 days, because the pore structure of the sol–gel undergoes complex changes for 3 weeks after preparation[61]. In order to prevent densification of the pore structure during ageing, which impairs the diffusion of glucose into the matrix, sorbitol was added to the sol-gel. Before the measurements, the sorbitol was washed out of the matrix leaving pores, which the glucose can easily enter. Without addition of sorbitol, the apparent activity of glucose is approximately 10-fold lower than the expected. The typical response of the sensor for different concentrations of glucose is shown in Fig. 15.17. The limit of detection was calculated to be 3 μM. The detection limit is given by the intercept value plus three times the standard deviation of this value. The reported limit of detection was 15 times more sensitive than the corresponding leaky waveguide sensor developed by Hulme[54]. Figure 15.17 also shows the resulting calibration graph for the sensor over the range of 1–50 μM glucose. The developed sensor showed a good reproducibility of fluorescence quenching of the ruthenium complex due to the liberated gluconic acid for a periodic exposure of 10×10^{-6} M glucose (R.S.D. = 1.7%, $n =$ 6). This indicates that there is no diffusion limitation of the glucose analyte or product accumulation due to the increasing porosity of the sol–gel caused by the sorbitol.

Polythiophene-CLW was used in the direct detection of paraoxon using pH sensitive, fluorescein isothiocyanate (FITC) covalently immobilized to the OPH enzyme entrapped in a porous sol–gel waveguiding layer[53]. The sensing scheme is based on the measurement of the quenching of the fluorescence of the fluorescein isothiocyanate due to the liberation of protons as a result of the hydrolysis of the OP pesticides[55,56]. Figure 15.18 shows the typical response of the OPH sensor as well as the calibration curve to different concentrations of OP pesticides. The detection limit of the sensor using the OPH enzyme for paraoxon and parathion was 2.3×10^{-7} M and 9.3×10^{-6} M for diazinon. This detection limit is set to the value of the

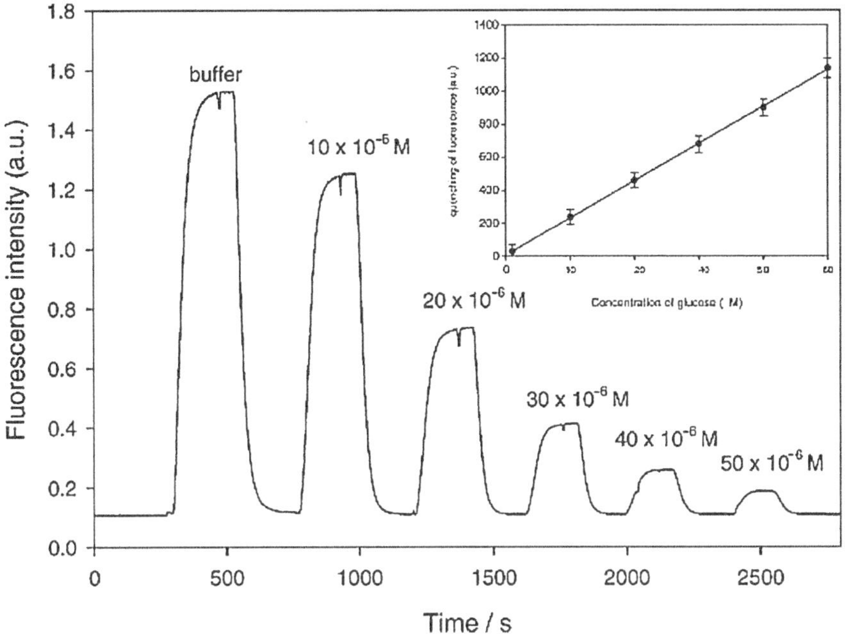

Fig. 15.17 Response of glucose oxidase immobilized in sol–gel using MCLW for detection of various concentrations of glucose and calibration curve. Reprinted from Ref. 52 with permission. © 2008 Elsevier

concentration that gave a signal three times the background noise. The sensor reproducibility of fluorescence quenching of the fluorescein isothiocyanate was very good (R.S.D = 3%, $n = 7$) due to the liberated protons for a periodic exposure of 10×10^{-7} M paraoxon.

15.6.2.4 Integrating the Sensor with the Electric field

The metal layer in the MCLW can be used to electrochemically control the binding of analytes to the immobilized recognition elements. This yields such effects as enhanced diffusion of different analytes through the immobilization waveguide matrix and accelerates the binding and reduces the non-specific binding. Zourob et al.[52] applied the same principle to enhance the diffusion of Janus Green B dye by application of an electric field and regeneration of the ligand by reversing the potential. Janus Green B is a cationic azine monoazo dye, which weakly interacts with the hydroxyl groups within the agarose by ionic and van der Waals forces[54]. Janus Green B was chosen as it is used in the staining of deoxyribonucleic acid (DNA) in agarose gels. The dye binds to the negatively charged phosphate group

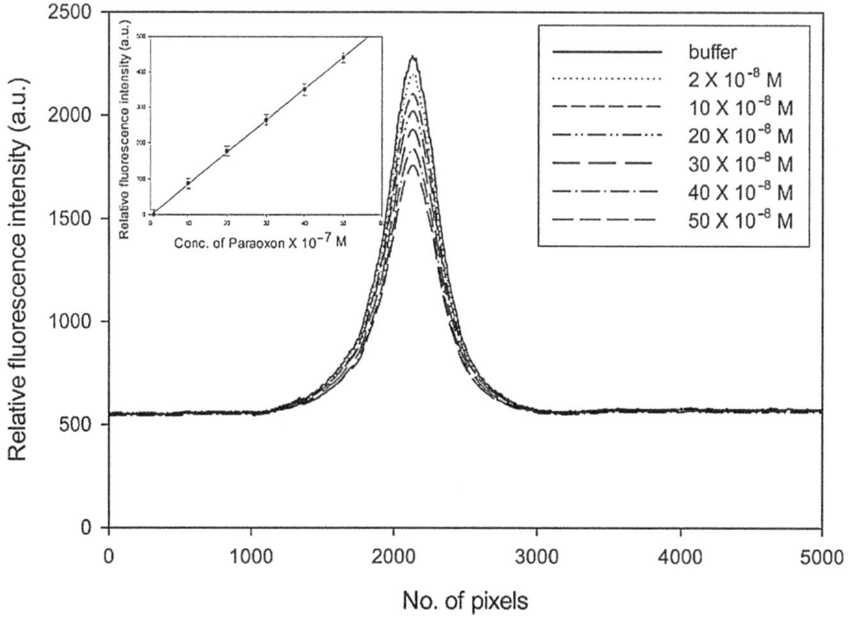

Fig. 15.18 The decrease in the fluorescence of FITC-labelled OPH TE_0 mode of PT-CLW and the calibration curve for different concentrations of paraoxon. *Error bars* represent ±1 standard deviation, $n = 4$. Reprinted from Ref. 53 with permission. © 2008 The Royal Society of Chemistry

on the helical backbone of the DNA[62]. It was reported by Hulme that this dye takes 30 min to bind to the agarose and it is difficult to remove it completely from the agarose matrix[54]. In this work a negative 1.3 V potential was applied to the metal layer with respect to the indium tin oxide (ITO) cover layer of the flow cell, to enhance the diffusion of the dye, so that the binding of the dye occurred in less than 2 min as shown in curve C in Fig. 15.19. Curve B shows that this process led to a decrease in the TE_0 mode intensity compared to that obtained when the dye was incubated for 30 min without application of the electric field. Reversing the potential resulted in the removal of most of the dye molecules from the matrix, allowing more than 98% of the original mode intensity to be recovered, as shown in curve D. The same principle was used to enhance the diffusion of biotin to the immobilized anti–biotin–FITC conjugate on an agarose matrix[52]. Similarly 1.3 V positive potential was applied to the metal layer with respect to the ITO flow-cell cover layer, resulting in an enhancement of the binding of biotin to the anti–biotin–FITC conjugate. This led to an increase in the TE_0 mode height of the MCLW due to the fluorescence emission. Contrarily, reversing the potential resulted in a negligible decrease in fluorescence emission. This indicates that the unbound biotin did not affect the fluorescence signal. Clearly, the electric field significantly reduces the analysis time.

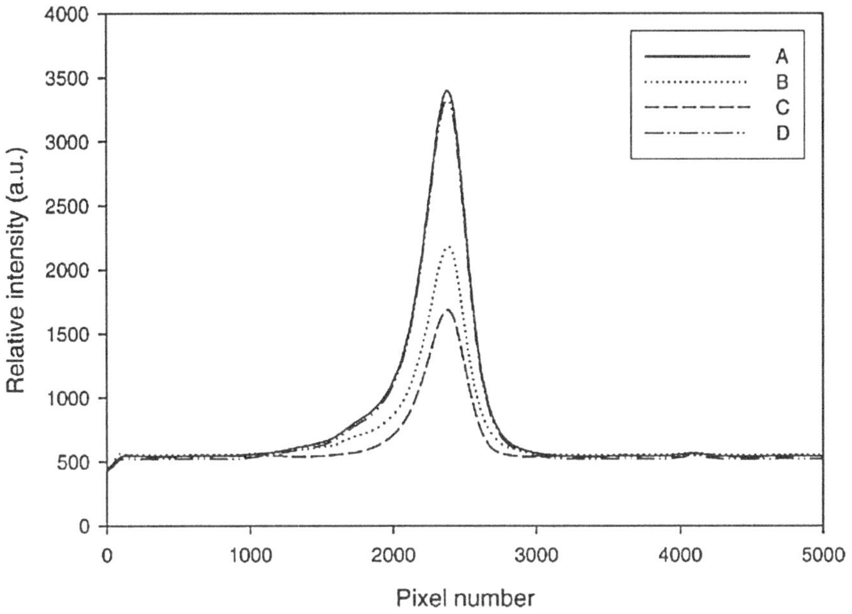

Fig. 15.19 Change in the absorbance of the TE_0 mode of the MCLW mode to Janus Green B at 610 nm when an electric field is applied. (A) Deionized water, (B) Janus green B after 30 min without applying electric field, (C) Janus green B after applying the electric field for 2 min and (D) Janus green B after reversal of the potential for 2 min. Reprinted from Ref. 52 with permission. © 2008 Elsevier

15.6.3 Bacterial and Cell Detection

Mammalian cell detection has been demonstrated using an Ag-CLW sensor operating in reflection mode, and a Si-CLW sensor has been applied for bacterial detection, where the sensor was operating in reflection- and scattering mode.

15.6.3.1 Reflection Mode: Detection of Bacterial Cells

The measurement of RI is a favourable technique to study biomolecular interactions, allowing real-time analysis of biospecific reactions without the need for biomolecular labelling. Zourob et al. employed MCLW sensors[20–23] and polythiophene clad leaky waveguide (PT-CLW) sensors for bacterial detection using RI changes[24]. The chips were modified by anti-BG antibody using glutaraldehyde on silanized sensor surface. Then the sensor was exposed to different concentrations of BG spores (10^3–10^8 spores per mL). Figure 15.20 shows the PT-CLW sensor response for different concentrations (1×10^5, 3×10^5 and 7×10^5 spores per mL). The limit of detection was estimated to be 8×10^4 cells per mL when 5 mL of BG spores suspension was pumped through for 20 min. This value is given by the

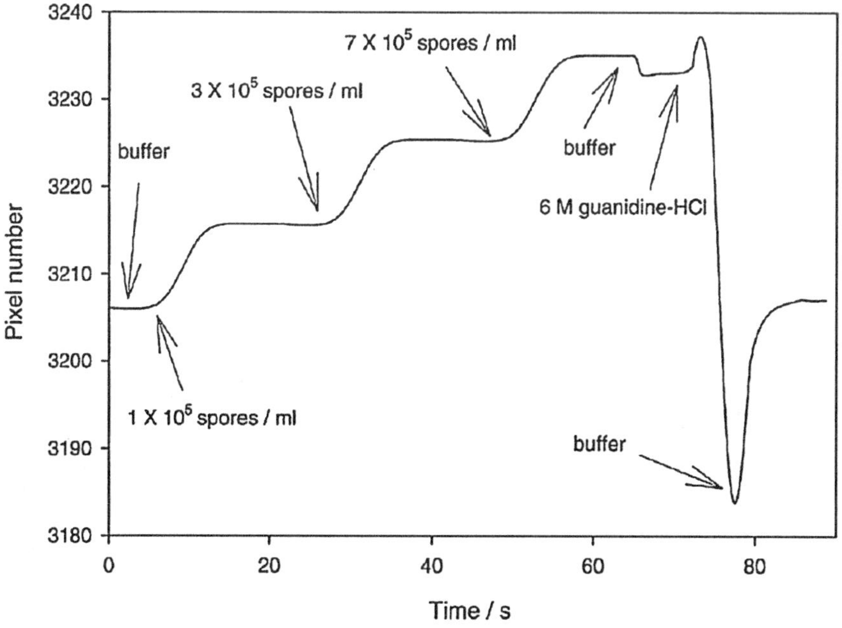

Fig. 15.20 RI sensorgram for 1, 3 and 7 $\times$ 10^5 spores per mL of bacteria using PT-CLW sensor. *Arrows* indicate the times at which various solutions were added. Reprinted from Ref. 24 with permission. © 2008 Elsevier

concentration corresponding to a response of the intercept plus three times the standard deviation of this value. It can be concluded from the result that extending the evanescent field to the whole volume of the cell improves the detection limit for bacterial monitoring in comparison with short penetration depths of the evanescent field sensors such as SPR and RM.

Skivesen et al.[27] have performed experiments detecting skin cells using both a dip- and a peak-type MCLM. The experiments compare the peak- and dip-type MCLW using two different types of skin cells: normal human dermal fibroblast (NHDF) and human ceratinocyte cell line (HaCaT). The waveguides structure were in case of the dip-type MCLW glass substrate ($n = 1.517$), silver cladding (60 nm), followed with a spin-coated polystyrene waveguiding layer ($n = 1.59$, $d = 330$ nm) and in the peak-type MCLW glass substrate ($n = 1.517$) coated with titanium cladding (5 nm), followed by silica waveguiding layer ($n = 147$, $d = 240$ nm). The sensors were operated in reflection mode and the two types of MCLW sensors both showed very sharp and distinct resonances. Both sensors have a large probing depth of the evanescent field in the cover media. The dip-type MCLW can be tuned to obtain either a low or a high penetration depth by adjusting the film parameters and thus, the sensor can be used for measurements on both thin adlayer and micron-scale objects. The dip-type sensor can be operated in multimode operation or supporting both a TE- and TM-mode resulting in modes with different penetration depths in the cover, giving a sensor that can measure at different depths in the cover medium. Thus, the dip-type MCWG sensor has the same advantages as the reverse symmetry dielectric waveguide. Contrarily, for the peak-type MCWG sensor it is not possible to tune the penetration depth. Detection limits of 1 cell per mm^2 are achievable with both MCWG sensor types but the dip-type shows a slightly improved detection limit compared with the peak-type sensor. Only the results of the peak-type experiment are shown here in Fig. 15.21. Figure 15.21a shows three sensorgrams taken at different times during the cell settlement. It is seen that the peaks tend to (1) shift towards larger illumination angles and (2) decrease in intensity, as cells settle at the surface. The increase in angle is due to the fact that the refractive index of a cell is larger than water, typically 1.35 (http://www.olympusmicro.com), whereas the reduced peak intensity is caused by the fact that the cell diameter is larger than the wavelength of the light. Hence, the cells are not perceived as a homogeneous layer with uniform thickness and refractive index, but rather a collection of particles, which is generally known to deform waveguide resonances[37].

The detected peak-angle vs. time is shown in Fig. 15.21b, in which a total peak shift of 0.2°, or $\Delta N' = 1.4 \times 10^{-3}$, is achieved. If this is compared to the observed surface cell density σ of 200 cells per mm^2, obtained from the microscope image in Fig. 15.21c, we arrive at a surface cell sensitivity ($\Delta N'/\Delta\sigma$) of 7×10^{-6} (cells per mm^2)$^{-1}$. The raw signal-to-noise ratio was determined to be 23 leading to a detection limit of 8.7 cells per mm^2. This is slightly above the detection limit of the dip-type sensor. However, by using point averaging over several data points, this figure could be reduced by at least an order of magnitude. The dip-type Ag-CLW sensor shows a slightly lower detection limit of 8 cells per mm^2. The cell detection limits derived above are based on the microscope image in Fig. 15.21c. However, the rounded shapes of the cells suggest that they might not be completely adherent, indicating that the cell membranes are probably not as close to the surface as they would have been, had the cells been adherent. Hence, one could expect larger sensitivities in the case of well-spread, adherent cell bodies.

15.6.3.2 Scattering Mode: Detection of Bacteria

Peak-type Polythiophene-CLW (PT-CLW) chips with a sol–gel guiding layer have been used for bacterial detection in scattering mode. Here, the sensor structure is illuminated at resonance angle for the peak resonance[24]. Antibody for BG spores were immobilized via a glutraldehyde activated surface. The sensor surface was exposed to different concentrations of BG spores (10^2–10^8 spores per mL) to find the detection limit of the device. The PT-CLW sensor response after applying a concentration of 1×10^6 spores per mL is shown in Fig. 15.22, which shows a typical response of scattered light from captured BG bacteria spores. The sharp spikes represent individual particles on the surface. The plots were analysed to count the number of particles present on the sensor surface. The MCLW was found to generate a lower background signal than the PT-CLW, due to the smooth surface of the vacuum-deposited silica on the MCLW, PT-CLW produce a rougher surface

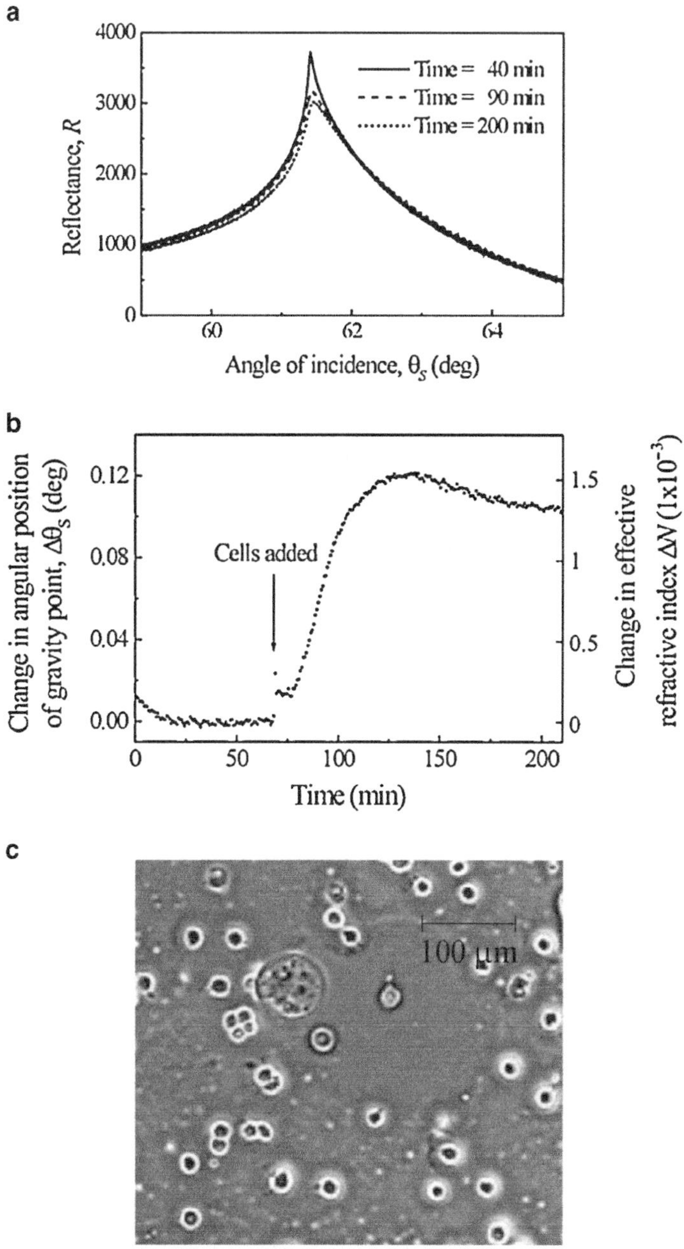

Fig. 15.21 Peak-type detection of HaCaT cells settling onto the sensor surface. (**a**) Three sensorgrams obtained after 40, 90 and 200 min; (**b**) angular position of the peak vs. time and (**c**) microscope image of the exposed sensor surface taken right after the measurements. The area shown is $460 \times 400\ \mu m^2$. Reprinted from Ref. 27 with permission. © 2008 Elsevier

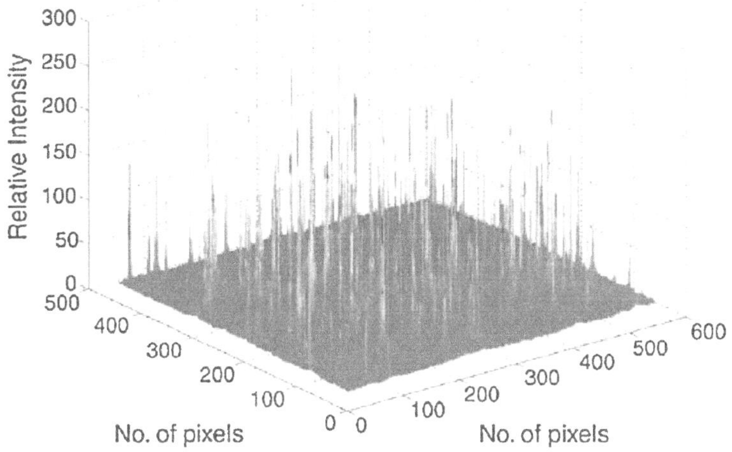

Fig. 15.22 Images of scattering of BG spores using PT-CLW. Reprinted from Ref. 24 with permission. © 2008 Elsevier

due to spin coating. The limit of detection was estimated to be approximately 1×10^4 cells per mL when 5 mL of BG spores suspension was pumped through for 20 min; this was based on the minimum response shift that could be distinguished on the instrument.

As the BG spores flowed across the surface of the sensor chip, the spores appeared either as a diffuse area of moving light or as small and bright points moving slowly with occasionally instantaneous stop, presumably showing spores captured by the immobilized antibody. No such attachment was observed with BG spores at control surfaces coated with BSA or FCS, or with *E. coli* strain on anti-BG antibody coated surfaces. Upon stopping the flow, the diffusely emitting particles appeared to settle onto the PT-CLW chip where they became brighter and more sharply defined. Capture of the spores at the surface was indicated by cessation of movement and the bacteria remaining in place when the flow was restarted[24].

Figure 15.23 shows the normalized area occupied on a functionalized silica surface by captured bacterial spores, as determined from surface imaging, plotted against bacterial spore concentrations (1×10^3–1×10^{10} cells per mL). Three different tests have been performed: BG spores binding to immobilized anti-BG spores, BG spores binding to immobilized non-specific protein BSA (control surface) and *E. coli* binding to immobilized anti-BG spores (control surface). The BG/anti-BG assay shows significant BG capturing, while there is no significant BG capturing for the other two assays. The curve in Fig. 15.23 shows a clear relationship between the BG bacterial spores captured by the immobilized anti-BG spores and the concentration of the tested solution. The dose-response was found to be linear over five decades of bacterial concentration (regression coefficient, $r > 0.97$; the probability of the linear fit, $p < 0.001$). The standard curve for the BG/anti-BG assay reaches a plateau at cell concentrations above 8×10^7 cells per mL

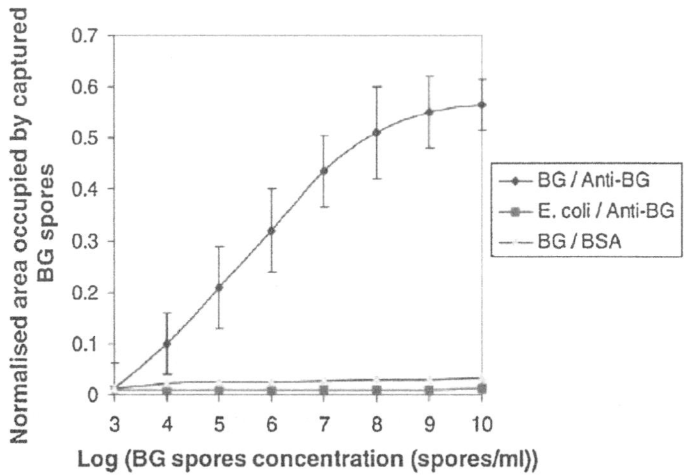

Fig. 15.23 Normalized area occupation of BG and anti-BG bacterial spores vs. solution concentration of three different assays: (*filled triangle*) BG/BSA functionalized control surface, (*filled diamond*) BG/Anti-BG immobilized surface and (*filled square*) *E. coli*/Anti-BG immobilized surface. The surfaces were exposed for 60 min to the respective solutions. *Error bars* represent ±1 standard deviation, $n = 4$. Reprinted from Ref. 21 with permission. © 2008 American Chemical Society

corresponding to a surface coverage of bacterial spores of 50–60%. This plateau indicates a certain degree of saturation of possible binding events in the close vicinity of the surface. This may be due to cells already at the surface sterically hinder other spores interacting with the immobilized antibody. In addition, negative charges on the waveguide silica surface and cell-to-cell repulsive forces may prevent close packing of cells. Indeed, as cell surfaces consist of many different charged groups both short- and long-range interactive forces between cells would be expected[63].

15.6.3.3 Immobilization Matrices

The type of matrix used for immobilization of the recognition element for bacterial cell detection is crucial to achieve high sensitivity. Two important conditions should be considered specifically for bacterial detection (1) the accessibility of the recognition elements in the immobilization matrix for bacteria binding on the sensor surface and (2) to obtain the binding of the analytes within the most sensitive region of the evanescent field, immediately adjacent to the sensor surface.

Zourob et al.[21] performed a comparative study of a number of surface immobilization matrices for preparation of bacteria immunosensors to find the best-suited immobilization matrix. The matrices are evaluated by operating the MCLW sensor in scattering mode and measuring the bacteria coverage on the sensor surface along

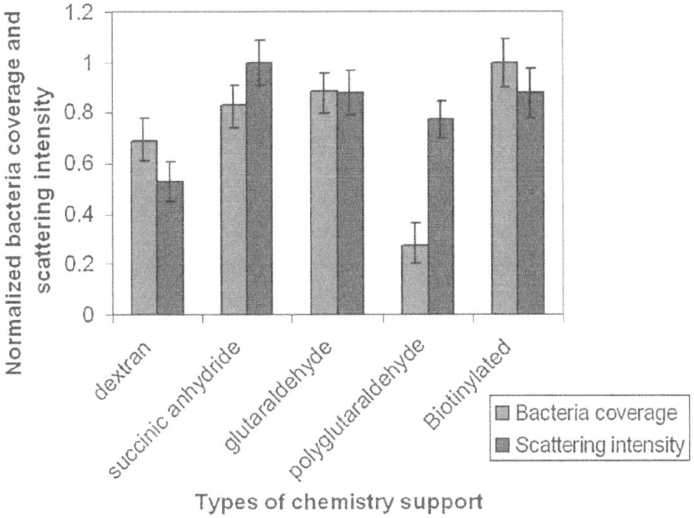

Fig. 15.24 Normalized spore coverage after 1-h incubation and scattering intensity using different supports. Reprinted from Ref. 21 with permission. © 2008 American Chemical Society

with the scattering intensity after bacteria binding for each of the matrices. The results are shown in Fig. 15.24.

Five immobilization matrices have been tested and compared. These include a dextran layer on the sensor surface, succinic anhydride modified surface, monomer glutaraldehyde activated surface, polyglutaraldehyde activated surface and streptavidine/biotinylated antibody coated surface.

The amount of bacteria captured by the immobilized antibody (bacteria coverage) by the different matrices and the resulting scattering intensity varied greatly depending on the surface modification and immobilization chemistry used. In general, the bacterial coverage using dextran and polyglutaraldehyde showed low bacteria coverage; while the succinic acid modified surface, monomer glutaraldehyde and the streptavidine/biotinylated coated surfaces resulted in the highest bacteria coverage.

The scattering intensity for captured bacteria to the immobilized antibodies element was found comparable to most of the chemistries except that for the dextran layer. The glutaraldehyde (monomer and polymerized) coated surface, succinic acid and the streptavidine/biotinylated antibody coated surface showed the highest scattering intensity due to the placement of captured cells very close to the sensor surface. However, dextran showed the poorest scattering intensity. Dextran coating is a 3D-hydrogel, commonly used for detection of small proteins. However, for bacterial cell detection the applicability for the dextran layer is limited. The pore sizes in the dextran hinder the diffusion of bacteria into the dextran layer, and also dextran is very compact at the sensor surface due to the binding mechanism of dextran and thus this hydrogel will waste the most sensitive area of the evanescent field.

15.6.3.4 Enhancing Performance of Deep-Probe Optical Waveguides Systems

Processes that enhance bacterial interactions with a defined immobilized matrix on solid surfaces are of particular relevance for enhancing the sensor performance. Generally, when working with bacterial cell detection one of the main features to consider is the diffusion of these large analytes to the immobilized recognition element on the sensor surface. In order to enhance the sensor performance, the analytes of interest in the sample should preferably approach the immobilized recognition elements rapidly and be captured on the sensor surface. Apart from the time issue, the sensor performance can for example be limited due to (1) mass transport of the large analyte such as bacteria, which limits the binding of targets to the immobilized recognition receptors and (2) non-specific binding.

The best approach to enhance the diffusion of the analyte towards the immobilized recognition element is to use a physical force field that directs the cells in a continuous flow-through system onto an activated sensor surface where they can be captured by the recognition element. In this section, the sensor systems, based on deep-probe MCLW sensors, will be integrated with an electric field and ultrasound standing waves (USW), respectively. These fields exert physical force fields, which concentrate and enhance capturing of bacteria spores into immobilized antibodies on the sensor surface[22,23]. The integration improves the detection limit by a few orders of magnitude and shortens the analysis time.

Integration with Electric Field

The isoelectric point of bacteria can be exploited to enhance the sensor performance by integrating an electric field across the sample volume into the sensor system. This results in a physical force pushing the cells towards the sensor surface. The isoelectric point for most bacterial cells is ≤ 4[64]. Hence, bacteria have a net negative charged surface in deionized water (pH = 7.0) and a net positive charge at low pH levels (≤ 3). Thus, the bacteria can be directed either towards the sensor surface or to the top cover of the flow channel by applying a negative voltage to the metal layer of the waveguide depending on the pH of the buffer solution or the direction can be changed by applying a positive voltage. The advantage of coupling the waveguide sensor chip with an electric field is the capability to concentrate and enhance the bacteria capturing from a diluted sample to the immobilized antibodies achieving a shorter analysing time.

Zourob et al.[22] constructed a flow-cell incorporating the MCLW and ITO electrode as shown in Fig. 15.25. Initial experiments were conducted after treating the surface with BSA overnight. BG bacterial spores were introduced to the sensor system at a constant flow rate of 200 μL min^{-1} in 50 mM histidine buffer, and the sensor system was operated in real-time scattering mode using a CCD camera.

Different concentrations of BG bacterial spores (10^3–10^7 spores per mL) were introduced into the electric field sensor chip-based flow cell and an appropriate

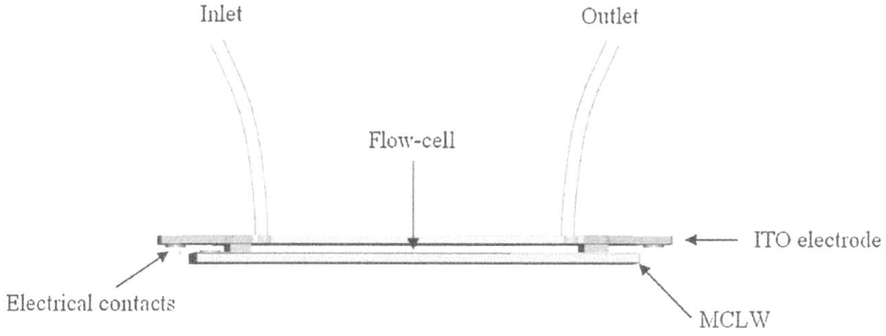

Fig. 15.25 A schematic diagram for MCLW sensor chip integrated with electric field. Reprinted from Ref. 22 with permission. © 2008 The Royal Society of Chemistry

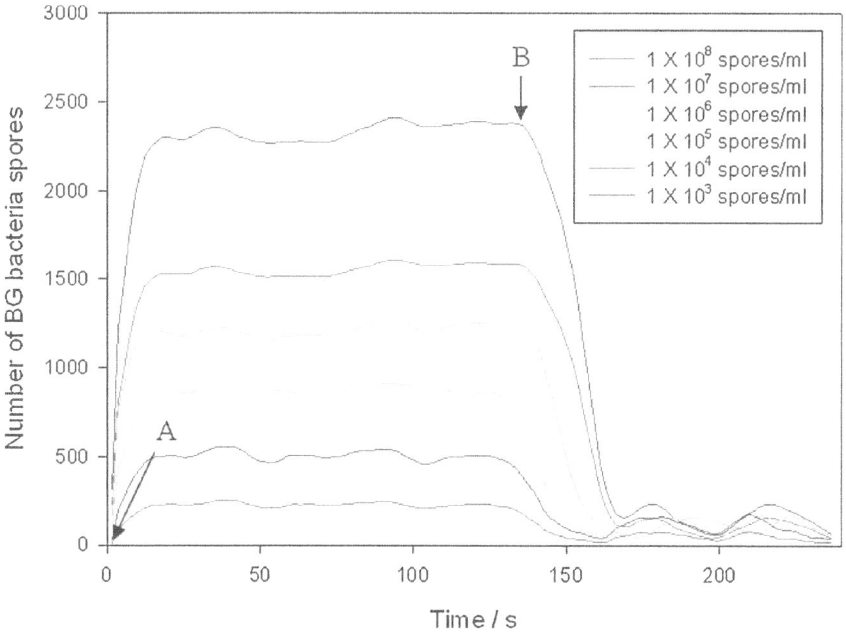

Fig. 15.26 The effect of applying positive (A) and negative (B) potential to the metal layer of the MCLW sensor on deposition and repulsion of different concentrations of BG bacteria spores. Reprinted from Ref. 22 with permission. © 2008 The Royal Society of Chemistry

DC potential was applied so that BG bacterial spores were either collected onto or repelled from the MCLW sensor surface. Figure 15.26 shows the detected number of bacterial spores on the sensor surface vs. time. At time A, a negative potential was applied to the ITO electrodes and a positive potential to the metal layer of the MCLW sensor, resulting in an increase in the number of BG bacterial spores driven to the sensor surface. At time B, the potential is reversed and a decrease in the number of BG bacterial spores on the sensor surface is observed, hence the bacterial spores are repelled away. The diagram shows the results for six different concentrations of BG bacterial spores. The results indicate that for each of the bacteria suspension approximately the same number of BG bacterial spores is deposited onto the sensor surface and repelled away from the sensor surface when the potential is reversed.

The same experiments were repeated with immobilized antibodies on the MCLW sensor surface before depositing the BG bacterial spores. Figure 15.27 shows the number of BG bacterial spores concentrated on the MCLW sensor surface, with a positive potential applied to the metal layer with respect to the ITO layer at time A, and subsequently when reversing the potential at time B. The remaining bacterial spores on the MCLW sensor surface after reversing the field are captured by the immobilized antibodies as shown in Fig. 15.27. The number of bacterial spores captured was estimated by calculating the number of scattering points on the surface and this was confirmed by direct counting immediately after terminating the experiments.

The number of cells captured using the MCLW with the electric field applied for 2 min was found to be equal to the number of bacteria captured in 1 h using the same concentration of bacteria but with no applied electric field. Thus, the electric field gave a 30-fold faster response time. The limit of detection was found to be 1×10^3 cells per mL when the electric field was applied for 2 min. This value has been

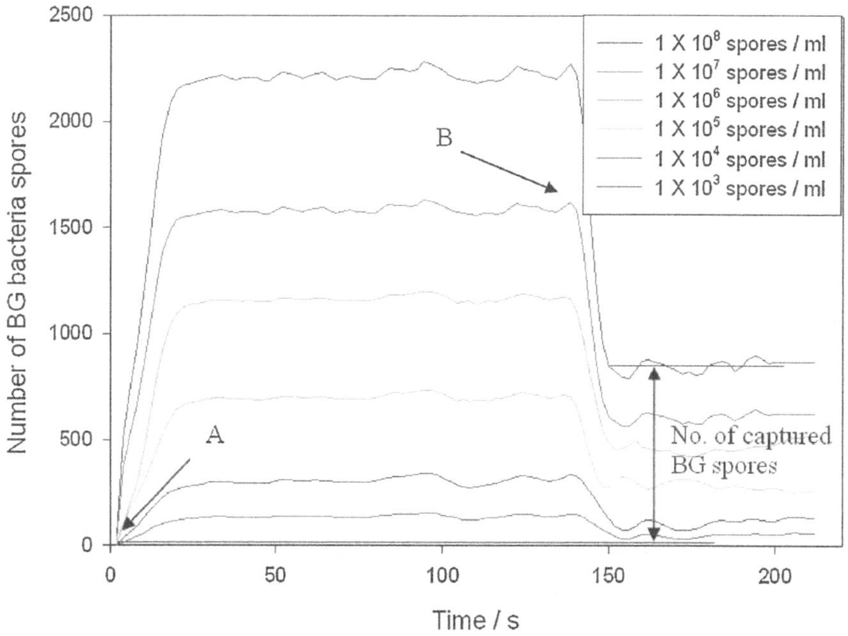

Fig. 15.27 The use of an electric field potential to capture/repel BG spores and demonstration of the number of bacteria spores captured on the sensor surface. Reprinted from Ref. 22 with permission. © 2008 The Royal Society of Chemistry

taken as the concentration that gave a signal three times the background noise. The significant enrichment of captured bacteria spores in time indicates a practical way to substantially improve the detection limit and shorten the response time of future waveguide sensors for bacterial agents.

Integration with Ultrasound Standing Waves

Instead of using electric field to concentrate the bacterial cells to the MCLW sensor, detection system using ultrasonic standing waves across the flow cell has been proposed. When particles suspended in a fluid are exposed to an ultrasonic standing wave, they experience an acoustic radiation pressure. This drives them towards the (usually pressure) nodes or antinodes of the acoustic field, where they aggregate to form bands of particles in localized clumps, bounded by clarified medium. This is because the mass density and the speed of sound in the particles are greater than the respective values of the medium. This phenomenon has been used to develop a system with a region of standing wave radiation. The bacteria are moved to a central nodal position located on the sensor surface to enhance the bacteria capture rate of biosensor surfaces, by driving the bacteria onto the sensor surface coated with immobilized antibodies. USW are appropriate for working with biological cells in aqueous medium, since they do not have deleterious effect on the particles. USW systems are amendable to miniaturization and can be applied to different particle sizes. Furthermore, the USW systems can be applied to environmental samples without problems unlike the electric field, which suffers problems from the presence of the salts.

The sensor system with USW has been tested for the detection of BG bacterial spores using a MCLW sensor coated with immobilized antibodies operated in scattering mode. A schematic diagram of the chamber construction used in integrating the MCLW and USW is shown in Fig. 15.28. The basic construction of the flow chamber is based on the description by Hawkes et al.[65] It comprises a 30 mm wide and 0.66-mm thick (i.e. 3-MHz fundamental resonance thickness) PZ26 polished ceramic ultrasonic transducer (Ferroperm, Krisgard, Denmark) with a 1.5-mm thick (i.e. equivalent to a half wavelength in the metal at 3 MHz) stainless steel coupling plate glued to the transducer, separating it from the water layer, a spacer to define the channel dimensions and a 1-mm thick ($\lambda/4$ wavelengths at 3 MHz) MCLW chip acoustic reflector. Generation and control of the frequency and voltage was applied to the transducer as described before[23,65,66]. The ceramic transducer was driven by an amplifier, with a sinusoidal signal from a function generator (Hewlett Packard 3326A) and a RF voltmeter was used to detect the voltage across the ceramic.

The integrated system has been tested for enhancement of bacterial capturing by introducing a solution of bacteria spores (concentration of 106 cells per mL) at 200 μL min^{-1} flow rate into the flow cell. Here, a small number of spore "clumps" were observed forming in the central region of the chamber when the ultrasound (0.5 V at 1.92-MHz frequency) was turned on. The clumps were held stationary

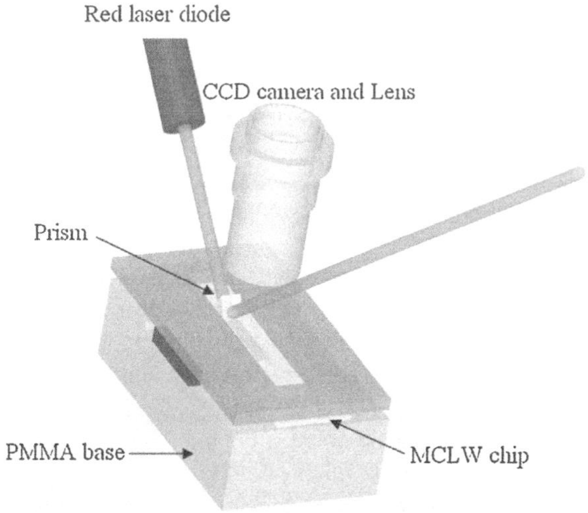

Fig. 15.28 Schematic of the ultrasound-MCLW instrumental setup. Reprinted from Ref. 23 with permission. © 2008 American Chemical Society

against the flow and increased in size with exposure time as more spores flowed into the system. It was found that the ultrasound deposited over 96% of BG bacteria spores from the bulk solution onto the MCLW sensor surface. Thus, the integration of USW enhances the capturing of BG bacteria spores to the immobilized antibodies and shortens the analysis time. The number of bacteria spores captured during ultrasound application over 3 min exceeded that of captured spores in 1 h when no ultrasound was applied to the system, using the same concentration and flow rate[23].

An additional experiment has been performed with the sensor system (0.4 V, 2.92 MHz) where the ultrasound was applied over a limited area of the sensor surface. After the experiment, the sensor area was imaged on CCD by ensuring propagation of the waveguide mode along the line of the flow of the solution at the sensor surface. Two images were taken as shown in Fig. 15.29 for two different positions of incident light into the waveguide structure. In Fig. 15.29a, the point of incident light was chosen to be on the area of the sensor surface that was exposed to the bacteria detection but before the ultrasound depositing centre. Figure 15.29b shows the scattered intensity from the sensor surface, when the point of incident light was at a point of the area that had been exposed to the ultrasound but after the ultrasound. It was possible to monitor the effect of the ultrasound on the surface of the MCLW sensor, where the difference in depositing in relation to specific regions of the MCLW is clearly observed in Fig. 15.29. Little or no depositing occurred before the spores reached the ultrasound field (Fig. 15.29a), whereas maximum deposition occurred within the area exposed to ultrasound (Fig. 15.29b). A small number of BG bacteria spores were deposited in the region beyond the field, after

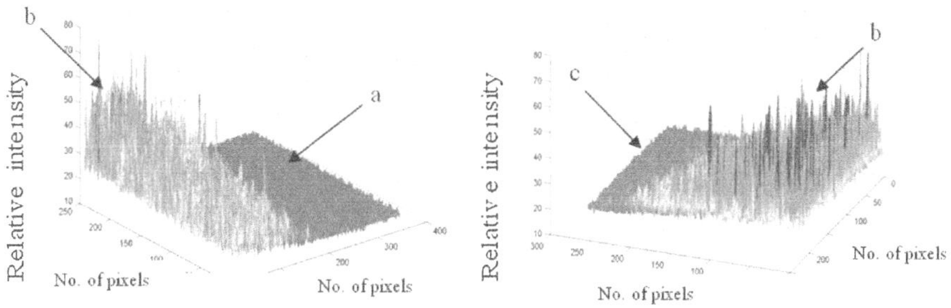

Fig. 15.29 Comparison of bacteria deposited before (area a), within (area b) and after (area c) the ultrasound electrode area. Reprinted from Ref. 23 with permission. © 2008 American Chemical Society

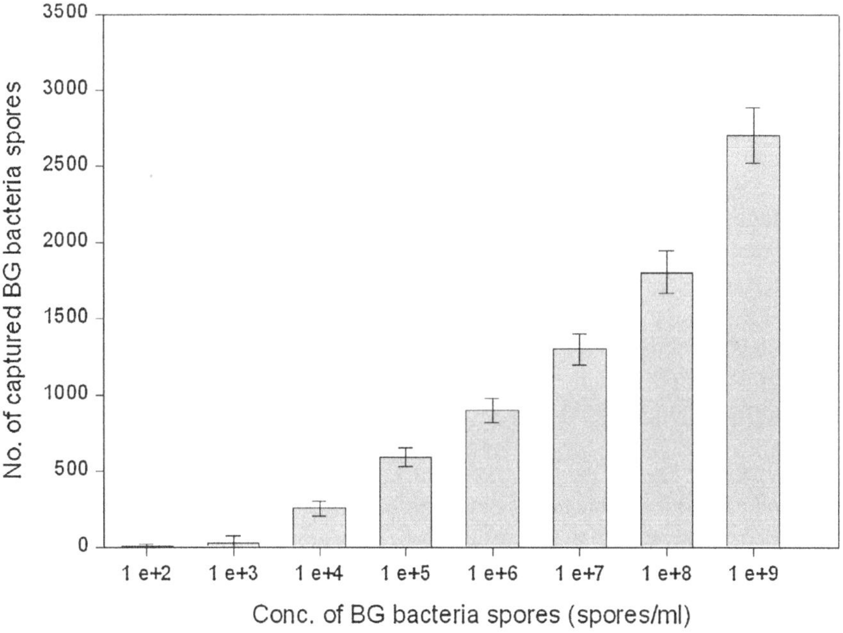

Fig. 15.30 The relationship between the number of BG captured and concentration of BG suspensions. Reprinted from Ref. 23 with permission. © 2008 American Chemical Society

the flow had carried the spores past the ultrasound-exposed area (Fig. 15.29c). Spore binding approaches saturation at the region of maximum attachment (Fig. 15.29b). The reduced amount of attachment further downstream (Fig. 15.29c) may be accounted for by the reduced concentration of spores near the surface due to earlier spore removal from suspension[23].

The general position of the spore clumps in the chamber did not change during the frequency scan from 2.910 to 2.920 MHz. This enabled the effect of changing the spore concentration to be examined using the MCLW sensor, while ensuring that the same areas of the chamber could be compared to each other. The regions of maximum spore adherence, at the centre of the clumping patterns, were compared using different concentrations of bacteria spores. Figure 15.30 shows the relationship between the numbers of BG bacteria spores captured onto the immobilized anti-BG over a concentration range of 10^2–10^9 spores per mL. This was achieved by counting the number of BG bacteria spores captured from the images taken in real time over the area of mode propagation for 3 min.

15.7 Concluding Remarks

In this chapter, we described the development of different configurations of deep-probe optical evanescent wave sensors, such as reverse waveguides based on low index substrates and metal-clad leaky waveguides, for different chemical and biosensing applications. It was shown that increasing the penetration depth of the evanescent field increases the sensitivity of such sensors for large micron scale biological objects like bacteria and living cells. Waveguide configurations were presented where the light is guided in a thick sensing layer and where the amount of immobilized recognition receptors within the sensing layer is increased which results in an enhanced sensitivity. The metal layer in the MCLW was assigned an additional role when the sensor system is integrated with electric fields. By applying an electrical potential to the metal layer the diffusion of the analyte was accelerated towards the immobilized recognition receptors. This resulted in a shortened analysis time and a reduction in non-specific binding. In addition, ultrasonic standing waves have been integrated with MCLW sensors to concentrate and enhance entrapment of bacterial spores onto immobilized antibodies on the sensor surface. The integration improves the detection limit by a few orders of magnitude, shortens the analysis time significantly and reduces non-specific binding, which causes false positive results.

References

1 Harrick, N. J., Internal Reflection Spectroscopy, Wiley, New York, NY, 1967
2 Buckle, P. E.; Davies, R. J.; Kinning, T.; Yeung, D.; Edwards, P. R.; Pollard-Knight, D.; Lowe, C. R., The resonant mirror: A novel optical biosensor for direct sensing of biomelcular interactions. Part II: Applications, *Biosens. Bioelectron.* **1993**, 8, 355–363
3 Cush, R.; Cronin, J. M.; Stewart, W. J.; Maule, C. H.; Molloy, J. O.; Goddard, N. J., The resonant mirror: A novel optical biosensor for direct sensing of biomolecular interactions. Part 1: Principle of operation and associated instrumentation, *Biosens. Bioelectron.* **1993**, 8, 347–354.

4 Stamm, C.; Lukosz, W., Integrated optical difference interferometer as refractometer and chemical sensor, *Sens. Actuators B* **1993**, 11, 177–181
5 Huber, W.; Barner, R.; Fattinger, C.; Hubscher, J.; Koller, H.; Muller, F.; Schlatter, D.; Lukosz, W., Direct optical immunosensing (sensitivity and selectivity), *Sens. Actuators B* **1992**, 6, 122–126
6 Nellen, Ph.; Tiefenthaler, K.; Lukosz, W., Integrated optical input grating couplers as biochemical sensors, *Sens. Actuators* **1988**, 15, 285–295
7 Ramsden, J. J., Review of new experimental – techniques for investigating random sequential adsorption, *J. Stat. Phys.* **1993**, 73, 853–877
8 Lukosz, W.; Tiefenthaler, K., Sensitivity of integrated optical grating and prism couplers as (bio) chemical sensors, *Sens. Actuators B* **1988**, 15, 273–284
9 Horvath, R.; Lindvold, L. R.; Larsen, N. B., Reverse-symmetry waveguides: Theory and fabrication, *Appl. Phys. B* **2002**, 74, 383–393
10 Tien, P. K., Integrated optics and new wave phenomena, *Rev. Mod. Phys.* **1977**, 49, 361–420
11 Goddard, N. J.; Pollard-Knight, D.; Maule, C., Real-time biomolecular interaction analysis using the resonant mirror sensor, *Analyst* **1994**, 119, 583–588
12 Horvath, R.; Pedersen, H. C.; Cuisinier, F. J. G., Guided wave sensing of polyelectrolyte multilayers, *Appl. Phys. Lett.* **2006**, 88, 111102–111104
13 Goddard, N. J.; Singh, K.; Bounaira, A. F.; Holmes, R. J.; Baldock, S. J.; Pickering, L. W.; Fielden, P. R.; Snook, R. D., Anti-resonant reflecting optical waveguides (ARROWS) as optimal optical detectors for μ-TAS applications. In Micro Total Analysis Systems μ-TAS/ Proceedings; **1998**, 97–100
14 Goddard, N. J.; Hulme, J.; Malins, C.; Singh, K.; Fielden, P. R., Asymmetric anti-resonant reflecting optical waveguides (ARROW) as chemical sensors, *Analyst* **2002**, 127, 378–382
15 Horvath, R.; Pedersen, H. C.; Larsen, N. B., Demonstration of reverse symmetry waveguide sensing in aqueous solutions, *Appl. Phys. Lett.* **2002**, 81, 2166–2168
16 Horvath, R.; Pedersen, H. C.; Skivesen, N.; Selmeczi, D.; Larsen, N. B., Monitoring of living cell attachment and spreading using reverse symmetry waveguide sensing, *Appl. Phys. Lett.* **2005**, *86*, 071101
17 Horvath, R.; Pedersen, H. C.; Skivesen, N.; Svanberg, C.; Larsen, N. B., Fabrication of reverse symmetry polymer waveguide sensor chips on nanoporous substrates using dip-floating, *J. Micromech. Microeng.* **2005**, 15, 1260–1264
18 Horvath, R.; Skivesen, N.; Larsen, N. B.; Pedersen, H. C., Reverse symmetry waveguide for optical biosensing, In Frontiers in Chemical Sensors. Novel Principles and Techniques; Orellana, G.; Moreno-Bondi, M. C., Eds.; Springer Series on Chemical Sensors and Biosensors; Springer, Berlin, **2005**, Vol. 3, 279–301
19 Zourob, M.; Mohr, S.; Fielden, P. R.; McDonnell, M. B.; Goddard, N. J., Small-volume refractive index and fluorescence sensor for micro total analytical system (μ-TAS) applications, *Sens. Actuators B* **2003**, 94, 304–312
20 Zourob, M.; Mohr, S.; Fielden, P. R.; Goddard, N. J., The development of a metal clad leaky waveguide sensor for the detection of particles, *Sens. Actuators B* **2003**, 90, 296–307
21 Zourob, M.; Mohr, S.; Treves-Brown, B. J.; Fielden, P. R.; McDonnell, M. B.; Goddard, N. J., An integrated metal clad leaky waveguide sensor for detection of bacteria, *Anal. Chem.* **2005**, 77, 232–242
22 Zourob, M.; Mohr, S.; Treves-Brown, B. J.; Fielden, P. R.; McDonnell, M. B.; Goddard, N. J., An integrated optical leaky waveguide sensor with electrically induced concentration system for the detection of bacteria, *Lab Chip* **2005**, 5, 1360–1365
23 Hawkes, J. J.; Coakley, W. T.; Mohr, S.; Treves-Brown, B. J.; Fielden, P. R.; McDonnell, M. B.; Goddard, N. J., Optical leaky waveguide sensor for detection of bacteria with ultrasound attractor force, *Anal. Chem.* **2005**, 77, 6163–6168
24 Zourob, M.; Mohr, S.; Treves-Brown, B. J.; Fielden, P. R.; McDonnell, M. B.; Goddard, N. J., Bacteria detection using disposable optical leaky waveguide sensors, *Biosens. Bioelectron.* **2005**, 21, 293–302

25 Skivesen, N.; Horvath, R.; Pedersen, H. C., Optimization of metal-clad waveguide sensors, *Sens. Actuators B* **2005**, 106, 668–676

26 Skivesen, N.; Horvath, R.; Pedersen, H. C., Peak-type and dip-type metal-clad waveguide sensing, *Opt. Lett.* **2005**, 30, 1659–1661

27 Skivesen, N.; Horvath, R.; Pedersen, H. C.; Thinggaard, S.; Larsen, N. B., Deep-probe metal-clad waveguide, *Biosens. Bioelectron.* **2007**, 22, 1282–1288

28 Born, M.; Wolf, E., Principles of Optics, 7th edn; Cambridge University Press, Cambridge, **1999**

29 Zourob, M.; Elwary, S.; Turner A. A. F., Principles of Bacterial Detection; Biosensors, Recognition Receptors and Microsystems; Springer science + business media, ISBN: 978–0–387–75112–2, **2008**

30 Tiefenthaler, K.; Lukosz, W., Sensitivity of grating couplers as integrated-optical chemical sensors, *J. Opt. Soc. Am. B* **1989**, 6, 209–220

31 Horvath, R.; Pedersen, H. C.; Skivesen, N.; Selmeczi, D.; Larsen, N. B., Optical waveguide sensor for on-line monitoring of bacteria, *Opt. Lett.* **2003**, 28, 1233–1235

32 Horvath, R.; Lindvold, L. R.; Larsen, N. B., Fabrication of all-polymer freestanding waveguides, *J. Micromech. Microeng.* **2003**, 13, 419–424

33 Skivesen, N.; Horvath, R.; Pedersen, H. C., Multimode reverse-symmetry waveguide sensor for broad-range refractometry, *Opt. Lett.* **2003**, 28, 2473–2475

34 Horvath, R.; Cottier, K.; Pedersen, H. C.; Ramsden, J. J., Multidepth screening of living cells using optical waveguides, *Biosens. Bioelectron.* **2008**, 24, 799–804

35 Horvath, R.; Skivesen, N.; Pedersen, H. C., Measurement of guided lightmode intensity: An alternative waveguide sensing principle, *Appl. Phys. Lett.* **2004**, 84, 4044–4046

36 Horvath, R.; Voros, J.; Graf, R.; Fricsovszky, G. Textor, M.; Lindvold, L. R.; Papp, E., Effect of patterns and inhomogeneities on the surface of optical waveguides, *Appl. Phys. B* **2001**, 72, 441

37 Horvath, R.; Cottier, K., Imageless microscopy of surface patterns using optical waveguides, *Appl. Phys. B* **2008**, 91, 319–327

38 Rohrbach, A., Observing secretory granules with a multi-angle evanescent wave microscope, *Biophys. J.* **2000**, 78, 2641–2654

39 Perkins, E.; Squirrell, D., Development of instrumentation to allow the detection of micro-organisms using light scattering in combination with surface plasmon resonance, *Biosens. Bioelectron.* **2000**, 14, 853–859

40 Lenney, J. P.; Goddard, N. J.; Morey, J. C.; Snook, R. D.; Fielden, P. R., An electro-osmotic flow system with integrated planar optical waveguide sensing, *Sens. Actuators B* **1997**, 39, 212–217

41 Jakeway, S. C.; de Mello, A. J., Chip-based refractive index detection using a single point evanescent wave probe, *Analyst* **2001**, 9, 1505–1510

42 Wang, S.-L.; Huang, X.-J.; Fang, Z.-L.; Dasgupta, P. K., A miniaturized liquid core wave-guide-capillary electrophoresis system with flow injection sample introduction and fluorometric detection using light-emitting diodes, *Anal. Chem.* **2001**, 73 4545–4549

43 Salamon, Z.; Brown, M. I.; Tollin, G., Plasmon resonance spectroscopy: Probing molecular interactions within membranes, *TIBS* **1999**, 24, 213–219

44 Salamon, Z.; Tollin, G., Graphical analysis of mass and anisotropy changes observed by plasmon-waveguide resonance spectroscopy can provide useful insights into membrane protein function, *Biophys. J.* **2004**, 86, 2508–2516

45 Zdzislaw, S.; Gordon, T., Optical anisotropy in lipid bilayer membranes: Coupled plasmon-waveguide resonance measurements of molecular orientation, polarizability, and shape, *Biophys. J.* **2001**, 80, 1557–1567

46 Zdzislaw, S.; Goran, L.; Gordon, T., Plasmon-waveguide resonance and impedance spectroscopy studies of the interaction between penetratin and supported lipid bilayer membranes, *Biophys. J.* **2003**, 84, 1796–1807

47 Tollin, G.; Salamon, Z.; Cowell, S. M.; Hruby, V. J., Plasmon-waveguide resonance spectroscopy: A new tool for investigating signal transduction by G-protein coupled receptors, Life Sci. **2003**, 73, 3307–3311

48 Tollin, G.; Salamon, Z.; Hruby, V. J., Techniques: Plasmon-waveguide resonance (PWR) spectroscopy as a tool to study ligand-GPCR interactions, *Trends Pharmacol. Sci.* **2003**, 24, 655–659
49 Alves, I. D.; Salamon, Z.; Varga, E.; Yamamura, H. I.; Tollin, G.; Hruby, V. J., Direct observation of G-protein binding to the human δ-opioid receptor using plasmon-waveguide resonance spectroscopy, *J. Biol. Chem.* **2003**, 278, 48890–48897
50 Alves, I. D.; Cowell, S. M.; Salamon, Z.; Devanathan, S.; Tollin, G.; Hruby, V. J., Different structural states of the proteolipid membrane are produced by ligand binding to the human δ-opioid receptor as shown by plasmon-waveguide resonance spectroscopy, *Mol. Pharmacol.* **2004**, 65, 1248–1257
51 Alves, I. D.; Ciano, K. A.; Boguslavski, V.; Varga, E.; Salamon, Z.; Yamamura, H. I.; Hruby, V. J.; Tollin, G., Selectivity, cooperativity, and reciprocity in the interactions between the δ-opioid receptor, its ligands, and G-proteins, *J. Biol. Chem.* **2004**, 279, 44673–44682
52 Zourob, M.; Goddard, N. J., Metal clad leaky waveguides for chemical and biosensing applications, *Biosens. Bioelectron.* **2005**, 20, 1718–1727
53 Zourob, M.; Simonian, A.; Wild, J.; Mohr, S.; Fan, X.; Abdulhalim, I.; Goddard, N. J., Optical leaky waveguide biosensors for the detection of organophosphorus pesticides, *Analyst* **2007**, 132, 114–120
54 Hulme, J., Optical Waveguide Biosensors, PhD thesis, UMIST, Manchester, UK, **2002**
55 Rainina, E.; Simonian, A.; fremenco, A.; Varfolomeyev, S.; Wild, J., The development of a new biosensor based on recombinant *E. coli* for the direct detection of organophosphorus neurotoxins, *Biosens. Bioelectron.* **1996**, 11, 991–1000
56 Mulchandani, A.; Chen, W.; Mulchandani, P.; Wang, J.; Rogers, K. R., TITEL, Biosensors for direct determination of organophosphate pesticides, *Biosens. Bioelectron.* **2001**, 16, 225–230
57 Mayes, A. G.; Blyth, J.; Millington, R. B.; Lowe, C. R., A holographic sensor based on a rationally designed synthetic polymer, *J. Mol. Recognit.* **1998**, 11, 168–174
58 Marshall, A. J.; Young, D. S.; Blyth, J.; Kabilan, S.; Lowe, C. R., Metabolite-sensitive holographic biosensors, *Anal. Chem.* **2004**, 76, 1518–1523
59 Marshall, A. J.; Blyth, J.; Davidson, C. A. B.; Lowe, C. R., pH-Sensitive holographic sensors, *Anal. Chem.* **2003**, 75, 4423–4431
60 Malins, C., Glever, H. G., Keyes, T. E., Vos, J. G., Dressick, W. J., Mac- Craith, B. D., Sol-gel immobilised ruthenium(II) polypyridyl complexes as chemical transducers for optical pH sensing, *Sens. Actuators B* **2000**, 67, 89–95
61 McDonagh, C.; Sheridan, F.; Butler, T.; McCraith, B. D., Characterization of sol-gel-derived silica films, *J. Non-Cryst. Solids* **1996**, 194, 72–77
62 Marquette, C. A.; Blum, L. J., Luminol electrochemiluminescence-based fibre optic biosensors for flow injection analysis of glucose and lactate in natural samples, *Anal. Chim. Acta* **1999**, 381, 1–10
63 Watts, H.; Lowe, C.; Pollard-Knight, D., Optical biosensor for monitoring microbial cells, *Anal. Chem.* **1994**, 66, 2465–2470
64 DeFlaun, M. F.; Condee, C. W., Electrokinetic transport of bacteria, *J. Hazard. Mater.* **1997**, 55, 263–277
65 Hawkes, J. J.; Long, M. J.; Coakley W. T.; McDonnell, M. B. Ultrasonic deposition of cells on a surface, *Biosens. Bioelectron.* **2004**, 19, 1021–1028
66 Spengler, J. F.; Jekel, M.; Christensen, K. T.; Adrian, R. J.; Hawkes, J. J.; Coakley, W. T., Observation of yeast cell movement and aggregation in a small-scale MHz-ultrasonic standing wave field, *Bioseparation* **2000**, 9, 329–341
67 Salamon, Z.; Macleod, H. A.; Tollin, G., Coupled plasmon-waveguide resonators: A new spectroscopic tool for probing proteolipid film structure and properties, *Biophys. J.* **1997**, 73, 2791–2797

Section III
Microfluidics Enabled Photonic Sensing Systems

Chapter 16
Optically Resonant Nanophotonic Devices for Label-Free Biomolecular Detection

Julie Goddard, Sudeep Mandal, and David Erickson

Abstract Optical devices, such as surface plasmon resonance chips and waveguide-based Mach–Zehnder interferometers, have long been successfully used as label-free biomolecular sensors. Recently, however, there has been increased interest in developing new approaches to biomolecular detection that can improve on the limit of detection, specificity, and multiplexibility of these early devices and address emerging challenges in pathogen detection, disease diagnosis, and drug discovery. As we describe in this chapter, planar optically resonant nanophotonic devices (such as ring resonators, whispering gallery modes, and photonic crystal cavities) are one method that shows promise in significantly advancing the technology. Here we first provide a short review of these devices focusing on a handful of approaches illustrative of the state of the art. We then frame the major challenge to improving the technology as being the ability to provide simultaneously spatial localization of the electromagnetic energy and biomolecular binding events. We then introduce our "Nanoscale Optofluidic Sensor Arrays" which represents our approach to addressing this challenge. It is demonstrated how these devices serve to enable multiplexed detection while localizing the electromagnetic energy to a volume as small as a cubic wavelength. Challenges involved in the targeted immobilization of biomolecules over such a small area are discussed and our solutions presented. In general, we have tried to write this chapter with the novice in mind, providing details on the fabrication and immobilization methods that we have used and how one might adapt our approach to their designs.

D. Erickson (✉)
Sibley School of Mechanical and Aerospace Engineering, Cornell University, Ithaca, NY 14853, USA
e-mail: de54@cornell.edu

X. Fan (ed.), *Advanced Photonic Structures for Biological and Chemical Detection*, Integrated Analytical Systems,
DOI 10.1007/978-0-387-98063-8_16,

16.1 Introduction

The global market for biosensors and bioelectronics has grown from $6.1 billion in 2004 to an expected $8.2 billion in 2009 (roughly the time of the publication of this book) with approximately 40% of this representing demand within USA. Within this larger market optical sensors and biosensors are expected to grow the fastest with the large medical/diagnostic market continuing to offer the best opportunities (note that the above figures are compiled from a variety of commercially available market surveys that I will not specifically reference here). Readers interested in details of existing commercially available optical biosensors and application areas are suggested to consult a recent review paper by Rich and Myszka[1], who provide an excellent survey of the 2006 literature.

This enlarged interest is largely driven by recent advancements in proteomics and genomics, which have yielded a larger number of more specific biomarkers that can be associated with specific disease states[2–5] and pharmacological responses.[6] Point-of-care sensing platforms that can detect these biomarkers at very low concentrations among an enormous number of background interferants in a blood or saliva sample could lead to earlier stage diagnosis of complex diseases like cancer[7–10]. The second broad application area relates to the growing need for military, government, and other civilian organizations to be able to rapidly detect and identify pathogenic threats. The detection of water-[11], food-[12], or air-borne bacterial pathogens[13,14] requires devices with high sensitivity and specificity but introduce other engineering challenges such as larger sample volumes to be processed and often more complex background media. While greater sensitivity leading to earlier detection is critical, it is perhaps more important to minimize false alarm rates since the economic and social consequences of a detection event can be significant. Viral detection adds further complications to detection platforms in that they must be able to account for high rates of mutagenic drift and therefore detect what amounts to a moving target.

16.1.1 A Brief Overview of Biomolecular Detection

Though much of this book covers various methods of biomolecular detection and thus it can be assumed that the reader has a certain familiarity with the concept, we thought we would begin this chapter with a high-level overview of the different methods to help place our work in context. Uninterested readers can skip ahead to Sect. 16.1.2. In the most general sense, there are two ways in which a biomolecular detection reaction can be carried out. The first is homogeneously, where both the target (e.g., a blood-borne antibody created by the immune system in response to the presence of a tumor) and probe (e.g., an antigen that will specifically bind to the target antibody) are dispersed in the solution phase. Binding between the probe and the target is detected by observing changes in bulk solution properties such as optical absorbance, fluorescence, or electrical conductivity. Quantification is typically done then by relating the change in the interrogated bulk property to the concentration of target in the sample through a predetermined calibration curve. If no change is detected the target concentration is said to be below the limit of detection (LOD). The slope of this calibration curve is referred to as the "internal sensitivity" (see Erickson et al.[15] for more details on the use of these terms). The advantage of homogeneous techniques is that they tend to have higher overall reaction rates since both the target and probe are mobile in solution and therefore often the time required for analysis is lower. Additionally since all the chemistry is carried out in solution, the overall detection reaction can be easier to conduct and may be more robust. The major disadvantage of homogeneous reactions is that they are much harder to multiplex since binding is not spatially localized within the reaction volume. To increase the parallelity of such reactions, it is common to use either arrays of tiny reaction wells (such as is done in 96 or 384 well plates) or to spectrally multiplex using, for example, fluorescent tags that emit at different colors.

Heterogeneous detection reactions involve first immobilizing a probe on a surface and then introducing the test sample to the surface. The presence of a target is determined by observing changes in a particular surface property (e.g., fluorescence, refractive index, or mechanical compliance). Heterogeneous reactions are much easier to multiplex since the probes and detection signal can be spatially localized. They can also have fundamentally greater limits of detection since one is essentially concentrating the detected analyte from a dispersed 3D phase to a 2D one. The disadvantage is that the analysis time can be longer since the targets must diffuse to the probe site. The incorporation of microfluidic elements to confine the targets closer to the surface (thereby decreasing the diffusion distance) can help to mitigate this problem.

Heterogeneous detection reactions are commonly further subdivided into what are known as labeled or label-free techniques. Labeled techniques are those that require at least one more reaction step, either prior to or after the probe–target reaction, to attach a secondary molecule to the probe that can be more easily detected than the reaction itself. The most commonly used labels are fluorescent tags (as is done in most modern nucleic acid and immunological microarrays) though other tags such as nanoparticles and radiolabels can also be used. Though the sensitivity and specificity of labeled detection can be very high (particularly if the label is also specific to the target), the added complexity of the labeling step is often undesired for point-of-care, portable, or rapid result diagnostic/detection applications. Label-free techniques seek to eliminate this secondary step and exploit some intrinsic changes in surface properties caused by the binding of the target to the surface immobilized probe. Techniques for accomplishing this are classified by the sensor transduction mechanism as either electrical, mechanical, or optical. We have discussed the advantages and disadvantages of these different modalities in a recent review paper[15] and so will not go into in detail here. Briefly, however, mechanical techniques rely on detecting changes in the mechanical behavior of a surface due to the accumulation of mass on it. This can manifest itself in either a change in the natural frequency of a cantilever or membrane, additional surface stress induced by electrostatic repulsion/attraction between bound molecules, or changes in the transmission properties of surface guided acoustic waves. Electrical techniques sense changes in surface conductivity or capacitance due to the accumulation of charged molecules. Since often only an impedance measurement is required, electrical methods represent likely the simplest transduction mechanism but they have additional complications such as sensitivity to background electrolytes and relatively complex device fabrication.

16.1.2 Advantages of Label-Free Optical and Nanotechnological Techniques

Optical techniques represent perhaps the most pervasive method of label-free biomolecular detection. Though a number of different properties can be probed, the most common methods look for changes in the local refractive index near the surface as the amount of bound molecules increases. The primary advantages of optical techniques over analogous mechanical or electrical label-free methods are the relative ease with which devices can be fabricated and the broad range of fluids and environments in which they can be used (e.g., gas, water, and serum). Though numerous different architectures have been developed, surface plasmon resonance (SPR) biosensors have been the most commercially successful to date due to their inherently high sensitivity and lack of any complex assembly or fabrication steps. There is also considerable knowledge in the use of SPR including the extraction of kinetic data, which presents a more robust picture than might be available with a simple qualitative test. The serial nature of the traditional SPR approach, however, has proved to be a limitation of the technology in the past, though the advent of SPR-imaging techniques will likely help to resolve this.

As mentioned above, one of the major strengths of heterogeneous array-based technologies is in their ability to provide very high degrees of multiplexing. In general, however, the relatively low sensitivity of such approaches limits the types of targets, which can be reasonably expected to be interrogated, and places stricter requirements on the amount of sample processing and detection infrastructure that is required. Emerging nanotechnologies such as nanoparticles[16], nanowires and nanotubes[17,18], nanomechanical resonators[19], and (as will be discussed primarily here) nanophotonics[20–22] are of interest largely to address this failing. While many of these devices tend not to have much greater internal sensitivity (i.e., slope of the sensor response curve in response to changes in bulk properties such as conductivity or refractive index) than traditional techniques, their inherent advantage is that the total surface area or volume that is probed tends to be much smaller. As a result, the total mass required to impart a measurable transduction signal is significantly lower and therefore the potential for greatly improved LOD is significant. In general, however, the extension of these technologies to the extreme parallelity of the 2D microarray format is complicated by the challenges involved in functionalization of individual sensor and 2D optical or electrical addressing of reaction sites with submicrometer spacing.

16.1.3 Overview of This Chapter

The goal of this chapter will be to provide an overview of the use of planar, optically resonant nanophotonic devices for biomolecular detection. Nanophotonics[23,24] represents the fusion of nanotechnology with optics and thus it is proposed that sensors based on this technology can combine the advantages of each as discussed above. Although many of the issues are the same, we focus here on optical resonance rather than plasmonic resonance (such as is used in emerging local SPR and surface-enhanced Raman spectroscopy-based biosensors).

In Sect. 16.2, we present a noncomprehensive overview of the state of the art in nanophotonic approaches to biomolecular detection. The goal of this section will be to provide sufficient background to place our work in the context, rather than providing an exhaustive review of the current literature. Following this, we attempt to take a high-level overview of the field and propose a series of overall challenges for advancing it. Sections 16.4 and 16.5 describe some of our general approaches to solve these challenges, namely, the development of nanoscale optofluidic sensor arrays (NOSA) and techniques for targeted immobilization of probes within optical cavities.

16.2 Review of Nanophotonic Devices for Biomolecular Detection

16.2.1 Brief Overview of Evanescent Field-Based Optical Sensing

While most of the optical energy is confined within the structure itself, solid-core dielectric waveguides have an exponentially decaying tail of the guided optical mode, referred to as the evanescent field, that impinges a small distance (typically on the order of the wavelength of light) into the surrounding medium. Binding of a target at a probe site within this evanescent field causes a change in the local refractive index in that region, imparting a slight phase shift to the propagating optical mode. Likely the simplest way to detect this phase shift is through interferometry with the simplest practical on-chip implementation useful for biosensing being the Mach–Zehnder interferometer (MZI) (see Refs.25–27). These designs typically consist of an input optical waveguide, which is split into two arms of equal length and then recombined to form the output optical waveguide. A section of one of these arms, called the sensing arm, is functionalized with the desired probe agent. The second arm is referred to as the reference arm and is generally left without modification. In the absence of any surface modifications to either of the arms, the light recombining at the output port remains in phase, giving rise to constructive interference and maximal light intensity at the output port. When binding occurs at the surface of the sensing arm, it changes the local refractive index and the resulting phase shift causes the output power to drop due to destructive interference effects. An analogous approach using a Young's interferometer is discussed in Chap. 10.

16.2.2 Microring Resonators

The drawback of traditional devices such as the interferometers discussed above is that the interaction length between the optical field and the binding sites necessary for producing detectable phase shifts can be very large, often in the order of a centimeter. The major disadvantage of this is that it therefore requires a relatively large amount of actual bound mass to make an appreciable change in the transduction signal. Therefore, while the sensitivity to bulk properties, such as refractive index or mass per unit area, can be quite high, the response to absolute measures, like bound mass, tends to be relatively low (in comparison with some of the technologies to be discussed below).

Microring resonators consist of a ring waveguide adjacent to a bus waveguide. Light from a laser travels down a bus waveguide and evanescently couples into the ring resonator. When the optical path length around the ring is equal to an exact multiple of the wavelength of the excitation light, the conditions for constructive interference are met and the ring is said to be on-resonance. Under such conditions, the intensity of the light will build up in the rings until such a time as the intrinsic losses are equivalent to the input power. By exciting the bus waveguide with a tunable laser or broadband source, these resonance conditions can be detected at the output end of the bus waveguide as sharp dips in transmission. Binding events along the surface of the microring increase the refractive index in the evanescent field, effectively optically lengthening the ring and causing the resonant dips to red shift to longer wavelengths. The shift in the resonant wavelength can be related to the amount of bound mass on the surface, which in turn can be related to the solution phase concentration by empirical calibration. The degree to which this energy is trapped within the ring is reflected in a quantity called the Q-factor with higher Qs leading to greater intensity trapped in the ring (to be precise, the definition of the Q-factor is the ratio of the energy stored to the amount of energy lost in a round trip).

As an example of such a device, Chao et al.[28] demonstrated polymer microring resonators of 45μm radius having a Q-factor of 20,000. They were able to detect an effective refractive index change of 10^{-7} RIU and had a detection limit of approximately 250 pg mm^{-2} mass coverage on the microring surface, which translated into ~160 fg of actual bound mass. In comparison with the interferometric devices presented above, the major advantage of ring-type devices is that because the light circulates within the ring each photon interacts with each bound molecule multiple times thereby potentially increasing the effective path length of the device.

As such while such devices may have similar sensitivity to bulk measures like mass per unit area, the actual required surface area is much smaller and thus the actual amount of mass required to impart a detectable signal is much smaller. This translates into a lower total mass LOD. In addition, it is relatively easy to multiplex such devices by placing rings of different sizes (and therefore exhibiting resonant peaks at different wavelengths) along a single bus waveguide.

16.2.3 Whispering Gallery Mode Devices

Operationally similar to ring resonators, microcavities sustaining whispering gallery modes (WGMs)[29,30] have recently been demonstrated for label-free biosensing. WGMs correspond to light being confined along a circular orbit along the edge of a sphere-, disk-, or cylinder-shaped structure. WGMs have been extensively studied in liquid droplets and fused silica spheres[31], both of which have nearly atomic scale smoothness. In such microcavities, optical losses are significantly lower than in other optical resonators and the Q-factor can exceed a hundred million. A number of variations on such devices have been demonstrated as biosensors for example that of Armani et al.[32] who demonstrated label-free biomolecular detection of a variety of different analytes with a solution phase LOD as low as 5 aM for Interleukin-2.

16.2.4 Planar Photonic Crystal-Based Biosensors

Photonic crystals[33] are composed of periodic dielectric structures. One of the features this periodicity gives rise to a range of wavelengths, which are not allowed to propagate within the structure, is referred to as the photonic bandgap. The size of the bandgap and its spectral position can be tuned by varying the refractive index contrast of the dielectric materials and/or the periodicity of the structure[34]. These properties of photonic crystals make them extremely useful in a number of applications, including biosensing. Of the different architectures that have been developed, high Q-factor 1D and 2D photonic-bandgap microcavity[35] sensors are particularly interesting as the probed volume is shrunk down to the size of the optical cavity, which can be on the order of λ^3. As alluded to above, since the mode volume is so small the total amount of mass required to result in a measurable change in the refractive index (similarly reflected by a change in the wavelength of the resonant peak) can also be very small. Examples of such systems include that of Lee et al.,[36] who demonstrated a 2D resonant photonic crystal-based biosensor for protein detection and Schmidt et al.,[22] who demonstrated a nanoparticle sensing in a unique 1D cavity. The drawback of these designs is that the large photonic bandgap prohibits having multiple sensing sites along the same waveguide. As such, the number of targets, which can be screened for at once, is relatively small. Ideally, one would like an architecture that combines the high Q-factor and low mode volume sensing of the above devices with the ability to multiplex multiple detection sites along a single waveguide.

16.3 Our Approach to Improving on the State of the Art

There are numerous challenges involved in creating better biomolecular sensors, a number of which we have outlined in a recent review[15] and discussed throughout this book. Here we focus on how new device approaches can address the other emerging challenges such as the ability to simultaneously exhibit low LOD and high sensitivity, how to provide strong multiplexing capabilities, and how to simplify construction/assembly.

The technical challenge in addressing the former two of these is illustrated in Fig. 16.1. If we regard the goal of the sensor element itself as being able to provide as large of an output transduction signal as possible for the smallest amount of

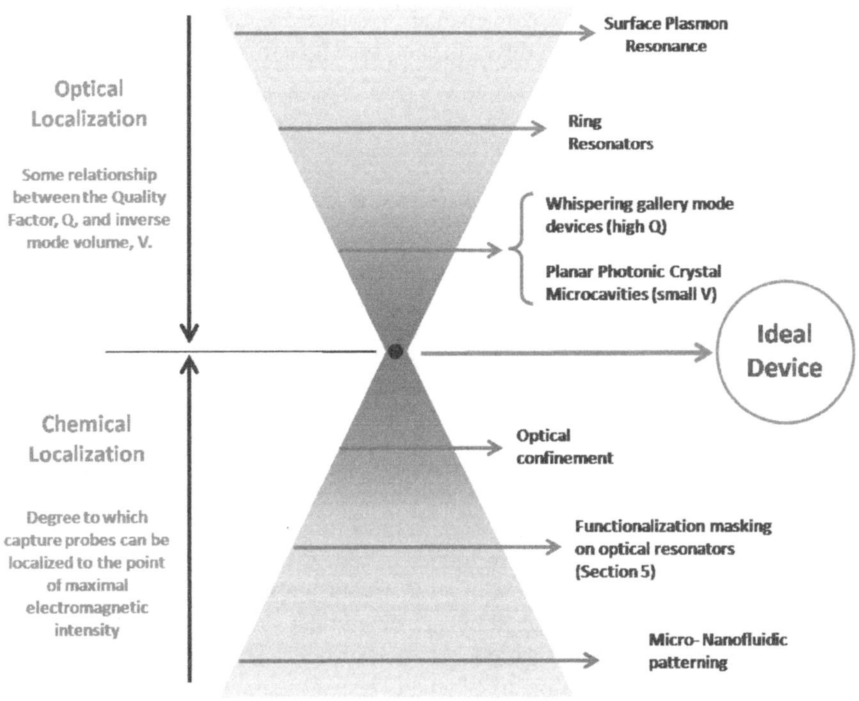

Fig. 16.1 Convergence of optical and chemical localization to improve biosensor performance. For label-free optical detection, ideally one would like to develop a system that maximizes the overlap between the optical localization (where the electromagnetic energy is concentrated) and the chemical localization (where the biomolecular target will be immobilized)

bound mass, it is clear that the optical sensor system must simultaneously confine the electromagnetic energy and the binding events to the same spatial location and that the energy density in this location should be as high as possible. Confining the electromagnetic energy down to a very small volume (such as in a photonic crystal microcavity) decreases the amount of functionalized area that is optically probed, but if the immobilization site is not also minimized, the same amount of mass is bound simply less of it is "seen." Therefore, without a chemical localization technique the improvement is likely to be marginal. We have placed broad groups of existing technologies at different locations along the axis in Fig. 16.1, to provide the reader with a general idea of how they might rank relative to each other. It is important to note, however, that because of the wide range of technologies within a particular subgroup, it is likely that different implementations could be at vastly different locations along these axes.

In the remaining two sections of this chapter, we discuss our efforts to address the challenges discussed above, through the development of an evanescently coupled linear optical resonator array architecture, which we have termed "Nanoscale Optofluidic Sensor Arrays." Section 16.4 outlines our technique for localizing electromagnetic energy in a device format that enables high quality factor, extreme multiplexibility, and relatively simplistic fabrication (approaching the apex in Fig. 16.1 from the top). Section 16.5 outlines our technique for spatial localization of the immobilization chemistry to ensure that the great majority of the binding events are "seen" by the sensor (approaching the apex in Fig. 16.2 from the bottom). Our goal given below will be to demonstrate the general design principles behind the approach and outlining how the techniques can be implemented on other designs. For further details, readers are referred to some of our recent papers namely Mandal and Erickson[37] and Goddard et al.[38]

16.4 Nanoscale Optofluidic Sensor Arrays

As mentioned above, the advantage of 1D photonic crystal cavities as biosensors is that they are likely the simplest possible optical devices able to concentrate electromagnetic energy down to spatial volumes as little as λ^3. The drawback of these designs is that the large photonic bandgap prohibits having multiple sensing sites along the same waveguide. Here we describe our technique for overcoming this limitation using evanescently coupled linear optical resonators. In the subsections below, we describe our approach, the fabrication and integration techniques used to construct the devices and some preliminary experimental/numerical results characterizing the technique. As will be demonstrated, this technique has a potential label-free LOD in the order of tens of attograms, while allowing for 2D multiplexing at reaction site densities at least equivalent to those found in a standard microarray.

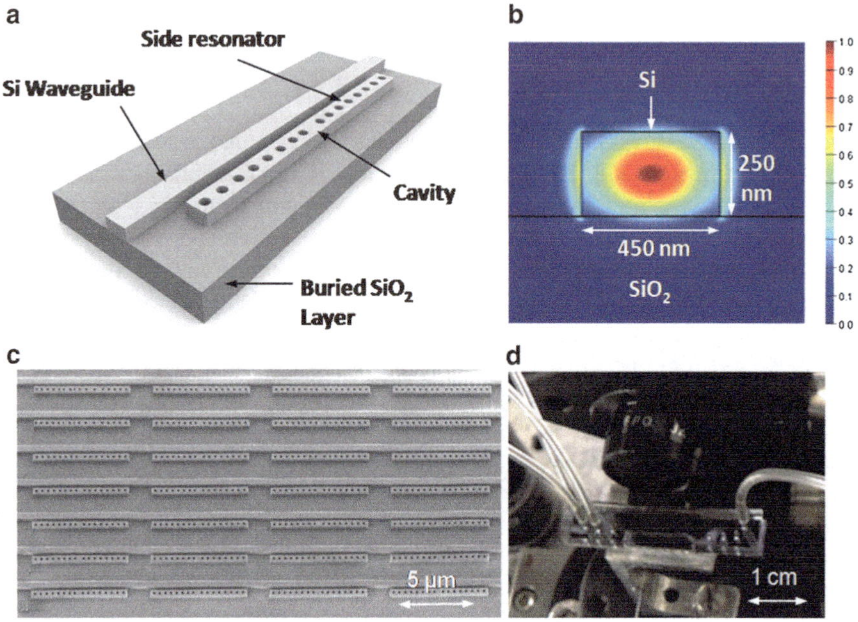

Fig. 16.2 Nanoscale optofluidic sensor arrays (NOSA). (**a**) 3D illustration of a NOSA sensing element. It consists of a 1D photonic crystal microcavity, which is evanescently coupled to a Si waveguide. (**b**) The electric field profile for the fundamental TE mode propagating through an air-clad Si waveguide on SiO_2. (**c**) SEM of a NOSA device array. It illustrates how this architecture is capable of two-dimensional multiplexing, thus affording a large degree of parallelism. (**d**) Actual NOSA chip with an aligned PDMS fluidic layer on top. Reprinted from Ref. 37 with permission

16.4.1 Design Overview

Figure 16.2a shows a 3D illustration of our sensor design. It consists of a silicon (Si) waveguide with a 1D photonic crystal microcavity (side resonator) that lies adjacent to the waveguide. The side resonator consists of a central defect cavity with eight holes on either side, which form the 1D photonic crystal. The Si waveguide was designed to be 450-nm wide and 250-nm tall to make it single mode. The low index silicon dioxide (SiO_2) layer which lies beneath the high index Si waveguide helps confine the light within the waveguide core, preventing optical losses into the lower substrate. Figure 16.2b shows the fundamental quasi-TE mode for this waveguide geometry.

A central defect cavity in the 1D photonic crystal gives rise to a defect state in the photonic bandgap. By varying this defect cavity spacing, we can tune the resonant wavelength of this state across the bandgap of the side resonator. Analogous to the ring resonators and WGM devices described above, light corresponding to the resonant wavelength couples evanescently into the side resonator and is sustained within it. This results in a dip in the output spectrum of the waveguide at the resonant wavelength. Because the resonant structures lie to the side of the waveguide, the bandgap does not interfere with the light transmission outside of that which lies in the resonant peak. Thus our unique design allows multiplexing along a single waveguide by placement of a large number of side resonators along the waveguide, each of which is fabricated to have a slightly different resonant wavelength. Figure 16.2c shows an example 4 by 7 NOSA array and Fig. 16.2d shows the final device bonded to PDMS fluidics.

Figure 16.3 shows the steady-state electric field distribution in a given device (computed using 3D FDTD simulations) for both the (a) on resonance and (b) off resonance conditions. As can be seen, there is a significant amount of light amplification within the resonator. Relative to the evanescent field at the sidewalls and top of the resonator, we observe the inner most holes of the side resonator to have a stronger optical field. This causes the resonators to be very sensitive to refractive index changes within these holes due to the large degree of light–matter interaction inside them. Figure 16.3c demonstrates the typical output spectrum of a device with four evanescently coupled resonators (again computed using FDTD simulations). In this particular simulation, each resonator consists of four holes on either side of the central cavity as opposed to the eight shown in the above figures. The inherent advantage of our sensor design is apparent from this graph. By tuning the input light across a range of wavelengths and having a large number of side resonators (each designed to possess a unique resonant wavelength within this tunable range) placed alongside a single waveguide, the output spectrum will consist of a large number of sharp dips in an otherwise flat spectrum wherein each dip corresponds uniquely to one of the resonators. Any shift in one of the resonances indicates a change in the refractive index of the local environment

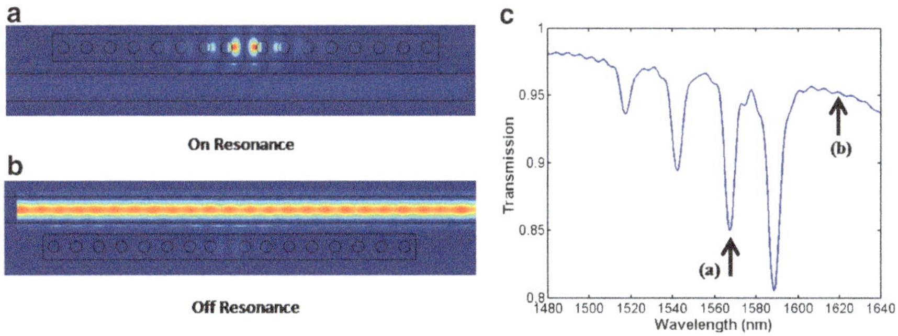

Fig. 16.3 Simulation of transmission spectrum for a four-resonator array. FDTD simulation showing the steady-state electric field distributions when the device is excited at the (**a**) resonant wavelength and (**b**) nonresonant wavelength. Note that the color levels in this image are scaled to the maximum field intensity in each image not to each other. The field levels in (**b**) are roughly of 20 times greater magnitude than those shown in (**a**). (**c**) Output spectrum for a device consisting of a waveguide with four evanescently coupled side cavities adjacent to it. Here each resonator consists of a cavity with four holes on either side. Reprinted from Ref. 37 with permission

around the corresponding resonator. In this manner, a large number of detections can be done in parallel on a single waveguide.

16.4.2 Device Fabrication and System Integration

Figure 16.4 shows a schematic illustrating a version of the current fabrication method for our device. Briefly, the photonic devices were fabricated on SOI wafers having a device thickness of 250 nm. XR-1541 e-beam resist (HSQ, Dow-Corning Corporation) was spun on the wafer and the devices were patterned using a Leica VB6-HR electron beam lithography system. The Si device layer was then etched vertically down using a chlorine-based inductively coupled plasma etching system. The remaining XR-1541 was dissolved in a dilute 100:1 HF solution. To increase coupling efficiencies, nanotapers[39] were fabricated and a layer of hard baked SU-8 (Microchem) photoresist was used as a cladding to the nanotapers.

The top microfluidic architecture was fabricated using a multilayer soft lithography technique with polydimethylsiloxane (PDMS). In the figure, we illustrate a

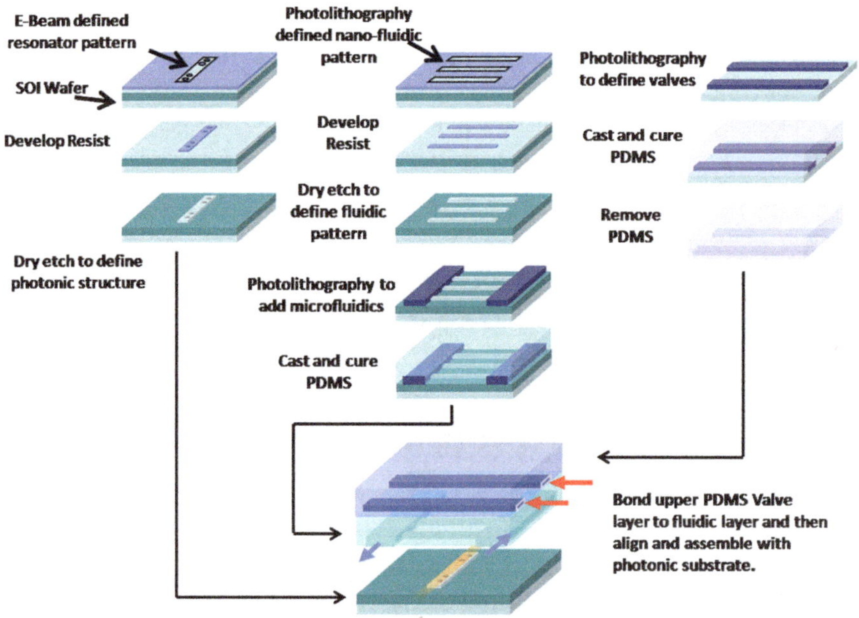

Fig. 16.4 Fabrication and assembly of the NOSA platform with PDMS microfluidics. The three elements of the fabrication are shown with the *left column* showing the steps involved in fabrication of the photonic structure, the *middle column* showing the fabrication of the fluidics, and the *right column* the fabrication of the valve layer. The *lower image* shows the assembly of the three elements into an integrated device similar to that shown in Fig. 16.2d

technique whereby nanofluidic channels are first defined in positive relief on an SOI wafer (analogous to how the photonic structures were fabricated) and then a microfluidic structure photolithographically patterned over them also in positive relief. In the majority of cases, however, the nanofluidic structures are not necessary and we forgo this step and simply pattern the microfluidics. In either case, liquid PDMS is spun coated over the fluidics layer and allowed to cure for approximately 90 min. Once cured the upper PDMS layer containing the pneumatic valves is bonded to the top side of the fluidic layer. To fabricate the complete device, the PDMS assembly was then aligned using a home-built rig and bonded reversibly to the chip. Precise alignment of the channels with the resonators during bonding was ensured by using a modified overhead optical microscope setup.

16.4.3 Experimental Demonstration

A typical, experimentally obtained, output spectrum of a waveguide with five resonators of differing sizes is shown in Fig. 16.5a. In this first case, all the five resonators had water as the surrounding medium. As can be seen, each resonator contributes a sharp dip to the output spectrum of the device. We observe that each 1D resonator possesses a large Q-factor varying from 1,500 to 3,000 and a full

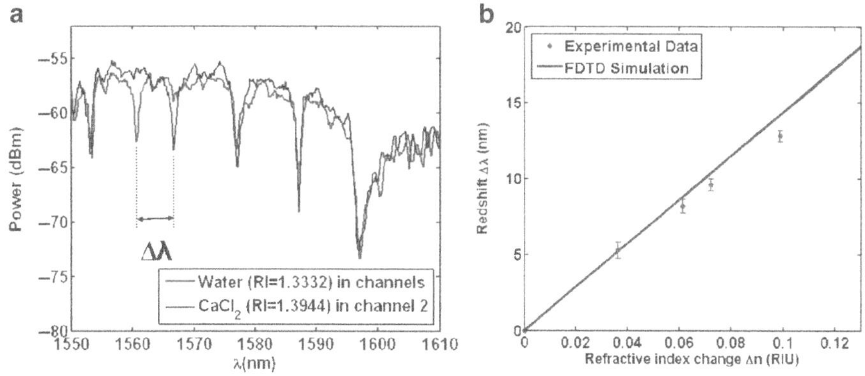

Fig. 16.5 Response to refractive index interrogation of a single NOSA waveguide. (**a**) Output spectrum for a NOSA where one of the five resonators is fluidically targeted, first with water and then with a $CaCl_2$ solution. The resonance of the targeted resonator shifts toward the red end of the spectrum due to the higher refractive index of the $CaCl_2$ solution. (**b**) Experimental data (with *error bars* indicating inter-device variability) showing the redshifts for various refractive index solutions. The *solid line* is the theoretically predicted redshift from FDTD simulations. The experimental data is in excellent agreement with the theory. Reprinted from Ref. 37 with permission. © 2008 Optical Society of America

width at half maxima of less than a nanometer. This is important for two reasons. First, higher Q-factors make it easier to detect very small shifts in the resonances. Equally as important, however, is that as the peaks get narrower it allows us to pack the output spectrum with a larger number of closely spaced dips and thus allows us to multiplex a larger number of resonators onto a single waveguide. Given the operational range of a standard 1,550-nm tunable laser (such as the one used here) and the linewidth of the observed resonances, we expect that 50 such side resonators could be incorporated on a single waveguide allowing us to perform as many as 50 detections in parallel on a single waveguide.

To investigate the refractive index sensitivity of the device we performed an experiment wherein one of the resonators was targeted by a fluidic channel, which was initially filled with water. When a higher refractive index solution of calcium chloride is passed through the channel, it changes the resonance condition of the resonator and pushes its unique resonant dip toward the red end of the spectrum as also shown in Fig. 16.5a. It is important to note that the other peaks are unaffected. In this way, one can confirm positive binding events occurring at any one of the resonators since only their corresponding resonances would show a shift in the output spectrum. Resonators with no binding occurring will show no shift in their output resonance.

The RI sensitivity of the NOSA devices was characterized by flowing fluids of different refractive indices through a channel targeting a particular resonator. We used deionized (DI) water as well as various concentrations of calcium chloride solution. The molar concentration of the $CaCl_2$ solutions varied from 1 to 5 M. The refractive indices of all the liquids were initially measured using a commercial refractometer. Figure 16.5b shows a plot of the shift in the resonant peak as a function of the change in refractive index of the fluid flowing through the channels. We observe an excellent match between the experimental data and the theoretically predicted redshifts. The slight discrepancy between the experimentally observed redshifts and those predicted by theory can possibly be attributed to optical losses in the waveguides and resonators arising due to surface roughness at their walls. The device exhibits a bulk refractive index sensitivity of over 130 nm for a unit shift in refractive index. Assuming a spectral resolution of 10 pm, we estimate the bulk refractive index detection limit of this device to be approximately 7×10^{-5}. Thus, while the refractive index LOD of this device is not as good as techniques like SPR, the ability to confine drastically the detection volume by targeting the holes allows us to lower the mass LOD.

16.4.4 Estimates of Device Performance

It is important to note that for applications such as biosensing, the device does not measure changes in the bulk refractive index of the surrounding medium, but rather respond to local changes in the refractive index at the surface of the sensor. As a result, the magnitude of the resonant shift will be dependent on a combination of factors such as the biolayer thickness and the effective change in refractive index of the bound targets. To model this here, we performed detailed 3D FDTD simulations wherein we studied the sensitivity of this sensor design and determined how to achieve the lowest mass LOD using this architecture. We assumed that the resonator was initially functionalized by a 50-nm thick single-stranded DNA monolayer (ssDNA). When a detection event occurs, the complementary ssDNA strand of the target would hybridize with the functionalized capture probes forming double-stranded DNA (dsDNA). The ssDNA monolayer and the dsDNA monolayer were assumed to possess refractive indices of 1.456 and 1.53, respectively, and a binding density of 1.49 pmol cm^{-2}.[40] The molecular weight of the nucleic acids used in our simulations is 57,000 Da. We varied the number of holes being functionalized to study the mass sensitivity of the device as a function of the number of functionalized holes. Simulations were performed for the cases of two holes (the innermost holes on either side of the cavity), four holes (the inner two holes on either side), and so forth, up to 16 holes (eight holes on either side) being functionalized with an ssDNA capture probe. We have calculated the term $\Delta\lambda/\Delta m$ in all these cases where $\Delta\lambda$ is the shift in the resonant wavelength of the device caused due to positive binding events resulting in the formation of dsDNA and Δm is the mass of the bound target. This $\Delta\lambda/\Delta m$ term is indicative of the mass sensitivity of the device. We use nucleic acids as are model species in this case due to the availability of data relating the change in local refractive index with surface concentration of immobilized probes and bound targets. For very large nucleic acid targets, however, there may be steric effects, which precludes transport into the resonator holes potentially

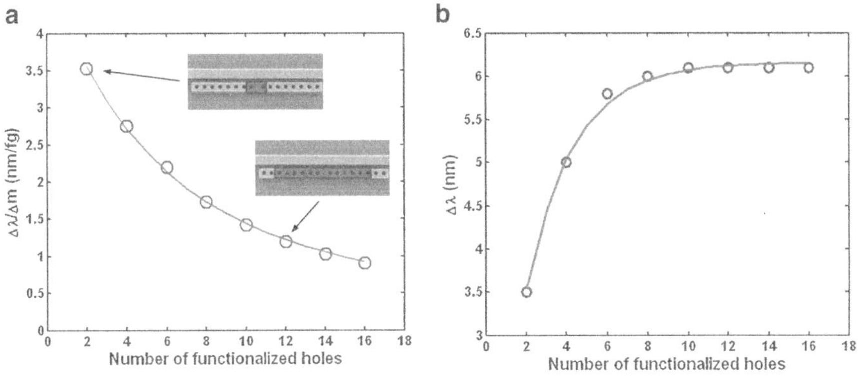

Fig. 16.6 Estimates of device performance in response to nucleic acid binding. (**a**) FDTD simulation showing the mass sensitivity of the device plotted as a function of the number of functionalized holes. The *circles* indicate the sensitivity values calculated from the simulations. The *solid curve* shows a least squares fit using an analytical model for the device sensitivity, which is described below. (**b**) Plot illustrating the dependence of the shift in resonant wavelength of a resonator on the number of functionalized holes. The *blue circles* indicate the data obtained from 3D FDTD simulations. The *solid curve* is a best-fit curve of the form $a(1-e^{-bN})$ where a and b are arbitrary constants. The values of a and b used here are 6.159 nm and 0.4273, respectively. Reprinted from Ref. 37 with permission. © 2008 Optical Society of America

degrading the overall sensitivity. For smaller nucleic acids or antibody/antigen systems we do not expect this to be a problem.

The circles in Fig. 16.6a show the calculated sensitivity, $\Delta\lambda/\Delta m$ for these different cases. As can be seen, the innermost holes are the most sensitive to any refractive index changes in the local environment as opposed to the holes that are further away from the cavity. These results can be explained by noting that the evanescent field is largest inside the innermost holes and decreases inside holes that are situated further away from the cavity. This is important to note because targeting only the inner most holes for functionalization allows for the lowest possible limit of mass detection for this device. In the case where only the inner two holes are functionalized we find that the resonance shifts by 3.5 nm when 1 fg of DNA binds to the resonator. Therefore, a mass change of 10 ag would result in a mass surface density of 0.84 ng cm^{-2} and an approximate shift of 0.01 nm, which can be experimentally detected. We therefore take this as the potential LOD of the device.

Figure 16.6b is a plot illustrating the dependence of the wavelength shift $\Delta\lambda$ on the number of functionalized holes N. We observe that an exponential function of the form $a(1-e^{-bN})$ (shown on the chart), where a and b are arbitrary constants that approximate this dependency quite well. To reiterate from above, although the resonant shift is larger for the greater number of holes, which are functionalized, more bound mass is required to impart this change. Thus increasing the number of holes, which are functionalized, tend to negatively affect the mass LOD.

16.5 Immobilization of Biological Recognition Elements Over Nanophotonic Biosensors

As alluded to above, as the design and fabrication of nanophotonic biosensors becomes more and more sophisticated, the difficulty in limiting immobilization of biomolecular recognition elements to the regions of highest optical intensity increases. Design of the optimal device requires maximal overlap between the concentrated electromagnetic energy and the biomolecular interaction to be detected. Binding events that occur outside this region of highest optical intensity are essentially lost (i.e., not detected by the device), negatively affecting the LOD of the device. An additional challenge is that as the probed area becomes smaller and optical intensity higher, the device is less able to average out surface heterogeneity and contaminants. Therefore, uniformity of immobilization chemistry and cleanliness of the final device are essential. In this section, we discuss our techniques for targeted immobilization of biomolecular capture probes over silicon nanophotonic devices. In targeting this chapter to the researcher more likely to be familiar with photonics than surface chemistry, we have focused here on providing practical advice on probe immobilization rather than detailed surface analysis. Readers interested in the latter should consult our recent paper[38].

16.5.1 Importance of Surface Cleanliness

In order to achieve uniform surface functionalization and subsequent biomolecule immobilization, the nanosensor surface must be free of any processing or environmental contaminants. Particular care must be taken when handling the finished devices, as well. Ungloved hands will of course contaminate the surface with various oils and salts, but even gloves can be a source of contamination, from the variety of oils and anticaking powders used to facilitate manufacturing of the gloves. The initial step in functionalizing the surface of a nanophotonic sensor often includes the generation of a silane monolayer. The stability of these monolayers depends on the uniform deposition of the silane and the ability of the silane molecules to cross-link with each other and with the hydroxylated surface. Surface contaminants will prevent the formation of well-defined monolayers. In addition, residues that interrupt the continuity of the cross-linked molecules allow moisture to penetrate the monolayer, resulting in hydrolysis of the siloxane bonds and loss of surface functionality.

Selection of an appropriate initial cleaning protocol will depend on the sensor substrate material. In the case of polymers, sonication in a range of solvents (in which the polymer is insoluble) followed by drying under nitrogen is often sufficient. For silicon-based nanosensors, a more thorough cleaning can be obtained by an ultraviolet (UV)/ozone treatment or oxygen plasma etching, both of which work by fragmenting the surface contaminants into lower molecular weight molecules that can be flushed away under vacuum or dissolved in solvents. In the case of plasma etching, it is important to remember that a dirty chamber often results in deposition of additional contaminants. A 10–20 min empty chamber etch is a prudent precleaning step. Wet chemical methods such as Piranha, MOS cleaning, and RCA cleaning methods provide even more aggressive cleaning. *Caution must be exercised when handling these wet chemical solutions, as many are strong bases, acids, or oxidizers, and can be explosive.* Analysis of the nanosensor surface following cleaning will ensure that the chosen cleaning protocol meets the needs of the device.

16.5.2 Biopatterning Techniques

For reasons described above, it is also critical that the patterning technique used to localize the biorecognition elements does not leave any residues. Microarray and inkjet printing are widely used biopatterning techniques, but their resolution is limited to a spot size between 10 and 100 μm. Additionally, the potential for capillary flow along topography makes these techniques less suitable for multiplexed optical nanostructures. Microcontact printing offers biopatterning on the scale of several micrometers, but the stamp can foul the surface, affecting device sensitivity[41–43]. Parylene biopatterning has proven to be an effective means to immobilize biomolecules onto surfaces on the micrometer scale or smaller[44–47]. Unfortunately, it has been observed that a grassy parylene residue often remains on the substrate after standard lithography processing. This residue may be a result of contamination by metal particulates during the parylene deposition process and has been observed to be unaffected by longer etch times. Metal particulates embedded within the parylene thin film can form stable complexes with oxidized parylene[48], which are then resistant to complete etching. Because parylene is biocompatible, such residue may not pose an issue in applications such as fluorescence detection or cell patterning. However, in the development of optical nanosensors, surface contamination reduces signal intensity, which adversely impacts device sensitivity[49]. Therefore, when utilizing this technique for biopatterning onto optical nanostructures, it is necessary to ensure that the patterned substrate is free of any residues in order to maintain device sensitivity and high optical Q-factor.

A number of published reports investigating the long-term stability of parylene thin films have suggested that exposure to UV irradiation results in oxidative degradation[48,50,51]. On the basis of these reports, we have developed a simple method to eliminate the residues that often remain after standard parylene patterning, while retaining the bulk integrity of the patterned parylene film. In this technique, we demonstrate that UV irradiation followed by rinsing in an alkaline solution removes 98% of this parylene residue and demonstrated it on the device from Sect. 16.4. In order to prepare the surfaces for capture probe immobilization, sensor substrates were submerged in Piranha solution (3:1 mixture of concentrated sulfuric acid and 30% hydrogen peroxide) for 30 min, followed by rinsing in copious Milli-Q water in order to clean the surface and generate surface silanol (Si–OH) groups. A 1-μm film of Parylene C was vapor deposited onto clean silicon substrates, after which Microposit S1818 positive photoresist (Rohm and Haas, Marlborough, MA) was spun on to a thickness of approximately 2 μm. After exposure through a photomask and development of the exposed regions of the photoresist, the exposed parylene was etched in a Plasmatherm SLR-720 reactive ion etcher (30 sccm O_2, 60 mTorr, 150 W).

Grassy parylene residue was removed by exposing the etched substrates to 10 mW cm^{-2} UV irradiation at 254 nm for 1 min. The photooxidized residue product was then dissolved by rinsing the patterned nanosensor in deionized water adjusted to pH 12 by sodium hydroxide. Various irradiation and alkaline rinse conditions were tested from which it was determined that this combination provided the desired result without impacting the bulk parylene. Following UV/base treatment, substrates were rinsed in ethanol to remove residual photoresist. SEMs of control and UV/base-treated etched parylene substrates were taken to illustrate the effect of parylene residues on optical nanodevice performance, as well as the ability to target nucleic acid probes over the parallel 1D photonic crystal resonators, as shown in Fig. 16.7. Since nanometer-scale coatings have been reported to cause significant red-shifts in optical devices[52], device sensitivity will be affected by any residue. Our UV/base treatment effectively removed the residual parylene as shown in Fig. 16.7b. After UV/base treatment and immobilization of four nucleic acid probes

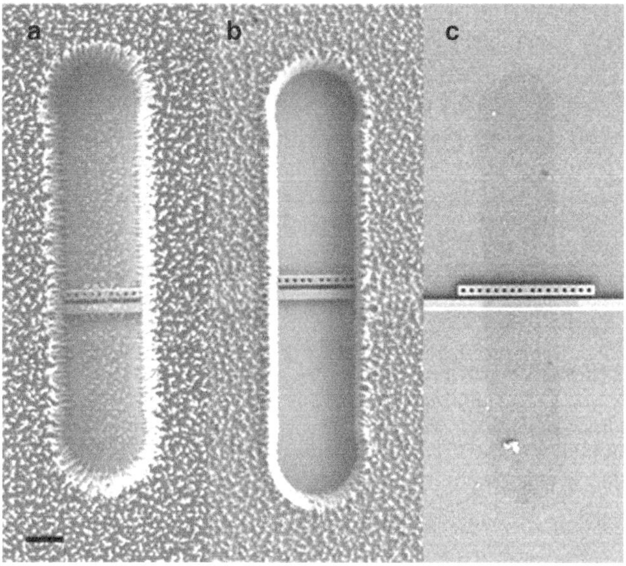

Fig. 16.7 SEM of parylene etched optical nanostructures (**a**) without UV/base treatment and (**b**) with UV/base treatment. (**c**) SEM image after removal of parylene. *Darkened spot* shows location of probes. Scale bar represents 2 μm

over parallel resonators (details on probe immobilization follow), the parylene remained intact and was fully removed (Fig. 16.7c). Immobilized capture probes appear slightly darker due to differences in charging effects during SEM imaging between the biopatterned region and the unfunctionalized SiO_2 substrate. Again, for further details on the analysis, readers are referred to our recent paper[38].

16.5.3 Biopatterning for Multiplexed Detection

It is often desirable to immobilize different biomolecules on different sensing elements in close proximity on the same nanophotonic sensor in the development of a multiplexed sensor. This is the case in the example of parallel 1D photonic crystal resonators described in Sect. 16.4. Cross-contamination of biomolecules must be avoided in order to preserve high specificity. We have found that a combination of parylene biopatterning and polydimethylsiloxane (PDMS) microfluidics is a convenient means to immobilized multiple biomolecules in close proximity without cross-contamination as shown in Fig. 16.8. Parylene biopatterning is first used to expose only the regions of highest optical intensity of the nanosensor for functionalization. Second, a set of PDMS microfluidics is applied to the parylene-patterned nanophotonic sensor, and the biomolecules to be attached

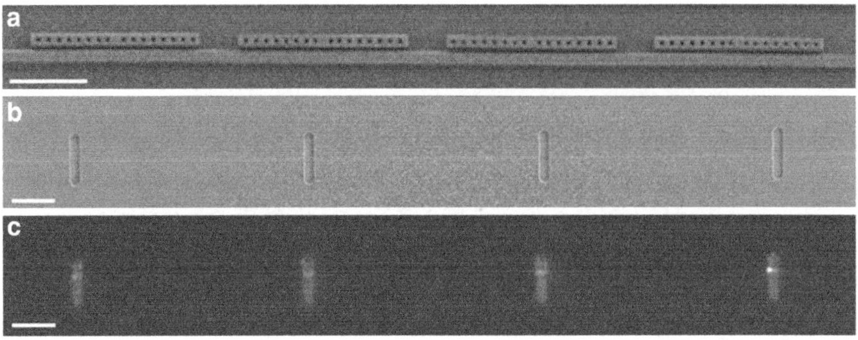

Fig. 16.8 Multiplexing immobilization on paralell resonators. (**a**) SEM of parallel resonators (scale bar is 3 μm). (**b**) SEM of parallel parylene patterned 1D photonic crystal resonators (scale bar is 20 μm). (**c**) Fluorescence micrograph of patterned capture probes after parylene removal (scale bar is 20 μm)

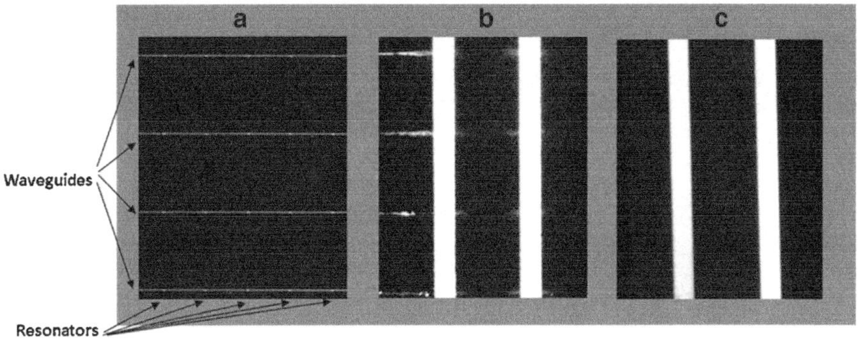

Fig. 16.9 Micrographs illustrating microfluidic targeting over structured substrate. (**a**) Optical micrograph of optical nanostructure. (**b**) Without parylene patterning, fluidic channels leak along waveguides within 5 min. (**c**) With parylene patterning, channels maintain integrity for full hour

are drawn through the microfluidic channel. In using this combined technique, we have observed an additional benefit to using parylene biopatterning over topographical substrates such as our photonic crystal resonators and waveguides. Although parylene deposition is reported to be a conformal coating, it provides sufficient planarization to prevent leaking during fluidic targeting of biomolecules over structured nanodevices. Figure 16.9 illustrates the fluidic integrity of structured nanodevices with, Fig. 16.9c, and without parylene, Fig. 16.9b. In both images, alternating channels of fluorescein isothiocyanate in water and water alone were withdrawn for 60 min across a nanostructured optical device. Leaking was evident on the uncoated substrates after less than 5 min; whereas parylene-coated devices exhibited no leakage through the course of the study.

16.5.4 Chemistry of Biomolecular Immobilization

In addition to selecting an appropriate biopatterning technique, consideration must be given to the method of immobilizing the biorecognition element. The major methods of immobilizing biological recognition elements include adsorption, electrostatic interactions, ligand/receptor pairing, and covalent binding. Adsorption is the simplest, and is suitable for single-use devices; however, it provides no specificity with regards to biomolecule orientation, and it is unlikely to withstand rigorous rinse steps. Electrostatic interactions can be long lasting, provided the pH and ionic strength of the sensor environment remains within the appropriate isoelectric points of the biomolecule and substrate. Ligand receptor pairing, such as the biotin–streptavidin interaction, provides the strongest noncovalent bond. The commercial availability of numerous biotinylated bioconjugation reagents (antibodies, oligonucleotides, enzymes, etc.) has made this method of immobilization a convenient option. A covalent bond provides the strongest linkage between the biomolecule and the substrate, and provides the greatest potential for a reusable device. By targeting specific reactive functional groups on both the biomolecule and the substrate, one is also better able to control the orientation of the immobilized biomolecule. This is of particular importance in the case of antibodies and oligonucleotides, in which proper orientation greatly impacts analyte binding efficiency.

As mentioned previously, the initial step in functionalization is often achieved by use of a silane cross-linker. Use of a mixed monolayer, in which multiple silane components are cross-linked together to form the silane monolayer, may help space out the reactive functional groups to prevent overcrowding of biorecognition elements, which adversely affects analyte binding efficiency. Silane monolayers can be formed by submersion in a solution of silane in organic solvent or by vapor phase deposition. Exposure to water must be carefully controlled to prevent polymerization of the silane and formation of multilayers. The deposited monolayer should be allowed to cure (overnight at room temperature or 1–2 h at 60–80°C is sufficient) to allow formation of siloxane cross-links. Biological recognition elements can then be directly linked to the nanophotonic sensor substrate by conjugation to the silane's terminal functionality; however, it is often beneficial to link the biomolecule via a tether molecule. Tether molecules such as poly(ethylene glycol) reagents and dendrimers can improve analyte binding efficiency by shielding the biomolecule from adverse surface-induced effects and by lessening steric hindrance[53–56]. Poly(ethylene glycol) offers additional advantages by reducing nonspecific adsorption of nontarget molecules within the sample matrix[57] and, in the case of sensors that incorporate micro- or nanofluidics, by providing a hydrophilic surface that improves fluid flow. Another consideration in selecting an appropriate functionalization chemistry is that in order to avoid high background signal from nonspecific adsorption, there should be little affinity between the target analyte and the tether molecule.

16.5.5 A Brief Practical Example for the Immobilization of Nucleic Acid Probes

To close off this section, we will briefly overview a surface functionalization chemistry scheme we have developed for immobilization of four separate nucleic acid capture probes over our NOSA resonators described in Sect. 16.4. The goal of this technique is to maximize hybridization efficiency while minimizing nonspecific adsorption of noncomplementary target. This is achieved by using carboxylic acid terminated generation 4.5 dendrimers (Dendritech, Inc., Midland, MI) as a tether molecule. Figure 16.10 illustrates the overall reaction scheme for immobilization of DNA capture probes onto patterned SiO_2 substrates. We have used this functionalization chemistry in combination with the residue-free parylene biopatterning and microfluidic targeting techniques described previously to limit the immobilization of fluorescently tagged nucleic acid capture probes over the regions of highest optical intensity within 1D photonic crystals, as described above. Utilizing carboxy-terminated dendrimers to tether nucleic acid capture probes to surfaces offers several advantages. Their branched, semispherical structure distances the immobilized capture probes from the solid surface, thus improving immobilization density and hybridization yields when used as a linker molecule[53–56,58]. When the dendritic linker is carboxyterminated, electrostatic interactions between target analyte and silicon substrate are minimized, thus reducing nonspecific binding vs. the more commonly used amine-terminated PAMAM dendrimer. We thus designed our surface functionalization chemistry to take advantage of these unique properties, while also limiting the overall height of the biopatterned region to the region of highest optical intensity within the evanescent field of the resonators.

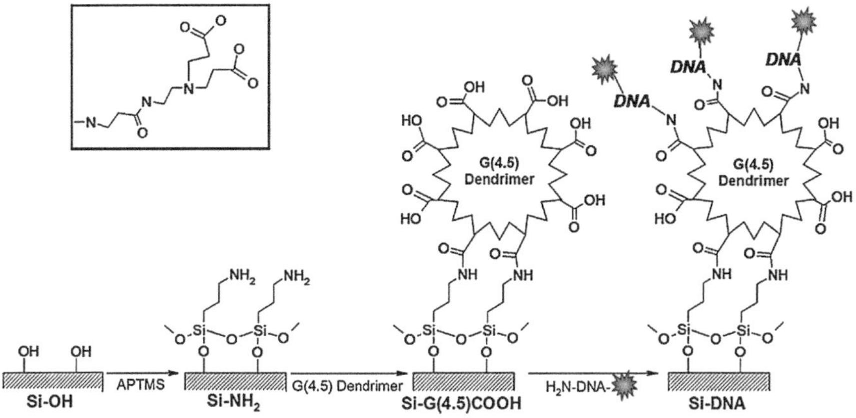

Fig. 16.10 Reaction scheme for immobilization of DNA onto functionalized SiO_2 substrates. Plasma-treated SiO_2 substrates are denoted as Si-OH, APTMS functionalized substrates are denoted as Si-NH_2, dendrimer functionalized substrates are denoted as Si-G(4.5)COOH, and substrates to which DNA capture probes have been immobilized are denoted as Si-DNA. *Inset*: Repeat unit of PAMAM dendrimer possessing terminal carboxylic acid functionality

Following parylene patterning and UV/base treatment as described above, patterned substrates were shaken in a solution of aminopropyltrimethoxysilane (APTMS, TCI America, Portland, OR) in ethanol for 5 min, followed by ethanol and water rinses, and curing for 1 h at 80°C. Generation 4.5 carboxylic acid-terminated dendrimers were covalently linked to the amine-functionalized substrates using water soluble carbodiimide chemistry[59]. Substrates were shaken for 2 h at room temperature in a solution of 10 μM dendrimer in 0.067 M phosphate buffer, pH 7.4, containing EDC and NHS, followed by rinsing in deionized water and drying under air. Subsequently, amine-terminated DNA capture probes (Operon Biotechnologies, Huntsville, AL) were bound to the carboxyl-functionalized SiO_2 surfaces via two-step water-soluble carbodiimide chemistry[59]. Briefly, surfaces were first shaken for 15 min in 0.1 M MES buffer, pH 6, containing EDC and NHS, followed by rinsing in water and drying under air. In order to separately address parallel resonators such that a different capture probe could be immobilized on each resonator, a PDMS microfluidics were applied to the substrate as described above. Amine-terminated DNA probes were diluted to 150 μM in 0.1 M MES buffer, pH 5.0 (final conjugation buffer was pH 7.0). Probes were withdrawn through microchannels for 2 h at room temperature, followed by in-channel water rinsing. Nucleic acid functionalized substrates (Si-DNA) were then removed from the plexiglass housing, PDMS microchannels and parylene film were peeled off the substrate, and the substrate was rinsed in deionized water and air dried.

16.6 Summary

In this chapter, we have attempted to describe broadly the advantages available from the use of planar nanophotonic devices as biomolecular detectors. We have reviewed the state of the art in these devices and described a few technical challenges involved in improving these devices. In the context of these challenges, we have introduced our "Nanoscale Optofluidic Sensor Arrays" which represents our attempt to address them.

References

1 Rich, R. L.; Myszka, D. G., Survey of the year 2006 commercial optical biosensor literature, *J. Mol. Recognit.* **2007**, 20, 300–366
2 Sander, C., Genomic medicine and the future of health care, *Science* **2000**, 287, 1977–1978
3 Srinivas, P. R.; Kramer, B. S.; Srivastava, S., Trends in biomarker research for cancer detection, *Lancet Oncol.* **2001**, 2, 698–704
4 Srinivas, P. R.; Verma, M.; Zhao, Y. M.; Srivastava, S., Proteomics for cancer biomarker discovery, *Clin. Chem.* **2002**, 48, 1160–1169
5 Growdon, J. H., Biomarkers of Alzheimer disease, *Arch. Neurol.* **1999**, 56, 281–283

6 Ross, J. S.; Schenkein, D. P.; Kashala, O.; Linette, G. P.; Stec, J.; Symmans, W. F.; Pusztai, L.; Hortobagyi, G. N., Pharmacogenomics, *Adv. Anat. Pathol.* **2004**, 11, 211–220
7 Hernandez, J.; Thompson, I. M., Prostate-specific antigen: A review of the validation of the most commonly used cancer biomarker, *Cancer* **2004**, 101, 894–904
8 Ward, A. M.; Catto, J. W. F.; Hamdy, F. C., Prostate specific antigen: Biology, biochemistry and available commercial assays, *Ann. Clin. Biochem.* **2001**, 38, 633–651
9 Sidransky, D., Emerging molecular markers of cancer, *Nat. Rev. Cancer* **2002**, 2, 210–219
10 Wulfkuhle, J. D.; Liotta, L. A.; Petricoin, E. F., Proteomic applications for the early detection of cancer, *Nat. Rev. Cancer* **2003**, 3, 267–275
11 Straub, T. M.; Chandler, D. P., Towards a unified system for detecting waterborne pathogens, *J. Microbiol. Methods* **2003**, 53, 185–197
12 Rasooly, A.; Herold, K. E., Biosensors for the analysis of food- and waterborne pathogens and their toxins, *J. AOAC Int.* **2006**, 89, 873–883
13 McBride, M. T.; Masquelier, D.; Hindson, B. J.; Makarewicz, A. J.; Brown, S.; Burris, K.; Metz, T.; Langlois, R. G.; Tsang, K. W.; Bryan, R.; Anderson, D. A.; Venkateswaran, K. S.; Milanovich, F. P.; Colston, B. W., Autonomous detection of aerosolized *Bacillus anthracis* and *Yersinia pestis*, *Anal. Chem.* **2003**, 75, 5293–5299
14 Stetzenbach, L. D.; Buttner, M. P.; Cruz, P., Detection and enumeration of airborne biocontaminants, *Curr. Opin. Biotechnol.* **2004**, 15, 170–174
15 Erickson, D.; Mandal, S.; Yang, A.; Cordovez, B., Nanobiosensors: Optofluidic, electrical and mechanical approaches to biomolecular detection at the nanoscale, *Microfluid. Nanofluid.* **2008**, 4, 33–52
16 Seydack, M., Nanoparticle labels in immunosensing using optical detection methods, *Biosensors Bioelectron.* **2005**, 20, 2454–2469
17 Zheng, G. F.; Patolsky, F.; Cui, Y.; Wang, W. U.; Lieber, C. M., Multiplexed electrical detection of cancer markers with nanowire sensor arrays, *Nat. Biotechnol.* **2005**, 23, 1294–1301
18 Li, C.; Curreli, M.; Lin, H.; Lei, B.; Ishikawa, F. N.; Datar, R.; Cote, R. J.; Thompson, M. E.; Zhou, C. W., Complementary detection of prostate-specific antigen using In(2)O(3) nanowires and carbon nanotubes, *J. Am. Chem. Soc.* **2005**, 127, 12484–12485
19 Majumdar, A., Bioassays based on molecular nanomechanics, *Dis. Markers* **2002**, 18, 167–174
20 Ouyang, H.; Striemer, C. C.; Fauchet, P. M., Quantitative analysis of the sensitivity of porous silicon optical biosensors, *Appl. Phys. Lett.* **2006**, 88, 163108
21 Chow, E.; Grot, A.; Mirkarimi, L. W.; Sigalas, M.; Girolami, G., Ultracompact biochemical sensor built with two-dimensional photonic crystal microcavity, *Opt. Lett.* **2004**, 29, 1093–1095
22 Schmidt, B.; Almeida, V.; Manolatou, C.; Preble, S.; Lipson, M., Nanocavity in a silicon waveguide for ultrasensitive nanoparticle detection, *Appl. Phys. Lett.* **2004**, 85, 4854–4856
23 Pollock, C.; Lipson, M., Integrated Photonics, Kluwer, Norwell, MA, **2003**.
24 Prasad, P., Nanophotonics, Wiley, Hoboken, NJ, **2004**.
25 Luff, B. J.; Wilkinson, J. S.; Piehler, J.; Hollenbach, U.; Ingenhoff, J.; Fabricius, N., Integrated optical Mach-Zehnder biosensor, *J. Lightwave Technol.* **1998**, 16, 583–592
26 Prieto, F.; Sepulveda, B.; Calle, A.; Llobera, A.; Dominguez, C.; Abad, A.; Montoya, A.; Lechuga, L. M., An integrated optical interferometric nanodevice based on silicon technology for biosensor applications, *Nanotechnology* **2003**, 14, 907–912
27 Heideman, R. G.; Lambeck, P. V., Remote opto-chemical sensing with extreme sensitivity: Design, fabrication and performance of a pigtailed integrated optical phase-modulated Mach-Zehnder interferometer system, *Sensors Actuat. B-Chem.* **1999**, 61, 100–127
28 Chao, C. Y.; Fung, W.; Guo, L. J., Polymer microring resonators for biochemical sensing applications, *IEEE J. Sel. Top. Quantum Electron.* **2006**, 12, 134–142
29 Matsko, A. B.; Ilchenko, V. S., Optical resonators with whispering-gallery modes – Part I: Basics, *IEEE J. Sel. Top. Quantum Electron.* **2006**, 12, 3–14

30 Ilchenko, V. S.; Matsko, A. B., Optical resonators with whispering-gallery modes – Part II: Applications, *IEEE J. Sel. Top. Quantum Electron.* **2006**, 12, 15–32
31 Arnold, S.; Khoshsima, M.; Teraoka, I.; Holler, S.; Vollmer, F., Shift of whispering-gallery modes in microspheres by protein adsorption, *Opt. Lett.* **2003**, 28, 272–274
32 Armani, A. M.; Kulkarni, R. P.; Fraser, S. E.; Flagan, R. C.; Vahala, K. J., Label-free, single-molecule detection with optical microcavities, *Science* **2007**, 317, 783–787
33 Joannopoulos, J. D.; Meade, R. D.; Winn, J. W., Photonic Crystals: Molding the Flow of Light, Princeton University Press, Princeton, NJ, 1995
34 Erickson, D.; Rockwood, T.; Emery, T.; Scherer, A.; Psaltis, D., Nanofluidic tuning of photonic crystal circuits, *Opt. Lett.* **2006**, 31, 59–61
35 Foresi, J. S.; Villeneuve, P. R.; Ferrera, J.; Thoen, E. R.; Steinmeyer, G.; Fan, S.; Joannopoulos, J. D.; Kimerling, L. C.; Smith, H. I.; Ippen, E. P., Photonic-bandgap microcavities in optical waveguides, *Nature* **1997**, 390, 143–145
36 Lee, M. R.; Fauchet, P. M., Two-dimensional silicon photonic crystal based biosensing platform for protein detection, *Opt. Express* **2007**, 15, 4530–4535
37 Mandal, S.; Erickson, D., Nanoscale optofluidic sensor arrays, *Opt. Express* **2008**, 16, 1623–1631
38 Goddard, J.; Erickson, D.; "Bioconjugation Techniques for Microfluidic Biosensors" *Analytical and Bioanalytical Chemistry* **2009**, 394, 469–479
39 Almeida, V. R.; Panepucci, R. R.; Lipson, M., Nanotaper for compact mode conversion, *Opt. Lett.* **2003**, 28, 1302–1304
40 Elhadj, S.; Singh, G.; Saraf, R. F., Optical properties of an immobilized DNA monolayer from 255 to 700 nm, *Langmuir* **2004**, 20, 5539–5543
41 Barbulovic–Nad, I.; Lucente, M.; Sun, Y.; Zhang, M. J.; Wheeler, A. R.; Bussmann, M., Bio-microarray fabrication techniques - A review, *Crit. Rev. Biotechnol.* **2006**, 26, 237–259
42 Xu, L. P.; Robert, L.; Qi, O. Y.; Taddei, F.; Chen, Y.; Lindner, A. B.; Baigl, D., Microcontact printing of living bacteria arrays with cellular resolution, *Nano Lett.* **2007**, 7, 2068–2072
43 Mannini, M.; Bonacchi, D.; Zobbi, L.; Piras, F. M.; Speets, E. A.; Caneschi, A.; Cornia, A.; Magnani, A.; Ravoo, B. J.; Reinhoudt, D. N.; Sessoli, R.; Gatteschi, D., Advances in single-molecule magnet surface patterning through microcontact printing, *Nano Lett.* **2005**, 5, 1435–1438
44 Ilic, B.; Craighead, H. G., Topographical patterning of chemically sensitive biological materials using a polymer-based dry lift off, *Biomed. Microdevices* **2000**, 2, 317–322
45 Moran-Mirabal, J.; Tan, C.; Orth, R.; Williams, E.; Craighead, H.; Lin, D., Controlling microarray spot morphology with polymer liftoff arrays, *Anal. Chem.* **2007**, 79, 1109–1114
46 Orth, R. N.; Kameoka, J.; Zipfel, W. R.; Ilic, B.; Webb, W. W.; Clark, T. G.; Craighead, H. G., Creating biological membranes on the micron scale: Forming patterned lipid bilayers using a polymer lift-off technique, *Biophys. J.* **2003**, 85, 3066–3073
47 Atsuta, K.; Suzuki, H.; Takeuchi, S., A parylene lift-off process with microfluidic channels for selective protein patterning, *J. Micromech. Microeng.* **2007**, 17, 496–500
48 Majid, N.; Dabral, S.; McDonald, J. F., The parylene-aluminum multilayer interconnection system for wafer scale integration and wafer scale hybrid packaging, *J. Electron. Mater.* **1989**, 18, 301–311
49 Byun, K. M.; Yoon, S. J.; Kim, D.; Kim, S. J., Sensitivity analysis of a nanowire-based surface plasmon resonance biosensor in the presence of surface roughness, *J. Opt. Soc. Am. A: Opt. Image Sci. Vis.* **2007**, 24, 522–529
50 Fortin, J. B.; Lu, T. M., Ultraviolet radiation induced degradation of poly-*para*-xylylene (parylene) thin films, *Thin Solid Films* **2001**, 397, 223–228
51 Pruden, K. G.; Sinclair, K.; Beaudoin, S., Characterization of parylene-N and parylene-C photooxidation, *J. Polym. Sci. Part A: Polym. Chem.* **2003**, 41, 1486–1496
52 Lee, M.; Fauchet, P. M., Two-dimensional silicon photonic crystal based biosensing platform for protein detection, *Opt. Express* **2007**, 15, 4530–4535

53 Le Berre, V.; Trevisiol, E.; Dagkessamanskaia, A.; Sokol, S.; Caminade, A. M.; Majoral, J. P.; Meunier, B.; Francois, J., Dendrimeric coating of glass slides for sensitive DNA microarrays analysis, *Nucleic Acids Res.* **2003**, 31, e88
54 Pathak, S.; Singh, A. K.; McElhanon, J. R.; Dentinger, P. M., Dendrimer-activated surfaces for high density and high activity protein chip applications, *Langmuir* **2004**, 20, 6075–6079
55 Benters, R.; Niemeyer, C. M.; Wohrle, D., Dendrimer-activated-solid supports for nucleic acid and protein microarrays, *Chembiochem.* **2001**, 2, 686–694
56 Benters, R.; Niemeyer, C. M.; Drutschmann, D.; Blohm, D.; Wohrle, D., DNA microarrays with PAMAM dendritic linker systems, *Nucleic Acids Res.* **2002**, 30, e10
57 Harris, J. M., Poly(Ethyelene Glycol) Chemistry: Biotechnical and Biomedical Applications, Plenum, New York, NY, **1992**, 385
58 Caminade, A. M.; Padie, C.; Laurent, R.; Maraval, A.; Majoral, J. P., Uses of dendrimers for DNA microarrays, *Sensors* **2006**, 6, 901–914
59 Hermanson, G. T., Bioconjugate Techniques, Academic, New York, NY, **1996**, 785

Chapter 17
Droplet-Based Cavities and Lasers

Kristian Mølhave, Anders Kristensen, and Niels Asger Mortensen

Abstract The self-organized and molecularly smooth surface on liquid microdroplets makes them attractive as optical cavities with very high quality factors. This chapter describes the basic theory of optical modes in spherical droplets. The mechanical properties including vibrational excitation are also described, and their implications for microdroplet resonator technology are discussed. Optofluidic implementations of microdroplet resonators are reviewed with emphasis on the basic optomechanical properties.

17.1 Introduction

Freely suspended liquid droplets are characterized by their shape determined by surface tension leading to ideally spherical shape and smooth surface at the subnanometer scale. These properties suggest liquid droplets as optical resonators with extremely high quality factors, limited by material absorption. Liquid microdroplets have found a wide range of applications for cavity-enhanced spectroscopy and in analytical chemistry, where small volumes and a container-free environment is required – for example for protein crystallization investigations. This chapter reviews the basic physics and technical implementations of light-matter interactions in liquid-droplet optical cavities.

K. Mølhave (✉)
DTU Nanotech, Technical University of Denmark, Ørsteds Plads, 2800 Kongens Lyngby, Denmark
e-mail: kristian.molhave@nanotech.dtu.dk

X. Fan (ed.), *Advanced Photonic Structures for Biological and Chemical Detection*, Integrated Analytical Systems,
DOI 10.1007/978-0-387-98063-8_17, © Springer Science+Business Media, LLC 2009

17.2 Optical Modes

Liquid droplets of spherical shape constitute a particular simple class of optical resonators, where the spherical symmetry allows for a simple classification of whispering-gallery modes (WGMs) in terms of their radial quantum number p as well as by their angular momentum quantum number l and the azimuthal quantum number m which can have $(2l + 1)$ values, meaning that the resonance frequency ω_{pl} has a $(2l + 1)$ degeneracy. For more details we refer to Ref. 1. For the present purpose we focus on the lowest-order radial modes ($p = 1$) with Fig. 17.1 illustrating one particular example. The resonances are to a very good approximation given by $n\,\chi \sim l$ where n is the refractive index of the liquid while $\chi = ka$ is a dimensionless size parameter with a being the radius of the droplet and $k = \omega/c = 2\pi/\lambda$ being the free-space wave number. The modes are consequently equally spaced in frequency with a corresponding spacing in wavelength given by

$$\Delta\lambda = \left|\frac{\partial\lambda}{\partial\chi}\right|\Delta\chi \cong \frac{\lambda^2}{2\pi a}\,\frac{\arctan\left(\sqrt{n^2-1}\right)}{\sqrt{n^2-1}}. \tag{17.1}$$

By careful expansion of the last fraction, the expression is found to have a $1/n$ asymptotic dependence for a large refractive index significantly exceeding the index of the surrounding medium, i.e., when $n >> 1$. To give an impression of a typical mode spacing, droplets with a volume of $V = 100$ pL, corresponding to a radius of $a \sim 30$ μm, will in the visible have a mode spacing of around $\Delta\lambda \sim 1$ nm.

17.3 Mechanical Modes

The equilibrium shape of the droplet may be addressed within the framework of Gibbs free-energy arguments. Analysis and estimates are significantly eased by

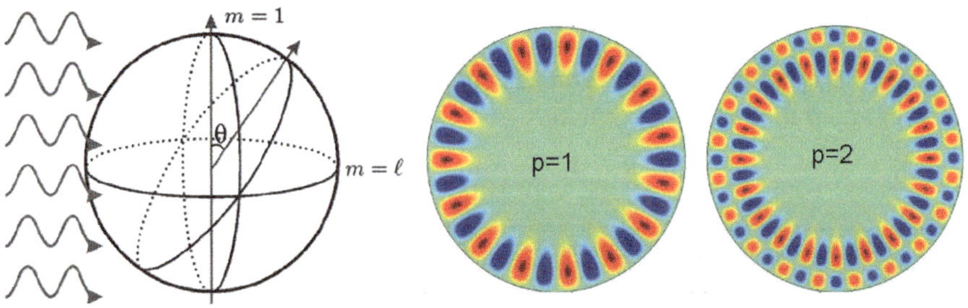

Fig. 17.1 Illustrations of whispering-gallery modes (WGM) in a spherical optical resonator. The WGM modes are classified in terms of their radial quantum number p as well as by their angular momentum quantum number l and the azimuthal quantum number m which can have $(2l + 1)$ values, meaning that the resonance frequency ω_{pl} has a $(2l + 1)$ degeneracy

introducing the characteristic length scale, usually referred to as the capillary length[2]

$$\xi = \sqrt{\frac{\gamma}{\rho g}}. \tag{17.2}$$

Here, γ is the surface tension while ρ is the liquid mass density and g is the strength of the gravitational field. The characteristic length scale allows for dimensional analysis with the result that droplets with a characteristic size $a \ll \xi$ will have a shape strongly determined by the surface tension. Obviously, a spherical shape will then be in favor when minimizing the Gibbs free energy. Using typical numbers for water droplets in air we find that $\xi \sim 3$ mm. For typical experiments employing droplets with $a < 500$ μm, the spherical shape is not perturbed significantly by body forces such as gravitation.

When the droplet is no longer studied under equilibrium conditions, a levitated or suspended droplet is in principle a complicated dynamical system as discussed in, e.g., Ref. 3. However, as already noted by Lord Rayleigh in 1879,[4] the vibrational spectrum of a liquid droplet can be characterized according to two different classes of modes. The first class is that of surface-tension driven surface modes while the second class consists of compression-driven compressional modes. Since for most practical cases the liquid can be considered incompressible, the latter class of compressional modes is necessarily in the high-frequency regime. Likewise, the low-frequency response is due to surface-shape oscillations conserving the volume. Quite interestingly, the surface vibrational modes are sound waves with modal properties similar to the optical WGMs and consequently they are characterized by their angular momentum quantum number l_{vib}. For low amplitude oscillations and $p_{\mathrm{vib}} = 1$, the spectrum is particularly simple[4]

$$\omega_{\mathrm{vib}} \cong \sqrt{\frac{\gamma l_{\mathrm{vib}}(l_{\mathrm{vib}} - 1)(l_{\mathrm{vib}} + 2)}{\rho a^3}}, \tag{17.3}$$

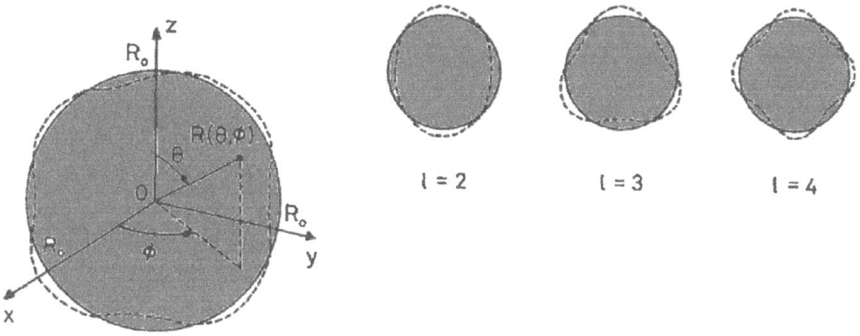

Fig. 17.2 Illustration of low-frequency acoustical whispering-gallery modes (WGM) in a liquid droplet

with l_{vib} taking integer values larger than unity. Figure 17.2 illustrates the low-frequency acoustical WGMs. Consequently, there is a lower bound and a critical frequency (corresponding to $l_{\mathrm{vib}} = 2$)

$$\omega_{\mathrm{vib}}^{*} \cong \sqrt{\frac{8\gamma}{\rho a^3}}, \tag{17.4}$$

below which external vibrational perturbations cannot excite liquid-droplet oscillations and thus the droplet will attain its equilibrium shape with a spherical symmetry.

Water droplets with a radius below 10 μm remain inert to acoustic vibrations and noise up to the 100-kHz regime, while above this frequency vibrations may be excited.

17.4 Implementations

Spherical optical resonators, as realized by liquid microdroplets, and their interaction with light have been studied in a large range of implementations. As a common trend, surface-energy driven self-organization into highly spherical and smooth surface topography is a key element in forming the system. The individual implementations mainly differ in strategies to suspend and stabilize the self-organized droplets.

17.4.1 Free-Falling Droplets

The most basic implementation to exploit the attractive optical properties of liquid-droplet resonators is illustrated in Fig. 17.3, where a stream of free-falling droplets is passing the experimental apparatus. In the example in Fig. 17.4, the droplets are formed of a solution of laser dye – rhodamine 590 or 640 – in ethanol, where each individual droplet forms an optically pumped microlaser as it is passing a green pump laser beam from an Nd:YAG laser. In the work by Qian et al.,[5] this was used to highlight the liquid-air interface for studies of morphology resonances, as illustrated in the right panel of Fig. 17.3. Controlled droplet generation is obtained by a piezoelectric dispensing apparatus oscillating at a frequency of 30–70 kHz, whereby also mechanical vibrations are excited, leading to a rich, but controlled ensemble of droplet shapes for investigations. The lifetime of the actual mechanical oscillations are sufficiently stable to be investigated in the free-fall configuration.

To estimate the available interaction time in a free-fall droplet experiment the steady-state balance between gravity and Stokes drag can be analyzed. A free-falling droplet will be subject to both gravitational and drag forces. Assuming that

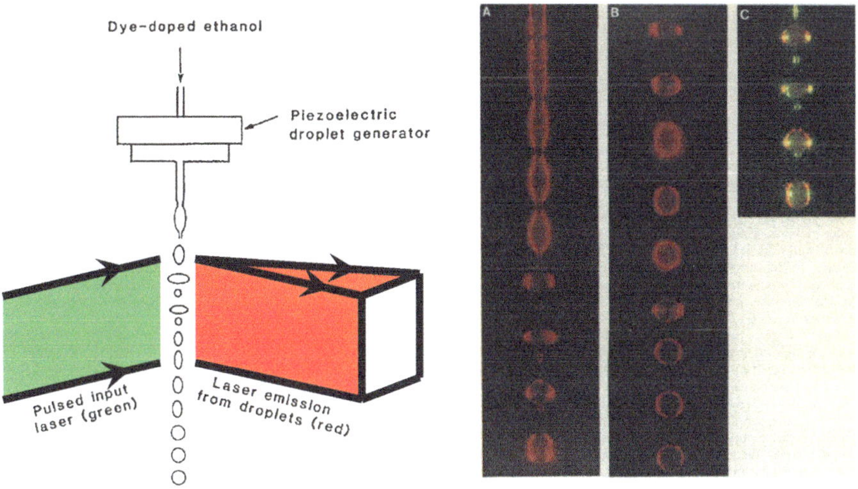

Fig. 17.3 Lasing of free-falling dye-doped ethanol droplets. Laser dye-doped ethanol droplets of radius 20–40 μm are ejected form a piezoelectric aerosol generator and excited by pulsed *green laser*. Reprinted from Ref. 5 with permission. © 2008 American Association for the Advancement of Science

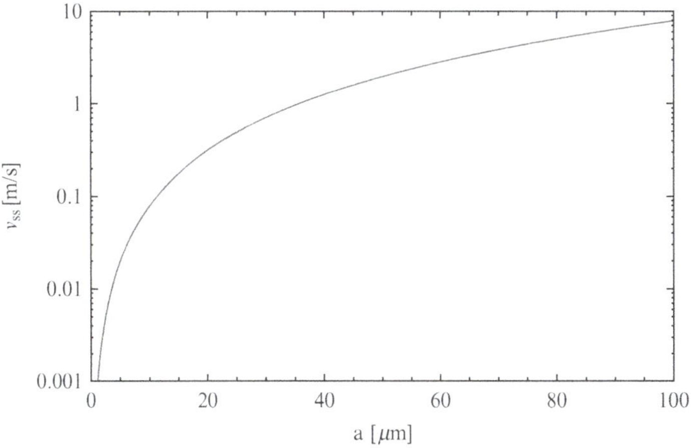

Fig. 17.4 Steady-state velocity of freely falling microdroplets as function of droplet radius calculated from the balance between gravitation and Stokes drag

$a \ll \zeta$, so that the droplet is of spherical shape, the drag is given by the Stokes result.[2] In the steady-state limit where the drag and gravitational forces balance we get the characteristic velocity

$$\left.\begin{array}{l} F_{\mathrm{g}} = \frac{4\pi}{3} a^3 \rho_{\mathrm{liquid}} g \\ F_{\mathrm{d}} = 6\pi \eta_{\mathrm{gas}} a v \end{array}\right\} \Rightarrow v_{\mathrm{ss}} = \frac{3}{9} \frac{\rho_{\mathrm{liquid}} g}{\eta_{\mathrm{gas}}} a^2. \tag{17.5}$$

Here, η_{gas} is the viscosity of the gas surrounding the liquid droplet and ρ_{liquid} is the mass density of the liquid. Figure 17.4 shows the steady-state velocity of a water droplet in air as a function of the droplet radius. The quadratic dependence on the droplet radius gives rise to a dramatic slow down, thus making visualization of falling microdroplets practical.

17.4.2 Ultrasonically Levitated Droplets

Gradients in the pressure field, p, associated with acoustic waves can serve to balance gravitation

$$F_{\mathrm{g}} = \frac{4\pi}{3} a^3 \rho_{\mathrm{liquid}} g = -\nabla p \tag{17.6}$$

and thereby levitate solid as well as liquid objects, and in particular droplet of radius a smaller than the capillary length ζ. For standing acoustic waves, the droplets can be captured in positions slightly below nodal points. Obviously, both the frequency and intensity of the sound wave play a central role in the position and stability of the captured droplet.

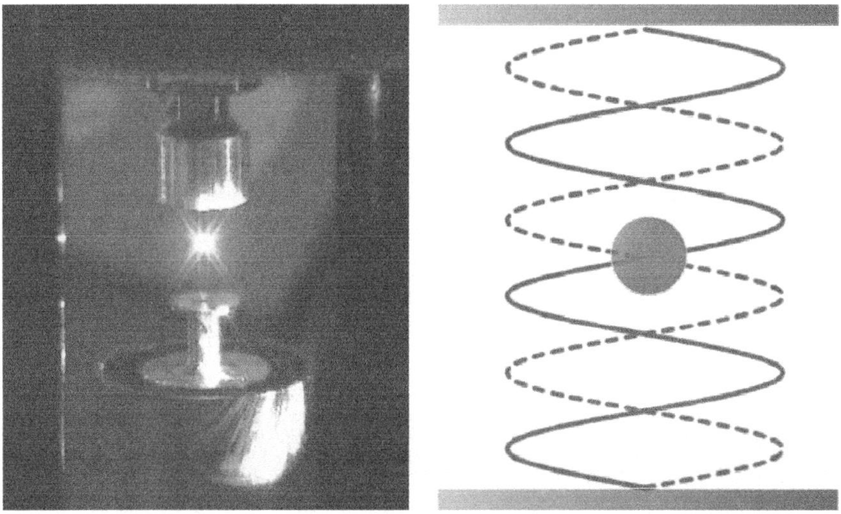

Fig. 17.5 Ultrasonically levitated microdroplet dye laser. *Left*: Photograph of a lasing levitated microdroplet. *Right*: Schematics of ultrasonic field with the microdroplet being trapped at a node in the ultrasonic field. Reprinted from Ref. 11 with permission. © 2008 Optical Society of America

The ultrasonic levitation technique was introduced in the 1930s and does not rely on any specific properties of the sample except size and mass, and has been used in bioanalytical and analytical chemistry applications[6–10]. The acoustic levitator consists of an ultrasonic transducer and a solid reflector supporting standing waves (see Fig. 17.5).

In experiments by Azzouz et al.,[11] a levitated droplet dye laser was realized and analyzed, with perspectives of harvesting the ultra high-Q microdroplet optical resonators for laser intracavity spectroscopy. In this work, the levitator was operated at a frequency of $\Omega_{\mathrm{vib}}/2\pi \sim 58$ kHz corresponding to a wavelength of $\Lambda \sim 6$ mm. The levitator can hold particles of radius $a < \Lambda$ and droplets with $a \sim \Lambda/6$ require a minimum of ultrasonic power. Large droplets may be deformed by the body forces (gravity and ultrasonic pressure gradients), which in practice limits the droplet size to $a < \xi$, the capillary length, which is on the order of 1 mm for a water droplet. Furthermore, the droplet shape may also be spheroidally deformed by applying large ultrasonic pressure amplitude. Rhodamine 6G (Rh6G) laser dye was dissolved in ethylene glycol (EG). The liquid sample was placed in a nodal point of the levitator by means of a computer-controlled piezoelectric microdispenser. Typically, a total volume of $V = (4\pi/3)a^3$ corresponding to a volume of 700 nL was introduced by repeated addition of pL drops.

The levitated laser dye droplet was optically pumped by a pulsed (pulse length 5 ns, repetition rate 10 Hz), frequency-doubled Nd:YAG laser ($\lambda = 532$ nm) in free-space optical setup. Droplet light emission was collected by a multimode optical fiber placed at an angle of approximately 50° relative to pump laser beam. Collected light was analyzed in a fixed-grating spectrometer with a resolution of FWHM 0.15 nm.

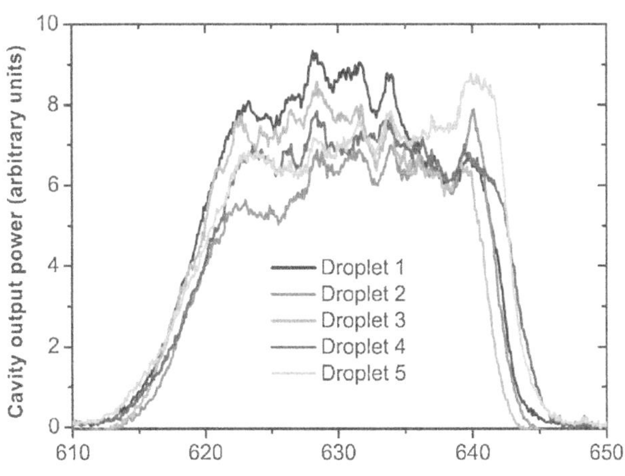

Fig. 17.6 Reproducible lasing spectra from dye doped microdroplets. Each spectrum is obtained in a fixed setup by a well-controlled loading of a 750-nL EG droplet with a Rh6G dye concentration of 2×10^{-2} mol L^{-1} which is subsequently pumped with an average pump power of 1.20 mW. Reprinted from Ref. 11 with permission. © 2008 Optical Society of America

The performance and applicability of the levitated droplet liquid dye laser are challenged by droplet evaporation and bleaching of the laser dye. In order to minimize these effects, Azzouz et al.[11] used a measurement scheme, where nominally identical droplets were loaded consecutively, and each droplet was at

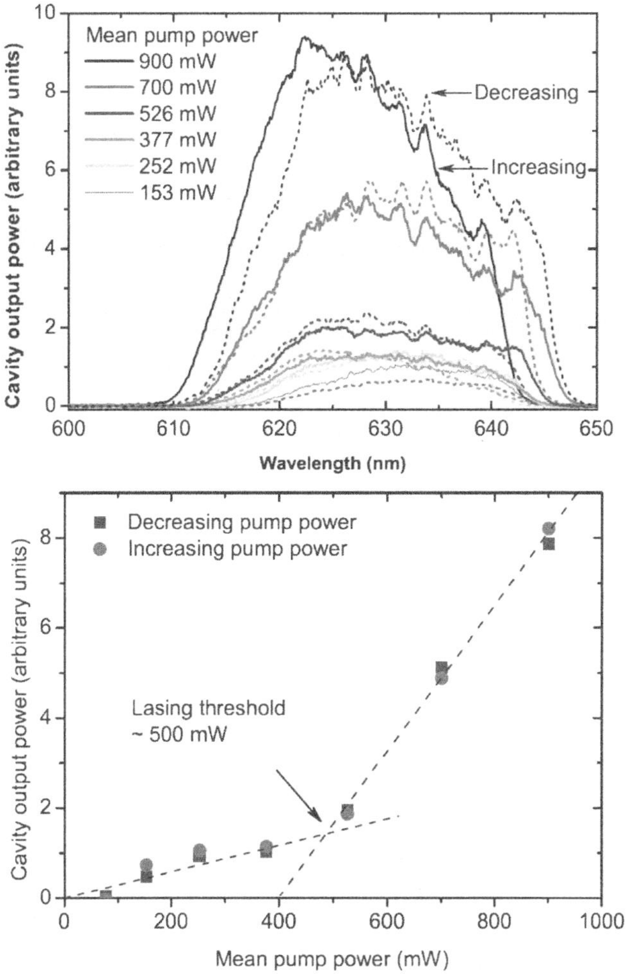

Fig. 17.7 *Top*: Cavity output power for increasing and decreasing average pump power. Each spectrum is obtained in a fixed setup by pumping a 0.75 μL EG droplet with a Rh6G dye concentration of 2×10^{-2} mol L^{-1}. The pump power is first increased from zero up to level above 1,000 mW (*dashed curves*) and subsequently again lowered (*solid curves*). The overall small differences between *dashed* and *solid curves* indicate the weak influence of evaporation dynamics during the measurement series. *Bottom*: Cavity output power vs. mean pump power. Data is obtain from the peak power of output spectra obtained in a fixed setup by pumping a 0.75-μL EG droplet with a Rh6G dye concentration of 2×10^{-2} mol L^{-1}. The *dashed lines* are guides to the eyes indicating a lasing threshold of around 500 mW in the mean pump power. Reprinted from Ref. 11 with permission. © 2008 Optical Society of America

maximum exposed to 100 pump pulses from the Nd:YAG laser – corresponding to 10 s of operation – before it was replaced by the next droplet.

Figure 17.6 shows five emission spectra obtained from normally identical $V =$ 750 nL EG droplets, with Rh6G concentration of 2×10^{-2} mol L^{-1}. An average pump power of 1.20 mW was used. The observed variation in output power was attributed to fluctuations in the pump power.

A second measurement series illustrates the lasing action of the dye droplet, where consecutively loaded droplets are pumped at different average pump power. The measured spectra shown in Fig. 17.7 show the dye droplet output power vs. pump power. In the measurement sequence, the pump power was first increased from 150 to 900 mW and subsequently decreased again. The reproducibility of the obtained spectra and the lasing threshold are seen from Fig. 17.7, respectively. The experiments reveal no significant indication of neither bleaching nor evaporation.

The levitated droplets and droplet dye lasers may conveniently be operated with acoustic frequencies below the critical for excitation of droplet vibrational modes, (17.4), to facilitate stable and highly spherical optical resonators.

17.4.3 Electrostatically Levitated Droplets

Charged microdroplets formed by, e.g., electrospraying or from charged piezoelectric droplet dispensers[12] can be trapped in an ion trap, such as a Paul trap[13]. The ion trap levitation principle makes it possible to trap particles with a wide range of charge to mass ratios, q/m, depending on the trap design and applied electrical fields,

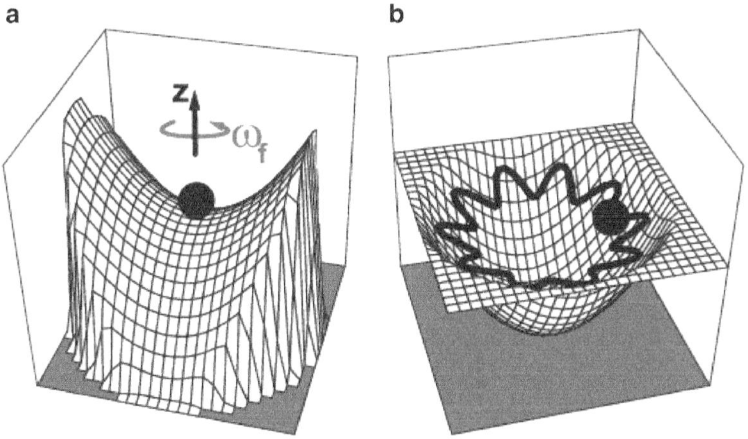

Fig. 17.8 Illustration of the trapping principle in an ion trap. The effect of rotating the saddle potential in part (**a**) is a pseudopotential well illustrated in part (**b**). For particles with suitable mass (and charge) the particle motion in the pseudopotential is indicated by the *black line*. The motion is a combination of a secular motion in the pseudopotential well and a small amplitude micromotion at the frequency of rotation ω_f. If the particle motion is damped, the particle can come to rest at the bottom of the well

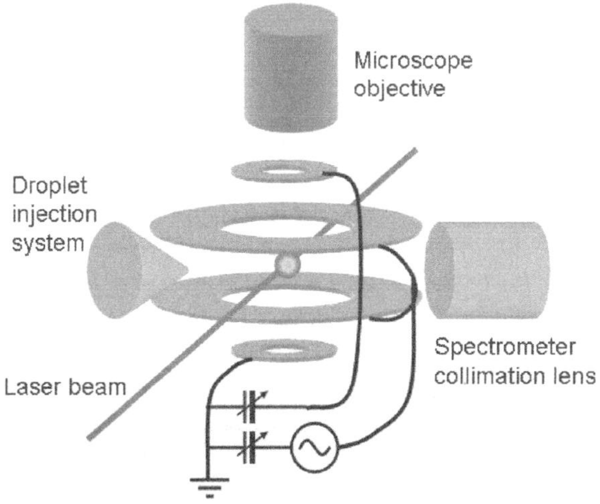

Fig. 17.9 Sketch of a typical setup for ion trap experiments on lasing microdroplets. The oscillating field between the inner and outer ring electrodes forms the trapping potential, and gravitational forces can be opposed by static electrical fields to move the droplet to the trap center with no micromotion

thereby circumventing the instability problems observed in for instance ultrasonic levitation for small droplets in the range interesting for single-mode observations.

Ion traps have been used in a number of studies of optical properties of microdroplet lasers[14,15] and the ion trap itself can be used as a useful mass spectrometer for fundamental studies of trapped particles and micrometer to nanoscale aerosols[16].

A static electrical field in free space cannot have potential minima and hence cannot trap a charged particle. By using oscillating fields, a pseudopotential can be formed as illustrated in Fig. 17.8, where a mass with a suitable charge to mass ratio can be trapped.

For the microdroplet experiments[14,15,17], small traps as sketched in Fig. 17.9 that allow high-resolution microscopy of the droplet have been used. The traps are generally formed by ring electrodes to make a cylindrically symmetric trapping region with an electrical saddle potential-driven AC electrical fields to form a three-dimensional pseudopotential well and DC fields to compensate for gravitational forces and imperfections in the trap geometry.

The ion trap methods allow measurements over extended periods of time of individual microdroplets[14] and can also be used as a high precision mass spectrometer for characterizing the droplet by its secular motion in the trapping potential[17]. For instance Tona et al. have done extensive studies[18] of the detailed mode structure[19] in droplets and the dependence on polarization[20] and dye concentration.[21]

Recently, even practically, single-mode lasing of a droplet has been reported for quantum dot doped droplets[14], when the droplet diameter was sufficiently small and pump laser intensities were high enough as shown in Fig. 17.10. Interestingly, the

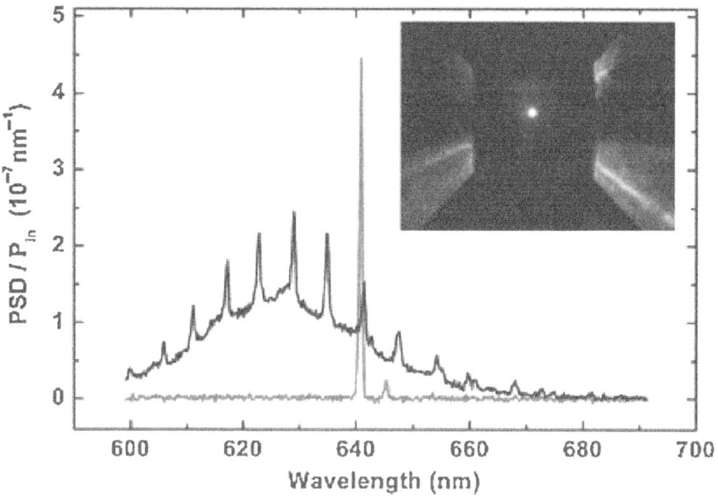

Fig. 17.10 Schäfer et al.[14] have observed single mode lasing from core-shell CdSe/ZnS nanocrystal quantum dots in a glycerine-water mixture. The fluorescence spectrum (*black line*) showed clear peaks of WGM and single-mode lasing (*grey line*) was observed for sufficiently small droplets (~10 μm) and high pump laser intensities (53 mJ cm^{-2} in 10-ns pulses at 532 nm). *Insert* shows the droplet trapped between the electrodes. Reprinted from Ref. 14 with permission. © 2008 American Chemical Society

observed lasing threshold was much lower than anticipated, tentatively explained as a sign of the quantum dots being attracted to the surface charge of the trapped droplet.

17.4.4 Droplets on Super Hydrophobic Surfaces

In order to achieve stationary and highly spherical microdroplets, the possibility to use superhydrophobic nanostructured surfaces has also been explored to make lasing[22] and Raman lasing microdroplets[23], where the high contact angle makes it possible to make long-term measurements on nearly spherical microdroplets at rest.

Superhydrophobic surfaces can be made by a variety of methods, where nanostructured surfaces provide the highest super hydrophobicity and contact angles. The nanostructured surfaces can be made by spin-on coating of particles[22] or nanotubes/nanograss surfaces[24] that can be both hydrophobic and oligophobic, thereby widening the range of solutions that can be used.

Evaporation of the droplets is an issue on the surfaces, since the vapor pressure of the liquid increases as the droplet radius decreases, thereby making the droplets evaporate even in a saturated vapor environment. The droplet volume can be stabilized by using the WGM size-dependent absorption peaks in the droplets in a supersaturated environment, where droplets increase in size until absorption at a WGM resonance increases the temperature and the droplet achieves a steady-state volume with femtoliter precision[25].

17.4.5 Liquid Suspension and Optically Trapped Droplets

Optical trapping[27] has for years been a tool for manipulating particles in gasses and liquids[28], and can conveniently be used to avoid the Brownian motion of droplets in dispersion. Lasing has been observed from optically trapped droplets[29]. Aerosol droplets can also be trapped in air[30]. By trapping the droplet in air the microdroplets' mass can be measured by observing their fall when the trapping laser has been turned off[28]. Instead of having the dye embedded in the microdroplet cavity, it can also be in the media surrounding the cavity and be used as laser media by the evanescent-wave-coupled gain from the internal WGM. This has been used to make WGM lasers in optical fibers immersed in dye solutions[31].

In optical tweezer experiments, the optical scattering force is used to trap particles, but the force can also be used to control the shape of liquid droplets[26]. An infrared laser with 43-mW power focused onto a microdroplet on a superhydrophobic surface enabled up to 40% reversible tuning of the equatorial diameter of the droplet[26]. Such effects must naturally also be taken into account when exciting laser modes in droplets in experiments with levitated drops.

17.4.6 Phase Matched and Critical Coupling to Droplet Resonators

An efficient optical coupling to the WGMs is instrumental in order to harvest the full potential of the high-Q droplet resonators. In most reported experiments, the droplet resonators are probed by free-space excitation, where, e.g., a Gaussian laser beam excites resonator modes and scattered light or fluorescence is detected. This approach

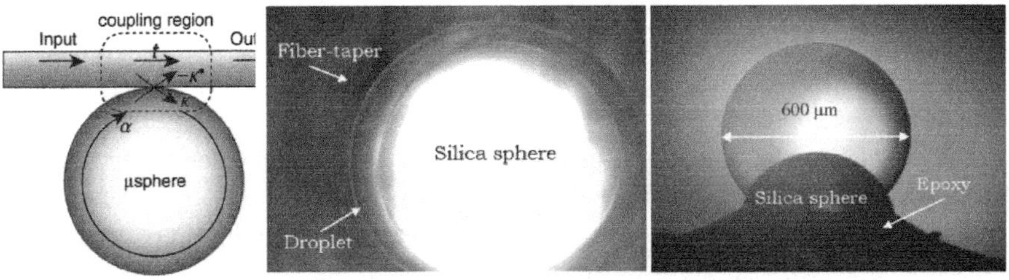

Fig. 17.11 *Left*: Outline of fiber taper coupling scheme to microsphere resonator. Reproduced from Ref. 33. *Right*: Fiber taper coupling configuration for liquid embedded microdroplet. Reprinted from Ref. 35 with permission. © 2008 Optical Society of America

suffers from a very small spatial overlap and a phase mismatch between the free-space Gaussian modes and the WGMs in the droplet, resulting in a very inefficient coupling.

Efficient, phase-matched excitation of WGMs in spherical cavities can be obtained by fiber-taper coupling[32–34]. A section of a single-mode telecommunication optical fiber is tapered down by heating and stretching[32] to form a thin waist of diameter down to 1 μm. The guided mode has a large evanescent tail around the thin fiber in the tapered region whereby the propagation constant can be matched to that of the relevant WGM in the microsphere resonator[34]. The silica microspheres are formed by melting a fiber tip, and the remaining fiber subsequently is used to hold the microsphere. When the fiber is placed in close vicinity to the microsphere resonator, as outlined in Fig. 17.11, the matched propagation constants and a large spatial overlap enables a highly efficient coupling between fiber waveguide- and microsphere WGMs. The system is probed by measuring the transmission through the fiber, which will exhibit dips (and reflections) when energy is coupled to WGMs propagating in the equatorial plane of the microsphere resonator.

The fiber-taper coupling scheme was applied to liquid-immersed water-droplet microsphere resonators by Hossein-Zadeh and Vahala[35]. The water droplets of diameter 0.5–1 mm were generated by a syringe and immersed in a low refractive immiscible cladding liquid - an index matching oil with a refractive index of 1.3, and trapped on a silica sphere which was fixed to the bottom of a liquid container

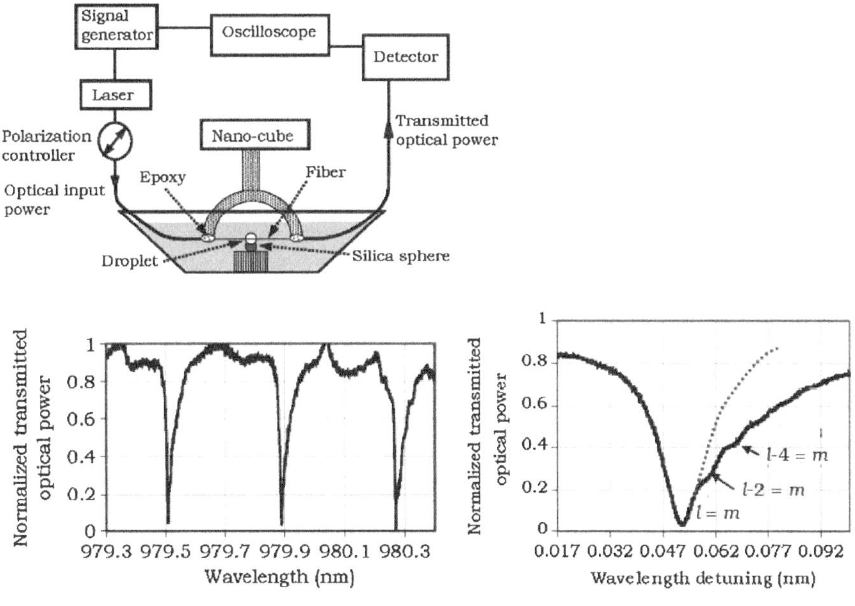

Fig. 17.12 *Top*: Outline of experimental setup for fiber taper coupling to liquid embedded microdroplet resonators. *Bottom*: Measured optical transmission spectra form a fiber taper coupled 600-μm diameter liquid embedded microdroplet resonator. Reprinted from Ref. 35 with permission. © 2008 Optical Society of America

(see Fig. 17.11). The liquid-cladding immersion stabilizes the water droplet by preventing evaporation and by damping mechanical vibrations. In addition, the low refractive index of the index matching oil serves to increase the evanescent tails of the guided modes in the fiber and droplet, and thereby enhances the special mode overlap. However, the low index contrast between droplet and cladding liquid of 0.029 also leads to large radiation losses for smaller droplets. The measured quality factor Q of the resonator has three contributions:

$$\frac{1}{Q} = \frac{1}{Q_{\text{ext}}} + \frac{1}{Q_{\text{abs}}} + \frac{1}{Q_{\text{rad}}}, \tag{17.7}$$

where Q_{ext} is the external quality factor associated with the waveguide coupling and Q_{abs} is the absorption loss limited quality factor, which is 1.98×10^5 for water around the applied vacuum wavelength, 980 nm. With the applied index matching liquid of refractive index 1.297, the measured quality factor Q of the water droplet resonator is estimated to be absorption limited for droplet diameters above 250 μm.

Figure 17.12 shows the outline of the experimental setup and the measured optical transmission spectrum for a fiber-taper coupled 600-μm diameter water droplet immersed in an index matching liquid cladding of refractive index 1.297. Critical coupling, where all energy is coupled to the resonator and the measured transmission is zero, is observed at vacuum wavelengths around 980 nm, with a measured quality factor of the fundamental mode of 7.5×10^4, in very good agreement with the calculated value of 7.79×10^4.

17.5 Outlook

As demonstrated by the many implementations, microdroplet lasers provide a unique combination of features:

- Individual droplets are cheaply and easily generated on demand by, for instance, piezodispensers, electrospray, or similar techniques.
- Droplets can be trapped and kept for long-time observations by several methods as described above, though evaporation can be an issue in some implementations.
- The WGM laser mode structure is determined by the droplet size and shape (which can be influenced by optical trapping forces[26]), laser polarization, and dye concentration[18].
- Droplet size and contents can be nondestructively characterized by many methods[36].

These features enable fast generation and characterization of micro- and nanoscale volumes with high sensitivity for changes in volume and chemical content. Chemical reactions can be followed by Raman spectroscopy, and even individual bacteria have been enclosed in drops and detected in droplet spectra[37]. For high sensitivity, high Q-factors are often considered important and extremely high Q-factors up to 10^8 have been achieved in hollow core droplets with gas bubbles surrounded by dye solution[38]. Microdroplet lasers seem to have several interesting future applications at the intersection of optofluidics with chemical and biotechnological methods for sensitive measurements on small liquid volumes as well as the more fundamental studies of high-Q cavity lasing.

Acknowledgments This work is supported by the Danish Technical Research Council (grant no. 26-02-0064) and the Danish Council for Strategic Research through the Strategic Program for Young Researchers (grant no. 2117-05-0037).

References

1 Symes, R.; Sayer, R. M.; Reid, J. P., Cavity enhanced droplet spectroscopy: Principles, perspectives and prospects, *Phys. Chem. Chem. Phys.* **2004**, 6, 474–487
2 Bruus, H., Theoretical Microfluidics, Oxford University Press, Oxford, **2008**
3 Yarin, A. L.; Weiss, D. A.; Brenn, G.; Rensink, D., Acoustically levitated drops: Drop oscillation and break-up driven by ultrasound modulation, *Int. J. Multiphase Flow* **2002**, 28, 887–910
4 Landau, L. D.; Lifshitz, E. M., Fluid Mechanics, Butterworth Heinemann, Oxford, 1987
5 Qian, S. X.; Snow, J. B.; Tzeng, H. M.; Chang, R. K., Lasing droplets – highlighting the liquid-air interface by laser-emission, *Science* **1986**, 231, 486–488
6 Omrane, A.; Santesson, S.; Alden, M.; Nilsson, S., Laser techniques in acoustically levitated micro droplets, *Lab. Chip.* **2004**, 4, 287–291
7 Santesson, S.; Nilsson, S., Airborne chemistry: Acoustic levitation in chemical analysis, *Anal. Bioanal. Chem.* **2004**, 378, 1704–1709
8 Santesson, S.; Ramirez, I. B. R.; Viberg, P.; Jergil, B.; Nilsson, S., Affinity two-phase partitioning in acoustically levitated drops, *Anal. Chem.* **2004**, 76, 303–308
9 Santesson, S.; Andersson, M.; Degerman, E.; Johansson, T.; Nilsson, J.; Nilsson, S., Airborne cell analysis, *Anal. Chem.* **2000**, 72, 3412–3418
10 Santesson, S.; Degerman, E.; Johansson, T.; Nilsson, J.; Nilsson, S., Bioanalytical chemistry in levitated drops, *Am. Lab.* **2001**, 33, 13–18
11 Azzouz, H.; Alkhafadiji, L.; Balslev, S.; Johansson, J.; Mortensen, N. A.; Nilsson, S.; Kristensen, A., Levitated droplet dye laser, *Opt. Express* **2006**, 14, 4374–4379
12 Schafer, J.; Mondia, J. P.; Sharma, R.; Lu, Z. H.; Wang, L. J., Modular microdrop generator, *Rev. Sci. Instrum.* **2007**, 78, 066102
13 Paul, W., Electromagnetic traps for charged and neutral particles, *Rev. Mod. Phys.* **1990**, 62, 531–540
14 Schafer, J.; Mondia, J. P.; Sharma, R.; Lu, Z. H.; Susha, A. S.; Rogach, A. L.; Wang, L. J., Quantum dot microdrop laser, *Nano Lett.* **2008**, 8, 1709–1712
15 Tona, M.; Kimura, M., Parallel-plate ion trap useful for optical studies of microparticles, *Rev. Sci. Instrum.* **2004**, 75, 2276–2279
16 Wang, S. Y.; Zordan, C. A.; Johnston, M. V., Chemical characterization of individual, airborne sub-10-nm particles and molecules, *Anal. Chem.* **2006**, 78, 1750–1754
17 Schlemmer, S.; Illemann, J.; Wellert, S.; Gerlich, D., Nondestructive high-resolution and absolute mass determination of single charged particles in a three-dimensional quadrupole trap, *J. Appl. Phys.* **2001**, 90, 5410–5418
18 Tona, M., Study on Spherical Microlasers Levitated in an Ion Trap, Kochi University of Technology, Kochi, **2002**

19 Tona, M.; Kimura, M., Novel lasing modes observed in a levitated single dye-doped microdroplet, *J. Phys. Soc. Jpn* **2000**, 69, 3533–3535

20 Tona, M.; Kimura, M., Polarization effects in both emission spectra and microscopic images of lasing microdroplets levitated in an ion trap, *J. Phys. Soc. Jpn* **2002**, 71, 425–428

21 Tona, M.; Kimura, M., Dependence of lasing modes of microdroplets on dye concentration, *J. Phys. Soc. Jpn* **2003**, 72, 1238–1243

22 Kiraz, A.; Sennaroglu, A.; Doganay, S.; Dundar, M. A.; Kurt, A.; Kalaycioglu, H.; Demirel, A. L., Lasing from single, stationary, dye-doped glycerol/water microdroplets located on a superhydrophobic surface, *Opt. Commun.* **2007**, 276, 145–148

23 Sennaroglu, A.; Kiraz, A.; Dundar, M. A.; Kurt, A.; Demirel, A. L., Raman lasing near 630 nm from stationary glycerol-water microdroplets on a superhydrophobic surface, *Opt. Lett.* **2007**, 32, 2197–2199

24 Gjerde, K.; Kumar, R. T. R.; Andersen, K. N.; Kjelstrup-Hansen, J.; Teo, K. B. K.; Milne, W. I.; Persson, C.; Molhave, K.; Ruabahn, H. G.; Boggild, P., On the suitability of carbon nanotube forests as non-stick surfaces for nanomanipulation, *Soft Matter* **2008**, 4, 392–399

25 Kiraz, A.; Kurt, A.; Dundar, M. A.; Yuce, M. Y.; Demirel, A. L., Volume stabilization of single, dye-doped water microdroplets with femtoliter resolution, *J. Opt. Soc. Am. B Opt. Phys.* **2007**, 24, 1824–1828

26 Kiraz, A.; Yavuz, S. C.; Karadag, Y.; Kurt, A.; Sennaroglu, A.; Cankaya, H., Large spectral tuning of liquid microdroplets standing on a superhydrophobic surface using optical scattering force, *Appl. Phys. Lett.* **2007**, 91, 231102

27 Ashkin, A., Optical trapping and manipulation of neutral particles using lasers, *Proc. Natl Acad. Sci. USA* **1997**, 94, 4853–4860

28 Kaiser, T.; Roll, G.; Schweiger, G., Investigation of coated droplets in an optical trap: Raman-scattering, elastic–light-scattering, and evaporation characteristics, *Appl. Opt.* **1996**, 35, 5918–5924

29 Kitamura, N.; Kitagawa, F., Optical trapping – chemical analysis of single microparticles in solution, *J. Photochem. Photobiol. C Photochem. Rev.* **2003**, 4, 227–247

30 Hopkins, R. J.; Mitchem, L.; Ward, A. D.; Reid, J. P., Control and characterisation of a single aerosol droplet in a single-beam gradient-force optical trap, *Phys. Chem. Chem. Phys.* **2004**, 6, 4924–4927

31 Moon, H. J.; Chough, Y. T.; An, K., Cylindrical microcavity laser based on the evanescent-wave-coupled gain, *Phys. Rev. Lett.* **2000**, 85, 3161–3164

32 Knight, J. C.; Cheung, G.; Jacques, F.; Birks, T. A., Phase-matched excitation of whispering-gallery-mode resonances by a fiber taper, *Opt. Lett.* **1997**, 22, 1129–1131

33 Cai, M.; Painter, O.; Vahala, K. J., Observation of critical coupling in a fiber taper to a silica-microsphere whispering-gallery mode system, *Phys. Rev. Lett.* **2000**, 85, 74–77

34 Little, B. E.; Laine, J. P.; Haus, H. A., Analytic theory of coupling from tapered fibers and half-blocks into microsphere resonators, *J. Lightwave Technol.* **1999**, 17, 704–715

35 Hossein-Zadeh, M.; Vahala, K. J., Fiber-taper coupling to whispering-gallery modes of fluidic resonators embedded in a liquid medium, *Opt. Express* **2006**, 14, 10800–10810

36 Ray, A. K.; Souyri, A.; Davis, E. J.; Allen, T. M., Precision of light-scattering techniques for measuring optical-parameters of microspheres, *Appl. Opt.* **1991**, 30, 3974–3983

37 Tanyeri, M.; Kennedy, I. M., Detecting single bacterial cells through optical resonances in microdroplets, *Sensor Lett.* **2008**, 6, 326–329

38 Huston, A. L.; Lin, H. B.; Eversole, J. D.; Campillo, A. J., Effect of bubble formation on microdroplet cavity quality factors, *J. Opt. Soc. Am. B Opt. Phys.* **1996**, 13, 521–531

Chapter 18
Single Molecule Analysis with Planar Optofluidics

Holger Schmidt and Aaron R. Hawkins

Abstract Integrated optofluidics uses integrated optical elements for building a new generation of analytical devices that feature planar architectures for both fluidics and optics. Liquid-core optical waveguides that can simultaneously guide both liquids and light through micron-scale hollow channels are essential components of such a platform. Here, we review advances in single particle detection, manipulation, and analysis in integrated optofluidic chips based on liquid-core antiresonant reflecting optical waveguides (ARROWs). Starting from a discussion of physical principles and fabrication methods, we discuss demonstrations of liquid-core waveguiding, fluorescence and Raman detection, single molecule analysis, and all-optical particle manipulation. Special emphasis is placed on the on-chip implementation of fluorescence correlation spectroscopy for analysis of biological particles, including liposomes and viruses. An outlook on future opportunities and challenges is given.

18.1 Introduction

Contemporary developments in chemical and biological sensor technology include high-throughput detection, analysis of minute trace amounts of liquid analytes, and sensitivities down to the single molecule level at which ensemble averaging is eliminated. These capabilities have enabled measurements of the intrinsic properties of molecules and access to the details of reaction kinetics. Optical sensing methods have a long tradition in bio/chemical analysis as they provide great versatility coupled with high sensitivity. Examples include detection via absorption,

H. Schmidt (✉)
School of Engineering, University of California Santa Cruz, 1156 High St., Santa Cruz CA 95064, USA
e-mail: hschmidt@soe.ucsc.edu

A.R. Hawkins
ECEn Department, Brigham Young University, 459 Clyde Building, Provo, UT 84602, USA

X. Fan (ed.), *Advanced Photonic Structures for Biological and Chemical Detection*, Integrated Analytical Systems,
DOI 10.1007/978-0-387-98063-8_18,

emission (fluorescence), scattering, refractometry (index changes), polarimetry, and wavelength-dependent spectroscopy[1]. Numerous powerful implementations of these techniques based on both basic and highly specialized optical microscopes have been developed and demonstrated.

The need for improved sensor performance has led to the emergence of micro and nanofluidics. These fields seek to develop miniaturized analysis systems that combine the desired attributes in a compact and cost-effective setting. These platforms are commonly labeled as labs-on-chip or micro total analysis systems (μTAS)[2], often using optical methods to realize a desired functionality. The preeminent role that optics play has recently led to the notion of optofluidics as an independent field that deals with devices and methods in which optics and fluidics enable each other[3]. Most of the initial lab-on-chip advances, however, occurred in the area of fluidics, while the optical components continued to consist largely of bulk components such as polarizers, filters, lenses, and objectives.

The pursuit of complete miniaturization of the sensing systems provides strong incentives to draw from the progress in the field of integrated optics and create fully planar, integrated optofluidic devices. Historically, integrated optics was driven by the telecommunications industry, and numerous optoelectronic elements including waveguides, sources, detectors, polarizers, modulators, and others have been developed. In recent years, efforts to combine these elements with fluidic environments have substantially increased, and the development of suitable planar waveguides is being pursued vigorously. Although initially most devices relied on evanescent field coupling from a solid-core waveguide to liquid surroundings[4,5], liquid-core waveguides are now receiving increasing attention. They have the inherent advantage of exposing the liquid analyte to the full optical power as both light and liquid are being guided through the same microscopic channels. A number of clever approaches to realizing a liquid-core waveguide have been developed, including nanoporous waveguides[6], liquid-core waveguides (LCW) with Teflon claddings[7,8], metal-clad waveguides[9,10], liquid-liquid waveguides[11,12], slot waveguides[13,14], hollow- core photonic crystal fibers (HC-PCF)[15–17], and Fresnel fiber[18].

Here, we review the status of optofluidic sensor devices based on liquid-core antiresonant reflecting optical waveguides (LC-ARROWs). After a brief discussion of the history and salient characteristics of ARROWs, we give an overview of a silicon-based microfabrication technology used for their construction, followed by a description of optical sensing devices based on a planar LC-ARROW platform. The demonstration of single bioparticle detection sensitivity and an outlook on future challenges and opportunities illustrate the potential of this approach.

18.2 ARROW Waveguides

18.2.1 Principle and History

It is well known that the low refractive index of water ($n = 1.33$) causes problems for the construction of optical waveguides that guide light efficiently through liquids over appreciable distances. Since all solid materials that are commonly used in microfluidic systems (silicon, glass, silicon nitride, PDMS, and others) have a higher index than water, they cannot be simply placed as cladding materials around the liquid core to achieve conventional index guiding that works so successfully for optical silica fiber. Teflon is an exception with an index of around 1.3 but it is very difficult to construct micro-fluidic structures using this material. One of the best solutions to this problem is to reduce the refractive loss at the core-cladding interface by increasing the portion of light reflected back into the core using wave interference. This is the idea behind the antiresonant reflecting optical waveguide (ARROW). ARROWs were first proposed and demonstrated for creating SiO_2 ($n = 1.45$) waveguides on silicon ($n = 3.5$)[19]. A side view of this original structure is shown in Fig. 18.1a.

In a ray picture, the wave vector $\boldsymbol{k}$ can be decomposed in a transverse component $\boldsymbol{k}_x = :\boldsymbol{k}_T$ and a propagation vector $|\boldsymbol{k}_z| = :\beta$. By simultaneously invoking an antiresonance condition for the round-trip phase shift $\Delta\Phi$ in the cladding material ($\Delta\Phi = 2t{\cdot}\boldsymbol{k}_T + \Phi_{\text{interface}} = (2N - 1)\pi$, $N =$ odd) and a mode condition in the core ($\boldsymbol{d}_C = \pi/\boldsymbol{k}_T$; $\boldsymbol{d}_C$: core thickness), one can show that the required thickness $\boldsymbol{t}$ for efficient light confinement in the core for a given set of materials is given by:

$$t = \frac{N\lambda}{4\sqrt{n^2 - n_C^2 + \frac{\lambda^2}{4d_C^2}}}, \qquad (18.1)$$

Equation (18.1) is known as the antiresonance condition for ARROW waveguides. For typical core dimensions of several microns, the last term under the square root can be neglected unless the cladding and core indices are approximately equal.

Using this design principle, Duguay et al. were able to fabricate solid-state ARROWs with losses as low as 0.4 dB/cm at $\lambda = 1.3$ μm[19]. In 1995, Delonge and Fouckhardt first proposed the use of ARROWs in the context of chemical sensing[20]. Alternating ARROW layers of TiO_2 and SiO_2 were used to confine light horizontally in a liquid core for improved capillary electrophoresis and liquid chromatography. Bernini et al. proposed building liquid-core ARROWs with hexagonally shaped cross sections using a wafer bonding approach[21]. Although these

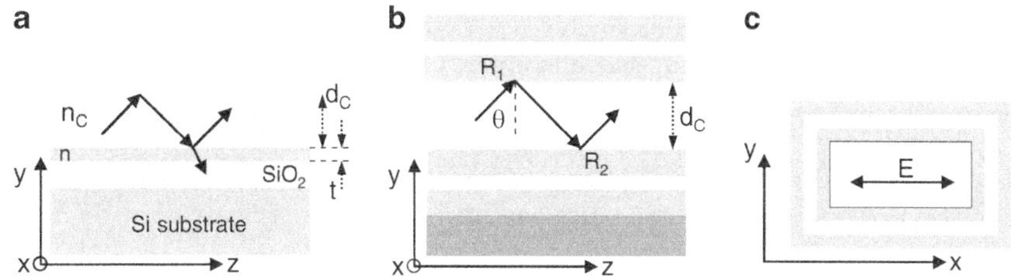

Fig. 18.1 ARROW waveguides. (**a**) Side view of light ray propagating in original solid-state ARROW[19]; (**b**) side view of ray propagating in hollow-core ARROW; (**c**) hollow-core cross section with light polarization

structures proved challenging to fabricate, waveguides with simpler, rectangular cross sections, and core dimensions on the order of 100–150 μm were built and successfully used as optical sensors based on refractometry (detection of refractive index changes)[22,23].

ARROW-based optofluidic devices for chemical and biological sensing have several desirable optical attributes, including:

- Low waveguide loss
- Dispersion control (wavelength dependent loss)
- Efficient coupling between solid and liquid-core waveguides
- Polarization control
- Single mode propagation (for interferometric measurements[1])

Among these, the first three can be considered as the most important ones. Therefore, we will discuss these aspects of LC-ARROWs in more detail in the following sections.

18.2.2 Waveguide Loss

To understand the essence of the waveguide loss characteristics of the LC-ARROW, we consider the cross section of a waveguide with multiple high-index cladding layers shown in Fig. 18.1b. Again keeping with the intuitive ray picture, we see that a light ray experiences two loss events per transverse round trip in the core because of the finite reflection coefficients $\boldsymbol{R_1}$ and $\boldsymbol{R_2}$ at the top and bottom core-cladding interfaces. One can then show that the power loss coefficient α of this one-dimensional waveguide is given by

$$\alpha = \frac{2 - R_1 - R_2}{2d_C \tan\theta}, \tag{18.2}$$

where θ is the angle of incidence of the ray as determined by the mode condition and shown in the figure. We note that the loss depends on both wavelength λ and polarization $\boldsymbol{p}$ through $\boldsymbol{R_{1,2}}(\lambda, \boldsymbol{p})$. The reflectivities for a multilayer structure can be determined from a matrix formalism[24]. Archambault et al. derived closed-form expressions for the minimum and maximum values of the waveguide loss in an ARROW with $\boldsymbol{M}$ cladding layers[25]. It is instructive to first consider the case of a simple leaky waveguide ($\boldsymbol{M} = 0$, low-index core surrounded by high-index cladding), which had been treated extensively by Marcatili and Schmelzer[26]. In this case, the loss is given by

$$\alpha = \frac{\lambda^2}{n_C d_C^3 \sqrt{n^2 - n_C^2}} \tag{18.3}$$

and reveals the characteristic dependence on the square of the wavelength and the inverse of the third power of the core thickness. In practice, ARROWs have typically been made from alternating layers of two different cladding materials (indices $\boldsymbol{n_1}$ and $\boldsymbol{n_2}$). In this case, Archambault's expressions for the loss can be simplified for $\boldsymbol{M}$ ARROW layers to

$$\alpha = \varepsilon \frac{\lambda^2}{n_C d_C^3 \sqrt{n_1^2 - n_C^2}} \sqrt{\frac{n_2^2 - n_C^2}{n_1^2 - n_C^2}}^{\,M} . \tag{18.4}$$

We see that the essential dependencies of the leaky waveguide are preserved, but the loss is reduced by the square root factor for each additional antiresonant layer. For example, in the case of water surrounded by silicon nitride and oxide layers ($\boldsymbol{n_C} = 1.33$, $\boldsymbol{n_1} = 2.1$, $\boldsymbol{n_2} = 1.46$), each additional layer reduces the loss by a factor of 0.37. Equation (18.4) also shows that the loss is determined by two main factors: the index difference between the core and the first cladding layer, and the index difference between the cladding layers. Finally, ε is a polarization dependent factor that illustrates the importance of this property for the ARROW loss. $\varepsilon = 1$ for TE polarization (s-polarized wave) and $\varepsilon = \boldsymbol{n_1}^2/\boldsymbol{n_C}^2\ (\boldsymbol{n_1}^2/\boldsymbol{n_2}^2)\boldsymbol{M}$. For the same example system, this means that the TM loss is higher by a factor of $2.5 \times 2.07\boldsymbol{M}$. This has important practical consequences as fluidic channels have two-dimensional cross sections. The example of a rectangular cross section is shown in Fig. 18.1c. It is evident that x-polarized light appears TE-polarized along the y-direction and TM-polarized in the x-direction. Calculating the loss of this two-dimensional structure requires solving Maxwell's equations in two dimensions, which can be done numerically using commercial software packages (e.g., Photon Design). Fortunately, the total loss is given in good approximation by the sum of the one-dimensional losses that can be calculated with (18.4) for the two directions[27]. Since the TM loss is larger, light polarized along the wider edge of the rectangle as shown in Fig. 18.1c experiences lower loss.

18.2.3 Dispersion Control

Many chemical and biological sensing applications are spectroscopy based, i.e. they involve several discrete wavelengths or spectral ranges. A prime example is fluorescence sensing in which molecules are excited by light at a short wavelength and fluorescence is reemitted at a longer wavelength. Other examples include Raman scattering or fluorescence resonance energy transfer (FRET) where fluorescence from two different molecules (donor and acceptor) is compared. A typical experiment requires the discrimination between different signals at wavelengths, using optical elements such as spectrometers, bandpass or edge filters. The requirements for these devices can be rather stringent, often requiring $\geq$50 dB discrimination in the case of separating weak signals from strong excitations.

In the context of complete device miniaturization, it would be highly desirable to replace conventional dispersive optical elements with a method that is capable of doing the required wavelength-dependent optical processing on the chip. Interference-based waveguides such as LC-ARROWs are ideally suited for this task as the interference phenomenon that allows for waveguiding is intrinsically dispersive. Figure 18.2 illustrates the design potential for LC-ARROWs by displaying substantial differences between the one-dimensional loss of three different LC-ARROW structures, each consisting of six top and bottom layers of alternating silicon nitride and oxide.

All three are designed for low loss at $\lambda = 690$ nm, i.e., the layer thicknesses fulfill (18.1). The solid line corresponds to the case where $N = 1$ for all layers. The dashed line corresponds to the case where only a single layer (the first bottom layer) was changed to $N = 3$. The loss minimum at 690 nm remains, but the behavior at other wavelengths changes substantially. Changing both the first and second layer to $N = 3$ results in the dotted line, which differs from both previous spectra.

With a suitable choice of the ARROW layers, spectra can be custom-designed, for example with low loss at a fluorescence wavelength and high loss at the excitation wavelength to implement on-chip filtering. In this particular case, creating a high-loss peak corresponds to making the layer resonant (N even) at the excitation wavelength and a closed form design rule for choosing the ARROW layer order N for the fluorescence wavelength λ_F can be derived as

$$\frac{\Delta\lambda}{\lambda_F} = \frac{1}{N+1}, \tag{18.5}$$

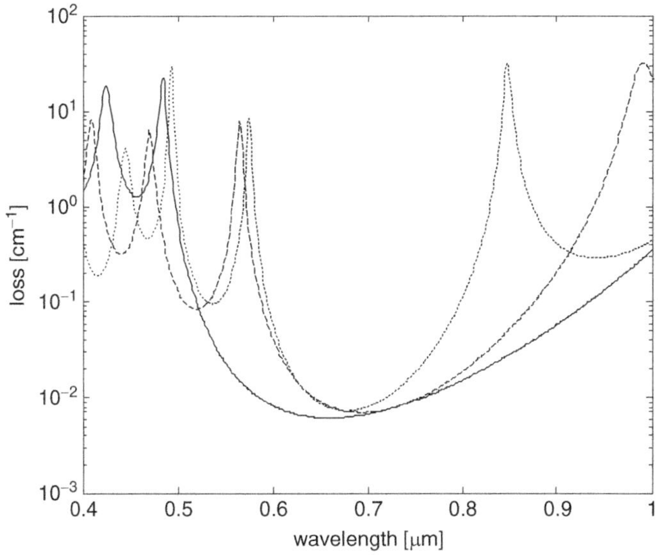

Fig. 18.2 Wavelength dependence of loss in ARROW waveguide for different antiresonant orders of cladding layers (see text)

where $\Delta\lambda$ is the wavelength difference between fluorescence excitation and emission. This rule also applies to separating Rayleigh from Raman lines in a Raman scattering experiment. Similar design rules can be derived for the more complicated case of FRET that involves multiple excitation and emission wavelengths[28], and transmission discrimination in excess of 35 dB has recently been demonstrated in solid-core ARROWs[29].

18.2.4 Solid-Liquid Interface

A final essential consideration for designing and constructing integrated LC-ARROW devices is an interface between solid-core and liquid-core waveguides. Extending the liquid-core waveguides to the edges of the sensor chip leads to analyte evaporation and leakage. In addition, this would mean that the fluid supply would enter through the same opening used for coupling light into the waveguide, a problem that plagues HC-PCF[30]. Consequently, ways have to be found to couple light from solid-core waveguides into and out of the liquid channels. The most straightforward solution is to vertically align a solid-core waveguide at the end or side of a liquid-core channel as shown in Fig. 18.3a. The cross section in Fig. 18.3b shows that the problem to solve then is to design the vertical ARROW layers such that they transmit light from the solid into the liquid core while still guiding along the liquid channel at the desired wavelengths.

Fortunately, there is substantial design flexibility. The rays corresponding to the lowest order mode propagate near grazing incidence ($\theta \sim 90°$) inside both the liquid and solid cores. This means that a ray propagating in the solid core hits the vertical ARROW layers that enclose the liquid core nearly perpendicularly ($\theta \sim 0°$) at the solid-hollow intersection. To first approximation, this interface can, therefore, be treated as a one-dimensional transmission matrix problem[24]. Since the material is identical at both ends, perfect transmission can be achieved if all intermediate layers are $\lambda/2$ (latent) layers. This leaves us with the following two conditions for a typical fluorescence experiment that requires efficient coupling of excitation light

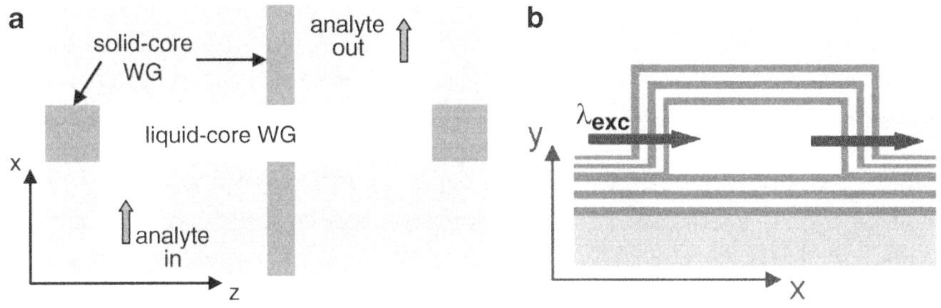

Fig. 18.3 (**a**) *Top view* of LC-ARROW chip; (**b**) Cross-sectional view of liquid-core section showing light coupling between liquid and solid cores

at λ_{exc} across the solid–liquid interface and low-loss guiding at the fluorescence wavelength along the liquid core:

$$t_1 = M\frac{\lambda_{\mathrm{exc}}}{2n_1} \quad ; \quad M : \text{integer}, \tag{18.6}$$

$$t_1 \approx \frac{N\lambda_{\mathrm{F}}}{4\sqrt{n_1^2 - n_{\mathrm{C}}^2}}. \tag{18.7}$$

To fulfill both conditions for the first cladding layer simultaneously, one can either optimize for the material index or the desired emission wavelength. The two solutions are given by[27]

$$n_1 = \frac{n_{\mathrm{C}}}{\sqrt{1 - \frac{N^2}{M^2}\frac{\lambda_{\mathrm{F}}^2}{4\lambda_{\mathrm{exc}}^2}}}, \tag{18.8}$$

$$\lambda_{\mathrm{F}} = \frac{2\sqrt{n_1^2 - n_{\mathrm{C}}^2}}{n_1}\frac{M}{N}\lambda_{\mathrm{exc}}. \tag{18.9}$$

We have shown that antiresonant dielectric layers can be used to design low-loss liquid-core waveguides that are suitable for implementing planar sensor device geometries. The following sections will describe in more detail how the design principles laid out here were implemented in silicon-based LC-ARROW chips and used for optical sensing and detection of a wide variety of substances.

18.3 ARROW-Chip Fabrication and Loss Characterization

18.3.1 Thin-Film Planar Fabrication and Structures

We have purposely pursued making waveguides with small hollow cross-sectional areas (approximately 10 μm × 5 μm). Using waveguides with such small areas highlights the utility of fluid filled ARROWs by (1) limiting the total fluid contained in the waveguides and allowing for analysis of small analyte volumes with single-particle sensitivities and (2) maintaining high optical intensities down the length of a fluid channel in tightly confined optical modes. Although ARROWs of larger cross sections (100 μm widths) had been made through wafer bonding[22], a new fabrication approach had to be developed for these smaller waveguides. To do so, we have relied on surface micromachining, thin film deposition, and sacrificial etching.

Fabricating waveguides on the surface of a wafer provides ready access to fluidic channels without the need for through-wafer etching as required in a wafer bonding process. Surface waveguides can also take advantage of an air interface that can be used for total-internal-reflection waveguiding. This is especially useful for nonhollow waveguides used to direct light on and off a chip (see Fig. 18.3). Surface micromachining also allows for the addition of metallic lines and pads to a chip for routing electrical signals, and is also closely compatible with standard procedures used in the microelectronics industry – very important for low cost manufacturing.

Figure 18.4 outlines the surface micromachining procedure we have developed for creating hollow ARROWs. The heart of the process involves surrounding a sacrificial core with silicon dioxide and/or silicon nitride layers and then removing the sacrificial layer with acid etching. First, a substrate is coated with silicon dioxide or nitride layers using plasma-enhanced chemical vapor deposition (PECVD) – see Fig. 18.4a. These layers are grown to specific thicknesses (see (18.1)) to achieve ARROW-based optical confinement. This process takes place at approximately 250°C. A thin layer of sacrificial material is then deposited and defined into a line using photolithography and etching – see Fig. 18.4b. A variety of sacrificial materials may be used, including photosensitive polymers and metals. Layers of PECVD oxide and nitride are then grown over the sacrificial material – see Fig. 18.4c. The conformal nature of this process is important to ensure that the sacrificial layer is completely enclosed. Other common methods of thin film deposition such as vapor phase deposition or sputtering can produce conformal coatings but the films are very weakly joined around corners. PECVD depositions are done at temperatures of at least 250°C to produce dense films with as little optical material loss as possible. It should be noted, however, that depositions over 300°C adversely affect most sacrificial materials, either by carbonizing them in the case of polymers or causing migration in the case of metals.

The final step of the procedure is to expose the sacrificial material to an acid etch from either end of the patterned line – see Fig. 18.4d. Upon etch completion, we are

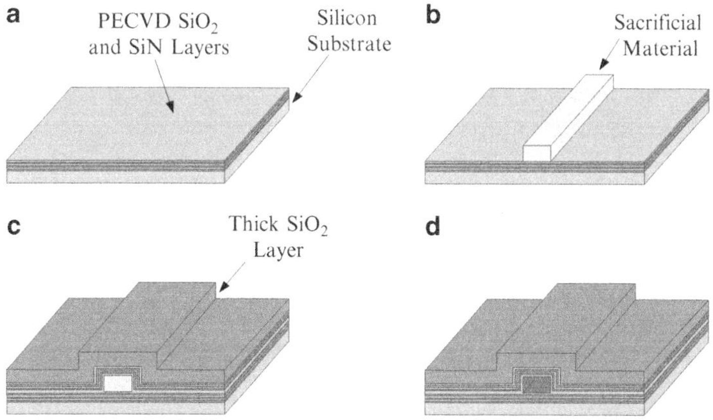

Fig. 18.4 Fabrication steps used to make ARROW chips using sacrificial etching and chemical vapor deposition

left with a hollow tube with walls composed of silicon dioxide and silicon nitride layers. Certain polycarbonate sacrificial materials[31] can simply be vaporized by heating them to a sufficiently high temperature. We have chosen to avoid these so that the sacrificial layer maintains its shape throughout the PECVD deposition process and to avoid any residues that the vaporization process might leave on the inner walls of the waveguides.

A number of sacrificial layers have been investigated in the context of the fabrication process outlined above: aluminum[32], SU8 (a photosensitive epoxy)[33], and photoresist[34]. Aluminum is most quickly removed using a nitric and hydrochloric acid etching solution (aqua regia), while SU8 and photoresist are removed using a sulfuric acid and hydrogen peroxide solution (Nanostrip). The different sacrificial materials result in differently shaped hollow core cross sections as illustrated in Fig. 18.5, providing flexibility when designing microfluidic devices. The alternating silicon dioxide and silicon nitride layers are evident in the images in Fig. 18.5. The layers nearest to the hollow core are between 100 and 200 nm thick. The topmost layer is a silicon dioxide film grown to around 3 μm thick. This relatively thick top layer gives the waveguide mechanical strength.

In order for such structures to be useful for fluid and light guiding, the channels must have smooth inner walls, be of reasonable length, and mechanically strong. The first criterion minimizes optical scattering or interruptions in fluid flow and is met by the conformal PECVD coating as evident in scanning electron microscope (SEM) micrographs shown in Fig. 18.5. AFM scans over PECVD films have shown that the surface roughness is around 8 nm rms. Mechanical strength of the hollow structures has been investigated using finite-element modeling software (ANSYS 6.0). To provide the necessary stress and strain constants for the software model, a set of experiments was performed. The results of the model indicate that the critical failure pressure for a hollow channel can be given by the simple expression

$$P_c = 2S_t\left(\frac{t_h}{w}\right)^2, \tag{18.10}$$

where S_t is the tensile strength of the overcoat material, t_h is the thickness of the overcoat layer, and w is the width of the channel. This simple equation reveals the functional dependence of the pressure on the width and thickness, and agrees within

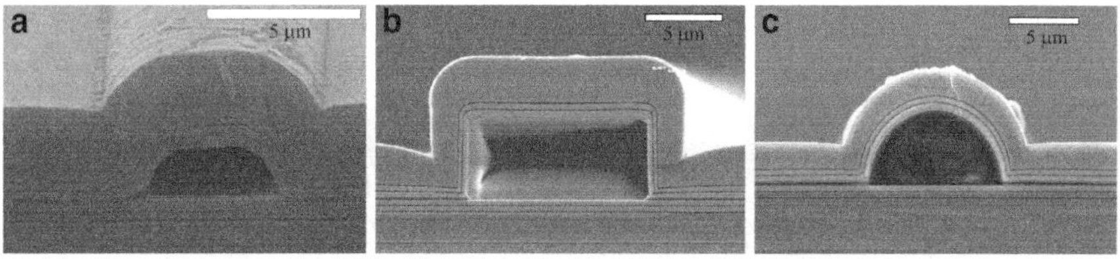

Fig. 18.5 Scanning electron microscope images showing cross sections of ARROWs with (**a**) trapezoidal geometry formed from an aluminum core, (**b**) rectangular geometry formed from an SU8 core, and (**c**) arched geometry formed from a reflowed photoresist core

10% of the values calculated using the finite-element simulation when $t_h/w < 1/10$. Tests were done on real structures by varying their core width and overcoat thickness to confirm this expression.

From experiments and simulations, we have found that the following design considerations provide optimal channel stability: For rectangular microchannel geometries, a ratio of $t_h/w > 1/10$ is needed; for a trapezoidal channel structure, the ratio must be $t_h/w > 1/25$; and for arch shapes the ratio should be $t_h/w > 1/35$[35–17]. These design rules are dependent on the cross-sectional geometry and also the etch solution used to remove the sacrificial core. Acid etches that produce gaseous by-products can create very high pressures within the channel, leading to breakage if the channel walls are not strong enough.

Etch times were investigated for aluminum and SU8 sacrificial cores patterned on silicon. Samples were periodically removed from the acid etch solution, and the amount of sacrificial core that was removed was measured using an optical microscope. We found that the etch length as a function of time follows the equation

$$l(t) \approx \sqrt{c_0 t} \tag{18.11}$$

where $l(t)$ is the length of the channel etched in a given time and c_0 is a constant related to the concentration of the etch solution, temperature, and channel geometry. Equation (18.11) indicates the diffusion-limited nature of the etch mechanism. Under optimized etch conditions, aluminum cores clear at a rate of around 10 mm in 24 h, while SU8 cores clear at a rate of around 1.5 mm per 24 h. Using cores that could be dissolved in solvents could potentially increase the etch rate significantly.

18.3.2 Solid and Liquid Core Waveguides on the Same Substrate

To utilize integrated optofluidic waveguides in a compact lab-on-a-chip platform, it is important to be able to interface between fluid-filled waveguides and standard solid waveguides. The solid waveguides serve the purpose of directing light on and off the chip and into and out of a fluidic medium. An example optofluidic platform is shown in Fig. 18.6. Using solid waveguides allows for a T-configuration in which light is injected down the length of a fluid filled waveguide from one arm of the T while fluid enters the waveguide through another arm. Solid waveguides also allow for perpendicular excitation of the fluid-filled waveguides, critical for devices such as fluorescence sensors[38]. Closeups of both features are shown in the scanning electron images in Fig. 18.7a, b, respectively.

Our thin-film ARROW fabrication process lends itself well to the simultaneous creation of solid and hollow core waveguides. To form the solid core waveguides, the hollow core ARROWs are defined on a silicon substrate as described previously. Then, the solid waveguides are added using photolithography and CF_4-based reactive ion etching (RIE). Light is confined in the solid waveguide in the transverse

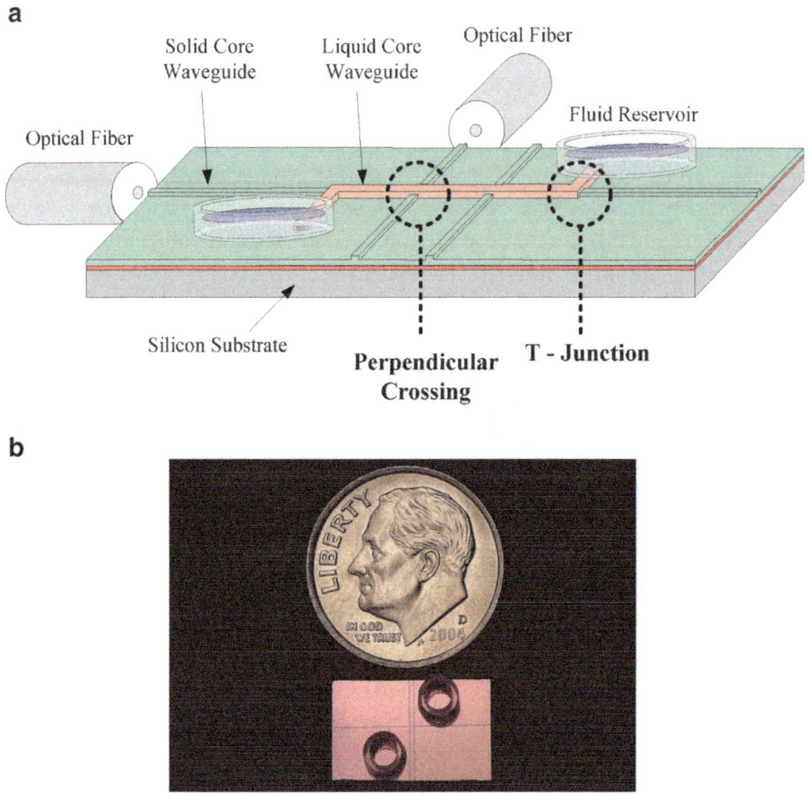

Fig. 18.6 (**a**) ARROW-based chip utilizing both solid and liquid core waveguides. The intersections of these two types of waveguides are highlighted. (**b**) Photograph of an ARROW-based chip with fluid reservoirs attached

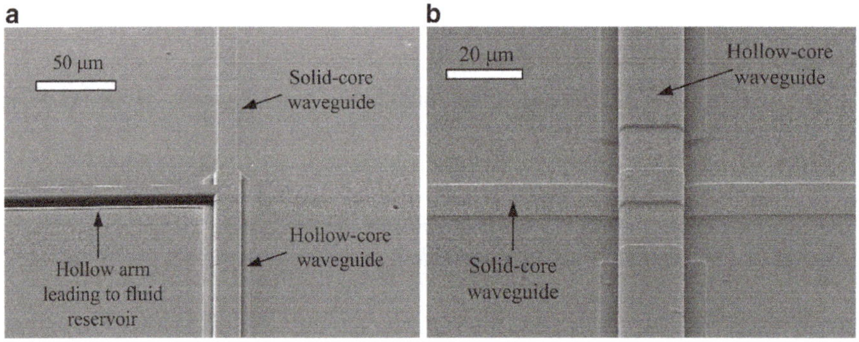

Fig. 18.7 Scanning electron microscope images showing interfaces between solid core and hollow core waveguides. (**a**) *Top view* of the interface at the end of a hollow core waveguide. (**b**) *Side view* of a solid core waveguide intersecting a hollow core waveguide

direction by the ARROW layers and air above the waveguide. In the lateral direction, confinement is provided by etching into the top layer. To create an efficient interface between the solid-core and hollow-core waveguides, the optical modes should be vertically aligned. This constrains the choice of core height and dielectric layer thickness[39]. Because the waveguide widths are determined by photolithography, tapers can be easily added to engineer interfaces between the hollow waveguide and off chip elements like optical fiber.

18.3.3 Transmission Losses in Liquid-Filled Waveguides

Optical losses in the liquid-filled ARROWs are typically measured using the cutback method. This measurement technique assumes optical transmission in the waveguide can be described according to

$$\text{Transmission} = \kappa e^{-\alpha L}, \qquad (18.12)$$

where κ is a constant related to the coupling of light into and out of the waveguide, L is the length of the waveguide, and α is the loss coefficient. By measuring the optical power transmitted through waveguides of different lengths, the loss coefficient α can be determined.

Loss coefficients were measured for both rectangular and arched-shaped ARROWs as seen in Fig. 18.5. The waveguides were filled with ethylene glycol during testing and 633 nm light was coupled into them from an optical fiber. Ethylene glycol was used because it evaporates very slowly, allowing sufficient time for testing. Rectangular core waveguides 15 μm wide and 3.5 μm tall had a loss coefficient of 0.33 cm^{-1}.[40] Arched-shaped waveguides 15 μm wide at their base and 3.5 μm tall had a loss coefficient of 0.26 cm^{-1}.[41] These measured loss coefficients are higher than predicted by (18.4) due to scattering and material losses in the PECVD grown layers.

18.4 Optical Spectroscopy with LC-ARROWs

In addition to being able to guide light with low propagation loss, one also needs to be able to excite and detect characteristic optical signals such as fluorescence or scattering inside the liquid-core waveguides. As a first step in this direction, straight LC-ARROWs as described in the previous section were filled with fluorescent Alexa 647 dye molecules in ethylene glycol using capillary force at the open ends. Fluorescence from the Alexa molecules was generated by sending light from a HeNe laser ($\sim$1 mW) through a single mode fiber end-coupled to the ARROW facet. Transmitted light was collected with an objective lens at the output facet, the copropagating excitation beam was removed by an interference filter, and the remaining signal was recorded with a silicon photodetector. The main outcomes of these experiments were the ability to achieve a detection sensitivity of ~10 pM or ~500 molecules inside the excitation volume of 57pL. In addition, a collection efficiency of ~15% of the emitted fluorescence signal for each facet could be extracted in good agreement with a theoretical model[40].

This successful demonstration of fluorescence detection in LC-ARROWs paved the way for using this concept for optical sensing, manipulation, and analysis of molecules on an optofluidic chip. One main goal of this work was to evaluate the detection limits that can be achieved in this approach and, if possible, to push this limit to the single particle regime. In the following, we will describe some of the recent accomplishments made by our groups in high-sensitivity particle analysis on a chip. Unless otherwise noted, the experiments will use the generic two-dimensional chip layout of interconnected solid and hollow-core waveguides shown in Fig. 18.6.

18.4.1 Raman Scattering in Linear Waveguides

Raman detection is a very attractive detection mechanism because it is label-free and the inelastic scattering process results in molecule-specific spectral signatures. The inherently low Raman scattering efficiency can be mitigated by attaching the sample molecules to metallic nanoparticles that provide surface enhancement of the signal (SERS). The waveguide arrangement for the SERS measurements is shown in Fig. 18.8a. In this case, the liquid-core ARROW had an L-shape, i.e., while one end was still open, the other was now connected to a 10 μL fluidic reservoir that was attached on the top of the chip and used to fill the channel with analyte solution. Excitation light was coupled into the liquid core via an SiO_2 solid core ARROW

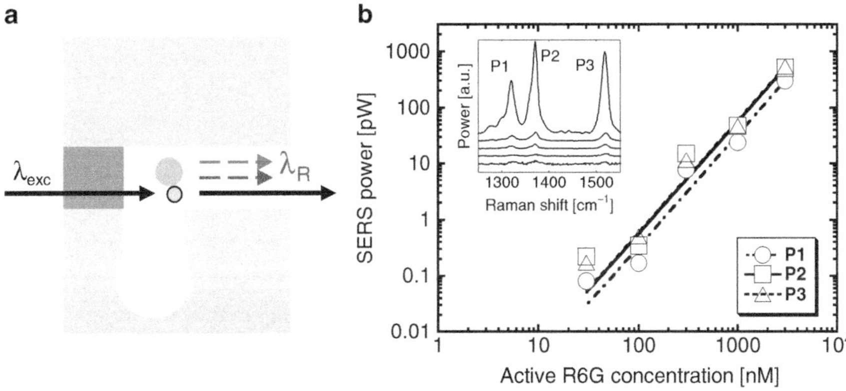

Fig. 18.8 SERS detection in LC-ARROW. (**a**) *Top view* of experimental beam geometry exciting rhodamine 6G molecules bound to silver nanoparticles (λ_{exc}: excitation beam, λ_R: Raman signals); (**b**) R6G concentration-dependent SERS power for three representative Raman peaks P1-P3, inset: spectra at various excitation powers

waveguide as described in Sect. 18.3.2, thereby demonstrating a transition between optofluidic and conventional solid-state ARROWs. SERS from rhodamine 6G (R6G) molecules bound to silver nanoparticles was then detected as a function of concentration and excitation power[42]. The concentration dependence is shown in Fig. 18.8b along with the observed characteristic Raman spectrum of R6G. A detection sensitivity of 30 nM of freely diffusing R6G molecules demonstrated highly sensitive Raman detection on an optofluidic chip with planar optical beam paths.

18.4.2 Single Molecule Fluorescence Sensing

Single molecule fluorescence detection generally requires optical excitation/detection volumes on the order of picoliters or below[28], which corresponds to a cube with edge length 10 μm or less. The small mode areas in our LC-ARROWs provide the necessary optical confinement in two dimensions, but without additional confinement in the third dimension, single molecule detection sensitivity would remain elusive as shown by the results from fluorescence in linear waveguides[40]. The key step to push the excitation volume into the subpicoliter volume was the introduction of a perpendicularly intersecting excitation waveguide as shown in Fig. 18.7b. The sample geometry for single molecule detection experiments is sketched in Fig. 18.9a. As in the earlier fluorescence experiments, Alexa 647 dye molecules in ethylene glycol were used to fill the liquid channel. The new fully planar geometry of a perpendicularly intersecting excitation beam has two highly beneficial effects. First, the physical separation of excitation and fluorescence beam paths led to a substantial reduction in the number of background counts on the detector. Second, the excitation volume is now lithographically confined on the micron-scale

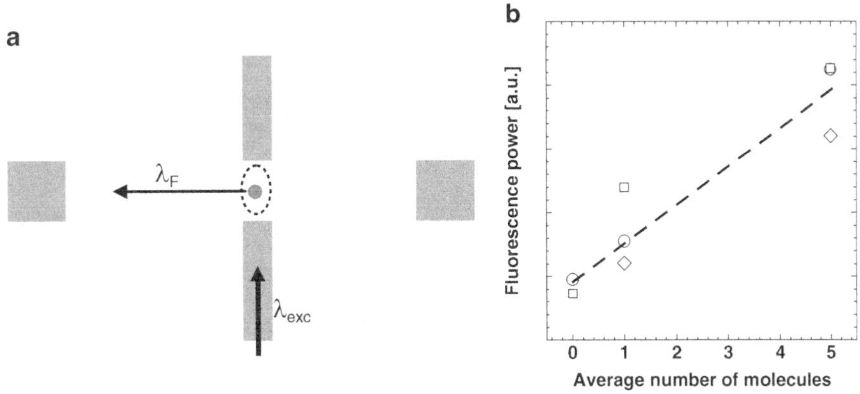

Fig. 18.9 Single molecule fluorescence detection in LC-ARROW chip. (**a**) Top view of experimental beam geometry of dye molecule in sub-picoliter excitation volume (dotted ellipse) (λ_{exc}: excitation beam, λ_F: fluorescence signal); (**b**) fluorescence signal as function of molecules in excitation volume; *symbols*: different experimental runs, *dashed line*: linear fit

in all three dimensions by the LC-ARROW cross section and the solid core waveguide width. This results in a subpicoliter optical excitation volume of 85fL that is symbolized by the ellipse in Fig. 18.9a[38]. Figure 18.9b shows the detected fluorescence signal as a function of the number of molecules in the excitation volume. Single molecules could clearly be resolved on this chip without the need for expensive microscopy techniques. It was also demonstrated that fluorescence can be excited at different points along the liquid channel using four solid-core waveguides in parallel[38], which suggests that two-dimensional networks of liquid and solid-core waveguides are feasible.

Subsequently, the devices were expanded to the full layout shown in Fig. 18.6a in which solid-core ARROWs are used for all paths to the chip edge and the fluidic channels are terminated with reservoirs. Electrodes can be placed into these reservoirs, enabling the application of an electric field across the channel. The utility of this setup was demonstrated by introducing Alexa dye molecules into one reservoir and driving them through the channel electrokinetically[43]. From the delay between the turn-on of the drive voltage and the detection of an optical signal from the waveguide intersection a dye velocity of 0.28 mm/s could be extracted. This initial experiment showed the feasibility of combining planar ARROW waveguide detection with electrical and simple microfluidic control.

18.4.3 Single Bioparticle Analysis with Fluorescence Correlation Spectroscopy

The results of Yin et al. showed that it is possible to detect single molecules in LC-ARROWs using integrated fluorescence, i.e., by detecting the fluorescence signal ***F(t)*** and averaging it over a sufficiently long time to achieve an acceptable signal level above the background that results from detector dark counts and spurious fluorescence and Raman signals from solvents and waveguide layers. In practice, however, more sophisticated fluorescence analysis methods are often used to improve sensitivity or access various molecular properties. These methods include fluorescence resonance energy transfer (FRET), fluorescence lifetime imaging (FLIM), fluorescence recovery after photobleaching (FRAP), fluorescence correlation spectroscopy (FCS), and others.

FCS, in particular, is a very attractive method that works best at low concentration levels, i.e., in the few to single particle regime. It is also extremely versatile and is used to detect binding processes, cell structure, conformational changes, and in high-throughput drug screening. In FCS, the fluctuations $\delta \boldsymbol{F(t)}$ of the fluorescence signal are the basis of the measurement, which is usually obtained from the autocorrelation function

$$G(\tau) = \frac{\langle \delta F(t) \delta F(t+\tau) \rangle}{\langle F(t) \rangle^2}, \tag{18.13}$$

where the brackets denote a temporal average. In the simplest case of conventional confocal microscopy and a single species of molecules of interest (concentration **c**) that diffuse in and out of the ellipsoidal excitation volume V_{exc}, one can calculate $G(\tau)$ explicitly to find[44]

$$G(\tau) = \frac{1}{c \cdot V_{\mathrm{exc}}} \frac{1}{\sqrt{(1+\tau/\tau_{\mathrm{Dx}})(1+\tau/\tau_{\mathrm{Dy}})(1+\tau/\tau_{\mathrm{Dz}})}}, \tag{18.14}$$

where τ_{Dx}, τ_{Dy}, and τ_{Dz} are the diffusion times for diffusion out of the excitation volume along the three dimensions (typically $\tau_{\mathrm{Dx}} = \tau_{\mathrm{Dy}}$). Figure 18.10a shows a generic FCS curve obtained from (18.14). The most important features are the $\boldsymbol{G}(0) = 1/(\mathbf{cV_{exc}}) = 1/\langle N \rangle$, which corresponds to the average number of molecules in the excitation volume, and the rollover behavior at longer τ from which the diffusion times and hence the particle diffusion constant $\boldsymbol{D}$ can be extracted by fitting (18.14) to the data.

We have analyzed FCS in our ARROW-chip geometry[45]. In principle, a similar analysis as for a bulk microscope can be applied. The situation in the waveguide is different, however, because of the (generally) finite reflectivity of the vertical ARROW layers (see Sect. 18.2.1). This can result in a standing wave pattern of the optical intensity across the width of the waveguide, which changes the excitation volume for the FCS measurement[46]. This cavity effect can be included via a correction factor in the equation for $\boldsymbol{G}(\tau)$[45]. Calculated autocorrelation curves for different ARROW layer reflectivities are shown in Fig. 18.10b along with an inset of the intensity distribution in the liquid core. When compared with Fig. 18.10a, one notices the emergence of a second shoulder in the FCS trace at shorter time scales. This initial rolloff is due to a particle diffusing in and out of the smaller volumes that are created by the standing wave. Clearly, the design of the optofluidic chip

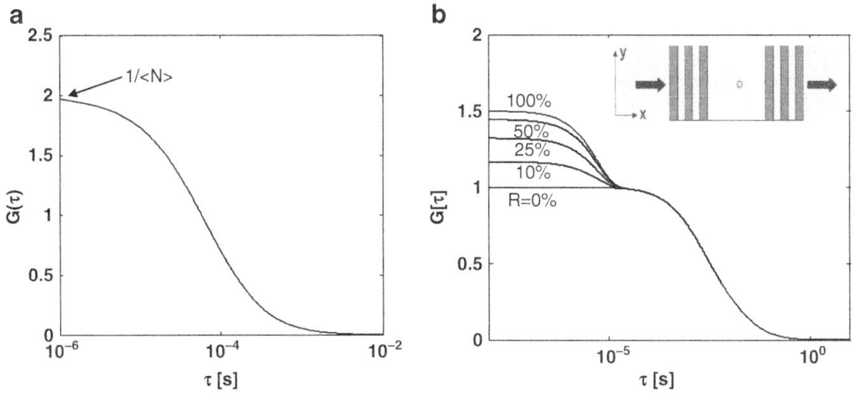

Fig. 18.10 (**a**) Generic FCS autocorrelation trace for $\langle N \rangle = 0.5$ particles in excitation volume; (**b**) FCS traces in LC-ARROW for different reflectivities $\boldsymbol{R}$ of vertical ARROW layers, inset: cross-sectional view of standing wave pattern for $\boldsymbol{R} > 0$

determines the microphotonic structure and can therefore be used to tailor the FCS response to the experimental needs.

FCS was measured using Alexa 647 dye at various concentrations in LC-ARROW chips. Clean FCS signals could be obtained for concentrations as low as 10 pM, corresponding to an average number of only 0.35 dye molecules in the excitation volume in excellent agreement with integrated fluorescence measurements[38]. In addition, the dye diffusion coefficient was extracted from a fit to the data. A value of 250 $\mu m^2/s$ was found in good agreement with a model assuming spherical particles. Other FCS derivatives such as two-photon or two-focus FCS also lend themselves perfectly to the integrated waveguide environment[44]. Having shown the utility of the LC-ARROW chip for advanced fluorescence analysis, we investigated different types of biomolecules in this integrated setting: liposomes and bacteriophages. The experimental procedure was the same for all cases.

Liposomes are lipid bilayer vesicles that are used for drug delivery, for example in cancer or asthma treatment. Here, we used 200-nm diameter liposome nanoparticles filled with Alexa dye[47]. Even though the dye fluorescence is detected, the resulting FCS curves represent the properties of the much larger and heavier liposomes. In addition to optical excitation, the liposomes could also be moved electrokinetically through the liquid channels by applying a bias voltage across the fluid reservoirs. The electrokinetic movement affects the FCS signal as the liposomes' residence time in the optical excitation volume decreases with increasing flow velocity. This qualitative notion is borne out by the experimental autocorrelation curves shown in Fig. 18.11a that are taken at applied voltages ranging from $V =$ 0V to $V = 180$V. As expected, the characteristic rollover is shifted to shorter delay times with increasing voltage. The solid lines show one-parameter fits to the data that include the additional drift via a standard exponential factor exp $(-(v_z\tau)^2/\ (w_z^{\ 2} + 4D\tau))$ in $G(\tau)$[44,45], where v_z is the drift velocity along the ARROW channel, w_z is the full width at half maximum of the excitation mode along z, and D is the diffusion coefficient of the liposomes. Excellent agreement

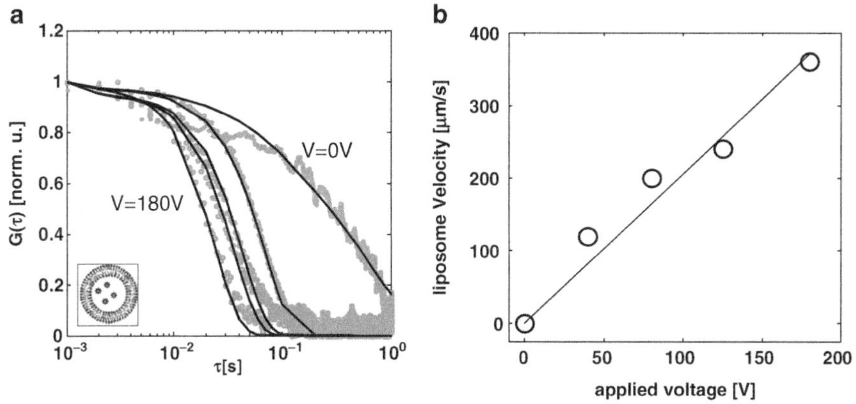

Fig. 18.11 FCS of liposome nanoparticles in LC-ARROW. (**a**) *Symbols*: data for various voltages applied across channel, *lines*: fits to ARROW FCS model, inset: dye-filled liposome; (b) extracted liposome velocity versus applied voltage (*symbols*: data, *dashed line*: linear fit)

between experiment and the model is found and allows for extraction of the diffusion coefficient at $V = 0$ ($\boldsymbol{D} = 3\,\mu m^2/s$) and the voltage-dependent drift velocities. These are depicted in Fig. 18.11b as a function of applied voltage and show a linear dependence in agreement with a field-independent mobility. As few as $1/\boldsymbol{G}(0) = 0.6$ liposomes could be detected, again demonstrating the high sensitivity afforded by the ARROW-based optofluidic device.

Finally, we extended our detection capabilities to bacterial viruses that are almost an order of magnitude smaller than the liposomes. Q-β bacteriophages (26-nm diameter, labeled with Alexa dye molecules) were provided by M.G. Finn's group (Scripps Institute, La Jolla). A sketch of these biomolecules is shown in the inset to Fig. 18.12a. The results of an FCS experiment (excitation wavelength 633 nm) are shown in Fig. 18.12a. Fluorescence bursts from virus particles passing the waveguide intersection are clearly visible. Figure 18.12b depicts the corresponding FCS signal (dots) along with a fit to the ARROW FCS model (line). Excellent agreement is found once again, and allows for extraction of the phage diffusion coefficient ($\boldsymbol{D} = 17\,\mu m^2/s$) and an average number of detected viruses of 0.8. This demonstrates that LC-ARROW chips can be used for detection and analysis of biomolecules on the single particle level[48].

18.4.4 All-Optical Particle Control and Manipulation

Molecular and particle analysis using well-established optical methods such as fluorescence and Raman scattering is but one of the possibilities that LC-ARROWs offer. A whole new research area based on the study and exploitation

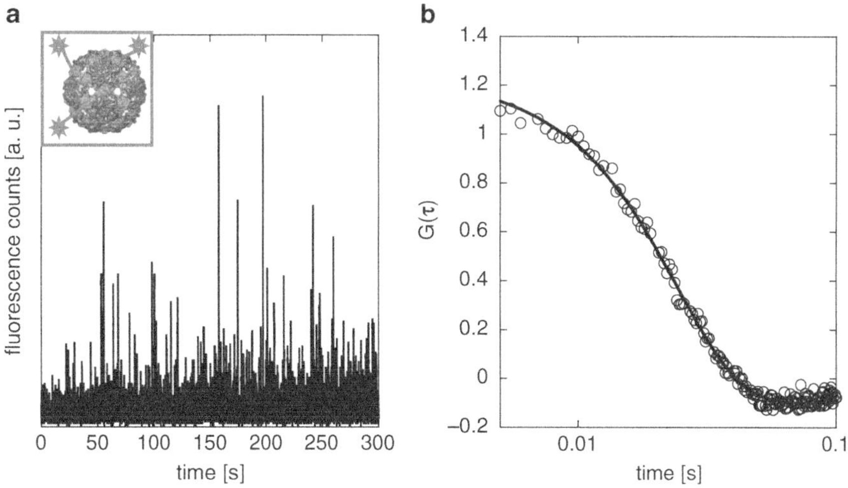

Fig. 18.12 FCS of Q-beta bacteriophage. (**a**) *Symbols*: data for various voltages applied across channel, *lines*: fits to ARROW FCS model, inset: dye-filled liposome; (**b**) extracted liposome velocity versus applied voltage (*symbols*: data, *dashed line*: linear fit)

of all-optical particle control has recently been opened. It is based on the notion that light can not only be used to elicit a detectable optical response from particles of interest, but can also be used to mechanically control the position and movement of the particles.

The potential to optically change and control the position or velocity of a particle becomes immediately apparent when one considers the fact that each photon carries a momentum $p = h/\lambda$ determined by its wavelength λ and the Planck quantum h. Reflection or absorption of a photon then corresponds to a momentum change that has to be compensated for by the particle itself. This radiation pressure has been used extensively since the early 1970s to push and trap particles[49]. It is the basis for optical dual-beam traps and laser tweezers that have proved very useful for single particle studies in molecular biology and other areas[50,51], Figure 18.13a shows a schematic view of a particle in a laser beam and the optically induced forces it experiences. The scattering force $\boldsymbol{F}_S$ pushes the particle along the direction of the beam, while gradient forces $\boldsymbol{F}_G$ move the particle to regions of high intensity.

Optically induced particle movement in liquid-core waveguides was shown in hollow fibers[52] and photonic crystal fibers[53], and particle trapping in a liquid core using light propagation in solid core waveguides was observed using both divergent beams[54] and evanescent fields[55]. We have recently demonstrated the use of optically induced particle movement to implement new ways for investigating not the properties of the particle, but the optical properties of the liquid-core waveguide itself[56].

The first possible application is a novel method to measure the liquid-core waveguide loss. Typical methods to determine waveguide loss in our ARROW-chips are either destructive (cutback method[57]) or inaccurate (measuring light scattered from the top of the waveguide). Direct access to the liquid-core loss is, however, possible by analyzing the scattering force a particle experiences when exposed to a beam in the liquid core. This situation is shown in Fig. 18.13b. The forward scattering force $\boldsymbol{F}_S$ depends on the local power, which is itself loss dependent. By equating the scattering force to the Stokes-Navier

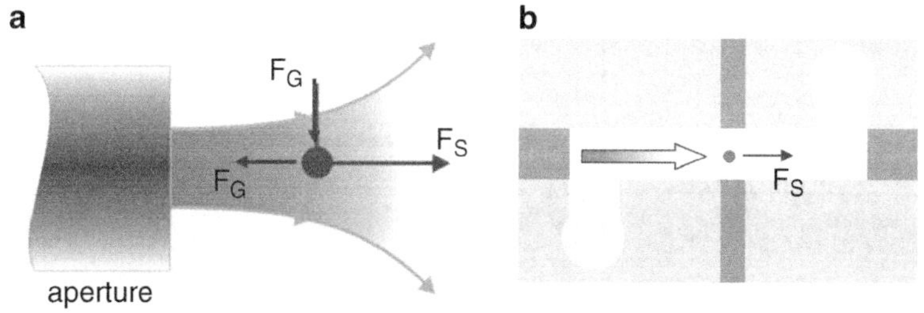

Fig. 18.13 (**a**) Radiation pressure from diverging beam on dielectric particle along with resulting scattering and gradient forces; (**b**) scattering force on particle in LC-ARROW due to beam propagating in waveguide mode

drag force, one can relate the waveguide loss a to the particle velocity $\boldsymbol{v} = \mathrm{d}z/\mathrm{d}t$ according to

$$F_S(z) = Q\frac{nP_0}{c}\mathrm{e}^{-\alpha z} = 6\pi\eta r\frac{\mathrm{d}z}{\mathrm{d}t}, \tag{18.15}$$

where $\boldsymbol{Q}$ is a dimensionless factor[49], $\boldsymbol{n}$ is the particle index, $\boldsymbol{P}_0$ is the optical power at $z = 0$ in the liquid core, η is the viscosity, and $\boldsymbol{r}$ the particle radius. One simply has to record the position (velocity) as it is pushed by the light along the waveguide to measure the waveguide loss quickly and nondestructively. Using polystyrene microbeads ($\boldsymbol{n} = 1.59$, $\boldsymbol{r} = 0.5$ μm), this method was successfully used to determine the waveguide loss in both single and multi-mode liquid-core ARROWs[56]. Figure 18.14a shows the recorded particle position and a fit based on the analysis of (18.15) in excellent agreement with the data.

A second application of optically induced particle motion lies in the possibility to explore the mode structure inside the liquid core. Realizing that the lateral position of a particle is determined by the gradient force, which tries to keep it at the point of highest intensity, one can map out the lateral (along x) intensity profile of the waveguide as the particle is moved along the liquid core by the scattering force. We demonstrated this method experimentally by recording the full trajectory of a polystyrene microbead optically pushed along a multi-mode ARROW waveguide (liquid core width 12 μm and height 5 μm) by a frequency-doubled Nd/YAG laser ($\lambda = 532$ nm, $\boldsymbol{P}_0 \sim 12$ mW). Figure 18.14a shows the frequency histogram of the particle obtained from all recorded lateral (x) positions. A multi-peak structure is clearly revealed that corresponds to the profile found from a commercial mode solver (Photon Design) when coupling into multiple lateral modes is taken into

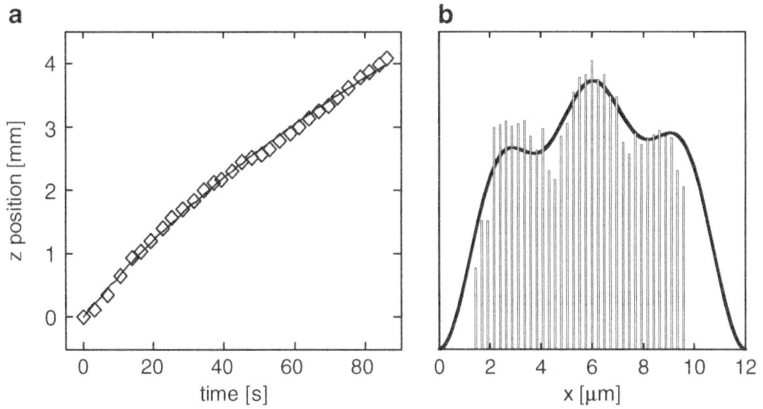

Fig. 18.14 LC-ARROW analysis based on radiation pressure. (**a**) Time-dependent microbead position for extraction of waveguide loss (*symbols*: data, *line*: fit); (**b**) lateral mode profile determination (*bars*: histogram of measured lateral particle position, *line*: multimode profile calculated with commercial mode solver

account (solid line). It was also shown that the particle trajectory indeed follows the high-intensity regions of the waveguide and that single lateral mode behavior can also be observed[56].

18.5 Conclusions

The preceding sections have shown that liquid-core ARROW waveguides are a viable technology for optofluidic sensing on a chip. By combining both liquid and solid-core waveguides, optical and fluidic geometries can be defined with great flexibility. The fact that both optical and fluidic paths are confined to the chip plane allows for the elimination of microscopy in three dimensions and is a promising path toward compact and robust biomedical analysis instruments. The deliberate choice of liquid-core waveguide dimensions of only a few microns results in the implementation of subpicoliter optical excitation volumes that enable single-molecule fluorescence sensitivity and are precisely defined by microfabrication steps. The implementation of advanced fluorescence sensing methods on a chip allows for the detection, analysis, and (electrical) manipulation of extremely small quantities of biological nanoparticles. Future refinements in both fabrication methods as well as optical waveguide design will undoubtedly further improve the performance and commercial potential of these sensors.

LC-ARROWs can, however, have an equally big impact on fundamental science, in particular molecular biology and single particle measurements. In this case, the central issue is the ability to both detect and control a single particle of interest. For instance, while FCS boasts single particle detection sensitivity, it is based on stochastic processes of particles entering and leaving the optical excitation volume. If one were able to deliberately place a single particle in the excitation volume and hold it there for an extended time period, a whole new class of studies will be possible that provide insight into kinetic pathways not accessible by ensemble measurements.

One possible avenue to realize true single-particle studies is to combine nanopore gating with all-optical particle control. Figure 18.15a shows schematically how a nanoscopic opening (nanopore) etched into the ARROW top layers can act as a gate for sequentially introducing individual particles of interest into the LC-ARROW. To this end, the nanopore diameter has to match the particle dimension. We have already demonstrated nanometer control over nanopores in silicon nitride layers down to diameters of 11 nm. This is sufficient for studying a vast variety of macromolecules including the liposomes and phages discussed in Sect. 18.4. Additional capabilities could be gained by monitoring ionic current levels through the nanopore and analyzing current blockades induced by translocating particles[58]. Once a particle has entered the waveguide, it needs to moved in a controlled way to the waveguide intersection. This can be accomplished by building on the radiation pressure methods introduced in Sect. 18.4.4. In particular, optical traps can be used for this purpose. Figure 18.15b shows how a dual-beam trap can

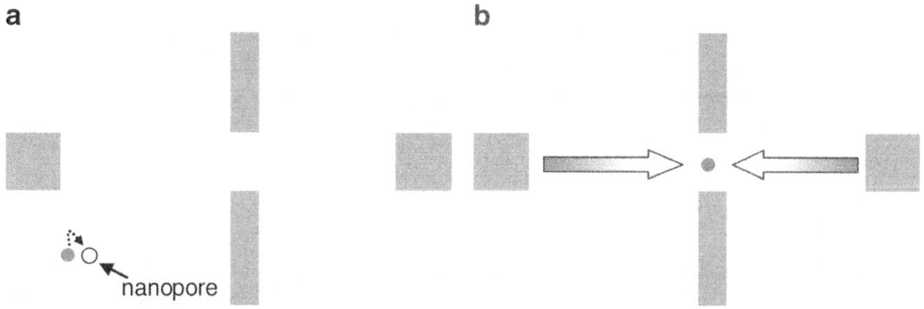

Fig. 18.15 Advanced methods for single particle analysis in LC-ARROW chips. (**a**) Nanopore added to reservoir for single particle entry into LC-ARROW; (**b**) Optical dual-beam particle trap based on balancing the scattering force due to counter-propagating beams

be implemented in an LC-ARROW simply by balancing the scattering force of two counter-propagating laser beams that lose intensity along the waveguide due to the propagation loss. This concept was successfully used to efficiently trap microbeads at the waveguide intersection for extended periods of time[59]. Other trapping methods are also conceivable and have been demonstrated, including an electro-optical method[60].

The simultaneous implementation of both single particle gating and control will result in a new experimental platform that can take advantage of the unique features of the microfluidic environment such as small analyte volumes, small thermal mass, etc. Combined with the nearly unlimited possibilities of integrated optical design, it can be expected that integrated optofluidic methods using LC-ARROWs and other approaches will play an increasing role in investigating chemical and biological processes at the most fundamental level.

Acknowledgments Many students and colleagues have contributed to the research described in the preceding sections. It is a great pleasure to acknowledge the contributions of D. Yin, J.P. Barber, D.W. Deamer, P. Measor, E.J. Lunt, S. Kühn, M.I. Rudenko, M. Smith, B.S. Phillips, M.R. Holmes, D. Ermolenko, H.F. Noller, L. Seballos, J. Zhang, M.G. Finn, U. Håkanson, and V. Sandoghdar. The different aspects of this work were enabled by financial support from the National Institutes of Health (grants R21EB003430 and R01EB006097), the National Science Foundation (grants ECS-0528730 and ECS-0528714), the W.M. Keck Foundation (National Academies Keck Futures Initiative Award NAKFI-Nano14), the California Systemwide Biotechnology Research & Education Program Training Program (grant UC-GREAT 2005–245), NASA (NASA/UARC Aligned Research Program (ARP) grant), the D. Huber Foundation, and Ted Goldstein.

References

1 Potyrailo, R. A.; Hobbs, S. E.; Hieftje, G. M., Optical waveguide sensors in analytical chemistry: Today's instrumentation, applications, and trends for future development, *Fresenius J. Anal. Chem.* **1998**, 362, 349–373

2 Reyes D. R.; Iossifidis, D.; Auroux, P.; Manz, A., Micro total analysis systems. 1. Introduction, theory, and technology, *Anal. Chem.* **2002**, 74, 2623–2636
3 Psaltis D.; Quake S. R.; Yang C., Developing optofluidic technology through the fusion of microfluidics and optics, *Nature* **2006**, 442, 381–386
4 Homola, J.; Yee, S. S.; Gauglitz, G., Surface plasmon resonance sensors: review, *Sensors Actuators B* **1999**, 54, 3–15
5 Sharma, A. K.; Jha, R.; Gupta, B. D., Fiber-optic sensors based on surface plasmon resonance: A comprehensive review. *IEEE Sensors J.* **2007**, 7, 1118–1129
6 Risk W. P.; Kim H. C.; Miller R. D.; Temkin H.; Gangopadhyay S., Optical waveguides with an aqueous core and a low-index nanoporous cladding, *Opt. Express* **2004**, 12, 6446–6455
7 Schelle, B.; Dress, P.; Franke, H.; Klein, K. F.; Slupek, J., Physical characterization of light-guide capillary cells, *J. Phys. D: Appl. Phys.* **1999**, 32, 3157–3163
8 Datta, A.; Eom, I.; Dhar, A.; Kuban, P.; Manor, R.; Ahmad, I.; Gangopadhyay, S.; Dallas, T.; Holtz, M.; Temkin, H.; Dasgupta, P., Microfabrication and characterization of teflon AF-coated liquid core waveguide channels in silicon, *IEEE Sensors J.* **2003**, 3, 788–795
9 Mohebbi, M.; Fedosejevs, R.; Gopal, V.; Harrington, J. A., Silver-coated hollow-glass waveguide for applications at 800 nm, *Appl. Opt.* **2002**, 41, 7031–7035
10 Grewe, M.; Grosse, A.; Fouckhardt, H., Theoretical and experimental investigations of the optical waveguiding properties of on-chip microfabricated capillaries, *Appl. Phys. B.* **2000**, 70, S839–S847
11 Wolfe, D. B.; Conroy, R. S.; Garstecki, P.; Mayers, B. T.; Fischbach, M. A.; Paul, K. E.; Prentiss, M.; Whitesides, G. M., Dynamic control of liquid-core/liquid-cladding optical waveguides, *PNAS* **2004**, 101, 12434–12438
12 Vezenov, D. B.; Mayers, B. T.; Wolfe, D. B.; Whitesides, G. M., Integrated fluorescent light source for optofluidic applications, *Appl. Phys. Lett.* **2005**, 86, 041104
13 Almeida, V. R.; Xu, Q.; Barrios, C. A.; Lipson, M., Guiding and confining light in void nanostructures, *Opt. Lett.* **2004**, 29, 1209–1211
14 Xu, Q.; Almeida, V. R.; Panepucci, R. R.; Lipson, M., Experimental demonstration of guiding and confining light in nanometer-size low-refractive-index material, *Opt. Lett.* **2004**, 29, 1626–1628
15 Mach, P.; Dolinski, M.; Baldwin, K. W.; Rogers, J. A.; Kerbage, C.; Windeler, R. S.; Eggleton, B. J., Tunable microfluidic optical fiber. *Appl. Phys. Lett.* **2002**, 80, 4294–4296
16 Erickson, D.; Rockwood, T.; Emery, T.; Scherer, A.; Psaltis, D., Nanofluidic tuning of photonic crystal circuits, *Opt. Lett.* **2006**, 31, 59–61
17 Yan, H.; Gu, C.; Yang, C.; Liu, J.; Jin, G.; Zhang, J.; Hou, L.; Yao, Y., Hollow core photonic crystal fiber surface-enhanced Raman probe, *Appl. Phys. Lett.* **2006**, 89, 204101
18 Martelli, C.; Canning, J.; Lyytikainen, K.; Groothoff, N., Water-core Fresnel fiber, *Opt. Exp.* **2005**, 13, 3890–3895
19 Duguay, M. A.; Kokubun, Y.; Koch, T.; Pfeiffer, L., Antiresonant reflecting optical waveguides in SiO_2-Si multilayer structures, *Appl. Phys. Lett.* **1986**, 49, 13–15
20 Delonge, T.; Fouckhardt, H., Integrated optical detection cell based on Bragg reflecting waveguides, *J. Chromat. A* **1995**, 716, 135–139
21 Bernini, R.; Campopiano, S.; Zeni, L.; Sarro, P. M., ARROW optical waveguides based sensors, *Sensors Actuator B* **2004**, 100, 143–6
22 Campopiano, S.; Bernini, R.; Zeni, R.; Sarro, P. M., Microfluidic sensor based on integrated optical hollow waveguides, *Opt. Lett.* **2004**, 29, 1894–1896
23 Bernini, R.; DeNuccio, E.; Minardo, A.; Zeni, L.; Sarro, P. M., Liquid-core/liquid-cladding integrated silicon ARROW waveguides, *Opt. Comm.* **2008**, 281, 2062–2066
24 Yeh, P., Optical Waves in Layered Media, 2nd edn.; Wiley-Interscience, New York, 2005
25 Archambault, J. L.; Black, R. J.; Lacroix, S.; Bures, J., Loss calculations for antiresonant waveguides, *J. Lightw. Tech.* **1993**, 11, 416–423
26 Marcatili, E. A. J.; Schmeltzer, R. A., *Bell Syst. Tech. J.* **1964**, 43, 1783

27 Schmidt, H.; Yin, D.; Barber, J. P.; Hawkins, A. R., Hollow-core waveguides and 2D waveguide arrays for integrated optics of gases and liquids, *IEEE J. Sel. Topics Qu. Elec.* **2005**, 11, 519–527

28 Lakowicz, J. R., Principles of Fluorescence Spectroscopy, 3rd edn.; Springer, New York, **2006**

29 Measor, P.; Lunt E. J.; Jones C.; Hawkins, A. R.; Schmidt, H., Characterization of optofluidic ARROW rejection filter devices, In Technical Digest, IEEE Summer Topical Meetings, Acapulco, MX, July 21–23, **2008**

30 Cordeiro, C. M. B.; de Matos, C. J. S.; dos Santos, E. M.; Bozolan, A.; Ong, J. S. K.; Facincani, T.; Chesini, G.; Vaz, A. R.; Brito Cruz, C. H., Towards practical liquid and gas sensing with photonic crystal fibres: Side access to the fibre microstructure and single-mode liquid-core fibre, *Meas. Sci. Tech.* **2007**, 18, 3075–3081

31 Metz, S.; Jiguet, S.; Bertsch, A.; Renaud, P., Polymide and SU-8 microflu-idic devices manufactured by heat-depolymerizable sacrificial material technique, *Lab Chip.* **2004**, 4, 114–120

32 Yin, D.; Barber J. P.; Hawkins A. R.; Schmidt H., Low-loss integrated optical sensors based on hollow-core ARROW waveguides, *Proc SPIE* **2005**, 5730, 218–225

33 Yin, D.; Barber, J. P.; Hawkins, A. R.; Deamer, D. W.; Schmidt, H., Integrated optical waveguides with liquid cores, *Appl. Phys. Lett.* **2004**, 85, 3477–3479

34 Barber, J. P.; Lunt, E. J.; George, Z. A.; Yin, D.; Schmidt, H.; Hawkins, A. R., Integrated hollow waveguides with arch-shaped cores, *IEEE Photonic Tech. Lett.* **2006**, 18, 28–30

35 Hubbard, N. B.; Howell, L. L.; Barber, J. P.; Conkey, D. B.; Hawkins, A. R.; Schmidt, H., Mechanical models and design rules for on-chip micro-channels with sacrificial cores, *J Micromech Microeng* **2005**, 15, 720–727

36 Barber, J. P.; Conkey, D. B.; Lee, J. R.; Hubbard, N. B.; Howell, L. L.; Schmidt, H.; Hawkins, A. R., Fabrication of hollow waveguides with sacrificial aluminum cores, *IEEE Photonic Tech Lett.* **2005**, 17, 363–365

37 Lee, J. R.; Barber, J. P.; George, Z. A.; Lee, M. L.; Schmidt, H.; Hawkins, A. R., Microchannels with different core shapes fabricated using sacrificial etching, *J. Micro/Nanolith. MEMS MOEMS* **2007**, 6, 013010

38 Yin, D.; Barber, J. P.; Deamer, D. W.; Hawkins, A. R.; Schmidt, H., Single-molecule detection using planar integrated optics on a chip, *Opt. Lett.* **2006**, 31, 2136–2138

39 Barber, J. P.; Lunt, E. J.; Yin, D.; Schmidt, H.; Hawkins, A. R., Monolithic fabri-cation of hollow ARROW based sensors, *Proc. SPIE* **2006**, 6110, 61100H

40 Yin, D.; Barber, J. P.; Hawkins, A. R.; Schmidt, H., Highly efficient fluo-rescence detection in picoliter volume liquid-core waveguides, *Appl. Phys. Lett.* **2005**, 87, 211111

41 Yin, D.; Barber, J. P.; Lunt, E. J.; Hawkins, A. R.; Schmidt, H., Optical char-acterization of arch-shaped ARROW waveguides with liquid cores, *Opt. Express* **2005**, 13, 10564–10570

42 Measor, P.; Lunt, E. J.; Seballos, L.; Yin, D.; Zhang, J. Z.; Hawkins, A. R.; Schmidt, H., On-chip Surface-enhanced Raman scattering (SERS) detection using integrated liquid-core waveguides, *Appl. Phys. Lett.* **2007**, 90, 211107

43 Yin, D.; Barber, J. P.; Hawkins, A. R.; Schmidt, H., Single molecule sensitivity and electrically controlled fluorescence detection in integrated planar ARROW waveguides, In CLEO/QELS conference, Long Beach, CA, May 21–26, **2006**

44 Rigler, R.; Elson, E. S., Fluorescence Correlation Spectroscopy, 1st edn.; Springer, New York, **2001**

45 Yin, D.; Lunt, E. J.; Barman, A.; Hawkins, A. R.; Schmidt, H., Microphotonic control of single molecule fluorescence correlation spectroscopy using planar optofluidics, *Opt. Express* **2007**, 15, 7290–7295

46 Lenne, P. F.; Etienne, E.; Rigneault, H., Subwavelength patterns and high detection efficiency in fluorescence correlation spectroscopy using photonic structures, *Appl. Phys. Lett.* **2002**, 80, 4106–4108

47 Yin, D.; Lunt, E. J.; Rudenko, M. I.; Deamer, D. W.; Hawkins, A. R.; Schmidt, H., Planar optofluidic chip for single particle detection, manipulation, and analysis, *Lab Chip* **2007**, 7, 1171
48 Rudenko, M. I.; Kühn, S.; Lunt, E. J.; Deamer, D. W.; Hawkins, A. R.; Schmidt, H., Ultrasensitive Qβ Phase Analysis Using Fluorescence Correlation Spectroscopy on an Optofluidic Chip, *Biosensors and Bioelectronics* **2009**, 24, 3258–3263
49 Ashkin, A., Optical trapping and manipulation of neutral particles using lasers, *PNAS* **1994**, 94, 4853–4860
50 Ashkin, A.; Dziedzic, J. M.; Optical trapping and manipulation of viruses and bacteria, *Science* **1987**, 235, 1517–1520
51 Ashkin, A.; Dziedzic, J. M.; Yamane, T., Optical trapping and manipulation of single cells using infrared laser beams, *Nature* **1987**, 330, 769–771
52 Renn, M. J.; Pastel, R.; Lewandowski, H. J., Laser guidance and trapping of mesoscale particles in hollow-core optical fibers, *Phys. Rev. Lett.* **1999**, 82, 1574–1577
53 Mandal, S.; Erickson, D., Optofluidic transport in liquid core waveguiding structures, *Appl. Phys. Lett.* **2007**, 90, 184103
54 Cran-McGreehin, S.; Krauss, T. F.; Dholakia, K., Integrated monolithic optical manipulation, *Lab Chip* **2006**, *6*, 1122–1124
55 Schmidt, B. S.; Yang, A. H. J.; Erickson D.; Lipson, M., Optofluidic trapping and transport on solid core waveguides within a microfluidic device, *Opt. Express* **2007**, 15, 14322–14334
56 Measor, P.; Kühn, S.; Lunt, E. J.; Phillips, B. S.; Hawkins, A. R.; Schmidt, H., Hollow-core waveguide characterization by optically induced particle transport, *Opt. Lett.* **2008**, 33, 672–674
57 Hunsperger, R. G. Integrated Optics, 5th edn.; Springer, New York, **2002**
58 Deamer, D.; Branton, D., Characterization of nucelic acids by nanopore analysis, *Acc. Chem. Res.* **2002**, 35, 817–825
59 Kühn, S; Measor, P; Lunt, E.J.; Phillips, B.S.; Deamer, D.W.; Hawkins, A.R.; Schmidt, H., Loss-based optical trap for on-chip particle analysis, *Lab on Chip* **2009**, 9, 2212–2216
60 Kühn, S.; Measor, P.; Lunt, E. J.; Hawkins, A. R.; Schmidt, H., Particle manipulation with integrated optofluidic traps, Technical Digest, IEEE Summer Topical Meetings, Acapulco, MX, July 21–23, **2008**

Chapter 19
Optofluidic Ring Resonator Dye Microlasers

Siyka I. Shopova, Scott Lacey, Ian M. White, Jonathan D. Suter, Yuze Sun, and Xudong Fan

Abstract We discuss versatile, miniaturized optofluidic ring resonator (OFRR) dye lasers that can be operated regardless of the refractive index of the liquid. The OFRR is a piece of a thin-walled fused silica capillary that integrates the photonic ring resonator with microfluidics. In an OFRR dye laser, the active lasing materials (such as dyes) are passed through the capillary whereas the circular cross section forms a ring resonator and supports whispering gallery modes that provide optical feedback for lasing. Because of the high Q-factors extremely low lasing threshold is achieved (25 nJ/mm^2). The operation wavelength can conveniently be changed by using different dyes and fine-tuned with solvent. The OFRR laser is excited through direct excitation or through efficient energy transfer. The laser can be efficiently out-coupled through a fiber taper in contact with the capillary, thus providing easy guiding for the laser emission. Theoretical analysis and experimental results for OFRR lasers are presented.

19.1 Introduction

Optofluidic dye lasers have become an important research direction as they can be implemented in the design of a complete "lab-on-a-chip" system[1]. They combine the advantages of dye lasers with the convenience of microfabrication[2–7]. Optical feedback in these microfluidic dye lasers is provided by Fabry-Pérot-type cavities[3–5,7] or embedded distributed feedback gratings[2,6]. Ring resonator dye lasers based on evanescent wave coupled gain have also been demonstrated[8–12]. In a ring resonator, the whispering gallery modes (WGMs) form because of the light total internal reflection along the curved interface between the high and low

X. Fan (✉)
Department of Biological Engineering, University of Missouri, Columbia, MO 65211, USA
e-mail: fanxud@missouri.edu

X. Fan (ed.), *Advanced Photonic Structures for Biological and Chemical Detection*, Integrated Analytical Systems,
DOI 10.1007/978-0-387-98063-8_19,

refractive index (RI) media[13]. WGMs can have extremely high Q factors ($>10^9$), resulting in a significant reduction in lasing threshold. However, all of these design configurations impose some limitations. In microfluidic lasers, especially in the ring resonator-based lasers, selection of a liquid with an appropriate RI is critical. For example, high RI liquid is needed in DFB lasers and liquid waveguide lasers[6,7,14]. On the one hand, high index core is also necessary to achieve lasing in a capillary with a thick wall (tens to hundreds of micrometers in thickness), and no lasing is observed when the core RI is lower than that of the wall[8,15]. On the other hand, for the ring resonator lasers that rely on the evanescent gain in the surrounding medium, the low RI liquid is imperative, as no WGM would exist when the RI of the liquid is larger than that of the ring resonator[10,12]. The requirement of the liquid RI significantly limits the flexibility and practicality of microfluidic lasers.

Recently, we have demonstrated another type of micro-optofluidic dye laser[16–20] based on opto-fluidics ring resonators (OFRRs)[21,22]. The schematic of these resonators is shown in Fig. 19.1. The OFRR is made of a glass capillary with an outer diameter (OD) of a few hundreds of micrometers and wall thickness of a few micrometers. The OFRR serves as a ring resonator and as a microfluidic channel and requires no additional optical feedback components and alignment. It is mechanically robust and chemically inert and can accommodate various solvents and dyes. Laser emission can be efficiently out-coupled through multiple waveguides in contact with the OFRR, as shown in Fig. 19.1b. The OFRR exhibits great versatility in handling the liquids of various RIs[19,23]. When the core is filled with a low RI solvent, the cross section of the capillary forms a ring resonator and the evanescent field of the WGM in the core provides the feedback. When the core is filled with a high RI solvent, the WGM forms at the liquid-wall interface and provides the optical feedback for lasing. Theoretical analysis and experimental demonstration have shown that in both low and high RI cases, the high Q-factors can be maintained to achieve low-threshold laser oscillation and that the laser

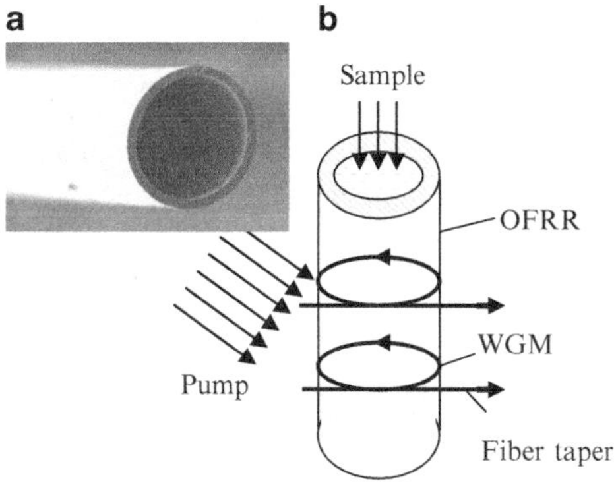

Fig. 19.1 An optofluidic laser based on the OFRR. (a) SEM picture of the OFRR cross section. The wall thickness is 5 μm. (b) Side view of the OFRR laser. The pump laser is focused on the capillary and the OFRR laser emission can be out-coupled through a waveguide. Reprinted from Ref. 18 with permission. © 2008 Optical Society of America

emission can be out-coupled via a waveguide or an optical fiber taper in contact with the OFRR exterior surface for easy light delivery[23].

In addition to direct excitation, we have demonstrated OFRR laser when the excitation is achieved through energy transfer. Optofluidic lasers in which the gain medium is directly excited impose several limitations. The selection of the dye is significantly limited by the availability of pump lasers. Furthermore, due to the small size of the microfluidic channel, the absorption of pump light is weak. Although the low absorption efficiency may be alleviated by using a higher dye concentration, the existence of dye self-absorption leads to an increased lasing threshold, severely compromising the laser performance. These issues can be overcome by using fluorescence resonance energy transfer (FRET) in two-component laser gain medium, in which the acceptor is pumped by the energy nonradiatively transferred from the donor.

19.2 Theoretical Considerations

19.2.1 Interaction of a WGM with the Gain Medium

In an OFRR, WGMs reside in both core and wall. Therefore, only a fraction of the WGM interacts with the gain medium. The lasing behavior of this partially interacting system has been analyzed by Moon et al.[10] For a given dye concentration, the minimum fraction of the excited molecules, $\gamma(\lambda)$, is given by:

$$\gamma(\lambda) = \frac{\sigma_a(\lambda)}{\sigma_e(\lambda)}\left[1 + \frac{Q_{dye}(\lambda)}{\eta_1 Q_{empty}(\lambda)}\right], \tag{19.1}$$

where σ_a and σ_e are the absorption and emission cross section, respectively. Q_{dye} is the Q-factor related to the dye absorption, i.e., $Q_{dye} = 2\pi n_2/\lambda\sigma_a\rho$, where n_2 is the RI of the wall, λ is the WGM wavelength, and ρ is the dye concentration. η_1 is the fraction of the WGM in the core. Q_{empty} is the empty cavity Q-factor when the ring resonator is filled with the solvent in the absence of dye. Q_{empty} is determined by:

$$Q_{empty}^{-1} = Q_{rad}^{-1} + Q_{wall}^{-1} + Q_{sca}^{-1} + \eta_1 Q_{sol}^{-1}, \tag{19.2}$$

where Q_{rad}, Q_{wall}, and Q_{sca} are the Q-factors determined by the radiation loss, loss in the wall medium, and loss resulting from the surface scattering. $Q_{sol} = 2\pi n_2/\lambda\alpha_{sol}$, where α_{sol} is the absorption coefficient of the solvent.

When $\eta_1 Q_{empty} << Q_{dye}$, the lasing threshold depends on $\eta_1 Q_{empty}$ and its wavelength is determined by $1/\sigma_e$. When $\eta_1 Q_{empty} >> Q_{dye}$, all WGMs have nearly the same lasing threshold with lasing spectral positions determined by σ_a/σ_e. In this case the lasing threshold and its spectral position are insensitive to a small change in Q_{empty}. Another special case is when Q_{empty} is limited by the solvent absorption,

$$\gamma = \frac{\sigma_a}{\sigma_e}\left[1 + \frac{\alpha_{sol}}{\sigma_a \rho}\right], \tag{19.3}$$

suggesting that the lasing threshold is independent of η_1 and that all modes should have the same lasing threshold, regardless of the value of Q_{empty} relative to Q_{dye}. This behavior is consistent with the experimental results shown later.

19.2.2 WGM Spectra of OFRR Containing Different Liquid Core RI

To understand the full potential of OFRR lasers, which are not limited by the choice of solvent, we have performed theoretical analysis on the versatility of the OFRR in

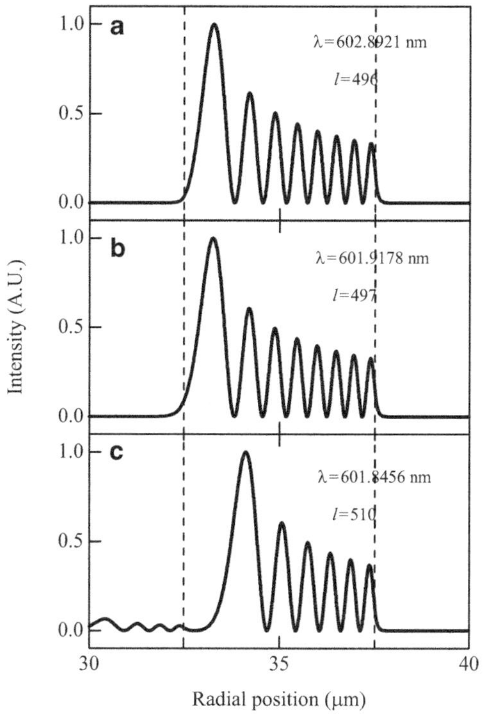

Fig. 19.2 WGM distribution for an OFRR with OD = 75 μm and wall thickness, $t = 5$ μm when the core RI is 1.36 (**a**), 1.445 (**b**), and 1.626 (**c**). The wall RI is 1.45 and the RI for the surrounding medium is 1.0. The WGM spectral position, λ, and the corresponding angular momentum, l, are labeled in the figure. All modes are polarized along the OFRR longitudinal direction. The fraction of light in the core (η_1) and outside the resonator (η_3) is: (**a**) $\eta_1 = 0.2\%$, $\eta_3 = 0.6\%$; (**b**) $\eta_1 = 0.9\%$, $\eta_3 = 0.6\%$; (**c**) $\eta_1 = 4.7\%$, $\eta_3 = 0.53\%$. *Dashed lines* indicate the OFRR inner and outer surface. Reprinted from Ref. 19 with permission. © 2008 Optical Society of America

handling liquids of different RI in the core[23]. The WGM spectral position and mode profile are calculated using a three-layer Mie scattering model[22]. Figure 19.2 plots the WGM intensity distribution for the core RI (n_{core}) of 1.36, 1.445, and 1.626. When $n_{\text{core}} = 1.36$ (Fig. 19.2a), an evanescent field exists both inside and outside the OFRR, which provides the feedback for lasing and enables the WGM out-coupling through a waveguide. It should be emphasized that the wall thickness is crucial in low-threshold lasers, as the thick wall prevents the low radial order WGMs, which have the highest Q-factors, from being exposed to the core. No lasing is observed when a thick-walled capillary is filled with ethanol dye solution[8,15]. When n_{core} is increased to 1.445 (Fig. 19.2b), the OFRR essentially becomes equivalent to a solid silica cylinder ring resonator and more light is pulled into the core. When the core RI becomes higher than the wall RI, the core itself also becomes a ring resonator and the WGMs exist at the core/wall boundary, in addition to the wall/air boundary, as shown by the ripple-like structure in Fig. 19.2c. Since the wall is thin, the WGMs in these two boundaries interact and form bonding and antibonding modes[23,24]. Figure 19.2c shows a new hybrid mode formed by the 4th order WGM in the core interacting strongly with the 6th order WGM in the wall. These new modes have large presence in both core and wall and also have sufficient exposure outside the ring for the coupling to the waveguide. The coupling can also be interpreted as resonant tunneling of the WGM through the potential well formed by the OFRR wall. Because of large phase mismatch between the core WGMs and the fiber or waveguide, the coupling strength decreases extremely rapidly with the increased wall thickness[23].

19.3 Direct Pumped OFRR Lasers

19.3.1 Experimental Realization

Figure 19.3 presents the experimental setup for the OFRR dye laser. The fused silica OFRR with OD = 75 μm and $n = 1.45$ is fabricated using an in-house CO_2

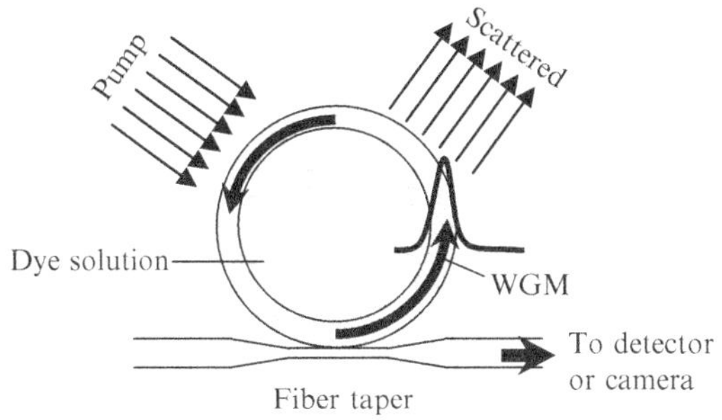

Fig. 19.3 Optofluidic laser setup (cross-section view) – fiber taper is used to evanescently couple out the laser emission

laser pulling station. The shape after pulling is a uniform cylindrical capillary with a wall thickness of about 5 μm. R6G dye in different solvents (ethanol ($n = 1.36$), chloroform ($n = 1.445$), and quinoline ($n = 1.626$)) is flowed through the OFRR with a peristaltic pump at a flow rate of 10 μL/min. A pulsed laser (Opolette, 532 nm, 10 ns pulse width, and 20 Hz repetition rate) is loosely focused onto the side of the OFRR through a cylindrical lens so that approximately a 3 mm portion of the OFRR is homogeneously pumped. The emission spectra are acquired by an Ocean Optics USB 4000 spectrometer with a spectral resolution of 3.7 nm. The laser emission is coupled out through a tapered fiber in touch with the OFRR for both low and high liquid core RI.

In our initial experiments, we used HF to etch the interior surface, as detailed previously[17,22]. The etching resulted in a decrease in the wall thickness of the resonator, so a larger fraction of the wall-confined WGMs is exposed to the laser dye solution in the core. However, etching also degrades the inner surface of the capillary, thus reducing the Q-factor of the resonator.

19.3.2 Characterization of the OFRR

For characterization of the empty cavity Q-factor, the OFRR is initially filled with ethanol in the absence of dye. We use a tunable diode laser at 690 nm, close to the dye emission wavelength (~600 nm). The laser light is transmitted via a fiber taper and is scanned in wavelength. When the laser wavelength coincides with the resonant wavelength of a WGM, the light is coupled into the OFRR, leaving a spectral dip at the fiber output. Figure 19.4a shows a spectral linewidth of 0.068 pm, corresponding to a Q-factor of 1.0×10^7, indicative of the high quality of the OFRR as a laser cavity, even after HF etching. The observed Q-factor has contributions from the scattering loss at the OFRR inner surface, ethanol absorption, and fiber coupling loss, i.e., $Q^{-1} = Q_{sca}{}^{-1} + Q_{abs}{}^{-1} + Q_{coupl}{}^{-1}$. Q_{abs}, related to the absorption losses, is estimated to be much larger than 10^8 because of the low ethanol absorption in the visible range. Q_{coupl}, related to the coupling losses through the tapered fiber, is larger than Q_{sca} (the modes are usually undercoupled). Therefore, at 690 nm, the Q factor is mainly determined by the surface roughness resulting from the HF etching. For a comparison, at 1,550 nm, a Q-factor of 6×10^6 and in excess of 10^7 are obtained when the OFRR is filled with ethanol and air, respectively, indicating that the ethanol absorption plays a more significant role in determining the OFRR Q factor in the near infrared regime.

In the OFRR, the wall thickness is important, as it determines the fraction of the light in the core that interacts evanescently with the gain medium when the solvent is of low RI. To precisely characterize the OFRR thickness noninvasively, we use the method developed previously[22], in which various concentrations of water-ethanol mixtures are passed through the OFRR, and the WGM spectral shift in response to RI changes in the core is plotted as demonstrated in Fig. 19.4b. By matching the experimental sensitivity results with those obtained from our

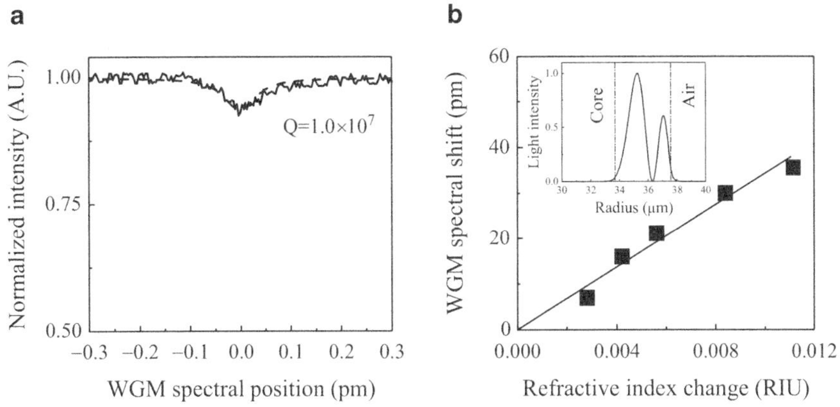

Fig. 19.4 (**a**) Q-factor measured at 690 nm. Lorentzian fit (*dashed line*) shows a linewidth of 0.068 pm. (**b**) Response of the WGM around 1,550 nm to the RI change in the core. A sensitivity of 3.5 nm/RIU corresponds to a wall thickness of 3.8 μm. Inset: radial distribution of the 2nd order WGM (~1,550 nm). Reprinted from Ref. 17 with permission. © 2008 American Institute of Physics

simulation based on Mie theory, we find a wall thickness of approximately 5 μm for the capillaries directly after pulling and about 3.5–4 μm for the etched capillaries.

19.3.3 Lasing and Out-Coupling Through a Tapered Fiber

Lasing is achieved at the wavelength with the largest gain, which depends on the modified Q-factor, $\eta_1 Q$. The observed lasing with the resonators that have been etched with HF is at 570–580 nm, as shown in Fig. 19.5a. Figure 19.5b shows the lasing emission from an OFRR that is pulled under heat but not etched. The Q-factor of these resonators is in excess of 10^8, 2 orders of magnitude better than the etched resonators. Because of such high Q-factors, the measured laser threshold is a record low 25 nJ/mm^2. In Fig. 19.6, we further show the dependence of the lasing wavelength on the dye concentration. With the decreased dye concentration, $Q_{dye} > \eta_1 Q_{empty}$ is satisfied. As a result, the lasing peak moves to a shorter wavelength, according to $1/\sigma_e(\lambda)$, as discussed previously.

The versatility of the laser resonator in handling dye solutions with different RIs has also been demonstrated (see Fig. 19.7)[19]. The lasing has been achieved with a larger core RI as the ethanol ($n_{core} = 1.36$) has been replaced with chloroform ($n_{core} = 1.445$) and quinoline ($n_{core} = 1.626$), respectively. Chloroform has absorption of 1.7×10^{-5} cm^{-1}, therefore, $\eta_1 Q_{empty}$ for chloroform is similar to that for ethanol (having absorption 1.55×10^{-5} cm^{-1}). For quinoline, our absorption measurement shows absorption of approximately 15.2 cm^{-1} at 600 nm. As a result, Q_{empty} is limited by the solvent absorption ($Q_{sol} = 10^6$). In addition, the γ value for quinoline is twice the value for ethanol in agreement with the higher lasing threshold observed experimentally for R6G in quinoline[19].

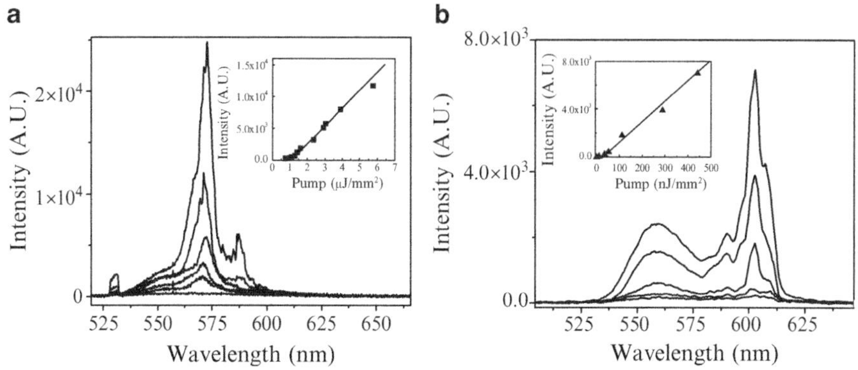

Fig. 19.5 The spectral position of the laser emission and the threshold curves (in the insets) for (**a**) an etched resonator with OD = 75 μm and the wall thickness of 3.8 μm having threshold at pump energy density of 1 μJ/mm^2 and (**b**) unetched resonator with OD = 75 μm and the wall thickness of 5 μm with a threshold of 25 nJ/mm^2. Reprinted from Ref. 20 with permission. © 2008 International Society for Optical Engineering

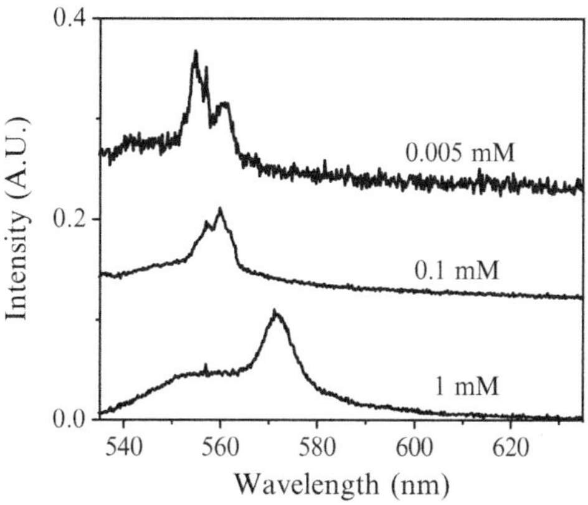

Fig. 19.6 Lasing peak of the etched laser cavity in different dye concentrations of ethanol solutions. Reprinted from Ref. 20 with permission. © 2008 International Society for Optical Engineering

The evanescent out-coupling of the laser emission through a tapered fiber is another important result of this work. A tapered optical fiber (approximately 2 μm in waist diameter) is brought in contact with the OFRR. Coupling of emission to the waveguide can be as high as 50%.[16] In addition, the tapered fiber only collects the light from the WGM and therefore also serves as an optical filter to remove the pump light and the broad spontaneous emission, as evidenced in Fig. 19.8.

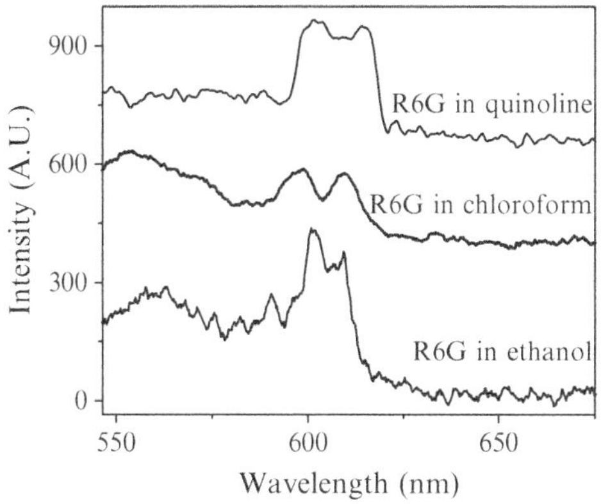

Fig. 19.7 Laser emission of R6G (2 mM) in different solvents using a high *Q*-factor OFRR; the spectra are acquired from the scattered light. Traces are shifted vertically for clarity. Reprinted from Ref. 20 with permission. © 2008 International Society for Optical Engineering

19.4 OFRR Lasers Pumped by Energy Transfer

19.4.1 FRET Laser Concept

Using energy transfer between suitable dyes provides great flexibility in laser wavelength without the need to change pump wavelengths. The concept of this OFRR FRET laser system is illustrated in Fig. 19.9. The donor and acceptor are coflowed through the OFRR, and the energy from the donor is transferred to the acceptor whose emission is coupled to the WGMs of high Q-factors. The donor and acceptor can be of dye-dye type and quantum dot (QD)-dye type. Multiple dyes or QDs can also be used for cascaded energy transfer. Furthermore, in an OFRR FRET laser, the donor and the acceptor can be homogeneously distributed in solution, or they can be placed together within the Förster distance (~10 nm) through some mechanisms like DNA hybridization, for example.

19.4.2 Characterization of FRET in Free Space

In our experiment, we first use R6G and LDS722 as the donor and acceptor. It is important to characterize the FRET signal of this pair in a linear regime below the lasing threshold to provide a performance reference for the OFRR FRET laser that we will investigate later. Figure 19.10 shows the characterization of their FRET behavior for varying acceptor/donor concentration ratios when the donor is excited with a low power CW laser at 532 nm. As shown in Fig. 19.10a, in the absence

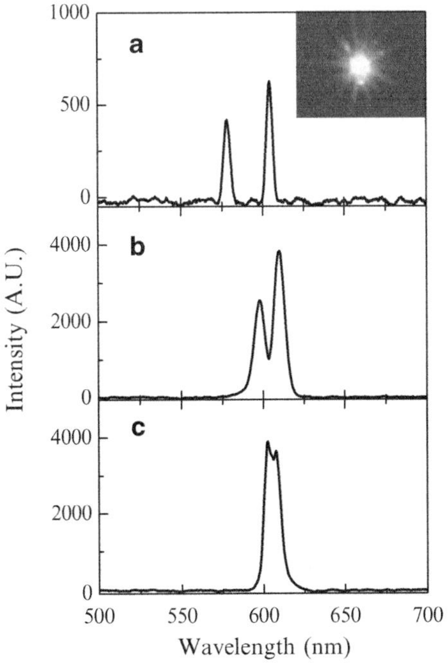

Fig. 19.8 Laser emission through the tapered fiber for different dye solvents – ethanol (**a**), chloroform (**b**), and quinoline (**c**). The inset shows the out-coupled light from the tapered fiber. Reprinted from Ref. 19 with permission. © 2008 Optical Society of America

of R6G, no LDS722 emission is observed. When 0.1 mM of R6G is added, the acceptor emission emerges with the peak at 700 nm, and increases with the increased acceptor concentration. The energy transfer efficiency for each acceptor-to-donor ratio (A/D) is plotted in Fig. 19.10b, which can be derived by: $\varepsilon = 1 - I_D/I_{D0}$, where I_D and I_{D0} are the donor intensity in the presence and absence of the acceptor, respectively. ε is related to the Förster distance, R_0, by:[25]

$$\varepsilon = 1 - \exp\left(-1.42\frac{c}{c_0}\right), \tag{19.4}$$

where c_0 is the critical concentration that corresponds to one acceptor molecule in a sphere of the radius R_0. Curve fit in Fig. 19.10b yields $c_0 = 1.7$ mM, which results in a Förster distance of 6.2 nm.[25] Transfer efficiency of 50% is obtained when the acceptor concentration is 0.83 mM (i.e., A/D = 8.3), reflecting a highly efficient energy transfer between the donor and acceptor. The quantum yield of the acceptor, ϕ, can also be estimated from:

$$\varepsilon = \frac{I_A/\phi}{I_A/\phi + I_D} = \frac{1}{1 + \phi I_D/I_A}, \tag{19.5}$$

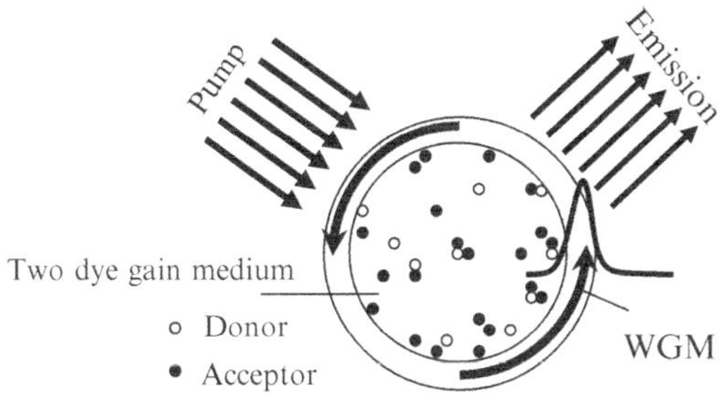

Fig. 19.9 Concept of the OFRR FRET laser

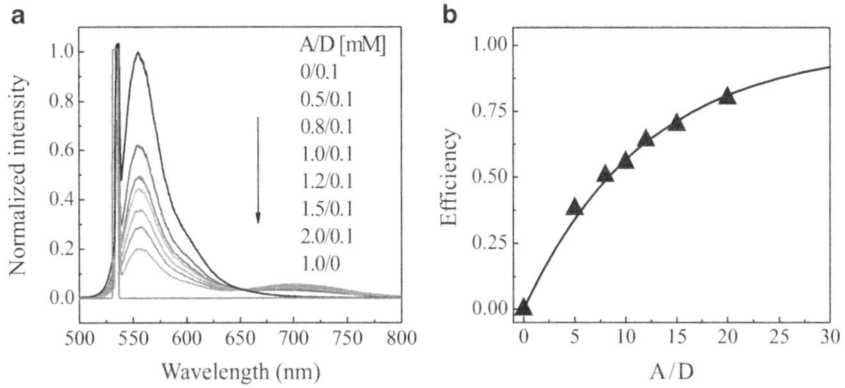

Fig. 19.10 (**a**) Normalized spectra of FRET in free space for R6G (donor) and LDS722 (acceptor). The bottom curve (A/D = 1.0/0 mM) is acquired in the absence of the donor. The donor concentration for the remaining curves is fixed at 0.1 mM. The CW pump wavelength: 532 nm. A/D: acceptor-to-donor ratio. (**b**) Energy transfer efficiency, ε, calculated from (**a**). *Solid line* is the theoretical curve for the efficiency with a parameter $c_0 = 1.7$ mM, corresponding to $R_0 = 6.2$ nm. Reprinted from Ref. 18 with permission. © 2008 Optical Society of America

where I_A is the acceptor intensity. For LDS722, $\phi = 7\%$, much lower than that for R6G.

19.4.3 Lasing Through FRET

The OFRR fabrication and characterization are the same as for direct pumped lasers. The dye mixture in methanol is passed through the OFRR at a flow rate of 10 μL/min. For OFRR FRET lasers, the donor is excited under the same conditions as in the direct pump case.

When pure R6G is passed through the OFRR, the lasing emission is observed at 565 nm. By adding an acceptor dye (LDS722), while the pump intensity remains

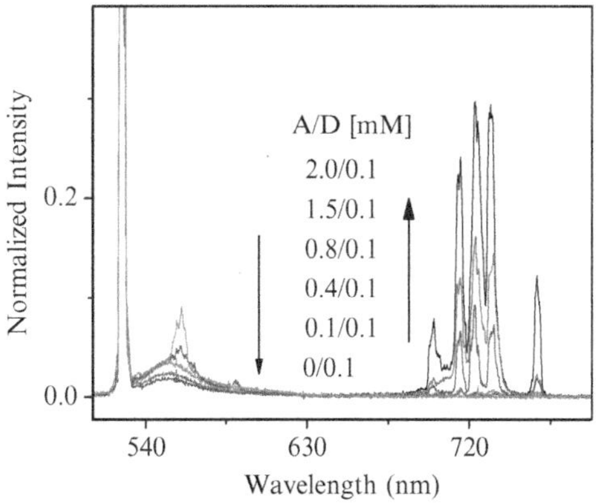

Fig. 19.11 OFRR-FRET laser spectra for various acceptor concentrations. Donor concentration is fixed at 0.1 mM. The pulsed laser wavelength is 526 nm. Reprinted from Ref. 18 with permission. © 2008 Optical Society of America

the same, the donor lasing decreases (Fig. 19.11). The prominent peaks start to emerge at the longer wavelength side of the acceptor emission spectrum (715, 723.5, and 733 nm). As the donor/acceptor (A/D) ratio increases, the donor lasing is completely quenched and acceptor lasing emission increases. Note that the lasing peaks result from donor's and acceptor's molecules that lie within 100 nm from the OFRR interior surface, as only the emission from those molecules can be coupled into the WGM. Therefore, the effective sample volume is approximately 20 picoliter, 0.3% of the total illuminated volume. The lasing threshold for the mixture with 2 mM LDS722 is estimated to be approximately 0.3 μJ/mm^2.[18] Given the low quantum yield of LDS722, a much higher lasing threshold would be expected if the dye were directly excited.

In addition to the short-ranged nonradiative Förster transfer, the energy transfer between donor and acceptor can also be achieved through the mediation of the WGM[26,27]. In the case of cavity-assisted radiative energy transfer, the donor emission is first coupled into the WGM that stores the photons for extended time for subsequent absorption by acceptors. To determine which effect is dominant in our OFRR FRET laser, we carry out a few control experiments. The first one is shown in Fig. 19.12a where the A/D remains constant and the pump power is set to be below, near, and well above the R6G lasing threshold. In all three cases, the acceptor lasing is achieved. The effective photon conversion efficiency is determined from $\varepsilon = (1 + I_D/I_A)^{-1}$. For donors and acceptors lying within 100 nm from the OFRR interior surface, below the lasing threshold of R6G, all the donor excited states are converted to lasing photons in the acceptors, which means that ε is nearly 100%. When the pump power increases, R6G starts to lase, at 562 nm. However, the acceptor lasing intensity remains nearly the same, making ε drop to 91%. When R6G intensity is well above the threshold, acceptor lasing intensity increases, however at a much lower percentage, and ε becomes 63%. This experiment

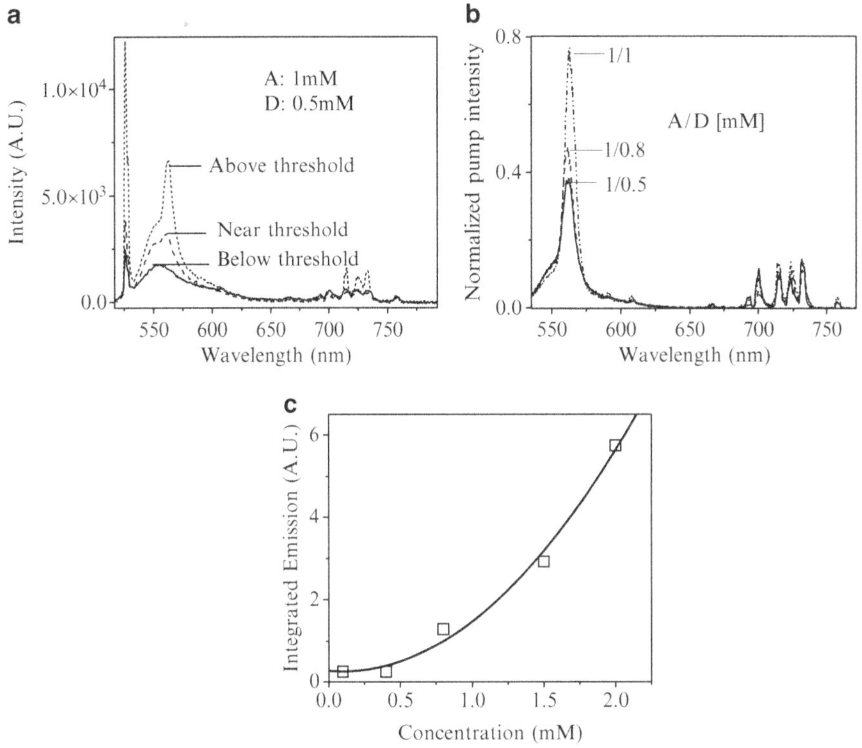

Fig. 19.12 Characteristics of the nonradiative Förster transfer. (**a**) FRET spectra show that $\varepsilon =$ 100%, 91%, and 63%, respectively, when the pump intensity is below, near, and above the R6G lasing threshold. (**b**) FRET spectra show a decreased ε (61%, 41%, and 34%) when R6G concentration increases while LDS722 concentration and the pump power remain the same. (**c**) LDS722 emission as a function of its concentration. Reprinted from Ref. 20 with permission.

can be explained by the fact that nonradiative transitions' lifetime is shorter than the fluorescence lifetime of R6G, which is 3 ns, and thus the probability for nonradiative transfer is higher. However, above the R6G lasing threshold the average excited state lifetime is reduced to that of nonradiative transitions and the R6G lasing process competes with the energy transfer process. This result agrees with the nonradiative Förster transfer model, but contradicts the radiative cavity-assisted energy transfer model, in which the acceptor intensity would be expected to increase proportionally when the donor lases. Another piece of evidence to confirm the dominant effect of the nonradiative Förster transfer is plotted in Fig. 19.12b, in which the R6G concentration changes while the pump intensity and the acceptor concentration remain constant. R6G lasing is achieved in all three R6G concentrations. Despite the increase in R6G emission intensity, the photon conversion efficiency decreases with the increase of R6G concentration. This experiment also shows that the nonradiative Förster transfer dominates over the cavity-assisted energy transfer.

Finally, Fig. 19.12c plots the integrated acceptor emission shown in Fig. 19.11 as a function of the acceptor concentration. As discussed in Ref. 25, the rate constant of the transfer process is $\eta_{D \to A} = \frac{1}{\tau_D}\left(\frac{R_0}{R}\right)^6$, where τ_D is the life time of the donor state and R_0 is the Förster distance. In terms of concentration, this means that the rate of nonradiative transfer is proportional to the square of the concentration. The nonradiative transfer rate is what determines the population of the upper level of the acceptor molecules and consequently the acceptor emission. The concentration dependence in Fig. 19.12c is quadratic, suggesting the energy transfer is predominantly nonradiative.

To further demonstrate the OFRR FRET capability, we show in Fig. 19.13a lasing achieved with a different dye pair, Rhodamine B and LD800. Another interesting energy transfer concept that is performed in the OFRR is illustrated in Fig. 19.13b. Here, we show cascade FRET, where the energy is transferred from Coumarin 480 to R6G, and then to LDS722. Strong acceptor laser emission emerges when the ratio between LDS722 and Coumarin is 10:1. R6G emission is nearly invisible, indicative of a highly efficient energy transfer between R6G and LDS722. As a control, no LDS722 emission is observed in the absence of R6G, showing that Coumarin and LDS are completely decoupled. Therefore, a high concentration of Coumarin can be used to enhance the pump light absorption efficiency without concern about the donor absorption at the acceptor lasing wavelength.

An OFRR FRET laser using orange CdSe/ZnS core/shell QDs as the donor and Nile blue dye as the acceptor is shown in Fig. 19.13c. QDs have much better chemical/photo stability, higher quantum yield, and higher absorption cross section than dyes. Moreover, all sizes of QDs can be excited by a single blue light source and their fluorescence can be tuned by changing the QD size to match the absorption band of various acceptors[28,29]. Therefore, using QDs as a donor enables the maximal pump flexibility and multiple-emission capability in microfluidic lasers. Because of the large QD absorption, only 200 nM QD is needed to achieve sufficient energy transfer, which corresponds to only 4 atto-mole QDs.

In addition to a flexible, low threshold, and compact microfluidic laser, the OFRR FRET laser can also be used for ultrasensitive intra-cavity bio/chemical sensing[30,31]. When compared with conventional FRET, the lasing action achieved from nonradiative transfer is a nonlinear effect. As shown in Fig. 19.11, in the presence of the acceptor, the gain in the donor is reduced, resulting in an exponential decrease in the donor emission. Meanwhile, when the gain in the acceptor is achieved through the energy transfer from the donor, stimulated emission takes place in the acceptor, producing significantly higher intensity than spontaneous emission. Therefore, the conventional Förster transfer is tremendously enhanced in this lasing system. As a result, the sensitivity of a FRET assay can be improved by utilizing the OFRR platform.

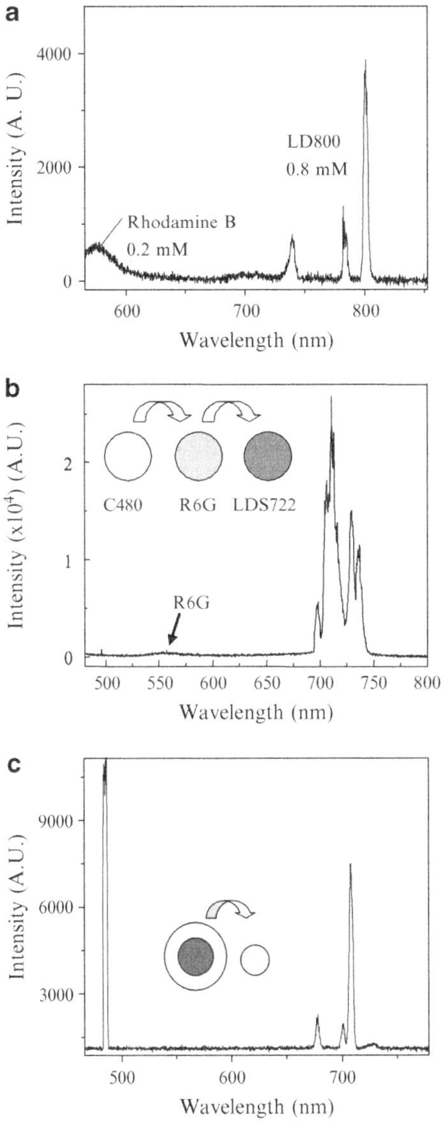

Fig. 19.13 (**a**) FRET between 0.2 mM Rhodamine B and 0.8 mM LD800. (**b**) Lasing of 2 mM LDS722 through cascade FRET from Coumarin 480 (0.2 mM) and R6G (0.4 mM). (**c**) Lasing of 0.125 mM Nile blue in methanol through FRET from 200 nM CdSe/ZnS core/shell QDs. The pump laser is at 485 nm. Reprinted from Ref. 18 with permission. © 2008 Optical Society of America

19.4.4 DNA Controlled FRET Laser

Recently, DNA has been studied as a robust material for photonics applications[32], and particularly as a polymer matrix in laser gain material[33]. This material can withstand high-intensity laser fields and relatively high temperatures.

Here, we demonstrate another use of DNA in laser gain materials. Single-stranded DNA molecules labeled with a donor and complementary DNA (cDNA) labeled with an acceptor can be used as a model system to lock the donor and acceptor together. DNA has a well-defined distance between two adjacent bases (approximately 0.34 nm)[34], thus providing an excellent nano-sized ruler to measure the donor-acceptor separation and to control the Förster energy transfer. Higher efficiency of energy transfer can be achieved by decreasing the distance between the donor and acceptor. As a result, lasing action from the acceptor dye can be achieved at significantly lower dye concentrations, when compared with free dye molecules in solution.

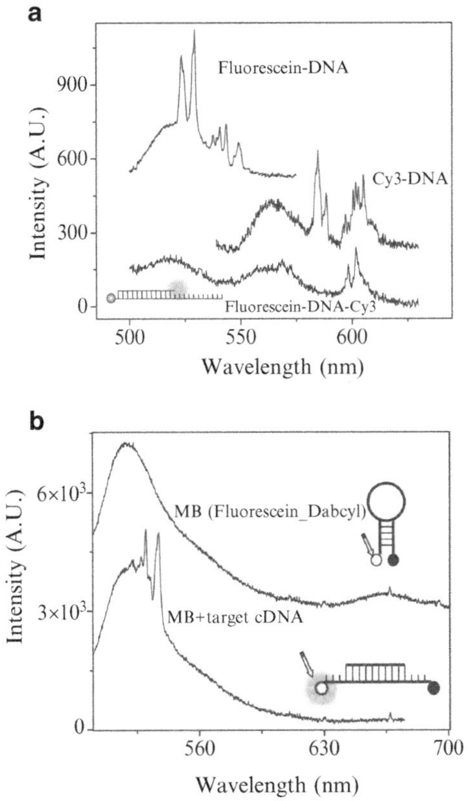

Fig. 19.14 (**a**) Emission from dye-DNA complexes: Fluorescein-labeled DNA at 60 μM (*Top*); Cy3-labeled DNA at 60 μM (*Middle*); After DNA hybridization at 30 μM for each cDNA sample (*Bottom*). (**b**) Emission from a probe MB (*Top*); Emission after adding a target cDNA (*Bottom*)

We have demonstrated lasing in dye-labeled DNA gain material in two different designs. In the first experiment, the two single DNA strands, each consisting of 15 base units, are labeled with a donor dye (Fluorescein) and an acceptor dye (Cy3), respectively. The actual sequences used in this experiment are: probe 5′-/FLC/ATTGTGACTGCGGTT and target 5′-/Cy3/AACCGCAGTCACAAT. The DNA is designed with no looping and folding so that the DNA is relatively rigid and sticks straight[35]. Each DNA-dye complex is flowed through an OFRR and excited with a nanosecond laser of wavelength close to the absorption maximum of the dye. Strong lasing peaks, shown in the top and middle trace in Fig. 19.14a, are observed in solutions of 60 μM DNA concentration. When the two DNA strands are mixed together and let to hybridize, the dye molecules are brought closer. The bottom trace in Fig. 19.14a shows lasing at the wavelength of the acceptor; the donor emission is reduced below the lasing threshold. With this donor-acceptor design, further reduction to submicromolar concentrations has also been achieved.

Figure 19.14b demonstrates another experiment in which a hairpin-shaped DNA, called a molecular beacon (MB)[36], is used. The MB is labeled with a fluorescent dye molecule (Fluorescein) on one end and a quencher (DABCYL) on the other end. The DNA molecule (5′-/FLC/CCA CGC TTG TAG GTC AAC CCC CGT GG/DAB/-3′) is designed such that the first and last five bases are complementary to each other, so it forms a loop and a stem hybrid that keeps the dye and the quencher close. Tris(hydroxymethyl)aminomethane (TRIS) solution (20 mM TRIS at pH 8 plus 1 mM $MgCl_2$) of 25 μM MB is flowed through an OFRR. Excitation with a nanosecond pulsed laser at 470 nm results in residual fluorescence from the donor that is not completely quenched by a relatively weak quencher of DABCYL. However, after adding an equal amount of unlabeled target DNA molecules, several distinctive sharp lasing peaks are observed on the top of the broad fluorescence. The target DNA is complementary to the loop part of the probe, so after the hybridization the hairpin-shaped DNA opens up, the dye and quencher are separated far enough to significantly decrease the energy transfer, which results in the increased gain for the acceptor to achieve lasing emission.

19.5 Conclusion

We have demonstrated low-threshold versatile optofluidic dye lasers based on the OFRR that are capable of utilizing dye solutions with any RI and that the laser light can be out-coupled efficiently with an optical fiber or a waveguide. The out-coupling fiber also serves as a filter to remove the pump light and the broadband luminescence. In addition to the direct excitation, we have also demonstrated a laser action achieved through energy transfer. Because the lasing of either donor or acceptor is highly sensitive to the energy transfer between the two types of carriers, it can be used for monitoring the interactions between larger molecules (DNA or proteins) to which dye molecules (or QDs) are attached. Our future work will focus on achieving single-mode lasing oscillation in the OFRR. CW pumped

OFRR lasers and phase-locked OFRR lasers will be investigated. Finally, applications of the OFRR in highly sensitive intracavity bio/chemical sensors will also be explored.

Acknowledgment This work is supported by the Wallace H. Coulter Foundation Early Career Award, NSF-CAREER (CBET-0747398), and NSF (ECCS-0853399)

References

1 Psaltis, D.; Quake, S. R.; Yang, C. H., Developing optofluidic technology through the fusion of microfluidics and optics, *Nature* **2006**, 442, 381–386
2 Balslev, S.; Kristensen, A., Microfuidic single-mode laser using high-order Bragg grating and antiguiding segments, *Opt. Express* **2005**, 13, 344–351
3 Galas, J. C.; Torres, J.; Belotti, M.; Kou, Q.; Chen, Y., Microfluidic tunable dye laser with integrated mixer and ring resonator, *Appl. Phys. Lett.* **2005**, *86*, 264101
4 Helbo, B.; Kristensen, A.; Menon, A., A micro-cavity fluidic dye laser, *J. Micromech. Microeng.* **2003**, 13, 307–311
5 Kou, Q.; Yesilyurt, I.; Chen, Y., Collinear dual-color laser emission from a microfluidic dye laser, *Appl. Phys. Lett.* **2006**, 88, 091101
6 Li, Z. Y.; Psaltis, D., Optofluidic distributed feedback dye lasers, *IEEE J. Sel. Top. Quantum Electron.* **2007**, 13, 185–193
7 Vezenov, D. V.; Mayers, B. T.; Conroy, R. S.; Whitesides, G. M.; Snee, P. T.; Chan, Y.; Nocera, D. G.; Bawendi, M. G., A low-threshold, high-efficiency microfluidic waveguide laser *J. Am. Chem. Soc.* **2005**, 127, 8952–8953
8 Knight, J. C.; Driver, H. S. T.; Hutcheon, R. J.; Robertson, G. N., Core-resonance capillary-fiber whispering-gallery-mode laser, *Opt. Lett.* **1992**, 17, 1280–1282
9 Moon, H.-J.; An, K., Interferential coupling effect on the whispering-gallery mode lasing in a double-layered microcylinder, *Appl. Phys. Lett.* **2002**, 80, 3250–3252
10 Moon, H.-J.; Chough, Y.-T.; An, K., Cylindrical microcavity laser based on the evanescent-wave-coupled gain, *Phys. Rev. Lett.* **2000**, 85, 3161–3164
11 Moon, H.-J.; Park, G.-W.; Lee, S.-B.; An, K.; Lee, J.-H., Laser oscillations of resonance modes in a thin gain-doped ring-type cylindrical microcavity, *Opt. Commun.* **2004**, 235, 401–407
12 Shevchenko, A.; Lindfors, K.; Buchter, S. C.; Kaivola, M., Evanescent-wave pumped cylindrical microcavity laser with intense output radiation, *Opt. Commun.* **2005**, 245, 349–353
13 Chang, R. K.; Campillo, A. J. Optical Processes in Microcavities, World Scientific, Singapore, 1996
14 Gersborg-Hansen, M.; Kristensen, A., Tunability of optofluidic distributed feedback dye lasers, *Opt. Express* **2007**, 15, 137–142
15 Moon, H.-J.; Park, G.-W.; Lee, S.-B.; An, K.; Lee, J.-H., Waveguide mode lasing via evanescent-wave-coupled gain from a thin cylindrical shell resonator, *Appl. Phys. Lett.* **2004**, 84, 4547–4549
16 Suter, J. D.; Sun, Y.; Howard, D. J.; Viator, J. A.; Fan, X., PDMS embedded opto-fluidic microring resonator lasers, *Opt. Express* **2008**, 16, 10248–10253
17 Shopova, S. I.; Zhu, H.; Fan, X.; Zhang, P., Optofluidic ring resonator based dye laser, *Appl. Phys. Lett.* **2007**, 90, 221101
18 Shopova, S. I.; Cupps, J. M.; Zhang, P.; Henderson, E. P.; Lacey, S.; Fan, X., Opto-fluidic ring resonator lasers based on highly efficient resonant energy transfer, *Opt. Express* **2007**, 15, 12735–12742

19 Lacey, S.; White, I. M.; Sun, Y.; Shopova, S. I.; Cupps, J. M.; Zhang, P.; Fan, X., Versatile opto-fluidic ring resonator lasers with ultra-low threshold, *Opt. Express* **2007**, *15*, 15523–15530
20 Shopova, S. I.; Lacey, S.; White, I.; Sun, Y.; Zhu, H.; Zhang, P.; Fan, X., Optofluidic ring resonator dye microlasers, In Proceedings of SPIE, 6872 (Laser Resonators and Beam Control X) San Jose, California, 68720W, **2008**
21 X.; White, I. M.; Zhu, H.; Suter, J. D.; Oveys, H., Overview of novel integrated optical ring resonator bio/chemical sensors, In Proceedings of SPIE, 6452 (Laser Resonators and Beam Control IX) San Jose, California, 6452M, **2007**
22 White, I. M.; Oveys, H.; Fan, X., Liquid core optical ring resonator sensors, *Opt. Lett.* **2006**, 31, 1319–1321
23 White, I.; Gohring, J.; Sun, Y.; Yang, G.; Lacey, S.; Fan, X., Versatile waveguide-coupled optofluidic devices based on liquid core optical ring resonators, *Appl. Phys. Lett.* **2007**, 91, 241104
24 Teraoka, I.; Arnold, S., Coupled whispering gallery modes in a multilayer-coated microsphere, *Opt. Lett.* **2007**, 32, 1147–1149
25 Forster, T., Transfer mechanisms of electronic excitation, *Disc. Faraday Soc.* **1959**, *27*, 7–17
26 Arnold, S.; Folan, L. M., Energy transfer and the photon lifetime within an aerosol particle, *Opt. Lett.* **1989**, 14, 387–389
27 Götzinger, S.; Menezes, L. d. S.; Mazzei, A.; Kühn, S.; Sandoghdar, V.; Benson, O., Controlled photon transfer between two individual nanoemitters via shared high-q modes of a microsphere resonator, *Nano Lett.* **2006**, 6, 1151–1154
28 Clapp, A. R.; Medintz, I. L.; Mattoussi, H., Forster resonance energy transfer investigations using quantum-dot fluorophores, *Chemphyschem* **2006**, 7, 47–57
29 Leatherdale, C. A.; Woo, W. K.; Mikulec, F. V.; Bawendi, M. G., On the absorption cross section of CdSe nanocrystal quantum dots, *J. Phys. Chem. B* **2002**, 106, 7619–7622
30 Rose, A.; Zhu, Z. G.; Madigan, C. F.; Swager, T. M.; Bulovic, V., Sensitivity gains in chemosensing by lasing action in organic polymers, *Nature* **2005**, 434, 876–879
31 Wun, A. W.; Snee, P. T.; Chan, Y.; Bawendi, M. G.; Nocera, D. G., Non-linear transduction strategies for chemo/biosensing on small length scales, *J. Mater. Chem.* **2005**, 15, 2697–2706
32 Steckl, A. J., DNA - a new material for photonics?, *Nature Photon.* **2007**, 1, 3–5
33 Yu, Z.; Li, W.; Hagen, J. A.; Zhou, Y.; Klotzkin, D.; Grote, J. G.; Steckl, A. J., Photoluminescence and lasing from deoxyribonucleic acid (DNA) thin films doped with sulforhodamine, *Appl. Opt.* **2007**, 46, 1507–1513
34 Alberts, B.; Johnson, A.; Lewis, J.; Raff, M.; Roberts, K.; Walter, P. Molecular Biology of the Cell, 4th edn.; Garland, New York, NY, **2002**
35 Zuker, M., Mfold web server for nucleic acid folding and hybridization prediction, *Nucleic Acids Res.* **2003**, 31, 3406–3415
36 Tyagi, S.; Kramer, F. R., Molecular beacons: Probes that fluoresce upon hybridization, *Nature Biotechnol.* **1996**, 14, 303–308

Index

Q

R

ERRATUM TO:

Single Molecule Analysis with Planar Optofluidics

Holger Schmidt[1] ✉ and Aaron R. Hawkins[2]

(1) School of Engineering, University of California Santa Cruz. 156 High St., Santa Cruz, CA 95064, USA, hschmidt@soe.ucsc.edu
(2) ECEn Department, Brigham Young University, 459 Clyde Building, Provo, UT 84602, USA

X. Fan (ed.), *Advanced Photonic Structure for Biological and Chemical Detection*, Integrated Analytical Systems,
DOI 10.1007/978-0-387-98063-8_18,

DOI 10.1007/978-0-387-98063-8_20

The original version of this article unfortunately contained a wrong legend to Fig. 18.12.

The correct legend is:
Fig. 18.12 FCS of Q-beta bacteriophage. (a) fluorescence signal showing bursts from individual virus particles, inset: dye-labeled phage capsid; (b) corresponding FCS signal (symbols) with fit to ARROW-based FCS model (solid line)

The original online version for this chapter can be found at DOI 10.1007/978-0-387-98063-8_18

GPSR Compliance

The European Union's (EU) General Product Safety Regulation (GPSR) is a set of rules that requires consumer products to be safe and our obligations to ensure this.

If you have any concerns about our products, you can contact us on ProductSafety@springernature.com

In case Publisher is established outside the EU, the EU authorized representative is:

Springer Nature Customer Service Center GmbH
Europaplatz 3
69115 Heidelberg, Germany

Batch number: 10789854

Printed by Printforce, the Netherlands